THEORY AND PRACTICE OF EARTH REINFORCEMENT

IS Kyushu '88

PROCEEDINGS OF THE INTERNATIONAL GEOTECHNICAL SYMPOSIUM ON THEORY
AND PRACTICE OF EARTH REINFORCEMENT / FUKUOKA KYUSHU / 5-7 OCTOBER 1988

Theory and Practice of Earth Reinforcement

Edited by
TOYOTOSHI YAMANOUCHI
Kyushu Sangyo University, Japan
NORIHIKO MIURA
Saga University, Japan
HIDETOSHI OCHIAI
Kyushu University, Japan

Under the auspices of the Japanese Society of Soil Mechanics and Foundation Engineering
A.A.BALKEMA / ROTTERDAM / BROOKFIELD / 1988

CIP-DATA KONINKLIJKE BIBLIOTHEEK, DEN HAAG

Theory

Theory and practice of earth reinforcement: proceedings of the international geotechnical symposium on theory and practice of earth reinforcement, Fukuoka Kyushu, 5-7 October 1988 / ed. by Toyotoshi Yamanouchi, Norihiko Miura, Hidetoshi Ochiai. – Rotterdam [etc.]: Balkema. – Ill.
ISBN 90 6191 820 0 bound
SISO 692.4 UDC 624.131.5
Subject heading: earth reinforcement.

The texts of the various papers in this volume were set individually by typists under the supervision of each of the authors concerned.

Published by
A.A.Balkema, P.O.Box 1675, 3000 BR Rotterdam, Netherlands
A.A.Balkema Publishers, Old Post Road, Brookfield, VT 05036, USA

ISBN 90 6191 820 0
© 1988 A.A.Balkema, Rotterdam
Printed in the Netherlands

International Geotechnical Symposium on Theory and Practice of Earth Reinforcement / Fukuoka Japan / 5-7 October 1988
© *1988 Balkema, Rotterdam. ISBN 90 6191 820 0*

Foreword

Studies with respect to earth reinforcement materials and their applications have been started historically by Henry Vidal in France since 1966. In addition, the first stage of their remarkable development was mainly in Europe where the First International Conference on Geotextiles was held in Paris in 1977. In Japan, the fundamental studies and applications of earth reinforcement commenced in laboratory in the middle of the 1960s to meet the extensive needs of practicing engineers who encountered many geotechnical problems caused by widely distributed soft ground. Thereafter, they have been very active not only in carrying out the fundamental studies of earth reinforcement, but also in applying them to civil engineering practices.

Among various districts in Japan, the geotechnical group in Kyushu is a pioneer in geotextile research, and in the beginning of the year 1985, an international symposium concerning the theory and practice of earth reinforcement was proposed by the Kyushu Chapter of the Japanese Society of Soil Mechanics and Foundation Engineering (JSSMFE). Its realization has been extensively discussed by those who were engaged in earth reinforcement researches. Such a movement has attracted the attention of the headquarters of JSSMFE in Tokyo, and the Society recognized the significance of having the symposium in Kyushu, the nearest island of Japan to Asian countries. It was decided that the International Symposium on Theory and Practice of Earth Reinforcement (IS Kyushu '88) be held in Fukuoka, under the auspices of JSSMFE.

The symposium aims at inviting more than two hundred participants from twenty countries to discuss various problems and/or topics with respect to earth reinforcement, for the benefit of collecting and exchanging a large amount of knowledge of the methods or techniques being developed recently and to spread these to all countries in the world for further development. Professor Bengt B.Broms and Professor Masami Fukuoka, the present and past presidents of the International Society of Soil Mechanics and Foundation Engineering, are invited to give special lectures.

The contents and themes of the symposium are as follows: 1) Tests and materials: strength and durability tests, corrosion problems of materials, contrivance and characterization of new material, and the new application of conventional materials involving natural fibres; 2) Shallow and deep foundations: the basic theory, development of the design methods and their applications in soft grounds and in road constructions; 3) Slopes and excavations: the theory of slope reinforcement, reinforcing techniques of natural and cutoff slopes, application for slope protection, the ground reinforcement accompanied by excavation, and the temporary reinforcement method in excavation works in soft grounds; 4) Embankments: the guideline for design, the reinforcement method of embankment foundations, the reinforcement at the boundary between soft ground and embankment, the damage of reinforcement during construction, and 5) Wall structures: the reinforcement theory of retaining walls, design methods for retaining wall reinforcement, the reinforcement of backfill materials, jointing techniques of reinforcement materials, and the monitoring system.

The symposium is sponsored by the Science Council of Japan. Also, the symposium receives support from fourteen societies and/or associations including the International Geotextile Society, the Japan Chapter of the International Geotextile Society and the Japan Society of Civil Engineers. It should be especially noted that helpful subsidies are granted to the symposium from four societies and/or foundations including the Japan Society for the Promotion of Science. I am most grateful for such valuable help, besides I would like to express my sincere gratitude to the members of the International Advisory Group who have given their support to the preparation of the symposium.

Toyotoshi Yamanouchi
Chairman of IS Kyushu '88

These proceedings have been partially supported by a grant from the Commemorative Association for the Japan World Exposition.

STEERING COMMITTEE

Prof. T.Yamanouchi (Chairman)
Prof. H.Fujimoto
Prof. T.Iseda
Prof. M.Ishido
Prof. K.Kaku
Prof. N.Miura (Secretariat)
Prof. T.Nishida
Prof. S.Ohara
Mr T.Sawayama
Mr Y.Sugita
Mr J.Taguchi
Prof. M.Takayama
Prof. H.Uehara
Mr T.Watanabe
Mr K.Yamaguchi
Prof. N.Yoshida

EXECUTIVE COMMITTEE

Prof. N.Miura (Chairman)
Prof. G.Aramaki (Head of Division)
Assoc. Prof. T.Esaki
Mr K.Fukami
Assoc. Prof. K.Hashiguchi
Assoc. Prof. S.Hayashi
Prof. T.Hirata
Assoc. Prof. M.Hyodo
Mr K.Ikebe
Assoc. Prof. S.Imaizumi
Mr M.Kimoto
Prof. R.Kitamura
Dr T.Kokusho
Mr K.Koreeda
Assoc. Prof. T.Koumoto (Head of Division)
Mr M.Kubota
Mr M.Kudo
Mr K.Matsui
Prof. H.Murata
Assoc. Prof. S.Murata
Prof. H.Ochiai (Head of Division)
Dr M.Ohtsubo
Prof. K.Onitsuka
Mr J.Otani
Mr A.Ozaki
Dr A.Sakai
Mr N.Sumita
Prof. A.Suzuki
Assoc. Prof. Y.Tanabashi
Assoc. Prof. F.Tatsuoka
Assoc. Prof. S.Yasuda
Prof. K.Yasuhara

INTERNATIONAL ADVISORY GROUP

Prof. T.Adachi
Prof. T.Akagi
Prof. L.R.Anderson
Prof. A.S.Balasubramaniam
Prof. J.R.Bell
Assoc. Prof. D.T.Bergado
Dr E.W.Brand
Prof. B.B.Broms
Assoc. Prof. M.B.Cao
Prof. K.Fujita
Prof. M.Fukuoka
Prof. K.Gamski
Dr J.P.Giroud
Prof. G.Gudehus
Prof. R.D.Holtz
Prof. M.Hoshiya
Prof. T.S.Ingold
Prof. C.J.F.P.Jones
Prof. S.K.Kim
Dr K.Kutara
Prof. S.L.Lee
Prof. T.Matsui
Prof. A.McGown
Dr Z.C.Moh
Prof. H.B.Poorooshasb
Prof. M.V.Ratnam
Prof. R.K.Rowe
Prof. F.Schlosser
Prof. C.K.Shen
Assoc. Prof. Y.Taesiri
Assoc. Prof. F.Tatsuoka
Prof. S.Valliappan

SPONSORSHIP

The Science Council of Japan
The Japan Society for the Promotion of Science
Commemorative Association for the Japan World Exposition
The International Geotextile Society
Architectural Institute of Japan
Japan Federation of Geological Survey Enterprises Associations
Japan Landslide Society
Japan Society of Civil Engineers
Japan Society of Engineering Geology
Japanese Committee for International Society for Rock Mechanics
The Japan Chapter of the International Geotextile Society
The Japanese Society of Irrigation, Drainage and Reclamation Engineering
The Mining and Metallurgical Institute of Japan
The Society of Materials Science, Japan
The Kajima Foundation
Shimadzu Science Foundation

International Geotechnical Symposium on Theory and Practice of Earth Reinforcement / Fukuoka Japan / 5-7 October 1988
© 1988 Balkema, Rotterdam. ISBN 90 6191 820 0

Contents

2. Shallow and deep foundations

3. Slopes and excavations

4. Embankments

5. Wall structures

Special lectures

International Geotechnical Symposium on Theory and Practice of Earth Reinforcement / Fukuoka Japan / 5-7 October 1988
© 1988 Balkema, Rotterdam. ISBN 90 6191 820 0

Fabric reinforced retaining walls

Bengt B.Broms
Nanyang Technological Institute, Singapore

ABSTRACT: The design of fabric reinforced retaining walls is reviewed in the article. Methods are presented to calculate the lateral earth pressure acting on vertical and sloping walls and the evaluation of the force, spacing and length of the fabric reinforcement. The stability of the wall when the length of the fabric either increases or decreases with increasing depth also has been investigated.

1 INTRODUCTION

Investigations of the possible applications of woven high strength synthetic fabrics as reinforcement in soil first started about 25 years ago in order to increase the bearing capacity of embankments and footings or to reduce the lateral earth pressure acting on sheet pile and retaining walls. In Fig 1 is shown a fabric reinforced retaining wall and an embankment constructed of L-shaped concrete elements. The fabric reinforcement has in this application a dual function in that it both reduces

the forces initiating failure and increases the stability of the wall. The main purpose of the L-shaped face elements is protect the fabric against vandalism and direct sunlight. They can be made very light since the lateral earth pressure acting on the wall is very low as indicated in the paper. It is also possible to shotcrete the face.

Several synthetic polymers are used as reinforcement in soil such as polypropylene, polyester, polyethylene, polyamide and glass fibres. A wide variety of both woven and non-woven fabrics are available where the fibres

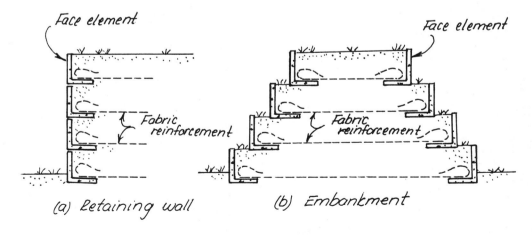

(a) Retaining wall   (b) Embankment

Fig.1 Applications of fabric as reinforcement in soil

are either bonded or interlocked. Also composite materials are used with a high resistance to environmental attacks.

Geofabric has several advantages as reinforcement in soil compared with steel or aluminium strips since the fabric has a high tensile strength and is flexible and thus not affected by large settlements. There are materials available with a tensile strength of up to 400 kN/m or more. Also the durability is high when the fabric is not exposed to direct sunlight. The surface area of the geofabric is large so that the force in the fabric can be transferred effectively to the soil. The required anchor length is thus short. The application of fabric as reinforcement in soil is, therefore, often a very economical method compared with other soil improvement methods since heavy equipment or skilled labour are not required for the construction. The reduction of the cost has in some cases been substantial, 40% or more compared with alternate methods.

The main disadvantages and limitations with geofabric when used as reinforcement in soil are the large deformations required to mobilize the tensile strength of the fabric, the lack of experience with the method especially about the long term performance as discussed in the paper and that the fabric had to be protected against direct sunlight which is particularly important for polypropylene. The deformations can be reduced by prestretching. There are also some questions about the durability (van Zanten, 1986) and that the fabric can be damaged when crushed rock or gravel is used as backfill. The sharp corners of the rock may puncture the fabric locally thereby reducing the tensile strength. The rupture strain can also be reduced but the stiffness is practically unchanged. It is therefore important that the fabric is protected against mechanical damage and chemical attacks. The material is also difficult to splice. The tensile strength of the seams can be only 50% of that of the unspliced material.

In this paper, the applications of geofabric as reinforcement in the backfill behind retaining walls and in embankments in order to reduce the lateral earth pressure and to increase the stability are reviewed. Design methods are proposed for retaining structures and high embankments reinforced with fabric based on the strength and deformation properties of the reinforcing material and on the interface friction and adhesion between the fabric and the soil.

2 STRENGTH AND DEFORMATION PROPERTIES OF GEOFARBIC

2.1 Stress-strain Relationship

Test methods have been developed to determine the strength and the stress-strain properties of geofabric used as reinforcement in soil. The factors which are of particular interest are :

* the short and long term tensile strength of the fabric when confined in the soil.

* the axial strain at working load and at failure and how it is affected by time (creep).

* the interface friction and adhesion between fabric and soil

The main uncertainties are the effect of time and the environment on the ultimate strength of the geofabric and on the bearing capacity of the soil and the increase with time of the lateral displacements and of the settlements caused by creep. Colin et al (1986) have investigated the tensile strength of fabric that has been buried in soil.

Uniaxial and biaxial tensile tests are commonly used to determine the short and long term tensile strength and the deformation properties of both woven and non-woven fabric materials (e.g. Myles, 1986). There are a number of short-comings with some of these testing methods. At, for example, strip load tests it is difficult to load the sample uniformly since the strain distribution is not uniform across the sample during the loading. Necking occurs at the centre when the ultimate strength is approached.

Accelerated creep tests (elevated temperature) are used to determine the creep rate and the long term tensile strength. However, the results from creep tests should not be extrapolated over more than two log-cycles of time as pointed out by Jewell and Greenwood (1988).

It is also well known that the confinement of the fabric in the soil improves its mechanical properties. For example, the axial strain in the fabric will be reduced considerably with increasing confining pressure. Confined tensile tests have been carried out by McGown et al (1981), El-Fermoui and Nowatzki (1982) Andrawes et al (1984) and by Chandrasekaran (1988).

A specially designed box (63.5 x 63.5 mm) was used by El-Fermoui and Nowatzki (1982) so that a normal pressure could be applied to the fabric during the testing. At the tests carried out by McGown et al (1981) and Andrawes et al (1984) a normal pressure was applied by means of rubber bellows on both sides of the fabric which was embedded in soil. The test results indicated that the stiffness and the tensile strength of the non-woven fabric increased with increasing confining pressure. The tensile strength of the woven fabric when confined was approximately the same as that of the unconfined samples. The confinement reduced somewhat the strain at failure while the stiffness of the fabric was increased. The main effect of the confinement was to reduce the lateral contraction of the fabric sample during the testing so that the stress distributed across the sample became much more uniform.

Both unconfined and confined direct tension tests have been carried out by Chandrasekaran (1988). Test results indicated that the tensile strength of the woven fabric was somewhat less than that of the single threads (yarn). Part of this difference could be attributed to the non-uniform strain distribution across the fabric samples during the testing. The confined tension tests were carried out inside a 300 mm diameter steel cylinder lined on the inside by a rubber membrane so that a confining pressure could be applied to the sand and thus to the fabric embedded in the sand. The strain distribution along the fabric was measured with strain gauges which were glued to the fabric. In Fig 2 is shown the stress-strain properties of the investigated woven polyester fabric at different confining pressures (100, 150, 200 and 300 kPa). It can be seen that the axial deformation of the fabric was reduced when the confining pressure was increased. At 30% to 50% of the ultimate strength of the fabric material, the working load, the axial strain was only 40% to 60% of the axial strain of the unconfined samples when the confining pressure was 300 kPa. The test data indicated that the ultimate strength of the fabric was not affected

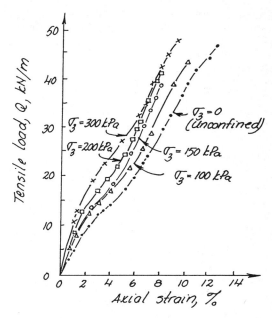

Fig.2 Tensile load versus axial strain curve for woven polyeter fabric at different confining pressures (after Chandrasekaran, 1988)

by the confining pressure.

Christopher et al (1986) investigated the effect of confinement with zero span tension tests where the distance between the two clamps is small. The test results from the zero span tension tests are very uncertain since the strain distribution is not uniform between the clamps. Slippage of the clamps affects the results as well.

2.2 Interface Friction Angle (ϕ_a')

The soil-fabric interface friction angle (ϕ_a') has been evaluated by several (e.g. Holtz, 1977; Myles, 1982; Miyamori et al, 1986; Chandrasekaran, 1988)

(1) by pull-out tests and

(2) by direct shear box tests.

A special 1.10 x 0.25 m direct shear box was designed by Holtz (1977) so that 0.15 m wide fabric strips could be tested. The normal pressure was applied with a pressure bag and the stress distribution along the fabric strip was determined with magnets glued to the fabric. The test data indicated that the interface friction angle (ϕ_a') of the investigated sand was the same as the angle of internal friction as determined by direct shear tests.

Myles (1982) used direct shear tests (100 x 100 mm) to investigate the soil-fabric interface friction. The fabric was located in the lower frame of the direct shear apparatus while sand was placed in the upper frame. The investigation indicated that the interface friction angle varied with the type of fabric material, woven and non-woven.

Miyamori et al (1986) investigated also the soil-fabric interface friction of non-woven fabric with direct shear tests. A relatively large shear box was used (316 x 316 mm). The interface friction resistance (ϕ_a') was found to be lower than the angle of internal friction of the investigated soil as determined by triaxial or direct shear tests. For dense sand the interface friction resistance was only 72% to 87% of the peak shear strength of the sand. A relatively large relative displacement was required to mobilize the peak resistance along the fabric because of the large thickness of the investigated

non-woven material.

Direct shear tests were also carried out by Chandrasekaran (1988) to determine the soil-fabric interface friction. The size of the direct shear box was 100 x 100 mm. It was found that the interface friction angle of the investigated angular sand decreased with increasing normal pressure from about 41° at a low normal pressure (50 kPa) to about 32° when the normal pressure was 500 kPa as shown in Fig 3. The angle of internal friction of the sand as determined by drained triaxial tests was 39°. The reduction of the interface friction angle with increasing confining pressure was thus substantial. The interface friction angle has also been found to be affected by size of the openings in the fabric. The openings of the investigated woven polyester fabric were about 0.1 mm while the average particle size of the sand was 1.2 mm. The corresponding interface friction angle was 34.3°. For cotton gauze where the size of the openings was about 1.0 mm, the average interface friction angle was about 40° which is approximately equal to the angle of internal friction as determined by triaxial tests (39°).

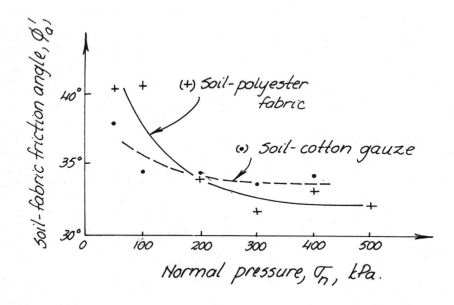

Fig.3 Variation of interface friction angle with normal pressure; woven polyester fabric and cotton gauze

3 FABRIC REINFORCED SOIL

The behaviour of the fabric reinforced soil can be demonstrated by triaxial tests as illustrated in Fig 4 where the fabric reinforcement is placed in horizontal layers in the soil as discussed by Broms (1977). This test method was first used by Schlosser and Long (1974) to investigate the effect of aluminium foil on the strength and deformation properties of sand. The spacing of the horizontal aluminium disks and the relative density of the sand were varied. The test results show that both the strength and deformation properties of the sand were improved noticeably. Two different failure modes were observed. At a low confining pressure, failure occurred by slippage along the reinforcement when the interface friction was fully mobilized (Fig 5). In this case the ultimate strength is governed by the interface friction angle. The shear strength of the reinforced soil was characterized by an apparent friction angle which is larger than the effective angle of internal friction of the soil (ϕ'). At

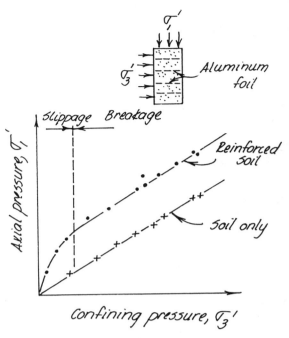

Fig.5 Failure curves for triaxial reinforced and unreinforced sand samples (after Schlosser and Long, 1974)

a high confining pressure, the ultimate strength of the samples is governed by the tensile strength of the disks. In this case, the ultimate strength of the reinforced soil is characterized by an apparent effective cohesion (c_r)

$$c_r = \frac{T}{D} \sqrt{K_p} \qquad (1)$$

where T is the tensile strength of the reinforcement (kN/m), D is the vertical spacing of the reinforcement and K_p is the coefficient of passive earth pressure. The friction angle corresponds to the effective angle of internal friction (ϕ') as illustrated in Fig 5.

Long et al (1983) investigated the stress distribution in triaxial samples reinforced with horizontal bronze discs. They found that the stress distribution along the discs was not uniform. The maximum interface friction resistance was mobilized near the edge of the disks. At the periphery of the sample the shear stress was significantly lower than the peak resistance. At the centre the resistance was zero since the

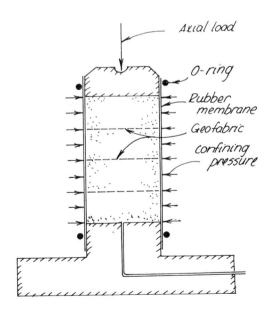

Fig.4 Triaxial tests with fabric reinforcement

relative displacement between the reinforcement and the soil was zero due to symmetry.

Broms (1977) investigated the stress-strain properties of triaxial samples reinforced with fabric disks. At these tests, the relative density of the sand and the spacing of the fabric layers were varied. The experiments indicated that the compressive strength of the cylindrical samples increased with decreasing spacing of the reinforcement and with increasing confining pressure. It was shown that the restraint offered by the reinforcement was equivalent to that of a confining pressure. The coefficient of lateral earth pressure K_b next to the fabric was estimated from the relationship

$$K_b = \frac{1}{1 + 2 \tan^2 \phi'_a} \qquad (2)$$

where ϕ'_a is the effective interface friction angle.

Halfway between two fabric layers the coefficient of active earth pressure K_a will govern since the shear stress along a horizontal plane through the sample is equal to zero due to symmetry.

$$K_a = \frac{1 - \sin \phi'}{1 + \sin \phi'} \qquad (3)$$

It was proposed to use an average value K_{av} on the coefficient of lateral earth pressure in the calculations of the equivalent confining pressure.

$$K_{av} = 1/2 \ (K_a + K_b) \qquad (4)$$

The vertical stress distribution along the fabric is then given by the equation

$$\sigma'_v = \sigma'_{vo} \ e^{\frac{2 \tan \phi'_a}{DK_{av}}(r_o - r)} \qquad (5)$$

where σ'_{vo} is the vertical stress at the perimeter of the sample ($r = r_o$), ϕ'_a is the effective interface friction angle, r is the distance from the centre of the sample and D is the spacing of the fabric layers. Eq (5) thus indicates that the vertical normal stress increases rapidly towards the centre of the sample.

It has thus been assumed in the derivation of this equation that the total tension force in the fabric at the periphery of the sample is equivalent to a uniformly distributed confining pressure and that the shear strength is fully mobilized along the soil-fabric interface. This is not the case at the centre of the sample where the relative displacement of the soil with respect to the fabric reinforcement is equal to zero. The increase of the bearing capacity depends mainly on the friction resistance along the reinforcement and thus on the interface friction angle ϕ'_a.

This friction angle is also affected by the stiffness of the reinforcement. It is lower for a non-woven than for a woven fabric.

The stress distribution in triaxial samples reinforced with aluminium foil, aluminium mesh and non-woven bonded fabric has been investigated by McGown et al (1978). They classified the reinforcement in the soil as inextensible and extensible depending on the failure mode of the samples. The reinforcement was classified as relatively inextensible when the reinforcement failed before the soil. In this case, the behaviour of the soil is brittle and the failure occurs suddenly without much warning. The term relatively extensible reinforcement was used when the sample failed before the reinforcement. The failure of the soil occurs in this case gradually. The behaviour is ductile and the reduction of the capacity beyond the peak strength is small.

Chandrasekaran (1988) has also investigated the behaviour of fabric reinforced soil using triaxial tests. The 100 mm or 200 mm diameter samples with angular medium coarse sand which were investigated were reinforced with horizontal fabric disks. Both woven and non-woven fabrics were tested. The spacing of the disks and the relative density of the sand were varied. The strain distribution along the fabric was measured with strain gauges attached to the fabric at different locations.

Typical stress-strain curves for the 100 mm diameter samples reinforced with woven polyester fabric disks are shown in Fig 6 at an applied confining pressure of 50 kPa. It can be seen that the ultimate strength increased with increasing number of layers and that the maximum increase was over 200% when

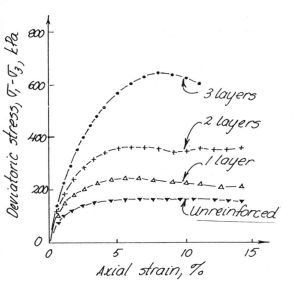

Fig.6 Stress-strain curves obtained for unreinforced and polyester fabric reinforced samples; confining pressure = 50 kPa (after Chandrasekaran, 1988)

three layers were used. Failure occurred by slippage along the fabric. It was found that stiffness of the sample was increased by the fabric as well as the deformation required to mobilize the peak strength.

The relative increase of the strength with decreasing spacing of the fabric layers has been plotted in Fig 7 at different confining pressures. It can be seen that the relative increase of the bearing capacity increased with increasing confining pressure.

The stress-strain relationships for the 200 mm diameter samples reinforced with woven polyester fabric disks are shown in Fig 8. The spacing of the layers was varied. The peak strength of the samples with three layers of reinforcement was governed by the tensile strength of the fabric as can be seen by the rapid drop of the capacity when the axial strain exceeded 12%. The ultimate strength of the sample reinforced by a single layer was governed by the interface friction angle (ϕ_a') rather than by the tensile strength of the fabric. The decrease of the resistance beyond the peak resistance was small.

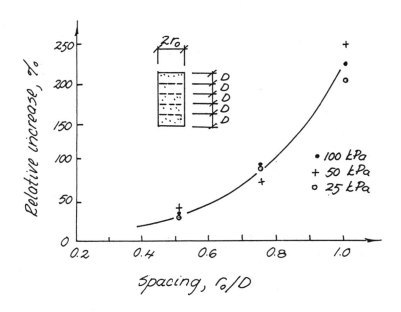

Fig.7 Relative increase in peak axial load of reinforced samples with decrease in fabric spacing

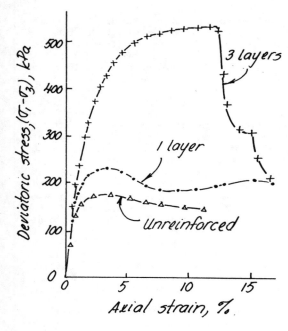

Fig.8 Stress-strain relationship for 200 mm diameter triaxial samples; confining pressure = 50 kPa

The strain distribution along the fabric reinforcement has been plotted in Fig 9. It can be seen that the axial strain in the fabric increased towards the centre of the sample. The gradient corresponds to the shear stress distribution along the fabric. The maximum shear stress and the maximum gradient were observed along the perimeter of the sample since the relative displacement there between the fabric and the soil was large. At the centre the shear stress is equal to zero because of the small relative displacement between the fabric and the soil.

The non-uniform shear stress distribution can be taken into account by assuming that the mobilized friction resistance along the fabric increases linearly from the centre of the sample toward the periphery and that the maximum mobilized interface friction angle is $(\alpha\phi_a')$ at the periphery of the sample. It has been observed that the mobilized friction angle $(\alpha\phi_a')$ for woven fabrics was less than the interface friction angle (ϕ_a') as determined by

e.g. direct shear tests. This reduction can be expressed by the coefficient α. Thus

$$\tau_a = \sigma_v' \frac{r}{r_o} \tan \alpha\phi_a' \qquad (6)$$

where r is the radical distance from the centre of the sample, r_o is the radius, ϕ_a' is the interface friction angle and α is a reduction factor which can be as low as 0.5 for non-woven fabrics. For woven fabrics test data indicate that $\alpha = 1.0$.

The vertical stress distribution along the fabric can then be estimated from the equation

$$\sigma_v' = \sigma_{vo}' \, e^{\left[\frac{\tan \alpha\phi_a' \, (r_o^2 - r^2)}{DK_{av} \qquad r_o}\right]} \qquad (7)$$

where σ_{vo}' is the vertical effective stress at the perimeter of the sample $(\sigma_{vo}' = \sigma_{ho}' \, K_a)$, D is the spacing of the fabric layers and $K_{av} = 1/2 \, (K_a + K_b)$.

A comparison between calculated and measured ultimate (peak) strengths as

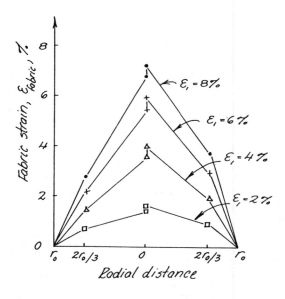

Fig.9 Typical plot of fabric strain gradient along the radius of the fabric disc; triple layered sample

evaluated from Eq (7) is shown in Figs 10a and 10b for samples with 100 mm and 200 mm diameter, respectively. The agreement with the modified calculation method has improved considerably compared with the original model (Eq. 5). The test results data thus indicate the importance of the relative displacement of the fabric with respect to the soil and of the mobilized interface friction angle. The results show that the relative displacement had to be considered in the design of fabric reinforced retaining walls.

An attempt has also been made to analyze the stress distribution in the fabric reinforced samples, using the finite element method (FEM) and assuming an hyperbolic stress-strain relationship for the soil. One major limitation of

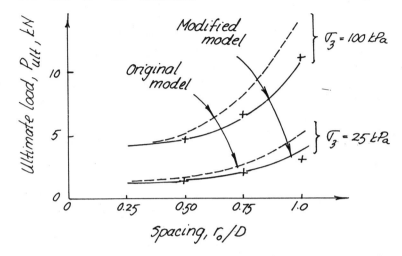

Fig.10a Comparison between experimental and computed peak axial loads; 100 mm diameter samples

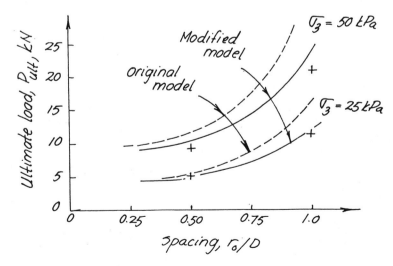

Fig.10b Comparison between experimental and computed peak axial loads; 200 mm diameter samples

the method is that the dilatancy of the soil cannot be considered. The calculated peak strength of the fabric reinforced samples by FEM was too high as well as the calculated stiffness of the soil.

The mobilized interface friction angle ($\alpha\phi_a'$) and thus the coefficient α were found to be affected by the fabric type, woven or non-woven. A much lower value (0.45 to 0.55) on the coefficient α was observed for the non-woven fabric due to the low stiffness compared with the investigated woven material. Also the spacing (D) of the layers affected the mobilized interface friction angle as well as the confining pressure.

Two different modes of failure was observed depending on the size of the samples, the confining pressure and the tensile strength of the fabric. Compression failure by slippage of the soil along the fabric occurred when the confining pressure was low and the spacing of the fabric layers was large. It was found that the axial strain increased gradually with increasing applied load. Tensile failure caused by tearing of the reinforcement occurred at the centre of the sample at a high confining pressure and a close spacing of the fabric layers. This type of failure occurred rapidly without previous warning. The capacity of the samples decreased suddenly when the sample was deformed further.

4 FABRIC REINFORCED RETAINING WALLS

4.1 Design Principles

Fabric can be used as reinforcement in the backfill behind retaining walls to increase the stability and to reduce the lateral earth pressure acting on the wall as discussed by e.g. Hausman (1976), Broms (1977), Ingold (1982) and Jones (1985). The spacing of the fabric layers is primarily governed by the long-term strength of the fabric, the interface friction or adhesion and by the total lateral earth pressure acting on the retaining structure. In the design of a fabric reinforced wall it is important to consider the variability of the strength of the fabric, the variability of the applied load, the consequences in case of a failure and the costs for the repair or reconstruction of the damaged wall. This can be done by using partial factors of safety.

The L-shaped elements shown in Fig 11

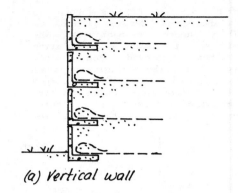

(a) Vertical wall

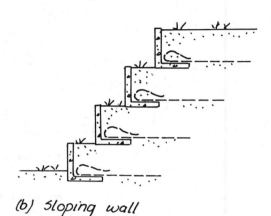

(b) Sloping wall

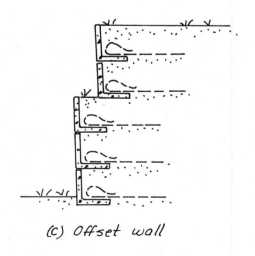

(c) Offset wall

Fig.11 Applications of geofabric as reinforcement for retaining walls and slopes

can either be placed just above one another (Fig 11a) or be staggered (Fig 11b). Another possibility is shown in Fig 11c where two or three elements are grouped together. The L-shaped wall elements can be made relatively light as discussed in the following. The main function of the elements is to protect the fabric against vandalism and direct sunlight. It is also possible to fold the fabric and to protect the surface with shotcrete reinforced with wire mesh.

Granular soil (sand, gravel or rock fill) is usually used as backfill material because of the high interface friction angle which corresponds approximately to the residual angle of internal friction of the soil. Also silt, marl and stiff clay have been utilized (e.g. Tatsuoka et al, 1986; Wichter et al, 1986).

One of the first application of fabric as reinforcement in soil was probably in France in 1971 where the method was used for a motorway (A15). In the United States, fabric was first used as reinforcement in soil in 1974 by the U.S. Forest Service in Oregon State. Bell and Steward (1977) have investigated both model and full-scale fabric reinforced walls. An experimental wall has been constructed and tested relatively early by the New York State Department of Transportation. Also the Colorado Division of Highways has built an experimental fabric reinforced wall (Bell et al, 1983). In Australia, a 20 m long fabric reinforced wall with a maximum height at 1.8 m has been constructed in Sidney by the N.S.W. Public Works Department. In Sweden, fabric reinforced model walls have been tested by Lindskog et al (1975), Broms

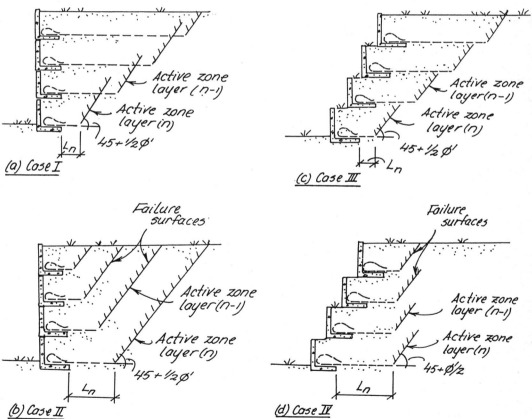

Fig.12 Fabric reinforced retaining wall (failure mechanisms)

13

(1975), Davidson and Ekroth (1975) and by Holtz and Broms (1977). In Germany Wichfer et al (1986) have investigated a 4.8 m high fabric reinforced wall. Extensive testing of fabric reinforced walls have also been carried out in England by e.g. John (1986).

There are four basic cases of fabric reinforced retaining walls as illustrated in Fig 12 depending on the length of the fabric reinforcement and the location of the L-shaped elements. In Fig 12a (Case I) the length of the geofabric behind the vertical wall decreases with depth. This variation of the length corresponds usually to the shape of the excavation required for the construction of a fabric reinforced wall. The volume of soil, the active, zone, which is affected when the wall fails by sliding along the base of the different layers is also shown in the figure. When sliding occurs along the bottom layer all the overlying layers will be affected. The failure surface will in this case be step-shaped and follow the underside of the overlying fabric layers as shown.

In Fig 12b (Case II), the length of the fabric increases with depth. In this case the volume of soil affected by the sliding will not be affected by the overlaying fabric layers. For the bottom layer the inclination of the failure plane that extends up to the ground surface is approximately $(45^{\circ} + \phi'/2)$. It should be noted that the shape and the inclination of the failure surface is affected somewhat by friction resistance along the wall and by the lateral deformations of the wall. It is proposed that this layer should be designed to resist the total active Rankine earth pressure for the full height of the wall neglecting the wall friction.

In Fig 12c and 12d (Cases III and IV) are shown sloping walls where the L-shaped elements are offset with respect to one another. The slope reduces considerably the lateral earth pressure acting on the wall. The length of the reinforcement decreases with depth in Fig 12c (Case III). In this case it is important to consider separately the stability of each row of elements. In Fig 12d (Case IV) the length of the fabric in the different layers increases with depth so that the slip surfaces generated by the different layers overlap.

4.2 Stress Distribution

A section through a fabric reinforced wall is shown in Fig 13. The lateral earth pressure acting on the wall elements will be small since part of the total lateral earth pressure is resisted by the fabric. The reduction of the lateral earth pressure by the fabric reinforcement which is affected by the interface friction angle or the interface cohesion along the fabric can be estimated from the equilibrium of the forces acting on a soil element located between two adjacent fabric layers as shown in the figure. For cohesive soils the lateral earth pressure in the backfill may change with time since it is affected by the porewater pressures in the soil. The undrained shear strength of the soil is normally used during the construction while the effective angle of internal friction (ϕ') as determined from e.g. drained triaxial or direct shear tests is applied for the long term conditions. A considerable increase of the force in the fabric reinforcement can occur during a heavy rainstorm when the porewater pressure in the soil increases (e.g. Tatsuoka et al, 1986).

The friction resistance along the fabric (f) will for a granular soil be proportional to the effective normal pressure (σ_v').

$$f = \sigma_v' \tan \alpha\phi_a' \qquad (8)$$

where $\tan \alpha\phi_a'$ is the coefficient of interface friction of the soil with respect to the fabric. It has thus been assumed in Eq (8) that the relative displacement between the fabric and the soil will be sufficient to mobilize fully the peak interface friction resistance $(\alpha\phi_a')$. For a woven fabric $(\alpha\phi_a')$ will correspond to (ϕ_d') the effective interface friction angle of the soil (eg. Holtz, 1977, 1985). For a non-woven material $\alpha\phi_a'$ can be less than ϕ_a' depending on the large deformation required to mobilize the peak interface friction (Chandrasekaran, 1988).

For a cohesive soil the resistance along the fabric is related to the undrained shear strength c_u as

14

determined by triaxial tests (UU–tests),
field vane tests or unconfined
compression tests.

$$f = \alpha c_u \qquad (9)$$

where α is a reduction factor. For
stiff clays a value of 0.4 to 0.6 is
proposed.

The stress distribution along the
reinforcement can then be estimated as
illustrated in Fig 13 as discussed by
Broms (1978). The friction resistance
(f) increases the lateral confining
pressure on a soil element located
between the two fabric layers from σ_h' at
the distance x from the L–shaped
elements to $(\sigma_h' + \Delta\sigma_h')$ at $(x + \Delta x)$. At
equilibrium

$$2 \ f \ dx = D \ d\sigma_h' \qquad (10)$$

where f is the friction resistance and D
is the spacing of the fabric layers. It
has thus been assumed that the resulting
friction force (2fdx) is distributed
uniformly over the height D between two
adjacent fabric layer. It should be
noted that the vertical and the lateral
earth pressure acting in the soil σ_v' and
σ_h', are related for granular soils

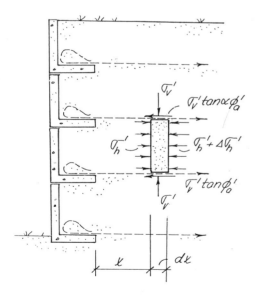

Fig.13 Stress distribution in fabric
reinforced soil

through the lateral earth pressure
coefficient K_{av} ($\sigma_h' = K_{av}\sigma_v'$) which is
somewhat larger than the Rankine
coefficient of active earth pressure
(K_a). The stress conditions in the soil
just at the fabric corresponds to the
earth pressure coefficient K_b [$K_b = 1/(1 + 2 \tan^2\phi_a')$] while at the centre halfway
between two layers the stress conditions
will be governed by the Rankine
coefficient of active earth pressure K_a.
It is, therefore, proposed to use an
average value K_{av} in the calculations,
$K_{av} = 1/2 \ (K_a + K_b)$].

From Eqs (8) and (10)

$$d\sigma_v' = \frac{2 \ \sigma_v' \ \tan \alpha\phi_a'}{DK_{av}} \ dx \qquad (11)$$

The solution of this differential
equation is

$$\sigma_v' = \sigma_{vo}' \ e^{\dfrac{2 \ x \ \tan \alpha\phi_a'}{DK_{av}}} \qquad (12)$$

where σ_{vo}' is the bearing capacity at the
backface of the wall and x is the
distance from the wall.

The bearing capacity of the fabric
reinforced wall thus increases rapidly
with increasing distance from the wall
as illustrated in Fig 14, where the
bearing capacity σ_v' has been calculated
for $\phi_a' = 30°$, $35°$ and $45°$. The
calculated relative bearing capacity
will increase ten times at a distance of
0.43D, 0.63D and 0.93D from the wall
when the interface friction angle ϕ_a' is
$40°$, $35°$ and $30°$, respectively. Thus a
relatively short distance is required to
develop the tensile resistance of the
fabric. The face elements can therefore
be designed to resist a lateral earth
pressure which is considerably smaller
than the Rankine active earth pressure.

The corresponding equation for a
cohesive soil is

$$d\sigma_v = \frac{2\alpha c_u}{D} \ dx. \qquad (13)$$

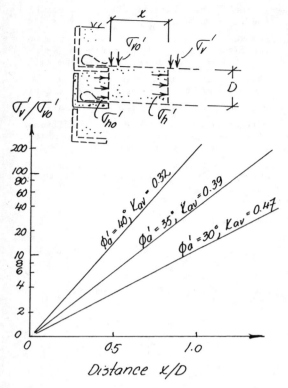

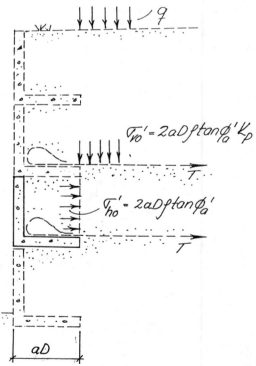

Fig.14 Increase of σ'_v/σ'_{vo} and σ'_h/σ'_{ho} with increasing distance x/D from the face of the retaining structure

The solution of this differential equation is

$$\sigma_v = \frac{2\alpha c_u x}{D} + \sigma_{vo} \qquad (14)$$

where σ_{vo} is the bearing capacity of the wall at the backface of the wall elements. Thus the bearing capacity increases linearly with increasing distance from the wall. This increase is less than that for a granular fill.

4.3 Design of Wall Elements

In Fig 15 is shown the stress conditions just at the backface of the L-shaped elements. For a granular material the equivalent confining pressure is there equal to $\sigma'_{ho} = 2aD\rho \tan\phi'_a$ due to the

friction resistance of the soil contained in the elements. At e.g. a = 1.0, D = 1.5 m, ρ = 20 kN/m^3 and ϕ_a = 35° the equivalent confining pressure σ'_{ho} is 42.0 kPa. The corresponding bearing capacity of the soil σ'_{vo} is $K_p\sigma'_{ho}$ or 77.5 kPa at K_p = 3.69. This bearing capacity corresponds approximately to the weight of a 8 m high fill at ϕ' = 35°. Thus the friction alone is not sufficient to prevent failure of the wall by slippage along the reinforcement if the height of the wall exceeds 5 to 10 m depending on the shear strength of the soil and the shape of the elements. It is therefore necessary to bolt or to glue the fabric to the face elements. Another possibility is to fold the fabric as illustrated in Fig 16a so that the force in the fabric can be transferred to the soil. The face elements can be made relatively light since the lateral earth pressure is low. It is proposed that the stem should be designed to resist a

Fig.15 Stress conditions at face elements

lateral earth pressure equal to ($K_a \sigma_v'$ - aρD tan ϕ_a').

The construction of the wall is greatly simplified if the fabric if folded as shown in Fig 16a and not connected to the wall. The main disadvantage with this method is the relatively large deformations required to mobilize the tensile strength of the fabric. Compaction of the granular material around the folded fabric is essential in order to reduce the lateral deformations as much as possible. The folds can be filled with coarse sand or

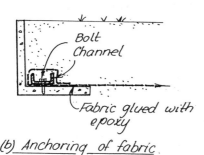

(a) Folding of fabric

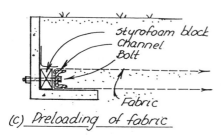

(b) Anchoring of fabric.

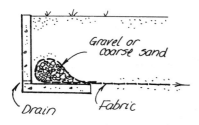

(c) Preloading of fabric

Fig. 16 Anchoring of fabric

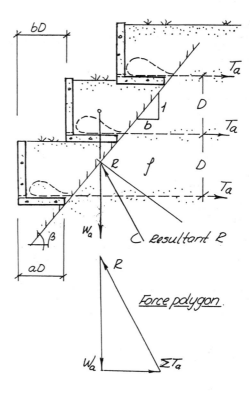

Fig. 17 Stability of face elements

gravel in order to improve the drainage since an excess pore water pressure will increase the lateral earth pressure.

The fabric has been glued and bolted to the face elements in Fig 16b. The deformations required to mobilize the tensile strength of the fabric is in this case small. A possible disadvantage with this method is that a damaged wall element is difficult to repair.

The fabric can be preloaded with the arrangement shown in Fig 16c in order to reduce the lateral displacement of the wall elements caused by creep in the fabric. This wall will be relatively easy to repair.

The design of the L-shaped face elements for a sloping wall is illustrated in Fig 17. The fabric will prevent sliding of the elements along an inclined failure surface which corresponds to the average slope (b) of the wall as shown. The average slope depends on the offset of the elements (bD) and on the spacing (D) of the fabric layers.

The average horizontal thickness of this zone is (aD + 0.5bD) where aD is the width of the base of the L-shaped elements. Each fabric layer had to prevent a block with a mass W_a from sliding which can be estimated from

$$W_a = (a + 0.5b) D^2 \rho \qquad (15)$$

The force T_a in the fabric which is required to prevent the L-shaped elements from sliding is then equal to

$$T_a = W_a \tan (\beta - \phi) \qquad (16)$$

This force decreases with increasing inclination of the slope as can be expected. At e.g. $\beta = 60^o$ and $\phi = 30^o$ then $T_a = 0.58 \, W_a$. The force in the fabric depends also on the height of the elements (D). At D = 1.5 m, $\rho = 20$ kN/m^3 and $W_a = 43.0$ kN/m then $T_a = 24.9$ kN/m. This force in the fabric had to be transferred to the L-shaped elements and to the soil.

The friction resistance (F_a) of the fabric within the elements is $2aD^2\rho\tan\phi_a'$. At, for example, aD = 1.0 m, $\rho = 20$ kN/m^3, D = 1.5 m and $\phi_a' = 30^o$ then $F_a = 34.3$ kN/m. The corresponding factor of safety (F_s) is 1.38 (34.3/24.9) which is not sufficient. It is desirable that the factor of safety should be at least 1.5 to 2.0. The factor of safety can be increased by folding or bolting the fabric to the elements (Fig 16a and 16b).

The L-shaped elements can even be replaced entirely by folding the fabric as illustrated in Fig 18. The different fabric layers will in this case function as a series of sandbags. The resulting tension in the fabric is governed by the radius (R) of the folds and by the total overburden pressure (Hρ+q) where H is the height of the wall and q is the surcharge load. The maximum tension in the fabric in the bottom layer can then be calculated from

$$T_b = K_a (H\rho + q)R \qquad (17)$$

where K_a is the coefficient of active earth pressure. At H = 10 m, $\rho = 20$

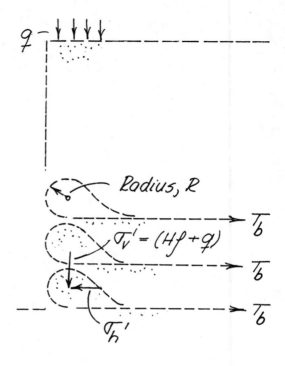

Fig.18 Design of folds

kN/m^3, $K_a = 0.3$ and R = 0.3 m then $T_b = 19.8$ kN/m. Thus a relatively high tension can develop in the bottom layer of the folded fabric of a high wall. It has been assumed in the derivation of Eq (17) that the shear resistance of the soil is fully mobilized. The tension in the fabric can be reduced by filling the folds with a granular material with a high angle of internal friction such as coarse sand or gravel or by decreasing the radius of the folds.

The fabric had to be protected in this case against direct sunlight by spraying the surface with shotcrete or by a precast concrete wall as shown in Fig 19. The required thickness of the shotcrete cover is 50 to 75 mm. Wire mesh can be used as reinforcement. Experience indicates that shotcrete cover can withstand relatively large differential settlements when reinforced. The precast concrete elements can be made very light since their only function is to protect the fabric against direct sunlight and vandalism. The face elements can be anchored to the fabric with steel or

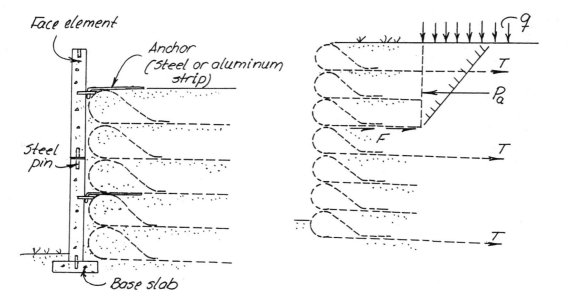

Fig.19 Light concrete slabs as face elements

Fig.20 Required length of fabric reinforcement

aluminium straps. The space between the wall and the face elements could be filled with loose sand to eliminate any voids behind the wall. The sand will also prevent loss of material from the folds in case the fabric ruptures or tears locally. The resulting lateral pressure acting on the face elements will be low due to arching.

It is not necessary to extend all the fabric layers the full length if the tensile resistance of the remaining layers is sufficient especially in the upper part of the wall as illustrated in Fig 20. However, the stability of each level had to be checked with respect to a failure surface extending below the fabric reinforcement as shown.

4.4 Design of Fabric Layers

The axial force in the geofabric depends mainly on the inclination of the wall and on the length of the fabric reinforcement (Cases I, II, III and IV). In Case I, where the length of the fabric reinforcement decreases with depth and the wall is vertical the stress distribution in the geofabric

will be as shown in Fig 21a. For a vertical wall the inclination of the critical failure surface is usually assumed to be $(45 + \phi'/2)$. In France, a spiral shaped failure surface is often used which can be approximated by two straight lines. The difference between the two methods is small.

The force in the fabric will increase with increasing distance from the two ends and reach a maximum approximately where the potential failure surface from the bottom element intersects the different layers. For a vertical wall the tension in the fabric reaches a maximum at a section that is inclined at approximately an angle $(45 + \phi'/2)$ with the vertical. For a sloping wall, the critical section is located further back into the slope.

It is proposed to calculate the axial force in the fabric as shown in Fig 21b. The indicated constant earth pressure distribution for a vertical wall is similar to that proposed by Terzaghi and Peck (1967) for braced excavations. Model tests and observations on instrumented fabric reinforced walls indicate that the force in the fabric

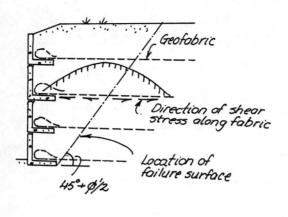

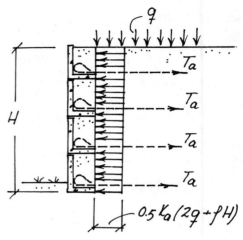

(a) Stress distribution along fabric

(b) Stress distribution along wall

Fig.21 Design of retaining wall
(Case I)

does not increase linearly with depth and that it is approximately constant (e.g. John, 1979, 1986). It is thus proposed that the fabric should be designed for a lateral earth pressure $\sigma_h' = 0.5\ K_a\ (2q + \rho H)$ where H is the total height of the wall, K_a is the Rankine coefficient of active earth pressure and ρ is the unit weight of the backfill material. It is proposed to use the critical state angle of friction in the calculation of the active earth pressure because of the large deformations required to mobilize the resistance of the fabric reinforcement. This corresponds to the residual friction angle at large deformations of the soil. This is a conservative assumption. The total lateral earth pressure is equal to the total active Rankine earth pressure.

Gourc et al (1986) and others have suggested that the vertical displacement of the fabric at the critical failure surface in the backfill could also be considered in the calculations. This displacement will increase the stabilizing effect by the fabric. It is proposed that this effect should not be taken into account in the design since a relatively large displacement is required to develop the maximum

resistance of the fabric.

The total lateral force at a section though the backface of the L-shaped elements will be resisted by the fabric reinforcement and by the friction along the base of the wall. The required spacing of the fabric (D) for a vertical wall can be calculated from the relationship.

$$D = \frac{T_a}{0.50\ K_a\ (2\ q\ +\ \rho H)} \qquad (18)$$

where T_a is the allowable long term tension in the fabric and q is the unit load acting on the surface of the backfill.

The design of the fabric reinforcement when the length of the layers decreases with depth is shown in Figs 22a and 22b. The different fabric layers should be designed to resist a tension force that corresponds to the active Rankine earth pressure as illustrated in Fig 22a. Each fabric layer had to resist the total lateral earth pressure acting above that layer as shown in Fig 22b. The bottom layer should be designed to resist a lateral force that corresponds to the total active Rankine earth pressure for the full height of the wall.

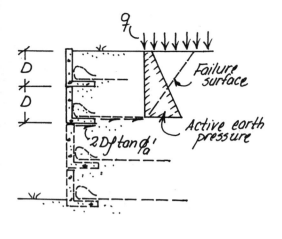

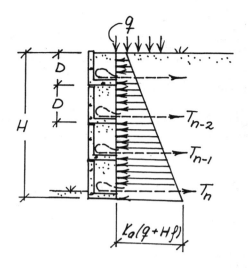

(a) Local failure

(b) Stress distribution along wall

Fig.22 Design   of   retaining   wall
(Case II)

The force in the fabric for a sloping wall is illustrated in Fig 23a and 23b (Case III). The fabric reinforcement should prevent failure along different possible failure surfaces in the soil. When the ground water level is low and the effective cohesion (c') is equal to zero then the soil located above an inclined plane steeper than φ' will be unstable. This plane thus governs the required length of the fabric.

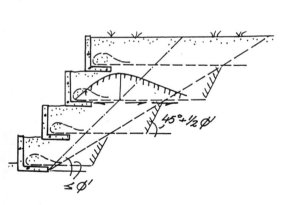

(a) Stress distribution along fabric

(b) Stress distribution along wall

Fig.23 Design   of   retaining   wall
(Case III)

The force in the fabric (T_n) will vary along the length and with the inclination of the failure plane. This force can be calculated from the weight of the soil mass (W_n) above the assumed failure plane and from the direction of the resultant (R) to the normal force and the shear force along the failure plane as shown in Fig 24. At equilibrium

$$\Sigma T_n = W_n \tan (\beta - \phi') \qquad (19)$$

where β is the inclination of the assumed failure plane. The total tension in the fabric $\Sigma T/H^2 \rho$ has been plotted in Fig 25 as a function of the inclination of the slope (b) for different values of the effective angle of internal friction ϕ' of the soil. It can be seen from the figure that the total force in the fabric reinforcement decreases rapidly with decreasing inclination of the slope (b) and with increasing angle of internal friction as shown in Fig 25. At e.g. b = 0.5 which corresponds to a slope of 63.4 degrees $\Sigma T_n/H^2 \rho = 0.03$. The total required

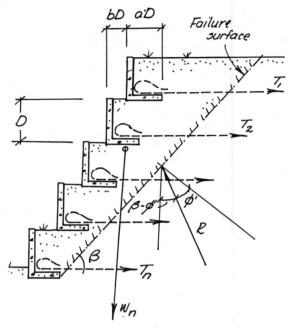

Fig.24 Stability of a sloping wall

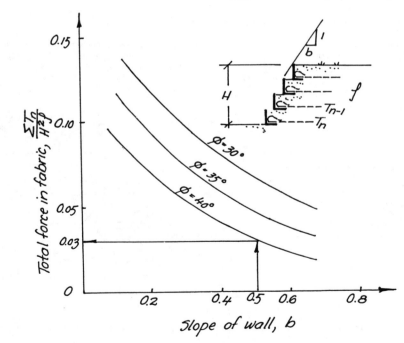

Fig.25 Force in fabric reinforcement for a sloping wall

22

anchor force ΣT_n is 60 kN/m at $\phi' = 40°$, H = 10 m and ρ = 20 kN/m³.

It is proposed to distribute the total lateral earth pressure uniformly between the different fabric layers. When one layer becomes overloaded a redistribtion of the applied load will occur due to the ductility of the fabric. At a vertical spacing of 2 m the maximum tension in each layer is 12 kN/m (60/5) since there are five layers in this case. At b = 0.3 which corresponds to a slope of 73.3 degrees, $\Sigma T_n/H^2\rho$ = 0.097. At $\phi' = 30°$, H = 10 m and ρ = 20 kN/m³, ΣT_n = 194 kN/m. The maximum tension in each layer is 38.8 kN/m (194/5).

The stress distribution along a sloping wall when the length of the reinforcement increases with depth is shown in Fig 26. Each layer should to be designed to resist a lateral earth pressure that corresponds to the height of the wall above that layer. For the second layer the height is 2D as shown in Fig 26. Fig 25 can be used to estimate the total lateral force. The height 2D should be used instead of H in this case.

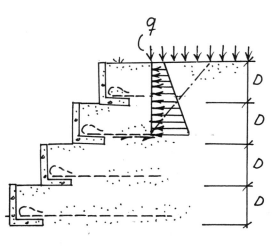

Fig.26 Required tensile strength of. fabric reinforcement-sloping wall

4.5 Length of Fabric Reinforcement

The required length of the fabric reinforcement can be determined as illustrated in Fig 27. The length should be sufficient so that the average slope of the failure surface extending beyond the fabric corresponds to the effective angle of internal friction of the soil (ϕ'). In Figs 27a and 27b are shown the required length for a vertical wall in the case the length of the fabric decreases or increases with depth, respectively (Cases I and II). The fabric reinforcement forces the soil and the wall located above the failure surface to move as a unit. This block of soil will be stable as long as the average slope of the failure surface is less than ϕ' when the effective cohesion of the soil (c') is small. This is a conservative assumption since the effective cohesion increases the stability. It is proposed to neglect the effective cohesion for permanent structures since the effective cohesion decreases with time and is very difficult to determine.

The face of the wall is sloping in Cases III and IV. The length of the fabric layers can either decrease or increase with increasing depth as illustrated in Figs 27c and 27d, respectively. The potential failure surface for the bottom layer is also indicated in the figure and the corresponding active zone. Failure will in this case occur along the base of the bottom layer. The failure surface will extend from the end of the fabric layers up to the ground surface. The inclination is ($45° + 1/2\ \phi'$) when the wall friction is small and can be neglected.

The length of the different layers is short in Cases II and IV compared with Cases I and III but the force in the fabric will be larger. Cases I and III will in general be more economical for cut slopes where the costs for the excavation of the soil is large compared with the cost for the fabric. Cases II and IV where the length of the fabric reinforcement increases with depth is normally more economical for fill slopes where excavation is not required. High strength fabric or several layers might be required in the bottom layer in order to resist the high lateral force caused by the total active earth pressure for the full height of the wall.

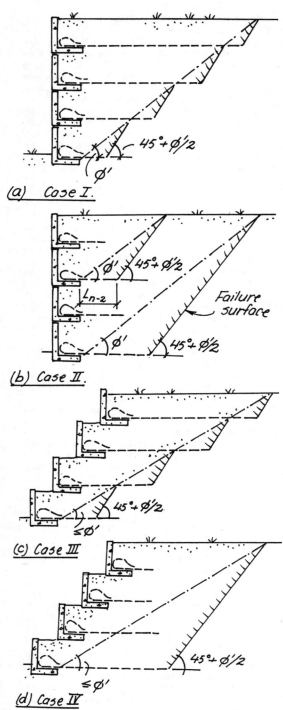

(a) Case I.

(b) Case II.

(c) Case III

(d) Case IV

Fig.27 Length of reinforcement

4.6 Bearing Capacity

It is also important to check that the bearing capacity of the soil below the wall is sufficient as illustrated in Fig 28a. The total weight (W_a) of the wall and of the soil contained in the elements will mainly be carried through the vertical stems of the L-shaped elements. The load will be vertical because of the fabric reinforcement that resists the lateral earth pressure acting on the wall.

The bearing capacity of the wall elements should also be checked when the wall is sloping as illustrated in Fig 28b. The load W_b which is transferred to the soil will be small in this case since the fabric has been designed to carry the weight of the L-shaped elements and the soil contained in the elements. It is also important to check the bearing capacity of the soil for the full height of the embankment. This can be done by assuming that the failure takes place along a circular slip surface below the elements. It should be noted that the fabric reinforcement improves the bearing capacity of the underlying soil also for this case.

4.7 Lateral Displacements

The lateral displacement of a fabric reinforced wall is often the largest at the top of the wall. The displacement at the bottom is usually small even close to failure. The lateral displacement is generally small at working loads. The maximum displacement increase rapidly as the ultimate capacity of the wall is approached.

The lateral displacements depend to a large extent on the type of fabric used, woven or non-woven. The maximum displacement is also affected by creep of the fabric material and thus by the ground temperature which is normally relatively constant except close to the wall. The lateral displacement caused by creep increases approximately with log (t). The creep of polypropylene fabric is usually larger than that of polyester but the initial deformation is smaller. The creep of polyaramid is very low. It is therefore important that creep tests are carried out over a range of temperatures.

The finite element method (FEM) has been used with some success to predict

24

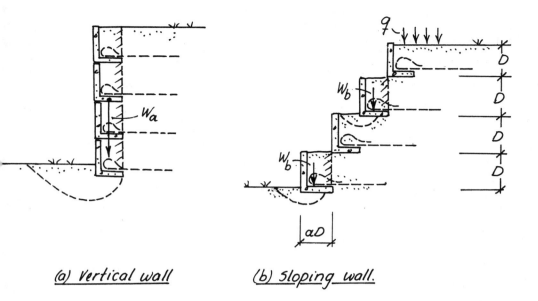

(a) Vertical wall   (b) Sloping wall.

Fig.28 Bearing capacity of wall
elements

the lateral displacements at working loads using a non-linear stress-strain relationship for the soil. The accuracy of the method depends mainly on how accurate the deformation properties of the soil (E and v) and of the fabric can be determined.

It is proposed to estimate the lateral displacement of a fabric reinforced wall from the load distribution in the fabric. The maximum tension in the fabric T_{max} can be estimated as

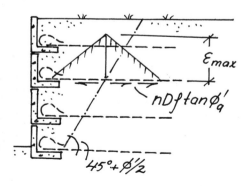

Fig.29 Strain distribution in fabric

described earlier by e.g. the Rankine earth pressure theory. This maximum force corresponds to an axial strain ϵ_{max} in the fabric which can be determined from confined tension tests. At working loads the axial strain may be only 40 to 60% of the strain determined from unconfined tension tests.

The tension in the fabric will decrease on both sides of the critical failure surface through the backfill. The reduction depends on the interface friction resistance ($\alpha\phi_a'$) and on the effective overburden pressure $\sigma_v'(= nD\rho)$.

The required length to reduce the tension in the fabric to zero is $T_{max}/nD\rho\tan\phi_a'$, where nD is the depth below the surface and ρ is the unit weight of the soil. At a linear variation of the strain distribution in the fabric the maximum lateral deformation of a vertical or sloping wall (δ_{max}) when the length of the fabric decreases with depth (Cases I and III) can be estimated from

$$\delta_{max} = \frac{\epsilon_{max} \cdot T_{max}}{nD\rho\tan \phi_a'} \qquad (20)$$

25

In the case the load distribution is uniform the lateral deformations will decrease approximately linearly with depth since the total overburden pressure $(nD\rho)$ increases with depth. Close to the surface δ_{max} = 144 mm at n = 1, D = 1.5 m, ρ = 20 kN/m^3, ϕ_a' = 30 degrees, T_{max} = 50 kN/m, ϵ_{max} = 5%. The corresponding lateral deformation at 6 m depth (n = 4) is 36 mm.

In the calculation of the lateral deflection of a vertical or an inclined wall when the length of the fabric increases with depth the lateral deformation of the soil behind the fabric reinforcement must also be considered. This deformation is governed by the lateral displacement of the underlying layers. The lateral displacement of the bottom layer $(\delta_{n,max})$ can be estimated from Eq (20). The corresponding lateral displacement for the following layer $\delta_{n-1,max}$ is

$$\delta_{n-1,max} = \delta_{n,max} + \frac{\epsilon_{n-1,max} \, T_{n-1,max}}{(n-1) \, D\rho \tan \phi_a'}$$

(21)

Since the force T_n increases linearly with depth when the spacing D is constant, then the lateral deformation as calculated by Eq (21) will also increase linearly with depth.

The lateral deformation required to develop the maximum resistance of a non-woven material is large compared with that of a woven fabric and the resulting deformation may be excessive for most structures. However, the behaviour of non-woven geotextiles is stiffer than indicated by unconfined tension tests when used as reinforcement in soil. The deformations can also be reduced by prestretching (Finnigan, 1977), by reducing the stress level in the fabric or by increasing the number of layers. The wall can often be inclined backwards, thereby reducing the lateral earth pressure acting on the wall. It is also possible to make allowance for the relatively large deformations of the fabric in the design. Fabric reinforced walls can often move considerably without any signs of distress.

4.8 Settlements

Relatively large settlements can normally be accepted, 100 to 200 mm, without any detrimental effects on the wall because of the large flexibility of the fabric reinforcement. The face elements can also tolerate large settlements depending on the size and shape of the elements. The settlements of a fabric reinforced wall is caused by

* the settlement of the fill which can be reduced by compaction and by

* the settlement of the soil below the fill and the wall. There may be soft clay or silt layers below the fill which can increase the settlements considerably.

The settlements below the fill and the wall can be reduced with embankment piles, stone or gravel columns or by replacing the compressible soil strata with granular material (sand or gravel).

4.9 Construction of Fabric Reinforced Walls

The following sequence is normally followed in the construction of a fabric reinforced retaining wall

a.   Before the beginning of the construction of the wall itself and the placement of the fill behind the wall the area should levelled and the soil compacted by e.g. a 3 to 5 tonne vibratory roller to at least 90% relative compaction as determined by the modified Proctor compaction tests (modified AASHO) in order to reduce the settlements of the soil below the fill. The area should be properly drained. Drain pipes might be required. Surface water should be channelled to collector drains.

b.   The bottom row of the prefabricated L-shaped elements can then be placed on the compacted surface and a 0.3 m thick layer of granular fill is spread out behind the concrete elements and compacted. The percentage of fines (< 0.06 mm) in the granular material should be less than 10 to 15%.

c.   Thereafter the first layer of fabric is rolled out on the compacted fill.

A string of gravel or coarse sand is placed on the fabric next to the wall for drainage. The fabric is folded over the gravel or sand layer as protection and to anchor the fabric. It is important that sufficient length of fabric is available for the folding.

d. Additional fill can then be placed on the fabric and compacted in 0.3 to 0.5 m lifts up to the top of the concrete elements. The maximum thickness of each layer depends on the available compaction equipment (plate vibrators or vibratory rollers). It is possible to use also a silty or a clayey material as fill except next to the fabric reinforcement. The thickness of the layers should then be reduced. The interface friction angle may otherwise be too low. The compaction should be done in the direction away from the concrete elements in order to pretension the fabric and to reduce the lateral deformations of the wall. The compaction next to the elements should be done in small lifts with a light plate vibrator. Otherwise the concrete element might be displaced during the compaction.

e. The next level of the L-shaped elements is placed on the compaction fill after the fill has been brought up to the top of the underlying elements. Additional fill can be placed behind the concrete elements after the next level of fabric has been rolled out in the same way as described above. Heavy equipment is not required to lift the concrete elements since they are light.

5 EXAMPLE

In the design of a fabric reinforced retaining wall it is important to consider

* the sliding resistance along different possible slip surfaces,

* the general stability of the wall with respect to a failure surface that extends behind the fabric reinforcement and below the structure (deep seated failure) and

* the bearing capacity of the soil below the wall

The resistance against sliding along the base of the 4 m high fabric reinforced wall shown in Fig 30 depends on the

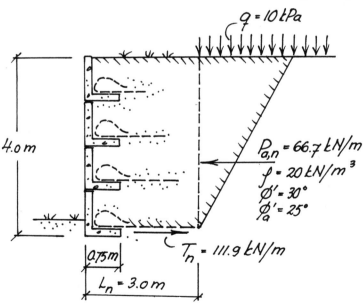

Fig.30 Design of a 4 m high fabric reinforced retaining wall – Sliding along base

27

length (3 m) of the fabric at the bottom of the wall. The force initiating failure $[P_{an} = 0.5 \, HK_a \, (2q + H\rho)]$ is governed by the coefficient of active earth pressure K_a, the height of the wall H and the surcharge load q. At H = 4.0 m, K_a = 0.33, q = 10 kPa and ρ = 20 kN/m³ the force initiating failure is 66.7 kN/m. The force resisting failure ($T_n = W_n \tan \phi'_a$) with respect to sliding along the base is 111.9 kN/m at L_n = 3.0 m, H = 4.0 m, ϕ'_a = 25° and ρ = 20 kN/m³. The corresponding factor of safety is 1.68 (111.9/66.7) which is considered to be satisfactory. A global factor of safety of 1.5 to 2.0 is usually required.

A high tension force will develop in the bottom fabric layer as the wall slides along the base. The maximum tension in the fabric depends on the friction resistance with respect to sliding along the bottom face of the fabric. Because the length of the fabric beyond the base of the L-shaped elements is 2.25 m, the maximum force in the fabric is 83.9 kN/m (2.25 x 4.0 x 20 x tan 25°)since ϕ'_a is 25 degrees. At a required factor of safety of 2.0 the minimum tensile strength of the fabric should be 167.8 kN/m (2 x 83.9). A maximum allowable load equal to 50% of the short term strength has been recommended for polyester fabric by Jewell and Greenwood (1988). A relatively high factor of safety is also necessary to limit the lateral displacement of the wall. Thus high strength fabric is required in this case in order to force the critical failure surface away from the wall.

The lateral displacement of the bottom layer can be estimated from Eq (20). The displacement thus depends on the maximum force in the fabric, which will be less than 66.7 kN/m (Pa) due to the friction resistance along the base of the L-shaped elements. This friction resistance is estimated to 28.0 kN/m (4 x 20 x 0.75 x tan 25°). At ϵ_{max} = 5%, T_{max} = 38.7 kN/m, nD = 4 m, ρ = 20 kN/m³ then $\delta_{n,max}$ = 0.051 m.

Also other possible slip surfaces should be checked as illustrated in Fig 31. The active earth pressure (P_a) initiating failure at the second level of the face element is 20 kN/m. The

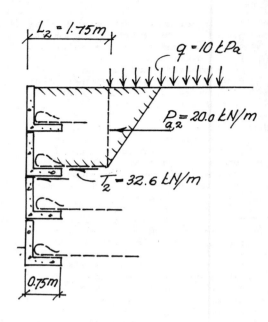

Fig.31 Local stability of retaining wall

force resisting sliding along the base is 32.6 kN/m (2 x 1.75 x 20 x tan 25°). The calculated factor of safety with respect to sliding along the base is 1.63 (32.6/20). This factor of safety is considered to be satisfactory.

The maximum tension in the fabric is estimated to 18.6 kN/m (1.0 x 2.0 x 20 x tan 25°). The required tensile strength of the fabric is thus 37 kN/m (2 x 18.6) at a minimum factor of safety of 2.0. It should be noted that the required tensile strength of the fabric decreases rapidly with decreasing distance below the ground surface.

The bearing capcity of the soil below the base of the L-shaped elements should also be checked. It should be noted that the fabric reinforcement will increase the bearing capacity since the lateral earth pressure is resisted by the fabric. Only the total over-burden pressure had to be taken into account. It is also important to consider the total stability of the wall with respect to a failure surface that extends behind the reinforcement and below the wall.

The fill behind the retaining wall should be carefully drained so that the excess pore water pressures in the soil will be small and can be neglected. The fabric reinforcement will also function as drainage layers which will reduce further the excess pore water pressures. The fill material should preferably be sand or gravel with a high permeability. It is desirable that the content of silt and clay size particles ($\leq$ 0.06 mm) should be less than 10 to 15%. The L-shaped concrete elements should be provided with drainage holes to relieve any excess pore water pressures in the fill behind the wall. If the pore water pressure increases in the fill during e.g. a heavy rain storm the lateral earth pressurs can be considerably higher than those indicated previously. This can be a problem in areas with a high annual rainfall such as Singapore.

Slotted PVC pipes can also be placed at the bottom of the backfill perpendicular to the wall in order to reduce the excess pore water pressures. The pipes should be connected to the gravel or sand layer just behind the wall.

6 SUMMARY

The application of geofabric as reinforcement in soil for vertical and inclined retaining walls have been reviewed. Geofabric can be used, for example, to reduce the lateral earth pressure on retaining walls or other retaining structures and to increase their stability. Calculation methods based on different possible failure mechanisms are presented in the paper so that the force in the fabric reinforcement can be estimated. The fabric reinforcement had to be designed to resist the total active Rankine earth pressure for the full height of the wall.

7 REFERENCES

Andrawes, K.Z., McGown, A. and Kabir, M.H., 1984. Uniaxial Strength Testing of Woven and Non-woven Geotextiles and Geomembranes, Vol.1, pp 41 - 56.

Bell, J. R. and Steward, J.E., 1977. Construction and Observations of Fabric Retained Soil Walls. Int. Conf. on the Used of Fabrics in Geotechnics, Paris, Vol. I, pp 123 - 128.

Bell, J.R., Barrett, R.K. and Ruckman, A.C., 1983. Geotextile Earth-Reinforced Retaining Wall Tests : Glenwood Canyon, Colorado, Transportation Research Record No 916, pp 59 - 69.

Broms, B.B., 1977. Polyester Fabric as Reinforcement in Soil. Int Conf on the Use of Fabrics in Geotechnics, Paris, Vol I, pp 129 - 135.

Broms, B. B., 1978. Design of Fabric-Reinforced Retaining Structures. Symposium on Earth Reinforcement, ASCE. Annual Convention, Pittsburg, pp 282 - 304.

Chandrasekaran, B., 1988. An Experimental Evaluation of Fabric Strength Properties and Behaviour of Fabric Reinforced Soil, Master of Engineering Thesis, National University of Singapore, 275 pp.

Christopher, B.R., Holtz, R.D. and Bell, W.D. 1986. New Tests for Determining the In-Soil Stress-Strain Properties of Geotextiles. Proc. 3rd Int. Conf. on Geotextiles, Vienna, Austria, Vol. 3, pp 683 - 688.

Colin, G., Mitton, M.T., Carlsson, D.J. and Wiles, D.M., 1986. The Effect of Soil Burial Exposure on Some Geotechnical Fabrics, Geotextiles and Geomembranes, Vol.4, pp 1 - 8.

Davidson, S. and Ekroth, N., 1975. Försök med vävförankrad stödmur, (Tests with retaining walls reinforced with fabric). Royal Institute of Technology, Stockholm, Unpublished Thesis, 43 pp.

Degoutte, G. and Mathieu, G. 1986. Experimental Research of Friction between Soil and Geomembranes or Geotextiles using a 30 x 30 cm Square Shearbox. Proc. 3rd Int. Conf. on Geotextiles, Vienna, Austria, Vol. 4, pp. 1251 - 1256.

El-Fermaoui, A. and Nowatzki, E., 1982. Effect of Confining Pressure on Performance of Geotextiles in Soil. Proc. 2nd Int. Conf. on Geotextiles, Las Vegas, Nevada, Vol. 3, pp 799 - 804.

Finnigan, J.A., 1977. The Creep Behaviour of High Tenacity Yarns and Fabrics Used in Civil Engineering

Applications. Proc. Int. Conf. Use of Fabrics in Geotechnics, Paris, Vol. 2, pp 305 - 309.

Fowler, J., 1982. Theoretical Design Considerations for Fabric Reinforced Embankments. Proc 2nd Int Conf of Geotextiles, Las Vegas, Vol 3, pp 671 - 676.

Gourc, J.P., Ratel, A. and Delmas, P., 1986. Design of Fabric Retaining Walls : The "Displacement Method". Proc. 3rd. Int. Conf. on Geotextiles, Vienna, Vol. 2, pp 289 - 294.

Hausmann, M.R., 1976. Strength of Reinforced Soil. Proc 8th Aust. Road Res. Conf., Vol 8, Sect. 13, pp 1 - 8.

Hausmann, M.R. and Vagneron, J.M., 1977. Analysis of Soil Fabric Interaction. Proc. Int. Conf. on the Use of Fabric in Geotechnics, Paris, Vol 3, pp 139 - 144.

Holm, G. and Bergdahl, U., 1979. Fabric Reinforced Earth Retaining Walls - Results of Model Tests C.R. Coll. Int. Reinforcement des Sols, Paris, pp 53 - 58.

Holtz, R.D., 1977. Laboratory Studies of Reinforced Earth using a Woven Polyester Fabric. Proc. Int. Conf. on the Use of Fabric in Geotechnics, Paris, Vol 3, pp 149 - 154.

Holtz, R.D. and Broms, B.B., 1977. Walls Reinforced by Fabrics - Results of Model Tests. Proc Int Conf. on the Use of Fabrics in Geotechnics, Vol 1, pp 113 - 117.

Holtz, R.D., 1985. Soil Reinforcement using Geofabric. Proc 3rd Int. Seminar on Soil Improvement, Singapore, pp 55 - 74.

Ingold, T.S., 1982. Reinforced Earth. Thomas Telford Ltd., London, England.

Jewell, R.A. and Greenwood, J.H., 1988. Long Term Strength and Safety in Steep Slopes Reinforced by Polymer Materials, Dept of Engineering Science, Univ. of Oxford, U.K., Soil Mechanics Report No 083/88, 31 pp.

John, N. W. M., 1986. Geotextile Reinforced Soil Walls in a Tidal Environment. Proc. 3rd Int. Conf. on Geotextiles, Vienna, Vol 2, pp 331 - 336.

John, N.W.M., 1977. Some Considerations on Reinforced Earth Design in the U.K., Proc. Int. Conf. on Soil Reinforcement, Paris, Vol. 1, pp 71 - 76.

John, N., Johnson, P., Ritson, R. and Petley, D., 1982. Behaviour of Fabric Reinforced Soil Walls. Proc 2nd Int. Conf. on Geotextiles, Las Vegas, Vol.3, pp 569 - 574.

Jones, C. J. F. P., 1985. Earth Reinforcement and Soil Structures. Butterworth and Co (Publishers) Ltd., London, England.

Lee, K.L., Adams, B.D. and Vagneron, J.M.J., 1973. Reinforced Earth Retaining Walls. Journ. of the Soil Mech. and Found. Div., ASCE, Vol. 99, No.SM10, pp 745 - 763.

Lindskog, G. and Eriksson, A., 1974. Rapport angående fältförsök med polyesterväv som jordarmering. (Report over Field Tests with Polyester Fabric as Reinforcement in Soil). Report No 50436, Swedish Geotechnical Institute, 32 pp.

Long, N.T., Legeay, G. and Madani, C., 1983. Soil-Reinforcement Friction in a Triaxial Test. Proc 8th European Conf. on Soil Mech. and Found. Engrg., Improvement of Ground, Helsinki, Vol. 1, pp 381 - 384.

McGown, A., Andrawes, K.Z., and Al-Hasani, M.M., 1978. Effect of Inclusion Properties on the Behaviour of Sand. Geotechnique, Vol. 28, No. 3, pp 327 - 346.

McGown, A., Andrawes, K.Z., Wilson-Fahmy, R.F. and Brady, K.C., 1981. A New Method to Determine the Load-Extension Properties of Geotechnical Fabrics. Transport and Road Research Labroatory, Department of the Environment, Department of Transport, Report no. SR 704.

Miyamori, T., Iwai, S. and Maiuchi, K., 1986. Frictional Characteristics of Non-Woven Fabrics. Proc. 3rd Int. Conf. on Geotextiles, Vienna, Austria, Vol. 3, pp 701 - 705.

Murray, R.T., 1980. Fabric Reinforced Earth Walls: Development of Design Equations. Ground Engrg., Vol. 13, No. 7, Oct 1980, pp 29, 31 - 16.

Myles, B., 1982.   Assessment of Soil
Fabric Friction by Means of Shear.
Proc. 2nd Int. Conf. on Geotextiles,
Las Vegas, Nevada, Vol 3, pp 787 –
792.

Myles, B. and   Carswell, I.G.,   1986.
Tensile Testing of Geotextiles.   Proc.
3rd Int. Conf. on Geotextiles, Vienna,
Austria, Vol. 3, pp 713 – 718.

Saxena, S.K. and   Budiman, J.S., 1985.
Interface Response of Geotextiles.
Proc. 11th Int. Conf. on Soil Mech.
and Found. Engrg., San Francisco, Vol
3, pp 1801 – 1804.

Tatsuoka, F., Ando, H., Iwasaki, K. and
Nakamura, K. 1986.   Performances of
Clay Test Embankments Reinforced with
a Non-woven Geotextile.   Proc. 3rd
Int. Conf. on Geotextiles, Vol. 2, pp
355 – 360.

Wichter, L., Risseeuw, P. and Gay, G.,
1986. Large-scale Test on the Bearing
Behaviour of a Woven-Reinforced Earth
Wall.   Proc. 3rd Int. Conf. on
Geotextiles, Vienna, Austria, Vol 2,
pp. 301 – 306.

van Zanten, R.V. (Editor),   1986.
Geotextiles and Geomembranes in Civil
Engineering, Balkema, 658 pp.

International Geotechnical Symposium on Theory and Practice of Earth Reinforcement / Fukuoka Japan / 5-7 October 1988
© 1988 Balkema, Rotterdam. ISBN 90 6191 820 0

Earth reinforcement – West and east

Masami Fukuoka
Science University of Tokyo, Noda, Japan

ABSTRACT: Natural reinforcing materials have been used in both west and east since long ago. Modern reinforcing materials are steel and geosynthetics. Much modern technology for reinforcing earth were mainly invented and developed in the west, but development of the technology has been carried out in the east as well. The past and present circumstances in the east are introduced in connection with the west.

1 INTRODUCTION

The reason for holding the International Symposium on Soil Reinforcement in Kyushu may be attributed to the fact that the level of engineering in this area has been high in both research and practical works. It is very well known that Professor Toyotoshi Yamanouchi, Chairman of the Steering Committee, Professor Emeritus of Kyushu University and Professor of Kyushu Sangyo University, has been a powerful leader for many years. An investigation committee for soil reinforcement has been established in the Kyushu Branch of the Japanese Society of Soil Mechanics and Foundation Engineering under the chairmanship of Professor M. Ishido, Kyushu Sangyo University, and the report, "Case Records on Soil Improvement, June, 1987", was published as the results of their activities. The report contains 3 chapters. Chapter 1: Reinforcing of soft ground and embankments. Chapter 2: Reinforcing of natural ground. Chapter 3: Literature. The number of items of literature collected is 881, 372 of these are from Japan and 509 from other countries, all of which are written in English. One can see that number of papers published in Japan is not so small, but it is questionable what percentage of papers has been translated and introduced to other countries.

Materials used for reinforcing earth have mainly been steel and petroleum products. As for the petroleum products, the development of new avenues of use has been remarkable, and international technology exchange has become active after the establishment of the international society. The International Geotextile Society was established in 1983, and the Japanese group of geotextiles began to work as early as in 1982 just after the 2nd International Conference in Las Vegas and has been active since then. The number of papers **published** has reached a considerable figure, but most of the papers have not been translated into English, like those of other countries in the world. Even valuable and interesting papers have not been translated into English. Dr. J. P. Giroud, President of the IGS, attended the Japanese Symposium in December of 1986 and showed a special interest in the presentations, and suggested to publish them for the journal, Geotextiles and Geomembranes. We Japanese must try to publish our papers in English as often as possible. The author assumes that it will not be any use to introduce how imported technology for earth reinforcement has been developed in Japan.

2 NATURAL REINFORCING MATERIALS

Throughout the world, natural reinforcing materials were used from old time. Babylonian constructed ziggurats made of soil mixed with plant stems more than 3,000 years ago. More than 2,000 years ago the Chinese constructed tide embankments by the use of fascine. Romans used mats made of ditch reed. The gabion was introduced to Japan in 6th century. Dutch used the fascine widely to treat soft ground around the 14th century. Shingen Takeda used various kinds of "ushiwaku" (ox frame), "ryogyu", "daiseigyu" and syakumokugyu", which were combinations of bamboo gabions and logs for river improvement. Earth reiforced

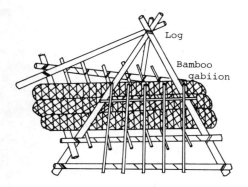

Fig. 1 "Ryogyu" for river work by Shingen Takeda (after K. Aki)

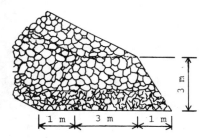

Fig. 2 Embankment reinforced by logs and bambooes for river work by Shingen Takeda (after K. Aki)

with timber, straw and reeds, etc., were used from old time in the east and west. Wooden piles were used to prevent a landslide mass in Japan 200 years ago. Originally, straw bags were used to keep rice in them 2,000 years ago, but straw bags filled with earth were used for forming embankments with steep slopes in Japan. The "sumo" wrestling ring began to be constructed using such sand bags since about 400 years ago.

3 STEEL

Steel wires began to be used to substitute bamboo nets for gabions about 100 years ago in Japan.

3.1 Steel nets

An artificial island was constructed for the purpose of constructing a vertical shaft to transport coal in the Ariake Bay in Kyushu from 1949 to 51. Steel nets were placed on the sea bed to reinforce the soft foundation and to separate the soft clay and the stone mound . Dr. Sadaichi Morita was in charge of this construction work. The thickness of the soft subsoil was about 14 m. Unit weight 14.8 kN/m³, specific gravity 2.45, void ratio 2.27, angle of internal friction 9°, and cohesion 5 kN/m². Height of the fill was about 9 m and its unit weight 20 kN/m³. The total settlement was estimated to be 2.7 m. The artificial island was round in shape. Its diameter was **178 m** at the foot of slope and 120 m at the top of the island. The island was made of reclaimed sand in the middle and surrounded by rocks. The inclination of the slope was 1 : 2.9 on the average, gentle at the lower part and steep at the upper part. A general cross section is shown in Fig. 3. Steel nets named Kawasaki steel nets, Type No. 6 (diameter 4.2 mm) with 18 cm mesh, were placed on the sea bed and filled with rocks of 30 - 50 kg in weight. Morita pointed out the reasons why the nets were used.

(a) The nets embedded in mud under the sea bottom will not rust as they are not exposed to fresh air.

(b) Unequal settlement can be prevented by the placement of nets.

(c) Gently sloped revetments are easy to repair after natural disaster.

(d) Construction cost is the lowest.

The sea bed runs dry about 3 hours for large low tide. The nets were laid by man power, and rocks were transported by boats at high tide, and placed in the right position during low tide. The total area of the nets was 13,000 m² and man power needed was 1,957 person days. This construction work was completed without any trouble and the island is still in good condition even after about 40 years. Steel nets have been used **occasionally** since then.

There was an example of retaining wall on a soft ground with steel nets of ordinary steel bars in the backfill.

Steel nets were used successfully for the construction of a retaining wall for temporary use in the USA. Steel nets are easier to have the stresses on the nets measured than are geotextile nets, as they deform elastically. The author of this steel wire net retaining wall measured

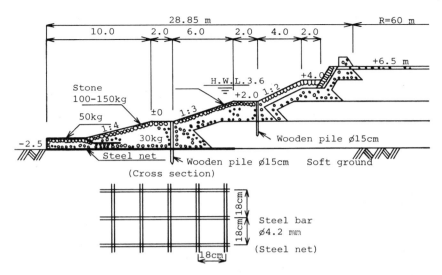

Fig. 3 Steel net reinforcement placed on soft ground, Miike **artificial** island for colliery, constructed in 1949-1951 (after S. Morita)

strains and stresses of the members and calculated the earth pressure against the front wall. The distribution of the design earth pressure was **trapezoidal.**

3.2 Reinforced earth (method of using steel strip and friction)

A modern method of reinforcing earth was invented by Henri Vidal of France. His method was named "tèrre armèe" in French. The first real retaining walls began to be constructed in 1964 and the first paper was published in 1966. A strip inserted in cohesionless soil restrains the relative movement of the soil by friction. This is the basic mechanisum of this method. Curved thin steel plates were laid horzontally one by one at the front wall of the retaining wall and steel strips connected to the plates to restrain the horizontal movement of the plates. A large retaining wall was constructed in southern France in 1968. Many researchers and engineers including engineers at the LCPC investigated both theoretically and experimentally. Later the concrete facing was replaced by the steel facing and the quality of strip was improved. This type of retaining wall has been used all over the world. Vidal's contribution to the area of earth reinforcement should be enormous.

Vidal came to Japan around 1966 and lectured to us at the Japanese Society of Soil Mechanics and Foundation Engineering. The Japanese National Railways (JNR) was interested in his retaining wall, and Hiroshi Uezawa of the JNR introduced his retaining wall in the **journal** of the Japanese Society of Soil Mechanics and Foundation Engineering (JSSMFE). Uezawa made a prompt report of his research work on the reinforced earth. Technical introduction on commercial base was completed, and an 18 m high retaining wall was constructed by the Kagoshima Toll Road Public Corporation, Kyushu. The author was asked to judge the construction of that high retaining wall in 1967. It was the highest retaining wall of this type in the world, it was said. The JNR decided to adopt it widely, and prepared standard for design and implementation. The author was chairman of the committee for that. A model retaining wall reinforced with geonets was placed on a vibration table to examine the stability during earthquakes, at the Technical Institute of the JNR. The JNR published a manual for the design and execution on the earth reinforcement including the reinforced earth in 1983. The Japan Road Public Corporation constructed a reinforced earth retaining wall at the Hokuriku Motorway at the suggestion of the author, in 1975. The Public Works Research

Institute (PWRI), Ministry of Construction, drafted a manual on reinforced earth retaining walls, and the Civil Engineering Research Center published a design manual to unify the design and execution in order to obtain a more reliable quality of the retaining walls.Many papers were written by many authors in the journals. A special volume for earth reinforcement was published by the JSSMFE, and Yoshiaki Hashimoto wrote a paper titled "Friction between soils and ribbed strips in reinforced earth walls". A book titled "Methods of reinforcing earth", was published by the JSSMFE in 1986, 14 % of the total 426 pages were allocated for the reinforced earth walls. The method of design against earthquake was described in this book.

Only examples of development in Japan are written here, but similar events occured in other countries.

3.3 Earth reinforcement with steel bars and anchor plates

The author was involved in countermeasures against landslide at the Narugo Dam in 1959. A small scale landslide was found at the left bank of the dam site. A tunnel was excavated horizontally, and the anchor concrete block was placed in a firm ground behind the sliding surface. Then a wire rope was attached to the block, and a concrete slab was made covering the surface of the sliding soil as shown in Fig. 4. Thus the movement of the slide was controlled. This was the origin of the earth reinforcement with steel bars and anchor plates. A high road embankment having 40 m in height and 1 : 1.5 in inclination was constructed at Kokanzawa of Sagami Lake, Kanagawa, in 1663. Artificial soft ground in the lake was

strengthened by sand compaction piles, but the slope seemed to be unstable even by a careful compaction. Therefore a method of earth reinforcement with steel bars and concrete plate anchors were inserted into the embankment as shown in Fig.5. The

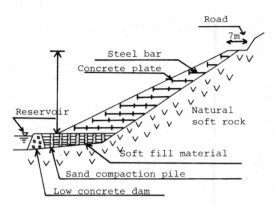

Fig. 5 High road embankment reinforced by steel bars with concrete plates, 1963

plates restrained the horizontal movement of the embankment body. Back of the steel bars were anchored in the rock by excavating drill holes if it was possible. It can be said that a combination of earth reinforcement and simple anchor was effectively used. Figure 6 shows a 20 m high embankment with the slope of 45° which was constructed by the use of this method in Toyama in 1982. The material of the slope was gravelly soil which collapsed during heavy rain.

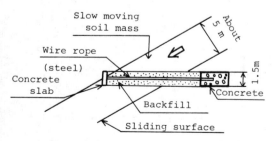

Fig. 4 Steel wire rope to check slow moving soil mass at Narugo Dam in Japan

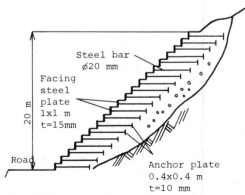

Fig. 6 Slope reinforced by steel bars with anchor plates

An experimental work aimed at applying
this method to retaining walls was
performed by the PWRI in 1965, but an
actual retaining wall was not constructed.
A high steel faced retaining wall was
constructed by the author at the Noda
Campus of the Science University of Tokyo
in 1979. A 5 m high concrete and fabric
faced retaining wall was constructed in
1980, and the earth pressure during
earthquake time was measured with this
retaining wall. A private company
constructed a concrete faced retaining
wall in 1983, and they constructed a 15 m
high steel faced retaining wall for the
JNR in Kyoto in 1985.

The steel bars with steel anchor plates
were laid between the soft subsoil and the
high embankment to prevent horizontal
movement of the embankment at Hayashima
Interchange in 1985.

The earth reinforcement method by the
use of steel bars and plate anchors have
advantages over that by the use of metal
strips because the round bars are more
resistant against rust, and because the
plates restrain any kind of soils, even
very soft clay or coarse gravel with sharp
edges. Ordinary soils excavated at the
construction site can be used without
using sand from other places. Ordinary
soils increase their strength by compaction, and moreover, the amount of
reinforcements can be saved by using their
characteristics of ordinary soils.

Strictly speaking the earth reinforcement method used for retaining walls with
rigid facing may not be called an earth
reinforcement method, but a wall between
the earth reinforcement method used for
retaining walls with rigid facing and
other ordinary reinforcement method was
removed since reinforced earth retaining
walls with concrete facing appeared.
Therefore, multiple anchored retaining
walls having concrete facing can be
regarded as a member of the earth
reinforcement family.

Wu Xiao-Ming of China invented the
retaining wall with concrete slab anchors
independently around 1976, and many
retaining walls of this type were constructed in China.

A sliding device between the wall and
anchor rod considering settlement of
backfill was invented in Japan and Prof.
Masaru Ho hiai et al made experiments for
this type of retaining wall.

3.4 Retaining walls in the UK

Collin J.F.P. Jones wrote a book titled
"Earth Reinforcement and Soil Structures"
which was translated into Japanese in
1986. There are many different kinds of
anchors in his book. R.T. Murray and M.J.
Irwin wrote a paper concerning the
cheapest type of anchors. They are a
double reverse bend of "Z" shape with a
short length at one end, and triangle
shape and incorporated a short length of
weld to prevent the anchor being distorted
when subjected to tension forces.

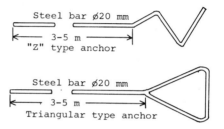

Fig. 7 Types of TRRL anchors(after Jones)

3.5 Reinforcement of natural ground

3.5.1 NATM

NATM stands for New Austrian Tunnelling
Method. This name seems to be used in
Austria, Japan and some developing
countries. The principle of this method is
to protect the inside wall of the tunnel
by positively utilizing the arch or ring
action of natural ground with rockbolts
and shotcrete. There are 3 main points
with NATM. They are rockbolts, shotcrete
and measurements of deformations. The
first record of rockbolts was of that A.
Bush used them at the coal mine at Frieden
in 1912. The rockbolts were used for
reinforcing hard rock in the 1930's, and
they came to be used even for soft rocks
and hard soil. When they are used for soft
rocks or hard soil, a drill hole in the
natural ground is filled with cement
mortar and a steel bar inserted into the
drill hole. The number of cases to which
this method was adopted increased since
around 1950. This method began to be
applied to civil works since L. v.
Racewicz of Austria suggested that the
rockbolts restrain natural ground around
the wall of tunnel with a circular section
to reduce the area of looseness. As for
shotcrete, the first application of

shotcrete to a tunnel was at the Maggia hydroelectric tunnel of Switzerland in 1951-55. It was found that the thin concrete layer had flexibility and behaved in harmony with the surrounding natural ground. L.v. Racewicz published a new method of tunnelling by writing a paper on the theme "The New Austrian Tunnelling Method (NATM)" from the November number of 1964 to the January number of 1965 of the Journal of Water Power. The NATM began to be used since around 1965 in Japan. The rockbolts were of the expansion type and used experimentally for the inclined shaft at Yoshioka in Hokkaido, which was a part of Seikan Tunnel of the JNR. An article concerning rockbolts and the shotcrete was included in the standard tunnel specifications by the Japanese Society of Civil Engineering. A genuine NATM was applied at Nakayama Tunnel of Jyoetsu Shinkansen by the full use of the rockbolts, shotcrete and measurements of deformations in 1976, and the technology of this method was completely established. This method was used for road tunnels since 1978. The shield tunnelling method had been used for tunnels in cities to avoid the ground subsidence caused by tunnelling works. The technology of NATM was improved by the use of auxiliary measures such as steel supporting members, mini-pipes, simple steel sheet piles, deep wells, etc. The amount of land subsidence caused by tunnelling works was therefore reduced remarkably. The cost of tunnel construction was reduced by using a more economical NATM than the shield tunnelling method. Even the depth of the tunnel is small, the thin layer above the tunnel is reinforced by the vertical rockbolts or concrete piles with reinforcements. As the result, the ground subsidence is kept to a minimum. The soil constants are determined by the soil survey before the design of the NATM tunnel. Deformation and the factor of safety against rupture of the ground are calculated by the use of the FEM etc. The behaviour of the earth is measured by the use of extensiometers, earth pressure gauges, pore pressure meters, reinforcement gauges, etc. as the construction work is carried out. The amount of rockbolts and supporting members are adjusted by interpreting the results of measurements. Kuriyama Tunnel of Narita Shinkansen, Kokubu River Tunnel, and Washuzan Tunnel of the Honshu-Shikoku Connecting Road are remarkable examples of the NATM tunnels, where the loose sandy soils of thin layer cover the tunnels.

3.5.2 Reticulated root piles

Reticulated root piles are a method of reinforcing natural earth on the basis of principles similar to the reinforced concrete, and developed by FONDEDILESPA of Italy. A similar method has been used to reinforce old buildings of brick and stones since long ago in Europe. The method was first applied to reinforce a natural slope in the USA in 1975. The first paper was published in 1978. The pile is composed of cement mortar and steel reinforcement, and is inserted into a hole excavated in the natural earth. This method was introduced to Japan in 1980. When a shield tunnel was excavated near an observatory in Tokyo, the foundation was consolidated by the use of the reticulated root piles. The observatory did not sink or tilt. The success of this work proved the effectiveness of those methods. Research work has been carried out in connection with the real construction work. Figure 8 shows a sketch of the cross-section of the reinforced slope for field test by Prof. T. Matsui et al.

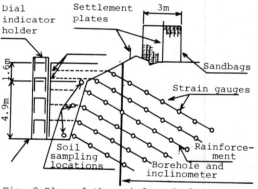

Fig. 8 Plan of the reinforced slope (after T. Matsui et al.)

3.5.3 Excavated slopes reinforces with steel bars

It seems that the method of reinforcing excavated slopes with steel bars was derived from the rockbolting method for tunnels. This method is used in both Europe and Japan. There are several methods for driving steel bars into the natural earth. Method of driving steel bars by breaker, method of making holes by a simple drill and method of making holes by drilling machine, etc. It is very common to drive deformed bars of 25mm in diameter and 3 m in length without using expensive drilling machine. The ordinary steel bar method differs from rockbolting, because it does not use any grouting. The former is cheaper than the latter, but the pullout resistance increases if the grout is carried out through a small hole in the steel bar. The slope excavated from the top of the ground is reinforced by steel bars step by step before the newly excavated portion of the slope becomes loose as time passes. Slopes are inclined more or less in general, but there is an example of vertical cut in uniform loose fine sand of 19 m high. However rockbolts 25 mm in diameter and 4 m in length were used in this case (Fig. 9).

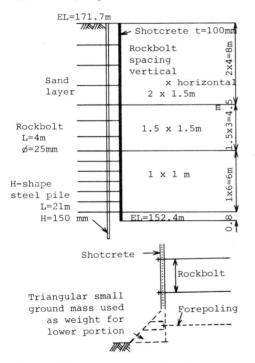

Fig. 9 Silty sand vertical cut reinforced by rockbolts (after Kitahara et al.)

Steel bar reinforcements have been used widely by the Japan Road Public Corporation and case records were analyzed by Teruki Kitamura et al in 1985. It was applied most frequently to soft rocks or hard soils. The most common lengths of steel bars covered from 2 to 4.5 m and the direction of driving bars was mainly downwards. The most common spacing was 1 to 1.75 m.

3.6 Other steel materials

There are many reinforcing materials using steel. One example is called "expand metal", which is perforated and expanded metal. This material is used to reinforce the earth. Steel bars for reinforced concrete are used to reinforce the earth in the form of lattice. Steel frames with rock inside is widely used as a Sabo dam (check dam) in Japan.

4 GEOTEXTILES AND RELATED MATERIALS

4.1 Geotextiles

The first geotextiles used for the purpose of reinforcing earth were woven textiles. A high tide occured in the North Sea in 1953, and the flood water covered 1,5000 km^2 of land and about 2,000 people were killed in the Netherlands. The Dutch government made the Delta plan, and decided to reconstruct the broken dikes over 22 years. Geotextiles were adopted instead of traditional fascines, because the former was difficult to fabricate in a short time. Nylon bags were used at the begining. The materials, machines, and methods of execution were revolutionally improved year by year, and total area of geotextiles used reached 10 million m^2 by the middle of 1980.

The PVC monofilament woven geotextile was first used at the base of the riprap under the sea dikes of Florida, USA, in 1958.

The woven vinyl fabric was first used instead of straw bags in early 1950's in Japan, but it was not widely used. A high tide accompanied by the Ise Bay Typhoon hit Nagoya Area, and about 5,000 people were killed by the sea water flooding through the broken sea dikes. The woven geotextiles were laid under the riprap for the purpose of separation and reinforcing. It is interesting that both the Netherlands and Japan used very expensive material for the reconstruction works. A new construction material may receive an opportunity to be used in times of emergency.

Ryuji Fukuzumi was the first person who used the woven geotextiles commercially to reinforce very soft clay ground along the sea coast. His report said, "The author, from around 1964, had been studying the application to civil works of synthetic chemical fibers. The first to be developed by the author was the mat method. This consisted of taking two

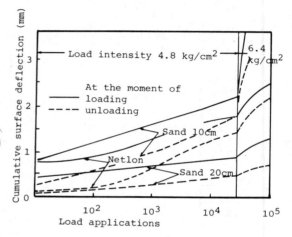

(a) Simplified model pavement sections

Model No. 1 — 4% Soil-cement / Crushed stone / Netlon

Model No. 2 — 4% Soil-cement / Crushed stone / Sand

Model No. 3 — 4% Soil-cement / Crushed stone / Sand

Subgrade (Kuroboku): CBR 1.3% Moisture 115%

(b) Relation between number of load applications and surface deflection for model pavement. Unimmersed case.

Fig. 10 Test results on structural effect of restraint layer on subgrade of low bearing capacity in flexible pavements (after T. Yamanouchi)

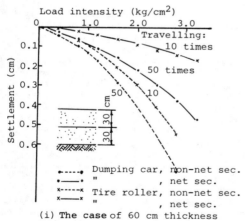

(i) The case of 60 cm thickness

(a) Results obtained from the plate bearing tests.

- ●----● Dumping car, non-net sec.
- ●——● " , net sec.
- ×----× Tire roller, non-net sec.
- ×——× " , net sec.

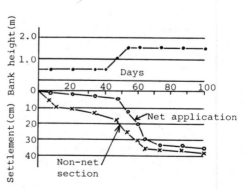

(b) Settlements observed using settlement meters.

Fig. 11 Test results obtained from test road for the work of Tohoku Expressway (after T. Yamanouchi)

sheets and sewing them together to make tubes. Sand was packed in these tubes in succession to make a mat. It was found there was a considerable problem about this method being practical from the standpoints of efficiency and economics.

In 1966, at the time of construction of the Sakai Factory of Ube Industries at reclaimed land, good-quality soil was directly spread as a part of surface-layer treatment work, but the spread soil sunk to the bottom of old sea bottom and the project had to be stopped for practical purposes. Application of the mat method was therefore planned, but since the area was as much as 50,000 m^2, while moreover, the construction period was short, the sheet method was thought of counting only on tensile strength, one of the effects of the mat method. --- Sand of thickness of 35 cm was banked, and on top of this good quality soil of 35 to 65 cm was spread. --- at the sheet method area using sheets made of vinylon of tensile strength of 3 kg/3-cm-width, even settlement was seen as a whole although there were some differences, and stabilization was achieved. --- At the stage of the test works, a Japanese patent was applied for in June 1966, which patent was granted in April, 1971.".

Shunske Sakai et al invented a method to reinforce geotextiles to cover soft ground, and prevent the swelling caused by pushing the filling soils by bulldozers.

This is a case of embankment using a weak cohesive soil with a high moisture content, but Toyotoshi Yamanouchi invented a method to reinforce the soil by laying a drain. The drain is lime covered with cardboards made of polyvinyl chloride. The method is called the "multiple-sandwich method".

4.2 Geonets

Geonets were invented in Britain to meet new demand. F.B. Mercer, inventer, wrote the following in his paper. "This process is a method for integrally extruding polymetric mesh structures in one step, from the molton polymer to the finished product, through a multiplicity of manufacturing steps. In the production of Netlon, molten polymer is extruded through two sets of opposing die orifices which are relatively displaced transversely to the direction of extrusion. The intersections are created when the orifices are in register and the mesh-forming strands are divided from the intersections with a shear action as the

dies move from the registration position. Geonets (Netlon) was invented in 1958. Prof. Yamanouchi carried out laboratory tests for geonet, and used it at the real construction site. He recognized that the geonet had the effect of interlocking and abutment of grids for stabilizing soils, compared with plane faced materials such as geotextiles. His paper was published in 1967. As the results of model tests of static and dynamic loading, he reached the following conclusions.(Fig. 10).

"From the above-mentioned tests it has been proved that pavement restrained by such a low pressure polyethylene net as Netlon is strengthened in structures under repeated loading. The direct effect of restrained layer in this case is as if the subgrade soil itself underwent improvement, and the author believes that the effect may be applied to pavement design by using the equivalent bearing capacity of subgrade or equivalent depth of sand blanket." He performed a static and dynamic tests to check the effect of the geonet for stabilizing soft ground, and obtained the following results. "As a result of the forgoing experiments, it has been shown that improvement of bearing capacity of soft soil can be effected by resinous net for static loading, but not always for repeated loading. These conclusions are thought to suggest the proper use of the net method, as well as proving useful for interpreting some of the field trials that have been already carried out." As the result of laboratory tests, the use of the net was useful only for the static load. However, the results of field tests proved that the net was effective under dynamic loading. Many tests were performed to construct the Tohoku Motorway more economically, and tests were made for resinous net that was selected as one of the promising materials. The test results were summarized as follows by Professor Yamanouchi. "This road test showed the additional informative data even in Fig. 11 (a) in which the effect of a resinous net application was distinctly advantageous for the case of placing it in a relatively thin sand layer. This result is not unlike that obtained from a model test by the author as noted in the introduction in this paper. Moreover, the test gave the significant information shown in Fig.11(b) that the settlement after a long time apparently become close to the settlement in the untreated section. This means that resinous nets elongate to follow the settlement to the bank in some cases."

Reclaimed land with soft clay cannot be

filled up directly with good soils. It was recognized that the surface of the soft ground could be strengthened by using geo-textiles but they were **occasionally** broken during construction. Therefore, the method of strengthening the geotextiles by covering rope nets, and filling with a layer of sand by the use of sand pumps or pipe lines was adopted. Methods of analysing the rope net method were present-ed by several authors. Shimizu proposed to use catenary taking into account of the strength of the ropes.

When the circumstances, which demand some new methods, are reached, many persons invent new good methods inde-pendently and almost at the same time. The use of nets for the purpose of strengthen-ing earth was also the case. S. Morita used steel nets at the Miike artificial island in 1949, and Kazuo Horimatsu of the Japanese National Railways (JNR) used fish nets to reinforce an embankment soil in 1964, and received a patent in the name of "Reinforcing body embedded in soil structure". He also received patents in following years and now has a total of 21. He also wrote many papers. Komei Iwasaki and colleagues of the JNR succeeded to reinforce embankments with newly developed resin nets in 1965. This was the first field test to strengthen an embankment by placing nets on a soft subsoil in Japan. The purpose of the test was to evaluate several methods of improving the soft subsoil for the vibration of trains. As for the effectiveness by static loading, there were 5 check points.

Initial settlement: average.
Long-term settlement: fairly good.
Lateral movement: excellent.
Increse of strength for subsoil: average.
As for the effectiveness by dynamic loading, there were 4 check points.
Up and down movement of the embankment: not clear.
Left and right movement of the embankment: fairly good.
Movement of ground near the foots of the embankment: bad.
In conclusion, the use of net was effective for vertical settlement and lateral movement, but exaggerated the vibration of the neighboring ground.
The JNR constructed 2.5 m high embankment on a soft ground, and examined the methods of safe execution, reducing the permanent settlement and the unfavorable effect of the old embankment at the time of increas-ing number of tracks. The methods tested were sand drains, lime piles ("chemico piles"), sand compaction piles ("composer"), gravel piles, and resinous nets. In

conclusion, the deformation of the ground was very little because the net distributed the local load over the wide area and the lateral movement of the ground was also small. The total amount of settlement and the increase of strength of subsoil were same as in the case of no net.

4.3 Geogrids

Mercer of the UK invented the high strength polyethylene net and also invented the extended polymer grid which had an extremely high elastic constant ("Tenser"). The material was imported to Japan and research and technical development were carried out for it. There were many examples of applications. The first report on the use of a geogrid for the purpose of reducing earth pressure against a retaining wall was the one by Naozo Fukuda et al. An example of use for the purpose of preventing unequal settlement was reported by Katsuyuki Kutara. An example of the use for reinforcing a long slope was reported by the Japan Road Public Corporation. An example of the use for constructing a vertical wall was reported by Kagoshima Developing Association in Kyushu. The two axially extended polymer grids were laid over the pile nets. The polymer grid has been used in the form of a mattress for foundations of roads, embankments, buildings, etc.

The micro-net was invented in the UK and introduced in Japan, but is not widely used at the moment.

4.4 Vertical drains

O.J. Porter of the US invented the sand pile around 1935. Walter Kjellman of Sweden invented the cardboard wick to drain saturated weak subsoils for the purpose of consolidation. A heavy machine was constructed to drive many vertical drains into the subsoil at Arlanda new International Airport around 1950. The machine was sold out to the Franky Pile Co. after completing the soil improvement of the airport. The author helped the introduction of the technique for card-board drains into Japan in 1963 and sixteen sets of heavy machines were manufactured by a Japanese company after improving upon the original Swedish design. The Japanese machines were used at the reclaimed land for NKK in Fukuyama City. The report from this construction work was made by Yoshiharu Watari. Theory, testing methods such as permeability test, estimation of effective diameter and the method of design proposed at that time are still used. Some of the machines constructed in Japan were exported to Europe. A method of soil improvement, which uses both cardboard drains and a vacuum, was invented by W. Kjellman. This was applied to consolidate soft tailings with fine grained coal at Fukuoka City. Thus the methods born in Sweden were applied to actual large construction sites and technically developed to a remarkable degree. The cardboard material was made from pulp in Sweden. Geotextiles and plastics began to be used instead of the pulp and it is said that high quality drain materials are produced by Japanese companies. When the vertical drains were driven through geotextiles laid on the soft subsoil, the geotextiles were broken, and the soft subsoil blew up through the broken parts. This trouble was overcome by the use of the rope net method. The big national project for Kansai International Airport is now being implemented. The thickness of the soft ground is about 20 m, and the water depth is about 20 m, and the height of the embankment 25 m. Many boats equipped with a huge machines which can drive 8 sets of drains at the same time are working effectively. About 300 thousand vertical drains will be driven in a very short period of construction.

4.5 Texsol

A method of reinforcing sand by mixing long continuous yarns was invented in France around 1979. E. Leflaive who invented the method said that he obtained the hint from the roots of the plants. This method was patented by the French government. It was introduced to Japan in 1987, and experimental works have been carried out extensively.

4.6 Method of analysis

Method of analysis has been developed by many engineers. The first group assumes a rupture surface. The reinforcing material across the surface is pulled and the tensile force on the material is compared with the tensile strength of the material. At the same time, whether the material is pulled out after the friction between the soil and the material exceeds the ultimate value of the friction or not. The second group divides the whole body into small elements of soil and reinforcing material. The relationship between stress and strain

is calculated by the FEM, etc. The deformation of the whole body and the possibility of rupture are examined while stress and strain is accumulated on the elements. Detailed discussion on this point is avoided because of space limitations and only the papers by Fumio Tatsuoka et al are introduced here, because they are important but not very well known. Everyone believes that the reinforcing materials are **effective,** because their Young's moduli are much higher than those of soils, but their Poisson's ratios are also important. The lower the Poisson's ratio the better. The Poisson's ratio of nonwoven geotextiles is zero according to the plane strain test by Tatsuoka et al. If nonwoven geotextiles are inserted into an embankment of cohesive soil, the soil around the geotextiles will not be saturated. It means that the soil maintains suction, which prevents the reduction of strength of the soil. This is a great merit of using non-woven geotextiles for the purpose of reinforcing embankments.

Analysis of deformation of geotextiles in earth was performed by using the FEM. The model proposed by Kutara et al is shown in Fig. 12 as an example. The geotextile was replaced by a plane truss element, and the joint elements were placed on discontinuous planes between the geotextile and soils respectively. The plane truss elements were connected with pins. Figure 13 shows the deformation characteristics of the joint element. It is assumed that the square elements of soils cannot resist tension. The pull out test was performed to determine the constants involved. A large scale model test of a retaining wall reinforced by geotextiles was performed, and the test results were compared with those of analysis.

There are a dozen centrifuge apparatuses in Japan. Those are used for analysing behavior of soils reinforced with geotextiles etc. The mechanical behavior of reinforced ground with high elongation characteristics was analyzed by Saito et al. As the result of centrifuge testing, it was recognized that the geotextile restrained well deformation near the ground surface. Friction between the ground and the geotextiles played an important role. Deformation in the ground caused by loading was reduced by the geotextiles, and a drain effect of geotextiles was observed.

The stress-strain relationships of soil structures changes with construction methods, non uniformity of soils, flexibility of supporting structures,

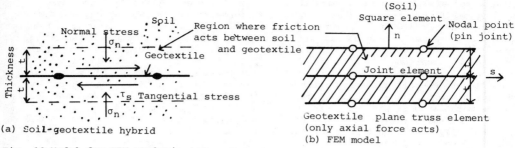

Fig. 12 Model for FEM analysis (after K. Kutara et al.)

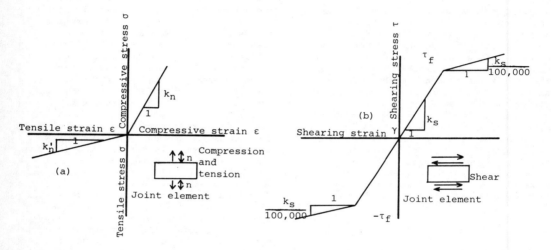

Fig. 13 Deformation characterics of joint element (after K. Kutara et al.)

foundations, percolating water from rain and snow, earthquakes, etc. Soil structures with reinforcements are more complex than those without reinforcements and so the accumulation of well documented case records is very important so that there is not only analysis to rely upon to make a safe and economical design.

4.7 Resistance against earthquake

Severe earthquakes occur frequently in Japan. Therefore, earthquake resistant design should be applied to the earth structures with reinforced materials in Japan. Some papers presented to the 3rd I S in Vienna discussed the earthquake resistant design using the conventional seismic coefficient method. Geotextiles are effective for preventing the liquefaction of saturated sands during earthquakes. This fact was revealed by Y. Mochizuki et al. The reinforcing effect is greater in the case of placing materials to direction of principal stress. If the reinforced materials are placed in the other direction, the reinforcing effect depends on the stiffness of the reinforcing materials.

Small scale shaking table test for the embankment reinforced by geotextile were conducted by Y. Koga et al. Various parameters such as the length of the reinforcements, the number of reinforcing layers, the slope of the face and the sort of the reinforcing materials. As the result, the reinforced part was acted as one block and the block resisted to the earth pressure like the gravity-type retaining wall. Figure 14 shows some results of the test. The model shaking table test of geotextile reinforced embankments on inclined ground was

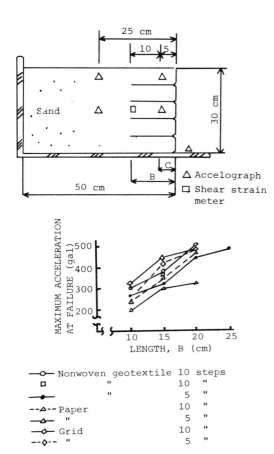

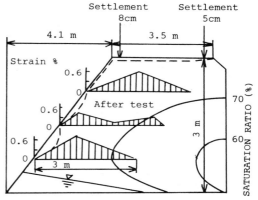

Construction (PWRI) and the private companies since 1985. Figure 15 shows one example of the large scale model test. The total amount of rainfall was 420 mm over 3 days. The saturation ratio and the deformed surface after the test are illustrated in the figure. The tensile strains are less than 1 %, which is much small compared to the strain at failure of 10 % of the polymer grids. Erosion gradually occurred in the surface soil between polymer grids. K. Kutara et al investigated the method of protecting slopes by the use of polymer grids while H. Miki et al performed the test using the embankments reinforced by nonwoven geotextiles having two different thicknesses. Only the thicker of the two was effective as the drainage system.

Fig. 14 Model shaking test of geotextiles reinforced embankment (after Y. Koga et al.)

Fig. 15 Settlement, strain and saturation ratio after 420 mm of rainfall(after H.Miki)

5 CONCLUSIONS

conducted by Y. Koga et al. As the result, it was revealed that earthquake resistance increases when geotextiles are thickly included and when geotextiles with large stiffness are used, and the slope surface is rolled up with geotextiles. Stability analyses were also performed.

4.8 Resistance against heavy rains

Resistance against heavy rains is important in areas which have heavy rain, because embankments and retaining walls may collapse during heavy rainfall. Prototype experiments were performed out of doors. Large scale model tests applying artificial rainfalls have been conducted by the research group including the Public Works Research Institute, Ministry of

Methods of soil reinforcement have been used in the west and east since very long ago. Methods of soil reinforcement using the modern materials such as steel and geosynthetics are made remarkable progress since the 1950's. Vidal's invention of reinforced earth and the development of the petrochemical industry had great influence on the promotion of these techniques. Many of the inventions and developments had been achieved in the west, and transfered to the east. However, some of discoveries and improvements were made in the east and some of them were developed remarkably through the extensive application to the real construction works. In this paper, the situation in Japan was emphasized, because the author thought that it may be necessary to introduce the papers written in Japanese and not yet public in other parts of the world. The content of this paper was divided into the two parts of steel and geosynthetics. At the end of this paper, research works for earthquakes and rains were described as it seemed that papers in this area have not been presented before in the west.

ACKNOWLEDGEMENTS

The author would like to express his sincere thanks to the authors of papers which were cited in this paper. Especially, he would like to thank those engineers who assisted greatly by sending their documents. In particular, these are Profs. T. Yamanouchi, N. Miura, T. Kimura, T. Matsui, S. Morita, F. Tatsuoka, and Messrs. K. Iwasaki, K. Horimatsu, Y. Watari, et al.

REFERENCES

Aboshi, H., Kawamoto, I and Inaba, A. 1965. On the paper drain method. Tsuchi-to-kiso No. 417. (in Japanese)

Broms, B.B. 1987. Stabilization of soft clay using geofabric and preloading. Proc. Int. Sym. Geo., Kyoto: 113-124.

Fukuoka, M. and Imamura, Y. 1983. Researches on retaining walls during earthquakes. Proc. World Conf. of Earthquake Eng., Vol.3, Sanfransisco: 501-508.

Fukuoka, M. and Morita, S. 1987. Failure of a retaining wall caused by heavy rain. Proc. Int. Conf. on Structural Failure, Singapore.

Fukuoka, M. and Goto, M. 1987. Embankment for interchange constructed on soft ground applying new methods of soil improvement. Proc. Int. Sym. on Prediction and Performance in Geot. Eng., Calgary: 169-176.

Giroud, J.P. 1986. From geotextiles to geosynthetics: a revolution in geotechnical engineering. Proc. Int. conf. Geo. Vienna Vol.1: 1-18.

Horimatsu, K. 1965. Material for reinforcing earth structures. Japanese patent.

Horimatsu, K. 1967. Method of reinforcing gravel layer of subgrade. Jap. patent.

Horimatsu, K. 1970. Method of reinforcing steep slopes using nets. Jap. patent.

Horimatsu, H. 1971. Method of reinforcing slopes using sandbags. Jap. patent.

Iwabuchi, S. & Arai, M. 1983. Application of root pile method to natural soils. Tsuchi-to-kiso 308, 31-9:29-33 (in Jap.)

Jones, C.J.F.P. 1978. The York method of reinforced earth construction. Proc. Sym. on Earth Reinforcement, ASCE Annual Con.

Pittsburg, Pennsylvania: 501-527.

Japanese National Railways 1978. Design Standards for earth structures:72-74.

JSSMFE, Kyushu Branch 1987. Examples of case histories of soil reinforcement. (in Japanese)

JSSMFE 1986. Soil Reinforcement(in Jap.)

Kamon, M., Ohashi, Y., Mizuhara,Y.,Tsujii, Y., Fukumori,I. & Kikuta,H. 1986. Drainage effect of band-shape drain material. 1st J.S.G., JC,IGS:95-100.

Kitahara, S., Ueno, M. & Uematsu, S. 1987. Study on the failure mechanism of vertical or steep cutting face and on the economical earth retaining method. Proc. Int. Sym. on Geomechanics, Bridges and Structures, Lanzhou, China: 397-406.

Kitamura, T., Yamada,N.,Okuzono,S. & Sano, N. 1987. Theory and practice on reinforced slopes with steel bars. Tsuchi-to-kiso 358, 35-11: 57-62.

Koga, Y. et al.1986. Small scale shaking test of the embankment reinforced by geotextiles. 1st. J.S.G., JC,IGS: 57-60.

Kurose, M. & Kimura, M. 1983. Effect of insertion of steel bars on slope stabilization. Tsuchi-to-kiso 308, 38-9:47-53.

Kumada, T., Sakata, Y. & Hirayama, H. 1986. Design, construction and cost of terre armee wall on soft ground. Kisoko 14-42: 86-92 (in Japanese)

Kutara, K., Gomado, M. & Takeuchi, T. 1986 Deformation analysis of geotextiles in soils using the FEM. Geotextiles and Geomembranes, 4: 191-205.

Lizzi, F. 1977. The in-situ reinforced earth, practical engineering in structurally complex formations. Proc. Int. Sym. on Geomechanics of Structurally Complex Formations Vol.1:327.

Leflaive, E. & Liausu,Ph. 1986. The reinforcement of soils by continuous threads. Proc. 3rd ICG Vol. 2: 523-528.

Matsui, T. & San, K.C. 1987. Reinforcement mechanism of cut slope with tensile inclusions.Proc. 8th Asian Reg. Conf. on SMFE, Vol.1, Kyoto:185-188.

Miki, H.,Kudo,K.,Tamura,Y.,Ikegami,M., Sueishi,T. & Fukuda,N. 1986. Large scale experiments on behaviour of embankment reinforced with polymer grids (rain). 1st. J.S.G. JC,IGS: 77-82.

Miyako, J., Watanabe, S., Iwasaki, K. & Suzuki, Y. 1969. Stabilization of soft subsoils,Report 2-Comparison of vibrational response of embankment with different treatments. Railway Technical Research Report 679: 1-36, 684:1-69(Jap).

Miura, N., Sakai, I. & Mouri, K. 1986. Model tests on reinforced pavement on soft clay ground. 1st J.S.G., JC,IGS:1-6.

Morita, S. 1952. Design and Construction of artificial island. Jour. of JSCE

37-6: 175-193. (in Japanese)

Morita, S. 1952. Subsidence of artificial island of Miike colliery. Jour. of J. society of Civil Engineering 37-8: 349-353.

Nishibayashi, K. 1982. Surface layer stabilization of soft ground using synthetic chemical fiber sheet. Proc. of Recent Developments in Ground Improvement Techniques, Short Course and Symposium on Soil and Rock Improvement Techniques Including Geotextiles, Reinforced Earth and Modern Piling Method, Bangkok: 239-254.

Ochiai, H. & Sakai, A. 1987. Analytical method for geogrid-reinforced soil structures. 8ARC, ISSMFE: 483-486.

Okuzono, S., Noritake, K. Terashima, H. Yasukawa, M. 1983. Model tests and practical examples of excavated slope reinforced with steel bars. Tsuchi-to-kiso 308, 31-9:55-62.

Saito, M., Miyako, J., Muromachi, T., Uezawa, H., Watanabe, S., Iwasaki, K., Kobayashi, S., Kurosawa, A. & Kotani, T. 1967. Reinforced embankment-plastic net is used as reinforcing subgrade. Railway Technical Research Report 595:73-78(Jap).

Stocker, M.F., Korber, G.W., Gassler, G. & Gudehus, G. 1979. Soil nailing. Proc. Colloque Int. sur de Renforcement des Sols, Paris (2): 469-474.

Schlosser, F. & Vidal, H. 1969. Reinforced earth. Bulletin de liaison des Laboratoires Routiers Ponts et Chausses, 41, Paris.

Taniguchi, E., Koga, Y. & Yasuda, S. 1987. Centrifugal model tests on geotextile reinforced embankments. Proc. 8ARC, ISSMFE: 499-502.

Tatsuoka, F. Ando, H., Iwasaki, K. and Nakamura, K. 1987. Performance of clay test embankment reinforced with a non-woven geotextile. Proc. Post Vienna Conf. on Geotextiles, Singapore:87-92.

Tatsuoka, F., Nakamura, K., Iwasaki, K. & Yamauchi, H. 1987. Behaviour of steep clay embankments reinforced with a non-woven geotextile having various face structures. Proc. Post Vienna Conf. on Geotextiles, Singapore: 387-403.

Uezawa, H., Nasu, M., Komine, K. & Yasuda, Y. 1972. Experimental research on embankment against earthquake by the use of a large scale vibration table. Report of Railway Engineering Institute.(in Jap)

Van Zanten, R.V. 1986. Geotextiles and Geomembranes in Civil Engineering. A.A. Balkema.

Vidal, H. 1966. La Terre Armee. Annales de l'Institut Technique de Batiment et des Travaux Publics, Nos. 223-229, Paris: 888-938.

Watari, Y., Higuchi, Y. 1986. Behaviour and analysis of geotextiles used on very soft ground for earth filling works. Geotextiles and Geomembranes 4:179-189.

Yamauchi, H., Tatsuoka, F., Nakamura, K., Tamura, Y. & Iwasaki, K. 1987. Stability of steep clay embankments reinforced with a non-woven geotextile. Proc. Post Vienna Conf. on Geotextiles, Singapore: 397-403.

Yamanouchi, T. 1967. Structural effect of restraint layer on subgrade of low bearing capacity in flexible pavement. Proc. 2nd Int. Conf. Structural Design of Asphalt Pavements, University of Michigan: 381-389.

Yamanouchi, T. 1967. Multiple-sandwich method of soft-clay banking using cardboard wicks and quicklime, Proc. 3rd Asian Reg. Conf. SMFE, Haifa:256-260.

Yamanouchi, T. 1970. Experimental study on the improvement of the bearing capacity of soft ground by laying a resinous net. Proc. Sym. Foundations on Interbeded Sands, Div. Applied Geomech. CSIRO, and Western Group, Aus. Geomech. Soc., Perth: 102-108.

Yamanouchi, T. 1975. Resinous net application in earth works. Proc. of the Conf. on Stabilization and Compaction, The School of Highway Engineering in Conjanction with Unisearch Ltd., The University of New South Wales, Sydney: 5.1-5.16.

Yamanouchi, T. and Miura, N. 1976. Soft clay banking using sandwich layers in situ made of wicked cardboard and quicklime. New horizons on construction materials, Vol. 1, Edited by H.Y. Fang, Envo Publishing Co., Inc. : 211-222.

Yamanouchi, T., Miura, N., Matsubayashi, N. and Fukuda, N. 1982. Soil improvement with quicklime and filter fabric. The Journal of the Geotechnical Engineering Division, Proc. ASCE, Vol. 108, No. GT7: 953-965.

1. Tests and materials

International Geotechnical Symposium on Theory and Practice of Earth Reinforcement / Fukuoka Japan / 5-7 October 1988
© 1988 Balkema, Rotterdam. ISBN 90 6191 820 0

Mechanics of orthogonally reinforced sand

T.Adachi
Kyoto University, Kyoto, Japan

H.B.Poorooshasb
Concordia University, Montreal, Canada

F.Oka
Gifu University, Gifu, Japan

ABSTRACT- The response of a cohesionless granular medium reinforced with a uniformly spaced orthogonal network of reinforcement to the action of external stresses is studied. The medium is treated as a continuum. The sand phase is assumed to behave according to the constitutive equation proposed by Poorooshasb and his colleagues and the reinforcement phase according to a set of rules that are described fully in the paper. The constitutive equation derived for the two phase soil (reinforced soil) is objective and hence may be used directly in any analysis involving such media.

INTRODUCTION

The material reported in this paper describes some preliminary results of joint research conducted in Kyoto and Concordia Universities regarding reinforced sand. The sand phase is assumed to behave as an elasto strain hardening plastic continuum obeying the constitutive relation proposed by Poorooshasb et al (1966,67) modified later by Poorooshasb and Pietruszazck (1986). The reinforcement phase is also treated as a continuum capable of withstanding large tensile but limited shearing stresses for bounded values of deformations. The final constitutive equation for the composite medium is derived by harmonising the deformation of the two phases of the medium and equating the sum of internal forces in the two phases to the external forces acting on the element. The relation obtained between stress and strain rates are objective with a non singular matrix relating the stress and strain rate spaces.

Rectangular Cartesian tensors and small deformation theory is used throughout.

STRESS DEFORMATION PROPERTIES OF THE REINFORCEMENT PHASE

The reinforcement is assumed to be orthogonal and to consist of a series of units placed at equal distances and mutually normal to each other. Fig.(1) shows two such cases.

Although the three dimensional reinforcement, Fig.(1,a), is unlikely to be of great practical interest the analysis presented here will include its treatment for the sake of completeness. The reinforcement scheme shown in Fig. (1,b) is in comon use and is sometimes referred to as sheet reinforcement.

Let the spacial axis of reference be denoted by $x_i(=x_1, x_2, x_3)$ such that they coincide with the directions of the reinforcements. Let $n_i=(n_1, n_2, n_3)$ be the fraction area of the reinforcements in the x_i directions. Then if the stress and strain tensors in the reinforcements are denoted by r_{ij} and

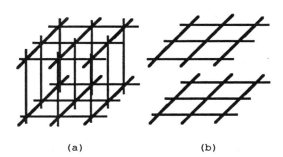

(a)   (b)

Fig.(1)-Orthogonal reinforcement
(a) Three dimensional and (b) Two dimensional reinforcements.

ε_{ij} respectively and remembering that the reinforcement phase is being treated as a continuum then the following set of relations exist

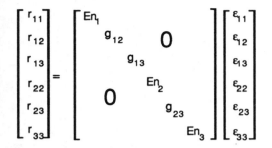

$$\begin{bmatrix} r_{11} \\ r_{12} \\ r_{13} \\ r_{22} \\ r_{23} \\ r_{33} \end{bmatrix} = \begin{bmatrix} En_1 \\ & g_{12} \\ & & g_{13} & & 0 \\ & & & En_2 \\ & 0 & & & g_{23} \\ & & & & & En_3 \end{bmatrix} \begin{bmatrix} \varepsilon_{11} \\ \varepsilon_{12} \\ \varepsilon_{13} \\ \varepsilon_{22} \\ \varepsilon_{23} \\ \varepsilon_{33} \end{bmatrix}$$

Eq. (1)

where E is Youngs modulus of the material of the reinforcement and g_{12}, g_{13} and g_{23} are equivalent shear moduli of the frame. They are not equal to the shear modulus of the reinforcing material rather to reinforcing structure. For example if the reinforcing sheet, Fig.(2),has a pitch of L (i.e. the distance between adjacent members is L) and the breadth (width) of a typical member is b then using elementary theory of structures the equivalent shear modulus is calculated to be $g=E(b/2L)^3$. Thus for a b/L ratio of 10 say,the value of g is evaluated to be E/8000! The above relation is with reference to a special frame of reference (i.e.when the x_i axis are directed along the reinforcement directions.) To obtain an objective constitutive relation they must be expressed in an arbitrarily chosen frame.To this end note that since both r_{ij} and ε_{ij} are treated as second order tensors then the coefficients relating them to each other must be a fourth order tensor which shall be denoted by C_{ijkl}.Stated

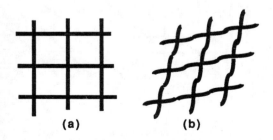

Fig.(2) Rigid grid (a) before shear (b) after shear.

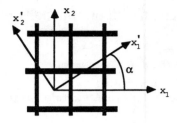

Fig.(3) Key figure.

otherwise C_{ijkl} is a fourth order tensor with principal values given in Eq.(1). The evaluation of the components of C_{ijkl} is straight forward and for the simpler two dimensional case is carried out in detail below. Referring to Fig.(3) let x'_i (i=1,2,3) be a new system of coordinates.It is required to obtain the components of the tensor C in the new system x'_i. The fourth order tensor C has $16(=2^4)$ components in two dimensions but only three constants would appear in the constitutive matrix.When the axis of reference are codirectional with the axis of reinforcements these are;

$$C_{1111}=En_1$$
$$C_{1212}=C_{2121}=g$$
$$C_{2222}=En_2$$

all other components being zero.In the x'_i system of reference and after symmetrizing the matrix the coeficients are obtained as;

$$C'_{1111}=\kappa_1+ \kappa_2\cos2\alpha- \kappa_3\sin^2 2\alpha/2$$
$$C'_{1112}= C'_{1211}=-(2\kappa_2\sin2\alpha+\kappa_3\sin4\alpha)/4$$
$$C'_{1122}= C'_{2211}=\kappa_3\sin^2 2\alpha/2$$
$$C'_{1212}=\kappa_2-\kappa_3\cos^2 2\alpha$$
$$C'_{1222}= C'_{2212}=-(2\kappa_2\sin2\alpha-\kappa_3\sin4\alpha)/4$$
$$C'_{2222}=\kappa_1-\kappa_2\cos2\alpha-\kappa_3\sin^2 2\alpha/2$$

where $\kappa_1=(En_1+En_2)/2$, $\kappa_2=(En_1-En_2)/2$, $\kappa_3= (En_1+En_2)/2-g$ and

$$\begin{bmatrix} r'_{11} \\ r'_{12} \\ r'_{22} \end{bmatrix} = \begin{bmatrix} C'_{1111} & C'_{1112} & C'_{1122} \\ C'_{1211} & C'_{1212} & C'_{1222} \\ C'_{2211} & C'_{2212} & C'_{2222} \end{bmatrix} \begin{bmatrix} \varepsilon'_{11} \\ \varepsilon'_{12} \\ \varepsilon'_{22} \end{bmatrix}$$

Note that if $n_1=n_2$ then $\kappa_2=0$,

$C'_{1111}+C'_{1122}= \kappa_1$, $C'_{2211}+C'_{2222}= \kappa_1$ and $C'_{1211}+C'_{1222}=0$. Thus for a state of pure compression whereby $\varepsilon'_{11}= \varepsilon'_{22}=\varepsilon$, $\varepsilon'_{12}=0$ the above relations yield $r'_{11}=r'_{22}=\kappa_1\varepsilon$, $r'_{12}=0$ as expected.

Finally since the deformation response of the reinforcing grid to stress is assumed to remain linear during the loading process (i.e. E and g are assumed to remain constant) then it is rational to state

$$\dot{r}_{ij} =C_{ijkl}\dot{\varepsilon}_{kl} \qquad (2)$$

where $\dot{r}_{ij}$ and $\dot{\varepsilon}_{kl}$ are the stress rate and the strain rate tensors respectively.

Having obtained the coefficients of the deformation matrix for the reinforcement the constitutive relation of the sand phase is examined next.

STRESS DEFORMATION PROPERTIES OF THE SAND PHASE

The constitutive relation proposed for sand assumes the existence of a global plastic potential ϕ and a local potential ϕ' which is derived from the global plastic potential. During virgin loading the plastic strain rate tensors are derived from ϕ and during stress reversals from ϕ'. Thus the strain rate tensor is related to the stress rate tensor by the relation;

$$\dot{\varepsilon}_{ij}= \dot{\varepsilon}_{ij}^{elastic} + \dot{\varepsilon}_{ij}^{plastic}$$
$$= E_{ijkl}\dot{s}_{kl}+ \dot{\lambda} \frac{\partial\phi}{\partial s_{ij}} \qquad (3)$$

where s_{ij} is the effective stress tensor associated with the sand phase, E_{ijkl} are the elastic moduli and λ' is the loading index. Its magnitude depends on whether the sand is loading, unloading (in which case its magnitude is zero) or reloading (in which case its magnitude is related to a conjugate quantity associated with the bounding surface, Dafalias(1982)).Specifically the loading index is given by the relation

$$\dot{\lambda} =h \frac{\partial f}{\partial s_{ij}}\dot{s}_{ij} \qquad (4)$$

where f is the yield function and h is a parameter that determines the magnitude of plastic strain increment tensor. It is related to the history of loading of the element.

Denoting by I_1, J_2 and J_3 the first invariant of the stress tensor and the second and third invariants of the stress deviation tensor respectively (i.e. $I_1=s_{ii}$, $J_2=\sqrt{S_{ij}S_{ij}}$ and $J_3= S_{ij}S_{jk}S_{ki}$ S_{ij} being equal to $s_{ij}- I_1\delta_{ij}/3$) the yield function F may be expressed through the relation

$$F= J_2-\eta(\varepsilon) I_1 g(\theta)=0$$

during the virgin loading process. In this last equation $\eta(\varepsilon)$ records the history of loading in terms of the total plastic strain ε, and θ is a function of the first and the third invariants (it is equivalent to Lode's angle). The function $g(\theta)$ has a certain symmetrical form about $\theta=n\pi/3$ for an isotropic sand. If $g(\theta)=$constant then the extended von Mises yield surface would obtain.
It must be emphasized that only during virgin loading the yield function f coincides with the bounding surface F. In general (e.g. during stress reversals) such a relation does not exist and indeed the kinematics of the yield surface within the bounding surface follows certain rules which can not be presented here.The interested reader may refer to the papers by the second author and his colleauges on stress deformation of sand.

STRESS DEFORMATION PROPERTIES OF THE REINFORCED MEDIUM.

Combination of Eqs.(2) and (3) results in the equation

$$\dot{r}_{ij}= C_{ijkl} (E_{klpq}\dot{s}_{pq}+\dot{\lambda} \frac{\partial\phi}{\partial s_{kl}})$$

which upon substitution from Eq.(4) for λ reduces to

$$\dot{r}_{ij}= C_{ijkl} (E_{klpq}+h \frac{\partial\phi}{\partial s_{kl}} \frac{\partial f}{\partial s_{pq}})\dot{s}_{pq} \qquad (5)$$

Equation (5) is the first equation relating the stress tensor in the reinforcement r_{ij} to stress tensor in the sand phase s_{ij}. In its derivation it has been tacitly assumed that the sand and the reinforcing grid deform harmoniously i.e. no slippage take place between the two phases of the composite medium.

A second equation to relate the stress tensors r_{ij} and s_{ij} may be obtained noting that their sum must equal the

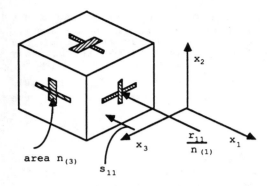

Fig.(5) **Schematic representation** of unit volume of reinforced sand

external applied stress σ_{ij}. Before this is done however it is worth stating that the r_{ij} tensor is a tensor of apparent stresses; see Fig.(5). The actual stress acting in the reinforcing bars is $r_{ij}/n_{(i)}$ where $n_{(i)}$ is the area fraction of the reinforcement in the ith face of the control volume of the medium. The situation is quite similar to flow of fluid through porous media where the apparent velocity is used in the constitutive equation of D'Arcy and is related to the loss of energy. The relation between apparent velocity and the actual velocity of flow is precisely the relation stated above i.e. $V_i = v_i/n_{(i)}$.

Thus the relation
$$r_{ij} + s_{ij}(1 - n_{(i)}) = \sigma_{ij} \qquad (6)$$
exists. Now if $n_{(i)}$ is small compared to unity then
$$r_{ij} + s_{ij} = \sigma_{ij} \qquad (6,a)$$
and a similar expression may be written for the rates of r_{ij}, s_{ij} and σ_{ij}.

Let for reason of convenience;
$$D_{ijpq} = C_{ijkl}\{E_{klpq} + h(\partial\phi/\partial s_{kl})(\partial f/\partial s_{pq})\}$$

Then Eq.(5) may be written as
$$\dot{r}_{ij} = D_{ijpq}\dot{s}_{pq} \qquad (7)$$

and when combined with Eq.(6) yields
$$\dot{s}_{ij} + D_{ijpq}\dot{s}_{pq} = \dot{\sigma}_{ij} \qquad (8)$$
Equation (8) relates the stress in the sand phase to the externally imposed stresses.

But $s_{ij} = s_{pq}\delta_{ip}\delta_{jq}$ where δ_{ij} is the Kronecker's delta. Thus Eq.(8) may be restated as
$$(\delta_{ik}\delta_{jl} + D_{ijkl})\dot{s}_{kl} = \dot{\sigma}_{ij} \qquad (9)$$

Before proceeding further it is worth noting that the fourth order tensor D_{ijkl} is, in all likelihood, a singular tensor (i.e. the inverse of the associated matrix may not exist). Such singularities may be the results of the reinforcement constitutive matrix (e.g. when no reinforcement is present in one of the directions x_i) or could be introduced if the sand is assumed to be a rigid plastic (rather than an elastic plastic) material.

The tensor $K_{ijkl} = D_{ijkl} + \delta_{ik}\delta_{jl}$ is, however, a non singular tensor. It is also an antisymmetric tensor since it is the sum of a symmetric tensor (the unit tensor $\delta_{ik}\delta_{jl}$) and the product of a symmetric (the C_{ijkl}) tensor and an antisymmetric tensor (the tensor of the elasto-plastic coefficients in the constitutive relation for sand).

Since K_{ijkl} is deemed to be non singular then Eq.(9) may be rewritten as
$$\{\dot{s}\} = [K]^{-1}\{\dot{\sigma}\} \qquad (10)$$

But from Eqs.(3) and (4)
$$\{\dot{\varepsilon}\} = [H]\{\dot{s}\}$$
where $[H] = E_{ijkl} + h(\partial\phi/\partial s_{ij})(\partial f/\partial s_{kl})$. Thus,
$$\{\dot{\varepsilon}\} = [H][K]^{-1}\{\dot{\sigma}\} \qquad (11)$$

which is the desired relationship between the stress and the strain tensor for the reinforced sand medium.

The various operations outlined above will be demonstrated by means of a simple example in the next section.

AXIAL LOADING OF AN ELEMENT REINFORCED IN A DIRECTION NORMAL TO THE AXIS OF THE MAJOR PRINCIPAL STRESS

Consider a sample of reinforced sand subjected to a state of axial loading whereby $\sigma_2 = \sigma_3$ remains constant with σ_1 increasing (triaxial compression test). The reinforcement is assumed to be of the sheet type with its plane normal to the direction of action of σ_1, Fig.(6).

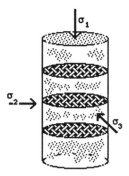

Fig.(6) Transversely reinforced sand in triaxial loading state

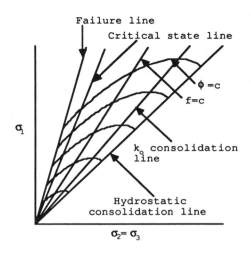

Fig.(7)- Yield loci and plastic potential curves for sand in triaxial compression.

Here $n_1 = 0$ and it is assumed that $n_2 = n_3$. This assumption is made only in view of the fact that otherwise it would be necessary to deal with a general state of stress for the sand phase: a discussion which is outside the space limitations of the present paper.

For the reinforcement phase the C matrix assumes the simple form;

$$\{\dot{n}\} = [C]\{\dot{\varepsilon}\}$$

$$[C] = \begin{bmatrix} 0 & 0 & 0 \\ 0 & En & 0 \\ 0 & 0 & En \end{bmatrix}$$

referring to principal directions.

The sand is assumed to be rigid plastic and for convenience the functions f, ϕ and h shall be expressed in terms of σ_1 and $\sigma_3 (= \sigma_2)$. In Fig.(7) are shown the f=const. and ϕ=const. curves associated with compression loading ($\sigma_1 > \sigma_3$).

If the state of stress experienced by the element is within the zone bounded by th two radial lines marked "critical state line" and the "hydrostatic consolidation line" then the sample contracts upon loading. Furthermore if the state of stress is to the right of the "k_o consolidation line" the strain components $\varepsilon_2 (= \varepsilon_3)$ would be positive (equivalent to a negative Poisson's ratio). This zone (i.e the zone bounded by the radial lines marked k_o and hydrostatic consolidation) is absent in preloaded or compacted samples and presents some peculiar behaviour.For

example it may be shown that a solution to certain type of problems may not exist at all when working in this zone of the stress space and assuming rigid plasticity.

If the state of stress is within the zone bounded by the radial lines marked "failure" and "critical state" then a loading of the sample produces expansion.

Now the matrix relating the strain rate tensor to stress rate tensor [[**H**] of Eq.(11)] for the sand is;

$$h \cdot \begin{bmatrix} f_{,1}\phi_{,1} & f_{,2}\phi_{,2} & f_{,3}\phi_{,3} \\ f_{,1}\phi_{,1} & f_{,2}\phi_{,2} & f_{,3}\phi_{,3} \\ f_{,1}\phi_{,1} & f_{,2}\phi_{,2} & f_{,3}\phi_{,3} \end{bmatrix}$$

Multiplying the above two matrices , adding the unit matrix [**1**] results the matrix [**K**] which has an inverse;

$$\frac{1}{det} \begin{bmatrix} 1+ \beta(f_{,2}\phi_{,2}+f_{,3}\phi_{,3}) & 0 & 0 \\ -\beta f_{,1}\phi_{,2} & 1+ \beta f_{,3}\phi_{,3} & -\beta f_{,3}\phi_{,3} \\ -\beta f_{,1}\phi_{,3} & -\beta f_{,2}\phi_{,3} & 1+ \beta f_{,2}\phi_{,2} \end{bmatrix}$$

where for convenience $\partial f/\partial\sigma_1$ has been shown by $f_{,1}$, $\partial\phi/\partial\sigma_1$ by $\phi_{,1}$ and so on, $\det=1+\beta(f_{,2}\phi_{,2}+f_{,3}\phi_{,3})$ and $\beta=Enh$. In a conventional triaxial test

$$\dot{\sigma}_2=\dot{\sigma}_3=0 \; ; \; f_{,2}=f_{,3} \text{ and } \phi_{,2}=\phi_{,3} \; .$$

Therefore the the principal components of s_{ij} are obtained from the relations

$$\left.\begin{aligned} \dot{s}_{11}&=\dot{\sigma}_1 & \text{(a)} \\ \dot{s}_{22}&=\dot{s}_{33}=-\frac{\beta\phi_{,3}}{\sigma_3\det}\dot{\sigma}_1 & \text{(b)} \end{aligned}\right\} \quad (12)$$

remembering that for the particular type of loading envisaged $f=\sigma_1/\sigma_3$ and hence $\partial f/\partial\sigma_1=f_{,1}=1/\sigma_3$.

From Eq.(12,a) it is seen that the rate of increase of sand stress in the axial direction is equal to the axial stress imposed on the soil. This is indeed as expected since no reinforcements exist in the axial direction.The rate of increase of s_{33} on the other hand is a function of the position of the stress point s_{ij} in the stress space, Fig.(7). If the stress point is in the zone bounded by k_o and hydrostatic lines the component $\phi_{,3}$ is positive and hence from Eq.(12,b) there will be a decrease in the magnitude of s_{33}.This is tantamount to saying that when loading in this zone   ,as far as the sand is concerned,a decrease in the confining pressure is experienced. This would lead to larger axial strains that when the soil is tested in the virgin state.This point whilst of some theoretical interest is unlikely to be of great practical value since as mentioned before preloaded or compacted soils do not exhibit such behaviour. Once the state of stress in the sand passes to a position on the left of the k_o line $\phi_{,3}<0$ and hence the rate of change of s_{33} is positive, Eq. (12,b).The soil is progressively getting stiffer (by virtue of the increase in the confining pressure s_{33}) and hence the magnitude of both axial and lateral strains would fall below the values measured for the non reinforced sand.

International Geotechnical Symposium on Theory and Practice of Earth Reinforcement / Fukuoka Japan / 5-7 October 1988
© *1988 Balkema, Rotterdam. ISBN 90 6191 820 0*

Dilatancy and failure of reinforced sand

R.R.Al-Omari, A.K.Mahmood & F.J.Hamodi
Building Research Centre, Baghdad, Iraq

ABSTRACT: The dilatancy characteristics and failure mechanism of reinforced sand are studied. It has been shown that a reinforced specimen has two options of failure. The first is to follow the minimum energy option as described by Rowe for sand alone which is termed here "underreinforced failure". The sand dilates stretching the reinforcement and thereafter a slip plane develops and yields the reinforcement. The effect of reinforcements may be taken as an enhanced confining stress and then the minimum energy lines and the dilatancy angles of reinforced and plain sand almost coincide. However the effect of stress level may be accounted for by utilizing the empirical equation of Bolton. The second option of failure termed here "overreinforced" is associated with rupture of sand-reinforcement bond and thereafter the bulging between layers. An equation is presented to estimate the position of the critical stage which separates between the two failures. The study is supported by an experimental investigation.

1 INTRODUCTION

1.1 Unreinforced sand

Shearing strength of cohesionless materials may conventionally be approached using Mohr-Coulomb formula based on a continuous material

$$\sigma 1 / \sigma 3 = \tan^2 (45 + \phi max / 2) \text{-------(1)}$$

where $\sigma 1$ and $\sigma 3$ are the major and minor principal stresses respectively, and ϕ_{max} is the maximum angle of shearing resistance.

According to this criterion the slip plane of failure is inclined at $(45 - \phi_{max} / 2)$ to the direction of σ_1 (Fig.1a). Rowe (1962) made his attempt to deal with sand as a particulate system. In his work the dilatancy occuring in the pack of particles in deformations to peak was considered. For a cubic pack of uniform spherical particles subjected to $\sigma 1, \sigma 3$, and intermediate principal stress $\sigma 2 = \sigma 3$, the energy ratio, E, which represent the ratio of the work done per unit volume on the assembly of particles by $\sigma 1$ to work done on $\sigma 3$ by the assembly during an increment of expansion was expressed as:

$$E = \sigma 1 /[\sigma 3 (1 + dV / V \epsilon_1)] = \tan(\phi_\mu + \beta) / \tan \beta \text{ ---(2)}$$

where dV is the incremental change in the volume, V, during the strain $\epsilon 1$ in the direction of $\sigma 1$. The angle ϕ_μ is the true angle of friction between the

mineral surfaces of the particles and β is the deviation of the tangent at the contact points from the direction of $\sigma 1$.

However, Rowe pointed out that for a pack of irregular particles, the principle of least work can be applied by taking $dE/d\beta = 0$ which yields $\beta = (45 - \phi_\mu /2)$ and the following equation becomes valid:

$$\sigma 1 /[\sigma 3 (1 + dV / V \epsilon_1)] = \tan^2 (45 + \phi_\mu / 2) \text{-------(3)}$$

Based on experimental results Rowe suggested replacing ϕ_μ in equation(3) by a frictional angle, ϕ_f, which approaches ϕ_μ and ϕ_{cv} for dense and loose packings respectively where ϕ_{cv} is the angle of shearing resistance at constant volume. The angle of dilatancy ψ can be calculated as :

$$\psi = \phi max - \phi f \text{ ----------------(4)}$$

Hanna and Youssef (1987) quoted a theoretical relationship between ϕ_{cv} and ϕ_μ by Horne, 1965. Koerner (1970) quoted the following relationship by Ladanyi, 1960 between ϕmax and its frictional component ϕf :

$$\frac{\sin \phi_f}{\cos^2 \phi_f} = \frac{\sin \phi_{max}}{\cos^2 \phi_{max}} + \frac{K}{3 - K} \left(\frac{3 - \sin \phi_{max}}{2 \cos^2 \phi_{max}} \right) \text{ ----(5)}$$

in which $K = d(\Delta V / V)/d \epsilon 1$

Though, the dilatancy effects were accounted for in equation (3), subsequent research highlighted the dependence of ϕmax on the stress level. Bolton(1986) correlated enormous results of past published work into the

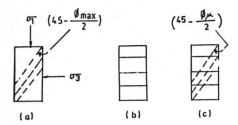

Figure-1.-Failure of plain and reinforced sand samples.

following empirical formula, for triaxial strain.

$$\emptyset_{max} - \emptyset_{cv} = 3 I_R \quad --------------(6a)$$

where I_R is a relative dilatancy index given by :

$$I_R = I_D (10 - \ln p) - 1 \quad -------------(6b)$$

I_D is the relative density and p is the mean principal stress, $(\sigma_1 + 2\sigma_3)/3$.

1.2 Reinforced sand

In reinforced sand ,though a considerable number of investigations have been carried out using triaxial (Gray and Al-Refeai,1986)or plane (McGown et al,1978) strain devices, the emphasis was mainly on the improvement of strength and secant modulus .In the present work a conceptual and experimental study of the dilatancy characteristics of reinforced sand is presented . These characteristics are correlated to the strength and failure mechanism and compared with those of the sand alone. The case of reinforcement rupture at or before peak is excluded from the study as it is impractical.

2 MECHANISM OF DILATANCY IN REINFORCED SAND

When a foreign body is included inside sand (Fig.1b) and the specimen continually loaded, the mechanism of mobilization and progress of the conventional slip plane may be altered according to the inclusion stiffness ,S, defined as the force per unit width per unit strain.

2.1 Underreinforced sand

If the stress level is high relative to S , the frictional or interlocking bond with reinforcement would be sufficient to extend the reinforcement during dilatancy. Failure would also occur through the development of the slip surface at $(45-\emptyset/2)$ to the direction of the major principal stress,σ_{1r} ,as this

mechanism provides the minimum energy ratio ,E, for failure (Fig.1c). Therefore, the lateral expansion for the sample which is a condition for such failure is expected to be almost equal to the unreinforced case and the value of Ψ is not expected to vary.

This kind of failure would be called "underreinforced failure" and it should be distinguished from that occuring in concrete as it is not associated with reinforcement yield before or at peak however, as in concrete it provides a less catastrophic collapse.

The tension resistance of reinforcements may be deemed to represent an increase in the value of σ_3 by an amount $\Delta\sigma_3$ (Ingold,1982). This amount is thus represented as:

$$\Delta\sigma_3 = S\epsilon_h N / H = \sigma_{3r} - \sigma_3 \quad ----------(7)$$

where ϵ_h is the lateral strain in the sample which is assumed not to significantly change through the hight,H,N is the number of reinforcing layers and σ_{3r} is the modified σ_3 for the reinforced sample.

The assumption of the equivalent confining stress,$\Delta\sigma_3$,implies that the value of $\emptyset_{max}$ is not altered and the failure envelope passes through the origin. Hence σ_{1r}/σ_{3r} is also not altered and equation (3) of Rowe may still be applied. The minimum energy lines of Rowe (σ_1/σ_3 vs $1+dV/V\epsilon$) for plain and reinforced sand may coincide if no abrasion or crushing occurs due to the increase in the value of p which becomes $(\sigma_{1r} + 2(\sigma_3 + \Delta\sigma_3))/3$.

If a considerable number of reinforcing layers is used then equation (6) which accounts for the stress level is efficient in the estimation of Ψ as far as the minimum energy option of failure is followed .

Post peak , the slip plane would pass through and yield the reinforcements unless their modulus is too low which is not often used. Therefore, a constant value is added to the residual strength which depends on the ductility of reinforcement. A reduction in the principal stress ratio even at high vertical strains is anticipated if the reinforcements is completely broken during the slip.

2.2 Overreinforced sand

When the reinforcement stiffness is high compared with the stress level,the assembly of particles would not be able to follow the minimum energy option described above. Thus the improvement in the capacity resistance of the composite system should be higher in this case which would be termed "overreinforced".

The new option available to the

assembly may be understood by considering the case of a sand layer squeezed between a strip footing and a rigid rough bed as reported in AL-Omari (1984).The bed may simulate a rigid rough (Glass paper) reinforcement. Frequent drops in the stress before peak were noticed in the three tests performed using a thin layer of height half the footing breadth. The slip surface initiated and then passed at the contact plane with the bed and the peak stress immediately and catastrophically dropped. A rational interpretation to that is the tendency of soil to fail through sliding at the underlying boundary,an option which requires either overriding of particles over the serrated face of the glass paper or the crushing of some particles to ease sliding,as long as dilation is limited. Actually, both took place,overriding caused the frequent drops and then crushing caused the immediate slip.The subsequence of these actions is affected by the grains toughness and the reinforcement roughness.A recent stereophotogrammetric measurement (AL-Omari and AL-Taweel,1988) of internal displacements indicated that dilation before peak,which usually takes place in deep layers to open the way for the progress of the slip surface,did not occur in that case.

It is therefore expected that reinforced triaxial specimens would not significantly dilate up to peak depending on the spacing between layers. Conventional dilatancy theories are not applicable in this case.

Assuming the boundary stresses remain principals and based on Mohr-Coulomb criterion , Hausmann(Ingold,1982) derived the following equation for the maximum friction angle of the reinforced sample, ϕ_r :

$$\sin \phi_r = \frac{K_a - (0.25 \, FNd) - 1}{(0.25 \, FNd) - K_a - 1} \quad \text{---(8)}$$

where F is the interface coefficient of friction, d is the sample diameter, and $K_a = \tan^2(45 - \phi_{max}/2)$.

2.3 The critical stage

It is known that a break in the failure envelope of reinforced sand appears at a critical value of σ_3 (Gray and AL-Refeai,1986). In fact, this break marks a change in the dilatancy characteristics leading to the underreinforced failure. The critical σ_3 may ideally correspond to a rise in the value $(1 + dU/V\varepsilon_t)$

A system failure by the rupture of sand-reinforcement bond is more catastrophic,particularly when a considerable number of layers is used,

the engineer should be able to manipulate the design so that failure would be through the minimum energy option.This could be approached by realizing that at the critical stage the value of the tensile stress required to develop a reinforcements strain ε_h necessary for minimum energy failure becomes equal to the frictional stresses mobilized at the interface of each reinforcement, thus:

$$S\varepsilon_h N \pi d \doteq F \sigma_{1r} \pi \frac{d^2}{4} \quad \text{------(9)}$$

which gives

$$S_c = \frac{F \sigma_{1r} d}{4 N \varepsilon_h} \quad \text{------(10a)}$$

or

$$N_c = \frac{F \sigma_{1r} d}{4 S \varepsilon_h} \quad \text{------(10b)}$$

where Sc is the critical stiffness if N is kept constant and Nc is the critical number of layers if S is kept constant.

The values Sc and Nc may be estimated by taking $\sigma_{1r} = (\sigma_3 + \Delta\sigma_3)$. Kp and $\Delta\sigma_3$ calculated from equation (7) using a trial value of Nc or Sc. Then this value changed until it becomes equal to the righthand side of equation (10). Similarly,the critical σ_3 for a constant S and N may be evaluated. The value of S or N in the design should be less than its critical value to ensure an underreinforced failure.

3 EXPERIMENTAL WORK

The sand used is sorted out from Karbala sand deposits located at the western part of Iraq. It has a particle size ranging from 0.425 to 1.18 mm with uniformity coefficient of 1.69.The value of D_{50} is 0.74 mm and the specific gravity is 2.75. The maximum and minimum porosities are 45% and 35% respectively.

Conventional triaxial apparatus was used in the investigation. The diameter of the sample was 100 mm and the length to diameter ratio was around 2. All the tests were carried out in the saturated condition using a relative density of 73% . Differences in the relative density were within ±3%. To eliminate the effect of varying σ_3 ,each specimen was overconsolidated to 690 kN/m^2 . However,the density was varied in the unreinforced case to determine the value of ϕ_{cv}. A burette was used to measure volume change in terms of the volume of water under atmospheric pressure displaced from the pore space of the sample.

A range of applied cell pressures was used. By changing this pressure, the relative stiffness of the same reinforcement is varied.

Two types of reinforcements were selected. A steel disc,2 mm thick,and a plastic mesh. The aperture size of the

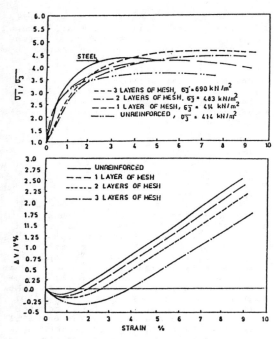

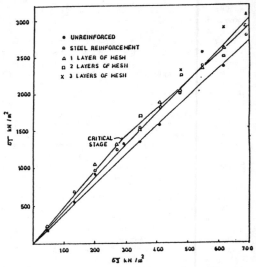

Figure-2.-Typical results of stress ratio and volume change VS axial strain.

Figure-3.-Failure envelopes of all the series of tests.

mesh is 9x7 mm which is appropriate to the value of D_{50} (AL-Omari et al 1987). The stiffness, S, of the mesh was 240 kN/m which enabled obtaining the two types of failure. The coefficient of interface friction between Karbala sand and each of the steel and mesh obtained using the shear box is 0.64 and 0.82 respectively which corresponds to 75% and 97% of the sand alone.

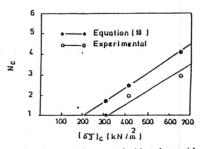

Figure-4.-Comparison of the hypothetical and experimental relationship between the critical confining stress and the critical number of reinforcing layers.

4 EXPERIMENTAL RESULTS

Typical results of stress ratio and volume change versus axial strain are shown in Fig.2. The failure envelopes for all the series of tests are plotted in Fig.3. A break is noticed only in the case of plastic mesh reinforcement and the position of the break agreed with visual observations of a transfer in the failure criterion from bulging between layers to formation of the slip plane. Overreinforced failure was maintained in the case of steel and the failure shape was noticed post peak as bulging of the top half of the sample. Underreinforced failure of mesh reinforced samples was according to minimum energy option and the slip plane yielded the reinforcements. The mesh reinforcements were examined after the tests and the yield was clearly noticed at positions were the slip plane has passed.

The critical confining pressure varied

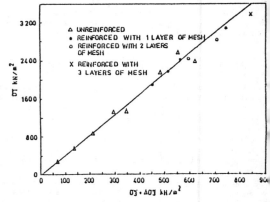

Figure-5.-Failure envelopes of unreinforced and underreinforced sand using the enhanced confining stress concept.

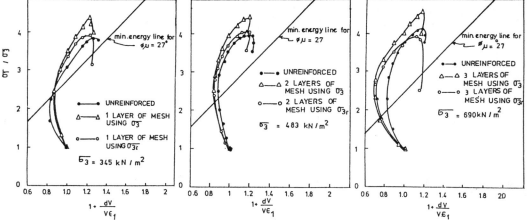

Figure-6.-Minimum energy lines of unreinforced and underreinforced Karbala sand

with the number of layers. This variation reasonably agreed with the prediction from equation (10) as shown in Fig.4. The difference is owing to the friction at the top and bottom platens. As mobilization of the kinetic angle of friction at the interface of a reinforcement takes place progressively, full mobilization over the total area of a reinforcement does not occur at peak but at the residual state. Thereby, it is found that the value of ϵ_h at the residual state should be used in the utilization of equation (10). The value used corresponds to 20% axial strain which in the three tests marked the start of the residual state.

The failure envelope of unreinforced and underreinforced samples is replotted in Fig.5 using the value $\sigma_{3r} = \sigma_3 + \Delta\sigma_3$ instead of σ_3. It is shown that a single envelope may reasonably fit the results. The minimum energy lines of selected number of these samples are shown in Fig.6. The experimental lines of plain and reinforced sand are very close using σ_{3r} which agrees with the argument given above.

Thus, the dilatancy characteristics of underreinforced sand are similar to those of sand alone and the theory of Rowe is applicable. The small difference is owing to crushing which not acounted for in Rowe's work.

It should be mentioned that the value of ϕ_{cv} was found to be 33.5° which gives $\phi\mu = 26°$ according to Horne's relationship. The value of Rowe's ϕ_f for a very dense packing was 28° The average of these two values was taken as $\phi\mu$ of Karbala sand.

The dilatancy component of $\phi\max$ taken as $\phi_{max} - \phi_f$ according to Rowe and Ladanyi, and $\phi_{max} - \phi_{cv}$ according to Bolton is plotted in Fig.7 against the stress level P. The dilatancy rate $(1+dV/V\epsilon_1)$ at peak is drawn against the confining

stress in Fig.8. It appears that there is no sudden change in this rate corresponding to the position of a break in the failure envelope.

The difference in the dilatancy rate between plain and reinforced sand is higher in the overreinforced case. The mechanism of strength enhancement may be through the restriction of dilation and hence increasing the interface coeficient of friction at which slippage may instanteously start at the interface of all layers. However, it should be realized that once the stress level approached the value of zero dilation, further increase of N may not increase the strength. The value of this stress may be estimated by taking $\phi_{max} - \phi_{cv} = 0$ using equation (6). This concept is different from that upon which equation (8) was derived when it was assumed that the enhancement directly dependent on the frictional area (number of layers). Thereby the value of ϕ_f determined from this equation did not agree with the experimental results for N larger than one. However, further experimental evidence are required to establish this point.

It should be mentioned here that at great axial strains reinforced samples suddenly start to contract. The state of a constant volume is thus inexistent. It was checked that this phenomenon is not due to a leak through the rubber membrane.

CONCLUSIONS

According to the failure mechanism, reinforced sand is classified to underreinforced and overreinforced.

Underreinforced failure is that which follows the minimum energy option as described by Rowe for plain sand. The sand stretches the reinforcements during

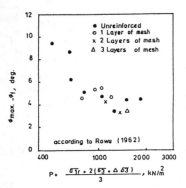 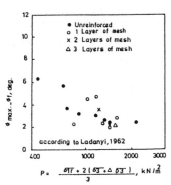 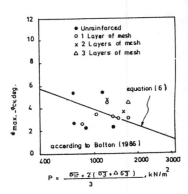

Figure-7.-Coincidence of the dilatancy angles of unreinforced and underreinforced sand as determined using different theories.

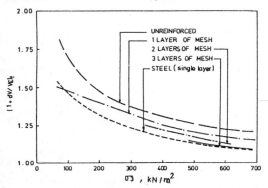

Figure-8.-Variation of dilatancy rate at peak with the confining stress.

dilation and eventually the conventional slip plane develops and yields the reinforcements. The effect of reinforcement may be considered as an enhanced confining stress, $\Delta\sigma_3$, and then the dilatancy characteristic of underreinforced sand becomes similar to that of plain sand and they can be approached using the conventional theories.

Overreinforced failure is characterized by the rupture of sand-reinforcement bond and the post peak bulging between layers, an option which the conventional dilatancy theories were not made for.

As both types of failure may be achieved for the same stiffness and amount of reinforcement by varying the confining stress. An equation is presented for estimating the position of the critical stage which separates between these failures.

Experimental results reasonably supported the above argument.

REFERENCES

AL-Omari,R.R.1984.Strip foundation on a sand layer overlying a rigid stratum. Ph.D. thesis,University of Strathclyde, U.K.

AL-Omari,R.R.,AL-Dobaissi,H.H.,and AL-Wadood,B.A. 1987. Inextensible geomesh included in sand and clay. Proceedings of International Symposium on prediction and performance in Geotechnical Engineering, Canada : 155-160.

AL-Omari,R.R.,and AL-Taweel,H.1988 Deformations in sand layers overlying a rigid rough stratum.Proceedings of 2nd Iraqi Engineering Conference, Mosul, Iraq .

Bolton,M.D.1986. The strength and dilatancy of sands. Geotechnique 36,No.1: 65-78 .

Gray,D.H.,and AL-Refeai,T.1986.Behaviour of fabric-versus fiber-reinforced sand. Journal of Geotechnical Engineering,ASCE, No.8 Vol.112:804-820 .

Hanna,A.M.,and Youssef,H.1987. Evaluation of dilatancy theories of granular materials. Proceedings of international Symposium on Prediction and Performance in Geotechnical Engineering,Calgary Canada:227-236 .

Ingold,T.S. 1982. Reinforced earth Thomas Telford Ltd,London.

Koerner,R.M.1970. Effect of particle characteristics on soil strength. Journal of the soil Mechanics and Foundations Division, ASCE,No.SM4,Vol.96:1221-1234.

McGown,A.,Andrawes,K.Z.,and AL-Hasani,M.M.1978. Effect of inclusion properties on the behaviour of sand. Geotechnique,28,No 3:327-346 .

Rowe,P.W.1962.The stress-dilatancy relation for static equilibrium of an assembly of particles in contact . Proceedings of Royal Socity of London,series A , Vol. 269:500-527.

Evaluation of the effects of construction activities on the physical properties of polymeric soil reinforcing elements

D.I.Bush
Netlon Limited, Blackburn, UK

ABSTRACT: This paper describes a site trial which reproduced UK practice for the construction of reinforced soil structures. The polymeric soil reinforcing geogrid was carefully recovered from the site trial and visual inspections and short and long-term tensile tests were carried out. The effects of construction activities are evaluated and recommendations for design are given.

1 INTRODUCTION

Before new polymeric geogrids were used in the Civil Engineering industry a controlled site trial was carried out to verify their effectiveness in withstanding the stresses and practices of reinforced soil construction. The paper describes standards for reinforced soil construction within the UK and organisation of the trial to reproduce construction practice.

The performance of 'Tensar' SR80 geogrid is assessed under two fill gradings and under single and multi-layer constructions. On recovery, the geogrids were inspected and tested and these results are presented. The effects of construction are evaluated and trends indentified. Suggestions are given for the application of these results to reinforce soil design.

A similar type of site trial was used as part of the assessment of fitness for purpose of 'Tensar' SR2 geogrids by the British Board of Agrément (Roads and Bridges Agrément Certificate 86/27, 1986) and also by the West German Institut fur Bautechnik in their Approval Certificate (Z20.1 -102, 1986).

2 TEST METHOD

In the UK, reinforced fill is selected and compacted according to the Department of Transport's "Specification for Highway Works" (1986). The compaction requirements are given in the form of a method specification for which the Engineer

1. classifies the fill type and appropriate method for compaction.

2. selects a compactor type and chooses the appropriate size of plant for the construction project.

3. reads from the specification the compacted layer thickness and number of passes.

For the proposed site trial of 'Tensar' SR80 geogrid:

1. A subangular limestone frictional fill was selected because it is very widely used in the UK.

2. A towed vibratory roller of mass 3300 kg/m width of roll was chosen as being typical of construction plant used on a large scale reinforced soil project.

3. The specification indicated that a 200mm thick layer of the correct density could be achieved with four passes of the selected roller.

3 SITE TRIAL

The site trial was carried out in a limestone quarry near Clitheroe in Lancashire, England. The trial area was organised into bays, shown in Figure 1. Two fill gradings are reported (Figure 2). One, designated fine fill, has 100% passing a 10mm sieve and 85% retained on a 5mm sieve. This is used in bays A and B. The other, designated medium fill, has 100% passing a 75mm sieve and 92% retained on a 37.5mm sieve. It is used in bays C and D.

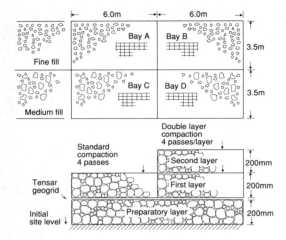

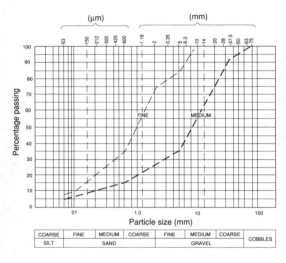

Figure 1 Schematic layout of test bays

A preparatory layer of each bay's fill was compacted on the quarry floor to produce a level surface on which the 'Tensar' SR80 geogrid reinforcing elements were placed. Typical dimensions of 'Tensar' SR80 are shown in Figure 3.

Fill was dropped from an excavator bucket onto the geogrid as shown in Figure 4. This is not good practice but unfortunately fill is often tipped from delivery wagons directly onto the reinforcement, so poor site practice was reproduced. The fill was dozed into a level layer and compacted with a towed vibratory roller as shown in Figure 5.

In bays B and D a second layer was spread and compacted to reproduce multi-layer construction practice.

Figure 4 Excavator dropping fill

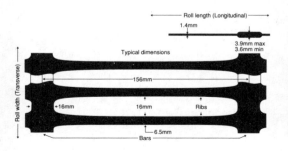

Figure 2 Particle size distribution curves

Figure 3 'Tensar' SR80 geogrid

Figure 5 Compaction

Table 1. Test bay data and visual classification of damage.

Fill Type	Bay code	Mean layer thickness		General abrasion	+ Bruises	+ Splits	+ Cuts
		First	Second				
Fine	A	171mm		25%	0	2	0
Fine	B	225mm	195mm	25%	1	0	0
Medium	C	149mm		63%	40	3	0
Medium	D	224mm	205mm	50%	27	16	0

+ Incidence of damage on six large size test specimens

The mean compacted layer thicknesses in the bays are shown in Table 1.

The 'Tensar' SR80 geogrid then had to be recovered from the trial bays. The bulk of the compacted fill was loosened and removed using the excavator back hoe and the remainder was manually removed using shovels and hand brushing. Any damage caused by recovery operations was clearly marked so that this damage would not be attributed to the trial.

4 SAMPLE PREPARATION

The geogrids were washed to remove any loose aggregate. Large size specimens of 'Tensar' SR80 geogrid, 15 ribs wide by 5 bars long, i.e. approximately 350mm wide by 610mm long, were cut at random from the lengths of recovered material. The specimens were conditioned for 24 hours at 20°C before visual inspection and tensile tests were carried out. Control specimens from the same batch of material were prepared and tested in the same manner.

5 VISUAL ASSESSMENT OF DAMAGE

The 'Tensar' SR80 geogrid specimens were inspected for the following categories of damage:-
  1. General abrasion of the geogrid surface - a subjective assessment.
  2. Bruising or flattening of the ribs or bars.
  3. Splitting of the ribs or bars allowing the passage of light.
  4. Cutting or severance of the ribs.

Six large size specimens from each bay were inspected and the total incidence of damage in each category was recorded as shown in Table 1.

6 SHORT-TERM INDEX TENSILE TESTS

Short-term Index tensile tests were carried out following the procedures developed by McGown et al (1984) for large size geogrid specimens. For this test, the specimens were placed in an Instron 1170 tensile test machine at 20°C, and extended at a constant rate of strain of 2% per minute.

In Figure 6 the load/strain curves show the range of results of test specimens from bay C, medium fill, single layer, against the mean load/strain curve of the control specimen. Figure 7 shows data from bay B, fine fill, double layer.

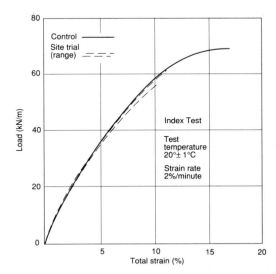

Figure 6   Load/strain curves for 'Tensar' SR80 from bay C and control

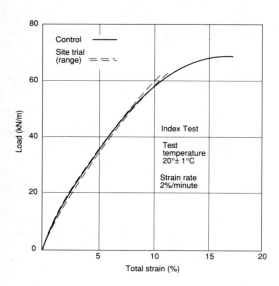

Figure 7   Load/strain curves for 'Tensar'
SR80 from bay B and control

The secant modulus at 5% strain was
calculated and the ratio of the secant
modulus of the site trial specimens to the
control specimens is shown in Table 2.

Table 2. Mean secant modulus at 5% strain
of site damaged specimens expressed as a
percentage of control specimens.

Product	Fill Type	Level of compaction (Number of passes)	
		Standard (4)	Double layer (4/layer)
Tensar SR80	Fine	96.3	100.4
	Medium	99.6	99.1

The peak load was recorded and the ratio
of the peak load of the site trial
specimens to the control specimens is shown
in Table 3.

Table 3. Mean peak tensile strength of site
damaged specimens expressed as a percentage
of control specimens.

Product	Fill Type	Level of compaction (Number of passes)	
		Standard (4)	Double layer (4/layer)
Tensar SR80	Fine	95.7	91.8
	Medium	83.4	88.2

7 LONG-TERM SUSTAINED LOAD TENSILE TESTS

Long-term sustained load tests were carried
out in accordance with the recommendations
of Andrawes et al (1986) and Murray and
McGown (1987). For these tests a load of
36.4 kN/m was applied within 5 seconds to
the large size specimens of 'Tensar' SR80
geogrid and sustained whilst measurements
of elongation with time were recorded. All
tests were carried out at 20°C.

Figure 8 shows the strain/log time curve
of specimens from bay D, medium fill,
double layer and control. Sherby-Dorn
curves of strain/log strain rate for these
specimens, shown in Figure 9, were
constructed from the strain/time data.
These specimens had not reached failure
when long-term testing stopped, but had
passed the performance limit strain of 10%,
used in the design of reinforced soil
structures.

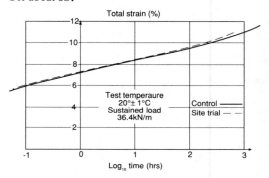

Figure 8   Strain-time curves for 'Tensar'
SR80 from bay D and control

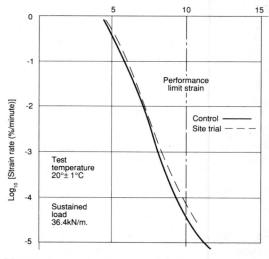

Figure 9   Sherby-Dorn curves for 'Tensar'
SR80 from bay D and control

8 DISCUSSION

The site trial reproduced typical conditions for extensive reinforced soil constructions within the UK and the heavy towed vibrating compactor was similar to those used on projects such as basal reinforcement or steep embankments. However, specifications for reinforced soil walls limit the mass of the compactor within 2 metres of the face in order to reduce compaction stresses on the wall face. Close to a retaining wall face the mass per metre roll of vibratory compactor would be limited to 1300kg compared to the 3300kg of the site trial vibratory compactor, so data derived from this site trial would be conservative when applied to reinforced soil retaining walls.

The trial also reproduced some typical examples of bad construction practice, such as dropping fill directly onto reinforcing elements rather than spreading fill by cascading it forward from a stockpile with a bulldozer.

Since similar construction equipment and techniques are used throughout the world, the results are not restricted to U.K. conditions.

During the recovery of the geogrid it was observed that when the excavator back hoe was loosening compacted fill, the free end of the length of geogrid was also moving. This clearly demonstrated the interlock and interaction between fill and 'Tensar' geogrid.

Visual inspection showed that the fine fill caused very little surface abrasion to the 'Tensar' SR80. The medium fill caused a little more surface abrasion and bruising to the geogrid.

From short-term Index test load/strain curves, examination of the change in secant modulus at 5% shows that there is little difference between any of the four bays and the control material. Working strains of up to 5% could be expected for steep embankment or soft foundation reinforcement and of up to 2% for retaining walls. Thus, neither of the fills used, nor the layered constructions compacted to the specification, alters the physical properties of 'Tensar' SR80 under working conditions.

The short-term Index tests did, however, show that there is a reduction in peak load carrying capacity of 'Tensar' SR80 geogrid. The medium fill influenced this trend more than the fine fill, double layers more than single layers and thinner layers more than thicker layers. Long-term strength of 'Tensar' SR geogrids is quoted for a performance limit strain of 10%. The short-term tests show that strain at rupture exceeds this peformance limit strain, so suggests there is no need to apply a partial factor of safety to account for change in rupture behaviour. This can be examined further with long-term tests.

The long-term sustained load tests have been carried out on geogrids in isolation, so there is no reliance on the effect of soil restraint. This is a lower bound condition and can be considered safe for all sites.

The effects of construction on the long term performance of 'Tensar' SR80 geogrids are found from examination of the strain/log time and Sherby-Dorn curves. These show a slightly higher strain than the control, but this variation is considered within the limits of experimental error.

The Sherby-Dorn curve is used to interpret the change in rate of strain with time and shows that the rate of strain continues to decrease beyond the 10% performance limit strain of 'Tensar' SR80. Furthermore, there is no suggestion of rupture in this range. This is confirmed by the short-term Index tensile tests.

9 DATA FOR DESIGN

The designer of reinforced soil structures needs to be able to assess the effects of construction on creep and rupture performance of the reinforcing elements.

UK practice has been to limit long-term strain (creep) in reinforced soil retaining walls and bridge abutments to the amounts specified in Technical Memorandum BE3/78 (revised 1987), i.e. 1% for retaining walls and 0.5% for bridge abutments. For the range of fill sizes reported, and construction to the UK specification, the experimental work described in this paper demonstrates that the "in isolation" test data of manufactured geogrids can be used with confidence for 'Tensar' SR geogrids to assess the effects of creep.

To guard against rupture of the reinforcing elements the designer uses a characteristic strength, above which the material will fail in tension from peak loading during the design life, and applies an overall factor of safety to take account of variations in loading and material properties.

The characteristic strength of 'Tensar' SR80 for a design life of 120 years and a performance limit strain of 10% is taken from manufacturer's literature (Netlon Limited, 1988), and is given in Table 4.

Table 4. Characteristic strength of 'Tensar' SR80 for a design life of 120 years.

In soil temperature	Characteristic strength of 'Tensar' SR80
10°C UK conditions	32.5 kN/m
20°C	30.5 kN/m

In the overall factor of safety there is a partial factor of safety, γ_{m2} which is intended to cover loss of strength due to site damage and non-uniform stress distribution across the reinforcement due to construction errors such as mis-alignment of the reinforcement and undulation of the compacted fill. Suggested values of γ_{m2} for 'Tensar' SR80 based on loss of peak strength are given in Table 5.

Table 5. Partial factors of safety γ_{m2}

Fill type	Well-graded fill of maximum particle size (mm)	Partial factors of safety γ_{m2} for 'Tensar' SR80
	——— 125 ———	
Coarse grained soils & crushed rocks		1.40
	——— 75 ———	
		1.30
	——— 20 ———	
		1.20
	——— 2 ———	
Sand, clay, PFA		1.10

However, this design approach for rupture appears to be unnecessarily conservative since it has now been shown that the long-term and short-term properties of 'Tensar' SR80 geogrid, under the range of fills compacted according to the specification, remain unaltered up to a performance limit strain of 10%.
It is suggested that for fills up to 75mm maximum size, compacted according to the specification, no partial factor of safety need be applied automatically to the long-term strength of the geogrid, but that the designers should consider what construction errors may arise outside the specified conditions and, if necessary, make a small allowance only for those conditions.

REFERENCES

Andrawes, K.Z., McGown, A. and Murray, R.T. 1986. The load-strain-time-temperature behaviour of geotextiles and geogrids. Proc. 3rd Int. Conf. on Geotextiles, Vienna, Vol. 3: 707-712.

Department of Transport, 1986. Specification for Highway Works, London: H.M.S.O.

Institut fur Bautechnik. Approval Certificate Z20.1 - 102, 1986. Reinforced soil structures with SR2 geogrids made of HDPE. Berlin.

McGown, A., Andrawes, K.Z., Yeo, K.C. and DuBois, D. 1984. The load-strain-time behaviour of 'Tensar' geogrids. Polymer Grid Reinforcement: 11-17, Thomas Telford Limited, London.

Murray, R.T. and McGown, A. 1987. Geotextile test procedures: background and sustained load testing, T.R.R.L. Application Guide 5, Dept. of Transport, UK.

Netlon Limited, 1988. Test methods and physical properties of 'Tensar' geogrids. Revised edition, Blackburn.

Roads and Bridges Agrement Certificate No 86/27 (1986). 'Tensar' SR2 polymer grid for reinforced soil walls. British Board of Agrement, Hertfordshire.

Technical Memorandum BE3/78 (revised 1987). Reinforced and anchored earth retaining walls and bridge abutments for embankments. Department of Transport, UK.

Natural materials for soil reinforcement

K.R.Datye
Bombay, India

ABSTRACT : The paper presents results of recent research on use of natural materials for soil reinforcement. The most promising protection methods are described e.g. coating by low density polyethylene (LDPE) melt of composite fabrics of natural fibre/ high density polyethylene (HDPE) slit tapes and ferrocement coating for small timber. Details of reinforcement are furnished.

1 NATURAL MATERIALS FOR SOIL REINFORCEMENT

In the past, natural materials have been looked upon as low cost substitutes for synthetics to be used in noncritical applications. The general belief has been that their strengths are uncertain and variable and the hazard of biodegradation renders them unsuitable for important structures.

Studies of material characteristics carried out by various researchers [Pama (1978), Fang (1981), Mwamila (1983), Yamanouchi (1986)] in the past have already established that fairly consistent characteristics are realised by judicious choice of natural materials.

Traditionally natural materials have been extensively used in the past for soil reinforcements. But these were not engineering applications i.e. no specific requirement regarding materials were stipulated nor was any scientific evaluation made of their performance.

Recent studies have brought out clearly that strength and deformation characteristics of several natural materials such as bamboo, timber and fibres compare very favourably with synthetics. In fact, with regard to the deformation modulus and creep, the natural materials are superiror to synthetics.

The main problems are the protection from biodegradation and attack from organisms. It is also necessary to develop suitable forms of reinforcement for various applications. These are illustrated in figure 1.

Various treatments were evaluated such as:

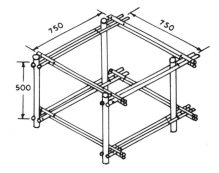

(a) Small timber crib frams

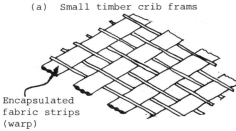

Encapsulated
fabric strips
(warp)
(b) Encapsulated natural fabric strips
(High performance geogrid equivalent)

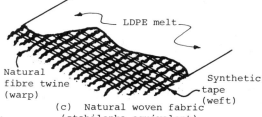

LDPE melt

Natural
fibre twine
(warp)

Synthetic
tape
(weft)

(c) Natural woven fabric
(stabilenka equivalent)
Fig. 1 : Forms of reinforcement

- Impregnation with water and oil borne preservatives.
- impregnation with synthetic polymers
- coating with cold setting liquid resins
- coating with synthetic melts
- encasement in concrete or cement composites e.g. ferro cement.

In the following paragraphs the most promising techniques are described and information presented regarding the techniques which have been tested on a small scale by fabricating trial pieces and using them in the field.

2 WOVEN FABRICS WITH NATURAL FIBRES

Tne fabric is made by weaving natural fibre twines in warp direction with HDPE slit tapes in weft direction. So far mechanised weaving on power looms has not been successful as the slit tapes have a tendency to get warped and twisted. Handlooms have been used successfully and the handloom weaving technique is cost effective in India. In fact, it helps to optimise the fabric design for specific use as range of applications is wide and a variety of products could be manufactured such as narrow tapes, wide strips and fabric rolls. Synthetic melt application is made possible by use of synthetic slit tapes in the weft direction. The melt provides a tough encapsulation, which would adequately protect the fabric against moisture ingress. Water soluble preservatives and fungicides successfully used in the past for treatment of timber, bamboo and thatch can effectively be used for treatment of the natural fibre. Further improvement can be achieved by coating the natural fibre with water repellants such as silicone and PVC emulsions.

Characteristics of the composite fabrics are compared with some of the high performance synthetic woven fabrics in table 1. The cost advantages are evident, the material characteristics are superior with regard to the deformation modulus and the composite fabric of the proposed design have good potential for manufacture in developing countries. It must be emphasised here that the manufacturing process involves a technology blend i.e. the natural fibres made in small units are combined with synthetics, woven again by small industrial units, but the final processing is done through an organised industrial unit where by necessary quality control and design optimisation can be achieved.

There is another avenue i.e. use of natural fibres coated by tough impermeable resins such as polyurethane. The manufacturing trial and experimental work is in an initial stage. The process seems to be promising but it is premature at this stage to arrive at a firm conclusion regarding the suitability of woven fabrics constituted of natural fibres coated by resins. There is an attractive prospect of using fabric of natural woven material for the geocells. The geocells by virtue of their higher modulus are expected to perform better than synthetic geocells as would be evident from table 2.

3 SMALL TIMBER FOR SOIL REINFORCEMENT FACING & CRIBS

The designs are based on use of pieces of 40 to 60 mm diameter small timber

Table 1. Comparison of costs and characteristics of synthetic and natural woven fabric for reinforcement

Description	Material	Weight (gm/m^2) & material price (Rs/Kg)	Ultimate tensile strength KN/M & strain at break (%)	Modulus at break (KN/M)	$Cost/m^2$ (Rs)
Synthetic woven (Stabilenka 200)	Polyester yarn	. 460 @ 120	200 (10)	2000	55 IE
Natural woven	. Sisal fibre . HDPE slit tape . LDPE melt	. 528 @17 . 50 @70 . 120 @60	200 (4)	5000	20 LE

IE - Estimated at current international prices
LE - Estimated at current local prices
1 US $ = Indian Rs. 12.8

Table 2.   Comparison of costs and characteristics of synthetic and natural geocells

Material	Geocell Description & seam type	Tensile stregnth (Kn/M) & strain (%)	Modulus (KN/M)	Material requirement (kg/m²) & unit price (Rs/kg)	Cost of geocell (Rs./m²)
Unreinforced HDPE (after Koerner-'86)	200 mm (d) x 230 mm cell size, Ultra-sonic welding	110¢ (13)	846	3.0 @ 70	210 IE
Woven Natural Fibre	100 mm (d) x 200 mm cell size, Mechanical stitching	110 (2)	5500	. Sisal-0.96 @ 25 * . HDPE- 0.06 @ 70 . LDPE- 0.04 @ 60	31 LE

¢ From Gundle H D specifications ;   * includes cost of polyurethane coating
IE - Estimated at current international prices, LE - Estimated at current local prices
1 US $ = Indian Rs. 12.8

in lengths of 1.5 to 3 m. The use of small timber is made possible by development of a shear connector design. This design has been tested in the field over the last two years. One overflow crib dam has been built which has been performing satisfactorily for one year. Numerous tests have been carried out on samples fabricated in small workshops. The joint strength in single shear is found to be at least equal to the yield load of bolt and thus an allowable joint strength of 1200 kg has been realised for 40 mm dia timber. As the bolts are fixed in oversized holes cracks are neither initiated nor propagated along the bolt holes. The geometry of the contacting surfaces where the load transfer is achieved is such that stress concentrations are avoided; there are no sharp corners or notches and the confining forces mobilised by the bolts ensure that minor cracks and defects would not compromise the joint strength. The studies have substantiated the feasibility of manufacturing the small timber reinforcing elements to get consistent performance, using very simple tools. The confirmatory tests are underway and the results would be presented at the International Conference on timber engineering, Seattle, U.S.A. in Sept. '88.

The small timber cribs would be cost effective as compared to concrete cribs. Potential application of small timber cribs are illustrated in Fig.2. By judicious design, a combination of small timber ties and crib facings would replace gabions. The small timber reinforcements are eminently suitable for anchored wall constructions such as those developed by Fukuoka et al 1986 (Fig. 3). The extensions would be small and the form of joints used would impart adequate flexibility. Timber reinforcements would compare very favourably with high performance synthetic grids such as Signode, table 3. They can also replace steel rods and short pile reinforcements for bases of embankments on soft soils such as those described by Broms (1987).

Protection of timber members by ferrocement is a very worthwhile and workable system. By using colloidal silica admixtures very tough and durable cement mortar coats of 6 to 8 mm thickness can be produced. Corrosion resistance can be achieved by use of galvanised iron, aluminium or high modulus synthetic meshes. By using a suitable coating and providing joints in the protective annulus, it could be ensured that the coating would be relieved of stress and therefore cracking would be avoided.

4   CONCLUDING REMARKS

Many researchers have recognised the superior deformation characteristics of natural reinforcement materials; their energy saving potential and economic benefits are self-evident. Yet engineering applications on a significant scale have not so far been possible. This paper describes application details and suggests cost effective preservation techniques which aim at overcoming the deficiencies of natural materials.

71

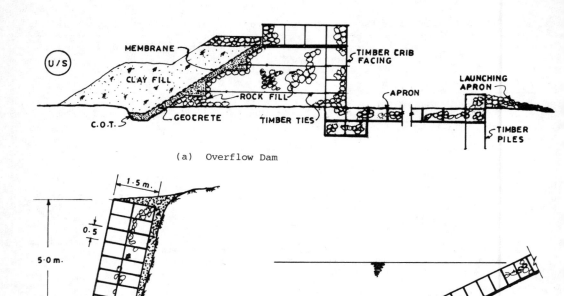

(a) Overflow Dam

(b) Retaining wall   c) Flexible revetment

Fig. 2 : Applications of small timber crib

Table 3. Comparison of costs and characteristics of synthetic and natural geogrids for reinforcement

Description	Material & price (Rs/kg)	Weight (gm/m^2)	Strength at 2% strain (KN/m)	Modulus at 2% strain (KN/m)	Approximate cost (Rs/m^2)	Cost at Equivalent strength (Rs./m^2)
Signode TNX-5001	Polyester (PETP) @ 164	550	46	2300	90 IE	90
Tensar SRI	HDPE @ 75	872	23	1150	65 IE	130
Timber (40 mm Ø)	Treated timber @ 4.0	1430	25 ¶	1250*	11.5¢	21

IE – Estimated at current international prices, LE – Estimated at current local prices
¶ – Allowable stress (100 kg/cm^2) governs
* – Strain at allowable stress is less than 0.1%
¢ – Including fixture & mortar coat which are 40% and 60% respectively of the cost of treated timber.
1 US $ = Indian Rs. 12.8

72

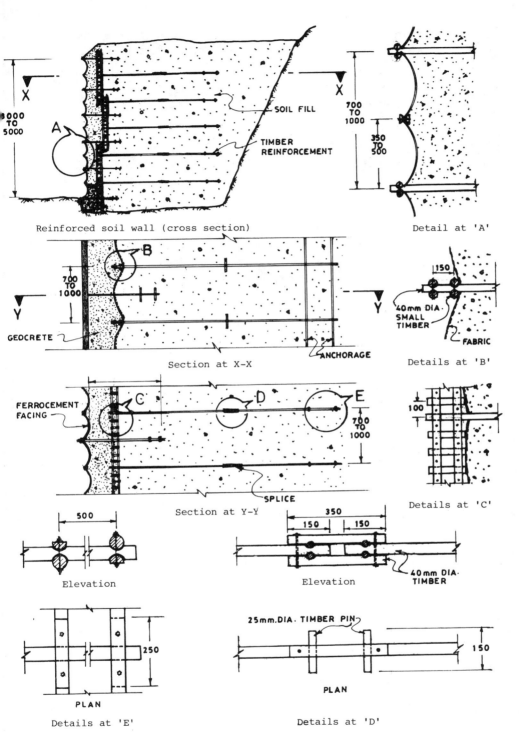

Reinforced soil wall (cross section)

Detail at 'A'

Section at X-X

Details at 'B'

Section at Y-Y

Details at 'C'

Elevation

Elevation

Details at 'E'

Details at 'D'

Fig. 3 : Timber reinforced soil wall

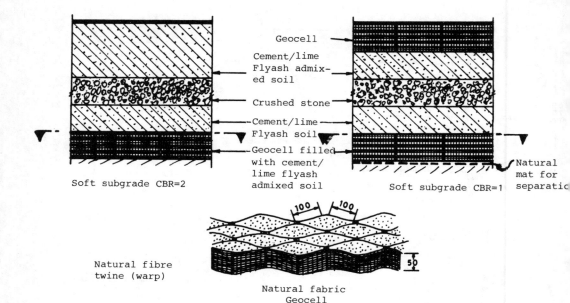

Fig. 4 : Geocell reinforced road base

It is suggested that the proposed protection methods should be subjected to extensive field evaluation. This could best be done without any risk, in applications such as timber cribs for erosion control, soil conservation structures (Fig. 2) and geocells for road bases (Fig. 4), where periodic replacement of the reinforcement is possible.

REFERENCES

Broms, B.B. 1987. Fabric reinforced soil. Proc. of the post vienna conference on geotextiles, Singapore.

Datye, K.R. 1987. Geotextile use in India - Recent experience and suggested developments. Proc. of the post vienna conference on geotextiles, Singapore.

Fang, H.Y. 1981. Low cost construction materials and foundation structures. Proc. of the second Australian conference on engineering materials, Sydney.

Fukuoka, M. etal. 1986. Fabric faced retaining wall with multiple anchors. Geotextiles and Geomembranes, Vol.4, Nos.3 & 4, special issue on geotextiles in Japan. England: Elsevier Applied Science Publishers Ltd.

Koerner, R.M. 1986. Designing with geo-synthetics. New Jersey: Prentice Hall.

Mwamilla, B.L.M. 1983. Reinforcement of concrete beams with low modulus materials in form of twines. Proc. of the symposium on appropriate building materials for low cost housing, Nairobi, Kenya.

Pama, R.P. etal. Development of bamboo pulp boards for low cost housing. mimeo (1978).

Yamanouchi, T. 1986. The use of natural and synthetic geotextiles in Japan. Proc. of IEM-JSSME joint symposium on geotechnical problems, Kuala Lampur, Malaysia.

International Geotechnical Symposium on Theory and Practice of Earth Reinforcement / Fukuoka Japan / 5-7 October 1988
© 1988 Balkema, Rotterdam. ISBN 90 6191 820 0

Dutch progress in the standardization of geotextile test methods

M.Th.de Groot
Delft Geotechnics, Netherlands

G.den Hoedt
Akzo Industrial Systems, Netherlands

A.H.J.Nijhof
Technical University Delft, Netherlands

ABSTRACT: Dutch progress is reported in the standardization of test methods for the determination of geotextile properties. Draft standards of permittivity and characteristic pore size tests are outlined.
Results are presented of an interlaboratory test programme on tensile strength test methods. The influence of specimen width and clamp system is discussed.

1 INTRODUCTION

The concept of functional design becomes more and more accepted in the geotextiles engineering world.
The general idea of functional design is well described by Koerner (1986) and takes as a starting point the functions that a geotextile has to fulfil in a specific application.
It goes without saying that standard methods for testing all possible properties of geotextiles which relate to those functions are vital for both industry, designers and last but not least, for the structure of which the geotextile becomes a part.

This paper deals with results that were obtained so far by the hydraulic and mechanical working groups of the Dutch Standardization Committee on Geotextiles (353 50). As yet both working groups confined themselves to so-called index tests. Performance tests will be dealt with later on.

The hydraulic group has drafted a standard on the determination of the permittivity and of the characteristic pore size of geotextiles, see 2.

Up to now the Working Group on Mechanical Properties has focused mainly on tensile properties.

Where it is believed appropriate the Dutch Standardization Committee seeks agreement with international developments.

2 HYDRAULIC PROPERTIES

2.1 Introduction

Hydraulic aspects are involved in the filter function, the separation function and the drainage function of geotextiles. Very often these functions are of primary interest, e.g. in erosion control.

In terms of functional design, the main properties of geotextiles to fulfil these functions are:
• permittivity, related to water permeability through the geotextile
• transmissivity, related to water permeability in the geotextile plane
• characteristic pore size, related to its capacity of soil retention.

2.2 Permittivity

The permittivity of a geotextile is defined as the ratio between the filter velocity and the hydraulic head over the geotextile specimen.

$$\psi = \frac{u_f}{\Delta h} \qquad (1)$$

where

ψ is the permittivity in s^{-1}
u_f is the filter velocity through the geotextile in m/s
Δh is the head in m

Groundwater flow generally is expressed in terms of permeability coefficients (m/s).

Taking a geotextile as a porous medium,

through the concept of the permittivity ψ one can arrive at a geotextile permeability coefficient k by assuming Darcy's law.

$$u_f = k \cdot \frac{\Delta h}{t_g} \qquad (2)$$

where t_g is the geotextile thickness in m

Comparison of (1) and (2) gives

$$\psi = \frac{k}{t_g}$$

Yet, this formula is too much simplified, since it does not take into account the nature of the flow regime.

With sufficiently low filter velocities flow through the geotextile will be laminar and there is a linear relation between filter velocity and hydraulic head, i.e. ψ is a constant.

With higher filter velocities, however, flow through the geotextile is no longer laminar, it becomes turbulent and the permittivity ψ becomes a function of the filter velocity. Therefore a standard filter velocity must be used.

A second reason why permittivity and permeability are not readily correlated concerns Darcy's law, which is valid for laminar flow conditions.
Following Forchheimer (1901) the non-Darcy relation between filter velocity and hydraulic head reads:

$$\frac{\Delta h}{t_g} = a \frac{u_f}{k_g} + b\left(\frac{u_f}{k_g}\right)^2 \qquad (3)$$

where a and b are dimensionless constants and k_g is a geotextile constant with dimension m/s.
Darcy's law follows from (3) with b = 0. Pure turbulent flow results if a = 0.
The permittivity ψ can now be obtained by rewriting (3). It follows

$$\frac{1}{\psi} = t_g\left(\frac{a}{k_g} + \frac{b\,u_f}{k_g^2}\right) \qquad (4)$$

again describing ψ as a function of the filter velocity.
Although the existing Dutch draft standard has not been based on the Forchheimer approach, it is believed that an ideal test method would determine the constants a and b and thus establish relation (4), which is valid for both laminar and turbulent flow conditions.
The flow type dependency of the permittivity has lead to the use of a standard filter velocity u_s of 10 mm/s. The alternative of a standard hydraulic head has been discarded since the criterion for the flow regime, the Reynolds number, depends on the filter velocity, not on the hydraulic head.

2.3 Permittivity test procedure

The Dutch draft standard method for the determination of the permittivity precisely describes the principles, terminology, test equipment, test execution including sampling and specimen preparation, calculations and logging of relevant data.
As an example Table 1 shows results of five measurements on one specimen.

From Table 1 it can be derived, that at the standard velocity u_s = 10 mm/s the head equals 36 mm.
The average temperature was 13 °C, giving a temperature corrected standard hydraulic head of 39 mm.
The standard permittivity ψ_s becomes

$$\psi_s = \frac{u_s}{\Delta h_s} = \frac{10}{39} = 0.26 \ s^{-1}$$

Table 1. Measurements on one specimen for permittivity calculation

test no.	head Δh	mass of passed water m_i	time interval t i	water temp.	filter velocity $u_i = \dfrac{1000\,m_i}{A\,t_i}$
	(mm)	(g)	(s)	(°C)	(mm/s)
1	10.5	161	20.0	13.4	4.1
2	4.5	113	24.0	13.2	2.4
3	27	215	15.0	13.1	7.3
4	63	316	10.0	13.0	16.1
5	149	529	10.0	12.8	26.9

A is the effective flow area = 1963 mm²

Table 2. Sieve test results for the characteristic pore size determination

specimen	sand fractions (μm)	D_m (μm)	amount of passed sand (g)	amount of sand on and in the geotextile (g)	percentage of sand on and in the geotextile (%)
1	250 - 300	275	8.01	41.99	83.98
	300 - 355	328	4.03	45.97	91.94
2	250 - 300	275	9.11	40.89	81.78
	300 - 355	328	3.96	46.04	92.08
3	250 - 300	275	7.84	42.16	84.32
	300 - 355	328	3.38	46.62	93.24
4	250 - 300	275	10.23	39.77	79.54
	300 - 355	328	5.53	44.47	88.94
	355 - 425	390	2.77	47.23	94.46
5	250 - 300	275	7.01	42.99	85.98
	300 - 355	328	2.63	47.37	94.74

2.4 Characteristic pore size test procedure

The draft Dutch standard method for the determination of the characteristic pore size is based on the Delft Hydraulics method of dry sieving with well specified sand fractions according to NEN 2560 (Veldhuijzen van Zanten, 1986: 176-213).

The pore size O(p) is defined as the pore size in the geotextile that equals the average grain size of a hypothetical sand fraction, of which p% remains on and in the geotextile after execution of the standard test.
The execution of the standard test is precisely described and includes principles, terminology, test equipment, sampling and specimen preparation calculations and logging of data. Subsequent sievings with 50 g of selected sand fractions is prescribed during 300 ± 2 s with a vertical amplitude of 0.75 mm and a frequency of 50 Hz.
As an example of the determination of the characteristic pore size consider table 2, where the relevant test quantities are listed. D_m is the average grain size of the selected sand fractions.
Mean value and standard deviation per sieved sand fraction of sand percentage that remain in and on the geotextile can now easily be determined. Finally the O(90) can graphically be derived giving O(90) = 314 μm.

3 MECHANICAL PROPERTIES

The purpose would be to standardize only one test method that applies for both woven and non-woven fabrics (ISO TC38/SC21). With this in mind the Strip Tensile Test and Wide-Width-Tensile-Test methods are discussed below.

3.1 Strip Tensile Test (S.T.T.)

The S.T.T. (e.g. ISO 5081, width/length = 50/200) has been used for many years to assess the load-elongation characteristics of textile fabrics.
For geotextiles (ranging from high-elongation non-wovens with a strength of 5 kN/m to coarse, sometimes high-modulus, woven fabrics with a strength of 1000 kN/m) the S.T.T. however has some deficiencies.
For geotextiles with a high lateral contraction (most non-wovens, some types of woven fabrics) the S.T.T. does not give a fair indication about the real strength of the geotextile used in full width (generally of the order of 5 m). The S.T.T. can yield either much too low values, or (even) too high values. Which one will be the case depends on the geotextile structure, and is related to the degree of freedom of the constituent fibres or threads to reorientate during the tensile process. It also depends on the clamping system (fixed or frictional). In principle the S.T.T. is suitable only for (light) woven fabrics with little crimp, i.e. almost without any lateral contraction.

3.2 Wide-Width-Tensile-Test (W.W.T.T.)

To overcome some of the drawbacks of the S.T.T. for geotextiles, the W.W.T.T. has been developed (e.g. AFNOR 38014, W/L = 500/100).
Here, at least in the central part of the specimen, the lateral contraction is nearly zero, so more or less the desired plane strain state is reached (i.e. deformations are limited to planes in tensile direction and perpendicular to the specimen).

At the edgings of the specimen, however, lateral contraction still develops, leading to stress concentrations and initiation of failure.
This can be observed especially at the corners of the specimen, where the discontinuity from the fixed, compressed state in the clamp to the free state outside the clamp has its influence.

Nevertheless, the use of stiff, fixed clamping systems is essential for the W.W.T.T., otherwise the state of plane strain will not be obtained anywhere. For capstan clamps the effective gauge length is increased so far that a high W/L ratio cannot be accomodated on existing tensile machines.

For strong woven fabrics the W.W.T.T. has comparable drawbacks, as with large specimen widths the load capacity of nearly all testing machines will appear to be insufficient. Furthermore, problems related to satisfactory clamp constructions will increase excessively.

To investigate acceptable compromises between the S.T.T. and the W.W.T.T. the Dutch Interlaboratory Test Programme has been set up (see 3.4).

3.3 Test procedure of load-elongation measurement

The edges of a specimen can be prepared in different ways. It is preferred to prepare the width of the woven specimen by ravelling and that of the non-wovens by cutting.
The very measurement of the force is technically not a problem.

A correct measurement of the extension is more difficult, however. In practice the displacement of the (stiff) clamps is often used as a measure for the extension of the specimen. Because of the slipping of the specimen this measurement is not correct.
At the same time the extension of the material, especially in inhomogeneous non-wovens, varies a lot from place to place.

For calibration reasons a pre-tension of the specimen shall be prescribed, e.g. 1 % of the nominal tensile strength.
The origin of the strain axis coincides with the pre-tensioned state. The so-called daylight point (BS 6906: Part 1) cannot be measured in a reproducible way and is unusable.
In the case of capstan clamps the use of an extensometer is a must.

Mostly the rate of extension is realized by a constant rate of the moving clamp, but should be expressed for comparison reasons in rate of strain (e.g. in %/min). Because various materials have different strains at break, it is important to prescribe an average time-to-break. The textile standard is either 20 or 30 s, but in view of the highly extensible non-wovens the according rate of extension can be too high. A time-to-break between 1 and 2 minutes is therefore suggested.

3.4 The Dutch Interlaboratory Test Programme

In the Netherlands an interlaboratory test programme was set up, in which 6 laboratories participated.
For this test 10 types of current geotextiles have been selected, together covering a wide range of production techniques and stress-strain behaviour (see figure 1 and table 3).

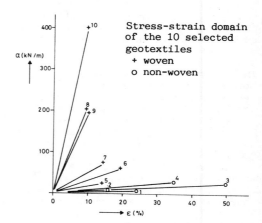

Fig. 1

All important geotextile structures and materials are represented in this scheme. Most difficulties were to be expected both with geotextiles exhibiting high elongation/high lateral contraction and with high-strength, high-modulus woven fabrics.

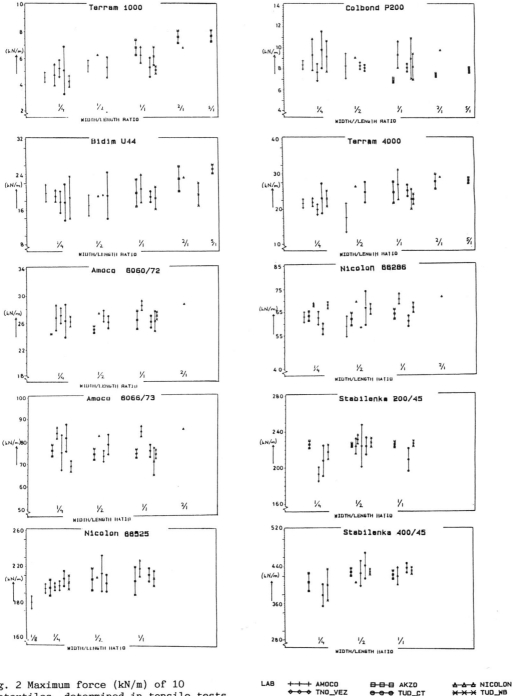

Fig. 2 Maximum force (kN/m) of 10 geotextiles, determined in tensile tests by 6 laboratories at different width/length ratio's.

LAB		
+–+–+ AMOCO	🔲–🔲–🔲 AKZO	▲–▲–▲ NICOLON
◆–◆–◆ TNO_VEZ	⊖–⊖–⊖ TUD_CT	✕–✕–✕ TUD_WB

Table 3: Description of the 10 geotextiles

	Type	structure	polymer	brand	nominal values		
					weight (g/m²)	α (kN/m)	ε_f (%)
1	NW	fil., TB	PP/PE	Terram 1000	140	8	24
2	NW	sf, NP, CB	PETP	Colbond P200	200	8	16
3	NW	fil., NP	PETP	Bidim U44	340	24	50
4	NW	tapes	PP/PE	Terram 4000	370	28	35
5	W	fil., TB	PP	Amoco 6060	140	24	14
6	W	multifil.	PA	Nicolon 66286	215	65	20
7	W	splitfibre	PP	Amoco 6066	520	74	15
8	W	multifil.	PETP	Stabilenka 200	450	200	10
9	W	splitfibre	PP	Nicolon 66525	780	200	10
10	W	multifil.	PETP	Stabilenka 400	820	400	10

Used abbreviations and symbols:

α	tensile strength per unit width (kN/m)	NP	needle punched
		CB	chemical bonding
ε_f	elongation at max. force	TB	thermobonding
NW	non-woven	PA	polyamide
W	woven	PE	polyethylene
fil.	filament	PETP	polyester
sf	staple fibre	PP	polypropylene

The six participating laboratories were:
- Akzo Industrial Systems, Arnhem
- Amoco Fabrics, Gronau
- Nicolon, Almelo
- TNO-Fibre Institute, Delft
- Techn. University Delft,
  Dept. of Civil Engineering
- Techn. University Delft,
  Dept. of Mech. Engineering

The laboratories agreed upon the following regulations for the interlaboratory test:
- registration of test conditions (normally T = 20 °C, RH = 65 %)
- tests only in length direction
- time-to-failure between 1 and 2 minutes
- number of specimens per sample per test width: 5
- test width (if possible): 50, 100, 200 and 500 mm

The results concerning the tensile strength are presented in Fig. 2, where vertical lines represent 95 % confidence intervals.
Results concerning elongations, lateral contraction and modulus are still in statistical evaluation, and cannot yet be presented.

4 PRELIMINARY CONCLUSIONS

1. In general the geotextile permittivity is a function of the flow regime and thus of the filter velocity. Therefore a standard filter velocity for an index test must be used.

2. The concept of permittivity may become a reliable design tool when properly related to both laminar and turbulent flow conditions. Forchheimers approach is suggested.
3. The Wide-Width-Tensile-Test offers a possible solution to some problems related to lateral contraction of non-wovens; for strong woven fabrics, however, a large specimen width increases the testing difficulties significantly.
4. For non-woven fabrics the use of stiff clamps can be recommended. For woven fabrics the use of capstan clamps is often favourable.
5. For woven fabrics a specimen width of 100 mm could be recommended. The width/length ratio can be chosen between 1/2 and 1/1.
6. For non-woven fabrics a specimen width of 200 mm could be recommended, in combination with a length of 100 mm, (W/L = 2).

REFERENCES

Veldhuijzen van Zanten, R. (ed.) 1986. Geotextiles and Geomembranes in Civil Engineering. Rotterdam: Balkema

Koerner, R.M., 1986 Designing with Geosynthetics. Englewood Cliffs: Prentice-Hall

Forchheimer, P. 1901. Wasserbewegung durch Boden. Zeit. Ver. Deutsche Ing., 45

International Geotechnical Symposium on Theory and Practice of Earth Reinforcement / Fukuoka Japan / 5-7 October 1988
© *1988 Balkema, Rotterdam. ISBN 90 6191 820 0*

Clay geotextile interaction in in-soil tensile tests

K.J.Fabian
Coffey and Partners, Darwin, NT, Australia
(Formerly: Queensland Institute of Technology)

A.B.Fourie
University of Queensland, Brisbane, Australia

ABSTRACT: In-soil tensile tests were carried out on woven and non-woven needle punched geotextile specimens in clay confinement. It was found that the geotextile modulus greatly increased due to the confinement. The increase of the modulus was inversely proportional to the geotextile strain. Due to the different mechanism of the clay geotextile interaction the increase on the non-woven geotextile was considerably larger (up to ten times) than on the woven geotextile (up to three times).

1 INTRODUCTION

The applicability of cohesive backfill in geotextile reinforced retaining walls is a economically important research topic in areas where good quality backfill is not readily available. Previous research has shown that those geotextiles which have high transmissivity and are able therefore to drain cohesive soil can be effectively used to increase the load bearing capacity of clay: Fabian and Fourie (1985), Ingold (1981), Tatsuoka and al (1986). Non-woven needle-punched geotextiles have high transmissivity as the fibres of the fabric are oriented randomly and relatively loosely, providing large void content and seepage paths in the fabric. These characteristics of the manufacturing of the non-woven geotextiles are also the cause of the large deformability which is an important mechanical property of non-wovens.

In certain applications, where the displacements and deformations are normally required to be kept low such as retaining walls, the large deformability is not advantageous.

In-soil tensile tests using granular soil have shown that the soil confinement decreases the deformability of both woven and non-woven geotextiles: McGown and al (1982), Leshchinsky and Field (1987).

It is important to investigate this phenomenon in the case of clay backfill with particular emphasis on non-woven

geotextiles which otherwise could be very effectively used in cohesive backfill.

2 MATERIALS USED IN THE TESTS

The soil used in the tests was a silty clay (CL) with the soil indices given in Table 1.

Table 1. Soil Indices of Silty Clay

Plastic Limit	PL = 14%
Liquid Limit	LL = 28%
Plasticity Index	PI = 14%
Linear Shrinkage	LS = 4%
Unified Classification = CL	

The particle size distribution curve of the soil is shown on Figure 1.

Standard compaction tests were carried out on the silty clay and the maximum dry density was obtained as MDD = $1.95/m^3$ at the optimum moisture content of OMC = 12%. The Standard compaction curve of the silty clay is shown on Figure 2.

In the tests the soil was prepared at 19% moisture content. The soil layers -300,500,700 mm long; 100 mm thick; and 500 mm wide, were compacted with 110,190 and 270 blows, respectively, using a 15 kg mass dropped from 0.5 m height to provide a compaction effort 57 kg.m/litre which is equivalent to the compaction effort of the Standard compaction. At 19% moisture content and using Standard compaction the degree of saturation of the soil was about 95%. Thus the soil was deemed to be saturated.

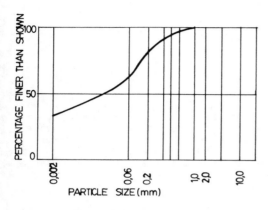

Figure 1: Particle Size Distribution Curve

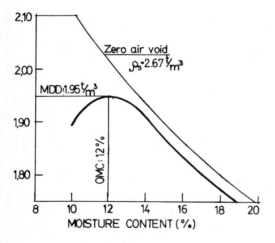

Figure 2: Standard Compaction Curve

Two geotextiles: a woven polypropylene and a non-woven, needle punched polyester were used in the tests. The most important parameters are summarised in Table 2. The results of tensile tests carried out on 200 mm wide strip specimens are collected on Figure 3.

Table 2.   Parameters of Geotextiles

Type	Non-woven	Woven
Thickness (mm)	1.8/0.8*	1.6
Weight (g/m²)	210	155
EOS (5m)**	60	250
Permeability (cm/s)	0.3/0.07*	0.003
Transmissivity (cm/s)	0.06/0.04*	N/A
Break load (kN/m)	10	25
Tangent modulus (kN/m)	28	197

* at 0.5,200 kPa normal stress, respectively

** equivalent opening size

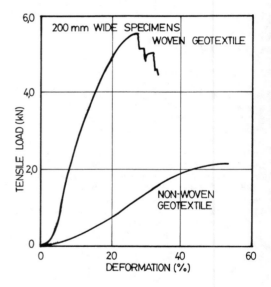

Figure 3: Results of Wide Strip Tensile Tests

3   TESTING PROCEDURE

To carry out the large scale pull-out (in-soil tensile) tests a large steel box was built. It allowed no relative lateral displacement between the soil layers on the two sides of the geotextile. The schematic diagram of the testing configuration is shown on Figure 4. The vertical load was applied from a load frame using a hydraulic ram and the load was distributed over the soil by a rigid steel plate.

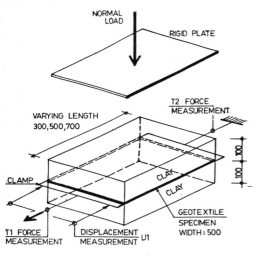

NORMAL LOAD

RIGID PLATE

T2 FORCE MEASUREMENT

VARYING LENGTH 300,500,700

100   100

CLAMP

CLAY

CLAY

GEOTEXTILE SPECIMEN WIDTH: 500

T1 FORCE MEASUREMENT

DISPLACEMENT MEASUREMENT U1

(ALL DIMENSIONS IN MM)

Figure 4: Schematic Diagram of Test Configuration

After the preparation and placement of the soil the geotextile sample was clamped on both ends. The in-soil tensile tests of this research were intended to investigate the effect of soil confinement on the geotextile modulus at low strains (max 10%) which are typical in retaining wall problems. The steel clamps were placed adjacent to the soil. Considering the width of the geotextile specimens (500 mm) and the small elongation (less than 35 mm) the aspect ratio of the unconfined section of the geotextile specimen was very large, the in-soil clamping was not regarded therefore important. No necking of the geotextile was evident in any of the tests carried out.

The clamps were attached to two proving rings to measure the load on both ends of the goetextile. The first proving ring (on the side of the application of the tensile force) measured the pull-out force (T1) exerted by a hydraulic pulling ram. The second proving ring anchored the other side of the geotextile specimen and measured the T2 anchoring force.

Two dial gauges were attached to the first clamp at equal distances from the point of application of the tensile (pull-out) force to measure the displacement-U1.

After preparation the normal load was applied using a hydraulic ram. Tests were carried out at 15,50 and 100 kPa normal

stresses. The dial gauges and proving rings were zeroed and then the lateral tensile force was applied. The rate of loading was 300 N/min. The tests were terminated at the failure of the geotextile specimen or at bond failure or at 4.2 kN tensile force, whichever was reached first. The loading rate was considered rapid enough to simulate undrained loading.

Before and after every test the length of the geotextile specimens were measured to determine whether any plastic deformation occured. No plastic deformation was detected on any of the specimens.

4 TESTING RESULTS

Typical measured load displacement curves are shown for the non-woven geotextile on Figure 5, and for the woven geotextile on Figure 6. From the load displacement curves the geotextile modulus can be determined using the method introduced by Leshchinsky and Field (1987). In this method it is assumed that the geotextile tensile strain has its maximum at the point of application of the tensile force and that it decreases linearly away from this maximum to zero at the other end of the textile specimen. The maximum geotextile strain may be calculated as:

$$\gamma = 2 * U1 / (L + U1) \qquad (1)$$

where L – length of geotextile specimen
U1 – displacement of geotextile at the point of application of the load

Assuming that the tensile force in the geotextile is linearly proportional to the geotextile strain, the geotextile modulus can be determined as:

$$E = (-L *(T1 + T2)/2U1 \qquad (2)$$

where E – geotextile modulus (kN/m)
T1 – force at the point of application of the load (kN/m)
T2 – force at the other end of the geotextile specimen (kN/m)

From the testing results obtained on the different length geotextile specimens the tensile moduli values were determined using Equation (2). The tensile modulus tensile strain relationships were then collected and summarised on Figures 7 and 8 for non-woven and woven geotextile specimens, respectively. Even though the above calculation technique may not be rigorously correct, i.e. the strain distribution along the geotextile may be non-linear, it provides a useful procedure for comparing the variation of the modulus with strain and normal stress.

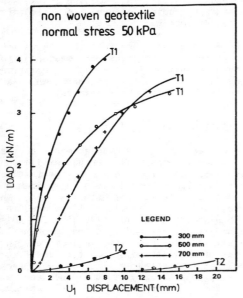

Figure 5: Load Displacement Curves

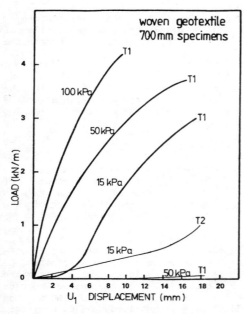

Figure 6: Load Displacement Curves

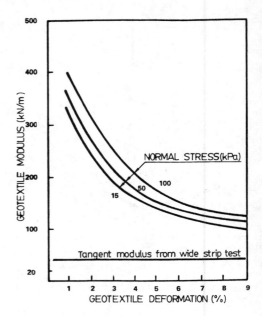

Figure 7: Geotextile Modulus vs. Geotextile Strain Non-Woven Geotextile

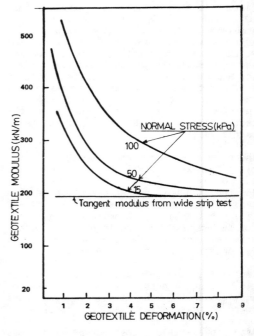

Figure 8: Geotextile Modulus vs. Geotextile Strain Woven Geotextile

84

5. DISCUSSION OF TESTING RESULTS

5.1 Observations

Studying the results of the tests carried out on geotextile specimens the following conclusions can be drawn:

1. The geotextile tensile modulus is inversely proportional to the geotextile strain in both woven and non-woven geotextiles. At low strains (1%) the tensile modulus can be considerably higher than the unconfined modulus. In the case of the non-woven geotextile this increase can be up to ten times, while in the case of woven geotextile the increase is just up to about three times.

2. The modulus of the non-woven geotextile was increased by soil confinement over the whole range of strains measured in the tests (0 to 9%) for all normal stress values.

3. The geotextile tensile modulus depends on the applied normal stress. An increase in the normal stress results in an increase in the value of the geotextile modulus.

4. In non-woven geotextile the normal stress affects the modulus only slightly. For the woven geotextile the normal stress has a more significant effect than for the non-woven geotextiles. The increase of the modulus is almost proportional to the normal stress. Furthermore at higher normal stresses the range of geotextile strain over which modulus increase can be expected, is wider.

5. A comparison of the load displacement curves in the in-soil tests (Figures 5,6) and in the wide strip tests (Figure 3) shows that the soil confinement reduces the offset and changes the shape and slope of the load displacement curve.

5.2 Mechanism of Clay Geotextile Interaction

The mechanism of the clay geotextile interaction in the in-soil tensile tests is governed by the following major factors:

1. During the application of the normal stress and tensile force, interlocking develops between the particles of the soil and the fibres of the geotextile. Due to this interlocking a high frictional resistance develops on the clay geotextile interface, which appears in the pull-out tests as a highly improved tensile response i.e. an improved tensile modulus.

2. The normal stress acting on the interface increases the frictional resistance between the clay particles and the geotextile fibres as well as the friction among the fibres inside the geotextile. This increased inter-fibre friction also contributes to the improvement of the tensile response of the geotextile. As the opening size of the non-woven geotextile is more akin to the particle size of the clay particles than that of the woven geotextile, the interlocking effect is more efficient on the non-woven geotextile. It is further enhanced by the disoriented nature of the fibres in the non-woven geotextile.

3. When applying the tensile force the point of application of the pull-out force is displaced and therefore strain develops in the geotextile. This results in relative displacement between the geotextile fibres and the clay particles, thus mobilising frictional resistance. At higher displacement (and thus strain) the relative displacement between the geotextile and the clay can be high enough to cause local slippage either through the breaking out of clay particles or through the unrevelling of the surface of the geotextile. This local slippage has two consequences.

1st: the load is transferred further away from the point of locad application and longer sections of the geotextile participate in the pull-out resistance. If the geotextile is not long enough full-scale bond failure occurs.

2nd: the fibres of the geotextile orient themselves in the direction of the pull-out force, hence reducing the frictional resistance inside the fabric itself. This leads to a deteriorated tensile response, or reduced tensile modulus.

It may be concluded that in woven geotextiles the improvement of the tensile modulus is caused fundamentally by the increased interfibre friction, and to a lesser extent by the mobilized bond strength. In non-woven fabrics the improvement of the modulus is caused primarily by the interlocking of the particles and the fibres of the geotextile, which is the source of the bond strength, and partly by the increased inter-fibre friction.

A further factor affecting clay geotextile interaction is the development of excess pore water pressure during the application of load. As the clay in this study was saturated, and the normal and tensile forces were applied rapidly an excess pore pressure developed in the clay.

As a hydraulic gradient is set up between the geotextile and the clay the pore water begins migrating towards the geotextile. Considering the length of the geotextile specimens and the rate of loading it can be assumed that only very small drainage can develop. Therefore a small increase of the bond strength results, which appears as a further increase of the tensile modulus.

Based on the results of previous research it can be assumed that there is no drainage in the plane of the woven fabric: Fourie and Fabian (1987). Consequently there is no increase in the bond strength on the woven geotextile clay interface during testing. The above observation that the effect of the normal stress on the tensile modulus of the geotextiles is stronger in the case of woven than non-woven geotextile verifies this hypothesis. The increase of tensile modulus is fundamentally caused by the improved interfibre friction-factor which is relatively unaffected by excess pore pressure.

6 CONCLUSIONS

The in-soil tensile tests carried out in the way as described gave valuable information on the effects of soil confinement on the geotextile modulus in the case of woven and non-woven geotextile specimens.

The tensile modulus of the non-woven needle punched geotextile increases in soil confinement due to

1. improved interfibre friction

2. interlocking between the soil particles and the fibres of the geotextile which results in the development of increased bond strength.

3. bond strength mobilized when the geotextile is in tension.

4. obstructed reorientation of the fibres when in tension.

The tensile modulus of the woven geotextile increases in soil confinement due to

1. improved interfibre friction and to a small extent due to

2. interlocking between the soil particles and the fibres of the geotextile which results in the development of

3. bond strength mobilized when the geotextile is in tension.

The confined modulus depends largely on the geotextile strain. Due to the difference in the mechanism of the clay geotextile interaction the soil confinement can significantly increase the modulus up to 10% strain on non-woven geotextiles, but only up to 3-4% strain on the woven geotextiles.

The extent of the increase is also different. At low strains the confined modulus of the non-woven geotextile was up to ten times higher than the unconfined modulus, while that of the woven geotextile was up to three times higher than the unconfined modulus.

The phenomenon the of modulus increase due to soil confinement should be taken into account in applications where the geotextile strain is relatively low.

7 REFERENCES

Fabian, K. and Fourie, A. 1986. Performance of geotextile reinforced clay samples in undrained triaxial tests . Geotextiles and Geomembranes, Vol. 4, p. 53-63.

Fourie, A. and Fabian, K. 1987. Laboratory determination of clay geotextile interaction. Geotextiles and Geomembranes, Vol 7 (in print).

Ingold, T.S. 1981. A laboratory simulation of reinforced clay walls. Geotechnique, Vol. 31, p. 399-412.

Leshchinsky, D. and Field, D.A. 1987. In-soil load elongation, tensile strength and interface friction of non-woven geotextiles. Proc. Geosynthetics '87 p. 238-249.

McGown, A, Andrawes, K.Z. and Kabir, M.H. 1982. Load extension testing of geotextiles confined in sand. Proc. 2nd Int. Conf. on Geotextiles, p. 793-798.

International Geotechnical Symposium on Theory and Practice of Earth Reinforcement / Fukuoka Japan / 5-7 October 1988
© 1988 Balkema, Rotterdam. ISBN 90 6191 820 0

Friction characteristics of polypropylene straps in reinforced minestone

T.W.Finlay, Mei-Jiu Wei & N.Hytiris
University of Glasgow, Glasgow, Scotland, UK

ABSTRACT: Tests to determine the friction characteristics of polypropylene strap reinforcement in minestone have been carried out using a large shear box, a laboratory pull-out test, and full-scale pull-out tests. The tests, results, and findings are reported, and differences in behaviour compared with ribbed steel reinforcement are discussed.

1 INTRODUCTION

In the search for new materials suitable for reinforced earth retaining wall construction, the University of Glasgow's Department of Civil Engineering has been investigating, with the assistance of a grant from British Coal's Extra Mural Research Committee, the possibility of using unburnt colliery spoil (minestone) as a fill in conjunction with polypropylene reinforcing straps, since minestone is a readily available fill material found in many areas of the U.K., while polypropylene offers better corrosion resistance than traditional galvanised steel straps.

This paper describes the properties of two minestones and the friction characteristics of two different types of polypropylene reinforcement obtained using shear box tests and pull-out tests. Similar tests on high adherence ribbed steel strips are reported for comparison.

2 MATERIALS

2.1 Minestone

Minestones from Wardley colliery and Wearmouth colliery were used as the fill material, Wardley exhibiting greater cohesion than Wearmouth. The properties of the materials are given in Table 1 and grading is shown in Fig. 1.

Table 1. Properties of minestone.

Property		Wardley	Wearmouth
Moisture content	%	9.7	5.6
Specific gravity	-	2.37	2.34
Liquid limit	%	31.0	-
Plastic limit	%	21.8	-
Plasticity Index	%	9.2	-
Bulk den.(loose)	kN/m^3	13.95	13.00
Opt. moist.cont.[*]	%	10.0	8.0
Max. dry den.[*]	kN/m^3	18.6	18.3
Angle of internal friction	deg.	39,46	52
Cohesion	kN/m^2	15,18	6
Uniformity coeff.		70	10
Permeability	cm/sec	2.8x10^{-3}	2.1x10^{-3}

[*] 2.5 kg Rammer

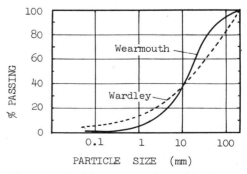

Fig. 1 - Grading curves for minestones.

2.2 Reinforcing straps

The reinforcing straps used consist of tendons developed by I.C.I. and made from ten bundles of high tenacity polyester fibres enclosed in a durable polyethylene sheath. Two types were used viz Paraweb (black) and Paralink (beige). The manufacturers claim that the straps are corrosion resistant against chemicals in the fill, have good resistance to abrasion and are unaffected by water. Paraweb is claimed to be superior to Paralink in terms of its resistance to ultra violet radiation and bacterial attack, while both straps are produced with a slightly roughened surface to enhance their frictional characteristics. Paraweb is 92 mm wide by 3.5 mm thick, while Paralink is 85 mm wide by 2.5 mm thick, and their stated breaking loads are 50 kN and 30 kN respectively. The ribbed steel straps used in some further tests were 40 mm wide by 5 mm thick with a breaking load of 98 kN.

2.3 Strap stiffness and creep

Before carrying out the main series of tests to determine the friction character-istics of the strap material in the fill, their tensile behaviour was studied by direct tension tests on sample lengths clamped at the ends. The results indi-cated that at loads greater than about 10 kN the polyester fibres began to slip within the sheath despite the clamping, and the ultimate load reached was only about 50% of the breaking load claimed by the manufactures. It appears that a breakdown in friction between the bundles of fibres and the outer sheath prevents the straps from reaching their full potential breaking load.

A point to note, however is that the direct tension tests were performed on samples only up to 300 mm long, whereas in practice longer straps are used, the straps are completely embedded in the fill and subjected to the surrounding earth pressure. The tension tests yielded values of modulus of elasticity of about 0.7 kN/mm² for the polypro-pylene straps compared with 170 kN/mm² for the steel straps.

Creep tests on the straps under a load of 2.2 kN gave rise to a creep of 0.12% for Paraweb and 0.23% for Paralink after 20 days, while at 60 days the Paraweb creep had increased to 0.17%.

3 TEST PROCEDURES

3.1 Shear box tests.

A 300 mm x 300 mm shear box was used to determine the coefficient of friction of each strap material. The lower half of the box was occupied by a hardwood block and the strap material was glued firmly to the surface of the block. Fill at natural moisture content and 96% maximum dry density, (corresponding to the measured field value) was placed in the top half of the box, and a normal load applied. The shear force required to cause sliding was then measured using a constant rate of strain of 1.05 mm/min. at normal stresses of 20, 60 and 120 kN/m².

3.2 Laboratory pull-out tests.

For this test, a steel box 2m long by 0.4m wide by 0.25m deep was used. The fill was compacted to mid-height in the box, a strap 1.5m long was then placed in position centrally and fed through a slot in the front face. Further fill was placed and compacted level with the top of the box. The top of the box, compris-ing a rubber membrane below a steel coverplate was then bolted in position. Overburden pressure similar to the normal stresses used in the shear box tests was simulated by air pressure introduced between the rubber membrane and the top of the box, and the strap was pulled at a constant rate of 50 N/minute until failure occurred.

3.3 Field pull-out tests

At Wardley colliery a large open-ended box 5m long by 2m wide by 4.5m high had been constructed by British Coal from rolled steel sections and timber railway sleepers. The box was used to accommo-date up to 18 straps at various levels and having different lengths. Placing and compaction of the fill was done via the open end of the box, the straps being fed through the front face. A patented pull-out device, supplied by British Coal, was used to pull out indi-vidual straps at a rate of 3 mm/min, with load and displacement being measured.

4 TEST RESULTS

4.1 Shear box

Table 1 presents the results of the shear box tests, not only for Paraweb and Paralink, but also for the high adherence ribbed steel straps.

Table 1 Shear box test results

Minestone	Sh.str. (fill)		Reinf. straps	c_a kN/m^2	δ deg
	c^1	ϕ^1			
Wardley	18	46	Paraweb	10	28
			Paralink	8	26
			Rib Steel	18	41
Wearmouth	6	52	Paraweb	10	29
			Paralink	4	26
			Rib Steel	10	46

The results appear to indicate that the strap friction is not greatly affected by the minestone type, and that the Paraweb is superior to the Paralink although both are inferior to the ribbed steel.

4.2 Laboratory pull-out

Again, ribbed steel straps have been included in the results for the purpose of comparison.

The test results are shown in Table 2. In the case of the laboratory pull-out tests, particularly under overburden pressures lower than about 80 kN/m^2 all

straps in Wardley minestone give higher pull-out loads than in Wearmouth. Paraweb gives a higher pull-out resistance than Paralink, both being less than the high adherence steel strap. This is also illustrated in the typical force v displacement curve of Fig. 2

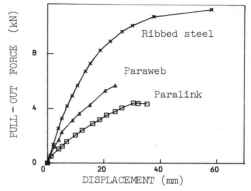

Fig 2 Pull-out v strap displacement from Wardley fill @ σ_n = 60 kN/m^2

4.3 Field tests

The field test results showed a high degree of scatter as can be seen in Fig. 3 for 4m long straps and the effects of different minestones and different strap types could not be easily separated. The results were not unexpected in view of the lower level of control of compaction and placing of the straps compared with laboratory conditions.

Table 2 Laboratory pull-out results

Reinforcing straps	Overburden pressure σ_n (kN/m^2)	Minestone fill			
		Wardley		Wearmouth	
		Displ.(mm) at pull-out	T_{max}(kN)	Displ.(mm) at pull-out	T_{max}(kN)
Paraweb	20	15	3.11	20	2.42
Paralink	20	33	3.25	39	2.13
Rib Steel	20	40	8.87	52	4.50
Paraweb	60	24	5.23	22	3.99
Paralink	60	31	3.96	29	3.41
Rib Steel	60	52	11.19	81	4.80
Paraweb	120	26	7.77	39	8.31
Paralink	120	42	5.62	35	7.23
Rib Steel	120	62	11.98	98	12.13

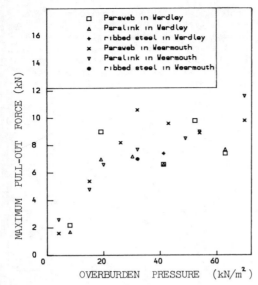

Fig 3 Field pull-out tests

5. DISCUSSION

The design of reinforced earth walls
requires a knowledge of the coefficient
of friction between the fill and the
reinforcement.

In the design method used in the U.K.
(Department of Transport 1978) the friction coefficient, μ, can be measured
directly in a shear box. Alternatively,
the friction coefficient can be taken as
$\mu = \alpha \tan \phi'$ where ϕ' is the effective
angle of internal friction of the fill
and α is a multiplying factor with a
value lying in the range 0.45 to 0.5
for metallic reinforcement.

Pull-out tests have been used as an
alternative to shear box tests to model
the behaviour of a strip subject to
tensile load and lead to the apparent
friction coefficient

$$f^* = \frac{T_{max}}{\sigma_n . \, 2.b.L.}$$

(Alimi et al 1977), where T_{max} is the
maximum pull-out load, σ_n is the overburden pressure, and b and L are the
strip width and length respectively.

It is possible to relate μ and f^* for a
cohesive frictional soil. At any
overburden pressure in the shear box,

$$\mu = \frac{c_a}{\sigma_n} + \tan \delta \qquad \text{where } c_a = \text{unit}$$

adhesion, and δ = angle of frictional
resistance. In terms of pull-out of a
thin rigid reinforcing strap, length L,
width b embedded in a fill under an
overburden pressure σ_n, the maximum
pull-out force

$$T_{max} = 2.b.L \; \sigma_n \left(\frac{c_a}{\sigma_n} + \tan \delta\right).$$

Hence $\qquad f^* = \frac{c_a}{\sigma_n} + \tan \delta = \mu.$

Although the relationship between f^* and
μ is theoretically valid, many investigators have shown that comparisons based
on shear box and pull-out tests lead to
much higher values of f^* compared with μ,
especially at low overburden pressures.

Dilatancy and arching have been given as
possible causes of the difference
(McKittrick 1978;Guilloux et al 1979),
and other work (Finlay et al 1984) has
shown that 'free' pull-out compared to
pull-out through a slot can result in a
reduction in f^* of approximately 28%.

The results obtained from the tests
described have been put in terms of μ for
the shear box tests and f^* for the
laboratory and field pull out tests and
are shown in figs 4 and 5 for Wardley
and Wearmouth fills. It can be seen from
these figures that the f^* values for the
polypropylene straps are very much lower
than the μ values. This is in direct
contradiction to the generally accepted
behaviour of metallic and other rigid
reinforcing material, and the behaviour
must therefore be related to the flexibility and/or elasticity of the
reinforcement.

This hypothesis is partly borne out by
comparing the laboratory pull-out with
the pull-out calculated from the shear
box results as shown in fig 6. Since
the reinforcing material is rigidly
fixed during the shear box test, the
pull-out force v displacement behaviour
reflects this and actually shows
Paraweb and Paralink as giving superior
pull-out capacities at lower displacements than the ribbed steel. The
actual pull-out behaviour reveals that
the polypropylene straps undergo up to
three times the displacement necessary
to mobilise the maximum pull-out force
in the shear box. The much more rigid
ribbed steel on the other hand under-

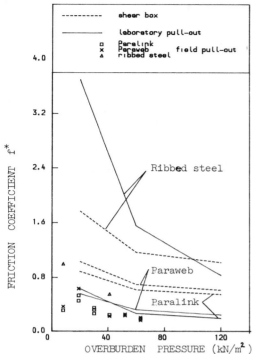

Fig 4 Comparison of friction coefficients in Wardley fill

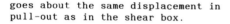

Fig 5 Comparison of friction coefficients in Wearmouth fill

goes about the same displacement in pull-out as in the shear box.

Following up this observation, some tests have been done in which the differential movement along the poly-propylene straps has been measured as a pull-out test progressed.

The original intention was to fix strain gauges to the polypropylene straps but it proved to be impossible to find an adhesive which would perform satisfac-torily and this approach was abandoned. The alternative approach used was to fix very thin high tensile steel wires to the 1.5m long straps at 0.5m spacing and feed them through holes at the rear of the box then over pulleys with tension-ing weights attached. The wire move-ment was then monitored by dial gauges.

The results from one test are shown in fig 7 and serve to illustrate the behaviour of the straps. At failure, the free end of the strap has only just begun to move although the pulled end has at that time moved through a considerable distance. This behaviour

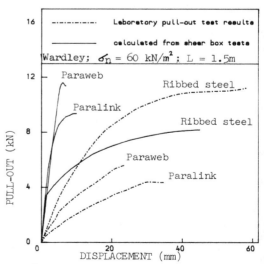

Fig 6 Comparison of laboratory pull-out

tests with pull-out calculated

from shear box tests

must lead to a departure from the assumption of uniform build-up of frictional force along the strap. The probable distribution is likely to be uniform towards the front end, dropping off to a low value at the free end, thus leading to a reduction in the pull-out force (and f^*) compared with the calculated value. This is confirmed by a calculation based on the limited results of the test shown in fig. 7 which indicates a reduction of about 25% in pull-out force compared with that obtained from the straight shear box values.

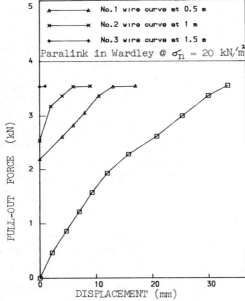

Fig 7 Differential strap movement on pull-out

Figs 4 and 5 also show that for the polypropylene straps the f^* values from field pull-out tests are in reasonable agreement with those from the laboratory pull-out tests whereas the ribbed steel field pull-out f^* values are much smaller.

In terms of design therefore it appears that the field pull-out behaviour of flexible reinforcement is more closely approximated to by laboratory pull-out tests than by shear box tests. The present tests also show that the field pull-out loads for rigid reinforcement are lower than predicted by laboratory pull-out or shear box tests.

Further work is now in progress to study the soil reinforcement interaction rigid and flexible reinforcement.

ACKNOWLEDGEMENTS

Thanks are due to British Coal's Minestone Services division for financing most of the work, and particularly to Dr. A.K.M. Rainbow, director, and Mr. S. Barnett, geotechnical engineer for their assistance in setting up the field tests. Also to Messrs I.C.I. for supplying the polypropylene straps, and to the Technicians in the Department of Civil Engineerng for their manufacturing skills.

REFERENCES

Alimi,I., Bacot, J., Lareal, P., Long, N.T. & Schlosser, F. 1977. Adherence between soil and reinforcement in-situ and in the laboratory. Proc. IX I.C.S.M.F.E. Vol.1: 11-14.

Department of Transport 1978. Reinforced earth retaining walls and bridge abutments for embankments. Tech. Memo. BE3/78

Finlay, T.W., Khattri, M.S. & Sutherland, H.B. 1984. The friction coefficient of metallic strip reinforcement. Proc. 6th Conf. on Soil Mech. and Found. Eng. Budapest 619-624.

Guilloux, A., Schlosser, F. & Long, N.T. 1979. Etude du frottement sable-armature en laboratoire. Proc. Int. Conf. Soil Reinforcement. Paris Vol.1: 35-40.

McKittrick, D.P. 1978. Reinforced Earth - Application of theory and research to practice. Proc. Symp. Soil Reinforcing and Stabilising Techniques. New South Wales Inst. of Technology/ New South Wales University.

International Geotechnical Symposium on Theory and Practice of Earth Reinforcement / Fukuoka Japan / 5-7 October 1988
© *1988 Balkema, Rotterdam. ISBN 90 6191 820 0*

Strength characteristic of reinforced sand in large scale triaxial compression test

S.Fukushima, Y.Mochizuki & K.Kagawa
Technical Research Division, Fujita Corp., Yokohama, Japan

ABSTRACT: A reinforcing mechanism, strength and deformation characteristics of the reinforced soil have been investigated by a laboratory element test using a small specimen. However, the properties of soil that reinforced with a full scale reinforcing material cannot be studied by the laboratory test as the size of specimen is very small. In this paper, a series of large scale triaxial compression tests were performed in order to investigate the strength and deformation characteristics of sand reinforced with full scale reinforcing materials (geo-grid, nonwoven fabric and metal strip for Terre Armee). The tests were carried out on a dry sand sample, the results were represented by comparing the stress-strain relationship of reinforced sand with that of unreinforced sand.

1 INTRODUCTION

The reinforced soil method reinforces the soil by arranging synthetic textile materials, metallic strip materials, and other materials (reinforcement) than soil having strong resistance to tension in the direction of the soil elongation. Friction generated between soil and reinforcement restrains the soil deformation by elongation and reinforces the ground. This method is now widely applied to construct comparatively small-scale bankings and cuttings. It is expected that the application range of this method will be further expanded, and the reinforced soil method will be applied to higher bankings, longer cuttings and other large-scale soil structures. Up to the present, the reinforcing mechanism and the effect of reinforcement have been examined by the shearing test of soil reinforced principally by model reinforcing materials. For this, the small-type element tests (triaxial compression and plane strain compression tests) with small specimen were used. It is impossible to study the effect of full scale reinforcement with these tests because of the smallness of the specimens. Particularly, to apply the reinforced soil method to construct a large scale soil structure, it is imperative to grasp the reinforcement when it is actually arranged in the ground. This report summarizes the results of experimental studies on the effect of full scale reinforcing materials when they are arranged in the sandy ground at same intervals.

2 TEST METHOD

2.1 Large scale triaxial compression testing system

A large scale triaxial compression testing system used for this study was developed to examine the stress-strain characteristics of the rock material for the rock fill dam. The size of the specimen was 120cm in diameter and 240cm in height. Figure 1 shows the sketch of this testing system. Highly compressed air pressure from a compressor is controlled to a specific level by the regulator. It is applied as lateral pressure to the specimen after being converted to water pressure inside the specimen volume change measuring tank (TA). A hydraulic unit feeds a fixed amount of fluid to the loading cylinder, which presses the loading plunger. The plunger compresses the specimen, and an axial load is applied to specimen. Axial displacement rate is changed by adjusting the flow from the hydraulic unit with the flow control

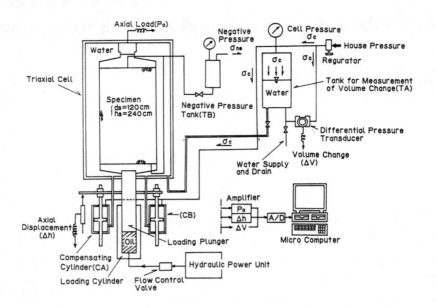

Figure 1.Ultra-large triaxial compression test

valve. The axial displacement rate was
determined at ε_a =0.3%/min in this test
program. The volume change of the specimen
was obtained by measuring the water level
inside the tank (TA) with a differential
pressure transducer. The water equivalent
to the volume change of specimen was led
to this TA from inside the triaxial cell.
If the loading plunger intrudes into the
triaxial cell, the water equivalent to the
volume of intrusion enters the compensating
cylinders (CA and CB) having one half of
the section of the loading plunger. This
stops the flow of the water from the
triaxial cell to the tank TA due to the
intrusion of the loading plunger, allowing
the water equivalent to the volume change
of the specimen to flow in and out. The
top and bottom ends of the specimen were
set so that these end faces would become
the principal stress plane. Friction was
eliminated by adhering a rubber sheet of
0.5mm to the cap and pedestal with silicon
sealant, then inserting silicon grease and
placing another rubber sheet as shown in
Figure 2.

2.2 Preparation of specimen

The sample used for the test was Toyoura
sand (Gs=2.64, e_{max} =0.977, e_{min} =0.605).

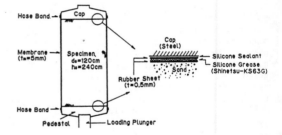

Figure 2.Specimen end face condition

The following describes how the specimen
was prepared. A loosely packed specimen
was obtained by allowing the sand in
air dry state to drop freely through a
multi-layered screen (four layered
combination of 9.54, 4.76, 4.76, and
4.76mm from the top respectively) from
the height of approximately 80cm (air
pluviating method). The void ratio of
specimen was approximately e=0.84-0.88
(Dr=34%). A densely packed specimen was
prepared by the air pluviating method like
the loosely packed specimen, then vibration
was applied by a vibrator to every sand
layer of 60cm. The void ratio of specimen

94

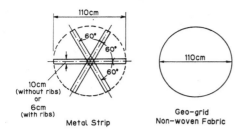

Figure 3 (a).Forms of reinforcement
and their dimensions

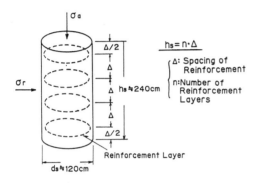

$$h_s = n \cdot \Delta$$

Δ: Spacing of
Reinforcement
n:Number of
Reinforcement
Layers

Figure 3 (b).Arrangement of
reinforcement inside the specimen

was approximately e=0.68-0.70 (Dr=70%).
The reinforcements used for the test are
the following three types. They are
actually applied to construction sites.
Figure 3 shows the reinforcement layers
inside the specimen.
(1) Geo-grid: Grid-type material made
of polypropylene. Applied in sliced disk
form with a diameter of 110cm. (Tenser SS-
2, grid knot dimensions: 28X40mm)
(2) Nonwoven fabric: Fabric-type
material made of continuous long filament
of 100% polypropylene. Applied in cut
circular form with diameter of 110cm.
(Toughnel construction matting U-90G)
(3) Strip material: Metal strip
materials used for Terre Armee method
(without rib, 10cm wide and with rib,
6cm wide). Material was cut in lengths of
110cm, and 3 pieces were crossed at the
center to form a layer.

Reinforced specimen was constructed as
follows. A sand was prepared as described
above. Reinforcement was arranged, as
shown in Figure 3(b), horizontally and at
equal intervals (reinforcement intervals:Δ
). Here the relation between the number of
reinforcement is shown by $h_s = n \cdot \Delta$.

3 TEST RESULTS

Figure 4(a) and (b) show the relationship
between the stress ratio and deformation
of sand reinforced by the geo-grids in
varied layer number of n=2, 4, and 8.
Figure 5 and 6 show the same relationship
of sand reinforced by nonwoven fabric and
strip material (with rib) respectively.
These figures show that the greater the
number of reinforcement layers or the
narrower the intervals, the greater the
effect of reinforcement. Of the three
types of materials, geo-grid reinforced
sand was particularly characteristic. It
had the greatest reinforcing effect among
the three, but the stress ratio sharply
dropped after the shearing deformation
advanced to certain extent. This trend
appeared stronger with the specimen having
the greater number of reinforcement
layers. The sharp drop of the stress ratio
is thought to be attributed to the
rupture of geo-grids. Figure 4(a) and (b)
show that the axial deformation at the
sharp drop of the stress ratio-deformation
curve varies by the sand density or the
denser the density of specimen, the
smaller the deformation. Judging from
these facts, when the ground is reinforced
using geo-grids, it is necessary to
consider the deformation when the rupture
of the reinforcing materials occurs.
Figures 7(a) and (b) show the maximum
stress ratios obtained from Figures 4, 5,
and 6, which are replotted against the
number of reinforcement layers to
investigate their effect. The ordinates of
these figures represent the rate of
increase of the maximum stress ratio by
the reinforcing material. They also show
the scale of the angle of internal
friction (cohesive component C=0) with
reinforcements. These figures show that
with the increase of the number of
reinforcing layers the effect of
reinforcement sharply increases, and that
the difference of effect becomes greater
by the types of reinforcements. However,
considering the fact that the reinforcing
materials are generally arranged at
intervals of 70-100cm in actual reinforced
ground, Figure 7(a) shows that there is
almost no difference of reinforcing effect

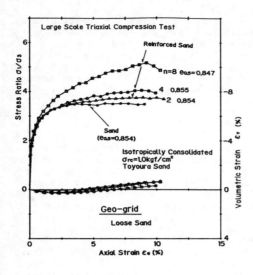

(a) Loose sand

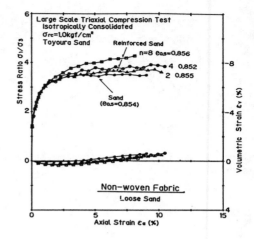

Figure 5. Relation between the stress and deformation of sand reinforced by nonwoven fabric

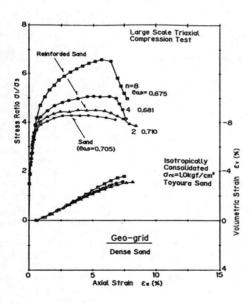

(b) Dense sand

Figure 4. Relation between the stress and deformation of sand reinforced by geo-grids

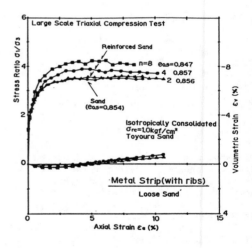

Figure 6. Relation between the stress and deformation of sand reinforced by strips

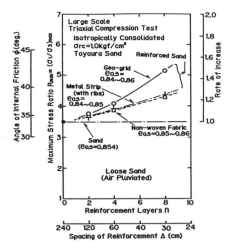

(a) Loose sand

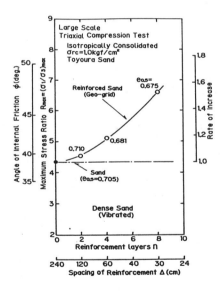

(b) Dense sand

Figure 7. Relation between the number of reinforcing layers and their effect

by the type of material, or nominal if any, in these intervals. The effect of reinforcement arranged in the ground at the intervals of 70-100cm is 10 to 20 % by the increment of the maximum stress ratio. This value is equivalent to 3-4 degrees by the angle of internal friction. Figure 8 shows the angle of internal friction (or the maximum stress ratio) of the tests conducted with the specimen reinforced with n=4 layers of geo-grids and strip material (with rib) under varying the confining pressure σ_{rc}. This figure shows that the relationship between the reinforcing effect and confining pressure differs between the geo-grids and strip materials. In other words, the reinforcing effect $\Delta\phi$ by the strip material remains almost constant irrespective of confining pressure, it becomes greater in the sand reinforced by geo-grids as the confining pressure becomes lower. This is thought to be caused by the difference of reinforcing mechanism between the two materials. That is, strip materials are effective in adding a fixed angle of friction to sands. On the other hand, geo-grids shows greater effect in the lower confining pressure range where the dilatancy of sand is strong. Therefore, the interlocking effect between the geo-grids and sand is thought to play an important role. Furthermore, the geo-grids increases its reinforcing effect as the confining pressure decreases. Its effect tends to become smaller than that of strip

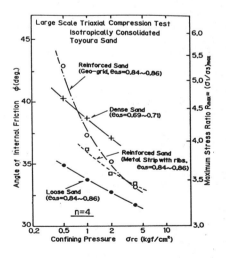

Figure 8. Relation between reinforcing effect and confining pressure

97

materials if the confining pressure increases. This indicates that this material is suitable for the construction of soil structures like embankments of considerably low height.

4 SUMMARY

Conventionally,the reinforcing mechanism and effect of the reinforcement material for sand reinforced with various materials were studied by the element tests with small specimens. The restriction of small-dimension specimen obliged the use of model reinforcing materials. This report has investigated, by experiments, the effect of actual reinforcement as arranged inside the sandy ground at likely intervals through the ultra large triaxial compression tests using large specimens. The descriptions so far made are confined only to the effect of soil deformation restraint by the reinforcing materials (deformation restraint by the friction that generated between soil and reinforcement), which has clear boundary conditions that can be examined by element tests (figure 9(a)). The effect of reinforcing materials in actually reinforced ground have, in addition to the above mentioned deformation restraint effect, another effect which disperses the load widely to the adjacent ground through the unification of portions which were subjected to the deformation restraint by the reinforcing materials (Figure 9(b)). The stress dispersion effect is determined by the boundary conditions of individual structures. The future problem would be how to incorporate this effect into the design for more rational construction of the reinforced ground.

REFERENCE

1)F.Tatsuoka,K.Kondo,G.Miki,O.Haibara, E.Hamada,T.Sato 1983. Fundamental study on tensile reinforcing of sand. Tsuchi-to-Kiso vol.31 No.9:11-19(in Japanese)
2)Japanese society of soil mechanics and foundation engineering 1986. Reinforced soil method. Part 2, Fundamentals of reinforcing mechanism and design of reinforced soil method:25-168(in Japanese)

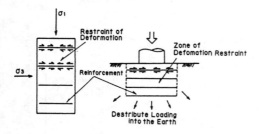

(a) Deformation restraint effect
(b) Deformation restraint effect and load dispersion effect

Figure 9.Effect of reinforcement

International Geotechnical Symposium on Theory and Practice of Earth Reinforcement / Fukuoka Japan / 5-7 October 1988
© *1988 Balkema, Rotterdam. ISBN 90 6191 820 0*

Functions and effects of reinforcing materials in earth reinforcement

S.Hayashi, H.Ochiai & A.Yoshimoto
Kyushu University, Fukuoka, Japan

K.Sato
Fujita Corporation, Japan

T.Kitamura
Japan Highway Public Corporation, Japan

ABSTRACT: When reinforcing materials are deposited in embankment or slope, not only the deformation conditions of soil but also the direction of reinforcing materials in soil influence the functions and effects of reinforcing materials. A new versatile shear test apparatus, which is developed in order to examine the reinforcing mechanism in soil, can demonstrate various types of strain conditions. In this paper, the usefulness and properties of this test apparatus are examined, and the results of simple shear test for sand specimen reinforced with two types of reinforcing materials are described.

1 INTRODUCTION

It is very important to clarify the reinforcing functions, material properties of reinforcement and interaction mechanism between soil and reinforcing materials for both slope reinforced with grouted steel bar and embankment reinforced with metal strips or geotextiles (Hayashi 1985). Besides, these data are needed for the purpose of establishing the design method of earth reinforcement. Many researchers have studied this subject (Jewell 1987, Tokue 1982), and it has become a common knowledge that the effects of the reinforcement in slope and embankment are influenced by the deformation conditions of soil and the direction of insert angle of reinforcing materials against slip surface.

In this paper, a new versatile shear test apparatus, whose specimen size is larger than that of the standard shear test apparatus, is developed in order to examine the reinforcing mechanism in sand. This apparatus can demonstrate various types of strain conditions such as uniform and nonuniform strain distributions. The purpose of this paper is to examine the functions and effects of different types of reinforcing materials in sand using this new apparatus. Here, the effective insert direction of reinforcing materials in soil and the reinforcing effects when the reinforcing materials are subjected to tensile and compressive forces are discussed using test results.

2 VERSATILE SHEAR TEST APPARATUS

2.1 Problems of using the standard direct shear test for reinforced soil

To examine the characteristics of reinforced soil, the test apparatus such as triaxial, torsional, direct shear and model test are generally used. Although direct shear test apparatus is most often used for examining the influence of the direction of reinforcing materials on earth reinforcement, the following problems are often generated:

1. It has been indicated that the strain condition in the specimen is by no means uniform as shown in Fig.1, so the functions of reinforcements in such strain condition are complicated.

2. The reinforcing materials deposited plurally are not in the same deformation conditions.

3. Vertical stress by rigid board is not uniformly loaded on the specimen in the ordinary case, much more in the case of deposited reinforcing materials, where the stress distribution is very complex.

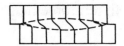

Fig.1 Deformation of specimen in direct shear test.

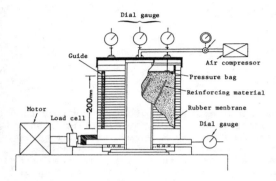

Fig.2 Versatile shear test apparatus.

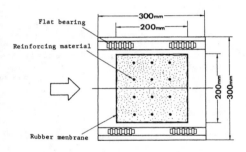

Fig.3 Plan of test apparatus.

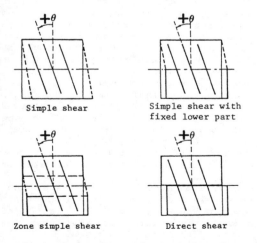

Simple shear

Simple shear with fixed lower part

Zone simple shear

Direct shear

Fig.4 Shear patterns by test apparatus.

4. From the results of model test on reinforcing mechanism using soil bin with bar-type reinforcing materials, it is clarified that the inside reinforcing material is different from the outside one. Therefore more than nine reinforcing bars have to be deposited in the specimen.

2.2 New versatile shear test apparatus

A new versatile shear test apparatus shown in Fig. 2 is developed to overcome the problems described above. The shear box consists of twenty elements made from aluminum. The friction between elements is reduced by flat bearing as shown in Fig. 3, and each element is restrained by four guided pins and rods. The shear box is designed to make the shear deformation uniform. The size of specimen in this apparatus is 200x200x200mm, and the specimen is covered with latex rubber membrane of 0.3mm in thickness.
 The top of shear box is fixed by rigid frames, and the shear stress is induced by applying displacement at the bottom.
The vertical stress is given by air pressure through rubber pressure bag at the top of shear box.
 This shear test apparatus can be demonstrated in several patterns of strain conditions as shown in Fig. 4 by substituting the frames at the top or bottom (or top and bottom) of shear box.
This means that the shear box is partially fixed by the frames and induces uniform and nonuniform strain conditions.

3   REINFORCING MATERIALS AND TEST
    CONDITIONS

The reinforcing materials used in this test are of bar and plane types. The bar-type material is a phosphor bronze bar whose size is 180mm in length and 3mm in diameter. Sand particles are applied on the surface of this bar to develop the skin friction with sand. A group of 4x3 bars is deposited in the sand as shown in Fig. 3. As demonstrated in Fig. 5, the gauges are attached to two bars at the center in group, so that the axial force and the bending moment of these two bars can be measured. On the other hand, the rectangular plane-type material is SS-type polymer grid whose size is 170mmx150mm, and three sheets are deposited in the specimen. All attachments to this material are set up in the same way as in the bar-type material.
 The tests are conducted by changing the

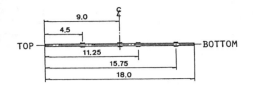

Fig.5 Position of attachment of strain gauge on reinforcing bar (phosphor bronze bar).

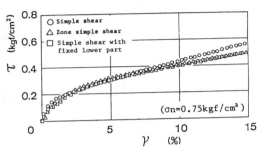

Fig.6 Stress - Strain relationship resulting from simple shear test, zone simple shear test and simple shear test with fixed lower part for unreinforced sand.

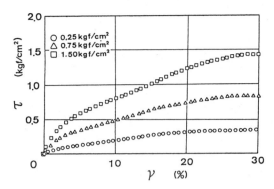

Fig.7 Stress - Strain relationship resulting from simple shear test with fixed lower part for unreinforced sand.

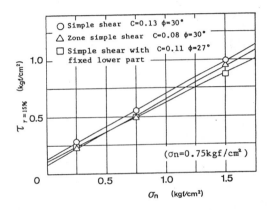

Fig.8 Shear stress - vertical stress relationship from simple shear test results for unreinforced sand.

direction angle θ of the reinforcing materials from $-20°$ to $+20°$.

The soil specimen is made of Toyoura Sand, whose properties are uniformity coefficient of 1.7, D_{50} of 0.18mm and specific gravity of 2.64. The specimen is prepared by multiple sieving pluviation method to keep the relative density of 82% constant.

In order to examine the usefulness and properties of this new apparatus, a series of shear tests of both reinforced and unreinforced sand specimens is performed.

4 UNREINFORCED SAND SIMPLE SHEAR TESTS

Figure 6 shows the shear stress-strain relationship of unreinforced sand specimens which are obtained from simple shear test, simple shear test with fixed lower part and zone simple shear test. Shear strain is defined as the ratio of the horizontal displacement of specimen against the height where simple shear deformation occurs. The stress-strain curves of these three tests are closed to each other until shear strain of 15%. The shear stress increases as shear strain increases. The clear peaks of the shear-strain curves do not appear until shear

strain of 15%, but the shear stresses settle down at a definite value around shear strain of 30%, as shown in Fig. 7. Strength coefficients obtained from the results at shear strain of 30% in the lower-part-fixed simple shear test have an observed cohesion of 0.12kgf/cm^2 and an angle of shear resistance of 41°. This result almost agrees with that (c=0.11 kgf/cm^2, φ=43.7°) of the plane strain compression test.

Figure 8 shows the shear stress plotted against the vertical stress of three shear patterns for unreinforced sand at shear strain of 15%. Results of the three shear patterns have nearly equivalent observed cohesion and angle of shear resistance. These results show that the test apparatus can demonstrate the three types of shear tests with a conservative accuracy.

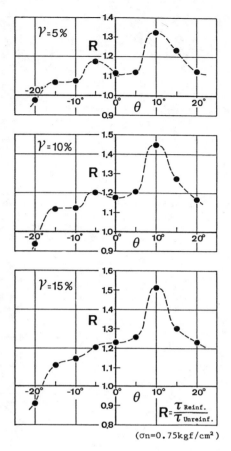

$R = \dfrac{\tau_{Reinf.}}{\tau_{Unreinf.}}$

$(\sigma n = 0.75 kgf/cm^2)$

Fig.9 Reinforcement ratio (R) - direction
       of reinforcing material (θ)
       relationship from simple shear
       test results for reinforced sand
       (phosphor bronze bar).

5 REINFORCED SAND SIMPLE SHEAR TESTS

5.1 Reinforced by bar-type material

The simple shear tests of sand reinforced
by using phosphor bronze bars are
performed by changing the insert direction
of reinforcing material, θ , from -20°to
+20° . Figure 9 shows the relationship
between the reinforcing effect
(reinforcement ratio : R) and the insert
direction of reinforcing material, θ , at
strain levels of γ=5, 10 and 15%. Here the
new parameter called reinforcement ratio R
is defined by the following formula.

$R = \dfrac{\text{Shear stress of reinforced specimen:} \tau_{Reinf.}}{\text{Shear stress of unreinforced specimen:} \tau_{Unreinf.}}$

At the angle of θ=-20°, this ratio R is
less than 1.0 for any strain level, so
that the reinforcing effect in this case
is negligible. For the case of θ>-15°, the
ratio R increases as the strain increases,
and the direction of reinforcing material
which gives the maximum reinforcing effect
is about +10°. As a whole, the more the
angle becomes positive,the higher the
reinforcing effect. Also, the increment of
reinforcement ratio R, which is about
0.11 ~ 0.20 for the case of θ=0° ~ +20°,
is enormously large, while that increment
for the case of θ=-15° ~ -5° is only about
0.02 ~ 0.05. Therefore, when using phosphor
bronze bars as reinforcing material,it
should be desposited at a positive value
of θ .

Figure 10 shows the distributions of
axial force of the reinforcing material
and Mohr's strain circles at strain levels
of 5, 10 and 15% when the reinforcing
material is deposited at θ=-10° and +10°.
In this figure, the reinforcing material
at the position of dotted line moves to
the position of solid line as shearing of
specimen increases. In the case of θ=+10°,
the distribution of the axial force
appears in tension in all strain levels
and becomes large as shearing increases.
As described from the Mohr's strain
circle, the reinforcing material is placed
in tensile strain region even at small
strain range (γ=5%). This means that the
distribution of axial force may be
explained from this point. On the other
hand, when the insert direction of
reinforcing material is θ=-10°, the
distribution of axial force is in
compression in the strain level of γ=5%,
and it becomes a tensile force as shearing
increases. It is considered that at the
beginning of shearing, the reinforcing
material is placed in the range of
compressive strain in Mohr's strain
circle, and by extending the tensile
strain range, this material moves
gradually into the tensile strain region.
In other words, the reinforcing material
initially functions as a compressive
material and gradually changes to tensile
material. In this case, the increase of
reinforcement ratio R is realized and the
effective reinforcement is performed. From
this test, it can be said that the
effective direction of reinforcing
material is when θ>0°.

5.2 Comparison between bar-type and
    plane-type materials

Figure 11 shows the relationship between

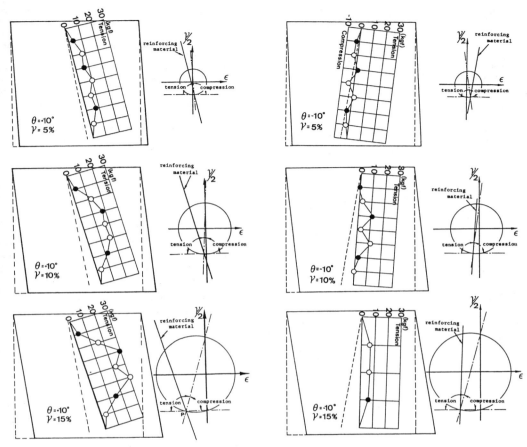

Fig.10 Axial force distributions and Mohr's strain circles ($\theta=+10°$, $-10°$).

reinforcement ratio R and direction of reinforcing materials θ, at the strain level of $\gamma=15\%$ for both phosphor bronze bars as bar-type reinforcing material and polymer grids as plane-type reinforcing material. Besides, the relationship between the volumetric strain and shear strain is given in Fig. 12. It is realized from Fig. 11 that the reinforcement ratio R for both cases are different from each other. In the case of negative θ, the ratio R of the plane-type is smaller than that of the bar-type, while in the case of positive θ, it is the opposite. This result is explained by the fact that the dilation for the case of polymer grid is larger than that for the case of phosphor bronze bar regardless of the value of θ, and the confining pressure in soil caused by the side ribs of polymer grid is larger compared with that of phosphor bronze bar shaft. In other words, when polymer grid

is placed at the negative value of θ, it reacts with the side ribs, and the ratio R becomes much small compared with that of phosphor bronze bar. On the other hand, as the reinforcing materials are placed at the positive value of θ, the polymer grid functions as a tensile material from the beginning of shearing, and it is considered that the ratio R becomes large, because the confining pressure against soil is high compared with that of phosphor bronze bar.

6 CONCLUSIONS

The main results of this paper are summarized as follows :
1. The new versatile shear test apparatus, which can model various types of strain conditions such as uniform and nonuniform strain distributions, is

103

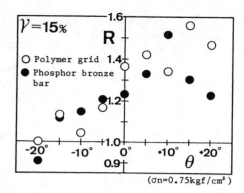

Fig.11 Reinforcing ratio (R)- direction
of reinforcing material (θ)
reletionship from simple shear
test results for reinforced sand.

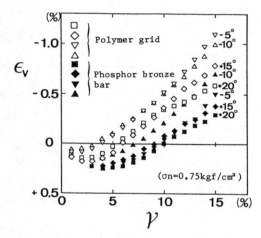

Fig.12 Volumetric strain - shear strain
relationship from simple shear test
results for reinforced sand(polymer
grid and phosphor bronze bar).

introduced. With the results of shear
tests shown in unreinforced sand, this
apparatus can therefore be useful for many
kinds of shear tests of reinforced soil.
2. The effects of reinforcement are
depicted when reinforcing material
functions as tensile material. On the
contrary, they are not expected when
reinforcing material operates as
compressive material. In the latter case,
the strength of reinforced soil goes down
sometimes.
3. The reinforcing effect of polymer
grids of the plane-type is larger than
that of phosphor bronze bars of the bar-

type in cases where the reinforcing
materials are deposited in the directions
of θ>0°. However the former is smaller than
the latter in cases where the reinforcing
materials are deposited in the directions
of θ<0°. Hence, it can be considered that
the influence of the direction of
reinforcing materials on the effect of
reinforcement appears clearly in the case
of deposited polymer grids rather than in
the case of deposited phosphor bronze
bars.

REFERENCES

Hayashi, S. et al 1985. Mechanism of pull-
out resistance of polymer grids in
soils, TSUCHI-TO-KISO, JSSMFE, Vol.33,
No.5, pp.21-26. (in Japanese)
Jwell, R.A. & Wroth, C.P. 1987. Direct
shear tests on reinforced sand,
Geotechnique Vol.37, No.1, pp.53-68.
Tokue, T. & Umetsu, K. 1982. Influence of
reinforcement direction on slope
stability, Proc. 17th Japan National
Conf. SMFE, pp.1165-1168. (in Japanese)

© 1988 Balkema, Rotterdam. ISBN 90 6191 820 0

Present state of knowledge of long term behaviour of materials used as soil reinforcements

Jean Marc Jailloux & Pierre Segrestin
Sté Terre Armee Internationale, Puteaux, France

ABSTRACT : Long term behaviour of the main metallic and synthetic materials used in soil reinforcement is examined, having regard to required service lives of the order of a century. Present-day knowledge of this area is summarized, and attention is drawn to major gaps where a particular and systematic research and experimental effort is still required.

1 - INTRODUCTION

1.1 Service life

There is a worldwide agreement that most civil engineering structures should be designed for a minimum service life of about a century. It is of course important to recall the difference between service life and lifetime : for as long as it is maintained in service, a structure must remain safe, i.e. an acceptable step away from the ultimate limit state of resistance or deformation, synonymous with failure.

Service life and durability are concepts which have become particularly important with the advent and increasing use of mechanically stabilized earthworks. For, unlike other civil engineering structures, their load bearing elements are difficult to inspect - and impossible to maintain. They are, in addition, buried in soil, a complex environment with physical and chemical characteristics which may vary greatly from site to site. Any guarantee to the effect that soil reinforcing materials have a service life of some one hundred years must be based on a sure knowledge of how their properties will evolve over the long term in extremely varied actual conditions of application.

1.2 Reinforcement materials

There are currently two large families of materials catering for the stabilization of backfill-based structures ; metallic reinforcements and plastic products. Metallic products are usually in the form of galvanized steel strips, but meshes may also be encountered and stainless steel or aluminium alloy strips have been used in the past. Plastics come in the form of polyester or polypropylene geotextile sheets, polyethylene grids, or polyethylene-coated polyester fibre belts.

2 - GALVANIZED STEEL

Up to now, galvanized steel has been the material used in the vast majority of mechanically stabilized embankments. Indeed, whenever major tensile forces are expected, steel continues to be the sole material used in civil engineering as a whole.

The prevalence of steel is due to its unique combination of qualities : tenacity, high modulus of elasticity, ductility, favourable economics. Its corrosion mechanisms and kinetics have been known for a long time ; it has been used in a wide variety of environments over very long periods, and can be protected to ensure that it can do its job for the time specified. Such experience is a major advantage.

In the reinforcing strips used by Terre Armée, durability is achieved by a combination of galvanization and a sacrificial thickness of steel. Galvanization is a technique which has been used for a good century now ; ever-improving, it protects 10% of world steel production. In soil reinforcement applications, a zinc coating has the additional merit of standing up well to rough treatment on construction sites during handling, backfilling and compaction.

2.1 Steel protected by zinc.

The corrosion of metals in an aqueous medium is solely electrochemical in nature : it is connected with the formation of micro-cells resulting from any heterogeneousness in the metal's surface or surroundings. A metal's tendency to corrode depends on the difference in potential it develops vis-à-vis an electrolytic medium. As this difference is greater for zinc, iron remains protected as long as there is any zinc nearby.

The behaviour of a galvanized steel strip in humid soil may be summarized as follows :

Phase 1 : Only the zinc, a tight and adherent coating, comes directly into contact with the soil. The zinc's oxidation reactions lead to the gradual formation of corrosion products which remain attached to the strip's surface and bind in adjacent soil particles. As the electrolyte changes, the reactions become slower.

Phase 2 : The zinc has completely disappeared in places, exposing the steel. Nearby zinc, however, continues to provide cathodic protection, and the steel does not corrode (Fig. 1). The soil round about becomes increasingly richer in zinc compounds (hydroxides, oxychlorides, carbonates ...) and forms an adherent gangue.

Phase 3 : The steel begins to dissolve in an environment which is very different from what it was at the outset, and much more slowly than if it had never been galvanized. The rate of the corrosion's advance continues to decline as time goes by.

The time spent in each phase will obviously depend on how aggressive the environment is ; indeed, the main aim of research has been to classify soils according to their chemical and electrochemical characteristics, in order to identify those most commonly encountered and estimate the maximum rate of corrosion to be expected.

2.2 Different soil-types

Data analysis techniques were used to assess the relative importance of corrosion-influencing factors : it was found that the aggressiveness of a backfill could be defined by reference to four parameters :
- its resistivity (measured after one hour's saturation with distilled water)
- its pH (as measured in water extracted from this mixture)
- chloride content
- and sulphate content.

The histograms below show the distribution of values found for these parameters in a representative population of 235 backfills. Borderlines were also drawn on these graphs to delimit common backfills of low aggressiveness (Fig. 2).

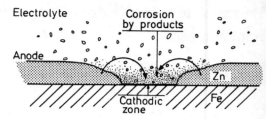

Fig.1 - Cathodic protection of steel by zinc

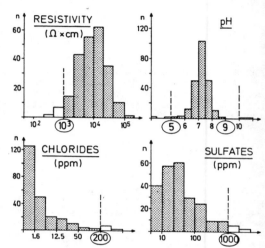

Fig. 2 - Electro/chemical characteristics of 235 backfills.

2.3 Previous research results

Valuable data on the order of magnitude of metal losses over long periods were initially obtained from publications concerned with the performance of galvanized steel culverts, forced conduits, and steel sheet piling or piles. But the most significant source from the point of view of soil reinforcement remains the report, written by M. Romanoff, on the experiments carried out by N.B.S. (U.S.A.) between 1910 and 1955. Thousands of samples of steel, galvanized or not, were buried at 128 different sites ; disregarding those placed near the surface in very clayey or organic soils, the first finding is that galvanized steel, after some ten years, has corroded four times more slowly than ordinary steel. Secondly, the average rate at which thickness is lost decreases over time ; the process may be described by the equation $P = AT^n$, where T is time, and A and n are constants representing soil characteristics, n always being < 1.

For example there are random backfills where the corrosion rate falls from 10μ per year at the end of the first year to 3μ after 10 years, tending towards 1μ per year after 100 years.

2.4 Research by Terre Armée.

Since 1970, Terre Armée Internationale has been pursuing its own programme of research aimed at confirming prior findings and carrying investigations further forward. Three main lines of experimentation are being simultaneously pursued :

1. A reference framework was provided by placing strip samples in boxes filled with selected soils which are kept in conditions identical to real backfills. Samples are removed and weighed at regular intervals. These tests show, for example, that the transition from zinc to iron has no effect on the slopes of weight loss curves (Fig. 3).

2. The same soils, and many others, were placed together with strip samples in a total of 200 cells (Fig. 4) where periodic measurements are made of instantaneous corrosion currents. Faraday's law is then applied to calculate the corresponding metal loss. Such measurements have been going on for twelve years now, agreeing well with the reference test findings.

3. Lengths of strip or durability samples are removed from structures currently in service (the oldest will soon be 20 years old) and here, too, average loss of thickness is measured.

All results for what might be considered routine-type backfills by reason of their physical and electrochemical characteristics and their water content – including those obtained during inspections of actual structures – are shown together in figure 5. The scale being logarithmic, a function of the type $P = AT^n$ would appear as a straight line. The first impression gained from this graph is that this is a

very coherent set of results, from the point of view both of numerical values and of the slopes of the curves. The points for field measurements are very much in line with the curves for experimental results and confirm the reliability of the laboratory measurements.

Straight lines can then be drawn on either side of this cluster of results and extrapolated over the period of service normally required, the extrapolation coefficient being of the order of 5 or 6 (trials extending over 12 to 18 years for a service life of 70 to 100 years). In the case of a rather aggressive backfill at the end of the range of common soils, the likely rate of corrosion is represented by $P = 25T^{0,65}$. This amounts to a combined loss of zinc and iron of 400µm per face at 70 years (in routine design calculations a notional 500µm of steel is deducted per face). In practice, though, most values will be situated towards the lower end of the range, as evidenced by the points plotted after inspections of actual structures.

3 – PASSIVABLE METALS

In the early days of Terre Armée's construction method, other metallic materials were also used for reinforcing strips, such as the aluminium alloy ASTM 5086 or the stainless steel AISI 430. These "passivated" metals are protected by a surface layer of oxide, expected to be extremely stable. And yet, some ten years later, problems were occuring in a number of cases. The experiments on the basis of which these materials had been specified,

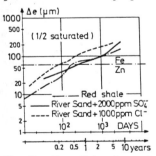

Fig. 3 - Transition from zinc to iron in aggressive soils.

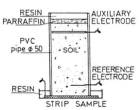

Fig. 4 - Corrosion cell

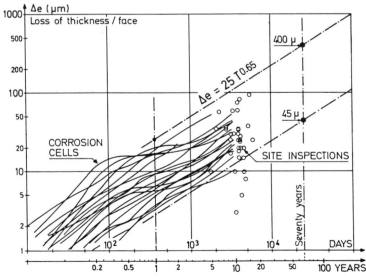

Fig. 5 - Actual loss of thickness in routine backfills.

while well designed, had been too short in duration and insufficient in number ; far too little is known, even now, of how these alloys corrode in soils.

Galvanized steel, on the other hand, has a "clean" accident-free record. Well established long before the development of the Terre Armée technique, it has since benefited from a further twenty years of operational applications and specific research.

4 - PLASTICS

The first retaining structure to be reinforced with plastic was built by Terre Armée at Poitiers (France) in 1970. Belts woven from high tenacity polyester yarns were incorporated in a structure forming part of a temporary development. When samples were first removed, in 1981, the yarns had lost a lot of their initial strength. In 1987, during a more thorough examination (the structure had since been taken out of service), the belts were found to have completely deteriorated where they entered the facing panels, as a result of hydrolysis in the vicinity of concrete. Many came away at the pull of a hand. In lengths of belt removed from the backfill, this time far from the facing, up to 35% of strength had been lost due to mechanical and chemical degradation, while chemical degradation alone accounted for a deterioration of some 4,5% in individual fibres. This experience further confirms the fact that when new materials, in particular synthetics, are used for soil reinforcement purposes, a variety of very real problems arise; these will be rapidly reviewed hereunder.

4.1 The desired mechanical properties of reinforcing materials.

A material must have a certain tenacity, the extent to which it may strain must be limited and, above all, any deformation must not progress over time : a material subject to creep poses problems, particularly when used with a relatively rigid facing.

A material must be hard and ductile enough to withstand impact shocks, puncturing by pebbles and the tracks of site equipment. Such mechanical aggression must not be able to seriously reduce a reinforcement product's strength, either sooner or later.

4.2 The degradation of polymers.

Degradation processes and their kinetics must first of all be identified. These follow a particular pattern in each individual material ; thus, the polyester rate

of hydrolysis accelerates over time, while that one of steel corrosion slows down. The morphology of degradation plays an important role : homogeneous dissolution promotes steady ductile behaviour, whereas localised cracking will result in brittle unpredictable behaviour.

Unlike rust on steel, deterioration in a plastic is difficult, even impossible, to see. The process may be started off by different factors :

- water, liquid or in the form of vapour, can penetrate into the molecular structure and act as a plasticizer sometimes greatly modifying the original mechanical properties. It also acts chemically by hydrolysis, a process which may be speeded up by salts and other substances dissolved in the water.

- light and UV rays have the effect of initiating and encouraging oxidation reactions in polymers. Prolonged exposure to light, prior to a "burial" in soil, may compromise its resistance to other forms of degradation.

-oxidation associated with increased temperature, aging mechanisms prompted by certain metallic ions present in soils, surface interactions leading to environmental stress cracking, the action of solvents, the gradual restructuring of molecules, etc ...all of these factors may be involved.

4.3 The effect of temperature

The vulnerability of polymers to the causes of degradation is largely dependent on raw materials, used additives, and manufacturing processes. Morever all degradation phenomena are accelerated by increased temperature : a rise from 10° to 20° C results, for example, in multiplying the rate of hydrolysis of polyester by a factor of 4.5, while creep in polyethylene increases tenfold. And, contrary to the commonly held view, soil temperatures vary to a depth of several meters ; near the surface, and near the facing (where, furthermore, stresses are high ...), diurnal and seasonal variations are even further accentuated by the effects of sunlight.

Since temperature affects degradation in an exponential way, following an Arrhenian law, it is not the average temperature at a site which determines a polymer's behaviour. Thus, near the facing, and in a temperate zone, hydrolysis in polyester advances 2,5 times faster, and the time to creep failure in polyethylene is 6 times shorter than it would be at a depth of some ten meters, at constant average site temperature ($\simeq$ 10°C).Temperature is therefore a significant variable which must not be overlooked when designing structures involving synthetic reinforcing materials. Such considerations are of uppermost importance in hot countries.

5 - THE CASE OF POLYETHYLENE

High density polyethylene (HDPE) is highly resistant to most aggressive chemicals, which makes it at first an attractive candidate for in-the-soil applications. Some of its weaker mechanical properties can be improved by drawing during manufacture, which promotes molecular orientation. Anyway over time and under a given load, HDPE continuously deforms until it fails : this is the phenomenon known as creep. Failure comes rapidly when the load is heavy, being preceded by major deformation : this is termed ductile failure. Under smaller loads, failure comes more slowly - but may be very sudden when it does occur, with no any preceding deformation phase : this is brittle failure, which is very difficult to predict (Fig 6).

We are aware of only one area of application of polyethylene, namely buried pipelines, where conditions are sufficiently similar and experience of long enough standing to be of relevance in the field of soil reinforcement. Such pipelines have been used for gas distribution for some thirty years now. In North America, cracks occurred in some pipes after just a few years, posing extremely serious problems for network operators. It was realized that mere extrapolation of the results of short term trials (about a year) did not allow an imminent brittle failure to be foreseen: the laws governing HDPE's ductile behaviour, with the pipe swelling prior to bursting, do not apply where the pipe splits suddenly in a brittle way (Fig. 7).

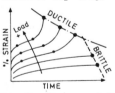
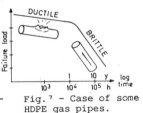

Fig.6 - Creep and failure modes of HDPE.

Fig.7 - Case of some HDPE gas pipes.

In addition, the process of brittle failure, to which polyethylene is intrinsically prone, may be greatly accelerated in certain environments which favour the development of cracks and existing defects, even if microscopic, especially in the zones subject to most stress. Thus, contact with water, or oil, may reduce a given resin's time to brittle failure in air by factors of, respectively, ten and one thousand. The result is a very considerable shift in the position of the boundaries of the brittle failure zone.

Given these uncertainties, HDPE gas pipes are used only in accordance with the following conditions, as laid down by the ASTM and the Plastic Pipe Institute :

1. Service life must never exceed 50 years.

2. Tensile stress must not, in practice, exceed 10% of short term breaking stress (and possibly less, depending on a maximum deformation criterion).

3. In addition, each product must have come through a series of qualifying tests carried out at different temperatures. Relying itself on certain correlations, this approach aims at ensuring that, in the conditions of use, there is no risk of the brittle failure zone being reached before the end of the service life.

In the present state of knowledge and not to mention the various aggravating factors encountered in soil reinforcement applications (installation, temperature..), it would seem prudent to refer to such specifications.

6 - THE CASE OF POLYESTER

The polyester (PETP) fibres used in many geotextiles pose few problems of creep, at usual temperatures. They are, however, vulnerable to hydrolysis, the chemical reaction process whereby a molecule of water causes a molecular chain to divide, creating new acid and alcohol chain extremities (Fig. 8). Molecular structure may be disrupted by other external factors such as certain chemicals, or UV rays. Fast-acting, the latter too cause chain scissions, and later substantially reduce resistance to hydrolysis.

Even in soils with a very low water content, the air is always saturated with water vapour, relative humidity being 100%. Some polyester products appear protected, encased in polyethylene ; this envelope is not, however, proof against water vapour, and within a few months the polyester fibres inside will be in a saturated environment.

Hydrolytic phenomena in polyester have been investigated above all by Mac-Mahon, (1959), though only in pure water and at temperatures in the 60-100°C range (Fig.9). In these conditions, molecular weight is halved in eight months at 80°C ; six years would be required at 60°C, and several hundred years, one might assume, at usual site temperature. However, no

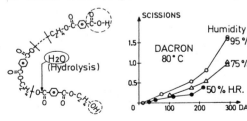

Fig.8 - Hydrolysis of a PETP molecula.

Fig.9 - Average number of scissions/molecula (Mac-Mahon).

systematic investigation of the effects of pH and certain catalytic phenomena has yet been carried out or published – though it is agreed that hydrolysis advances much faster in the presence of bases, particularly lime.

Mac-Mahon also studied the relationship between the kinetics of chain scissions (which accelerate over time) and changes in mechanical properties. He found that initially there was a relatively slow decline in mechanical strength, which fell in proportion to the average number of scissions in the macromolecular chains. The most striking discovery was, however, the existence of a transition zone beyond which the effect of hydrolysis on mechanical strength became catastrophic. The knee in the curve comes where the rate of degradation corresponds to about a halving of molecular weight (Fig. 10). Ultimately, then, accelerating hydrolysis in conjunction with the existence of a critical degradation threshold leads to a sudden collapse of mechanical strength.

There is little available experimental data on in-the-soil performance of polyesters over any appreciable period of time, except for a major report from the French Textile Institute, concerned with samples of geotextiles 3 to 11 years old. Most were used in drainage or separation functions, where mechanical characteristics are of less relevance. Unfortunately, little is revealed of the initial material characteristics or exact conditions of exposure, and the results are not very precise. Anyway, if loss of mechanical strength is compared with the number of chain scissions (derived from a titration of acid chain extremities), there is a certain consistency with Mac-Mahon's results (Fig 11), and therefore with presumptive effects of hydrolysis.

Moreover, the average rate of chemical degradation is about 50 times greater than might have been expected from the pure water results. This unfavourable outturn, probably due to chemical characteristics of the soils and aggravated by short term exposure to UV rays, prompts many questions about the life expectancy of structures incorporating polyester reinforcements.

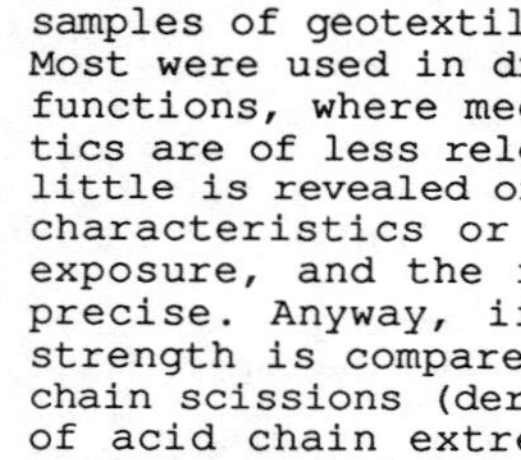

A tremendous amount of experimental work remains to be done, then, before sound specifications can be laid down for structures other than temporary.

7 – CONCLUSION

More than twenty years experience of the behaviour and aging of buried strips all points to the need for an open-minded, progressive and pragmatic approach to all new materials. Their individual degradation mechanisms must first be identified ; behavioural laws which take into account the various phenomena observed must then be derived from basic experimentation ; and, finally, long term tests must be embarked upon, in a variety of true-to-life environments, in order to determine reliable values for the different parameters of these laws. Given the service lives required in this field, and the magnitude of the risks, the duration of such trials is of paramount importance.

BIBLIOGRAPHY

Romanoff M – Underground Corrosion – NBS Circular 579 – April 1957.

Darbin M, Jailloux JM, Montuelle J – La perennité des ouvrages en Terre Armée. Résultats d'une expérimentation de longue durée sur l'acier galvanisé. Bulletin de liaison des LPC –141– janvier 86.

Verdu J. – Vieillissement des plastiques. AFNOR technique – Ed Eyrolles – 1984.

Segrestin P, Jailloux JM – Temperature in soils and its effect on the ageing of synthetic materials – Geotextiles and Geomembranes (IGS) – Vol 7 – 1988.

Mruk S.A. (PPI) – Validating the hydrostatic design basis (ASTM D 2837) of polyethylene piping materials – 9 th plastic fuel gas pipe symposium. New Orleans – November 1985.

Mac-Mahon W, Birdsall HA and al. J.chem.Eng.Data 4(1)57 – 1959.

Institut Textile de France – Contribution à l'étude du vieillissement des géotextiles (Contrat Ministère de l'Industrie) – Janvier 1983.

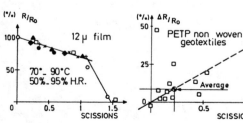

Fig.10 - Hydrolysis of PETP film. Residual strength (Mac-Mahon).

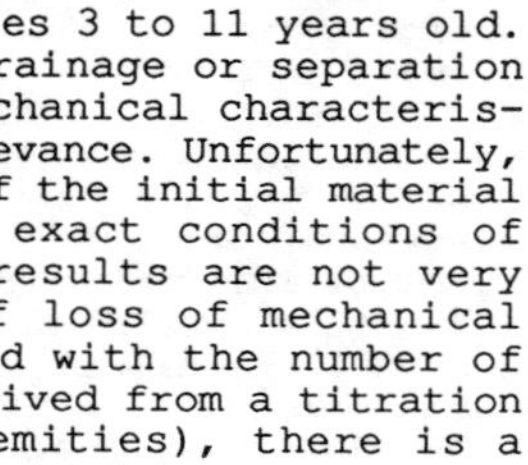

Fig.11 - Samples of buried geotextiles.Loss of strength (ITF).

International Geotechnical Symposium on Theory and Practice of Earth Reinforcement / Fukuoka Japan / 5-7 October 1988
© 1988 Balkema, Rotterdam. ISBN 90 6191 820 0

Creep behaviour of geotextiles

M.H.Kabir
Bangladesh University of Engineering and Technology, Dhaka, Bangladesh

ABSTRACT: A generalised nonlinear constitutive relation for creep behaviour of geotextiles is presented in this paper. Suitable testing technique, methods of data analysis and their presentation are suggested. The method of data presentation is based on isochronous load-strain curves and creep isochronous stiffness. Creep behaviour of four generic types of geotextiles is presented and practical significance of the work is discussed.

1 INTRODUCTION

Like all other polymeric materials geotextiles generally exhibit elasto-viscoplastic behaviour. Therefore, where performance data for geotextiles and related products are to be employed in Fundamental Analytical Designs (McGown and Kabir 1983) of permanent reinforced soil systems, creep test data should be used. For this, the development of suitable testing techniques, methods of data analysis and data presentation is of paramount importance. A rapid loading creep test methodology was developed for this purpose. The testing techniques developed are categorized into two main groups; in-isolation and in-soil tests. In-isolation test data may be used to obtain design parameters for those materials which are unaffected by in-soil confinement. However, in-soil tests should be used for obtaining design parameters for those materials whose properties are significantly altered when confined in-soil. The scope of this paper is limited to in-isolation creep behaviour of geotextiles only.

In terms of data analysis, two types of visco-elastic material behaviour, linear and nonlinear visco-elasticity have been taken into account. Many polymeric materials and so geotextiles, behave linearly at low levels of strain but become nonlinear at higher levels, hence the need for a generalised method of data analysis was recognised. A suitable method has been developed which is based on the multiple integral representation as described by Onaran and Findley (1965).

The suggested method of data presentation is based on isochronous load-strain curves and creep isochronous stiffness. To illustrate the load-strain-time behaviour using this technique, data analysis for four basic types of geotextiles are presented. These include a melt bonded non-woven (Terram), a needle-punched non-woven (Bidim), a woven (Lotrak) and a composite geotextile (Propex). Identity and properties of these geotextiles are presented later in Table 1. Finally, practical significance of the work undertaken is briefly discussed.

2 CREEP CHARACTERISTICS OF GEOTEXTILES

2.1 Background research

Little has so far been reported on creep behaviour of geotextiles. As geotextiles are generally made from polymers they are expected to demonstrate behaviour comparable to these when loaded over long periods of time.

Some of the empirical laws proposed to represent the creep behaviour of geotextiles may be described as follows:

$$\varepsilon = \varepsilon_o + A \log t \tag{1}$$

where ε = total strain, ε_o and A are functions of stress, temperature and nature of material. Finnigan (1977), Van Leeuwen (1977), Raumann (1981) reported use of this law for representation of short term creep with limited success.

Paute and Segouin (1977) used the three element rheological model, comprising a spring and a Kelvin model in series to portrary creep behaviour of geotextiles. However, they have presented test data only upto 8 hours duration.

Shrestha and Bell (1982) used the four element Berger model to portrary creep behaviour of geotextiles. They also used the so called "three parameter" equation and found that the rheological model offered a better fit to experimental data.

The main shortcommings of these investigations are that data from short duration tests (often only of several hours) were used. Moreover, these studies failed to present a generalised load-strain-time behaviour of geotextiles.

2.2 Developlment of generalized stress-strain-time behaviour.

It was observed by a number of researchers including Findley and Peterson (1958) that an empirical power function of time could be used to describe the creep of many polymers with reasonable accuracy over a wide span of time, and may be presented as

$$\varepsilon = \varepsilon_o + \varepsilon_t \cdot t^n \qquad (2)$$

where ε_o and ε_t are functions of stress and n is independent of stress and function of material only. Lai and Findley (1973) found n to be nearly independent of temperature and generally less than 1.0.

To compare the suitability of different time functions, (not stress functions), Findley and Peterson, presented the predictions from different equations with results of long time creep of plastic laminates. The data for the first 2000 hours were represented as well as possible by all the equations, then the tests were continued for about 10 years. It was observed by Findley and Peterson and confirmed later by Findley and Tracy (1974) that equation (2) gave a much more satisfactory prediction compared to those by other equations. This is due in part to the fact that creep of plastics, concrete and some metals under moderate stresses starts at a very rapid rate immediately after loading and progresses at a continuously decreasing rate. These are characteristics of equation (2).

Onaran and Findley represented creep of nonlinear visco-elastic materials as a series of multiple integrals. For uniaxial creep, the following expression results:

$$\varepsilon = R(t)P + M(t)P^2 + N(t)P^3 \qquad (3)$$

The Kernel functions R,M,N for constant loading for geotextiles and polymers are

expected to take the following form:

$$R(t) = \mu_1 + \omega_1 \, t^n \qquad (4)$$

$$M(t) = \mu_2 + \omega_2 \, t^n \qquad (5)$$

$$N(t) = \mu_3 + \omega_3 \, t^n \qquad (6)$$

where μ's and ω's are material functions and are dependent on temperature (T) and n is a function of material and may or may not be a function of temperature.

Substituting values of R,M and N from equations (4), (5) and (6) in equation (3), the following equation is obtained for constant temperature situation.

$$\varepsilon(t,p) = \varepsilon_o(p) + \varepsilon_t(p) \cdot t^n \qquad (7)$$

where $\varepsilon_o = \mu_1 \cdot p + \mu_2 \cdot p^2 + \mu_3 \cdot p^3 \qquad (8)$

and $\varepsilon_t = \omega_1 \cdot p + \omega_2 \cdot p^2 + \omega_3 \cdot p^3 \qquad (9)$

In the case of linear visco-elasticity the coefficients of second and third terms become equal to zero.

The constant μ's, ω's, and n are obtained by curve fitting the results of creep tests for at least three different loads, as described in subsequent sections.

3 TESTING TECHNIQUES

Test apparatuses and procedures developed to study the in-isolation creep behaviour of geotextiles, are described in the following sections.

3.1 Test apparatuses

Test specimens were selected and prepared in much the same way as described by Andrawes, McGown and Kabir (1984). Test set-ups each comprising a basic rig and a loading system, consisting either a direct loading or a lever loading device were designed and constructed to conduct one step, rapid-loading tests. A programmeable data logger capable of logging 100 channels per second, at a minimum logging interval of 0.2 second was used. This system was intended to measure and record loads and displacements under a very rapid loading situation. A schematic diagram of a test set-up developed for in-isolation testing is shown in Figure 1.

3.2 Test procedure

One step loading test using several different loads within the relevant operational range, was used to establish the load–strain –time relationship of the materials. The temperature of these tests was maintained at all times at $20^{\circ} \pm 1^{\circ}C$ and the relative humidity at 65 ± 2 percent.

The loads were applied as smoothly as possible and sustained for at least 1000 hours. During the initial part of the tests the loads applied and the resulting extentions of the test specimens were both monitored electronically. The load was measured using a load cell and the extension by the two L.V.D.T.s one on either side of the test specimen (Figure 1). After the rapidly varying initial part of the test, the deformations were measured using mechanical dial gauges.

4. DATA ANALYSES

To obtain the load–strain–time behaviour of geotextiles and to determine the load and time parameters mentioned in equations (7), (8) and (9) which constitute a maximum of seven parameters (for a three degree nonlinearity), the following steps are recommended.

1. Rewrite equation (7) in the form:

$$\log (\varepsilon - \varepsilon_o) = \log \varepsilon_t + n.\log t \qquad (10)$$

and plot $\log \varepsilon$ vs. $\log t$, using the equivalent zero time ($t_o=0$), where $2t_o$ is the time required for the ramped loading. Qualitative

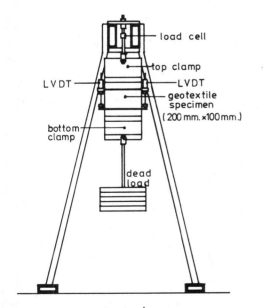

Fig. 1 The creep test rig

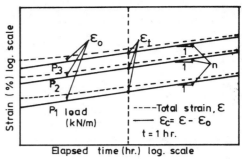

Fig. 2 Curve fitting for creep parameters

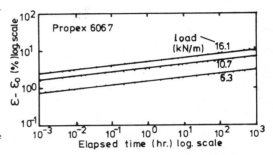

Fig. 3 Curve fitting for Propex

plots for different loadings (P) are shown by dotted lines in Figure 2. If for example it is assumed that $\varepsilon_o = 0$ in equation (10) then the plot will yield a straight line. However, this is not the case for most polymers and geotextiles. When a selected value of ε_o is subtracted from ε and the resulting reduced data ($\varepsilon - \varepsilon_o$) is again plotted against time, the lower set of curves result which are shown in Figure 2 as sharp lines. These ε_o values are not necessarily the equivalent of instantaneous strains, being only the values of strain required to make the straight lines fit the test data. The relevent plots for the composite geotextile Propex are presented in Figure 3.

2. The parameter n (equation (7)) is the common slope of the three sharp lines in Figure 2.

3. The three values of ε_o are then plotted against load and polynomials upto three degrees are fitted to obtain the parameters μ_1, μ_2 and μ_3 (equation (8)).

4. The three values ε_t (($\varepsilon - \varepsilon_o$) values at t = 1 hour) from Figure 2 are plotted against load P and polynomials upto three degrees are fitted to obtain the parameters ω_1, ω_2 and ω_3 (equation (9)).

The plots described in items 3 and 4 above appropriate for the geotextile Propex is presented in Figure 4. The value of n is also shown on this figure.

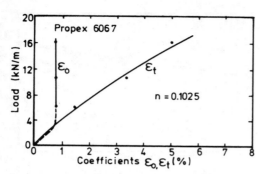

Fig. 4 Creep parameters of Propex

5 TEST RESULTS

5.1 Presentation of creep properties

It is suggested that the data obtained from
the foregoing analyses should be presented
in the form of three plots described as
follows:

PLOT (A): This is a plot of total strain
(ε) vs. time (t) for 1000 hours or more.

PLOT (B): After calculation, the isochronous load-strain relations may be plotted.
Using equation (7), the data may be extrapolated up to 10^6 hours.

PLOT (C): From the isochronous load-strain
plot, secant slopes at appropriate strain
levels may be computed for different times.
These values of secant slopes known as the
Creep Isochronous Stiffnesses (I_{sc}), may be
plotted against time (to a log scale) up
to 10^6 hours.

Typical equivalents of Plots (A), (B), and
(C) for the composite geotextile Propex are
presented in Figures 5,6 and 7 respectively.

5.2 Typical results

To illustrate the applicability of the method of data analyses, the load-strain-time
behaviour of the four types of geotextiles
are presented here. In all cases two creep
tests were performed for each load level
and the average was reported.

The results of creep parameters (μ, ω
and n in Figure 4), the isochronous load-
strain diagrams (Figure 6) and the diagrams
showing variation of Creep Isochronous
Stiffness, I_{sc} (at 5 and 10 percent strain
levels) with time for the composite geotextile Propex were presented earlier. The
relevant plots for the other three geotextiles, Terram, Bidim and Lotrak are presented in Figures 8,9 and 10. The values of
n and the expressions for ε_o and ε_t obtained for all the four geotextiles are also
presented in Table 1.

The agreement between test data and results from the mathematical model, in case
of all the four geotextiles possessing different structural and mechanical behaviour,
may be termed as very satisfactory. This
depict the versatility of the mathematical
model to portrary the load-strain-time behaviour of these geotextiles.

6 DISCUSSION AND CONCLUSIONS

In Fundamental Analytical Designs of geotextile reinforced soil systems,data from

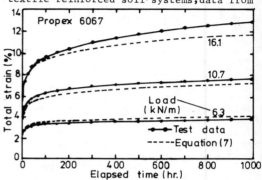

Fig. 5 ε-log t plot for Propex

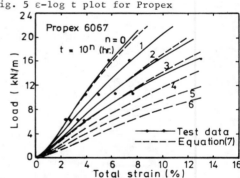

Fig. 6 Isochronous load-strain plots for
Propex

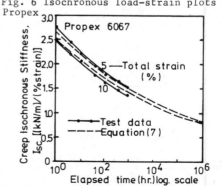

Fig. 7 Creep isochronous stiffness for Propex

114

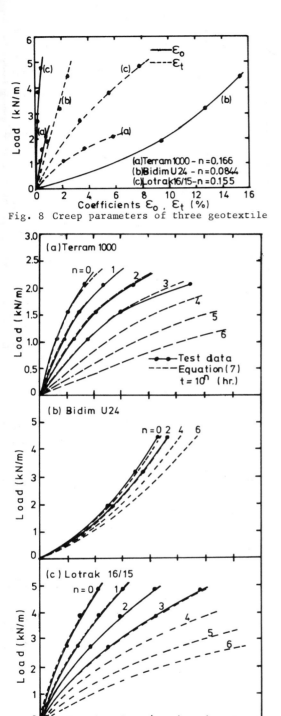

Fig. 8 Creep parameters of three geotextile

Fig. 9 Isochronous load-strain plots for three geotextiles

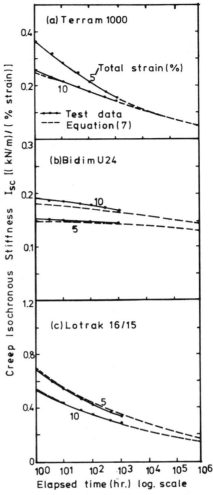

Fig. 10 Creep isochronous stiffnesses for three geotextiles

load-strain-time behaviour (using creep iso-chronous stiffness) of geotextiles must be used to evaluate the long term safety margins of the constituent soil and reinforcement. A soil which may have adequate safety margin under short term loading may loose that due to shedding of load by the reinforcement, undergoing long term deformation and vice versa. Therefore, an elaborate design should take into account the exact sharing of the load between soil and the reinforcement, throughout the design life of the structure. Apart from stability considerations, use of creep parameters is of vital importance in calculating deformations. Excessive deformations may impair the serviceability condition of the reinforced soil

Table 1. Identity and creep properties of geotextiles

Characteristics Geotextile	Method of constr.	Polymer compos.	Unit wt. (gsm)	Creep parameters		
				n	ε_o	ε_t
Terram 1000	Non-woven melt-bonded	*PP 67% PE 33%	140	0.166	$0.407P - 0.218P^2 + 0.087P^3$	$2.296P - 0.894P^2 + 0.553P^3$
Bidim U24	Non-woven	PES 100%	210	0.0844	$7.084P - 1.373P^2 + 0.125P^3$	$0.527P$
Lotrak 16/15	Woven tapes	PP 100%	120	0.155	$0.046P - 0.0118P^2 + 0.004P^3$	$0.436P + 0.346P^2 - 0.0215P^3$
Propex 6067	Composite wov+needl	PP 100%	650	0.1025	0.8 for $P > 0$	$0.2079P + 0.0103P^2 - 0.000161P^3$

*PP-Polypropylene, PE-Polyethylene, PES-Polyester.

structure itself or any structure supported by it.

On the basis of the work presented here, the following conclusions may be drawn.

1. The method of data analysis synthesised from the multiple integral representation of visco-elastic meterial behaviour was found to be a versatile technique which proved suitable for the four different types of geotextile tested.

2. The curve fitting technique to obtain the time dependent and load dependent parameters was also found to be versatile and suitable for the four types of geotextile.

3. The power function law of time dependency with the polynomial form of load dependency, were found to portray linear and nonlinear visco-elastic behaviour of geotextiles with reasonable accuracy.

4. The use of stiffness values from short term tests may be misleading, often unsafe, in the selection of geotextiles for long term use under load.

5. The creep isochronous stiffness (I_{sc}) should be used in design to obtain safe and allowable loads as well as limiting deformations in geotextiles used for reinforcement.

REFERENCES

Andrawes, K.Z., McGown, A. and Kabir, M.H. 1984. Uniaxial strength testing of woven and non woven geotextiles. Geotextiles and Geomembrane, Elsevier Applied Science Publishing 1:1:41–56.

Findley, W.N. and Peterson, D.B. 1958. Prediction of long time creep with ten year creep data on four plastic laminates. PASTM, 58:841.

Findley, W.N. and Tracy, J.F. 1974. 16 year creep of polythylene and PVC. polymer Engg. and Science, 14:577.

Finnigan, J.A. 1977. The creep behaviour of high tenacity yarns and fabrics used in civil engineering applications. Proc. Intl. Conf. on Use of Fabrics in Geotechnics, Paris, 2:305–309.

Lai, J.S. and Findley, W.N. 1973. Elevated temperature creep of polyurethance under nonlinear torsional stress with step changes in torque. Trans of the Soc. of Rheology, 17:1:129.

McGown, A. and Kabir, M.H. 1983. Specifications and testing of textiles for civil engineering earthworks. Proc. Conf. on Non Woven Fabrics UMIST, UK.

Onaran, K. and Findley, W.N. 1965. Combined stress creep experiments on a nonlinear viscoelastic material to determine the Kernel functions for a multiple integral representation of creep. Trans. of the Soc. of Rheology, 9:299.

Paute, J.L. and Segouin, M. 1977. Determination of strength and deformability characteristics of fabrics by dilation of a cylindrical sleeve. Proc. Intl. Conf. on the use of Fabrics in Geotechnics, Paris, 2:293–298.

Raumann, G. 1981. A plane strain hydraulic tensile test for geotextiles. Materials and structures, Research and Testing, RILEM 14:82:295–302.

Shrestha, S.C. and Bell, J.R. 1982. Creep behaviour of geotextiles under sustained loads. Proc. of Sec. Intl. Conf. on Geotextiles, Las Vegas, USA, 3:769–774.

Van-Leeuween, J.H. 1977. New methods of determining stress-strain behaviour of woven and non woven fabrics in the laboratory and in practice. Proc. Intl. Conf. on the use of Fabrics in Geotechnics, Paris, 2:299–304.

International Geotechnical Symposium on Theory and Practice of Earth Reinforcement / Fukuoka Japan / 5-7 October 1988
© *1988 Balkema, Rotterdam. ISBN 90 6191 820 0*

Long-term pull-out tests of polymergrids in sand

K.Kutara & N.Aoyama
Public Works Research Institute, Ministry of Construction, Tsukuba, Japan

H.Yasunaga & T.Kato
Technical Research Institute, Hazama Gumi Ltd, Yono, Japan

ABSTRACT: In order to establish the rational design methods for geotextile reinforced structures, it is necessary to clarify the long-term friction/adhesion characteristics between soil and geotextiles. Therefore, both the displacement-controlled short-term pull-out tests and load-controlled long-term pull-out tests were conducted respectively for the polymergrids embedded in sand. From the test results, it was found that, with a constant pull-out force applied to the polymergrid embedded in sand, each bar of the polymergrid is displaced with time but its free end is hardly displaced with time and that the increase with time in the deformation of polymergrid in sand is smaller than the increase obtained from the results of tensile tests.

1 INTRODUCTION

High tensile strength geotextiles are being utilized as reinforcing materials for steep slopes and reinforced earth structures. High effectiveness of the geotextiles is widely recognized, but the present design methods are unable to fully utilize their effectiveness. Particularly with respect to the friction/adhesion characteristics between soils and geotextiles, there are still many items which have not been clarified. Because of this, the current design methods require high strength and large lengths of geo-textiles. Thus, for establishing a rational design method for geotextile reinforced structures, it is necessary to clarify the friction/adhesion characteristics between soils and geotextiles.

Generally, for examining the friction/ adhesion characteristics between soils and geotextiles, pull-out tests or direct shear tests have been carried out. According to the researches conducted up to now, it was found that considerable deformation and displacement of geotextiles occur when pulling geotextiles embedded in soils, by which the tensile strains are mobilized in them. However, these tests were conducted by the displacement-controlled methods and were short-term tests, and long-term friction/ adhesion characteristics between soils and geotextiles were hardly studied.

Therefore, by using polymergrids embedded in sand, displacement-controlled short-term pull-out tests and load-controlled long-term pull-out tests were performed. Based on the test results, the short-term and long-term friction/adhesion between sand and polymer-grids (including the long-term deformation characteristics of polymergrids in sand) were examined.

2 TEST PROCEDURE

Two series of pull-out tests were performed, which were displacement-controlled tests and load-controlled tests. Through both tests, air-dried Toyoura sand (G_s=2.64, G_{50}=0.16mm, U_c=1.46, e_{max}=0.96, e_{min}=0.64) and polymergrids were used. An example of the polymergrids is indicated in Fig. 1. Both

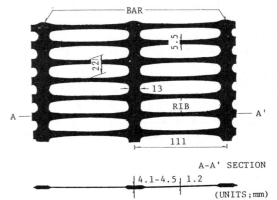

Fig. 1 An example of polymergrids

series of the tests were conducted under two conditions of the relative density, 20% and 65%, and under three conditions of the confining pressure, 2.0, 4.0, and 8.0tf/m². Each test piece of the polymergrids was 400mm wide and 700mm long in sand, and it had seven bars in sand. Each bar was named the survey point No.1-No.7(the bar closest to pull-out side was No.1 and the farthest bar was No.7). The pull-out force and the displacement of each bar were measured by means of load cells and displacement transducers.

2.1 Displacement-controlled pull-out tests

The apparatus of displacement-controlled tests is indicated in Fig. 2. It consists essentially of a steel box, 1200mm long, 600mm wide, and 600mm deep. The confining pressure was applied using rubber bags, with the same plan dimensions as the box, filled with air at the required pressure. The tests were performed by pulling the test piece of polymergrids out with a screw jack. The pull-out speed was adjusted to 1mm/minute.

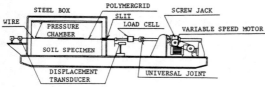

Fig. 2 Displacement-controlled pull-out test apparatus

2.2 Load-controlled long-term pull-out tests

The apparatus of load-controlled tests is indicated in Fig. 3. It consists essentially of a steel box, 1000mm long, 800mm wide, and 500mm deep. The confining pressure was applied by the same method as displacement-controlled tests. The tests were carried out by pulling the test piece of polymergrids out with a constant pull-out force. The pull-out force was to be applied in stages of 0.5tf/m each up. Each pull-out force stage had to be maintained for about twelve hours.

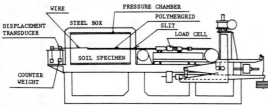

Fig. 3 Load-controlled pull-out test apparatus

3 TEST RESULTS

3.1 Tensile test results

Before pull-out tests, to review the stress and deformation characteristics of polymergrids alone, tensile tests were performed for the polymergrids by means of displacement control and load control.

As the test results, obtained were tensile force-strain relation shown in Fig. 4 and elongation-time relation by load-controlled long-term tensile tests shown in Fig. 5. By comparing two tests in Fig. 4, it is recognized that the strain in displacement-controlled test well coincided with the

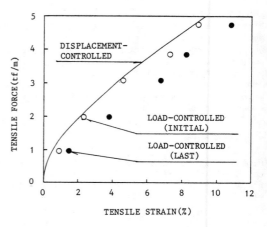

Fig. 4 Tensile test results: relation between tensile force and strain

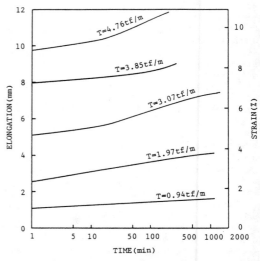

Fig. 5 Tensile test results: load-controlled long-term tensile tests

strain immediately after loading of the load-controlled test.

From Fig. 5, for the tensile force lower than 2tf/m, the strain increases almost in proportion to the time logarithmically indicated. When the tensile force further increases, the initial strain also increases and the increment of strain also increases with time. From this, when considering the long-term deformation, it is recognized to be necessary to review the long-term deformation characteristics of polymergrids alone.

3.2 Relation between pull-out force and pull-out displacement

When a pull-out force was applied to polymergrids embedded in sand, considerable displacement and deformation occurred. Typical examples of distribution of pull-out displacement are shown in Fig. 6 for displacement-controlled pull-out tests and in Fig. 7 for load-controlled pull-out tests. Even in the test under the same pull-

Table 1. Legend in Figs. 8-11

Confining pressure (tf/m^2)	Displacement-controlled	Load-controlled	
		Initial	Last
2.0	——————	○	●
4.0	—— · ——	△	▲
8.0	——— — —	□	■

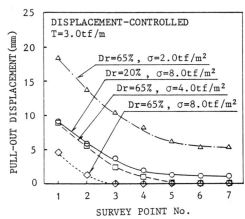

Fig. 6 Displacement-controlled pull-out test results: pull-out displacement distribution

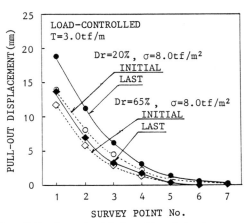

Fig. 7 Load-controlled pull-out test results: pull-out displacement distribution

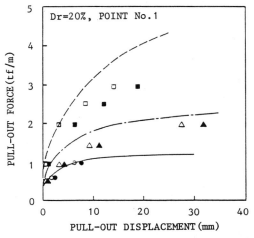

Fig. 8 Relation between pull-out force and displacement: Dr=20%, Point No.1

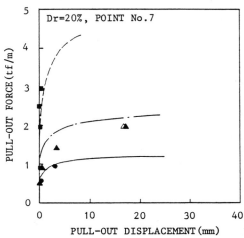

Fig. 9 Relation between pull-out force and displacement: Dr=20%, Point No.7

119

out force (T=3.0tf/m), the distribution of pull-out displacement varies between both the tests. However in both the tests, the displacement is large at the survey points near the pull-out side and decreases as the survey points become far from the pull-out side. Also, the pull-out displacement increases in the cases where the confining pressure decreases. It is also affected by the density of sand, and the pull-out displacement became larger as the relative density decreased even under the same confining pressure. From the above results, it was clarified that, when pulling out the polymergrids embedded in sand, the pull-out displacement of polymergrids was affected

not only by the magnitude of the pull-out force but also by the density and confining pressure of sand.

Typical examples of relation between pull out force and pull-out displacement are shown in Figs. 8 to 11. It was found that i the pull-out displacement of the free end (the survey point No.7) increases to a certain degree (about 10mm in the present experiment), then the pull-out force become. almost constant.

Now making comparison between the displacement controlled tests and the load-controlled tests, it was found that in the case of loosely compacted sand, the results of both the displacement-controlled and load-controlled tests are almost the same regardless of confining pressure and pull-out force. In the case of densely compacted sand, the displacement by load control is larger than by displacement control when comparing based on the time of application of the same pull-out force.

3.3 Long-term friction/adhesion characteristics between sand and polymergrids

From the results of load-controlled long-term pull-out tests, it was found that the pull-out displacement increases with time. Typical examples of the relations between pull-out displacement and time are shown in Figs. 12 to 14. In each case, at the survey point where the pull-out displacement is large, the increase in displacement with time was recognized to be also large, and the pull-out displacement at the free end (No.7) of polymergrids hardly increased with time. That is, it was found that region where the pull-out displacement increased with time is limited. This increase in displacement seems to be caused by the long-term deformation characteristics of the polymergrids itself in sand rather

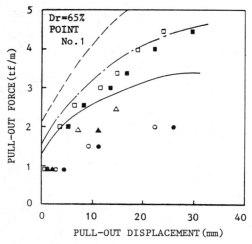

Fig. 10 Relation between pull-out force and displacement: Dr=65%, Point No.1

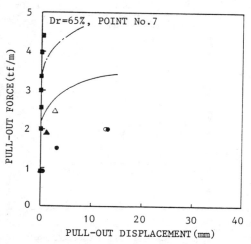

Fig. 11 Relation between pull-out force and displacement: Dr=65%, Point No.7

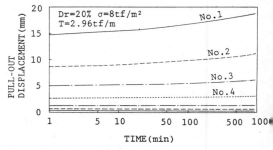

Fig. 12 Relation between pull-out displacement and time: Dr=20%, σ =8tf/m^2, T=2.96tf/m

120

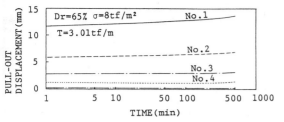

Fig. 13 Relation between pull-out
displacement and time: Dr=65%, σ=8tf/m^2,
T=3.01tf/m

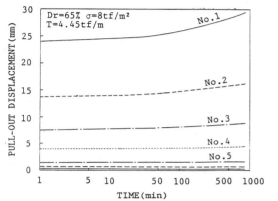

Fig. 14 Relation between pull-out
displacement and time: Dr=65%, σ=8tf/m^2,
T=4.45tf/m

than by the long-term characteristics of
shear resistance acting between sand and
polymergrids. Thus, when a pull-out force
acts to polymergrids embedded in sand and
the polymergrids are stable in short term,
then the long-term stability may be also
maintained.

3.4 Characteristics of the long-term
    deformation of polymergrids in sand

Based on the results of load-controlled
long-term pull-out tests, the long-term
deformation characteristics of polymergrids
in sand were reviewed. When a pull-out
force was applied to the polymergrids
embedded in sand, the polymergrids were
deformed and tensile strains were mobilized
between the bars. Mode of change with time
in the tensile strain under a constant pull-
out force was investigated as the long-term
deformation characteristics of the polymergrids in sand.
   Typical examples of relation between
strain and time are shown in Figs. 15 to 18.
In consequence, it was found that, as the
initial strain becomes larger, the strain

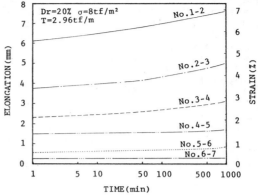

Fig. 15 Relation between elongation and
time: Dr=20%, σ=8tf/m^2, T=2.96tf/m

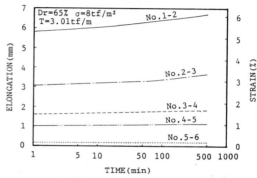

Fig. 16 Relation between elongation and
time: Dr=20%, σ=8tf/m^2, T=3.01tf/m

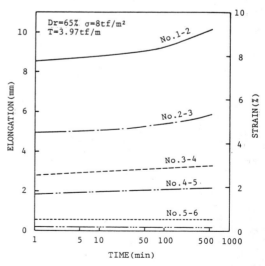

Fig. 17 Relation between elongation and
time: Dr=65%, σ=8tf/m^2, T=3.96tf/m

increases more thereafter. Also, by making a comparison based on the same initial strain, it was found that the increase in strain thereafter becomes smaller as the confining pressure increases. For instance, the increment in strain between the survey point No.2 and No.3 in Fig. 15 is larger than the increment in strain between the survey point No.3 and No.4 in Fig. 17.

Fig. 18 shows the comparison between tensile test results and pull-out test results. By making a comparison based on the same initial strain, the increase in strain is smaller in the pull-out tests than in the tensile tests. This was probably caused because the polymergerids were confined by sand and thus were not easily deformed.

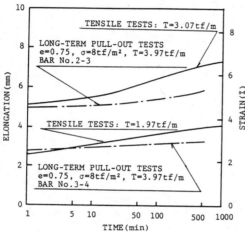

Fig. 18 Comparison between tensile test results and pull-out test results

4 CONCLUSIONS

The following conclusions can be drawn from the results of tests stated above:

1. When a pull-out force is applied to a polymergrid embedded in sand, considerable displacement and deformation occur. Also, the displacement and deformation vary depending on not only the density of sand, the confining pressure and the pull-out force but also the testing methods.

2. In the case of loosely compacted sand, the initial pull-out displacement immediately after loading by load-controlled pull-out tests well coincides with displacement-controlled pull-out displacement. On the other hand, in the case of densely compacted sand, the former is greater than the latter.

3. When polymergrids embedded in sand are pulled out with a constant force, the pull-out displacement increases with time. However the displacement at the free end of

polymergrids hardly increases with time. Thus, long-term stability can be maintained for the friction/adhesion between sand and polymergrids as long as they can provide short-term stability.

4. Cause of the long-term characteristics of the friction/adhesion between sand and polymergrids is the long-term deformation of polymergrids itself in sand rather than long-term shear deformation between sand and polymergrids.

5. The deformation of polymergrids embedded in sand increases with time in response to the initial strain occurred between each bar. And this increase varies depending on the density of sand.

6. When making a comparison based on the same initial strain, the increase with time in the deformation of polymergrids in pull-out tests is smaller than that in the tensile tests (see Fig. 19).

REFERENCES

Bergado, D.T. 1987. Laboratory pull-out tests using bamboo and polymer geogrids including a case study. Geotextiles and Geomembranes. Vol.5 No.3, pp. 153-189.
Ingold, T.S. 1983. Laboratory pull-out testing of grid reinforcements in sand. Geotechnical Testing Journal. Vol.6. No.3, pp. 101-111.
Jewell, R.A. 1985. Material properties for the design of geotextile reinforced slopes. Geotextiles and Geomembranes. Vol.2. No.2, pp. 83-109.
McGown, A., Andrawes, K.Z., Yeo, K.C., & DuBis, D. The load-strain-time behaviour of Tensar geogrids. 1984. Proceeding of Symposium on Polymer Grid Reinforcement in Civil Engineering. London.

International Geotechnical Symposium on Theory and Practice of Earth Reinforcement / Fukuoka Japan / 5-7 October 1988
© 1988 Balkema, Rotterdam. ISBN 90 6191 820 0

Modelisation and design of geotextiles submitted to puncture loading

F.Lhôte, J.M.Rigo & J.M.Thomas
University of Liege, Belgium

ABSTRACT : There are many uses of geotextiles. These geotextiles always serve as separation function even if they serve another function which is considered as the primary function. The application that best shows the use of geotextiles as separator is its placement between soft soil subgrades of low bearing capacity below it and fill above it (figure 1).
The geotextile is thus placed on the site and the stone dumped, spread and compacted above it. When studying the separation function, it appears that the required properties are the puncturing resistance, the burst resistance and the tear resistance. These properties can be studied from laboratory tests results but these results may be viewed as suitable criteria for later specification if they are considered in relation to site related stresses. Tests which have been carried out show that the puncturing resistance of geotextiles is a function of the subsoil bearing capacity.
This paper evaluates the relevant stress situations for geotextiles by means of puncture analysis of the mechanical properties of continuous filament needle punched polypropylene nonwoven geotextiles.

1. ON SITE BEHAVIOUR OF GEOTEXTILES

A geotextile's high elongation at break enables it to follow the unevenness and irregularity of the surface of the terrain and the shape of the stones without damage. Geotextiles with high elongation at failure are thus also unaffected by the specification criterion of tear propagation strength since even in the case of dynamic loading the occurence of perforations, which are a starting point for tear propagation stresses, is unlikely.

It is not easy to give precise value because it would depend of the way the geotextile is placed, risks of puncturing being lower if the geotextile is looser. According to Giroud {1}, an elongation at failure in the order of 100 % measured in a plane-strain tensile test is likely to prevent puncture of the textile in most cases.

Practically elongation of most geotextiles in plane-strain test is much lower than 100 %, so, this lack of deformability must be made up for by puncture resistance.

When placed in use, the geotextile must be able to undertake without damage :

- the local stress concentrations resulting from point loading by correspondingly high puncture strength (figure 2);
- the burst pressure occuring when a geotextile bridges gaps on one side and is subject to soil pressure in the other side (figure 3).

2. TEST PROCEDURE

So far the puncture strength of geotextile has been determined either by using a modified C.B.R. test with a pyramidal piston points (a 3-sided pyramid with sides 5,0 cm long and a height of 2,5 cm) or by an unmodified C.B.R. test according to DIN 54.307 E allowing for the grain form by means of a shape factor varying from 0.8 for round blunted shapes to 3.0 for pointed sharp-edged shapes {2}.

If these tests are easy to be carried out, they don't consider the subsoil bearing capacity. It is possible to take the subsoil into account by modifying the C.B.R. test (figure 4).

A cylinder 200 mm high is filled with a soil an the piston, 50 mm in diameter, is

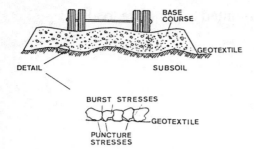

DETAIL   SUBSOIL

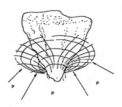

Figure 1 : Geotextile used as a separator

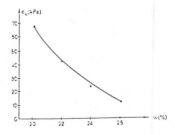

Figure 2 : Visualization of a stone puncturing a geotextile (after {5})

Figure 3 : Visualization of the burst pressure

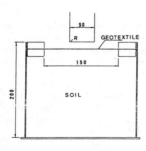

Figure 4 : Test device : puncture test taking into account the soil bearing capacity

pushed through the geotextile until failure. The force-elongation curve is recorded. Two ring diameters have been used for the experiments. The ring outside diameter is the cylinder inside one.

Wishing to study of the subsoil bearing capacity, the soil used for experimentation was an Hesbayan silt. By increasing the soil water content, it was possible to decrease the soil bearing capacity. The soil characteristics are given table 1.

Table 1. Characteristics of the soil used for experimentation

Atterberg limits	w_p = 21.1 %		
	w_ℓ = 31.2 %		
w %	γ/γ_w	C_u (kPa)	θ (°)
20	1.98	66.8	27
22	1.96	41.8	27
24	1.93	23.5	27
26	1.92	11.9	27

where : w is the water content by weight

γ/γ_w is the density

Cu is the undrained shear strength

θ the friction angle

The undrained shear strength decreases rapidly as the water content increases (figure 5).

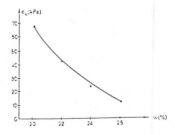

Figure 5 : Evolution of the soil undrained shear strength with its water content

3. TESTS RESULTS

It appears that the 120 mm diameter ring was too small for a puncture test and the results given here after are only the ones of the 150 mm diameter ring (tables 2 and 3).

124

Table 2. Puncture resistances obtained
with a 150 mm diameter ring (w)

μ (g/m2)	Cu = 66.8	Cu = 41.7	Cu = 23.5	Cu = 11.9	without soil
90	2240	2050	1840	1575	980
140	2960	2575	2185	1950	1360
200	3625	3250	2910	2675	1800
280	4655	4385	3975	3645	2425

Table 3. Piston displacement at break (mm)

μ (g/m2)	Cu = 66.8	Cu = 41.7	Cu = 23.5	Cu = 11.9	without soil
90	30	37	35	41	36
140	36	38	36	43	34
200	40	48	44	60	40
280	44	43	43	54	38

Tensile tests were also performed according to RILEM proposals. The results are
given table 4.

Table 4 : Tensile tests results

μ (g/m2)	R_1 (N/m)	R_2 (N/m)	ε_1 (%)	ε_2 (%)
90	6230	6800	80	34
140	9600	9840	83	42
200	12700	13800	110	56
280	19440	20600	111	69

where : R_1 is the tensile resistance in the
direction of production
R_2 is the tensile resistance in the
direction perpendicular to production
ε_1 is the strain at failure for the
tests in the direction of pro-
duction :

$$\varepsilon_1 = \frac{\ell_1 - \ell_0}{\ell_0} = 100 \text{ % (figure 6)}$$

ε_2 is the strain at failure for the
tests in the direction perpendi-
cular to the direction of production :

$$\varepsilon_2 = \frac{\ell_2 - \ell_0}{\ell_0} = 100 \text{ %.}$$

The soil above which the geotextile is
laid can absorb a big part of the puncture
energy. The higher is the soil undrained
shear strength, the higher is the puncture
force that the soil-geotextile system can
absorb.
The piston displacement at break is for a

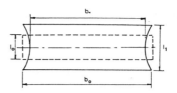

Figure 6 : Dimensions of the tensile test
specimen

given geotextile a constant.
Moreover the failure always occurs in the
direction perpendicular to the direction
of production. It seems then that the
rupture criteria is a function of the
strain ε_2. We also noticed that the displacement of a given point of the geotextile
during a puncture test follows a vertical
straight line. So it appears that there
wouldn't be any radial strain.
The shape taken by the geotextile sample
during the test may be assumed to be parabolic (figure 7). So for a given value of
the piston displacement r, it 's possible
to determine the length L^* of this parabole.
Indeed, if we assume (figure 7.b.) that
for :

$x = 0, y = 0$

$x = L \quad y = r \quad$ and $y' \dfrac{dy}{dx} = 0$

$y = ax^2 + bx + c$

then we have $y = -\dfrac{r}{L^2} x^2 + \dfrac{2r}{L} x$

and $L_1^* = \displaystyle\int_0^L \sqrt{1 + \left(\dfrac{d\,y}{dx}\right)^2}\ dx$

Moreover if we assume that the stress at
the ring is the ratio of the force at failure and the perimeter : $\sigma = F/2\,\pi\,r$, we
can, using the force-elongation curve, calculate the strain at the ring. We assume
too that the strain at break at the piston
is ε_2 and that the strains vary linearly
between the piston and the ring, then we
can also determine the length of L_2^* of
the parabole (figure 8).

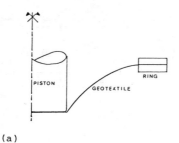

(a)

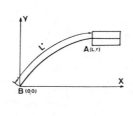

(b)

Figure 7 : Parabolic shape of geotextiles
index puncture loading

$$L_2 = \frac{\varepsilon_2 + \varepsilon_{ring}}{2} L$$

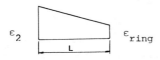

Figure 8 : Strain diagram

Table 5 : Results table

μ	ε_2	ε_{ring}	L_1^*	L_2^*	$\frac{L_1^* - L_2^*}{L_1^*} \times 100 \%$
(g/m2)	(%)	(%)	(cm)	(cm)	
90	34	17	6.4	6.4	0.0
140	42	14	6.6	6.5	1.5
200	56	19	7.2	6.9	4.2
280	69	19	7.2	7.3	-1.4

These comparison of L_1^* and L_2^* shows that
the rupture criteria seems to be the value
of the strain at break of the geotextile in
the direction perpendicular to the direction of production in a tensile test.

4. PUNCTURE ANALYSIS

Both puncture elongation and puncture
strength must be taken into account when
assessing the risk of puncture of a geotextile under static loading - whereby the
traffic loading from traffic movement on
the construction site is also assumed as a
static load.

Puncture strength

The forces acting at the contact points of
geotextile and rock must be determined in
order to evaluate the required puncture
strength. The required puncture strength
of a geotextile is generally expressed as :

$$F_{req} = \frac{\pi}{4} p' f_s d_m^2 \qquad (1)$$

Whereby should be considered that the magni
tude of the contact force depends upon the
number of contact points.
In this relation (1), we have :

F_{req} is the required strength as a function
of load and fill (kN)

d_m is the average diameter of the granular material (m)

p is the maximum pressure exerted on the
geotextile (kN/m^2)

f_s is a factor of safety.

The existing puncture strength is determined with a C.B.R. test with soil underneath the geotextile. The shape of the
grain can be taken into account by means
of a shape factor S_p.
According to {2} the shape factor varies
from 0.8 for round blunt shapes to 3.0 for
pointed, sharp-edged shapes.

5. ADAPTED DESIGN METHOD

It can be possible to define the right geotextile to use from what it is said before
(figure 9).
By formula (1) it is possible to find the
required resistance of the geotextile as a
function of d_m and p'.
Having the required resistance, it is possible to determine the geotextile to use.

Example :

For a truck with a tire inflation pressure
of 800 KPa in the stone base above the
fabric; p' the pressure on the geotextile
is according to {2} equal to 0,75 x tire
inflation pressure.
So p' = 0,75 x 800 = 600 kN/m^2.
So for stones of 8 cm in size, the required
resistance is, for a safety factor of 1
equal to :

126

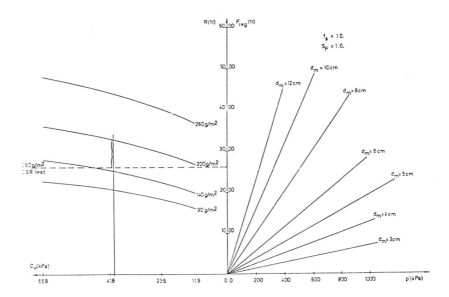

Figure 9 : design chart

$$F_{req} = \frac{\pi}{4} \times 600 \times 0,08^2 \times 1 = 3,02 \text{ kN}$$

For a soil with a Cu of 40 KPa, the required geotextile is, for a shape factor equal to 1, a 280 g/m² geotextile.
Let's note that according to DIN 54307 E, the puncture resistance of a 280 g/m² geotextile is only 2600 N so according to DIN 54307 E, the geotextile 280 g/m² wouldn't be sufficient (dotted line).
So the gain with soil beneath is obvious.

6. CONCLUSIONS

- It appears that the presence of soil beneath the geotextile has a great importance. In the case developed above, the gain was 27 %. Moreover this tests represents better the reality than the C.B.R. test.

- Without soil, the values obtained with the C.B.R. test or the modified C.B.R. test (with a 3 sided pyramidal piston) are very different. With a pyramidal piston with soil underneath the geotextile, there is probably a gain too, comparatively to the case without soil.
A bearing effect existing with the cylindrical piston will probably disappear to give place to a more local effect.

- The rupture criteria seems to depend only on the strain at break in the direction perpendicular to production.

7. REFERENCES

{1} Giroud J.P. Designing with geotextiles : Geotextiles and Geomembranes definitions, Properties and Design. I.F.A.I., 1984.

{2} Bell, J.R. & Koerner R.K. Designing with geosynthetics. Course notes, Drexel University, Philadelphia, 1984.

{3} Werner G. Design criteria for the separation of geotextiles on the basis of mechanical test procedures. Third International Conference on Geotextiles. Vienna, Austria ,1986.

{4} Lhote F. & Rigo J.M. Study of the Puncturing effect on Geotextiles. The Post-Vienna Conference on Geotextiles, Singapore, 1987.

{5} Koerner. Designing with Geotextiles. Prentice-Hall, Englewood Cliffs, New Jersey 07632, 1986.

International Geotechnical Symposium on Theory and Practice of Earth Reinforcement / Fukuoka Japan / 5-7 October 1988
© 1988 Balkema, Rotterdam. ISBN 90 6191 820 0

Mobilization of soil-geofabric interface friction

K. Makiuchi & T. Miyamori
Nihon University, Japan

ABSTRACT: In order to assess frictional properties of soil-geofabric system, laboratory investigations were made on several types of fabrics in contact with cohesive and non-cohesive soils using a friction testing apparatus. This paper presents the concept of displacement-induced mobilizing process of frictional resistance on the interface between fabrics and soils. Results indicate that the mobilized friction parameters of fabric are mostly lower than those of soil itself and affected by surface texture and thickness of fabrics, and type, density and moisture content of soils.

1 INTRODUCTION

The use of geofabric in earthworks is now widely accepted as an additional tool for reinforcing unstable soil-structures. The surface of a geofabric placed in a soil mass forms a discontinuous plane causing a slippage or a movement on the interface. In the theoretical analysis and the rational design of a reinforced earth, to estimate the effect of geofabric-soil friction is necessary for a soil reinforcement application, and friction parameters such as the coefficient of friction must be determined as an essential and indispensable factor. Emphasis is placed, however, on the lack of basic geotechnical knowledge about this interface friction mechanism.

The objective of this study, therefore, is to investigate fundamental behaviour of the soil-goefabric system. The frictional resistance is largely dependent upon a shear displacement. According to the Coulomb-Navier's shear law, the friction is obtained in terms of a friction angle and an apparent cohesion. These friction parameters can be described as a function of relative displacement.

Frictional characteristics between soils and geofabrics will depend on the geofabric's material composition, treatment, thickness and surface texture, and also the soil's type and condition. In this test, five different types of woven and non-woven geofabrics are used for cohesive and non-cohesive soils with their conditions varied.

There are several kinds of friction testing methods[1][2] such as a pulling-out test.

In this study an one-face test method was employed using a large shear testing apparatus.

2 APPARATUS

A large direct shear testing apparatus which consists of an upper and a lower boxes was used in measuring the friction between soils and geofabrics. The upper box has a cross-sectional area of 1000 cm² (31.6 x 31.6 cm), and its maximum available depth is 12 cm. The shearing is applied by pushing the upper box over the geofabric fixed to the larger lower box (47.8 cm long, 40.5 cm wide) with a height of 12 cm. Figure 1 shows the sectional view of the shear box.

The equipment is capable of giving a shear displacement of up to 62 mm without reducing the shear area. To apply horizontal shear loads, a variable speed electric motor is employed and the testing speed can be controlled 0 to 7 mm/min. Vertical

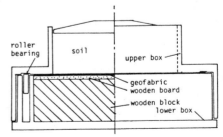

Fig.1   Cross-sectional view of the friction test apparatus

129

Table 1 Properties of geofabrics

	fabric	material	treatment	thickness (mm)	unit mass (g/m³)
non-woven	N - 1	polyester	needle-punched & heat bonded	0.72	95
	N - 2	polyester	needle-punched	2.2	135
	N - 3	polypropylene	needle-punched	4.0	500
	N - 4	polyester	needle-punched & heat bonded	0.8	110
	N - 5	polyester	needle-punched	4.0	420
woven	W - 1	polyester	plain weave	0.25	160
	W - 2	polyester	plain weave	0.35	235
	W - 3	polyester	twill weave	0.70	450
	W - 4	polyester	triple weave	3.00	1100

Table 2 Properties of river sand

Soil particle density (g/cm³)	2.67
Maximum grain size (mm)	2.0
Coefficient of curvature	1.75
Uniformity coefficient	0.72
Maximum void ratio	1.00
Minimum void ratio	0.65

Table 3 Properties of cohesive soil

Soil particle density (g/cm³)	2.74
Maximum grain size (mm)	4.76
Liquid limit (%)	130.7
Plasticity index	28.8
Natural water content (%)	80-126

confining loads are imposed through a loading frame and a double lever system.

A conventional standard shear testing apparatus with a diameter of 60 mm and a thickness of 20 mm was also used for estimating frictional properties of the sand and the cohesive soil.

3 MATERIALS

For this investigation, five types of non-woven geofabrics and four woven geofabrics of different material components, treatment and thickness were tested. Basic material characteristics of these geofabrics are listed in Table 1.

The granular soil used was a poorly graded and sub-angular river sand from which the coarse and the fine particles were removed. The properties of the sand are shown in Table 2. The sand was tested in both air-dried and wetted conditions. The moisture content of the wet sand specimen was controlled about 9 %.

The cohesive soil used for the test was a volcanic ash soil (so-called Kanto loam). The physical properties of the cohesive soil are given in Table 3.

4 TEST PROCEDURE

The lower shear box was filled with a spacer of rigid wood of 10 cm height, and a hard wooden board of 2 cm thickness on which a test specimen, geofabric, was glued with a commercial adhesive agent was rested. The upper box was then placed in position over the geofabrics.

In the test for loose sand, the air-dried sand was lightly poured in the upper box. When testing the dense sands, both the air-dried and wetted sands were put into the upper box in five layers, and the sands were compacted by tamping.

The cohesive soil was compacted in the conditions of water content of 102 % and 3 layers, 225 tampings per layer of which the compaction effort was determined by Proctor's optimum moisture content.

After the loading plate was placed on it, a normal load was applied to the soil-fabric system till the soil was consolidated. Application of the shear force commenced after the vertical settlement was completed.

The rate of shearing was about 0.5 mm/min. Measurements were made of the horizontal and the vertical displacements, and the friction force exerted in the system during shearing.

To determine the friction angle and cohesion component for each type of geofabrics, normal stress of 50, 100, 150 and 200 kPa were applied for each set of tests.

Standard direct shear tests of 6 cm in diameter and 2 cm in thickness were also performed with both the sand and the cohesive soil in order that comparisons were made between the soil-geofabric interface friction and the internal friction of soil alone.

5 FRICTION-DISPLACEMENT RELATION

Figure 2 shows typical results of the friction-shear displacement relationships between soils and woven fabrics.

It is found that those relationships are markedly affected by combination of soils and fabrics, moisture and compaction conditions of soils, and confining stress level. Similar tendencies of the relation are also observed in the cases of other soil-fabric interfaces.

Loose dry dands :
In the cases of loose dry sand (density index: $I_D = 0.46 - 0.57$) - woven fabric interfaces as shown in Figs.2(a) and (d), the friction increases gradually with the displacement and a high confining stress delay the appearance of its peak friction. Thin woven fabrics and all non-woven fabrics show monotonous smooth curves and do not have a well defined peak value, whereas thick woven fabrics develop somewhat a sawtoothed pattern in the relations.

Dense dry sands :
As shown in Figs.2(b) and (e), the friction curve between woven fabric and densely compacted dry sand ($I_D = 0.94 - 1.0$) has obviously a sharp peak and then the friction decreases to the residual stress level indicated by a steady value after the displacement in excess of 3-5 mm.

It is interesting to note that friction-displacement relations exhibit smooth curves in the case of thin woven fabrics and non-woven fabrics[3], while a friction curve of thick woven fabric-dense dry sand system fluctuates during shearing. This irregular variation of the curve becomes predominant with increase in thickness of geofabrics and stress level.

It is considered that this sawtoothed friction curve is caused by interlock-slip phenomena arising from the reiteration of forming and collapsing of bridging between the rough woven surface and sand particles, and of arching action of particles. Moreover, this phenomena seem probable from the performance of wave pattern occuring cyclically with a constant amplitude over a wide range of residual stress.

Several sharp mini-peal points of the friction curve develop over and over again until the friction reaches to its maximum value. From these figures it can be seen that each rising-up curve seems to have the same gradient, and the degree of fall of friction force is almost half of the mini-peak value.

Wet sand :
Figures 2(c) and (f) show the friction-displacement relation between woven fabric and dense wet sand of the moisture content of 9±1 % and the density index $I_D = 0.86 - 0.94$. These frictional resistance curves demonstrate the wave-shape pattern which is similar to those of the dense dry sands for both woven and non-woven fabrics.

Cohesive soil :
An example of friction curves for the woven fabric-cohesive soil system is illustrated in Fig.2(g). It may be seen that no obvious peak friction is given during shearing. Similar friction-displacement response is obtained for all geofabrics.

6 MOBILIZATION OF INTERFACE FRICTION

6.1 Determination of mobilized friction

Figure 3 shows a typical set of the confining normal stress and frictional stress relation. According to Coulomb-Navier's friction rule, a straight line through the plotted data can be obtained in good fitness, relevant to each displacement.

As parameters indicative of the frictional resistance of the system at a specified shear movement, both a friction angle, ϕ, and an apparent cohesion, c, are determined by the gradient, $\phi = \arctan(f/\sigma)$, and the intercept, respectively. It is observed from the test results that the intercept component, c, is negligible for all sand interfaces.

6.2 Mobilized friction parameters

The friction angle and the cohesion intercept are expressed as a function of displacement. In the cases of sand-fabric interfaces shown in Figs.4(a) and (b), it is observed that denser the sand, sharper the friction curve, and higher the peak friction. Further observation reveals that the residual friction are almost equal, and the similar pattern can be obtained for other goefabric-sand systems.

As shown in the examples in Fig.4(c), both the angle and the intercept of cohesive soil increase in proportion to a displacement, without any ultimate value. This tendency is the same as all cases of fabric-cohesive soil interfaces.

It is not reasonable to calculate the mobilized parameter of woven fabric-sand interfaces since their friction curves displays the sawtoothed fluctuation, therefore the friction angle at peak and residual states are illustrated in Fig.5. It is found that increasing of sand's void ratio, e = 0.80 to 0.65 reflects in increasing of the angle peak friction, $\phi = 5 - 10°$; however no significant difference in the angles of residual friction is observed. It is also found that thicker the fabrics, higher the friction angle.

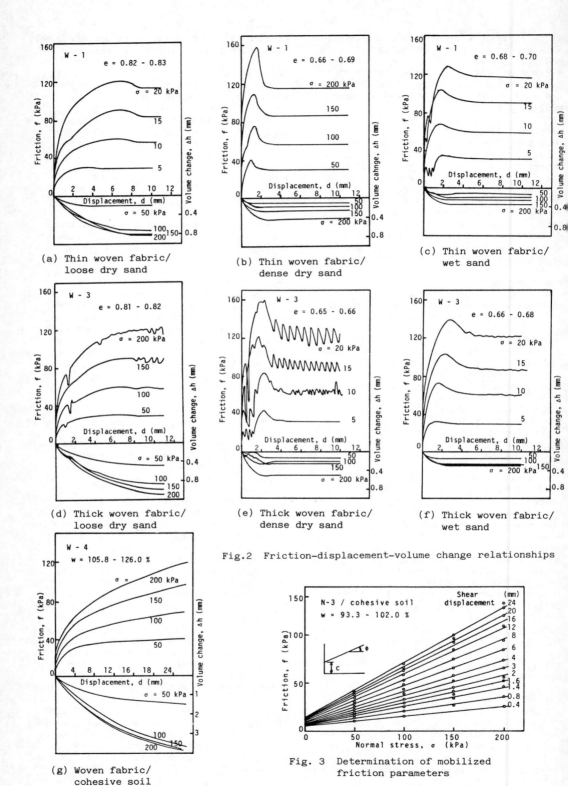

(a) Thin woven fabric/
    loose dry sand

(b) Thin woven fabric/
    dense dry sand

(c) Thin woven fabric/
    wet sand

(d) Thick woven fabric/
    loose dry sand

(e) Thick woven fabric/
    dense dry sand

(f) Thick woven fabric/
    wet sand

(g) Woven fabric/
    cohesive soil

Fig.2 Friction-displacement-volume change relationships

Fig. 3 Determination of mobilized
        friction parameters

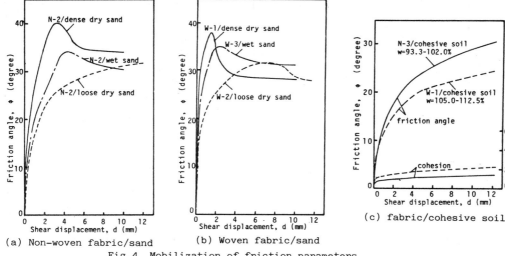

(a) Non-woven fabric/sand   (b) Woven fabric/sand
(c) fabric/cohesive soil

Fig.4 Mobilization of friction parameters

6.3 Ratio of friction parameters

In an attempt to evaluate the mobilized frictional resistance of geofabrics at the specified deformation, the ratios of the mobilized friction of soil-fabric interface of φ (soil/ fabric) to the mobilized internal friction angle itself φ (soil/soil) are calculated.

Figures 6(a) to (c) indicate the variation in the ratios for nonwoven fabric-sand interfaces. The ratio increases as a shear displacement develops. It appears that a dense sand or a thin fabric leads to an increase in the mobilizing ratio.

The ratios of friction parameters for woven and nonwoven fabrics adjacent to cohesive soil are illustrated in Fig.7. The ratios of friction angle are mobilized in the similar manner to the cases of sands. On the contrary, the ratio of cohesion intercept decreases markedly at the initial stage following the slight recovering with increase in deformation.

Figure 8 presents the ratio of friction angle at the peak and residual states for woven fabric-sand systems. The co-

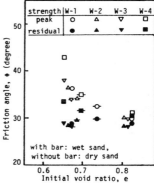

Fig. 5 Relation between void ratio and friction angle of woven fabric-sand interface

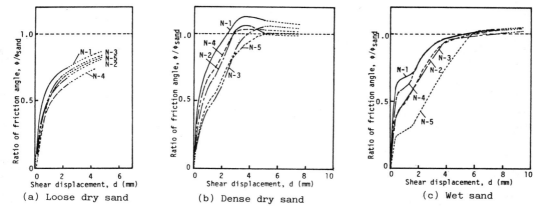

(a) Loose dry sand   (b) Dense dry sand   (c) Wet sand

Fig.6 Friction angle ratio of non-woven fabric – sand system

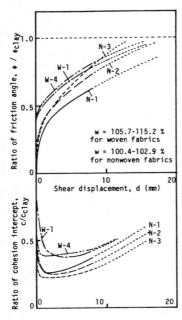

Fig. 7 Ratio of mobilized
friction parameters for
cohesive soil

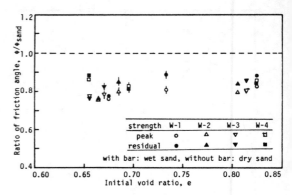

Fig.8 Relationships between void ratio and
friction angle ratio of woven fabric

hesion component is zero in these cases. It
is clearly demonstrated that the ratios are
not influenced from the sand's moisture content and void ratio, and provides the value
at range of 0.75 to 0.90. These values coincides with the reduction rate 80 to 90 % of
the coefficient of friction of geotextiles
which is now generally accepted in practice.
However, it should be recognized that the
ratio may be adopted under restrictions on
peak or residual friction.

7 CONCLUSIONS

Displacement-induced mobilization process
of frictional resistance was investigated
using a geofabric - soil interface friction
testing equipment. Results obtained in this
study are summarized as follows.

1) Mobilized friction on the fabric-soil
interface is much lower than that of soil
itself in a range of small displacement,
but increases as the displacement develops
and approaches finally toward the friction
value of the soil itself.

2) Denser the sand, higher the friction
angle for woven and non-woven fabrics.

3) Increasing in moisture content of sand
delays the appearance of peak friction of
the interface.

4) A friction-displacement curve of
woven-sand system displays an irregular
sawtoothed fluctuation caused by interlock-
slip phenomena. This behaviour is proportional to fabric's thickness, sand's density and confining stress level. This does
not occur in the cases of non-woven fabrics
and cohesive soils.

5) The ratio of mobilized friction angles
for a woven fabric-sand interface, at both
peak and residual states, has a constant
value of 0.75 to 0.90, regardless of void
ratio, moisture content and type of woven
fabric.

6) Friction curves of cohesive soil-
fabric systems demonstrates monotoneous
increasing curves without any peak, and
the ratios of mobilized friction parameters
of woven fabrics are slightly higher than
those of non-woven one.

REFERENCES

1) Myles, B. 1982. Assessment of soil fabric
   friction by means of shear. Proc. of 2nd
   Int. Conf. on Geotextiles, pp.787-792.
2) Nishigata, T. & I.Yamaoka 1986. Pull-
   out friction test of geotextiles. Proc.of
   1st Sym. on geotextiles, pp.41-46, Japan
   Chapter I.G.S. (in Japanese).
3) Miyamori, T., S.Iwai & K.Makiuchi 1986.
   Frictional characteristics of non-woven
   fabrics. Proc. of 3rd Int. Conf. on
   Geotextiles, pp.701-705.

International Geotechnical Symposium on Theory and Practice of Earth Reinforcement / Fukuoka Japan / 5-7 October 1988
© *1988 Balkema, Rotterdam. ISBN 90 6191 820 0*

Shaking table test on reinforced sand

Y.Mochizuki, S.Fukushima & K.Kagawa
Technical Research Division, Fujita Corp., Yokohama, Japan

ABSTRACT:Shaking table tests were conducted on saturated sand layer reinforced with wires in order to study the effectiveness of the reinforced earth method as a counter-measure for liquefaction.In this paper, the effects of arranging direction, spacing and area of reinforcement and rigidity of reinforcing materials on the liquefaction resist-ance are investigated.

1 INTRODUCTION

Liquefaction characteristics of reinforc-ed sand has so far been studied by the laboratory element tests (cyclic triaxial compression test and cyclic torsional simple shear test)[1)2)3]. As a result, it has been found that: If the reinforcing mate-rials are arranged in the direction of minimal principal stress, ample reinforce-ment effect can be obtained regardless of the rigidity of the materials, even with material like nonwoven fabric, and that if the reinforcement is arranged in the different direction of the principal stress, the reinforcement effect is affected by the rigidity of the materials.

During this study, the liquefaction phenomenon of actual ground (dissipation of pore water pressure from the upper layer), which is difficult to reproduce by the laboratory element tests, has been successfully reproduced using a small shaking table. The study has examined if wire, used as a model reinforcing material, will increase the ability of reinforced sand to resist liquefaction, and how the direction of reinforcement material arrangement and its rigidity affect the reinforcement effect.

2 TEST PROCEDURES

Figure 1 shows a small pneumatic type shaking table used for the test. The sand box dimensions were 110cm long, 30cm wide, and 60cm high. The walls in the direction of vibration were jointed to the bottom

plate by hinges to allow free shearing deformation of the model ground. The sand box was designed to be small since the purpose of the test was to examine the effect of reinforcement materials under various conditions.

The pneumatic cyclic loading system of this testing device was the same as a pneumatic type cyclic triaxial compression test[4]. Giant boosters were used so that the pressure could follow as in high frequency as possible.

The saturated sand layer was formed by pluviating dry Toyoura sand (Gs=2.64) into the water from the hopper (with 5mm wide opening) keeping the height of fall constant (50cm above the surface of the water). The water depth is 25cm from the bottom of the sand box. The void ratio was e=0.88. The side walls were hammered at 12 locations each for 10 times equally with a mallet to produce a state of void

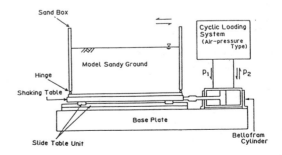

Figure 1. Small type shaking table

135

ratio of 0.80. Tests were made on two samples, void ratio of e=0.88 and 0.80. The water level was set at the ground level, and the sample was saturated for 12hours (this, although not perfect, was considered to raise the degree of saturation). The depth of the sand layer was arranged for 40cm. Figure 2 shows the arrangement of measuring equipment (such as pore water pressure gauge, accelerometer, and displacement gauge).

In this experiment, the sand layer having no reinforcement was subjected to vibration at several levels of acceleration. The number of loading cycles up to liquefaction (where the pore water pressure Δu equals to the initial effective earth pressure σ'_v) was arranged, and the mean curve of acceleration versus the number of loading cycles was obtained. A similar test was conducted with reinforced sand and the results were likewise arranged. The effect of reinforcement was examined by comparing the number of loading cycles up to the liquefaction under the same density. The reinforcement materials used for this experiment were based on the wire of 20cm long and 1.2mm diameter (wire length and diameter varied by the test). Sand was applied to the wire surface using Araldite and such wires were mainly arranged vertically in the area as shown in Figure 2.

The stress condition in the ground of this test was close to the case of torsional simple shear test. The direction of the reinforcement (vertical direction) under this test was inclined away from the direction of the principal stress, it failed to serve as a perfect tensile and compressive reinforcement. However, the arrangement of the reinforcements was based on vertical direction for the ease of actual construction work. The reinforced area in the model ground was limited to I =40cm and was located in the center of the sand box, this test was made different from the element test in that the entire specimen was reinforced at regular intervals. The reinforcement materials were arranged at regular intervals to one another (standard is $\Delta l=\Delta w=2.5$cm). If actual ground is assumed by applying only geometric scale, disregarding the law of similarity, the model ground of $\Delta l=\Delta w=2.5$cm, $h_R=20$cm will be equivalent at a scale of 1 to 25 to the case where the reinforcement materials of 5m long are arranged at intervals of I_R =10m, $\Delta l=\Delta w=62.5$cm to the sand ground of 10m deep.

Furthermore, in this test, the supply air pressure of the bellofrom cylinder for

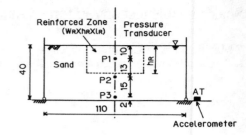

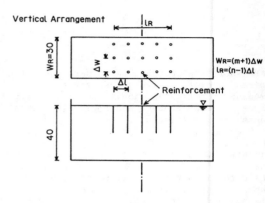

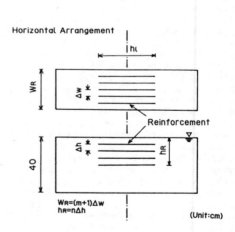

Figure 2. Measuring equipment and the arrangement of reinforcement

generating the specific table acceleration was checked by a gauge before application, and a testing frequency of 1.5HZ was applied.

3 TESTING RESULTS

Figure 3 shows the record of table acceleration and pore water pressure (P1) when the sand layer was vibrated with the void ratio e=0.80 and at a table acceleration of 156 gal. This figure shows that a specific acceleration was generated in the early stage, being kept almost at a constant value until liquefaction is reached.

Figure 4 shows the relationship between the acceleration and the number of loading cycles until the model ground reinforced by wire (d=1.2mm, 1=20cm, m=11, n=17, Λ1= Δw=2.5cm) reached liquefaction (e=0.88). Here, the number of loading cycles required till the liquefaction was measured at the location P1 inside the reinforced area. The solid line in the figure shows the relation between the acceleration and the number of loading cycles for unreinforced sand only. The curve of reinforced sand is located to the right of the one for the unreinforced sand. The figure shows that the reinforced sand required more loading cycles than the unreinforced sand at the same acceleration,hence the effect of reinforcement. Noting the acceleration that caused lique-faction at N=2, the scale of the reinforcing effect is an increase of about 20 gal.

The test was conducted by pluviating sand from 50cm above the surface into the water to keep constant density, but the sand was packed rather loosely, it was either liquefied immediately after the application of vibration or not liquefied at all. Therefore the test is considered imperfect for evaluating the effect of reinforcement.

Figure 5 shows the results of test with e=0.80 using the forementioned testing method. In this figure,α/g, $\sigma_d/2\sigma_c'$, and τ/σ_c' are commonly represented by the axis of ordinates and the number of loading cycles up to liquefaction by the axis of abscissas to make comparison with previously described element test (cyclic triaxial compression test: e=0.83, d=0.9mm, 8layers of wire, torsional simple shear test: e=0.77, d=0.9mm, arranged in vertical direction at intervals of =1.26cm)[5] The small shaking table test and element test cannot be simply compared because of the difference in drainage condition, saturation, confining pressure, preparing method of specimen, position of reinforcement insertion and spacing, and other conditions. However, as the trends of reinforcement effect, these two tests share a point that the smaller the acceleration ratio and stress ratio, the

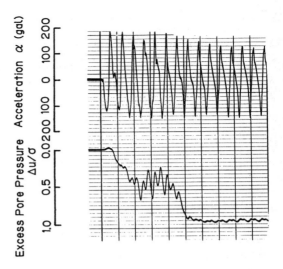

Figure 3. Record of table acceleration and pore water pressure

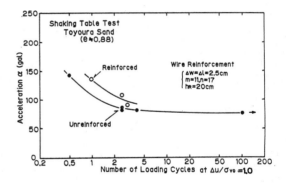

Figure 4. Acceleration versus number of loading cycles

greater the reinforcing effect. Noting the acceleration that produces liquefaction at N=10, the scale of the effect of the shaking table is an increase of about 40 gal. Supposing that the acceleration ratio and stress ratio are equal, the shaking table has the greatest effect. The following subsections describe the results of studies on various factors that affect the reinforcing effect of a small shaking table.

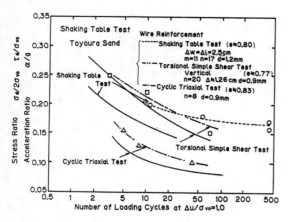

Figure 5. Comparison between shaking table test and element test

wires was about 1:8 or the wire of greater diameter had approximately eight times as great flexural rigidity (EI) as the wire of a smaller diameter. The tendency that the greater the diameter and flexural rigidity, the greater the reinforcing effect is same as the results of the torsional simple shear test. The degree of the reinforcing effect is remarkably greater than that shown by the latter, however.

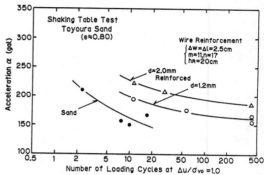

Figure 6. Effect of diameter (rigidity) of reinforcement

3.1 Effect of diameter (rigidity) of reinforcement

From the results of torsional simple shear tests of specimen reinforced with wires of different diameters or materials of different rigidity, it was found that the scale of reinforcing effect depended on the flexural rigidity of reinforcement material, and that the greater the rigidity, the greater the reinforcing effect, if the reinforcement was arranged in deviation from the direction of principal stress.

The state of stress of actual ground under the cyclic shearing stress during earthquake is close to that of the torsional simple shear test, where the direction of the minimum principal stress is inclined against the horizontal plane. Considering the actual construction method or the underground state, construction in the vertical direction is easier, and in this case, the reinforce-ment is arranged in deviation from the direction of the principal stress. Figure 6 shows the difference in the effect of reinforcement when two types of wires having different diameters (d=1.2mm and 2.0mm) were arranged vertically to model ground (m=11, n=17, $\Delta 1 = \Delta w = 2.5$cm). The figure shows that the reinforcement by the wires of greater diameter had a greater effect than the wires of smaller diameter. Since the two types of wires were of the same material, their rigidity E were equal. With regard to the geometrical moment of inertia I (πd^4 /64), the ratio between the two types of

3.2 Effect of direction of reinforcement arrangement

As stated in Subsection 3.1, reinforcement materials cannot always be arranged in the direction of principal stress. Therefore, the reinforcement must be arranged at the most effective location within the range feasible for construction. To study the effect of the direction of reinforcement arrangement, a comparison of the effect of reinforcement was made, as shown in Figure 7, between the reinforcement material arranged in horizontal direction (d=1.2mm, Δh= w=2.5cm, h_1=40cm) and in vertical direction (d=1.2mm, $\Delta 1 = \Delta w$=2.5cm, h_R=20cm). The figure shows that the reinforcement arranged horizontally had a greater effect than the one arranged in vertically. This trend coincides with the results of the torsional simple shear test. Furthermore, strictly speaking, in the case of model ground the shearing stress is applied horizontally in the Ko state, different from the state of stress in the torsional simple shear test. Therefore, the direction of minimum principal stress during vibration is closer to horizontal than to vertical, confirming the tendency of the test results.

138

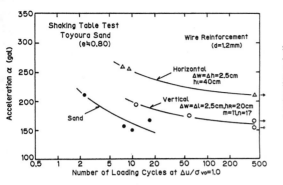

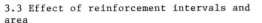

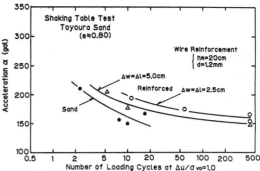

Figure 7. Effect of direction of reinforcement arrangement

Figure 8. Effect of reinforcement intervals

3.3 Effect of reinforcement intervals and area

The most important thing for actual construction work is the volume of materials and the depth they are buried to. Figure 8 ($\Delta l=\Delta w=2.5$cm, $\Delta l=\Delta w=5.0$cm) and Figure 9 ($h_R=20$cm, $h_R=35$cm) show the effect of reinforcement intervals and area. As previously stated, disregarding the law of similarity and noting only geometric scale, the reinforcement intervals of $\Delta l=\Delta w=2.5$cm are equivalent to 62.5cm in the ground, and $\Delta l=\Delta w=5.0$cm to 1.25m. Reinforcement depth of $h_R=20$cm is equivalent to 5m, and $h_R=35$cm to 8.75m, and both values are reasonable. Both figures show that the narrower the reinforcement interval, or deeper the area, the reinforcing effect becomes greater.

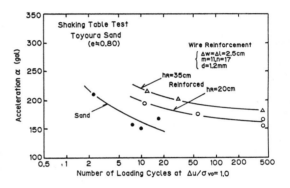

Figure 9. Effect of reinforcement area

4 CONCLUSION

The results of the shaking table experiments of saturated sand layer reinforced with wires are summarized as follows:

1) The ground will have much greater liquefaction resistance if the saturated sand layer is reinforced.

2) Similar to the cyclic triaxial compression test or torsional simple shear test, the reinfocing effect is great when the acceleration of shaking table is small.

3) The greater the rigidity of reinforement material, the narrower the reinforcement intervals, and the wider the reinforcement area becomes, the greater

reinforcing effect becomes.

Furthermore, in comparison between horizontal and vertical reinforcements, the former has a greater effect of reinforcement. Judging from the results thus far described, the reinforced earth method is considered as an effective and inexpensive measure to prevent liquefaction during earthquakes.

REFERENCE

1) Fukushima, Y. Mochizuki, K. Kagawa, & G. Kuno 1986. Application of reinforced earth method to prevent liquefaction. Tsuchi-to-Kiso. vol.35, No.4:5-10 (in Japanese)
2) Y. Mochizuki, S. Fukushima, & K. Kagawa 1986. Application of reinforced earth method as a preventive measure against liquefaction.

Proc. of the 1st Geotextile Symposium (in Japanese)

3)S.Fukushima,Y.Mochizuki,& K.Kagawa 1986. Development of torsional simple shear apparatus applicable to practical purpose. Tsuchi-to-Kiso. vol.35,No.7:61-65 (in Japanese)

4)Y.Mochizuki,S.Fukushima,& K.Kagawa 1986. Undrained cyclic triaxial compression test. Proc. of 21st annual conference of JSSMEF (in Japanese)

5)Y.Mochizuki,S.Fukushima,& K.Kagawa 1987. Application of reinforced earth method to prevent liquefaction (element tests and small shaking table tests). the Proc. of the 19th conference of Japan earthquake engineering.:257-260

International Geotechnical Symposium on Theory and Practice of Earth Reinforcement / Fukuoka Japan / 5-7 October 1988
© 1988 Balkema, Rotterdam. ISBN 90 6191 820 0

Evaluation of polymer grid reinforced asphalt pavement from the circular test track

T.Momoi
Nippon Hodo Research Laboratory, Tokyo, Japan

H.Tsukano, M.Matsui & T.Katsu
Nippon Hodo Research and Development Division, Tokyo, Japan

ABSTRACT : The structural behavior under traffic of asphalt pavement with strong polymer grid between surface course and base course, and between base course and subgrade was compared directly with control pavement of similar construction without grid.These tests were conducted by using the large scale circular test track.
The pressence of the grid did not affect the structural quality of the road as assesed by measuring the deflection and the vertical stresses acting on subgrade and base course. However the reinforcement of the base course reduced the developement of rutting and cracking on the road surface,and it was tentatively suggested that the pavement structure including combination of the grid reinforced thin asphalt surface course and the cement stabilised base course is effective for low traffic volume road and temporary road.

1 INTRODUCTION

This paper presents some results from a research program to investigate the effectiveness of a polymer grid [Tensar AR-1 and SS2. Mitsui] in improving the bearing capacity of bituminous pavement.

Many experiments [Brown(1985),Kennepohol (1985) and Haas(1984)] suggested that the considerable improvement in performance could be obtained by the application of grid placed between base course and sub-grade,and between surface course and base-course.

The present investigation has involved large scale tests conducted by using the circular test track, which has the arm of 12m and makes it possible to experiment on pavement built by using the same equipment as is used for actual pavement (see Fig-1).

This facility was completed in the field test area of Nippon Hodo located in Omiya city Saitama Pref. in Feburary 1986.

The goal of the research using this test facility is to develop the engineering records and criteria which can be used in design and construction of the grid rein-forced pavement as well as the improvement and maintenance of existing pavement by applying the grid.

As the first step of the research the structural behavior under traffic of a as-phalt pavement containing the grid was compared directly with control pavement of similar construction without the grid.

During loading, measurements were peri-odically made for the deflection,perma-nent deformation and extent of cracking on the pavement surface,and the transient vertical stress in each layer of pavement structure.

Suplemental investigations were also per-formed on the strain measurements acting on the surface of grid and the bituminous layer,and on the excavation of the pave-ment.

Fig-1 Circular test Track of Nippon Hodo Co.,Ltd.

2 CONSTRUCTION OF THE PAVEMENT IN THE TEST AREA

The test sections were constructed in the field test area in Saitama Pref.where subgrade soil is Kantoh loam and is known to have a low bearing capacity. Kantoh loam soil of 90cm was replaced with the pit run material of CBR 40. Prior to being replaced with this material,the Kantoh loam layer within a depth of approximately 30 cm below the surface of soil after the excavation,were stabilised with lime of 10% by weight.

The surface of pit run material refered to as the level of subgrade(Fig-2). Plate bearing tests and Benkelman deflection tests were carried out on the subgarde level and the surface of stabilised soil. These test results are given in Table-1.

The layout of completed trial sections is shown in Fig-3. The bituminous sections containing the grid are numbered 2-a,2-b and 4, and the control sections 1 and 3.

The bearing capacity on the subgrade in control section 3, estimated by the deflection and K-value, was higher than the remaining sections, as shown in Table-1. Then the datum, obtained from the control section 3, was excluded from a subject of the present analysis.

The grid used was made from polypropylene.The installation of the grid at the test section 2-a and 2-b was conducted in the following process. Before the tack coat was applied to the surface of the base course, the grid(Tensar AR-1) was unrolled. The end of grid was anchored to the surface of base course by nailing through a thin metal strip.Once the roll had been in position,a tension of 130 kg/m was applied as the another end anchored.The grid was then held in position on the surface by an application of a thin layer of asphalt concrete.

Asphalt concrete of about 2cm was spread on the grid manually and compacted by a hand controlled Bomag vibrating roller to avoid movement of the grid under delivery lorries or the asphalt paver.Immediatly after the thin layer of asphalt concrete was appllied, the dense graded asphalt mix was spread by the asphalt finisher and compacted to a total thickness of 5cm including the manually laid layer by a 15 ton pneumatic tired roller followed by an 8 ton vibraing roller. However the slight degree of disturbance of grid were observed during the asphalt paving operation.

Dense graded asphalt concrete used was designed to meet for request of surface course mixture shown in "The manual for design and construction of asphalt pavement" published by Japan Road Association using granite aggregate and filler,from Kuzuh hill quarry. The aggregate grading, binder content, mixing and compaction temperature were all within the ranges specified.

On the test section 4, the grid(Tensar SS -2) was placed on the subgrade.Immediatly

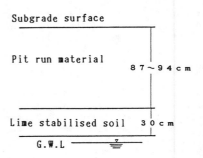

Fig-2 Soil strata in subgrade in field test area

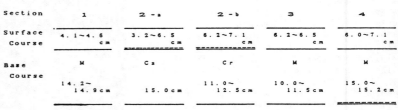

M : Mechanical stabilised crushed stone
C_s: Cement stabilised
C_r: Crusher run
------: grid

Fig-3 The layout of test section

after the grid was unrolled on the subgrade,grunular material was spread and compacted by a 15 ton pneumatic tired roller
and an 8 ton vibrating roller.The cement
stablised base course and the mechanically
stabilised base course were mixed to the
above mentioned specification .

3 NIPPON HODO CIRCULAR TEST TRACK

Nippon Hodo circular test track, which
has a total weight of approximately 10500
kg,is circular. Dual tire is located at
the end of special frame projecting from a
center core as shown in Fig.-1. Electric
motor, which is mounted on the frame,
drives dual tire.The arm is 12m long and
supported at its middle by a second rolling suport runing on the steel plate ring
rested on the surface of pavement to
prevent the weight of arm transmitting to
the dual tire and to serve the dynamic
stability of the arm.The parallel linkage
system were designed to connect the axle
of dual tire with the arm through the four
tiebars and eight hinges,to keep the dual
tire always perpendicular to the surface
of pavement.
  Both tires of dual tire were driven by
electric motor mounted on the arm.
Speed of 20km/hr was kept constant
throughout the testing period,
though this instrument has the
capacity to have maximum speed of
30 km/hr and make it possible to
change continuously.
The arm has the constant fixed
radiaus of 12 m but an additional
eccentric mechanism at the end
of the assembly was provided to
change the dual tire path
slowly back and forth across
the central wheel path.The
stretching and treching movement
of the arm was adjusted to
constant speed of 1.3 cm/sec.
Testing was continued by the
number of load applications,
which was approximately 200000 passes, to
attain until the rate of cracking of 60
% in the test section 1.As this facility
was operated only in the day time ex-
cluding the rainy or snowy days,testing
was initiated in February 1986 and
finished in January 1988.

4 INSTRUMENTATION AND FIELD MEASURE-
  MENT

  During construction, pressure gages to
measure the vertical stress were installed

in the subgrade and near the surface of
basecourse.Thermocouples were also installed in the middle point of the surface
layer. After construction,the holes of
diameter of 15 cm were made in the surface
course and or the base course by using
coreboring machine to expose the surface
of grid,then the strain gages were set on
to the surface of the transverse and
longitudinal members of the grid.Mesure-
ments by using these gages were conducted
periodically or after every 100000 passes.
  In additon to the testing above,measure-
ments of test track performance were con-
ducted.Surface deflection were determined
by using the Benkelmann beam.Rut depth was
mesured by profilometer. Surface cracking
was surveyed visually and mapped.

5 OBSERVATION OF PERFORMANCE

5.1 Permanent deformation and cracking

Fig-4,5 and 6 show the transverse surface
profile of test section,the transverse
distribution of dual tire versus mean radius of gyration and the cross section of
the rut depth in section 1,respectively.
  These figures show that the greater part
of the deformation were resulted from

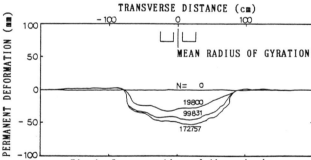

Fig-4 Cross section of the ruts in
section 1

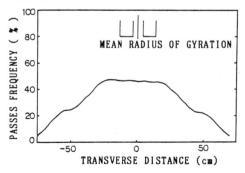

Fig-5 Transverse distribution of wheel
loads versus mean radius of gyration

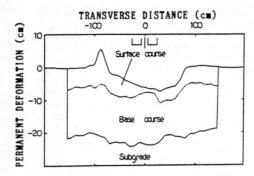

Fig-6 Cross section of pavement structure after number of two wheel passes
180000

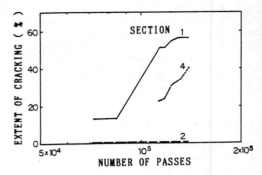

Fig-8 Variation of extent of cracking
versus number of passes

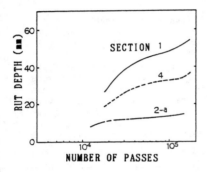

Fig-7 Measured rut depth versus number
of pasess

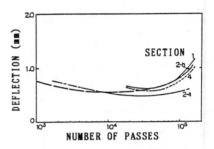

Fig-9 Benkelmann beam deflections

unbound layers and in particular from the
subgrade.A very small part of deformation
was resulted from the asphaltic mix and
this deformation was remarkable in the
position of greatest frequency of passes
of dual tire.

Fig-7 shows the relation between the
number of load applications and the rut
depth.The developement of rut depth in
test section 4 of which base course was
reinforced by grid,was less than that of
in the control section.

The relationship between the number of
load applications and the surface cracking
(Fig-8) shows that pressence of grid also
reduced the extent of cracking on the
pavement containing the unbound base
course.

In the case of test section 2-a,the
development of rut depth and craking was
very slow. However as the section 2-a has
the surface course under which was reinforced by the grid, and the cemented
stabilized base course,it is not clear

that these results were depended on
installation of the grid or not.

5.2 Deflection

The effect of the grid on the deflection
beneath a rolling wheel load was assesed
by comparing deflections measured at 3-5
reference positions in each section. As
noted above,considering that the test
section 2-a has the cement stabilized base
course,it should be concluded from Fig-9
that the deflection was not influenced by
the presence of grid. Radius of curveture,
which was estimsted from the deflection
curve,was unchanged by the grid.

5.3 Transient stress and strain

Fig-10 shows the transverse distribution
of the vertical stress in the control section 1 and the test section 4.Vertical
stress acting on subgrade concentrated
upon the center of wheel path and

144

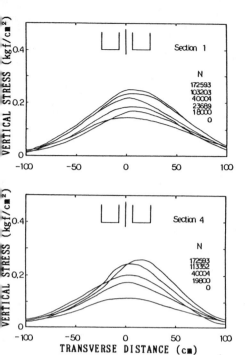

Fig-10　Variation of transient vertical
stress acting on the subgrade

increased as the number of load application is increases, however stress level
was unchanged.

The magnitude of the transient strain in
the members of grid which were installed
on the surface of base course and subgrade
were very high and were influenced on each
tire of dual tire.

6 COMPARISON WITH OBSERVATION

Many researches regarding to laboratory
experiments using full-scall model pavement structures have shown that reinforcement of base layer and asphalt layer improve the deformation resistance of the
pavement and dramatically extend the
service life of the road.The effectiveness of the reinforcement increase is as
the subgrade strength decreases.

The position of reinforcement is a significant variable and the existence of the
optimum location is suggested.Furthermore
reinforcement of the base layer is effective in distributing the load and reducing
the stress levels acting on the subgrade.

Present tests confirmed the observation
that the grid placed between base course
and subgrade surely improved the per-

formance of pavement assessed by measuring
the developement of rutting and cracking.
However the deflection on the road surface
and the stress acting on the subgrade was
not affected by the grid.

These results may be explained by considering that the strength of subgrade, on
which reinforced pavement was constructed,
was too high and the thickness of the surface course was too thin.

Furthermore considering that the deflection on road surface and the stress level
acting on the subgrade are unchanged,
though the performance of road are
improved,it has been tentatively suggested
that the effect of grid is not to delay
in appearing of initial failure of pavement such as rutting and cracking, but to
extend the period between initial failure
and complete deterioration of pavement.

The service life of pavement structure
of the section 2-a containing the reinforced surface course on the cement stabilised base course is very long and this
structure withstands the traffic load by
circular test track (N= 300000) and is
still in good condition like initial condition while the remaining test sections
were soon required　the overlay of asphalt
concrete of 5cm　with the grid.

7 CONCLUSION

The structural behavior of a bituminous
road of thin asphalt surface course
including granular base course with a
strong grid placed on the subgrade and
including cement stabilised bace course
with a grid under the surface course
have been investigated by using the
circular test track.The following
conclusion may be drawn.

1. The presence of the grid was surely
effective for the reduction of permanent
deformation and the craking on the pavement surface.

2. The structural quality of the road as
assesed by mesuring the deflection of
pavement beneath a rolling wheel load and
the transient stress acting on subgrade
was unaffected by the presence of the grid.

3. The service life of pavement
including of grid reinforced thin asphalt
layer on the cement stabilised base course
can be extended significantly.It is tentatively suggested that this pavement
structure is effective for the low traffic
volume road and temporary road.

Table-1   Bearing capacity on base and subgrade and dry density of base course in test are

	section No.	1	2	3	4
Bearing capacity K-value (Kgf/cm³)	On surface of :				
	base course	39	34	49	35
	subgrade	12	13	18	15
	lime stabilised soil	5	3	4	6
Deflection (mm)	On the surface of :				
	base course	0.35	0.47	0.28	0.35
	subgrade	0.70	0.63	0.64	0.60
Dry density (gf/cm³)	base course	2.209 ~2.199	2.000 ~2.075	2.207 ~2.19	2.209 ~2.199
Degree of compaction (%)		95~96	94~96	96~97	95~97

ACKNOWLEDGMENT

The works described in this report was
carried out by the members of Developement
and Research Division of Nippon Hodo Co.,
Ltd. The authors wish to thank the staff
of members concerned.

REFFERENCES

Mitsui Petrochemical Industrial Products
  Co.,Ltd. 1986. "Tensar Technical Note"
  in Japanese
S.F.Brown,J.M.Brunton,D.A.B Haghes and
  B.U.Brodrick.1985. "Polymer grid reinforcement of Asphalt". A.A.P.T.
G.Kennepohol,N.Kamel,J.Walls and R.C.G.
  Haas. 1985." Geogrid reinforcement of
  flexible pavements,design basis and
  field trials" A.A.P.T.
Ralph Haas.1984. "Structural behavior of
  Tensar reinforced pavement and some
  field applications".Symposium on
  polymer grid reinforce in Civil
  engineering

International Geotechnical Symposium on Theory and Practice of Earth Reinforcement / Fukuoka Japan / 5-7 October 1988
© 1988 Balkema, Rotterdam. ISBN 90 6191 820 0

Field pull-out test of polymer grid in embankment

H.Ochiai, S.Hayashi, J.Otani & T.Umezaki
Kyushu University, Fukuoka, Japan

E.Ogisako
Shimizu Corporation, Fukuoka, Japan

ABSTRACT: In order to examine the pull-out behavior of polymer grids in soils, a series of full scale field tests was performed in the way of pulling the polymer grids laid in a test embankment of 5.0m in height. Displacement of polymer grid in soil and pull-out resistance mobilized on grid junctions were discussed on the basis of a concept that the pull-out resistance concentrates and acts on each grid junction proposed by authors. As a result, it was concluded that the characteristic of pull-out resistance in actual size are basically very similar to the results from the laboratory model test.

1 INTRODUCTION

Most of the design method currently being employed for reinforced soil structures are based on the perfect plasticity theory which takes no account of displacement and deformation of reinforcing material in soil(Netlon Ltd. 1984). However, for some types of earth reinforcement such as polymer grid reinforced soil structures where the soil on upper and lower sides of the polymer grid is partially continuous, a pull-out resistance of the polymer grid in soil may be attributed to an interaction between reinforcing material and soil, thus depending on the displacement of polymer grid in soil. To make clear the mechanism, authors have conducted experimental studies in laboratory using a pull-out shear box apparatus and have emphasized the important role of grid junctions in the pull-out resistance(Hayashi et al. 1985, Ochiai et al. 1988).

In this paper, the results obtained from a full scale field pull-out test in an embankment of 5.0m in height are discussed in relation with the basic concept of pull-out resistance mechanism from the laboratory studies.

2 BASIC CONCEPT OF PULL-OUT RESISTANCE MECHANISM

When the polymer grid in soil is subjected to a pulling force, the polymer grid is pulled out in the soil as the grid itself deforms, and thus the pull-out resistance is mobilized on both grid junction and rib of the polymer grid. In the case of the polymer grid which the soil on either side of the reinforcing material is partially continuous, the resistance effect of the rib at right angle with the direction of pulling is assumed to be transferred to the grid junctions in a concentrated manner. Thus the resistance mobilized on each of grid junction plays a greater role than that mobilized on each rib. It is therefore considered that the pull-out resistance concentrate and act on each of grid junctions.

This behavior was observed visually by using a soil box with transparent plastic plate. To make clear the grid displacement in the observation, the plastic plate was marked by vertical red lines, and the marker rubber membrane strips were pasted with silicone grease on the plate. Applying constant confining pressure, the grid was pulled out with constant speed.

The results of observation are shown in Photo. 1 for two different displacement levels. It can be realized from these photographs that the grid junction plays very important role in the mechanism of mobilization of pull-out resistance. An elliptic slip field is formed in front of each grid junction and expanded with increasing the displacement level of grid junction. When the displacement reaches some large value, adjacent slip fields interact each other so that the pull-out resistance acting on each junction decreases and reaches the residual state. These results substanciate the validity of the concept which attaches importance to the role of grid junctions in the mobilization of pull-out resistance.

3 OUTLINE OF FIELD PULL-OUT TEST

A series of field pull-out tests was conducted in the way of pulling the polymer grids laid in a test embankment of 5.0m in height. The embankment was built whole of sand and part of decomposed granite(Masa soil) as shown in Fig.1. The steel sheet piles and H-section steel columns with penetration depth of about 1.0m were used for constructing the vertical embankment, and the steel columns of both sides were connected to each other by tie-rods. As shown in the figures, the polymer grids, which were 0.5m in width

and 2.0m and 4.0m in length for SR-2 type and 1.0m in width and 3.0m in length for SS-2 type, were laid at the depth of 1.0m, 2.5m and 4.0m from the top of embankment, respectively.

In the tests, the buried polymer grids were pulled at a constant pulling speed of 1.0mm/min by means of a center-hole type of oil pressure jack, and measurements were made of the pulling force at the forefront of the polymer grid and the displacement of each of grid junctions in soil. For the measurement of the displacement of grid junctions, thin stainless steel wires were attached to the grid junctions at the positions shown in Fig.3 and connected to dial gauges set up outside of the embankment. The wires passed through vinyl tube to avoid friction against the soil.

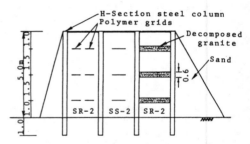

Fig.1 Test embankment (Front view).

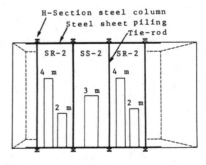

Fig.2 Test embankment (Plane view).

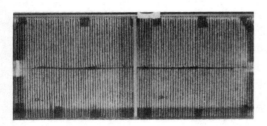

(a) Displacement level=5.0mm

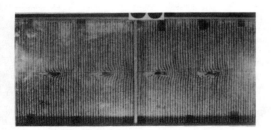

(b) Displacement level=30.0mm
Photo 1. Visual observation of mobilizing process in pull-out test.

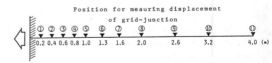

Fig.3 Position of measurement of displacement.

148

4 TEST RESULTS AND DISCUSSIONS

4.1 Displacement of polymer grid in soils

Examples of the actual measurements made of pulling force(Ft) and displacement of the grid junctions(Xi) during the tests are shown in Fig.4 and Fig.5. In the results, the SR-type of grid for reinforcing materials and sand and decomposed granite(Masa soil) for fill materials were used and the overburden pressure(σ_v) was 1.7tf/m^2 .

When a pull-out force is applied to the forefront of polymer grid in soil, the force is transmitted from forefront to back part, thereby the corresponding displacement of grid junction being caused. This behavior is the same as that observed in the laboratory test(Hayashi et al. 1985 Ochiai et al. 1988). For a design strength of polymer grid, the value of 40% of the tensile strength(Tf) is usually used as a non-creep stress level(Yamanouchi et al. 1985). At the condition of stress ratio Rf=Ft/Tf=0.4, distributions of displacement of grid junctions are summerized in Fig.6 and Fig.7 under taking note of effects of

overburden pressure (σ_v) and fill material. The results for the grid length of 4.0m is shown in Fig.6 and that of 2.0m is shown in Fig.7, respectively. Comparing both results, the more the grid length is long, the more the difference of fill materials under each overburden pressures is crucial. In the case of the grid length of 4.0m, the value of displacement of grid junction and the strained range of polymer grid in decomposed granite is larger than that in sand under the same overburden pressure. At the overburden pressure of σ_v =1.7tf/m^2 in both sand and decomposed granite, there exists the non-strained range in the grids of 4.0m in length, while the grids of 2.0m in length is strained in the whole length of the grids.

The effects of grid length and overburden pressure on the displacement distributions of polymer grid in soils are summarized in Fig.8 as a relation between the stress ratio Rf=Ft/Tf and the

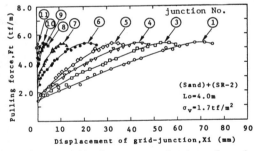

Fig.4 Force-displacement relation of each grid junction (sand)

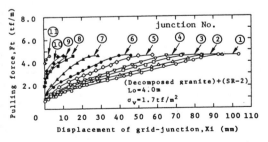

Fig.5 Force-displacement relation of each grid junction (decomposed granite)

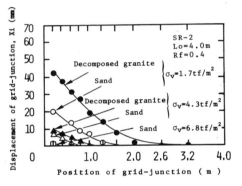

Fig.6 Distribution of displacement of polymer grid (grid length Lo=4.0m)

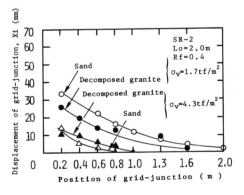

Fig.7 Distribution of displacement of polymer grid (grid length Lo=2.0m)

149

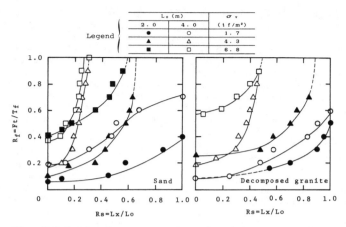

	Lo (m)		σ v
	2.0	4.0	(t f/m²)
Legend	●	○	1.7
	▲	△	4.3
	■	□	6.8

Fig. 8 Relation between ratios Rf=Ft/Tf and Rs=Lx/Lo
for sand and decomposed granite.

displacement ratio Rs=Lx/Lo, in which Lo is the original length of polymer grid and Lx is the length of strained range measured from the forefront of the grid. Pull-out behavior can be easily recognized from the figure. When both values of overburden pressure and grid length are small, the ratio Rs increases gradually with increasing the ratio Rf, and the curve finally reaches the line of Rs=1.0. This means that the grid is pulled out without non-strained range. On the other hand, when the value of overburden pressure increases, the trend of graphs shown in Fig.8 is quickly raised and the ratio Rf reaches 1.0. This means that even if the tensile strength(Tf) of polymer grid is applied, there is the non-strained range in back part of the grid. The tendency of rising of the curve for sand is more distinguished than that for decomposed granite because of the difference of the interlocking effect between soil and polymer grid.

According to these results, when the polymer grid is used as a tensile reinforcing material in an embankment on the expectation of mobilizing the interlocking effect, it is very important to decide the grid length with considering the properties of fill material. On the other hand, when it is expected to maintain the stability of embankment by the anchoring effect of polymer grid, the grid length has to be decided under considering the overburden pressure and the properties of fill material. Therefore, when the overburden pressure is

relatively small, the grid length has to be long enough, and on the contrary, the short length of grids may be used for the case of large value of overburden pressure.

4.2 Pull-out resistance mobilized on grid junctions

As described in Chap.2, the pull-out resistance of polymer grid in soil concentrates and acts on each grid junction. The resistance forces(Ti) mobilized on grid junctions are calculated by the analytical procedure using Ft-Xi curve as shown in Fig.4 and Fig.5(Ochiai et al.,1988). Fig.9 and Fig.10 show the relation between the resistance(Ti) and the displacement(Xi) at the grid junction for the case of Lo=4.0m and σ_v =1.7tf/m^2 without the non-strained range. At each grid junction, there exists a value of displacement which corresponds to a maximum value of mobilized resistance. And as the locations of grid junction is the back of grid, the value of the displacement decreases and that of the maximum resistance increases, thus the curve with a sharp peak being obtained.

However, when the polymer grid is pulled out without the non-strained range, a maximum value of the resistance is mobilized at a grid junction which is located at about half of total length of the grid in this example, and values of the resistances mobilized at grid junctions behind the location decrease.

150

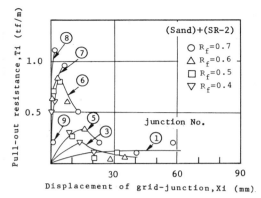

Fig.9 Relation between pull-out resistance (Ti) and displacement (Xi) of each grid junction (sand)

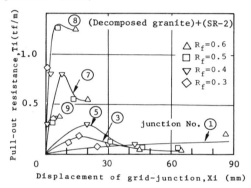

Fig.10 Relation between pull-out resistance (Ti) and displacement (Xi) of each grid junction (decomposed granite)

The reason why the resistance mobilized on each grid junction has a peak value at some displacement and decreases is that an elliptic slip field is formed in front of each grid junction and adjacent slip fields are affected each other as described in Chap.2.

It is also recognized from the comparisons of the results for sand and decomposed granite that the resistance curve mobilized on each grid junction depends on the properties of fill materials. The resistance curve with a sharp peak is obtained for fill material having a stress-strain relation with a peak value at relatively small strain range such as dense sand, while the curve with a flat peak corresponds to a stress-strain relation without a peak value as observed on decomposed granite.

All of discussions about the characteristics of pull-out resistance here are basically very similar to the results from the laboratory tests(Hayashi et al.1985, Ochiai et al.1988).

5 CONCLUSIONS

Results of field pull-out test using an embankment of 5.0m in height were examined to understand the pull-out mechanism of polymer grid in soil. The main conclusions aresummalized as follows :

1) The grid junctions play very important role in the pull-out resistance of polymer grid in soil. The resistance is considered to be mobilized at each grid junction in a concentrated manner.

2) When the polymer grids are used in soil as a tensile reinforcing material, it is important to decide the length of grid under considerations of the mechanical properties of soil and the magnitude of overburden pressure.

3) The pull-out resistance acting on the grid is not always uniform even at the same level of overburden pressure but varies with the displacement of grid junctions in soil.

4) The characteristics of pull-out resistance obtained from full scale field test are basically very similar to the results from the laboratory model test discussed in the previous paper.

REFERENCES

Hayashi, S. et al. 1985. Mechanism of pull-out resistance of polymer grids in soils, Tsuchi-to-Kiso, JSSMFE,Vol.33, No.5, pp21-26 (in Japanese).
Netlon Ltd. 1984. Guidelines for the design and construction of reinforced soil retaining walls using tensar geogrids.
Ochiai, H. et al. 1988. Pull-out behavior of polymer grid in soils and its analytical method, Memoirs of the Faculty of Eng., Kyushu Univ., Vol.48, No.2.
Yamanouchi, T. et al 1985. Tests for estimating of design strength of polymer grid, Proc. 30th. Symp. JSSMFE (in japanese).

International Geotechnical Symposium on Theory and Practice of Earth Reinforcement / Fukuoka Japan / 5-7 October 1988
© *1988 Balkema, Rotterdam. ISBN 90 6191 820 0*

Restraint effects on deformation of soft foundation with geotextile

B.K.Park & M.S.Lee
Chonnam National University, Kwangju, Korea

J.S.Jeong
Wonkwang University, Iri, Korea

ABSTRACT: This study aims at investigating the effects of geotextile on the reduction of two dimensional displacement and the addition of the ultimate bearing capacity in the model saturated clayey foundation (120 cm long, 25 cm wide and 47 cm high)by carrying out plate load test in the laboratory. Geotextile is placed on the surface of this model foundation. The rate of loading is controlled at 0.2mm/min to simulate the drained condition. When using geotextile, observational values of displacement and ultimate bearing capacity are compared with those calculated by the finite element method (FFM) and again with those observed in natural state (without use of geotextile).

1 INTRODUCTION

The use of geotextile has been increasing for the purpose of reinforcement of soft foundation, stabilization of soil structure, separation of adjacent different materials, etc. since the early 1980's(Zanten 1980:2).

It is reported that geotextile provides the function of filtration(Saitoh et al. 1985:6).

Settlement and lateral displacement are conspicuously reduced with the use of geotextile, which in turn means the reduction of stresses within the foundation(Yamaoka et al. 1985:31). Ultimate bearing capacity is increased in case of the construction employing geotextile. This is supposed to be due to the fact that geotextile makes the condition of local shear failure transfer to that of general shear failure in the foundation. The settlement is little affected depending on the physical properties of geotextile used(Yamaoka et al. 1985:31).

For the numerical analysis for adjacent different materials such as rock joints, the introduction of joint element is desired to obtain reasonable results(Goodman et al. 1968). In the FEM analysis for saturated clayey layers, the Biot equation is generally chosen as the governing equation. The numerical solution proves in good agreement with observational values by the use of the elasto-viscoplastic model(Sekguchi 1977)as constitutive equation(Kang 1988.) In this study, the FEM analysis on the deformation of the saturated clayey model foundation is

carried out in order to scrutinize the restraint effect on the settlement and lateral displacement including the increase of the ultimate bearing capacity of the two dimensional(which consists of large length compared with small width)model foundation (120 cm x 25 cm x 47 cm) covered with geotextile by performing plate load test (25 cm x 18 cm x 3.5 cm). Observation of horizontal and vertical displacement is also made for the above foundation during plate load test.

2 APPARATUS AND METHOD OF MODEL TEST

2.1 Loading and measurement

The measurements of two dimensional deformation, stresses and pore water pressure and the calculation of ultimate bearing capacity in saturated clay soft layer subjected to loading are carried out by the use of an apparatus, 120 cm long, 25 cm wide and 100 cm high of inside dimension. Uniformly reconsolidated samples are obtained with the apparatus.

Both stress and strain controls are carried out. The rate of loading is arbitrarily controlled depending on the simulation of drained and undrained conditions. In designing the apparatus for this study, the following points must be observed,

1. Two dimensional plate load test of model foundation and the production of

153

homogeneous sample for consolidation test.

2. Achievment of stress and strain controls.

3. Control of rate of loading ranged from drained to undrained condition.

4. Horizontally or vertically movable driving gear apparatus.

5. Auto self-recording system of stress, pore water pressure and earth pressure.

6. Preservation of water-proof and alleviation of side friction.

The apparatus is shown in Photograph 1.

Photo. 1 General view for test apparatus

2.2 Model soil tank

Model soil tank(120 cm long, 25 cm wide and 100cm high of inside dimension)made of steel plate and channel attached with acrylite transparent plate for observation through the front face and with another auxiliary steel channel to prevent lateral displacement of the frame was used for the preparation of the model foundation.

Those plates and channels are bolted for disassemblage which is necessary for plate load test after consolidation. The transparent observation window is slice-marked with grid-line of 5 cm spacing. The front face of the sample is also slice-marked with grid-line of 5 cm spacing in order to determine the displacement.

For preparation of reconstituted model foundation, The clay slurry passing #120 sieve mixed with water is put into model tank. The model tank is equipped with porous plates in the upper and lower parts.

The clay slurry is subject to uniform consolidation all over the horizontal area by filling the water into the water

bag through hose connected to water source over the frame.

Water head control is chosen in applying load for consolidation. The height of water source is arranged to obtain the required consolidation pressure. The pressure of 100kpa is secured by the difference of 10m in height between model soil tank and water source on the roof.

With the lapse of 30-60days, consolidation is almost completed. The soil sample obtained through the procedure is used for various soil laboratory tests such as model plate loading, consolidation, direct shear and triaxial test.

2.3 Lay out of geotextiles

Usually geotextiles is horizontally placed every certain interval of depth of foundation or banking. However in this experiment, only one layer of geotextile (A type) is covered on the surface of model foundation in order to examine the restrain effect on displacement and the increasing effect on ultimate bearing capacity. Another type of geotextile (B type) is used to determine the effect of the geotextiles used. The properties of geotextiles are shown in Table 1 while the layout of geotextile is shown in figure 1.

Table 1. Physical properties of geotextiles used.

Type	Nomen-clature	Weight (g/m^2)	Tension force(kg/m)
A	SM PP 300	220	75
B	SM PP 200	180	50

Type	Rate of elongation (%)	Coefficient of permeabil-ity (cm/sec)	Remarks
A	10-30	$\alpha^* \times 10^{-10}$	reinforce-ment
B	10-30	$\alpha^* \times 10^{-10}$	ment

α^* : coefficient ranged from 1 to 10

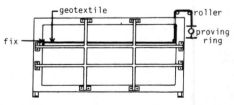

Fig. 1 Apparatus for measuring tension of geotextile

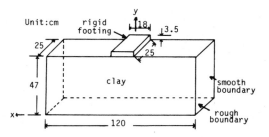

Fig.2 Test model

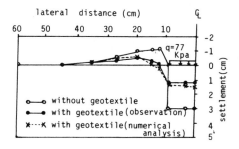

Fig.4 Settlement of surface

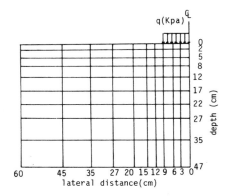

Fig.3 FEM grid

2.4 Test procedure

Loading is applied using the plate as shown
in figure 2 with the choice of strain
control of the rate of 0.2mm/min at which
pore pressure does not generate because of
drained condition.
During applying load, the stresses in soil
and pore water pressure were measured at
5-minute intervals while displacement was
measured every 30 minutes. The process was
continued while the foundation failed
completely.

3 NUMERICAL ANALYSIS

3.1 Program

In analysing results obtained from testings,
the Sekiguchis' elasto-viscoplastic model
(Sekeguchi 1977) was chosen as constitutive
equation and the joint element method
(Goodman et al. 1968) was introduced for the
adjacent different materials. The results of
the analysis were compared with observed
values to examine the accuracy of the
program developed by authors.

3.2 Boundary conditions

The FEM grid is shown in figure 3. As for
boundary conditions, the surface is free,
vertical sides are smooth so that vertical
displacement is allowed to occur along
these planes and the bottom is fixed. As for
geotextile the one end is fixed and the
other is connected to the proving ring to
measure the tension force.

4. THE RESULTS OF PLATE LOAD TEST AND FEM ANALYSIS

4.1 Settlement

The observed and numerical values of
vertical displacement on the surface
reinforced with (photograph 1) geotextile
are shown in figure 4 with satisfactory
results.
Also this figure shows observed values
without geotextile. Settlement and heave
are noticeably reduced with geotextile.

4.2 Lateral displacement

In natural state, at ultimate load, 77kpa,
a quite large lateral displacement develops
vertically at the edge of load plate as
shown in figure 5 & photograph 2. However
with the construction of geotextile the
constraint effect on displacement is
remarkable without large difference in
observed and approximate values.
Maximum displacement is located at depth of
a quarter of plate width from the surface.

4.3 Displacement vector and crack zone

Displacement pattern and crack zone are
shown in figure 6 and figure 7 respectively.

155

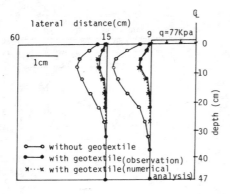

Fig.5 Lateral displacement at edge of model foundation

Photo.2 Displacements at failure

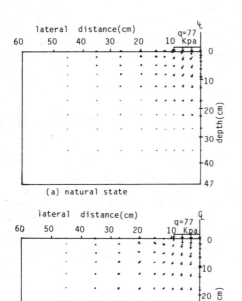

(a) natural state

(b) reinforcement of geotextile

Fig.6 Pattern of displacement

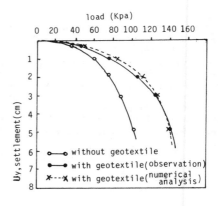

Fig.7 The schematic diagram of crack zone at shear failure

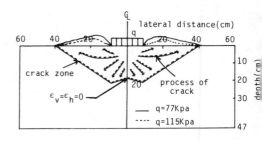

Fig.8 Load-settlement relation

Depending on the cases with or without geotextile, there appears large difference in magnitude but similar in pattern. With geotextile, displacements are conspicuously reduced, which in turn means a reduction in stresses. Crack zone shows the same pattern in both cases. But in natural state crack begins with as half a load as the reinforced state.

4.4 Ultimate bearing capacity

The ultimate bearing capacity depends greatly on geotextile. As shown in figure 8,

156

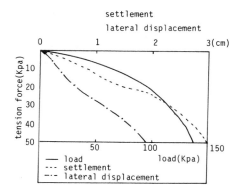

Fig.9 Settlement, lateral displacement and load-tension force relation

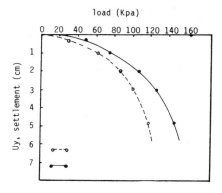

Fig.12 Load-settlement relation by FEM analysis

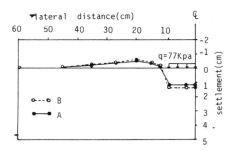

Fig.10 Settlement calculated by FEM analysis

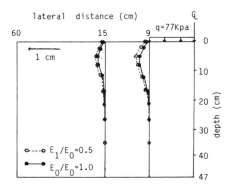

Fig.11 Lateral displacement calculated by FEM analysis

with geotextile the ultimate bearing capacity, 115kpa almost increases one and a half times over that of 77kpa in natural state. Previous study attributed this increase to the fact that geotextile

transfered local shear to general shear failure. This study shows the same tendency.

4.5 Load, settlement and lateral displacement-tension force of geotextile

The relation between tension-load, tension-settlement and tension-lateral displacement seems to be linear as shown in figure 9. The linearity seems due to the yield of foundation before the yield of geotextile. If geotextile yields before clay layer does the pattern of foundation failure will be very complicated.

4.6 The effect of different geotextiles on the behavior of foundation

In order to find out the behavior of foundation depending on the different geotextiles, FEM analysis is performed using 2 types of geotextle (A and B). Figure 10 to 12 show the results. In settlement, there is no significant distinction, which is very consistent with Yamaoka et al's result (1985).
   There is a little different in lateral displacements betwen case A and B.

5 CONCLUSIONS

Through plate load test on two dimensional saturated clayey model foundation, following results can be drawn from the comparisons between the FEM analysis and observed values from the start of loading to failure. The comparison is also done between the cases with/without geotextile covered.

1. Both settlement and lateral displacements are notably reduced due to the lay-out of geotextile.

2. With the construction of geotextile, the ultimate bearing capacity increases one and half times as compared with that in natural state.

3. The behavior of the model foundation shows similar magnitude and shape irrespective of the physical properties of geotextile used.

REFERENCES

Goodman, R.E.,R.L.Taylor & T.L.Brekke 1968. A model for the mechanics of jointed rock. Jour. of the Soil Mech. and Found. Divi., ASCE, Vol. 94, No. SM3: 637-654.

Kang, K.S. 1988. An Analysis of two-dimensional model foundation layer. Master's Thesis in Engineering, Graduate School, Chonnam National University, Kwangju, Korea.

Saitoh,K.,T.Kimura,S.Takahshi, Y. Nagano & E.Katoh 1985. Mechanical behavior of soft clay reinforced with qeotextile with high elongation. Proceedings of Symposium on Geotextile, JSSMFE: 1-6.

Sekiguchi,H.1977. Rheological characteristics of clays. Proc, 9th, Int. Conf., SMFE.

Yamaoka,I.,T.Nishigata & T.Yasuyuki 1985. Model test and numerical analysis results on reinforcement of soft subgrade with geotextile. Proceedings of Symposium on Geotextile, JSSMFE:27-32.

Zanten,R.V.(ed)1986. Geotextile and geomembrane in civil engineering. Jhon Wiley & Sons, New York, U.S.A.

APPENDIX

The physical and mechanical properties of soil sample produced in the model foundation apparatus described in 2.1 & 2.2 are shown in Table A.

Table A. Parameters of soil used in calculation.

λ	κ	M	C_α	ν	σ(Kpa)
0.146	0.023	1.34	0.048	0.374	100

G_0(Kpa)	K	q_u(Kpa)	σ_0(Kpa)	K_0	e_0
1500	0.597	45	100	0.597	0.972

γ (KN/m^3)	$\dot{\upsilon}_0$(sec^{-1})	λ	k_x(cm/sec)	k_y(cm/sec)	W_n(%)
18	0.1×10^{-6}	0.146	3.75×10^{-7}	3.75×10^{-7}	43

International Geotechnical Symposium on Theory and Practice of Earth Reinforcement / Fukuoka Japan / 5-7 October 1988
© 1988 Balkema, Rotterdam. ISBN 90 6191 820 0

Recent evolution in ISO standardization for mechanical testing of geotextiles

J.M.Rigo & F.Lhôte
State University of Liege, Belgium

D.Cazzuffi
Research Centre for Hydraulics and Structures, ENEL, Milano, Italy

G.den Hoedt
Akzo Industrial Systems, Arnhem, Netherlands

ABSTRACT : Nowadays geotextiles are worldwidely used. National Committees try to standardize tests on the basis of local experiences.
The Internationalization of the market of geotextiles makes essential the harmonization of tests procedures.
In 1985, I.S.O. (International Standard Organization) decided to create a sub-committee for the standardization of test procedures on geotextiles : I.S.O. TC 38/SC 21.
The working group 3 has the responsability of the study of the mechanical test procedures on geotextiles. The purpose of this paper is to present the recent evolution of the activity of this working group.

1 INTRODUCTION

In 1985, I.S.O. decided to set up a technical sub-committee for the standardization of test procedures on geotextiles :
ISO TC 38/SC 21. The first plenary meeting was held in Manchester in June 1985.
France held the chairmanship and the secretaryship of this committee.
It was decided to set up 5 working groups in order to prepare the technical bases for the plenary sessions. The 5 working groups and the chairmen are :
- WG 1 : terminology – J. Perfetti (France)
- WG 2 : sampling – M. Mortensen (Denmark)
- WG 3 : mechanical tests – J.M. Rigo
        (Belgium)
- WG 4 : hydraulic tests – B. Myles
        (United Kingdom)
- WG 5 : durability – to be appointed.
   The second plenary meeting of the ISO sub-committee TC 38/SC 21 was held in Paris in March 1987. After this second meeting, we can say that a lot of work has been done. So far, this conference is a good occasion to present the major results obtained by our working group.
   Very active experts from United Kingdom, West Germany, Austria, Denmark, Belgium, Canada, France, Italy, Netherlands, Switzerland and United States of America constitue the WG 3. The names of these experts are given at the end of this paper.
   The goals of WG 3 are to prepare ISO-standards for :

- conditioning atmosphere;
- tensile tests;
- tensile tests on geotextile junctions;
- tear tests;
- friction tests.

2 CONDITIONING ATMOSPHERE

WG 3 proposed and the plenary Assembly adopted (Paris 1987) that the conditioning atmosphere would include these three groups of conditions already recommended by different ISO standards :
- $20 \pm 2°C$; 65 ± 5 % of relative humidity;
- $23 \pm 2°C$; 50 ± 5 % of relative humidity;
- $27 \pm 2°C$; 65 ± 5 % of relative humidity.
These different conditioning atmospheres take into account the various local weather conditions in the world.

3 TENSILE TESTS

A great number of different tensile tests exists nowadays : the wish of WG 3 is not to elaborate new test procedures but to lead the experts to define a general philosophy for choosing a common test procedure on geotextiles. The best way in order to obtain this result is based on the principle of the value analysis; balancing the relevant factors and finally making a choice : it was presented by Leflaive at a R.I.L.E.M. 47-SM committee meeting on

geotextiles tests (Milano, November 1985). Following this presentation, each test procedure is estimated by a quotation coefficient, function of 8 criteria :
1. applicability to various uses of geotextiles;
2. direct relation between the test procedure and the geotextile's function;
3. applicability to all the different types of geotextiles;
4. accuracy;
5. reproductibility;
6. cost;
7. time before getting a response;
8. possible extension according to the market trend.

Each of these 8 criteria is graded from 1 to 5. Each grade is adjusted by a factor depending on the purpose of the test :
- quality control test;
- identification test (or acceptance test);
- design test.
The sum of these 8 factors is fixed to 20. By this way the final grades are rated up to 100.

A quality control test is a mean to check the production regularity of a type of geotextile. The producer is responsible in the choice of that method. The results of these control tests can be confidential. An external body might interact with this procedure :
1°) to recommend to follow through certain types of parameters because they seem to be significant; and this within the scope of a recommendation;
2°) for an approval procedure or quality attestation, in order to check wether or not the procedure has all the know-how and the materials needed for this qualification.

An acceptance test or an identification test is intended to identify the product if it meets the accepted specification
- the raw material;
- the production procedure;
- the physical characteristics of the components and final products.

These characteristics are given by the producer, under his responsibility and presented on technical sheets. These data may be checked easily by the contractors. Obviously a good accordance between the test procedures used by the contractors and the producers must exist.

A designing test is intended to the selection of a product for a given application. In each case, the choice of a test procedure depends on :
- the function(s) fulfilled by the geotextile. The test must give appropriate informations in relation with this (these) function(s);

- the design procedure (empirical, semi-empirical, analytical at different levels).
Experts found not less than ten different methods for the tensile test; these are :
- classical tensile test : 5 cm wide (ISO 5081);
- Grab test (DIN 53858);
- CBR test (DIN 54307);
- wide strip test : 50 cm wide (NF G 38-014);
- wide strip test : 20 cm wide (ASTM 4595-86);
- Sison test (SN 64 0550);
- St-Brieuc test (-);
- burst test (ASTM D 3786);
- biaxial test (-);
- tensile test with soil (-).

According to Leflaive's presentation described above, experts of WG 3 decided to adopt the following procedure as first step :
- the test procedure basis principle must be the wide strip tensile test;
- the 20 cm wide strip test is chosen as reference and basic test for ISO standard (ISO, first part);
- possibility is given if an improvement of the product characteristics is noticed, to use the 50 cm wide strip test (ISO second part);
- for all these tests the reference length is 10 cm;
- it's proposed for tests on geogrids (outside the centre of interest of ISO TC 38/SC 21, but viewed here for more convenience) to adopt a variable width for the samples and closed as possible to 20 cm width, including a whole number of elements. The length of these samples must also include a whole number of elements.

The tensile test on 50 cm wide samples calls for notion of equivalent strain to evaluate the strain under the maximum load. This is the strain of a piece of a geotextile that would be submitted at the moment of maximum load when lateral strains during the test are prevented.
This value of the strain is given by the following formula proposed by Rigo and Perfetti (1980) :

$$\varepsilon_R = \varepsilon_1 + \varepsilon_2 + \frac{\varepsilon_1 \, \varepsilon_2}{100} \qquad (1)$$

where ε_1 and ε_2 parameters are evaluated by the following expressions, according to figure 1

$$\varepsilon_1 = (\ell_r - \ell_o) / \ell_o \qquad (2)$$

$$\varepsilon_2 = (b_r - b_o) / b_o \qquad (3)$$

Recent investigations led by French experts (Leclercq, 1987 and Perrier, 1987) showed that it was possible to evaluate ε_R by measuring ε_1 in the sample central zone provided that there is no lateral contraction in that zone. Indeed (figure 2), it's possible to show that during a tensile test on wide sample, the value of ε_1 is not constant from one clamp to the other.

From the comparison (figure 3) between the values of ε_1 measured in the central zone and ε_R, it can be concluded that there is a good agreement in the case of the 50 cm wide samples.

This fact still has to be confirmed in the case of 20 cm wide samples. If the tests on 20 cm wide turn out to be convincing, the ISO test procedure could be simplified and consider only one width, namely 20 cm.

4 MEASUREMENT OF ε_R AND ε_1 IN THE CENTRAL ZONE

In order to contribute to the validation of this concept, tests were carried out at the Civil Institute of the State University of Liege. These results are given in Table 1.

The following products were considered : two thermobonded and two needle-punched non-woven geotextiles. Two widths were considered, namely 20 and 50 cm. The tests were carried out on the longitudinal direction and at the same rate of displacement, namely 50 mm per minute.

The following parameters were measured :
ε_1 = the longitudinal strains in the sample central zone
ε_1 is the sample strains determined from the displacement between the two clamps
ε_r is the lateral strain
ε_R was calculated by the formula (1).

For the measure of ε_1, an infra-red pulsed extensometer was used.

This apparatus was used to follow control markers 6.0 cm apart at the start.

These tests show that for non-woven geotextiles, no difference between the measure of ε_1 in the central zone and ε_R calculated by the formula (1) was observed for both sample widths of 20 and 50 cm.

The values of ε_1 obtained in both 20 and 50 cm wide samples are about the same.

The concept presented in the previous paragraph seems to be confirmed. More investigations are in progress.

Table 1. Comparison between ε_R and ε_1 from tensile tests on nonwoven geotextiles with different widths (20 and 50 cm) : two specimens for each width tested only in longitudinal direction at the same displacement rate

Type of product	μ (g/m2)	Specimen width (cm)	ε_1 (%)	ε_2 (%)	ε_R (%)	ε_1 (%)
NW TB PP CF	190	20	88 85	22 20	65 65	58 58
		50	77 83	10 10	67 73	63 60
NW TB PP CF	160	20	86 86	15 15	71 71	68 65
		50	80 55	6 5	70 43	71 40
NW NP PP CF	180	20	105 128	45 45	60 82	63 81
		50	109 86	17 17	92 69	48 52
NW NP PS CF	160	20	65 75	35 33	35 42	38 46
		50	75 65	16 15	47 40	50 40

NW : nonwoven   –   NP : needle-punched
TB : thermo-bonded

CF : continous filament
PP : polypropylene
PRS : polyester

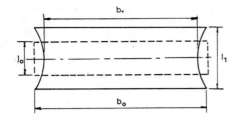

Figure 1 : Dimensions of a sample for a tensile test

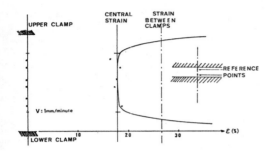

Figure 2 : Evolution of ε_1 in P.P. woven geotextile sample

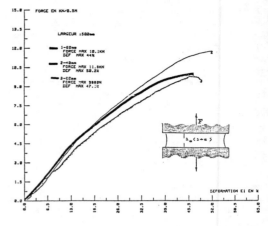

Figure 3 : Comparison between ε_1 measured in the central zone of the sample and ε_R (Perrier)

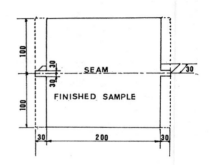

Figure 4 : Tensile test on geotextile junctions (from ASTM draft)

5 TENSILE TESTS ON GEOTEXTILE JUNCTIONS

This subject of work was appointed to WG 3 in Paris (March 1987) and must be solved easily and rapidly after the adoption of a final resolution for the tensile test procedure.

ASTL has, so far, adopted a tensile test procedure with 20 cm sample, presenting the junction to be tested in the central part of the sample.

The shape of the sample is given in figure 4. An over-width is necessary to make possible the sewing thread anchorage. The test results have to be compared to those obtained with the tensile test on 20 cm wide samples.

6 TEAR TEST

It's proposed to go closely into the different following aspects of the tear phenomenon :
- initiation of the tear process;
- tear propagation;
- residual resistance after damping.

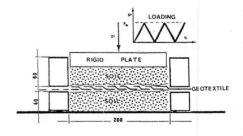

Figure 5 : Geotextile damaging before tensile test (Perrier, C.E.T.E.)

For the initiation of the tear process, future studies will be brought in order to examine the different possibilities of puncture tests on geotextiles. Information will be available in the new issue of the I.G.S. inventory on tests on geotextiles.

The following parameters will be taken into account :
- type of test (statical or dynamical);
- sample dimension;
- piston shape.

For the tear propagation, if the sample width for the tear propagation test. It is possible to continue the works on the basis of the Italian standard which already consider that width for the tear samples (Cazzuffi and al., 1986).

For the damaging effects on geotextiles, Perrier showed recently a method used and developed at C.E.T.E. laboratory. A geotextile (figure 5) is placed between two layers of soil in a box 200 mm wide, 600 mm long and 100 mm high. The soil is a 0/40 gravel, a mix of river and crushed gravels. A rigid metallic plate allows to apply a vertical force on the "soil-geotextile" system. Triangular loadings are applied 20 times with maximum value of 200 kPa. After that type of damaging, the geotextile is removed and is submitted to tensile tests, using 50 cm wide samples. This damaging method will be used as a reference for the works of the WG 3.

7 FRICTION

The subject of this study was intrusted to WG 3 in Paris in March 1987. No precise decision has been taken so far about the procedure and the test method. Contacts will be taken with ISO TC 183 (Soil Mechanics) having soil friction tests in its scope of duty.

8 CONCLUSION

The WG 3 is set up in Manchester in 1985, devoted to mechanical tests on geotextiles and has for goal to propose to plenary ISO sessions test methods in matter of conditioning atmosphere, tensile test, tensile test on junctions, tear and friction tests. These works make progress rapidly. The conditioning atmospheres were approved by the plenary session in Paris (1987) and the works concerning tensile tests and tensile tests on junctions are to be concluded.

The experts appointed by different national standardization institutes and actively participating in the activity of WG 3 are :

Mercer and Robinson	(U.K.)
Frobel	(U.S.A.)
Schneider	(A)
Mortensen	(DK)
Martin and Tonus	(CH)
Kristmann and Willemsen	(D)
Den Hoedt	(NL)
Delmas, Gourc and Leflaive	(F)
Rowe	(CND)
Cazzuffi	(I)
Rigo	(B)

REFERENCES

Cazzuffi D. - Venesia S. - Rinaldi M. - Zocca A. 1986. The mechanical properties of geotextiles : Italian standards and interlaboratory test comparison. Proc. Third International Conference on Geotextiles, Vienna, April.

Leclercq A. 1987. ITF Contribution to the discussion. ISO, WG 3 Meeting, unpublised Toulouse, November.

Leflaive E. 1985. Proposal of testing evaluation. RILEM, Technical Committee 47-SM Meeting. Unpublished, Milano, November.

Perrier. 1987. C.E.T.E. Contribution to the discussion. ISO, WG 3 Meeting, unpublished, Toulouse, November.

Rigo J.M. - Perfetti J. 1980. Nouvelle approche de la mesure de la résistance à la traction des géotextiles non-tissés. Bulletin de Liaison des Laboratoires des Ponts et Chaussées, n° 107.

International Geotechnical Symposium on Theory and Practice of Earth Reinforcement / Fukuoka Japan / 5-7 October 1988
© *1988 Balkema, Rotterdam. ISBN 90 6191 820 0*

Soil-reinforcement interaction determined by extension test

C.K.Shen, Oon-Young Kim, X.S.Li & Joonik Sohn
University of California, Davis, Calif., USA

ABSTRACT: This paper describes the use of a new technique to determine the soil-reinforcement interaction. The testing apparatus is essentially a triaxial cell fitted with the capability to house a hollow cylindrical sample. A hollow cylindrical sand specimen with a concentric layer of reinforcing material sandwiched in the middle is used in this investigation. The reinforcement is fastened at the base. The hollow specimen can be viewed as a "unit sheet" of a soil-reinforcement composite system of infinite horizontal extent. Axial load as well as inner and outer chamber pressures can be applied to perform a test. The specimen is first subjected to an isotropic stress state corresponding to the overburden pressure. Next, an extension test by reducing the axial load is carried out. Since the reinforcement is fastened at its lower end to the base, any tendency of relative movement between the reinforcement and the sand during an extension test can induce tensile force in the reinforcement thus forming a "reversed pull-out" test condition. Preliminary test results have demonstrated positively the feasibility of the new approach to study the soil-reinforcement interaction.

1 INTRODUCTION

It was almost a decade ago that the late Professor Kenneth Lee (1978) in his state-of-the-art report to the first soil-reinforcement conference held in Pittsburg said that one of the most important and yet least known parameters in working with reinforced soil is the bonding capacity describing the soil-reinforcement interface behavior. In the ensuing years much has been done in the laboratory and field to identify factors, such as dilation, compaction, grain size, type and rigidity of reinforcement, overburden, etc., affecting the mobilization of interface resistance in reinforced soil. However, the overall picture of the interface behavior is still unclear as ever due to the lack of a testing device that is capable to quantify the various factors influencing the mobilization of interface resistance.

Quantitative evaluation of soil-reinforcement interactive behavior in the laboratory is normally done by determining the soil-reinforcement bonding capacity using either the direct shear test or the pull-out test. The direct shear test measures the resistance to sliding of a soil over a planar surface of the reinforcing material. The quantitative measure of resistance is a friction coefficient obtained as the ratio of the peak shear stress to the normal stress. The direct shear test is considered to accurately determine the friction coefficient because the normal force and shear force are both measured quantities.

However, the loading conditions are not felt to adequately describe the in situ soil-reinforcement interaction. In the pull-out test, the reinforcement is embedded in a soil-filled box and the force and displacement are measured as the reinforcement is "pulled-out" of the box. The force required to initiate slippage of the reinforcement is known but, depending on the geometry of the various elements and the loading and boundary conditions of a test, the normal pressure on the reinforcement may not be known with much certainty (Johnston, 1985). The test, however, rather closely simulate the slippage failure that may occur in the field in that the axial force in the reinforcement varies from zero at the free end to a maximum at the face of the box.

Nonetheless, the complex boundary conditions and bulkiness of the apparatus limit the effective usage of the pull-out testing device. In view of the current laboratory testing capability, it is obvious that significant improvement in laboratory determination of soil-reinforcing element interactive behavior is much desired.

2 THE TESTING APPARATUS

It can be summarized that a desirable testing apparatus to study the soil-reinforcement interaction behavior should have the following capabilities:

a. Ability to test reinforced soil specimens to determine the interactive bonding capacity of a soil-reinforcement system.

b. The apparatus should properly simulate in situ loading conditions with minimum boundary effects.

c. It should be able to perform accurate tests under short and long loading conditions.

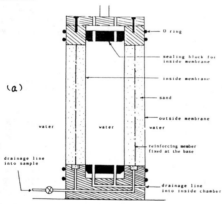

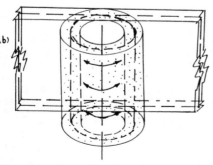

Fig. 1 A schematic diagram of a testing apparatus and a hollow cylindrical sample

A device that can satisfy the above is the triaxial testing apparatus with a hollow cylindrical sample (Kim, 1987). A schematic diagram of the apparatus and a sample is shown in Figure 1a. Briefly, a hollow cylindrical sand specimen with a concentric layer of reinforcing material sandwiched in the middle is formed. The reinforcement is fastened at the base. The hollow specimen can be viewed as a "unit sheet" of a soil-reinforcement composite system of infinite horizontal extent as shown in Figure 1b. To perform a test, the specimen is first subjected to an isotropic stress state corresponding to the overburden pressure. Next, an extension test (by reducing the axial load) is carried out and the load reduction vs. axial displacement readings recorded. The specimen is loaded in extension to failure by either the breakage of reinforcing material (tensile failure) or slippage which takes place at the soil-reinforcing element interface (i.e. the overcoming of the bonding capacity). In both cases necking, as a result of shear failure in the sand, occurs. The gradual reduction of axial load causes the specimen to elongate in the axial direction. Since the reinforcement is fastened at its lower end to the base, any tendency of relative movement between the reinforcing element and the sand can induce tensile force in the reinforcement thus forming a "reversed pull-out" test condition. This setup simulates closely the loading conditions existing at the soil-reinforcing element interface in reinforced soil structures.

The interactive behavior of a soil-reinforcement system is normally characterized by a Coulomb-type friction parameter, the apparent coefficient of friction, f. If the tensile capacity of the reinforcement is greater than the bonding capacity of the soil-reinforcement system, the reduction in axial load will lead to a "pull-out" failure; e.g. slippage between the soil and the reinforcement. This phenomenon can be detected by the presence of "necking" in the sand near the top of the specimen. The failure load is measured and the Coulomb-type friction parameter can be determined if the slipped length at the interface can be defined.

The apparatus described above satisfies most of the pertinent requirements needed for the study of interactive behavior of the various soil-reinforcement systems under a wide range of loading conditions anticipated in the field. The use of a hollow cylindrical specimen takes away most of the edge effects associated with

166

conventional pull-out test. The wall friction is completely eliminated because no such boundary exists in a hollow circular specimen. As a result the overburden pressure is applied evenly over the entire specimen. Furthermore, the end restraints at the base and top of a hollow specimen is far less severe than the constraints experienced in pulling the soil against the front panel of the pull-out box.

As stated earlier, an accurate measure of the length over which slippage takes place and the corresponding axial load are essential in determining the apparent coefficient of friction for the "pull-out" mode of failure. In an extension test with the reinforcement fixed at the base, the length over which slippage takes place is difficult to determine. If slippage occurs along the full length of the reinforcement and if the end constraint effect is ignored, referring to Figure 2, the horizontal force ΔP acting

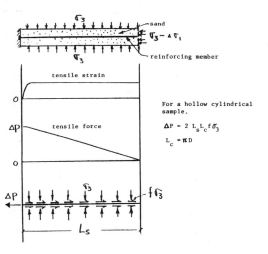

Fig. 2   Stress and strain developed along a reinforcing element

on the reinforcement prior to slippage can be calculated as

$$\Delta P = 2L_s L_c f \sigma_3 \qquad (1)$$

Where L_s is the slipped length, L_c is the circumferential dimension of the reinforcing element and σ_3 is the overburden pressure. Thus the apparent coefficient of friction,

$$f = \frac{\Delta P}{L_s} \left(\frac{1}{2L_c \sigma_3} \right) \qquad (2)$$

For a given test, $1/2L_c \sigma_3 = $ constant $= m$, Eq. 2 can be written as

$$f/m = \Delta P / L_s \qquad (3)$$

If slippage does not take place along the full length of the reinforcement, then the tensile strain and the tensile force distributions along the reinforcement may not be approximated as in Figure 2. On the other hand, it is understood that shorter reinforcements (shorter specimens) are more prone to have slippage failure along the full length of the reinforcement. The exact length of slippage for a given sample, nevertheless, is unknown. To circumvent this it is proposed that a number of identical samples of different heights be tested under identical loading conditions. The apparent coefficient of friction is then calculated according to eq. 3 for each sample assuming that L_s is the height of the sample (the total length of reinforcement). By plotting the corresponding ΔP vs. L_s values as shown in Figure 3, one can establish a line OA to

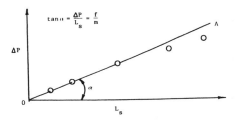

Fig. 3   Pull-out load vs. length of slippage

define the f/m ratio for the soil-reinforcement system tested. This approach is based upon the assumption that when friction is mobilized fully along the length of the reinforcement, the tensile force in the reinforcement is proportionate to its length. Furthermore, it is assumed that the fixation of the reinforcement at the base does not significantly affect the distribution of tensile force.

The testing device is operated with the automated control and data acquisition system originally developed for the automated triaxial testing system (Li, et al., 1988). The system has two feedback loops for axial and lateral actuators under software control; the two loops can work separately or synchronously. For the reinforced soil study, a uniform pressure in the inner and outer chambers representing the overburden pressure is applied to the hollow specimen. The specimen is initially subjected to a set of isotropic pressure (axial pressure equal to the chamber pressure) and followed by a gradual decrease in axial pressure leading to a failure of the specimen in extension.

3 PRELIMINARY TEST RESULTS

Described in this section is an exploratory laboratory investigation to demonstrate the feasibility of adopting the automated triaxial testing system and the hollow cylindrical sample to study the interactive behavior of reinforced soil systems. The dimensions of the hollow cylindrical sample were I.D. = 3.56 cm, O.D. = 6.35 cm, and L = 12.7 cm. The sand was the dry Monterey "0" sand placed at a density of approximately 1.5 ~ 1.6 gm/cm^3. A commercially available aluminum foil sheet was used as the reinforcement (ultimate strength = 1.0 to 1.5 N/mm at 2.5 to 2.7% elongation). The lower end of the aluminum foil cylinder was fastened at the base platen, whereas its upper end was free but in contact with the loading platen.

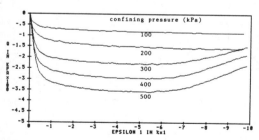

Fig. 4   Stress-strain curves - unreinforced

Figure 4 shows the extension test results of unreinforced samples under uniform chamber pressures ranging from 100 kpa to 500 kpa. The stress-strain curves (Q = deviatoric stress, epsilon 1 = axial strain) depict the behavior of a relatively loose sand; the angle of internal friction (ϕ) in the Mohr-Coulomb stress space is approximately 31°.   Figure 5

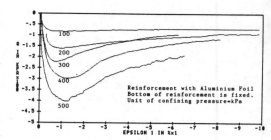

Fig. 5   Stress-strain curves - reinforced with aluminum foil

shows the corresponding stress-strain curves when a cylindrical aluminum foil reinforcement is placed in the middle of the sample. These curves exhibit the behavior of a dense sand, particularly under higher confirming pressures. The ϕ angle is measured at about 34°. It is also interesting to note that the rapid decrease in strength beyond the peak strain (1.0 ~ 1.5%) is associated with the detection of necking at the top of the sample (Fig. 6) indicating that slippage

Reinforced with Aluminium Foil, Hollow Cylinder Type

Confining Pressure= 500kPa

Axial Strain = -3.8%

Fig. 6 Necking due to slippage (pull-out)

between the sand and the reinforcing element did occur.   No breakage of the aluminum foil was observed in all the samples tested. Though the results presented are not sufficient to determine the apparent coefficient of friction for bonding capacity characterization, the increase in ϕ angle and the more brittle nature of the stress-strain behavior can be captured very well by this type of testing when a

168

Fig. 7 Breakage of reinforcement (plastic netting)

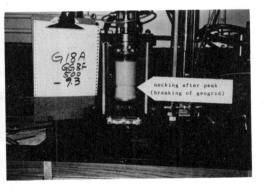

Reinforced with Geogrid(Fixed-Bottom), Hollow Cylinder Type

Confining Pressure = 500 kPa

Tensile Axial Strain = -9.3%

Fig. 8   Necking   due   to   reinforcement breakage

relatively   inextensible   aluminum   foil reinforcement is used. Additional tests using a very weak plastic netting reinforcement were also conducted. Figs. 7 and 8 show respectively the breakage of the reinforcement and the necking in the sample where breakage occurred.

4 SUMMARY AND CONCLUSIONS

This paper describes the development of a testing apparatus   for   the purpose of studying the interactive behavior of reinforced soil systems. The system can be used effectively to provide much needed basic knowledge pertinent to soil reinforcement technology for the emerging geofabric industry. The testing condition simulates closely the field loading conditions at the soil-reinforcement interface;   furthermore,   the   boundary constraints   around   the   specimen   are minimal enabling more accurate interpretation of test results.

An exploratory investigation to assess the   feasibility   of   the   approach   was discussed. The authors believe that the new device presents a viable alternative to the current methodology in reinforced soil testing. The authors also realize that concerns pertaining to the design and development of the system still exist. For instance, the development of a shear band   in   the   vicinity   of   the   soil-reinforcement interface is closely related to the grain-size and the density of the soil, the type of reinforcement, and the overburden   pressure;   therefore,   in deciding   the   thickness   of   the   hollow cylindrical sample, considerations should be given to the formation of the shear band. The importance of membrane effect in extension test is another factor to be considered   in   data   processing   and interpretation.

This new approach to conducting laboratory reinforced soil testing is albeit unconventional, the authors believe it will help to remove some of the limitations in our ability to obtain meaningful and quantifiable information on reinforced soils and thus benefit the geotechnical profession;   undoubtedly,   much   work   is still needed to develop the "ideal device" for reinforced soil testing. It is hoped that this paper may stimulate interest and attention to this very important issue.

5 ACKNOWLEDGEMENT

The idea of testing reinforcing sheet in the triaxial cell under extension was first suggested to the senior author by Barry Christopher.

6 REFERENCES

Johnston, R.S., Jr. 1985. Pull-out testing of tensar geogrids. Thesis, Department of Civil Engineering, University of California, Davis.

Kim, O.Y. 1987. Triaxial extension tests on hollow cylindrical samples of reinforced soil. Research Report, Department of Civil Engineering, University of California, Davis.

Lee, Kenneth L. (1978) STATE-OF-THE-ART-REPORTS, Mechanisms, analysis and design of reinforced earth. Symposium on <u>Earth Reinforcement</u>, ASCE, Pittsburgh, PA.

Li, X.S., Chan, C.K., and Shen, C.K. (1988) An automated triaxial testing system. To be published in ASTM special publication.

International Geotechnical Symposium on Theory and Practice of Earth Reinforcement / Fukuoka Japan / 5-7 October 1988
© 1988 Balkema, Rotterdam. ISBN 90 6191 820 0

The earth reinforcement effects of steel bar truss set in embankment

H.Yokota, T.Nakazawa & H.Fujimoto
Miyazaki University, Miyazaki, Japan

Y.Ono
Sumitomo Construction Co., Ltd, Tokyo, Japan

ABSTRACT: We investigated the reinforcing effects of full-size steel bar truss, which was welded to a steel band, on banking soil by pull-out and settlement tests in a soil chamber. The experimental results show that the pull-out resistance of the steel bar truss is more than 2.5 times approximately as large as one of the steel band plate used in Terre Armée and give the soil more effective stiffness.

1 INTRODUCTION

Recently, metal materials such as steel bar and steel band plate have been utilized (see Hashimoto(1979), Cartier and Gigan (1983) and Marczal(1983)) to increase the shearing strength of banking soil and to reduce the subsidence of embankment. The increase of shearing strength is caused by the pull-out resistance of these reinforcing materials. The resistance largely depend upon the friction and interlocking between reinforcing materials and soil. The subsidence of embankment may be reduced in the case that reinforcing materials and soil behave in one and the composite made of these two materials has large stiffness. It would be very meaningful to develop the reinforcing material, which has high pull-out resistance and gives large stiffness to soil.

We propose a steel bar truss welded to a steel band plate as the reinforcing material with these mechanical characteristics. This paper describes their mechanical characteristics in embankment, obtained from pull-out and deflection tests, and we investigate their reinforcing effects by comparing with ones of a steel band plate used in Terre Armée method.

2 EXPERIMENTAL METHOD

2.1 Reinforcing materials

Fig.1 shows the form of the reinforcing material used in the experiments and ex-

presses the full size used in banking. The band plate has the dimensional shape of 0.32cm × 10cm × 280cm (thickness × width × length). A truss of 5cm in height, made of steel bars of 6mm in diameter, was assembled at 20 cm pitch on the steel band plate. We use hereinafter the abbreviated word, PT, about this steel band plate mounted space truss of steel bars.

We made an experiment on the steel band plate used in Terre Armée method in order to obtain the reinforcing effects. This steel band plate (hereinafter abbreviated as TA) has the same shape as one of the PT. These reinforcing materials were set in a following soil chamber.

2.2 Soil chamber

A soil chamber is made of stiffened steel and has the dimensional shape of 3.0m × 3.0m × 0.3m as shown in Fig.2. We loaded to the reinforcing materials in soil as follows. A movable loading wall, to which reinforcing materials are attached, was pulled out horizontally by hydraulic

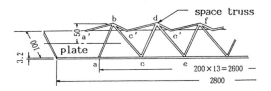

Fig.1 Form of reinforcing material (mm)

jacks. In the case of deflection tests, we used concrete blocks and H—section steel beams as surcharge load, which is transmitted through a loading plate to the reinforcing materials in the soil.

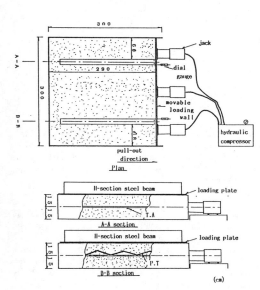

Fig.2 Soil chamber and reinforcing
material in soil

2.3 Banking soil

In the experiments, we used the disturbed Shirasu(sandy volcanic ash soil found in Southern Kyushu), having grain size accumulation curve as shown in Fig.3. Characteristics of the Shirasu are as follows.The specific gravity Gs=2.46, the optimum moisture content w_{opt} =26.8% and the maximum dry density ρ_{dmax} =1.27 g/cm^3. The density shows the variation of ρ =1.16–1.25g/cm^3, though the Shirasu was compacted by tamping and watering.

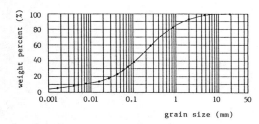

Fig.3 Grain size accumulation

2.4 Method of pull—out tests

The reinforcing materials PT and TA were placed in the soil chamber with the spacing of 1.5m each other, and were arranged in parallel in depth of 15cm from the soil surface as shown in Fig.2. Under vertical load of 1.15tf, we pulled out the reinforcing materials at the horizontal displacement speed of 2 mm/min.

In this case, we enable the attachment portion of the steel band plate to move correspondingly to the settlement of reinforcing materials, caused by soil compaction as in the case of actual construction works.

We measured the strains of steel band plates and obtained the pull—out force from the strains of the attached portion of the steel band plate to the movable loading wall.

2.5 Method of settlement tests

The reinforced soil has larger stiffness than soil itself and it is desirable that the reinforcing materials behave with soil in a body. We examined these stiffness increase and behavior in a body by two experiments.

We arranged the reinforcing materials PT and TA through the same procedure as in the pull—out tests. The reinforcing materials was placed in depth of 0cm, 10cm and 15cm from soil surface. We measured deflection of the reinforcing materials under the loads of 300–1500kgf. And we investigated the stiffness increase of the composite beams made of PT and soil cement by loading tests.

3 EXPERIMENTAL RESULTS AND DISCUSSION

3.1 Pull—out tests

Fig.4 shows the relation between the pull—out forces(F) and the horizontal displacements(δ). The pull—out force of TA shows peak value (F_{TA}) to small displacements. While, in the case of PT, the peak of the pull—out force (F_{PT}) occurs in the region of relatively large displacements. These peak values F_{TA} and F_{PT} are shown in Fig.5. It can be seen from this figure that F_{PT} is more than three times in maximum and 2.5 times in average as large as F_{TA}.

Fig.6 shows the axial strain distributions of TA and PT when their pull—out forces reached peak values. We consider that these distributions were obtained under same experimental conditions, though there was a little amount of scatter in the unit weight of soil.

We can consider from Fig.6 that the strain distribution of PT is expressed linearly as TA, and that, in other words, the shearing stress between PT and soil distributes uniformly along longitudinal axis of PT. Though the shearing stress at top and bottom sides are different from each other in the case of PT, we deal with the sum of the shearing stress at both sides, which distributes uniformly. Therefore, we propose the following apparent frictional coefficient f^* in the same way as TA,

$$f^* = \frac{F_{PT}}{2b\,\sigma_v\,L_e}$$

where
  b: width of reinforcing material PT,
  σ_v : average vertical stress acting on reinforcing material PT,
  L_e: length of reinforcing material PT.

Fig.7 shows the f^* values of PT and TA. It is reasonable that the value of TA is between 2 and 5. This may suggest the reliability of the f^* value about PT.

PT is composed of the portions of mountain and valley. The former is the "acbác" or "cedćé" and the latter is "bcdć" or "defé" in Fig.1. The soil

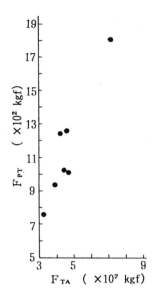

Fig.5 Pull-out resistance F_{PT} and F_{TA}

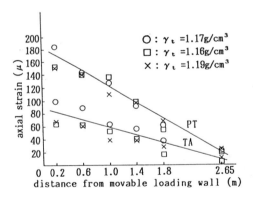

Fig.6 Axial strain distribution

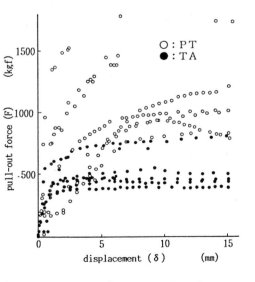

Fig.4 Relation between pull-out force and displacement

filled in the portion of mountain, for example "acbác", resists a pull-out force with the shearing strength of soil on the plane "abc" and "ábć". The portion of the valley, "bcdć", contributes to the resistance with the shearing strength on the horizontal plane containing bd-line. The pull-out resistance of PT is caused by these shearing strength except for the

173

frictional force between the steel band plate and the soil.

3.2 Deflection tests

Fig.8 shows the deflections produced at the center of PT and TA set in a soil, where the preceding load P is applied on the ground surface just above the center of PT and TA.

It is seen that the two reinforcing materials show same deflective behavior for each depth H in which the reinforcing materials are set,though PT has larger flexural rigidity than TA. Fig.8 may represent that PT behave with soil in a body, while we should consider that PT can not reduce the subsidence of embankment. The subsidence, however, may be reduced in the case that the volume ratio of PT to soil is larger than one in this experiment and soil is compacted more densely.

The composite beams, made of PT and soil cement (hereinafter abbreviated as CB), were tested to investigate the above mentioned subsidence problems. The soil-cement was made by mixing Shirasu with normal portland cement. The mixing rate of cement c_m is 5, 10 and 25 per cent of the dry weight of Shirasu, respectively. TA and PT beams,and soil-cement beam (hereinafter abbreviated as SB) were also tested to compare with the test results of CB.

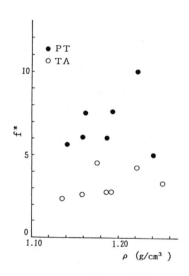

Fig.7 Apparent frictional coefficient

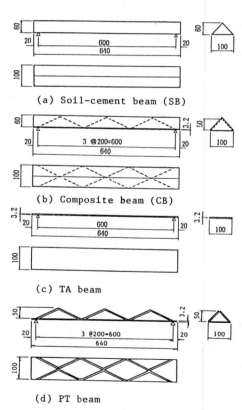

(a) Soil-cement beam (SB)

(b) Composite beam (CB)

(c) TA beam

(d) PT beam

Fig.9 Dimensions of beam specimens (mm)

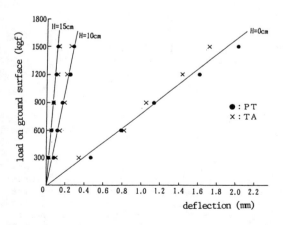

Fig.8 Deflections of reinforcing material

174

In these tests, the deflection was measured at the middle point of the beams when the concentrated load acted on the midpoint of the beam span. The configuration and dimension of these beam specimens are shown in Fig.9.

Compressive strength of soil-cement was obtained as follows;

$36.5 kgf/cm^2$ (c_m is equal to 25%)
$6.4 kgf/cm^2$ (c_m is equal to 10%)
$2.1 kgf/cm^2$ (c_m is equal to 5%)

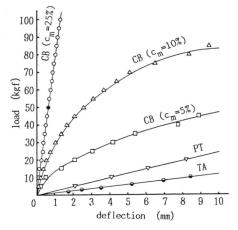

Fig.10 Load-deflection curves

The experimental load-deflection curves of the beams are shown in Fig.10. In this figure, blacken round symbol denotes the deflection when the soil cement beam of $c_m =25\%$ was broken in two. It can be seen that the flexural rigidity of the PT beam is two times as large as one of the TA beam. Then, it was found that the flexural rigidity of the CB of $c_m=25\%$ was about 50 times as large as one of the TA beam, and that the flexural rigidity of CB increased as c_m increased. For example, when load of 20kgf was placed on CB of $c_m= 5,10$ and 25%, it was estimated that the flexural rigidities of CB are in the ratio 4:13:26 if the one of PT beam was standardized. As seen from the results in the case of c_m=25%, the flexural rigidity of CB was nearly equal to one of SB for reason of small flexural rigidity of PT beam. The effect of composite beam appeared largely on increasing the failure load.

As it is clear from Fig.11 which shows the crack condition of CB at the ultimate state, only diagonal cracks appear along truss element. This may be due to the truss effect and the valley portions might play a role of compressive chord in the truss structure. This is the reason that

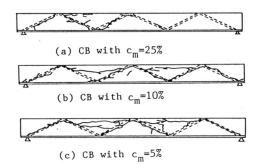

(a) CB with c_m=25%

(b) CB with c_m=10%

(c) CB with c_m=5%

Fig.11 Cracking patterns for CB with different mixing rate of cement

the toughness of CB is far larger than the total one of the PT beam and soil-cement beam. PT may be effective to reduce the subsidence of embankment, if this mechanism is made full use.

4 CONCLUSION

We proposed a reinforcing material which has high pull-out resistance and gives large stiffness to soil. The following characteristics became evident after examining the results of pull-out and deflection tests.

1. The pull-out resistance of the reinforcing material was more than 2.5 times as large as one of Terre Armée method.

2. The relatively large pull-out resistance was dependent upon the shearing stress of soil adjacent to the reinforcing material.

3. The reinforcing material deformed with soil in a body in spite of its relatively high flexural rigidity.

4. The experimental results of composite beam made of the reinforcing material and soil cement suggested that the soil filled in the valley portions of the reinforcing material might play a role of compressive chord in truss structure. This mechanism may be useful for examining the reduction of embankment subsidence by the reinforcing material PT.

ACKNOWLEDGMENT

The authors gratefully acknowledge the advice of Prof. Toshiaki Ohta of Kyushu University in the execution of this research. Thanks are extended to Mr. Shinsuke Kawano and Mr. Tamio Itoh of Miyazaki

University who provided assistance in the conduct of this experiment.

REFERENCES

Hashimoto,Y 1979. Behavior of reinforced earth embankment constructed with Kanto loam. International conference on soil reinforcement. Vol.2, pp.545-550.
Cartier,G & Gigan,J.P 1983. Experiments and observation on soil nailing structure. 8th European conference SMFE, Helsinki. Vol.2, pp. 473-476.
Marczal,L 1983. Measurements on reinforced soil structure. 8th European conference SMFE, Vol.2, pp.525-526.

2. Shallow and deep foundations

International Geotechnical Symposium on Theory and Practice of Earth Reinforcement / Fukuoka Japan / 5-7 October 1988
© 1988 Balkema, Rotterdam. ISBN 90 6191 820 0

Reinforcement of soft Bangkok clay using granular piles

D.T.Bergado, B.Panichayatum & C.L.Sampaco
Asian Institute of Technology, Bangkok, Thailand

N.Miura
Saga University, Saga, Japan

ABSTRACT: To study the effects of granular piles on the settlement of the soft Bangkok clay, a full scale embankment was constructed at the campus of the Asian Institute of Technology. Fully penetrating granular piles of 0.30 m in diameter, were arranged in triangular pattern at a spacing of 1.50 m and a 2.40 m high, well-instrumented embankment was constructed. To compare the settlement performance of the embankment with the nearby existing 4.0 m high test embankment on vertical drains, the height of the embankment was added up to 4.0 m after 345 days. Results show that after 900 days, the settlement of the embankment on granular piles was about 62% of the settlement observed on vertical drains. Comparison with past investigations indicates that granular piles reduced the settlement of soft clay foundations by as much as 20 to 40%. This implies that granular piles function as reinforcement to the clay rather than drains. The stress concentration factor measured in the granular pile foundation decreases with the increasing applied load.

1. INTRODUCTION

The use of granular piles as a soil improvement technique, is preferable when moderate increase in bearing capacity and/or moderate reduction, yet uniform rate of settlements, are required for foundations on soft clays. Bergado et al (1984b, 1987b) indicated that the bearing capacity of the soft Bangkok clay using granular piles increased by 4 times as much, the total settlements reduced by at least 30% and the slope stability factor of safety increased by at least 25%. Almost equal vertical strains were measured and the stress concentration varied from 2 to more than 5 between the pile and the surrounding clay.

To further study their effects on the settlement of a soft clay foundation, a full scale test embankment, 2.4 m high was constructed by Sim (1986) on a granular pile-improved foundation, and was raised to a height of 4.0 m by Panichayatum (1987) to provide a meaningful basis of comparison with the performance of the nearby test embankment constructed on a Mebra vertical drain-improved soil by Singh (1986). This paper presents the behaviour of the two test embankments to compare their effects on the settlement performance of the soft clay foundation and to assess the most suitable soil improvement techniques in the soft Bangkok clay. This study is the first of its kind in the subsiding environment in the Chao Phraya Plain in Thailand.

2. SITE LOCATION AND SOIL PROFILE

The test site is located inside the campus of the Asian Institute of Technology, 42 km north of Bangkok, Thailand. The soil profile is given in Fig. 1. The uppermost subsoil is divided into 3 distinct layers as follows:

(a) heavily overconsolidated, weathered clay layer of about 2.0 m thick with reddish-brown color;

(b) grayish soft clay layer underlying the weathered clay of about 6.0 m thick; and

(c) highly overconsolidated reddish-brown stiff clay layer of about 6.0 m thickness.

The groundwater table varies with the season and an average value of 1.50 m below

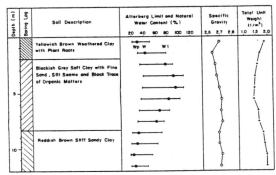

Fig. 1 Soil profile and basic soil properties at the site.

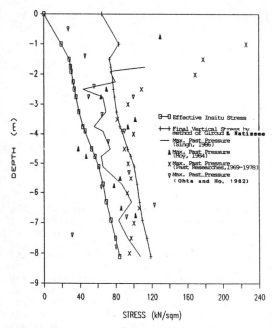

Fig. 2 Effective overburden and maximum
past pressure beneath the test
embankment.

the ground surface is assumed. From the borehole samples, the presence of fine sands and
silt lenses were observed in the soft clay
layer with few decomposed organic matters.
Bergado and Danzuka (1988) studied the macrofabric of AIT subsoil and obtained 2 domains,
namely: (a) almost homogeneous clay layer
with scattered fine sand and silt lenses
(mean = 0.2 % by area) from 2 to 7 m depth;
(b) almost heterogeneous clay layer containing higher density of fine sand and silt
lenses (mean = 20 %) from 7 to 9 m depth.

3. SOIL PROPERTIES

The index properties of the subsoil are
shown together with the soil profile in
Fig. 1. The maximum past pressure, p_c, are
plotted in Fig. 2, together with the values
obtained from previous investigations (Ohta
and Ho, 1982; Moy, 1984). The horizontal
coefficient of consolidation C_h was obtained
from consolidation tests with samples cut
perpendicular to the horizontal direction.
The variations of the vertical and
horizontal coefficients of consolidation, C_v
and C_h, respectively, are plotted in Fig. 3,
showing that C_h is about 3 to 4 times larger
than C_v. The resulting coefficients of
vertical and horizontal permeabilities were
found to be higher than the values obtained
from past studies (Ohta and Ho, 1982).

4. THE SUBSIDING ENVIRONMENT

The campus of the Asian Institute
Technology (AIT) is situated on a fla
deltaic marine deposit called the Ch
Phraya Plain in Central Thailand. The grou
subsidence of the Chao Phraya Plain
caused by the excessive groundwat
pumping for water supply through deep wel
(AIT, 1982). Presently, in AIT campus alon
1,500 cubic meters of water per day
pumped through deep wells from 200 m dep
(Bergado et al, 1984a). The ground subside
effects include differential settlemen
between structures on shallow foundatio
and on pile foundations, differenti
settlements in pavements and pathways, a
localized ground depressions in open fie
areas. Ground movements and buildi
settlements were monitored from precisi
levelling surveys (Bergado et al, 1987b).
average subsidence rate of 2.4 cm and
maximum of 8 cm per year were estimate
Nutalaya and Rau (1982) calculated
average subsidence rate of 6 cm per year a
a maximum of 10 cm per year in Bangkok are

5. SOIL IMPROVEMENT BY GRANULAR PILES

Granular piles are preferable when modera
increase in bearing capacity and/or
moderate reduction of settlements a
required for foundations on soft cl
subsoils. Moreover, their recharging effec
to minimize the problems caused
differential settlements such as cracking
concrete and asphalt pavements, as well
their effects of increasing the slo
stability factor of safety, has assum
position of greater significance a
interest for the past years. Barksdale a
Bacchus (1983) presented results of the
numerical models and case studies on vario
projects. Enoki (1987) presented the compa
ison of the experimental and analytic
results on the behavior of composite grou

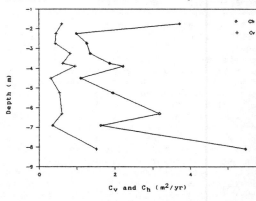

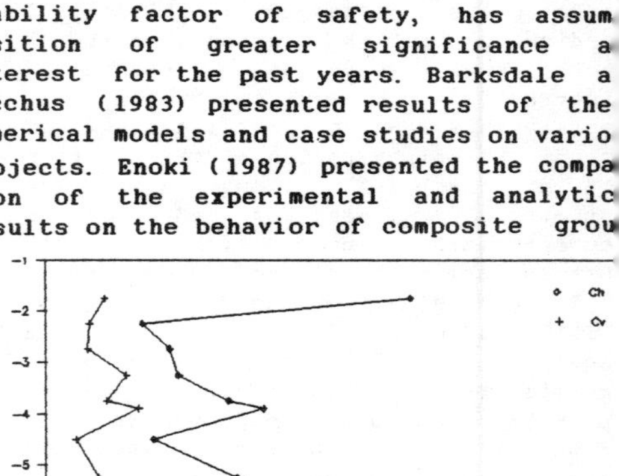

Fig. 3 Variation of C_v and C_h with depth.

consisting of granular pile and the
surrounding clay.

The gravel materials used to construct the
granular piles consisted of whittish-gray,
crushed limestones. The gravel was poorly
graded with a maximum size of 20 mm. The
compacted density varied from 1.70 to 1.81
t/m^3. The friction angles obtained from
direct shear tests varied from 39 to 45
degrees. The gravel materials and the
procedure of installation of the granular
piles were similar to that used by Bergado
and Lam (1987). The triangular pattern was
used. Full scale load tests were made by
Bergado et al (1987b) using rigid plate and
a stress concentration factor of more than 5
was calculated. However, in the test
embankment, a stress concentration factor of
2 was measured. Previous studies have shown
that the bearing capacity of the clay
subsoil increased by as much as 4 times in
the improved ground (Bergado et al, 1987a).

6. TEST EMBANKMENT ON GRANULAR PILES

A full scale test embankment was constructed
on granular pile foundation 10 days after
the installation of the last pile. The
granular piles were arranged in triangular
pattern with spacing of 1.50 m. The granular
piles have diameter of 0.30 m and length of
8.0 m fully penetrating the soft clay layer.
Prior to the construction of the test
embankment, the test site was instrumented
and the initial readings of all instruments
were recorded. The field instruments
consisted of piezometers, settlement plates,
inclinometers, etc. The embankment was
rectangular in plan with a first stage
height of 2.4 m and a second stage height of
4.0 m. It has dimensions of 13.7 m by 15.7 m
at the base and 1.3 m by 9.3 m at the top. A
drainage blanket of 0.25 m thick consisting
of clean sand was laid at the base on top of
the granular piles. The embankment was
compacted in layers by a light vibrating
plate tamper and was found to have an
average density of 2.0 t/m^3. The cross-
section of the test embankment is shown in
Fig. 4. For the second stage, the embankment
height was increased to 4.0 m after 345 days.

7. SOIL IMPROVEMENT BY VERTICAL BAND DRAINS

The principal alternative to large diameter
drain is the much smaller band-shaped drains
first employed by Kjellman (1948). A
prefabricated band drain consists of a
central core, whose function is to act as
free draining channel and to withstand
buckling stresses, enclosed by a thin
filter sleeve, which prevents the fine soil
particles from entering the central core but
allows free entry of pore water into the

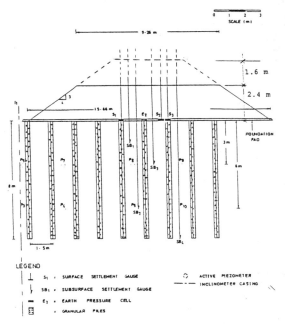

Fig. 4 Cross-section of test embankment
with granular piles and location of
field instruments.

core. The band drains are manufactured from
man-made polymer fabrics such as polyethel-
enes, polypropylenes, etc. Currently, there
are a number of available band drains. Mebra
drains were used in this study.

Barron (1948) presented a solution to
the problem of the consolidation of a soil
cylinder containing a central drain. The
problem formulated in terms of polar
coordinates yield the following differential
equation:

$$\frac{\partial u}{\partial t} = C_h \left[\frac{1}{r} \frac{\partial u}{\partial r} + \frac{\partial^2 u}{\partial r^2} \right] \qquad (1)$$

where C_h is the horizontal coefficient of
consolidation, r is the radial coordinate, u
is the excess pore water pressure and t
represents time. Barron (1948) suggested
two types of solutions based on either free
strain or equal vertical strain.
Jamiolkowski et al (1983) compared these two
solutions and indicated that almost similar
values of degree of consolidation were
obtained for the usual values of drain
spacing and soil compressibility. This
justifies the use of the less complicated
equal vertical strain solution. The average
degree of consolidation with respect to the
radial flow is given as:

$$U_h = 1 - \exp\left(-8 \, T_h/F(n)\right) \qquad (2)$$

181

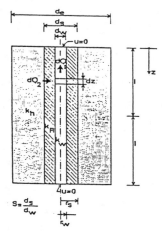

Fig. 5 Consolidation problem with well and smear effects (after Hansbo, 1979).

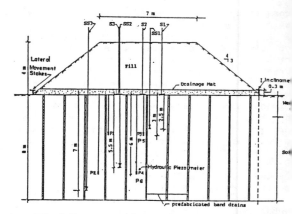

Fig. 6 Cross-section of test embankment with drains and location of field instruments.

In reality, the installation of the drain causes a disturbance around the drain depending on the size of the mandrel, the manner of installation, and the macrofabric of the clay which is called the smear effect. Also, headloss occurs in the flow along the drain due to the well-resistance. Solutions have been presented by Barron (1948) considering the effects of smear and well-resistance. Hansbo (1979) presented simplified solutions considering the effects of smear and well-resistance. The remoulding around the drain is included in the analysis by assuming an annulus of smeared clay with an internal diameter, d_s, surrounding the drain. Within the annulus, the soil is strongly remoulded and its coefficient of permeability, K_h', is lower than K_h, which characterized the original clay. Combining the effects of smear and well-resistance, F(n) takes the form:

$$F(n) = \ln(n/s) - (K_h/K_h') \ln(s) - \frac{3}{4}$$
$$+ Z(2\ell - Z) K_h/q_w \qquad (3)$$

where n is the ratio of the diameter of equivalent soil cylinder, d_e, to the equivalent nominal diameter of the drain, d_w, s is the ratio of smeared diameter, d_s, to the drain diameter, d_w, and the other terms are defined in Fig. 5.

8. TEST EMBANKMENT ON VERTICAL DRAINS

The test site is located inside the campus of the Asian Institute of Technology, 42 km north of Bangkok. The soil profile is as given in Fig. 1. The soil properties have been given in Figs. 2 and 3.

Before the installation of the drains, trenches of 0.6 m wide and 1.6 m deep were

excavated in the topmost weathered cla[y] layer to eliminate the arching effect[.] Mebra drains were installed in triangula[r] pattern at 1.5 m center to center spacing[.] The drains were installed by means of special mandrel down to 8.0 m depth t[o] fully penetrate the soft Bangkok clay laye[r] at the site. The size of the mandrel wa[s] minimized to reduce the smear effect. Th[e] rectangular shaped mandrel had inne[r] dimensions of 2.8 by 13.3 cm and oute[r] dimensions of 4.5 by 15.0 cm enough t[o] contain the 0.3 by 9.5 cm Mebra drains[.] Disposable shoes were installed at th[e] bottom of each drain for anchorage[.] Field instruments such as piezometers[,] surface and subsurface settlement gages[,] inclinometers, and lateral movement stake[s] were installed to monitor the behaviour o[f] the embankment foundation. Subsequently, [a] 4.0 m high test embankment with side slope[s] of 4H:3V was constructed. The base dimensio[n] consist of 14.6 by 16.6 m while the top dim[-] ensions were 5.0 by 7.0 m. The embankmen[t] construction lasted for one month. The ver[-] tical stress at the ground surface was calc[-] ulated to be 65.0 kN/m^3 due to the embankme[nt] loading. The detailed locations of the ins[-] trumentation and the typical section of th[e] embankment is shown in Fig. 6. The result o[f] this full scale loading test have bee[n] published elsewhere (Bergado et al, 1988).

9. COMPARISON OF SETTLEMENTS OF EMBANKMENTS ON GRANULAR PILES AND ON VERTICAL DRAINS

The observed settlements of test embankment[s] on granular piles and on vertical drain[s] after subtracting the effects of groun[d] subsidence are shown in Fig. 7.

At a height of 2.4 m, the maximum observe[d] settlements for embankment on granula[r] piles, 110 days after construction, wa[s]

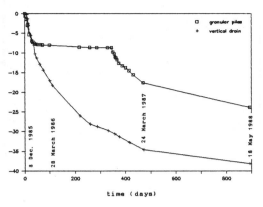

Fig. 7 Maximum observed settlements for both embankments after deleting subsidence effect.

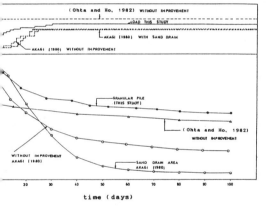

Fig. 8 Comparison of settlement records of embankment on granular piles with previous investigations.

Fig. 9 Test embankments on granular piles and plastic band drains.

about 44% of the maximum settlement observed for vertical drains. After 345 days, the settlement increment of the embankment on granular piles was closed to zero, with a maximum recorded settlement at the center surface settlement gauge of 8.65 cm, which is 29% of the maximum observed for the embankment on vertical drains.

After filling up of the embankment to reach 4.0 m height, it was observed that the settlements recorded, 470 days after the start of construction, was about 51% of the maximum observed settlement for vertical drains. A maximum surface settlement of 23.86 cm was then recorded about 900 days after construction for granular piles as compared to 38.20 cm for vertical drains. This amounted to about 62% settlement of granular piles embankment relative to the embankment on vertical drains.

Granular piles then, seemed to have reduced the settlement as compared to the vertical plastic band drains. This is probably due to the reason that granular piles are stiffer than the surrounding soil, such that their effect seems to reinforce the soil and support the embankment rather than the usual concept of providing drainage to the subsoil to accelerate consolidation.

10. COMPARISON WITH PAST INVESTIGATIONS

The time-settlement curves in this study from both test embankments were compared with the corresponding observations of the past studies of Akagi (1979) and Ohta and Ho (1982). As the settlements recorded in this study were made under different stress levels, it may not be appropriate to compare settlement magnitudes directly. However, relative comparisons can be made. As shown in Fig. 8 for instance, it can be seen that at 71 days after embankment construction, the embankment on granular piles with 2.4 m height settled nearly 40% less than that of the embankment of Akagi (1979) without improvement and about 20% less of the embankment of Ohta and Ho (1982). Thus, the granular piles seemed to function as reinforcement in the clay rather than drains. Figure 9 shows the two test embankments on improved soft Bangkok clay.

11. CONCLUSIONS

1) The settlement observed for the test embankment on granular piles was about 62% of the maximum observed for the embankment on vertical drains. Hence, granular piles seem to reduce settlements by acting as soil reinforcement to support the embankment, rather than providing drainage to accelerate the consolidation process.

183

2) Relative comparison with past investigations indicates that granular piles
decrease the settlement of soft clay
foundations by as much as 20 to 40%.
This confirms the idea that granular
piles function as reinforcement to the
clay rather than drains.

3) The stress concentration factor measured
in the granular pile foundation ranged
from 2 to 5 and was found to decrease
with the increasing applied load.

REFERENCES

AIT 1982. Investigation of land subsidence
caused by deep well pumping in the Bangkok
area (1978-1982). Final Report Submitted to
NEB. Bangkok, Thailand.

Akagi, T. 1979. Effect of displacement type
sand drains on strength and compressibility
of soft clays. Tokyo Univ. Publ. Japan.

Barksdale, R. D. & Bacchus, R. C. 1983. Design
and construction of stone columns: Vol. 1.
Re. No. FHWA/RD-83/026. NTIS, Virginia, USA.

Barron, R. A. 1948. Consolidation of fine-
grained soils by drain wells. Trans. ASCE.
113: 718-754.

Bergado, D. T., Balasubramaniam, A. S. & Apai-
pong, W. 1984a. Effects and investigations
of land subsidence in AIT campus, Bangkok,
Thailand. Proc. 3rd Int. Symp. on Land Subsidence, Venice.

Bergado, D. T., Rantucci, G. & Widodo, S. 1984b.
Full scale load tests on granular piles and
sand drains in the soft Bangkok clay.
Proc. Int. Conf. In-situ Soil and Rock
Reinforcements. Paris, France.

Bergado, D. T. & Lam, F. L. 1987. Full scale
load test of granular piles with different
densities and different proportions of gravel and sand in the soft Bangkok clay. Soils
and Foundations. 27: 1: 86-93.

Bergado, D. T., Huat, S. H. & Kalvade, S. 1987a.
Improvement of soft Bangkok clay using granular piles in subsiding environment. Proc.
5th Int. Geotech. Seminar Case Histories
in Soft Clay. Singapore.

Bergado, D. T., Khaw, L. G., Nutalaya, P. &
Balasubramaniam, A. S. 1987b. Subsidence
effects on infrastructures and settlement
predictions in the AIT campus, Chao Phraya
Plain, Thailand. Proc. 9th European Conf.
on Soil Mech. Found. Eng. Dublin, Ireland.

Bergado, D. T., Miura, N. & Danzuka, M. 1988a.
Reliability analysis of a test embankment by
variance reduction and nearest-neighbor methods. Proc. 6th ICONMIG. Innsbruck, Austria.

Bergado, D. T., Miura, N., Singh, N. & Panichayatum, B. 1988b. Improvement of soft
Bangkok clay using vertical band drains
based on full scale test. Proc. Int. Conf. Eng.
Problems of Regional Soils. Beijing, China.

Enoki, M. 1987. Consolidation characteristics

of composite ground. Proc. 8th As
Regional Conf. Kyoto, Japan.

Girroud, J. P. & Watissee, H. 1972. Stres
due to an embankment resting on a fir
layer of soil. Proc. 6th Aus. Road P
Board Conf. Paper No. 847. Canberra.

Hansbo, S. 1979. Consolidation of clay
band-shaped prefabricated drains. Gro
Eng. 5: 16-25.

Jamiolkowski, M., Lancellota, R. & Wolsk
1983. Precompression and speeding-up
consolidation. Proc. 8th European Co
Soil Mech. Found. Eng. Helsinki. 3: 1201-1

Kjellman, W. 1948. Accelerating consolid
ion of fine-grained soil by means of ca
board wicks. Proc. 2nd Int. Conf. Soil Me
Found. Eng. 2: 302-305.

Lam, F. L. 1985. Full scale test of granu
piles with different densities and diff
ent proportions of gravel and sand on s
clay. M. Eng. Thesis No. GT-84-19. A
Bangkok, Thailand.

Moy, W. Y. 1984. Properties of subsoil rel
to the stability and settlement of AIT
embankment. M. Eng. Thesis No. GT-83-
Bangkok, Thailand.

Nutalaya, P. & Rau, J. L. 1982. Quatern
geology and land use. Guidebook on P
Symposium Excursion. Ist Int. Symp. So
Geology and Landforms: Impact on Land U
Planning in Developing Countries. Bangk

Ohta, H. & Ho, Y. C. 1982. A trial embankm
on soft Bangkok clay, Phase I-IV. Resea
Report No. RR-146. AIT, Bangkok, Thaila

Panichayatum, B. 1987. Comparison of the
formance of embankments on granular pi
and vertical drains. M. Eng. Thesis
GT-86-8. AIT, Bangkok, Thailand.

International Geotechnical Symposium on Theory and Practice of Earth Reinforcement / Fukuoka Japan / 5-7 October 1988
© 1988 Balkema, Rotterdam. ISBN 90 6191 820 0

oad-carrying capacity of a soil layer supported by a geosynthetic overlying a oid

?Giroud
eoServices Inc. Consulting Engineers, Boynton Beach, Fla., USA

.Bonaparte, J.F.Beech & B.A.Gross
eoServices Inc. Consulting Engineers, Norcross, Ga., USA

BSTRACT: This paper presents equations and charts to design soil layer-geosynthetic
'stems overlying voids such as cracks, sinkholes, and cavities. These equations and
aarts were developed by combining tensioned membrane theory (for the geosynthetic) with
'ching theory (for the soil layer), thereby providing a more realistic design approach
aan one that considers tensioned membrane theory only.

INTRODUCTION

1 Description of the problem

a many practical situations, a load is
aplied on a soil layer-geosynthetic system
aat will eventually overlie a void. (In
ais paper, the term "void" is used
enerically for cracks, cavities,
epressions, etc.) Typical examples
aclude road embankments or lining systems
anstructed on foundations where localized
absidence or sinkholes may develop after
anstruction.

The design engineer has to verify that
ae load is adequately supported by the
ail-geosynthetic system, should the
absidence or sinkhole develop.

2 Scope of this paper

ais paper presents equations and charts
ar the case of a soil layer subjected to a
aiformly distributed normal stress and
verlying either an infinitely long void
olane-strain problem) or a circular void
axisymmetric problem). The parameters
ansidered in this paper are (Figure 1): b
width of the infinitely long void; r =
adius of the circular void; H = thickness
f the soil layer; γ = unit weight of the
ail; φ = friction angle of the soil (soil
ahesion is not considered); q = uniformly
istributed normal stress applied on the
op of the soil layer; y = geosynthetic
eflection; and α = geosynthetic tension
force per unit width) corresponding to the
eosynthetic strain, ε.

1.3 Prior work

The use of tensioned membrane theory to
evaluate the load-carrying capacity of a
geosynthetic bridging a void was presented
by Giroud (1981). Subsequently, Giroud
(1982) developed a design chart based on
tensioned membrane theory. This chart has
often been used to evaluate the load-
carrying capacity of a soil layer
associated with a geosynthetic. By doing
so, the internal shear strength of the soil
layer is neglected and this can be very
conservative. Therefore, Bonaparte and
Berg (1987) have suggested that arching
theory (for the soil layer) be combined
with tensioned membrane theory (for the
geosynthetic) to enable a more realistic
design approach.

This paper significantly extends the
earlier work of Giroud (1981 and 1982) and
Bonaparte and Berg (1987) and provides the
most extensive analysis yet of a soil-
geosynthetic system bridging a void.

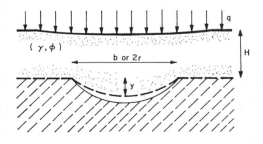

Fig. 1 Schematic cross section

1.4 Load carrying mechanism

The soil and underlying geosynthetic are assumed to initially be resting on a firm foundation. At some point in time, a void of a certain size opens below the geosynthetic. Under the weight of the soil layer and any applied loads the geosynthetic deflects. The deflection has two effects, bending of the soil layer and stretching of the geosynthetic.

The bending of the soil layer generates arching inside the soil, which transfers part of the applied load away from the void area. As a result, the stresses transmitted to the geosynthetic over the void area are smaller than the pressure due to the weight of the soil layer and applied stresses.

The stretching of the geosynthetic mobilizes a portion of the geosynthetic's strength. As a result, the geosynthetic acts as a "tensioned membrane" and can carry a load normal to its plane.

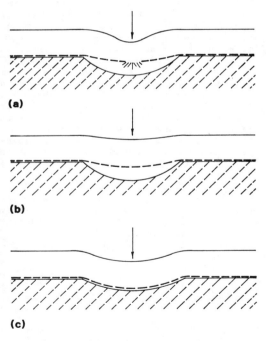

(a)

(b)

(c)

Fig. 2 Three design situations: (a) the soil -geosynthetic system fails; (b) the soil-geosynthetic system exhibits limited deflection and bridges the void; and (c) the soil-geosynthetic system deflects until the geosynthetic comes in contact with the bottom of the void

As a result of geosynthetic stretchi three cases can be considered: (i) soil-geosynthetic system fails (Figure 2 (ii) the soil-geosynthetic system exhib some limited deflection and bridges void (Figure 2b); and (iii) the so geosynthetic system deflects until geosynthetic comes in contact with bottom of the void (Figure 2c). In latter case, the mobilized portion of geosynthetic strength carries a portion the load applied normal to the surface the geosynthetic. The rest of the load transmitted to the bottom of the void.

2   ANALYSIS

2.1 Approach

The problem under consideration involve complex soil-geosynthetic interaction. problem can be greatly simplified, howev if the soil response (arching) is uncoup from the geosynthetic response (tensio membrane). Therefore, a two-step appro is used. First, the behavior of the s layer is analyzed using classical arch theory. This step gives the pressure the base of the soil layer on the port of the geosynthetic located above the vo Second, tensioned membrane theory is u to establish a relationship between pressure on the geosynthetic, the tens and strain in the geosynthetic, and deflection.

An inherent assumption in this uncoup two-step approach is that the s deformation to generate the soil arch compatible with the tensile strain mobilize the geosynthetic tension. T assumption has not been verified.

2.2 Arching theory

Terzaghi (1943) has established following equation for arching in the c of an infinitely long void, assuming t the lateral load transfer is achie through shear stresses along verti planes located at the edges of the void:

$$p = \frac{\gamma\, b}{2\, K\, \tan\phi} \left[1 - e^{-2\, K\, \tan\phi\, H/b} \right]$$
$$+ q\, e^{-2\, K\, \tan\phi\, H/b}$$

where: p = pressure on the geosynthe over the void area; K = coefficient lateral earth pressure; and other notati as defined in Section 1.2.

Using the same approach, Kezdi (1975) has established that Equation 1 can be used for a circular void if b is replaced by r (and not by 2r), which shows that arching is twice as significant for a circular void as compared to an infinitely long void.

Selection of a value for the coefficient of lateral earth pressure, K, is not easy since the state of stress of the soil in the zone where arching develops is not well known. Handy (1985) has proposed the following value:

$$K = 1.06 \ (cos^2\theta + K_a \ sin^2\theta) \qquad (2)$$

where: $\theta = 45° + \phi/2$, and $K_a = tan^2(45° - \phi/2)$.

Equation 2 was used previously by Bonaparte and Berg (1987). Another approach consists of using the coefficient of earth pressure at rest, expressed as follows, according to Jaky (1944):

$$K = 1 - sin\phi \qquad (3)$$

In Equation 1, K is always multiplied by $tan\phi$. Calculations carried out using Equations 2 and 3, show that $Ktan\phi$ does not vary significantly with ϕ, if ϕ is equal to or greater than 20°, which is the case for virtually all granular soils. The calculations show that a constant value of 0.25 can be conservatively used for $Ktan\phi$. As a result, Equation 1 becomes:

$$p = 2 \ \gamma b \ (1 - e^{-0.5H/b}) + q \ e^{-0.5H/b} \qquad (4)$$

Equation 4 is also valid for the circular void if b is replaced by r, and it was used to establish the chart given in Figure 3.

2.3 Tensioned membrane theory

The tensioned membrane theory has been used by Giroud (1981, 1984) to deal with the case of a geosynthetic overlying a void and subjected to a uniformly distributed stress normal to its surface.

In the case of an infinitely long void, the deflected shape of the geosynthetic is circular, the strain uniform, and the following relationship exist if y/b < 0.5:

$$1 + \epsilon = 2 \ \Omega \ sin^{-1} [\ 1/(2 \ \Omega) \ ] \qquad (5)$$

where: ϵ, y, and b are as defined in Section 1.2; and Ω = is a dimensionless factor defined by:

$$\Omega = (1/4) \ [2y/b + b/(2y)] \qquad (6)$$

As a result of Equations 5 and 6, there is a unique relationship between y/b, ϵ, and Ω, which is given in Figure 4.

Giroud (1981, 1984) has also shown that the tension in the geosynthetic, in the case of an infinitely long void, is given by:

$$\alpha = p \ b \ \Omega \qquad (7)$$

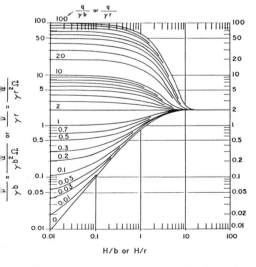

Fig. 3 Pressure, p, on the geosynthetic and geosynthetic tension, α

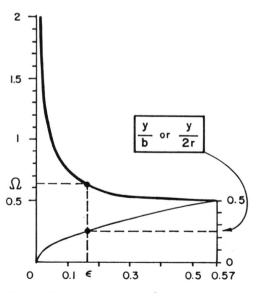

Fig. 4 Dimensionless factor Ω

187

As described by Giroud (1981), the deflected shape of the geosynthetic is not a sphere in the case of a circular void. As a consequence, incorporating 2r (diameter) instead of b (width) into Equations 5 and 6 gives only an approximate value of the average geosynthetic strain, ϵ.

Since the strain is not uniform, the tension, α, in the case of a circular void is not uniformly distributed in the geosynthetic and its average value is given approximately by Equation 7, with r substituted for b (Giroud, 1981, 1984).

It should be noted that Equation 7 can be used for a circular void only if the geosynthetic has isotropic tensile characteristics, i.e., the same tensile characteristics in all directions. If this is not the case, recommendations given in Section 4.1 should be followed.

3   SOLUTION OF TYPICAL DESIGN PROBLEMS

Typical design problems can be solved using the following equations which were obtained by combining Equations 4 and 7. In all the design cases considered below, the solution depends on the value of Ω which depends either on the allowable geosynthetic strain, ϵ, or on the allowable deflection, y.

In this section, the depth of the void is assumed to be such that the geosynthetic is not in contact with the void bottom. The case where the geosynthetic comes in contact with the bottom of the void is more complex and will not be addressed in this paper.

3.1 Determination of geosynthetic properties

The relevant equation for an infinitely long void is:

$$\alpha/\Omega = pb = 2\gamma b^2 (1 - e^{-0.5H/b}) + q\, b\, e^{-0.5H/b} \tag{8}$$

Equation 8 can be rewritten in a dimensionless form as follows:

$$\frac{\alpha}{\gamma b^2 \Omega} = \frac{p}{\gamma b} = 2(1 - e^{-0.5H/b}) + \frac{q}{\gamma b} e^{-0.5H/b} \tag{9}$$

This equation, which is related to the infinitely long void, was used to establish the chart in Figure 3.

Equations 8 and 9 can be used for circular void if b is replaced by r.

The above equations can be used to sol problems which consist of determining t required geosynthetic tension, α, for given strain, ϵ, when all other paramete are given (b or r, q, H, and γ Alternatively, the chart given in Figure can be used.

3.2 Determination of soil layer thicknes

The relevant equation for an infinite long void is:

$$H = 2\, b\, Log\, \frac{[q/(\gamma b)] - 2}{[\alpha/(\gamma b^2 \Omega)] - 2} \tag{1}$$

The same equation can be used for circular void by substituting r for b.

The above equation can be used to sol problems which consist of determining t required soil layer thickness, H, when ϵ other parameters are given (b or r, q, α, and ϵ). Alternatively, the chart giv in Figure 3 can be used.

3.3 Determination of maximum void size

There is no simple equation giving the v size (b or r) as a function of the otl parameters. In order to determine t maximum void size that a given soil lay geosynthetic system can bridge, it necessary to solve Equation 8 by trial error. To facilitate the process, a cha has been established (Figure 5) rewriting the two parts of Equation 9 i dimensionless form as follows:

$$\frac{p}{\gamma H} = \frac{2(1 - e^{-0.5H/b})}{H/b} + \frac{q}{\gamma H} e^{-0.5H/b} \tag{1}$$

$$\frac{p}{\gamma H} = \frac{\alpha}{\gamma H^2 \Omega} \frac{H}{b} \tag{1}$$

In Figure 5, Equation 11 is represent by a family of curves and Equation 12 represented by a family of straight lix at 45°. For a given set of parameters, t abscissa of the intersection between t relevant curve and the relevant straig line gives the maximum value of the widt b, of an infinitely long void, or radius, r, of a circular void.

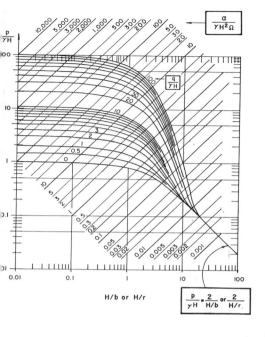

H/b or H/r

$$\frac{p}{\gamma H} = \frac{2}{H/b} \text{ or } \frac{2}{H/r}$$

Fig. 5 Chart for maximum void size determination

4 Determination of the maximum load

The relevant equation for an infinitely long void is:

$$q = 2\gamma b + \left\{ \frac{[\alpha/(\gamma b^2 \Omega)] - 2}{e^{-0.5H/b}} \right\} \gamma b \quad (13)$$

The same equation can be used for a circular void by substituting r for b.

The above equation can be used to solve problems which consist of determining the maximum uniform normal stress, q, which can be applied on the top of the soil layer, when all other parameters are given (b or r, H, γ, α, and ϵ). Alternatively, the charts given in Figure 3 or 5 can be used.

DISCUSSION

1 Anisotropic geosynthetic

Special precautions must be taken when using the equations and charts presented in this paper for anisotropic geosynthetics.

In the case of a long void, the geosynthetic should be installed with its stronger direction perpendicular to the length of the void since, theoretically, no strength is needed in the direction of the length of the void (according to the plane-strain model which corresponds to an infinitely long void). However, some strength is required lengthwise in places where the actual situation departs from a pure plane-strain situation, for instance near the end of the void.

In the case of a circular void, the tensioned membrane equation (Equation 7) is valid only if the geosynthetic has isotropic tensile characteristics. For practical purposes, Equation 7, and other equations as well as charts related to circular voids, can be used for woven geotextiles and biaxial geogrids that have similar tension-strain curves in two perpendicular directions. For woven geotextiles and biaxial geogrids that have different tensile characteristics in the two principal directions, two cases can be considered with circular voids, depending on the ratio between the geosynthetic tensions at the design strain in the weak and the strong directions: (i) if the ratio is more than 0.5, α should be taken equal to the tension in the weak direction; and (ii) if the ratio is less than 0.5, α should be taken equal to half the tension in the strong direction.

The rationale in the first case is conservativeness. The rationale in the second case is as follows. Comparison of Equation 7 written with b and the same equation written with r shows that, if the geosynthetic tension in one direction is less than half the tension in the other direction, the system placed over a circular void behaves as if it were on an infinitely long void with a width, b, equal to the diameter, 2r, of the actual void. Therefore, Equation 7 must be used, in this case, with 2r (instead of b) and α, or with r and $\alpha/2$, as recommended above.

There is another consideration when an anisotropic geosynthetic is used over a circular void. The complex pattern of strains in the geosynthetic resulting from different tensions in different directions may have a detrimental effect on the behavior of the geosynthetic. Therefore, it is recommended that, for holes which can be modeled as circular, isotropic geosynthetics (such as most nonwoven geotextiles) or "practically isotropic" geosynthetics (such as woven geotextiles or biaxial geogrids having similar tension-strain curves in the two principal directions) be used.

189

4.2 Influence of soil layer thickness

The influence of the thickness of the soil layer is illustrated in Figure 3. Three cases can be considered.

If the applied stress, q, is large (i.e., q > 2γb or 2γr), the pressure, p, on the geosynthetic, and consequently the required geosynthetic tension, α, decrease toward a limit as the soil layer thickness increases. In this case, it is beneficial to increase the thickness of the soil layer. For each particular situation, the amount by which the thickness should be increased can be determined using the chart given in Figure 3. This chart shows that it would be useless to increase the soil layer thickness beyond a limiting value of H = 20 b or 20 r.

If the applied stress, q, is small (i.e., q < 2γb or 2γr), the pressure, p, on the geosynthetic, and consequently the required geosynthetic tension, α, increase toward a limit as the soil thickness increases. In this case, it is detrimental, from the perspective of the design of the geosynthetic, to increase the thickness of the soil layer. (This is because the added load due to soil weight is not fully compensated by the effect of soil arching.)

If the applied stress, q, equals 2γb or 2γr, the pressure, p, on the geosynthetic remains constant and equal to q, regardless of the soil layer thickness.

The limit values for p and α are independent of the applied stress, q. The limit value for p is 2γb for an infinitely long void or 2γr for a circular void. The limit value for α is $2\gamma b^2\Omega$ for an infinitely long void or $2\gamma r^2\Omega$ for a circular void.

5   CONCLUSION

The analysis shows that the thickness of the soil layer associated with the geosynthetic plays a significant role. In contrast, the soil mechanical properties do not. It should not be inferred, however, that any soil will provide the same degree of arching. The equations used to prepare the tables and charts assume that the friction angle of the soil is at least 20°. Granular soils virtually always meet this condition. However, they should be well compacted to ensure arching because loose granular soils tend to contract when they are sheared or vibrated, which may destroy the arch.

Further refinements of the meth presented herein can be considered. F instance, it is possible that the degree soil arching depends on the geosynthet strain, whereas the method presented this paper does not consider the concept degree of soil arching. Also, the meth could be expanded to include cohesi soils. (The equations and charts present in this paper are essentially intended f granular soils; however, they can be us for saturated cohesive soils in the drain state, assuming that their cohesion is ze and provided that their drained fricti angle is greater than 20°.)

In spite of its limitations, the meth presented in this paper is believed to be useful tool for engineers designing soi geosynthetic systems resting on subgrad which may subsequently develop voids.

ACKNOWLEDGMENTS

The authors are indebted to G. Saunders, Mozzar, G. Kent, and A.H. Perry for assi tance during the preparation of this pape

REFERENCES

Bonaparte, R., and Berg, R.R. 1987. T use of geosynthetics to support roadwa over sinkhole prone areas. Proceedin of the Second Multidisciplina Conference on Sinkholes and t Environmental Impacts of Karst, Orland 437-445.

Giroud, J.P. 1981. Designing wi geotextiles. Materiaux Constructions, 14, 82: 257-272.

Giroud, J.P. 1982. Design of geotextil associated with geomembrane Proceedings of the Second Internation Conference on Geotextiles, 1, Las Vega 37-42.

Giroud, J.P. 1984. Geotextiles a geomembranes, definitions, properti and design. I.F.A.I. Publisher, S Paul, MN.

Handy, R.L. 1985. The arch in sc arching. Journal of Geotechnic Engineering, ASCE, 111, 3: 302-318.

Jaky, J. 1944. The coefficient of ear pressure at rest. Journal for Socie of Hungarian Architects and Engineer Budapest: 355-358 (in Hungarian).

Kezdi, A. 1975. Lateral earth pressur Foundation Engineering Handbook, Edit by Winterkorn, H.F. and Fang, H.Y., \ Nostrand Reinhold, New York: 197-220.

Terzaghi, K. 1943. Theoretical sc mechanics. John Wiley, New York.

International Geotechnical Symposium on Theory and Practice of Earth Reinforcement / Fukuoka Japan / 5-7 October 1988
© 1988 Balkema, Rotterdam. ISBN 90 6191 820 0

Prediction of bearing capacity in level sandy ground reinforced with strip reinforcement

Ching-Chuan Huang
University of Tokyo, Tokyo, Japan

Fumio Tatsuoka
Institute of Industrial Science, University of Tokyo, Japan

ABSTRACT: The bearing capacity of footing placed on level ground can be increased by placing horizontal layers of tensile reinforcing strips beneath the footing. A series of model tests were performed in order to investigate the reinforcing mechanism and develop a method for calculating the increase in the bearing capacity by reinforcing.

1 INTRODUCTION

The failure mechanism of level ground reinforced with long strips was assumed by Binquet and Lee (1975 a, b) as illustrated in Fig. 1 ; the active zone including a part of the strips was anchored by the remaining parts of strips located outside the active zone. It indicates that only the strips extending beyond the footing width can increase the bearing capacity.

The failure mechanism of reinforced level sandy ground loaded by a strip footing observed in the present study was very different from that mentioned above; at the peak footing load, shear bands were developed only in a limited area beneath the footing, with small strains outside the active zone. Further, by restraining possible strains in soil in the zone beneath the footing by means of short strips with the same length as the footing width, the bearing capacity was inceased remarkably. Based on these results, a new stability analysis method by using limit equilibrium method was developed.

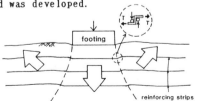

Fig. 1 Failure mechanism assumed by Binquet and Lee 1975 a, b.

2 MODEL TEST ARRANGEMENT

Model grounds were constructed by pluviating air-dried Toyoura sand through air at a controlled fall height in a sand box as shown in Fig. 2. By this method, homogeneous models having very similar dry densities γ_d were produced. Toyoura sand has a mean diameter of 0.16mm, a coefficiet of uniformity of 1.46 and a sub-angular to angular particle shape. The strength and deformation properties have been thoroughly studied by Tatsuoka et al. (1986).

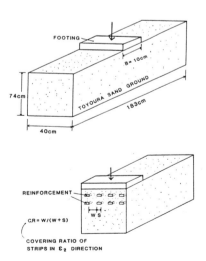

Fig. 2 Model test arrangement.

A rigid, rough, strip footing, which was 10 cm wide, guided against tilting was loaded at a constant axial displacement of 0.1mm-0.2mm/min. The side walls consisted of 3cm-thick transparent acryl plates with steel stiffeners outside. In order to reduce the effect of side-wall friction, the inside surface of each acryl plate was lubricated by means of a 0.05mm thick silicone grease layer and a 0.2mm thick membrane. Further, only the load at the central third of the footing was measured.

On the outside surface of membrane, 1cm-square grids were drawn. Strain fields on the lateral plane of model were constructed from the displacements of the nodes read to an accuracy of about 0.01mm using the pictures, which were taken occasionally during each test. From these pictures, shear bands also were identified.

Reinforcing strips, 0.5mm thick and 3mm wide, were made of phosphor bronze. Their surfaces were made rough by glueing model sand particles. The tensile forces were measured by means of strain gages attached to the strips.

The following four groups of tests were performed(see Fig.3):

Group-a: to study the reinforcing effect of short strips having the same length as the footing width B(=10cm). Additionally, three tests in which footing was placed on unreinforced ground were performed as reference tests.

Group-b: to study the effect of the length of reinforcement.

Group-c: to study the effect of the number of reinforcement layers.

Group-d: to study the effect of density of reinforcing strips(the effect of covering ratio CR as explained in Fig.2) .

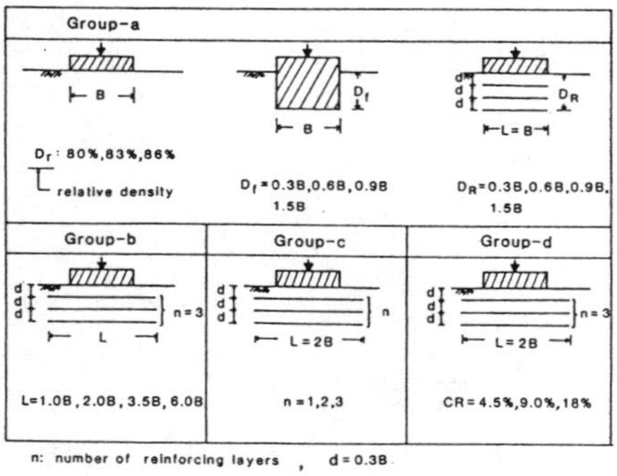

Group-a		
D_f: 80%,83%,86% relative density	D_f=0.3B,0.6B,0.9B, 1.5B	D_R=0.3B,0.6B,0.9B, 1.5B
Group-b	Group-c	Group-d
L=1.0B,2.0B,3.5B,6.0B	n=1,2,3	CR=4.5%,9.0%,18%

n: number of reinforcing layers , d=0.3B

Fig.3 Groups of test in present study.

3. FAILURE MECHANISM OF UNREINFORCED GROUND LOADED BY SURFACE OR DEEP FOOTING

The relationships between the normalized footing pressure N and s/B (footing settlement/ footing width)for Group-a are shown in Fig.4. Here $N=2*q/(\gamma_d*B)$, where q is the average footing pressure. It may be seen that the strength of reinforced ground increases in a very similar way to that by deeping the footing to the depth which is equivalent to that of the bottom layer of reinforcement(D_R).

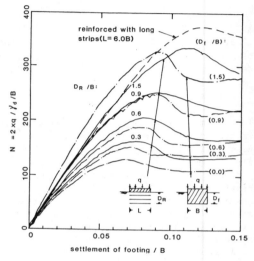

Fig.4 Results of Group-a .

Fig.5 shows the strain field in the unreinforced ground near the peak footing load. The lengths and directions of the two crossed segments represent the magnitudes and directions of the major and minor (extensional) principal strains developed for s/B=0.0 -> 0.07. The triangle beneath the footing is the wedge observed. Intensely strained bands may be seen only along the wedge(or active zone). Outside the active zone, only very limited strains can be seen.

This result suggests that the failure in such an unreinforced sand ground is rather progressive and the axial compressive strength of the active zone beneath the footing controls the bearing capacity of the ground.

Thus, it was assumed that the bearing capacity q_u for unreinforced ground loaded by a surface footing can be expressed by:

$$q_u = q_1 + q_2 \qquad (1)$$

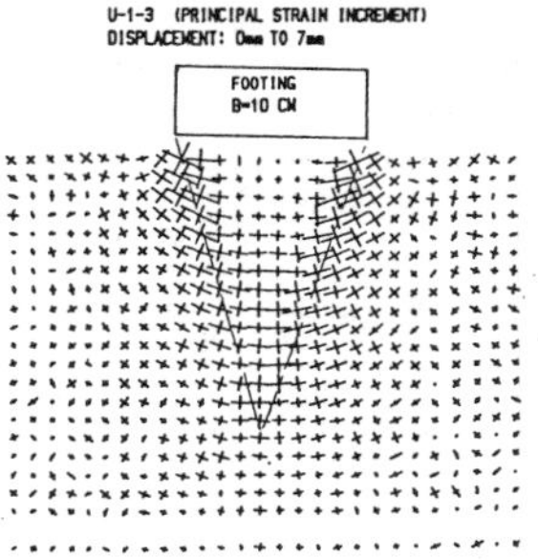

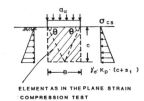

Fig. 5 Strain field for unreinforced ground at peak loading.

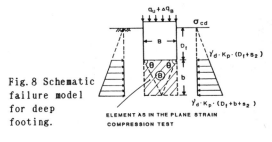

Fig. 8 Schematic failure model for deep footing.

Fig. 6 Schematic failure model for unreinforced ground.

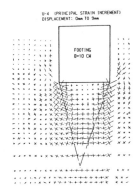

Fig. 7 Strain field for deep footing at peak loading.

Also in the unreinforced ground beneath a deep footing with a footing depth $D_f = 0.9B$, a strain field(see Fig. 7) similar to the one seen in Fig. 5 was observed. Thus, a similar equation to Eq. 1 was assumed for the bearing capacity q_B for a deep footing:

$$q_B = q_3 + q_4 \qquad (4)$$

q_3 = the compressive strength of a block (denoted by B in Fig. 8)beneath the deep footing, which is considered to behave like an element in a PSC test;

$$q_3 = K_p * \sigma_{cd} \qquad (5)$$
$$\sigma_{cd} = \gamma_d * (2D_f + b + 2s_2)/2 \qquad (6)$$

D_f, b, s_2: the depth of footing, the length of block and the settlement of footing at failure(see Fig. 8).

4 FAILURE MECHANISM OF REINFORCED GROUND LOADED BY A SURFACE FOOTING

Fig. 9 shows the strain field for the ground densely reinforced with short strips of L=B (Group-a) with a depth of the reinforced zone of $D_R = 0.9B$. It may be seen that this strain field is very similar to that for a deep footing with the same depth $D_f = 0.9B$ shown in Fig. 7. It may also be seen from Fig. 4 that each ground reinforced with short strips of L=B behave very similarly to the corresponding deep footing of $D_f = D_R$. This trend exists especially in the case of $D_R \leqq 0.9B$, and such a reinforcing effect may be called "the deep footing effect".

Fig. 10 shows the strain field for the ground densely reinforced with long strips of L=6B (Group-b) with a depth of the reinforced zone D_R equal to 0.9B. It may be seen that this strain field is similar to that for the case of short strips shown in Fig. 9 in the sense that the strain in the reinforced zone beneath the footing is effectively restrained and below the reinforced zone an intensely strained zone is formed.

q_1: the compressive strength of a block including the active zone, which is considered to behave like an element in the plane strain compression(PSC) test(Fig.6), and is assumed to be expressed by :

$$q_1 = K_p * \sigma_{cs} \qquad (2)$$
$$K_p = \tan^2(45° + \phi/2)$$

ϕ = the internal friction angle of sand in the corresponding PSC test

$$\sigma_{cs} = \gamma_d * K_p * (c + s_1)/2 \qquad (3)$$

γ_d = dry density of sand

c, s_1 = the length of block and the settlement of footing at failure (see Fig. 6).

q_2 is the bearing capacity component induced by the following factor, such as the friction at the lateral faces of the block element which do not exist in a PSC test.

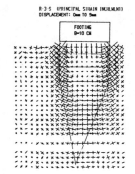

Fig. 9 Strain field for the ground densely reinforced with short strips.

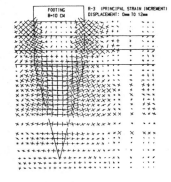

Fig. 10 Strain field for the ground densely reinforced with long strips.

As seen from Fig. 4, only small strength increase can be obtained by increasing the length of reinforcing strips from 1.0B to 6.0B. This result implies that the major part of the increase in the bearing capacity by reinforcing is due to "the deep footing effect", and a further increase is due to the side wall friction on the restrained zone.

The patterns of shear band observed at the peak footing load in all reinforced grounds can be claassified into the following two types, depending on the density of strips.

For densely reinforcing conditions (for either short or long strips), shear bands starting from the footing edges extends straightly downward approximately to the depth D_R, then form a wedge beneath the reinforced zone. This failure mode is called Failure mode-1. In this case, the bearing capaciy of reinforced ground is controlled by the strength of the zone including the wedge denoted by B (see Fig. 11 a).

For lightly reinforcing conditions, the shear bands starting from the footing edges form a wedge within the reinforced zone, but the apex of the wedge is deeper than that for the unreinforced ground (see Fig. 11 b).

This failure mode is called Failure mode-2 In this case, the bearing capacity of the reinforced ground is conrolled by the strength of the block A immediately beneath the footing.

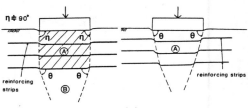

(a) Failure mode-1
(densely reinforcing)

(b) Failure mode-2
(lightly reinforceing)

Fig. 11(a), (b) Failure modes observed in the present study.

5 ANALYSIS OF REINFORCING EFFECT

The observed angles θ of shear bands defined from the horizontal line were slightly scattered as $\theta = 45° + \phi/2 + (3° \sim 7°)$ (ϕ is the angle of internal friction in the condition of σ_1-direction vertical to the bedding plane, refer to Tatsuoka et al. 1986. In the present study, ϕ is in the range of 48.8° — 50.2° . Since the effect of the small variation of θ (3° ~ 7°) on the results of analysis was found negligible, in the following, $\theta = 45° + \phi/2$ was assumed.

5.1 Increase in bearing capacity for Failure mode-1 (deep footing effect)

From Eqs. (1) and (4), the increace in the bearing capacity due to deeping a rigid footing is
$$\Delta q_B = q_B - q_u = q_3 - q_1 + (q_4 - q_2) \quad (7)$$
By assuming that $q_2 - q_4$ is relatively small and can be neglected, we obtain:
$$\Delta q_B = q'_3 - q_1 \quad (8)$$
For Failure mode-1, the increace in the bearing capacity due to reinforcing also is obtained by Eq. 8.

The results of the analysis are shown in Fig. 12, in which,
$$BCR_B (predicted) = (q_u + \Delta q_B)/q_u. \quad (9)$$
q_u is obtained from the reference test in Group-a for the same sand density as that for Δq_B by using the method of interpolation. In the following, the same method was used for obtaining q_u. In Fig. 12 the measured bearing capacities of deep footing in unreinforced grounds $q_u (D_f > 0)$ are

194

expressed by:

$$BCR(measured) = q_u (D_f > 0) / q_u \qquad (10)$$

It may be seen from Fig.12 that the values predicted by Eq.(8) agree fairly well with the measured values both for the deep footings and the reinforced grounds. Thus, it is concluded that the effect of densely reinforcing with short strips is very similar to that by deeping the footing to the depth of the bottom layer of reinforcement.

It is assumed that for the case of long strips (L>B), the tensile forces in the reinforcements outside the block A increase the normal forces on the side wall of Element A, and hence increase the upward side wall friction ΔS(see Fig.13). Thus, the bearing capacity increase for a ground reinforced with strips of L>B, Δq_c, is expressed by:

$$\Delta q_c = \Delta q_B + \Delta S \qquad (9)$$
$$\Delta q_B : \text{from Eq.(8)}$$
$$\Delta S = 2* \{ \sum_{i=1}^{n} T_{e.i} * \tan\phi * N_i \} /B \qquad (10)$$

n : the number of layers

N_i : the number of strips per unit length in the σ_2 direction at layer i.

B : the width of footing

$T_{e.i}$: the tensile force in strip i at the side faces of Element A. It is zero for L=B.

In the following analysis, the tensile fores measured at peak footing load are used In the case of L=B, ΔS is equal to 0, thus $BCR_B = BCR_C$.

Similarly, The predicted Δq_c can be expressed by:

$$BCR_c(predicted) = (q_u + \Delta q_c) / q_u$$

and the measured bearing capacity for reinforced grond is expressed by:

$$BCR(measured) = q_u(reinforced)/q_u$$

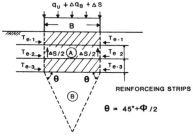

Fig.13 Schematic model showing the increase in side wall friction.

5.2 Increase in bearing capacity for Failure mode-2

It is assumed that the increase in the compressive strength of Element A (Δq_A) is due to the increase in the lateral confinement induced by tensile forces in the strips, and is expressed by:(see Fig.14),

$$\Delta q_A = K_p * \sigma_t \qquad (11)$$
$$\sigma_t = \{ \sum_{i=1}^{n} (T_{av.i} * N_i) \} /D_R \qquad (12)$$

N_i : the number of strips per unit length in σ_2 direction at layer i

$T_{av.i}$: averaged tensile force at layer i (= $(T_{max.i} + T_{e.i}) /2$) in Block A

$T_{max.i}$ and $T_{e.i}$ are the maximum value of tensile force at the center and at the side of Element A in strip i. $T_{e.i}$ is zero for L=B.

Again, the effect of reinforcing for Failure mode-2 can be expressed by :

$$BCR_A(predicted) = (q_u + \Delta q_A) / q_u$$

Base on this method, BCR_A and BCR_C for each test were calculated and the smaller one was used for evaluating the effect of reinforcing. It may be seen from Figs.15,16, and 17 that the prediction is satisfactory.

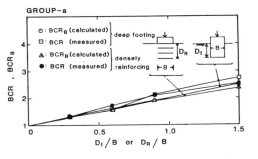

Fig.12 The results of Group-a.

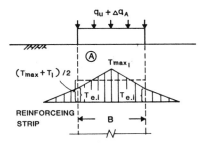

Fig.14 Schematic model showing the strength increase in reinforced zone A.

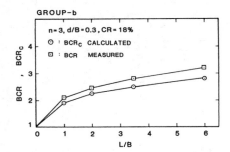

Fig. 15 The results of Group-b.

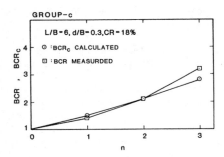

Fig. 16 The results of Group-c.

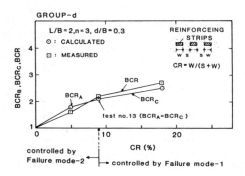

Fig. 17 The results of Group-d.

Theoretically, there exists an optimumn condition for designing when $BCR_A = BCR_C$. In the present study, Test No. 13 (L/B=2, D_R/B=0. 9 CR=9%, see Fig. 17) was found to be the optimum one.

6 CONCLUSION

The failure mechanism in a level ground reinforced with horizontal strips and the analysis of reinforcing effect on the bearing capacity have been presented. In practical conditions, the prediction of tensile forces in reinforcements is essential. An empirical method based on these tests has been developed and will be presented elsewhere in the future.

ACKNOWLEDGEMENT

This paper is a part of Master of Engineering thesis presented by the first author Appreciation is given to Japanese Rotary Yoneyama Foundation which offered a scholorship to the first author from April, 1986 to September, 1987.

REFFERENCES

1) J. Binquet and K. L Lee, 1975a. "Bearing capacity tests on reinforced slabs". Journal of Geotechnical Engineering Devision, ASCE, vol. 101, No. GT12, pp. 1241-1255

2) J. Binquet and K. L. Lee 1975b. "Bearing capacity tests on reinforced slabs". Journal of Geotechnical Engineering Devision, ASCE, vol. 101, No. GT12, pp. 1257-1276

3) Vito A. Guido, Dong K. Chang, and Michael A. Sweeney. "Comparison of geogrid and geotextile reinforced slabs". Canada Geotechnical Journal, vol. 23, No. 4, November 1986, pp. 435-440

4) Ching-Chuan Huang and Fumio Tatsuoka. "Deep-footing effect in bearing capacity of sand reinforced with tensile reinforcement". The 22nd Japan National Conference on Soil mechanics and foundation engineering, June, 1987

5) Kazuo Tani and Fumio Tatsuoka. "Effect of pressure level and model size on bearing capacity of model strip footing on sand". The 22nd Japan National Conference on Soil Mechanics and Foundation Engineering, June, 1987

6) Fumio Tatsuoka, Makoto Sakamoto, Taizo Kawamura and Shinji Fukushima . "Strength and deformation chacteristics of sand in plane strain compression at extremely low pressures". JSSMFE Soils and Foundations, Vol. 26, No. 1, pp. 65-84, March, 1986.

International Geotechnical Symposium on Theory and Practice of Earth Reinforcement / Fukuoka Japan / 5-7 October 1988
© 1988 Balkema, Rotterdam. ISBN 90 6191 820 0

Membranes in layered soils beneath pipelines

A.F.L.Hyde
University of Bradford, UK

K.Yasuhara
Nishinippon Institute of Technology, Fukuoka, Japan

ABSTRACT: Pipelines laid in reclaimed areas consisting of fill materials of varying strength often undergo unnacceptable differential settlements causing joint failure. Laboratory tests have been carried out to investigate a method of reducing settlements by placing a geosynthetic membrane below the pipe bedding material. It was shown that geotextiles particularly of the grid type when used as a separator between rounded aggregate pipe bedding and an underlying soft clay lead to reduced settlements and increased bearing capacities.

1. INTRODUCTION

A common form of failure of jointed pipes laid in trenches takes the form of differential settlement causing opening and sometimes fracture of the pipe joints. This type of failure often occurs in pipelines laid in reclaimed areas consisting of fill materials of varying strength. It is common practice to support pipes on beds of rounded pea gravel. Tests have therefore been carried out on the effects of a separating geosynthetic layer on the load deformation characteristics of a plate bearing on a rounded gravel overlying a soft clay subgrade. The purpose of the tests was to investigate a method of reducing pipeline differential settlements by placing a geosynthetic membrane below the pipe bedding material.

2. REINFORCED LAYERED SOIL

Considerable work has been carried out on the effects of geotextile layers in both bound and unbound road pavement systems and embankments on soft soils.

Many researchers studying the effects of a geotextile layer beneath the unbound granular layer in a road pavement have found that these materials lead to a reduction in the required design thickness of a granular layer. (Barenberg et al, 1975, Brown, 1978, Ruddock, 1977) While Brown (1978) also showed that

geotextiles induced enhanced stress distributing characteristics in pavement structures. In addition Jesberger (1977) studying a sub-base over silt and Jarret et al (1977) and Sorlie (1977) studying a sub-base over peat all showed an increased bearing capacity by the introduction of a geotextile layer. Clearly if geotextiles can reduce design thicknesses and increase bearing capacities then under similar conditions they should lead to reduced settlements under loading. Giroud et al. (1984) demonstrated this while Barvashov et al (1977) showed that pre-tensioning reduced displacement even further. In the case of embankments, Belloni and Sembenelli (1977) among others found that geotextile reinforcement led to reduced settlements.

Theoretical elastic analyses of geosynthetic layers by Brown (1978) seemed to suggest that the strains were reduced because of the increased modular ratio between the different layers. Using both empirical data and elastic analyses design charts have been produced for different types of geosynthetic layers. For example Giroud and Noiray (1981) present design charts for the thickness of a granular layer in a road structure as a function of the strength of the subgrade a geotextile modulus and number of load applications.

3. PIPE BED DESIGN

Pipe bedding is normally a single size granular material the purpose of which is to provide uniform support for pipes when laid and hence prevent structural failure due to surface loads. The field strength of a pipe is dependent on its position relative to the natural ground level, its depth, the shape of the trench, the density of the fill and the position of the water table. Walton (1970) defines the laboratory strength of a pipe as 80% of the ultimate force required to crush the pipe. A pipe bedding factor F_b is defined as the ratio of the field strength to the laboratory strength. Walton gives tables of bedding factor values for different loading conditions ranging from F_b =1.1 for a pipe laid directly onto the bottom of a trench and having a soil backfill to F_b =4.8 for a pipe having a 180° granular cradle with reinforced concrete laid over the top of the pipe and then a soil backfill. Waltons tables therefore give guidance to the designer on the choice of bedding material, thickness and placement geometry according to the loading conditions and position of the pipe.

4. EXPERIMENTAL METHODS

Assuming that load from a pipe is distributed to the underlying bedding material over an area given by a 60° cone from the centre of the pipe then a 350mm diameter pipe will apply a load over a strip 200mm wide. Since it was desired to model the settlement under a differentially loaded joint a loading plate 200mm x 200mm was chosen. Walton (1970) recommends a minimum granular bedding for a typical bedding factor F_b =1.5 of 100mm and therefore this depth of granular material was used in all tests.

The diameter of pipe considered is commonly manufactured in lengths of 1200mm. It is considered that a joint rotation of 5° would cause failure. The maximum allowable displacement under the joint is therefore 1200 x tan2.5 ° =52mm.

Materials

The subgrade consisted of Keuper Marl a silty clay (W_L =36%, W_P =19%, G_S = 2.7) A standard compaction test gave an optimum moisture content of 16.2% and an optimum dry density of 1.75 Mg/m^3. For testing purposes the subgrade was mixed to a moisture content W=21% in order to produce a

soft subgrade. This moisture content was well wet of optimum and core samples taken from the compacted subgrade gave quick undrained triaxial shear strengths C_u =15kPa. Shear vane tests carried out in the testing box agreed with this value. The bulk density in the test box was 2.07Mg/m^3 and the dry density 1.72Mg/m^3.

The single sized rounded aggregates were chosen to model typical pipe bedding materials. The gravels used were 5mm, 10mm and 20mm maximum sizes.

Three different types of material were used to separate the gravel and subgrade. These were Netlon a polyethylene grid, Terram a melded fabric and polythene sheeting. Four types of Netlon were used and these together with some of their properties are listed in table 1. The CEIII material was a low density polyethylene while all the others were high density.

The Terram used was a fibre mat comprising of a ramdomly oriented matrix of melded bicomponent filaments. The type used had a weight of 140 grammes/m^2 and while forming a continuous sheet was nonetheless water permeable. Terram is recommended for use in many situations including a separator for temporary roads.

Finally sheets of 1000 gauge polythene were used. This material being impermeable would not normally be used as a separator between the subgrade and bedding material but was tested for comparitive purposes as a low friction low strength material.

Table 1 Netlon properties

	CE111	CE121	CE131	CE152
Mesh aperture mm	8 x 6	8 x 6	27x27	74x74
Mesh thickness mm	2.9	3.3	5.2	5.9
Tensile strength KN/m	2.00	7.68	5.80	4.82
Strain at maximum load	41%	20.2%	16.5%	23.2%
Load at 10% extension KN/m	1.32	6.8	5.2	3.83

Procedure

Silty clay was compacted in a stiffened container having dimensions 0.95m x 0.71m x 0.5m deep. The marl was mixed to a moisture content of 21% and compacted in 100mm layers to a depth of 400mm. On top of this was laid the membrane as appropriate and a further 100mm of aggregate bedding material. (Figure 1)

198

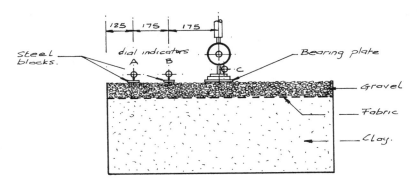

Figure 1   Testing arrangement

The layered system was then loaded by a hydraulic ram through a 200mm x 200mm square plate. It was estimated that the ratio of the size of the plate to that of the container would mean that the boundary effects were negligible. The tests were carried out under strain controlled conditions and each test was carried out in increments of 6.25mm penetration. Each value of penetration was maintained for ten minutes to allow creep and stress decay effects to stabilise. The penetration was limited to a maximum of 50mm which was regarded as the deflection value at which joint failure occurs. After the load was released the aggregate or bedding material was removed from the test container. Observations were made of any contamination of the aggregate and measurements were made of the final deformed surface profile.

5.   TESTING PROGRAMME

To obtain a comparison between the different fabrics and gravels used it was decided to test each type of fabric with the three different aggregates in turn as a three layer system except that the large mesh CE152 Netlon was not used with the 5mm gravel. In addition to this the clay was tested with the three different gravels without fabric as a two layer system.

6.   RESULTS

Tests were carried out to determine the load settlement characteristics of unreinforced layers of gravel overlying the clay. Results of these tests are presented in Figure 2. It is clear that increasing the size of

the bedding material has a beneficial effect. The settlement at a plate bearing stress of 60 kPa is reduced from 30mm for a 5mm gravel to 22mm for a 20mm gravel, a 27% reduction.

The introduction of a geosynthetic membrane reduces the settlement even further. For 5mm gravel, Figure 3, the Netlon grids have a more beneficial effect than the sheet separators. While polythene sheet reduces the settlement by 9% at a bearing stress of 60 kPa the corresponding reduction for Terram is 27% and for the grids from 37% to 55%. For 5mm gravel there is a marked and clear difference between the effects of each of the separators. However as the size of the gravel is increased to 10mm, Figure 4, the difference between each of the materials is not so clear and at some stress levels the sheet materials have as beneficial an effect as the grid separators. The reduction in settlement at 60 kPa is from 17% to 39%. Finally for 20mm gravel, Figure 5, although a membrane clearly reduces the settlement and increases the stiffness of the layered system the load deflection curves for the reinforced layers lie close to each other in a band. In this case the reduction in settlement at 60 kpa is from 18% to 32%.

The bearing capacity of all the layered systems was increased by the introduction of membranes and the percentage increase relative to the unreinforced bearing capacity at 50mm settlement is shown in table 2. It is clear that the stiffer high density polyethylene grids CE121 and CE131 have the greatest beneficial effect on bearing capacity although Terram worked well with the larger 20mm

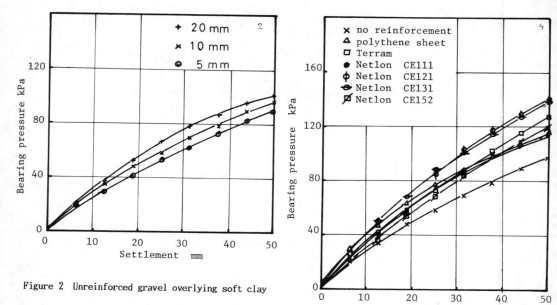

Figure 2 Unreinforced gravel overlying soft clay

Figure 4 10mm gravel underlain by geosynthetics

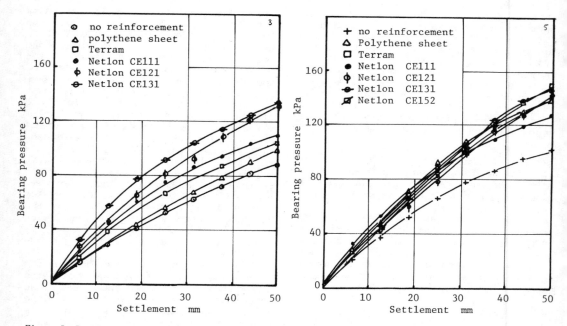

Figure 3 5mm gravel underlain by geosynthetics

Figure 5 20mm gravel underlain by geosynthetics

Table 2 Percentage increase in bearing capacity at 50mm settlement

	Size of Aggregate		
	5 mm	10 mm	20 mm
POLYTHENE	11	21	38
TERRAM 140	17	32	48
CE 111	24	22	26
CE 121	48	46	40
CE 131	39	44	45
CE 152		23	39

aggregate size.

Measurements of the clay profile after testing showed an increase of the depressed area of about 50%-60% for all reinforced sections. This would appear to indicate greater load spreading as the membranes take tensile stresses at the interface between the granular bedding and the soft clay.

7. CONCLUSIONS

Geotextiles particularly of the grid type when used as a separator between rounded aggregate pipe bedding and an underlying soft clay lead to reduced settlements and increased bearing capacities. These materials therefore seem to offer the potential of reducing severe differential settlements in pipelines laid in variable fill materials thus helping to prevent pipe joint failures.

8. REFERENCES

Barenberg, E. C., Dowland J. M. and Hales J. M. (1975), " Evaluation of soil-aggregate systems with MIRAFI fabric " The Highway Research Laboratory, Technical Report UILU-Eng-75, University of Illinois.

Barvashov, V. I., Budanov, V.G., Fomin A. N., Perkov J. R. and Pustikin, V.I. (1977), " Deformation of a soil foundation reinforced with prestressed synthetic fabric" Proc. 1st Int. conf. on The Use of Fabrics in Geotechnics, Paris, Volume 1, pp 67-69.

Belloni, L. and Sembenelli, P. (1977), " Road embankments on compressible soils constructed with the aid of synthetic fabrics" Proc. 1st Int. Conf. on The Use of Fabrics in Geotechnics, Paris, Vol. 1, pp 49-54.

Brown, S. F. (1978), " The potential of fabrics in permanent road construction", Proc. of Conf. on Textiles in Civil Engineering, UMIST, pp 1-30.

Giroud, J. P. and Noiray, L. (1981), " Design of geotextile reinforced unpaved roads" ASCE Proc. J. Geotechnical Engineering Div., 107, G.T.9, pp 1233-1254.

Giroud, J. P., Ah-Line, C and Bonaparte, R. (1984), " Design of unpaved roads and trafficked areas with geogrids", Proc. Polymer Grid Reinforcement Conf., London, Thomas Telford, pp 116-127.

Jarret, P. M., Lee, R. A. and Ridell, D. V. B. (1977), " The use of fabrics in road pavements constructed on peat", Proc. 1st Int. Conf. on The Use of Fabrics in Geotechnics, Paris, Vol 1, pp 19-22.

Jesberger, H. L. (1977), " Load bearing behaviour of a gravel sub-base – Non-woven fabrics on a soft subgrade system", Proc. 1st Int. Conf. on the Use of Fabrics in Geotechnics, Paris, Vol.1, pp 9-13.

Ruddock, E. C. (1977), " Fabrics and meshes in roads and other pavements: A State-of-the-art review", Technical Report, CIRIA, London.

Sorlie, A. (1977), " The effect of fabrics on pavement strength – Plate bearing tests in the laboratory" Proc. 1st Int. Conf. on The Use of Fabrics in Geotechnics, Paris, Vol.1, pp 15-18.

Walton, J. H. (1970), " The structural design of the cross section of buried vitrified clay pipelines", Clay Pipe Development Association Ltd. London.

International Geotechnical Symposium on Theory and Practice of Earth Reinforcement / Fukuoka Japan / 5-7 October 1988
© 1988 Balkema, Rotterdam. ISBN 90 6191 820 0

A few remarks on the design guide to pile net method

Tetsuya Iseda & Yoshihiko Tanabashi
Nagasaki University, Nagasaki, Japan

Kazuo Okayama
Ministry of Construction, Takeo, Japan

Eiji Arata
Ministry of Construction, Kurume, Japan

ABSTRACT: Pile net method has been used increasingly as one of the countermeasures for high organic soil deposits in Hokkaido and for high sensitive soft clay deposits in Kyushu, Japan. Pile net method may often be used in the future, because of its advantages which are a rapidity of construction, a facility of execution management, a large effect of restricting lateral flow and so on. However, at present, the rational design method based on its mechanism has not yet been established. This paper offers the useful knowledge for establishing the design guideline by synthetically examining the results of the in-situ tests executed on the Ariake clay ground, the model experiments and the finite element analysis.

1 OUTLINE AND FEATURES OF THE CONSTRUCTION METHOD

Pile net method is one of the pile works for reinforcing poor ground foundation. The pile net method is a relatively new which was developed in 1976 as the method of reinforcing the banking foundation in the peat deposit in Hokkaido, Japan. It is constructed by connecting the heads of the piles (wood, concrete, or steel) driven in a poor ground with the steel bars arranged in a net form as shown in Fig.2. And sand mats the permeable sheets or wire nets on them so that most of banking load is supported by the pile group as friction piles. The pile net method aims mainly at the increase of bearing capacity and the restriction of subsidence. When fill-loading is made on poor ground foundations, the restriction of ground displacement, the elimination of influence to surrounding structures, and the suppression of the ground vibration due to traffic load must be accounted for also (Sasaki, H. 1982).

The feature of this construction method is to transfer the banking load to deep strata by the flexible structure giving deflection to steel bars and enhanced the effect of pile group. Furthermore, such advantages as the rapid execution is

possible, and the maintenance management during and after the execution is easy.

The effect of restricting lateral displacement is large and the subsidence and deformation of the surrounding ground are minimized. It is expected that its adoption as the reinforcing method for the countermeasures to a poor ground will increase in future.

2 CASE RECORDS

The application of pile net method to Ariake clay ground has been carried out since 1981 at three districts in Saga plain for the following three purposes.

1. To study the effect of pile lengths (Ashihara district).

2. To study the effect of restraing lateral flow of the ground (Kutsugu district).

3. To illustrate the use short piles (Futamata district).

2.1 Geotechnical properties of the Ariake clay

The Ariake clay sedimented in Saga plain lies north of the Ariake Sea located in central Kyushu, is to be as one of the most soft clay in Japan. The Ariake clay

layer is sedimented in general 15-20m thick.
The geotechnical features of the Ariake clay are summarized from many investigations (Onitsuka, 1983).
The natural water content (W_n) is as high as 100-170%, and the liquidity index (I_L) is more than 1 and N-value is usually zero, therefore, the resistance to lateral flow is extremely small.
The sensitivity ratio (S_t) is above 16, and sometimes exceeds 100.
The geologic strata of the ground of Ashihara district, consist of the Ariake clay stratum, diluvial stratum that is the deposit of Quaternary and the base rocks (sandstone, shale) belonging to the Tertiary of Cenozoic era. The depth wide distribution of the soil characteristic values is shown in Fig.1. Besides, it has been confirmed that the groundwater level (borehole water level) is around 0.5m below the ground surface.

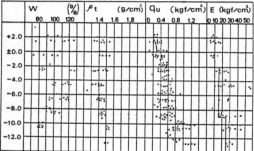

Fig.1   Soil properties at the site (Asihara)

2.2 Construction procedure

The construction procedure of pile net method in Ashihara district is as follows.
  1. Driving of piles
As for the piles, pine woods (diameter: 20cm at the tip, length: L=5,7 and 9m) were used, and those were driven in square arrangement at pile intervals of 1.1m, 1.3m and 1.5m, respectively, in the bank bed width of 22.5m. Further, preservative (creosote) was applied once from the pile tops to the depth of 3m.
  2. Connection of pile tops
After the piles were driven, by the clamping method shown in Fig.2, the steel bars (SR16, ϕ 16mm) joined by lap welding (one side, thickness 6mm) and pile heads were tied with clamps.
  3. Sand mats and the sheets
After pile tops were connected, the sand fill was spread all over by a bulldozer,

to a thick up 20cm over the pile heads. Then the sheets (tensile strength higher than 20kgf/cm^2) were laid.
  4. Execution of banking
Subsequent filling operation was done at 20cm lift thickness, and leveled and compacted by rolling with a bulldozer.
In the test construction in Ashihara district, the rapid execution in raising up to the final height of banking (3m in 40 days) has been carried out. As to the pile length, the construction for observation of not only 9m but also 7m and 5m, has been carried out continuously since July, 1981, for judging the effect of pile lengths.

2.3 Results of measurement

Taking Ashihara district as an example, the construction in this case used the pile of 9, 7 and 5m in length, and the difference of the construction effect due to the pile length was observed.
The thickness of layer is 16.5m. In this case, observation was carried out about the settlement under banking and the lateral displacement of the surrounding ground, the pore water pressure in the clay, the stress due to banking load, the force acting on the steel bars on pile tops. The arrangements of observational equipments are shown in Fig.3.
The settlement at February, 1987, in the case of the pile lengths of 9,7 and 5m were 22, 34 and 44cm, respectively, and the settlements decreased as pile lengths increased.
The stress acting in the steel bars changed according to pile intervals. As the pile intervals increased, the stresses also increased arose. Its maximum intensity was 200-400kgf/cm^2, equivalent to about 1/3-1/6 of the allowable tensile strength.
The lateral displacement on Feb. 1987 in the case of the pile length of 9. 7 and 5m were about 0.4, 1.0 and 1.2cm at the toe of the embankment slope.
As pile length increased, the lateral displacement decreased.
From the above discussion, we can summarize as follows.
  1. The settlement of the embankment is proportional to the thickness of comprresible soil strata below the tip of piles. The settlement between piles in the range where the piles were driven was small, and the settlement in the depth below pile tips was subjective.
  2. The lateral displacements tended to decrease as the pile length increased.

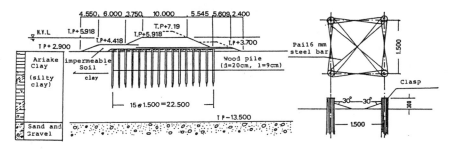

Fig.2 Outline of pile net method (Ashihara)

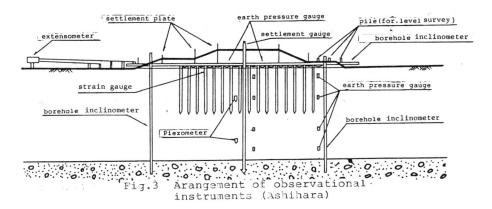

Fig.3 Arangement of observational instruments (Ashihara)

This is because banking load is transmitted to deeper ground through piles, but its relation to pile intervals has not yet been clarified.

3. The stress acting in the steel bars was about 200-500kgf/cm^2 in the case of embankment height less than 4m, amounting to only about 1/3-1/6 of the allowable tensile strength.

As mentioned above, in the pile net method, the effects has been recognized in the restriction of settlement, the prevention of lateral flow and the prevention of sliding failure.
However, as to such effect, particularly the mechanism of increasing the bearing capacity, the condition for the whole pile -net to act as one-body structure have not yet been clarified.

3 MODEL EXPERIMENT

3.1 Outline of experiment

A model experiment was carried out to clarify the bearing mechanism of the pile net method using the Ariake clay.
A soil tank for experiment is a plane strain two-dimensional model tank made

of acrylate plate and its dimension is 220*130*60cm.
The model ground and material for the experiment are shown in Table 1.

Table-1 Materials of model test

Model ground soil	Wn	112 ~ 123 %
	Gs	2.64
	Yt	1.34 t／m³
	c	0 ~ 10cm 0.10t／m²
		10 ~ 30 0.13
		30 ~ 50 0.17
		50 ~105 0.20
pile net	sheet mat	paper, tensile strength 0.027~ 33 kgf／3cm
	pile	wood φ8m Young's modulus 190000 ~200000 kgf／cm²
	steel bar	wire (#28 φ0.3m）

3.2 Experimental results

From the experimental results, it become obvious regarding bearing characteristics of the pile group without net and the pile group with net as shown in Fig.4.
The data obtained by Whitaker, T (1961) and Sower, F. (1961) are added in Fig.4.

205

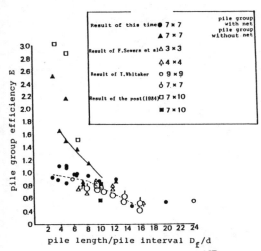

Fig.4 bearing characteristics of the pile group with and without net

4 CASE ANALYSIS

A slide joint is used to interface model considering the wall surface friction and the sliding. Also, FEM analysis with bi-linear anisotropic elasto-plastic model is carried out for the soil element (Iseda, T. and Tanabashi, Y. 1978). The analytical model is a plane strain condition as shown in Fig.5. The pile element is shown in this figure. The analytical conditions consist of three piling types such as non treated grouped pile, grouped pile and head connected piles by reinforcing bars. The step by step calculations are carried out with combinations of a pile length and a pile interval for embankment height from 0m to 6m.

(1) Failure type
The block failure where soil inside a pile and pile circumference fails/settle as a unit in a pile interval from 2 to 4 times of a pile diameter is occurred. All piles occur penetration failure that does penetrate in a pile interval over 4 times of a pile diameter.

(2) Pile group efficiency
A pile group efficiency (E) is defined by the bearing capacity ratio of pile group per unit pile to a single driven pile. Bearing capacity (a bearing capacity decrease by stress interference among the piles and also an increase of settlement) decreases by a pile group effect when pile intervals become narrow. A change of this pile group efficiency may also be a pile group that may not be over 1. The unique relation exists between pile length/pile interval (D_f/d) and pile group efficiency (E) as shown in Fig.4.

(3) Stress share between ground and pile top
The structure of a pile-net consists of pile tops connected by reinforcing bars as mentioned before. Because of the hammock effect generated by the reinforcing bar connection, a part of a load is transmitted to the piles. That proportional relation materializes in the interval of pile length/pile interval (D_f/d). As mentioned above, the bearing capacity characteristics of a pile-net can be explained by pile length/pile interval (D_f/d). Large bearing capacity is obtained, by connecting a pile head with reinforcement from this case.

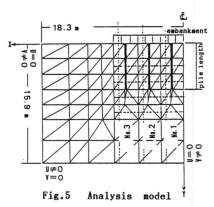

Fig.5   Analysis model

From the results of analysis, the piled zone may be considered as a integrated structure like a caisson. Incidentally, in case of 9m pile length with 3m embankment height, maximum horizontal stress increase between piles were $1.20tf/m^2$ as the measured value and $1.45tf/m^2$ as the calculated value. The maximum horizontal stress increase under piles were $3.80tf/m^2$ as the measured value and $5.85tf/m^2$ as the calculated value.

Attaching in the case of pile group construction, a relationship between expansion among three piles and embankment height was obtained by the parameter of the pile length to examine necessity of pile head connection by reinforcing bars in the pile net method (see Fig.6). Three piles shown in Fig.5. were designated No.1, No.2 and No.3 from one near a load center. Between the bank center and No.1 pile, the reinforcing bar connection is not necessary according to shrinkage having no connection with

206

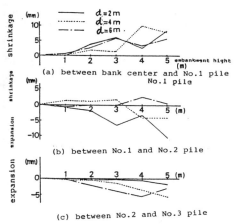

(a) between bank center and No.1 pile

(b) between No.1 and No.2 pile

(c) between No.2 and No.3 pile

Fig.6 Relation the extension of pile distance and embankment hight

banking height. Between No.1 and No.2 pile, this zone is a transient zone where the compression and expansion occur depending on the combination with embankment height and pile length. Between No.2 and No.3 pile, an expansion was entirely shown without being related to the embankment height as shown in Fig.6. If the hammock effect is not expected and any problem in trafficability has not occurred, it is not necessary to connect the pile heads by reinforcing bars for the total width of the considered piling area.

5 DESIGN GUIDELINES AND CONCLUSIONS

Conventional concepts for the pile net design are summarized in a flow chart as shown in Fig.7. From the results of various model experiments, field observations and case analysis, new design concepts against conventional concepts are presented as follows:

5.1 Pile interval and pile length

A conventional concept is based on Bierbaumer's formula, $d=1.5\sqrt{r D_f}$ where r is a pile radius, D_f is a pile length which is theoretically carried out to determine a pile interval without any stress interference between the piles. However, over and under bearing capacity values by pile group formula may be carried out due to the optional pile interval, pile length, and number of piles. Because of this, it is good to design the pile arrangement plan, by the

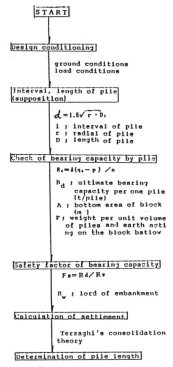

Fig.7 Design flow chart of pile net method

parameter (D_f/d) that is able to explain an actual phenomenon of bearing capacity characteristics.

A pile interval having a good efficiency in present conditions is within from 4m to 9m of D_f/d. In the case that D_f/d becomes small, the pile interval is extended when the pile length is constant. Whether or not this is the case, the structure that deform as a unit becomes a problem. Accordingly, if the pile length is assumed, pile interval may be determined.

5.2 Concept of bearing capacity

The present concept of bearing capacity is the one that regards an entire pile group of a pile-net as one block. With the ratio of a pile interval to pile diameter (5 to 7) usually used, it is clarified by an experiment that the soil between the piles does not integrally moved with piles. Also, the ratio of pile length (D_f) and pile diameter (D) may be indicated by $D_f/D>1$. From this case, in a pile-net, it is conceivable that plastic equilibrium state peculiar to general shear of shallow foundation will not happen.

207

It is adequate for a pile group of pile-net to handle as a short pile group that became independent individually.

Plastic equilibrium condition may be considered as a deep foundation case.

Bearing capacity of the pile-net is the bearing capacity of a single pile plus the ground reaction. Here, the ground reaction may be calculated by the relation with D_f/d as that shown in Fig.8. According to an existing experiment, the ground reaction magnitude is estimated to be from 10% to 60% of a pile bearing capacity in a range of $D_f/d=5$ to 7.

On the other hand, from results of the FEM analysis and measurements, stresses that act on the reinforcement reaches from 1/6 to 1/3 of the allowable stress of the reinforcement. Also, the netting width that influences the strength of circumferencial zone may be decreased in proportion to increasing of the ratio of pile length to the layer thickness (L/H). In the future, it is necessary to study the simplification of the pile arrangement and the application of new materials such as polymer grid.

REFERENCES

Iseda, T., Y. Tanabashi and T. Higuchi 1978. A finite element analysis considering wall surface friction. 14th JSSMFE Research Form: 989-992. (in Japanese)

Onitsuka, K. 1983. Ariake clay-Problem soils in Kyushu and Okinawa-. Kyushu Branch of JSSMFE: 23-39. (in Japanese)

Sasaki, H. 1982.Soft ground handbook. In M. Fukuoka (ed.), P.394-401. Tokyo, Kensetsu sangyo chosakai. (in Japanese)

Sowers, F. 1961. The Bearing capacity of friction pile group in homogeneous clay from model studies. 5th ICSMFE: 151-159.

Whitaker, T. 1961. Experiments with model piles in groups. Geotechnique. Vol.17, No.4: 147-159.

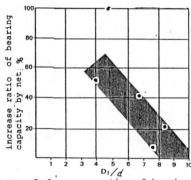

FIg. 8 Increase ratio of bearing capacity by pile net method

5.3 Settlement due to consolidation

The calculation methods of the settlement generated for the pile net method are as follows: Using overlap of the stress solutions as pile group settlement acting on the semi-infinite elastic body; and using one-dimensional consolidation theory that the stress acts at 1/3 of the pile length from pile point; and using FEM analysis. These methods have strong and weak points. However, as a simplified calculation method, one-dimensional consolidation calculation of Terzaghi with admissible accuracy on practical use may be cited.

5.4 Role of reinforcement

It is explained and clarified above that the ground reaction is added to the pile bearing capacity for the pile-net structure with pile head connection by reinforcing bars. A role of the reinforcement is to transmit the load between piles to the piles by hammock effect. Also, the reinforcements function to distribute stress by structure integration.

International Geotechnical Symposium on Theory and Practice of Earth Reinforcement / Fukuoka Japan / 5-7 October 1988
© 1988 Balkema, Rotterdam. ISBN 90 6191 820 0

The use of slip line fields to assess the improvement in bearing capacity of soft ground given by a cellular foundation mattress installed at the base of an embankment

C.G. Jenner & D.I. Bush
Netlon Limited, Blackburn, UK

R.H. Bassett
Kings College, London, UK

ABSTRACT: The paper describes the use of slip line fields in determining the increased bearing capacity of soft ground that can be mobilised when a cellular foundation mattress is installed at the base of an embankment. The analogies between the theories of plasticity used for metal pressings and extrusions, and the performance of soft cohesive soils under a relatively stiff, rough embankment base are described and developed. The possible failure modes of the soft foundations are discussed and the consequences examined (to demonstrate the development of the slip-line field), particularly when the soft foundation soil is relatively thin compared with the embankment base width. The effect of the cellular nature of the foundation mattress on the principle stresses in the granular fill in the mattress is discussed and the theory behind the determination of the required mattress strength described. A typical design using these theories is included to demonstrate the design method and the resulting mattress configuration.

1. INTRODUCTION

In many parts of the world a common construction problem is one of how to construct high embankments over soft foundation soils. The conventional methods of piling, soft soil replacement, phased construction over long periods of time or the use of very high strength geotextiles are not always practical or economic. The use of a cellular foundation mattress gives a solution that enables the maximum bearing capacity to be mobilised in the soft foundation whilst also forming a firm and stable working platform on which all construction plant travels thus leading to an increased rate of embankment construction.

Whilst the methods described and discussed can be applied to soft layers of any depth, the paper deals particularly with the condition of an embankment constructed over a relatively thin soft layer where the ratio of embankment base width to depth of underlying soft layer is greater than 4.

2. DESIGN CONCEPT

The incorporation of a cellular mattress creates an embankment foundation with the following characteristics:

(a) A perfectly rough interface between the mattress and the soft foundation due to the granular fill partially penetrating the base grid material.

(b) A stiff platform to ensure both an even distribution of load onto the foundation and the formation of a regular stress field within the soft foundation soil. The stiff platform is created by the high tensile strength of the polymer grid material used in the cellular construction to confine the granular infill.

These characteristics enable the cellular mattress to exert a degree of restraining influence on the deformation mechanism developing in the soft foundation material. This restraining influence in the mattress effectively rotates the principle stress direction compared with an embankment without these features from a near vertical direction in the embankment fill up to 45° inclined inward in the top of the soft foundation.

The corresponding rotation of the directions of maximum shear stress means that the surfaces on which failure is most likely to occur are also rotated to pass deeper into the foundation. The critical mode of failure for design thus becomes the plastic bearing condition rather than a slip circle and hence an enhanced bearing capacity can be developed with a full base friction situation (perfectly rough base) being the ultimate achievable value.

3. DEVELOPMENT OF THE SUPPORT MECHANISM AND PLASTIC YIELDED ZONES

For a properly designed mattress the soft underlying soil becomes critical before the granular mattress and this would deform laterally from under the embankment and the embankment would settle. As the overburden load increases during embankment construction the yielded area of underlying soil increases and the fully mobilised stress field moves progressively inwards with the central active zone, termed the 'rigid head', decreasing until the full embankment load is balanced (Fig. 1).

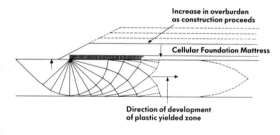

Figure 1.   Development of stress field

In the limit the two symmetrical stress fields meet under the centre of the embankment when complete plastic yield of the underlying soil is reached in a condition of limiting equilibrium. Whilst the stress fields are moving inwards the soil in the centre of the yielded zone of the soft layer begins to deform outwards at a greater rate than the lateral extension of the reinforced cellular layer. Work is done as the elements of soil within the soft layer deform and shear and this energy balances the loss of potential energy of the settled embankment. Once the stress condition at the rough interface reaches

C_u, slip would occur and thereafter remain unaltered by further deformation within the body of the soft layer.

The flexible, but still relatively stiff nature of a cellular mattress formed from polymer grid reinforcement enables the full foundation soil strength to be mobilised in the plastically deformed zones.

The ultimate bearing capacity for the condition of a soft layer being compressed between two relatively rough, rigid, parallel surfaces can be analysed using the same methods as those used for the pressing of metals. The analogies between the behaviour of soft soils and metal pressing are not in themselves new. In the same way as Prandtl used the theories of punching of metals as the basis of his foundation theories (Smith, 1982), the authors develop the approach of Johnson and Mellor (1983) who analysed the compression of a block between two rough, rigid plates.

The analysis assumes non-strain-hardening materials and does not allow for creep or strain rate effects in the compressed block. The soft foundation soils that would necessitate this type of design are generally very soft normally consolidated soils which, in the undrained condition can be expected to comply with these criteria.

3.1 Bearing Capacity from Slip-line Field

The bearing capacity of the soft soil layer can be quantified using a slip-line field as in Johnson and Mellor (1983), but with the addition of a fan field at the outer edge of the cellular mattress. The fan field enables the effect of the restraint applied by the continuing soft layer to be modelled.

The geometry of the resultant non-symmetrical slip-line field is used to define the length over which the constant bearing resistance acts at the toe of the embankment and over the 'rigid head' section in the centre of the embankment. The length at the toe is equal to the soft layer depth from the edge of the mattress to the point at which the first slip-line intersects the mattress (Fig. 2). The non-symmetrical nature of the slip-line field gradually decreases as the ratio of mattress width to soft soil depth increases. Above a ratio of approximately 9 the length from the centre line of the embankment to the edge of the 'rigid head' is constant at 1.25 times the soft layer depth.

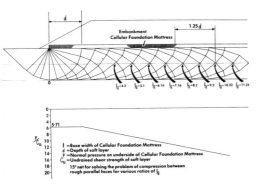
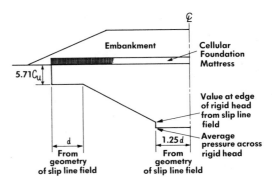

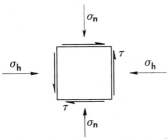

Figure 2. 15° slip line field and associated pressure diagram

A 15° slip-line field is used to determine the bearing resistance. The accuracy of the field is increased as the increment in the equiangular mesh is reduced but the authors have found the 15° field to be satisfactory for the range of conditions they have examined.

The ultimate bearing resistance from the foundation (area of the pressure diagram) is then compared with the imposed load from half the embankment to give a factor of safety against bearing failure. As consolidation of the soft material takes place its shear strength increases and the restraining requirement of the reinforcement becomes less critical. If consolidation and strengthening can occur during construction, then the full base friction value will be mobilised and the factor of safety will be greater.

4. STATE OF STRESS

If we examine an element of soil within the granular cellular mattress but interfacing with the soft layer, we have a stress condition on that element as below (Fig 3).

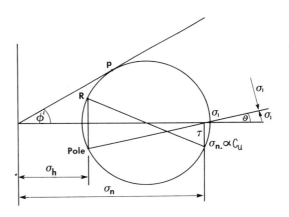

Figure 3. Soil element at interface

A Mohr circle construction can be drawn (Fig 4.) and point R represents the stress state on the vertical plane in the cellular mattress.

Figure 4. Mohr circle construction

In the limiting condition $\alpha = 1$ and $\tau = C_u$ the shear strength of the soft layer. The general expression for horizontal stress $\sigma h = \sigma_n - 2x$ where σ_n is the vertical stress on the element.

From the geometry of the circle

$$x = \{2\sigma n \sin^2\phi' \pm [4\sigma n^2 \sin^4\phi' - 4(\sin^2\phi'-1) (\sigma n^2 \sin^2\phi' - \tau^2)]^{\frac{1}{2}}\}/2(\sin^2\phi' - 1)$$

Hence the value of the horizontal stress σh (or confining stress) can be calculated. In the initial designs the value of horizontal stress σh was defined as $C_u/\sin\emptyset'$ which is the conservative limiting condition, point P, when both the granular material in the mattress and the soft soil are in a critical condition.

211

Figure 4 also indicates the angle to the horizontal through which the principle stress in the granular material must be rotated. The principle stress is rotated within the cellular mattress by the confining force that is developed by initial limited extension of the reinforcement as the foundation material deforms outward. Once this extension reaches a value to give the required confining force then the full shear strength of the soft soil is mobilised and no further extension occurs. The value of σ_1 caused by the embankment weight (i.e. the fill) and its active inclination can be found at each point moving in from the toe. The value of σ_1 sin θ can be calculated and summed $F_R = \Sigma \sigma_1$ sin θ for the required length over which the base friction must be mobilised to maintain stability to ensure that the component of the principle stress acting horizontally inwards at the base of the mattress is greater than the mobilised shear stress within the soft soil. The form of cellular mattress is chosen from the range of uniaxial geogrids (e.g. 'Tensar' SR grids) to provide the required horizontal resistance calculated from σh. For a 1m deep mattress the horizontal resistance equals the numerical value of σh.

5. DESIGN EXAMPLE

It is proposed to construct a 7m high embankment over a 4m thick layer of soft cohesive soil with an undrained shear strengh of 15 kN/m². A surcharge of 20 kN/m² will be applied.
Typical cross-section as below:-

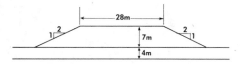

Figure 5. Embankment cross section

Width of Geocell Mattress
= Embankment base width - 6m
= 56 - 6 = 50m
This provides minimum cover of 0.5m to the mattress.

Ratio: $\dfrac{\text{Width of Geocell}}{\text{Depth of Soft Layer}} = \dfrac{50}{4} = 12.5$

Using the stress field we can now find $\dfrac{P}{C_u}$ at the edge of the 'rigid head'.

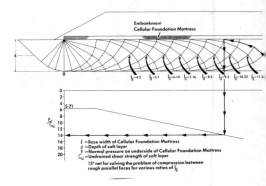

Figure 6. Evaluation of $\dfrac{P}{C_u}$

Therefore $P = 14\ C_u$ at the edge of the 'rigid head'.
The average pressure over the 'rigid head' is then calculated from

$$\frac{\bar{P}}{C_u} = \frac{2 \times (2 \times I + 0.5 \times d)}{2X} + \frac{P}{C_u}$$

(Johnson and Mellor, 1983, Eq. 12.11).

Where $\bar{P}$ = average stress over the rigid zone.

I = sum of (rotation of slip-line X horizontal chord length)
X = sum of horizontal chord lengths
d = depth of soft layer

In this case from the slip-line field:

Line	AB	BC	CD	
Chord length (units)	1.6	2.2	2.6	X = 6.4
Angle (radius)	.654	.395	.131	
I	1.05	0.87	0.34	I = 2.26

Table 1. Evaluation of X and I

Therefore
$$\frac{\bar{P}}{C_u} = \frac{2 \times (2 \times 2.26 + 0.5 \times 4)}{2 \times 6.4} + \frac{P}{C_u}$$

= 1.02

Therefore additional average pressure across the 'rigid head' zone is $1C_u$ added to the value determined at the edge of the zone.

The pressure resistance becomes

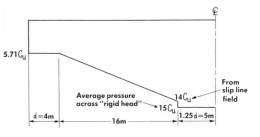

5.71C_u

Average pressure across "rigid head"

14C_u ← From slip line field

15C_u

d=4m — 16m — 1.25d=5m

Figure 7. Pressure diagram

Load from half the embankment including 20 kN/m² surcharge

$$= \frac{(14 + 28)}{2} \times 7 \times 19 = 2793$$

$$+ 14 \times 20 \qquad = 280$$

$$\overline{}$$

$$3073 \text{ kN/m}$$

Resistance from the pressure diagram:-

$$4 \times 5.71C_u \qquad = 23C_u$$

$$+ \frac{(5.71 + 14)}{2}C_u \times 16 \qquad = 158C_u$$

$$+ 5 \times 15C_u \qquad = 75C_u$$

$$\overline{}$$

$$256C_u$$

Therefore C_u required for equilibrium

$$= \frac{3073}{256} = 12 \text{ kN/m}^2$$

Actual C_u 15 kN/m² before any consolidation.

Therefore minimum FOS = 1.25. This figure will increase as consolidation takes place and is therefore satisfactory.

From the Mohr Circle Construction (Fig. 4), the horizontal stress

$\sigma h = \sigma n - 2x$

Where σn = vertical stress on the element and

$$x = \{2\sigma n \sin^2 \phi' \pm [4\sigma n^2 \sin^4 \phi' - 4(\sin^2 \phi' - 1)$$
$$(\sigma n^2 \sin^2 \phi' - \tau^2)]^{\frac{1}{2}}\}\Big/ 2(\sin^2 \phi' - 1)$$

τ = shear stress at the interface
 = C_u in the limiting condition

σn under the highest part of the embankment = 7 x 19 + 20 = 153 kN/m²

τ for FOS = 1 = 12 kN/m²

$\emptyset'$= 35° for the Geocell fill material

Therefore x = 55 kN/m²

Therefore σh = 153 - 55 x 2 = 43 kN/m²

The rotation of the principle stress occurs within the 1m mattress depth, therefore mattress strength required = 43 kN/m.
Using 'Tensar' SR80 grids a mattress formed in a 1m triangular chevron pattern has a long-term strength

$$= \frac{32.5}{1.1} (1 + \frac{1}{\sqrt{2}}) = 50 \text{ kN/m}$$

Characteristic strength of SR80 = 32.5 kN/m (Netlon Ltd, 1988a)

Without considering the 'Tensar' SS2 base.

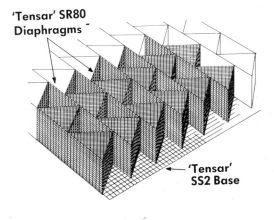

'Tensar' SR80 Diaphragms

'Tensar' SS2 Base

Figure 8. Cellular mattress configuration

213

Check $\sigma_1 \sin\theta < C_u$ for length of mobilised shear stress:

Distance from Mattress edge	Effective length	σ_1 (kN/m^2)	θ (Degrees)	$\sigma_1 \sin\theta$ x effective length
1m	2m	44	24	35.8
3m	2m	61	16.5	34.6
5m	2m	79	12.5	34.2
7m	2m	97	10	33.7
9m	2m	116	8	32.3
11m	2m	135	7	32.9
13m	2m	155	6	32.4
15m	2m	155	6	32.4
17m	2m	155	6	32.4
19m	2m	155	6	32.4

Table 2.   Evaluation of horizontal principle stress component   <u>333.1 kN/m</u>

Total C_u x L = 12 x 20 = 240 kN/m.   Therefore $\Sigma\sigma_1 \sin\theta > C_u$ x L   OK

CONCLUSIONS

A number of cellular foundation mattresses have been designed using the methods described in this paper and all are performing well. The practical and economic benefits of the cellular foundation mattress, both in developing the maximum bearing resistance of the soft soil and in preserving any natural surface 'crust' throughout construction, and in the long term, make this type of design and construction a very attractive solution.

REFERENCES

Edgar, S. (1987). The use of a High Tensile Polymer Grid Mattress on the Musselburgh and Portobello By-Pass. Proc. of Polymer Grid Reinforcement Symp. jointly sponsored by the SERC and Netlon Limited.

Johnson, W. and Mellor, P.B. (1983). Engineering Plasticity. Ellis Marwood Ltd, Chichester.

Netlon Limited (1988a). Test Methods and Physical Properties.

Netlon Limited (1988b). Tensar Geocell Mattress.

Robertson, J. and Gilchrist, A.J.T. (1987). Design and Construction of a Reinforced Embankment Across Soft Lakebed Deposits. Presented at the Int. Conf. on Foundations and Tunnels.

Smith, G.N. (1982). Elements of Soil Mechanics for Civil and Mining Engineers. Crosby Lockwood, London, Fifth Edition.

International Geotechnical Symposium on Theory and Practice of Earth Reinforcement / Fukuoka Japan / 5-7 October 1988
© 1988 Balkema, Rotterdam. ISBN 90 6191 820 0

An experimental study on the contribution of geotextiles to bearing capacity of footings on weak clays

S.I.Kim
Yonsei University, Seoul, Korea

S.D.Cho
Korea Institute of Construction Technology, Seoul, Korea

ABSTRACT: Geotextile effects as a reinforcement material were studied through laboratory model tests of a strip footing on weak clays. The geotextile effects on bearing
capacity and deformation of soil foundation were investigated in view of the distance
of footing from geotextile layer and the footing embedment ratio. Tests were carried
out under partially drainage condition in order to investigate closely bearing capacity,
settlement and sliding length of geotextile. From the experiments, it has been found
that the contribution of geotextile to the bearing capacity increase is high as the
distance of footing from geotextile layer reduces, as the embedment depth of footing
increases, and as the settlement of footing increases. And it has been also found that
the ratio of sand layer depth to footing width, H/B, which gives the greatest geotextile
effects falls between 0.5 and 1.0.

1 INTRODUCTION

The use of reinforcing materials to stabilize poor soil conditions predates the
Romans. However, the use of geotextile
as a reinforcement was not regularized before 1973 when a geotextile reinforcement
was applied in a bridge construction in
Sweden. Recently, the use of geotextile
as a reinforcing element in soils has
gained widespread use throughout the world.
Many researches for geotextile effect on
the bearing capacity of foundation soil
were performed.

Sorlie(1977) and Gourc et al.(1982) reported the reinforcing effect of geotextile
in unpaved road through laboratory model
tests. Giroud and Noiray(1981) presented
design charts for geotextile-reinforced
unpaved road. Sellmeijer(1982) developed
a analytical method for estimating the
bearing capacity increase of road foundation due to geotextile reinforcement.
Reinforced earth slabs have been studied
by: Binquet and Lee(1975), Fragaszy and
Lawton(1984) with a strip footing on sand
and aluminum foil as the reinforcement;
Akinmusuru et al(1982) with a square footing on sand and rope fiber as the reinforcement; Guido et al.(1985) with a square
footing on sand and geotextile as the reinforcement. This paper includes the geotextile effects on bearing capacity and
settlement through a series of laboratory
bearing capacity tests of a strip footing on
sand/clay layer reinforced with a geotextile.

2 EXPERIMENTAL WORK

2.1 Testing arrangement

Fig.1 shows the experimental set-up used
in the tests. The experimental model has
the plane dimension of 30cm x 110cm and
the height of the model is 70cm. The wall
of the model was made of 9mm thick Plexiglass in order to provide direct optical

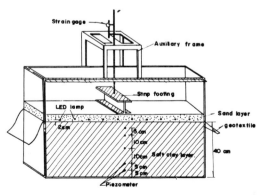

Fig.1. Experimental model set up

Table 1. Consolidation properties of clay

Before consolidation by self weight	water content, w(%)	60
	coefficient of consolidation, C_v(10^{-4}cm^2/sec)	7.4
	coefficient of permeability, k(10^{-8}cm/sec)	9.1
After consolidation by self weight	water content, w(%)	57
	initial void ratio	1.51
	compression index, C_c	0.305
	C_v in vertical direction (10^{-4}cm^2/sec)	4.73
	C_v in lateral direction (10^{-4}cm^2/sec)	5.12
	k in vertical direction (10^{-8}cm/sec)	6.70
	k in lateral direction (10^{-8}cm/sec)	7.35

Table 2. Properties of geotextiles

Item \ Trade name	K/M 8401	Bidim U_{64}
Structure	woven	non woven
Thickness (mm)	1.25	4.4
Weight (g/m^2)	350	550
Permeability (cm/sec : 0.02 bar)	2×10^{-2}	3×10^{-1}
Tensile strength (kg/cm)	90.6	-

observation of soil movement. Three numbers of steel strips were installed to reinforce the Plexiglass wall. Ten piezometers were installed symmetrically on both sides of the wall at different heights, 5, 10, 20, 30 and 35cm respectively from the bottom of the model.

Sample preparation was performed by placing a layer of geotextile over the 40cm deep clay layer, then sand was deposited until the desired level of height was achieved. The time dependent deformations were measured by utilizing a dial gauge on the footing and LED indicating lamps which had been inserted into the sand layer in a manner to form the square grid system of 10cm spacing. To simulate the rigid strip footing, a H-beam (10cmx30cm) was used and the vertical load was applied via loading plates. Dislocation of the model footing due to loading procedure was prevented by means of an auxiliary frame as shown in Fig. 1.

The clay used in all the tests has a liquid limit of 54% and a plasticity index of 21.9%.

The clay was mixed at slightly over the liquid limit, i.e water content of 58%-62% and smeared into the model box. On completion of each test the specimens were recovered for water content measurement and consolidation test. The consolidation properties of the clay are summarized in Table 1.

The sand used in all the tests has the mechanical properties of G_s=2.66, γ_{dmax}=1.65g/cm^3, γ_{dmin}=1.39g/cm^3 and has of uniform grain size ranging 0.59mm-2.0mm. The sand was placed above the clay layer and light compaction was applied manually by means of a wooden rod. Dry density of the sand layer was uniformly maintained between 1.47 g/cm^2 and 1.50 g/cm^2. The sand layer ranged in thickness up to 20cm. Geotextiles used in the experiments were a woven geotextile, K/M 8401 as reinforcement and a nonwoven geotextile, Bidim U_{64} as vertical drainage. The mechanical properties of geotextiles are shown in Table 2.

2.2 Vertical drain test using geotextile

Vertical drain test was performed before the bearing capacity test to investigate the effect of geotextile on consolidation. The nonwoven geotextile bands with a dimension of 43cm x 3.5cm were installed in the 40cm deep clay layer in square grid pattern to accelerate the consolidation. The spacing of them was 10cm. A plane view of inserted vertical drain is shown in Fig. 2.

The development and dissipation of the excess pore water pressure in the clay layer is shown in Fig. 3 when the overburden pressure from the 10cm deep sand was applied on the clay layer. Fig. 3 show that the primary consolidation of clay layer developed during one week owing to the geotextile effect.

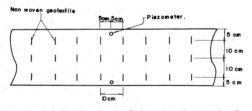

Fig.2. A plane view of inserted vertical drain

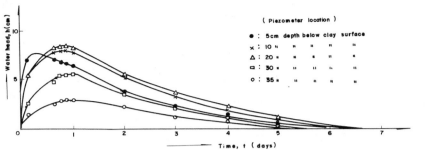

Fig.3. Development and dissipation of the excess pore water pressure

Therefore in this experiments, the bearing capacity tests were performed after one week under the condition of placing the sand layer on the clay layer with geotextile bands.

2.3 Bearing capacity test

The bearing capacity test was performed by acting loading plates on the strip footing placed on the sand/clay layer reinforced with a woven geotextile. Test loading is loaded incrementally by 0.1 kg/cm². The bearing capacity and settlement of foundation and sliding length of geotextile were investigated for each loading step.

Tests were carried out under the condition of partial drainage at 50 percent consolidation in order to investigate closely the geotextile effects on bearing capacity and settlement of foundation. Typical time dependent settlement is shown in Fig.4.

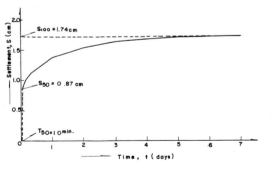

Fig.4. Time dependent settlement at the center of the sand surface for 0.144 kg/cm² of footing load

3. TEST RESULTS AND ANALYSIS

The bearing capacity of a strip footing on sand/clay layer reinforced with geotextile was studied in view of the following three parameters:

1. The depth below the footing of geotextile or the depth of sand layer, H. This was expressed as a dimensionless ratio H/B, where B is the width of the footing.

2. The embedment depth of footing, D. This was also expressed as D/B.

3. The settlement of footing, S. Geotextile effect on the deformation of foundation soil was also studied.

3.1 Effect of the depth below the footing of geotextile

The bearing capacity tests were performed with various H/B ratios to investigate geotextile effect on the bearing capacity of surface footing. Fig.5 shows the normailzed bearing capacity ratios to H/B=0.1 for various H/B ratios in the cases with and without geotextile, respectively. Fig.5 indicates that the effect of the depth of sand layer on the bearing capacity of reinforced soil is greater than the unreinforced soil. And the increase of the normalized bearing capacity ratio resulting from the use of geotextile to various H/B ratios is shown in Fig.6. From Fig.6, it was found that the increase of bearing capacity due to geotextile is high as the settlement of footing increases. The rate of measured increase of 37%, 41%, 19% and 12% for H/B of 0.5, 1.0, 1.5 and 2.0, respectively in the range of S/B between 0.05 and 1.0 was observed.

Binquet and Lee(1975) introduced a term, bearing capacity ratio, BCR for convenience in expressing and comparing test data:

$$BCR = q/q_o$$

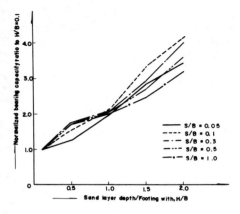

a) Case without geotextile

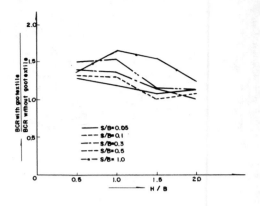

Fig.6. Increase of normalized bearing capacity ratio resulting from use of geotextile for various H/B ratios

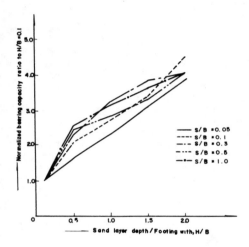

b) Case with geotextile

Fig.5. Normalized bearing capacity ratios for various H/B ratios

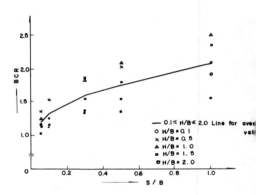

Fig.7. BCR-Settlement curves for various H/B ratios.

where q_o is the bearing pressure of the unreinforced soil and q is the bearing pressure of the reinforced soil, both measured at the same vertical settlement, S. Thus the BCR can be used to express the effect of geotextile for any desired settlement.

The BCR variation with settlement of footing for different values of H/B ratio is shown in Fig.7. It is found that the contribution of geotextile to the bearing capacity increase is high as the settlement of footing increases, for example the average value of BCR, BCR_{avg} is 1.186 for small settlement of S/B=0.05 and BCR_{avg}=2.088 for large settlement of S/B=1.0. As shown in Fig.7 the value of H/B which gives the greatest BCR is 0.5

for small settlements (S/B less than 0.3) and 1.0 for large settlement (S/B greater than 0.5). This indicates that H/B which gives the greatest geotextile effects falls between 0.5 and 1.0 for the settlements where S/B is less than 1.0.

From the experiments, it was found that when the depth of overlying sand layer is small, initial sliding of geotextile in sand/clay layer occurred at relatively low bearing pressure, which is due to small friction force between geotextile and soils. For this reason, the geotex - tile effect on the bearing capacity of strip footing is low for the small depths of sand layer where H/B is less than 0.5. In Table 3 are shown the bearing pressure for initial sliding and sliding length of geotextile with the variation of H/B ratios.

Table 3. Bearing pressure for initial sliding and sliding length with the variation of H/B ratios

	H/B				
Geotextile	0.1	0.5	1.0	1.5	2.0
Bearing pressure for initial sliding (kg/cm^2)	0.097	0.283	0.486	0.786	0.886
Sliding length (cm)	1.03	0.96	0.97	1.05	0.95

3.2 Effect of the embedment depth of footing

The bearing capacity tests were performed with various D/B ratios to investigate geotextile effect on the bearing capacity of strip footing under the condition of H/B=0.5. Fig.8 shows the normalized bearing capacity ratios to D/B=0.5 for various D/B ratios in the cases with and without geotextile, respectively. Fig.8 indicates that the increment rate of the normalized bearing capacity ratio varies in a similar fashion for the conditions with and without geotextile reinforcement.

The BCR variation with settlement of footing for different values of D/B ratio is shown in Fig.9. It shows that the contribution of geotextile to the increase of bearing capacity becomes high as the settlement of footing increases for the condition of 0.5≤D/B≤2.0. As shown in Fig.9 the value of D/B which gives the greatest BCR is 0.0 when S/B is less than 1.0. But the geotextile effects with regard to the embedment depth are similar in the range of 0.5≤D/B≤2.0 while the effect differs greater as the settlement of footing increases.

3.3 Geotextile effect on the deformation of foundation soil

The time dependent deformations of foundation soil were measured by utilizing LED indicating lamps inserted in the sand layer for each loading steps of the bearing capacity test. As shown in Fig.10, LED lamp location is typically changed with regard to the increase of vertical footing load. It shows that the lateral deformation of foundation soil corresponding to the increase of vertical load occurred to the outward direction receding from the footing for the case without geotextile

but to the inward direction being close to the footing for the case with geotextile. This indicates the geotextile response against the heaving and settlement of foundation soil. And the failure of the reinforced foundation soil occurred at high bearing pressure in a large deformation mode of circle due to the geotextile effect while the unreinforced foundation soil

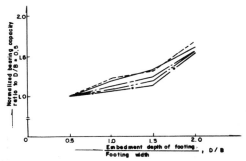

a) Case without geotextile

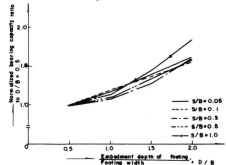

b) Case with geotextile

Fig.8. Normalized bearing capacity ratios for various D/B ratios

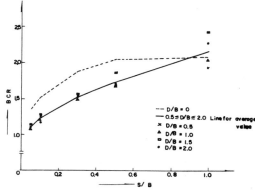

Fig.9. BCR variation with settlement of footing for different values of D/B ratio

219

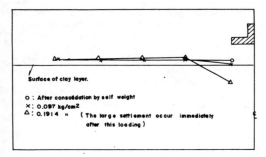

a) Case without geotextile

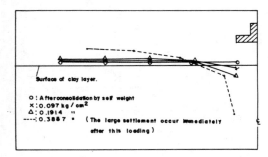

b) Case with geotextile

Fig.10. Typical changes of LED lamp location with regard to the increase of vertical load (in the case of H/B=0.5, D/B=0.0)

failed in a small deformation mode at low bearing pressure. It was also found that the vertical and lateral deformation of foundation soil becomes small as the distance of footing from geotextile layer increases and the embedment depth of footing increases regardless of the geotextile reinforcement.

4. CONCLUSIONS

This paper has reported a series of laboratory bearing capacity tests of a strip footing on sand/clay layer with and/or without reinforcing geotextile. The results from the experiments are summarized as follows:

1) The contribution of geotextile to the increase of bearing capacity becomes high as the distance of footing from geotextile layer is reduced. It also becomes high as the footing depth and the footing settlement increases.

2) The ratio of sand layer depth to footing width, which gives the greatest geotextile effect, falls between 0.5 and 1.0 for the settlements where S/B is less than 1.0.

3) The ratio of embedment depth of foot-

ing to footing width, which gives the greatest geotextile effect, is 0.0 for the settlements of S/B≤1.0 but the geotextile effect with regard to the embedment depth where D/B is between 0.5 and 2.0 is simillar for the settlements of S/B≤0.5.

4) Geotextile effect results in the bearing capacity increase over 100% for the conditions of 0.1≤H/B≤2.0, S/B≤1.0 and D/B≤2.0.

5) The failure of the reinforced foundation soil occurred at high bearing pressure in a large deformation mode of circle due to the geotextile effect while the unreinforced foundation soil failed in a small deformation mode at low bearing pressure.

REFERENCES

Akinmusuru, J.O., Akinbolade, J.A. & D.O. Odigie 1982. Bearing capacity tests on fiber-reinforced soil. Proceedings of Second International Conference on Geotextiles, Vol.3, Las Vegas, IFAI: 599-603.

Binquet, J. & K.L. Lee 1975. Bearing capacity tests on reinforced earth slabs. ASCE Journal of Geotechnical Engineering, 101: 1241-1255.

Fragaszy, R.J. & E. Lawton 1984. Bearing capacity of reinforced sand subgrades. ASCE Journal of Geotechnical Engineering 110: 1500-1507.

Giroud, J.P. & L. Noiray 1981. Geotextile-reinforced unpaved road design. ASCE Journal of Geotechnical Engineering, 107: 1233-1254.

Gourc, J.P., Matichard, Y., Perrier, H. & P. Delmas 1982. Bearing capacity of a sand-soft subgrade system with geotextile. Proceedings of Second International Conference on Geotextiles, Vol.2, Las Vegas, IFAI: 411-416.

Guido, V.A., Biesiadecki, G.L. & M.J. Sullivan 1985. Bearing capacity of a geotextile-reinforced foundation. Proceeding of 11th International Conference on Soil Mechanics and Foundation Engineering, San Francisco: 1777-1780.

Sellmeijer, J.B. & C.J. Kenter 1982. Calculation method for a fabric reinforced road. Proceedings of Second International Conference on Geotextile, Vol.2, Las Vegas, IFAI: 393-398.

Sorlie, A. 1977. The effect of fabrics on pavement strength; plate bearing tests in the laboratory. Proceedings of the International Conference on the Use of Fabrics in Geotechnics, Vol.1, Paris, Ecole Nationale des Ponts et Chaussees: 15-18.

International Geotechnical Symposium on Theory and Practice of Earth Reinforcement / Fukuoka Japan / 5-7 October 1988
© *1988 Balkema, Rotterdam. ISBN 90 6191 820 0*

Effect of installation methods on granular pile response

Madhira R.Madhav & C.Thiruselvam
Department of Civil Engineering, IIT, Kanpur, India

ABSTRACT: The effect of the method of installation—cased and uncased bore holes, number of lifts and compactive energy per lift given to granular piles and spacing, are studied in the field for single piles and in the laboratory for large groups. The paper describes the test methods, results, and the conclusions. The experimental results bring out the advantages of cased bore holes and larger compactive effort, in improving the behaviour of granular piles.

INTRODUCTION

Many sites which were considered to be uneconomical for development as stable and adequate support for conventional foundations and structures, can now be economically treated by the installation of granular piles (stone columns). Moderate increase in allowable bearing pressures and considerable reduction of settlement are the primary benifits of this method of ground improvement. In addition, by virtue of their high perviousness, granular piles act as free drains and permit rapid dissipation of excess pore pressures due to static and/ or seismic loading. Seed and Booker (1977) and Tatsuoka and Yoshimi (1980) recommend their use for improving resistance to liquefaction of alluvial deposits.

Granular piles are installed by the vibroflotation technique (Greenwood, 1970). In India, the rammed stone column and the preassembled stone column methods are popular (Datye, 1975). These two methods are labour intensive and are cost effective. No costly plant or equipment is required. Simple pile driving rigs can be adapted to install these piles. Mitchell (1982), Datye (1982a,b), Madhav (1982) and Barksdale and Bachus (1983) review the design, construction and analysis of the granular piles.

TEST DETAILS AND PROCEDURES

The soil that was used in both the in situ and the laboratory studies, is the local Kanpur silt available adjacent to the Geotechnical Engineering Laboratory in the academic area of IIT, Kanpur. The bore hole profile is given in Fig.1. The physical and the mechanical properties of the soil are Grain size : Sand 10-12 percent Silt 75-80 percent Clay 10-15 percent. Liquid Limit: 30 Plasticity Index 13-15; I.S. classification: CL or ML.
Undrained strength: $C_u = 0.35 \text{kg/cm}^2 \phi_u = 7.5°$.
$C_c = 0.33$; $p_c' = 0.65 \text{ kg/cm}^2$, $C_v = (0.6-1) \times 10^{-4} \text{ cm}^2/\text{sec}$. The granular pile material consisted of a mix of gravel and sand of the following grain size distribution.

Table 1: Grain size distribution of granular materials.

Percent Finer

Sieve size mm	6.25	2.0	1.0	0.6	.425	0.16
Gravel	95.60	17.4	1.7	1.6	1.4	1.4
Kalpi Sand	100	90.0	70.0	54.0	35.0	8.0

GL

	GL		SPT	q_c
				kg/cm^2
10	Silty clay (CL) LL 32	Sand 10	8	15
20	PI 16	Silt 69 Clay 21	10	30
30	Silty clay (CL) with		12	35
4·0	kankar			35
5·0	Silt (ML)	Sand 30% Silt 65% Clay 5%	6	38
	Non plastic			
6·0				

Figure 1. Soil Profile.

Angle of shearing resistance $\emptyset_s = 40°-45°$.

FIELD SET-UP

A test pit 1.5m x 1.5m by 1.0m deep was dug. Water was poured into the pit and allowed to stand for several days, to saturate the soil. For installing the pile, a bore hole 10cm dia and about 1m long was dug by gradual augering and removal of the soil. Prepared granular mix (2 gravel:1 sand) was poured into the hole in two or three lifts. Each lift was compacted by 5 or 10 blows from a standard penetration test hammer falling freely from a height of 35 cm. The set or the reduction in thickness of the pile length for each lift was measured. Care was taken to make the top surface of the pile as flat as possible. Loads were increased by 30 kg. each time, dial gauge readings being taken till there was no change. The piles were loaded to failure. The parameters varied were the method of installation of the pile-uncased and cased bore holes with angering, number of lifts, and compactive effort (number of blows per lift).

LABORATORY SET-UP

Keeping in view the constraints and limitations of small scale model tests, a C.B.R. mould (15 cm dia and 17 cm high) was used. Undisturbed soil samples were collected from the site adjacent to the Geotechnical Engineering Laboratory and kept in water for about a week for saturation. The mould was taken out of water, excess water drained and a bore hole made in the centre of the sample. The hole was filled with gravel-sand mixture in the ratio of 2:1, in two lifts, each lift being compacted by 5 or 10 blows from 2.5 kg. weight falling through a height of 25 cm.

The standard consolidometer frame was modified to accommodate the mould. The load from the lever was applied through a rigid plate. A dial gauge placed on top of the frame measured the settlement due to the loading. The loads were increased in stages and dial readings taken continuously till constancy. The test was continued till a stress level of 2.12 kg/cm² was reached. The parameters studied were the installation methods uncased and cased bore holes, diameter of the pile (5.1 cm, 6.1 cm, 7.62 cm or S/d = 3, 2.4 and 2.0), number of blows per lift(5 and 10).

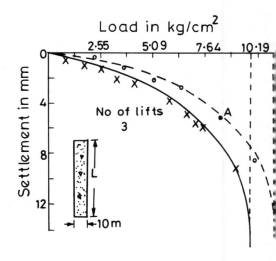

Figure 2. Load-Settlement Relation: In Situ Tests.

222

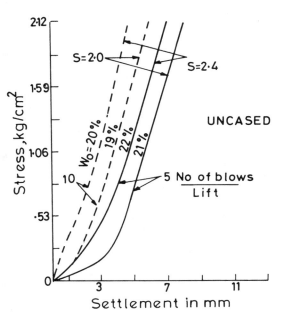

Figure 3. Stress-Settlement Relation: Model Tests; Uncased Hole.

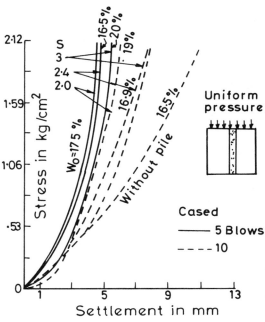

Figure 4. Stress — Settlement Relation: Model Tests; Cased Hole.

RESULTS

The load or stress versus settlement relation obtained from in situ tests is shown in Fig.2 for uncased bore hole with 2 and 3 lifts. The stress-settlement graph from the plate load test exhibits the typical local shear failure. The ultimate stress of about 1.15 kg/cm^2 corresponds to 6 C_u = 0.19 kg/cm^2, a value close to the one from undrained test. The load or stress-settlement graphs exhibit a linear trend at the lower stress level and become steep and nearly parallel to settlement axis at high stress levels. The ultimate loads in case of uncased hole are 800-825 kg if the pile is compacted in two lifts, 775-875 kg. for three lifts, and about 1450 kg. in case of cased borehole.

Cased bore hole causes densification of the in situ soil as the pipe is driven first, causing an increase in the strength of soil. Consequently the load carrying capacity is also very high. There is a possibility of loosening of soil in case of uncased bore hole and a reduction in strength resulting in

a smaller load carrying capacity. If the number of lifts is increased, the granular pile is compacted better and a higher load carrying is observed.

Table 2 compares the observed ultimate loads with those predicted based on pile type and bulging failure. The ultimate load based on pile formula is

$$Q_u = d \, l \, C_u + \frac{d^2}{4} \, 9 \, C_u \qquad (1)$$

while the ultimate load for bulging type failure according to Hughes and Withers (1975) is

$$Q_u = \frac{d^2}{4} \left(\frac{1+\sin\emptyset'}{1-\sin\emptyset'}\right) \, C_u \, N_c + \sigma_o \qquad (2)$$

where l and d are the length and diameter of the pile, C_u — undrained cohesion of in situ soil, and $\emptyset$ — the angle of shearing resistance of the pile material. It appears from the table that the pile formula predicts loads closer to observed ones indicating pile type failure rather than bulging failure. The length to diameter ratio of the piles is 8 to 9 and possibly bulging would govern failure for longer pile lengths.

223

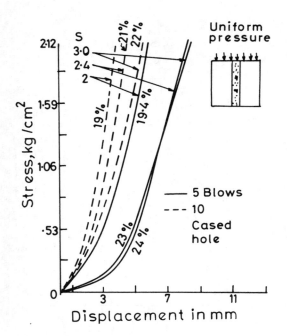

Figure 5. Stress—Settlement Rela-
        tion: Model Tests; Cased
        Hole.

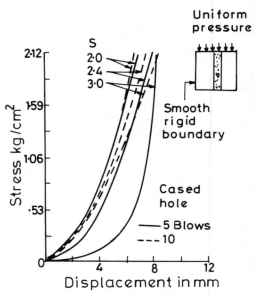

Figure 6. Stress—Settlement Rela-
        tion: Model Tests, Cased
        Hole.

LARGE PILE GROUPS

As mentioned earlier, a typical
unit is representative of the behaviour of granular pile reinforced
soil. It is observed that settlements terminate within 24 hours of
each stress increment.

The stress-settlement graphs
shown in Figs. 3-6, depict a flatter initial portion of the curves
at low stress levels and a steep
line at higher stress levels. An
yield stress, q_y, can easily be
established beyond which the granular pile reinforced soil behaves
as a very stiff material. The
effect of pile diameter and compactive energy on the unit cell behaviour is shown in Fig.3 for uncased
hole. The larger the diameter of
the pile closer is the spacing and
lesser will be the settlements. If
the pile is installed by heavier
compaction 10 blows per lift compared to 5 blows per lift, the settlement response is better i.e.,
lesser settlements are mobilised
due to higher unit weights achieved
for the granular pile material.
Thus with 5 blows per lift, the
unit weights are 1.4 to 2.55 g/cm^3

while with 10 blows they range
between 1.9 to 2.8 g/cm^3. A denser
granular pile is stiffer. The
settlements will naturally be
smaller. Similar results are observed for a cased bore hole (Fig.4).
The curve for untreated soil is
also shown in this graph for comparison.

The behaviour of granular pile
consisting of gravel alone is
shown in Fig.5. The pile is installed in a cased hole. If the angular gravel is replaced by peastones
and sand, the behaviour shown in
Fig.6 is observed.

The settlement reduction ratio,
S_R, is calculated as

$$S_R = \frac{\text{Settlement of treated ground}}{\text{Settlement of untreated ground}}$$

It is observed that the settlement
ratio is smaller for closer pile
spacing (larger pile diameter),
larger compactive effort (10 blows
per lift) and for angular material
than rounded pea gravel. The differences between the effects of uncased and cased methods of installation could not be established
clearly. Cased hole installation

however appears to be slightly better than the uncased method. The settlement reduction factors are in the range 0.34 to 0.74 a range on the higher side because of the small l/d ratio (2.3 to 3.45).

CONCLUSIONS

The effect of method of installation on the behaviour of granular pile reinforced soil is studied. The behaviour of a large group represented by a single unit consisting of a pile surrounded by soil influenced by it, is studied in the laboratory. Cased and uncased bore holes, amount of compactive energy, number of lifts, spacing of piles in case of groups, are the factors studied. The load carrying capacity of single piles and pile groups is more and the settlements are less in case of cased than uncased bore holes. Similarly larger the compactive energy, more the number of lifts, and closer is the spacing, the better will be response of the granular piles.

REFERENCES

1. Aboshi, H., Ichimoto, E., Enoki, M., and Harada, K., (1979) 'The Compozer – a Method to Improve Characteristics of Soft Clays by Inclusion of Large Diameter Sand Columns', Int. Conf. on Soil Reinforcement, Paris, Vol.1, pp. 211-216.

2. Barksdale, R.D. and Bachus, R.S., (1983), 'Design and Construction of Stone Columns', Vol.I, Rep.No. FHWA/RD-83/026, Federal Highway Adm., Washington.

3. Datye, K.R. (1982a), 'Simpler Techniques for Ground Improvement; 4th IGS Lecture, Hyderabad IGJ, Vol.12, No.1, pp.1-82.

4. Datye, K.R., (1982b), 'Settlement and Bearing Capacity of Foundation Systems with Stone Columns', Proc. Symp. on Soil and Rock Improvement Tech., Bangkok, Paper No.A-1.

5. Datye, K.R. and Nagaraju, S.S., (1975), 'Installation and Testing of Stone Columns', Proc. IGS Spec. Session, 5th ARC, Bangalore India, pp. 101-104.

6. Greenwood, D.A. (1970), 'Mechanical Improvement of Soils Below the Ground Surface', Proc. Ground Engrg. Conf., I.C.E., pp. 9-20.

7. Hughes, J.M.O., Wilhers, N.J., and Greenwood, D.A. (1975), 'A Field Trial of Reinforcing Effects of Stone Columns in Soil', Geotechnique, Vol.25, No.1, pp. 31-44.

8. Madhav, M.R. (1982), 'Recent Developments in the use and Analysis of Granular Piles', Proc. Symp. on Soil and Rock Improvement Tech., Bangkok, Paper No. 5.

9. Mitchell, J.K. (1981), 'Soil Improvement State of the Art Report', X ICOSMFE, Stockholm, Vol.4, pp. 509-566.

10. Seed, H.B. and Booker, J.R. (1977), 'Stabilization of Potentially Liquefiable Sand Deposits Using Gravel Drains', J. Geotech. Div., ASCE, Vol.103, No. GT7, pp. 757-768.

11. Tokimatsu, K. and Yoshimi, Y., (1980), 'Effects of Vertical Drains on the Bearing Capacity of Saturated Sand During Earthquakes', Int. Conf. on Engrg.

for protection from Natural Disasters, Bangkok, Jan. 1980, pp. 643-655.

APPENDIX I

The rammed stone columns and preassembled stone columns are two techniques developed by Datye (1975). They are described briefly below.

Rammed stone columns is a technique in which either closed end or an open pipe is driven into the ground. The hole may be created by boring also. Stone and sand are poured into the hole in two or three lifts and each lift is compacted by a heavy weight dropped through a fixed height a given number of times. The casing pipe is withdrawn gradually during the ramming phase.

Datye (1982) describes the preassembled stone columns through uncased bore holes. The casing if used is of short length only. An enclosure of bamboo strips into which stone is placed, is lowered, and the stone compacted. When the length of the granular pile to be installed is 8-10 m only, this method of installation is economic and rapid.

International Geotechnical Symposium on Theory and Practice of Earth Reinforcement / Fukuoka Japan / 5-7 October 1988
© 1988 Balkema, Rotterdam. ISBN 90 6191 820 0

Model and field tests of reinforced pavement on soft clay ground

N.Miura, A.Sakai & Y.Taesiri
Saga University, Saga, Japan

K.Mouri & M.Ohtsubo
Saga Prefectural Office, Saga, Japan

ABSTRACT: Model tests and numerical analysis of reinforced pavements by polymer grid are performed to investigate the effect of polymer grid in suppressing a non-uniform settlement of asphalt pavement constructed on the soft clay ground. The cyclic load tests on reinforced and unreinforced model pavements indicate that the polymer grid is useful for suppressing non-uniform settlement of pavement under cyclic loading. Field tests of 300 m length with six sections of different kinds of pavement are carried out in a road. The reinforcing effect of polymer grid is discussed and compared with a conventional pavement method.

1 INTRODUCTION

The Ariake clay, a highly sensitive and problematic marine clay, is deposited around the shores of Ariake Bay in northern Kyushu, Japan (Miura et al, 1987). Several geotechnical problems have been caused by this clay (Miura et al., 1986), and this study deals with one of them, non-uniform settlement which includes rutting in the pavement. This can be caused by soft spots in clay, insufficient compaction of base materials, traffic loads and also by subsidence due to pumping of ground water. Among techniques used to suppress such non-uniform settlement of the pavement, replacement of the soft clay with a good material and subgrade stabilization have been widely employed in Saga district. These conventional techniques, however, are generally expensive and during excavation, mixing, transportation of the clay, some problems are likely to occur both in engineering and environmental aspects. Moreover, the thicker and hence the heavier the pavement is, the greater the settlement occurs, and from this point of view the conventional methods are disadvantageous.

Recently, reinforcement technique for suppressing non-uniform settlement of pavement on soft ground, especially reinforcement in base course by using polymer grids, has been noticed and studied (Giroud et al., 1984). The mechanism of reinforcement by the polymer grid is considered as: tension membrane, interlocking and separation effects. Giroud et al. (1984) presented the design procedure of unpaved roads with polymer grid, in which they has taken into consideration the confinement and load distribution effect of polymer grid.

In order to examine the applicability of the polymer grid for reinforcement of the base on the very soft clay subgrade, a series of large scale laboratory tests were carried out and the results were analyzed by FEM. Based on the laboratory tests, test road of 300 m length was constructed on "Takeo – Fukudomi" route and the results obtained are discussed in this paper.

2 LABORATORY TESTS ON MODEL PAVEMENTS

2.1 Materials and testing method

Model pavements were constructed in a soil tank made of concrete (150cm x 150cm in area and 100cm in depth) as shown in Fig. 1. The clay used as subgrade is a very soft and sensitive marine clay, the Ariake clay, of which basic properties are: specific gravity = 2.625; natural water content = 129%; liquid limit = 117%; plastic limit = 39%. The clay was reconstituted by applying a pressure of 0.05 kg per sq. meter for 2 months into 60 cm thick subgrade, on which subbase and base course were compacted and then an asphaltic concrete layer of 5 cm thick was placed. Repeated load of 2 kg per sq. meter, frequency of 0.18 Hz (4 seconds load and 2 seconds unload) was applied through a loading plate of 20 cm in diameter. Vertical settlements of the surface

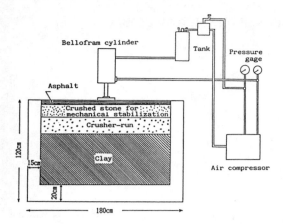

Fig.1 Model test of reinforced pavement

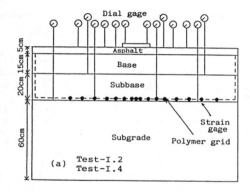

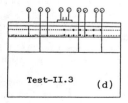

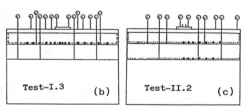

Fig.2 Schematic diagram of one and two
layers polymer grid reinforced
pavement

Table 1 Details of model tests

Test No.	Number of Polymer Grid Layer	Consolidation (kgf/cm²)	Polymer Grid	Location of Polymer Grid
Test-I.1	0			Unreinforced
I.2	1		SS2	On top of the subgrade
I.3	1	0.1	SS2	On top of the subbase
I.4	1		SS3	On top of the subgrade
Test-II.1	0			Unreinforced
II.2	2	0.05	SS1	On top of the subgrade and the subbase
II.3	2		SS1	On top of the subbase and in the base

and also that of each layer were measured by using dial gauges and differential transducers. Two types of model pavements were tested as shown in Fig.2. Test-I series are models of one-layer polymer grid system; Test-II series are models of two-layer polymer grid system. The details of these models are summarized in Table 1.

2.2 Effect of polymer grid

Modulus of subgrade reaction, K measured by using 20 cm in diameter loading plate and calculated equivalent K30 on each layer, before and after the cyclic loading, are summarized in Fig. 3 (Test-I.2). The ratios of modulus K of each layer to that of clay layer Kc, measured after cyclic loading tests, are indicated in Table 2, showing that the K/Kc value of reinforced pavement is 30% higher than that of the unreinforced one.

Figure 4 shows the surface settlement of model pavement with and without polymer grid, from which it can be seen that the polymer grid is very effective in suppres-

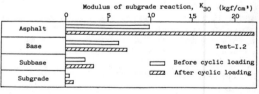

Fig.3 Modulus of subgrade reactions of
different layers from Test-I.2

Table 2 Ratio of modulus of subgrade
reactions of model tests

	Unreinforced (A) Test-I.1	Reinforced (B) Test-I.2	B/A
K_{30}(Subbase)/K_{30}(Subgrade)	2.8	3.7	1.32
K_{30}(Base) /K_{30}(Subgrade)	6.4	8.1	1.27
k_{30}(Asphalt)/K_{30}(Subgrade)	19.3	25.3	1.31

sing the settlement due to cyclic loads. The settlement curves for the models in two-layer system are comparably shown in Fig. 5, which suggests that the polymer grid can effectively suppress the settlement of the pavement, and in the case of Test-II.2 the accumulated settlement was generally less than 50% of that of the unreinforced pavement. From the above discussion, we can say that the polymer grid functions more effectively in suppressing the settlement when it is placed on the subgrade than when it is on the subbase.

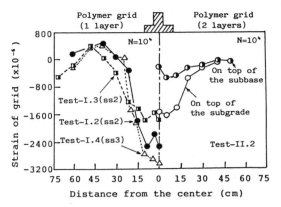

Fig.6 Strain distribution of polymer grids

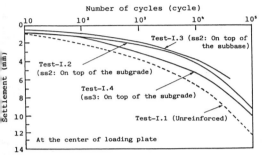

Fig.4 Surface settlements of one-layer reinforced and unreinforced pavements

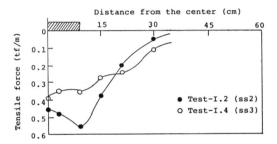

Fig.7 Calculated tensile force in polymer grids

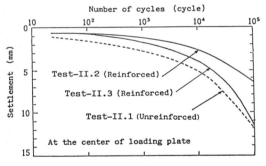

Fig.5 Surface settlements of two-layer reinforced and unreinforced pavements

Figure 6 shows the strain distribution of polymer grid under cyclic loading. This indicates that the magnitude of tensile strain of the polymer grid is maximum at the point just below the center of the loading plate. This figure also indicates that the maximum tensile strain is larger in Test-I.2 than in Test-I.3. This tendency is well consistent with the results in Figs. 4 and 5. It can be said that the polymer grid is better placed on the sub-

grade than on the subbase.

Comparison of tensile forces acted on polymer grids are shown in Fig. 7, in which the stress was calculated from the product of strain and Young's modulus. Since the Young's modulus of ss2 is larger than that of ss3, the tensile force of the former becomes larger than that of the latter, even though the magnitude of strain of ss2 is smaller than that of ss3 (see Fig. 6). The distribution of tensile force of polymer grid seems to be closely related to the distribution of the settlement of the corresponding layer, which suggests that one of the effects of polymer grid to suppress the settlement of pavement comes from tension membrane effect.

2.3 FEM analysis

For the deformation analysis of a reinforced soil structure with a heterogeneous material such as the polymer grid, analytical method capable of expressing the behavior of discontinuous plane should be used. The authors use the method in which the joint element representing the property of discontinuous plane is combined with the truss element transmitting the axial force only (Kutara et al., 1985). The joint element has two unit stiffness, a normal stiffness, kn, and a shear stiffness, ks. The polymer grid is modeled by the truss element whose ends are connected by the pin joint.

Deformation analysis by FEM was made for a model pavement shown in Fig. 2(a) reinforced by one layer polymer grid. We assume that all the materials composing the pavement are elastic. Young's modulus E and Poisson's ratio of these materials (asphaltic concrete, crushed stones for base course and subbase, clay for subgrade) are assumed as: E=500, 250, 100, 20 kg per sq. cm, ν = 0.38, 0.43, 0.43, 0.47, respectively. Also, stiffnesses of the joint element are assumed as ks = 1000 and kn = 100 kg per cu. cm, respectively. Detailed description on this analysis is appeared elsewhere (Sakai et al., 1988).

Figure 8 shows a comparison of the analytical and experimental results of surface settlement at 10,000 cycles. In this analysis, however, the interlocking effect of polymer grid is not taken into consideration. Experimental evidence indicates that the interlocking effect of polymer grid may play an important role in suppressing surface settlement of the pavement (Miura et al., 1986). Namely, in a previous experiment, we observed that a pavement reinforced with one layer of polymer grid settled less than the unreinforced pavement even though only the compression strain was measured in every parts of the polymer grid (The reason why the compression strain was measured instead of the tensile strain, contrary to the prospect, was considered that the surface of the base course on which a polymer grid was placed might be in

a slightly convex shape, and when vertical load was applied to the surface of the model pavement, the polymer grid was compressed downward resulting in the compressive strain). In order to consider the inter locking effect in the deformation analysis further investigation is necessary.

3 OBSERVATION IN TEST ROAD

3.1 Description of test road

Test sections were constructed on a public road "Takeo-Fukudomi route" in Shiroishi cho, Saga. The main purpose of constructing this test section is to check the function of the polymer grid as a reinforcing material for pavement on the very soft ground. The deposited clay in this area is the Ariake and its thickness is about 15 m. Ground subsidence due to pumping up of ground water also produces difficult geotechnical problems in this district. Pavements in this area are unavoidably experienced settlement caused not only by consolidation (in both static and repeated loadings) but also by subsidence. Thus, development of technique for suppressing non uniform settlement of pavement is very important.

Figure 9 illustrates the schematic diagram of test road of 300 m long which has 6 sections of 50 m each. On the subsoil of Ariake clay, tailings refused from coal mines had been placed in a thickness of about 60 – 80 cm, then the top 20 cm of this tailings was excavated and made into a flat surface of being approximately equal CBR at any point. Sections 1 and 3 were constructed by conventional methods; the former was by lime stabilization of 40 cm thick and the latter was by placing a subbase of 25 cm and a base course of 20 cm, each of them is 5 cm thicker than that of reinforcement sections. In sections 2, 4, 5 and 6 polymer grids (ss3 in sections 2 and 4, and ss2 in sections 5 and 6) were placed at the interfaces as shown in Fig. 9. The base and subbase were reduced by 5 cm each compared with unreinforced section 3. To investigate the effect of polymer grid on the function of stress distribution, earth pressure cells were placed in each section as indicated in Fig. 9.

The construction costs of sections 2 to 6 were almost the same, 3600 to 3800 yen per sq. meter, and that of section 1 was 5600 yen per sq. meter. The purpose of this study is to examine the effectiveness of polymer grid as reinforcement of the base course. Therefore, the performance of section 1 is omitted from the discussion.

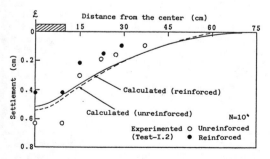

Fig.8 Comparison of measured and predicted settlements by FEM

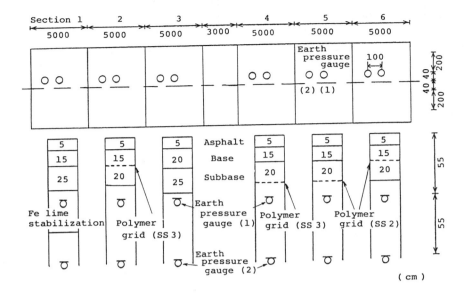

Fig.9 Plan and cross sections of test road

To evaluate bearing capacity of every layers of each section, plate load tests of 30 cm diameter (JSF T25-81) were carried out. Also Benkelman beam tests were performed, and during these tests vertical stresses were measured by the earth pressure cells.

3.2 Test results

Modulus of subgrade reaction, K30 are plotted in Fig. 10, which were obtained by loading tests immediately after the completion of the pavement. The experimental facts show that the modulus K30 measured on the surfaces of the reinforced sections (2, 4, 5 and 6) are lower than that of conventional method 3. Differences of vertical pressures between the five sections, shown in Fig. 11, are similar to that of the modulus, K30 in Fig. 10. Shown in Fig. 12 are results of Benkelman beam tests which were carried out immediately after construction, one and three months later, showing that the conventional method of construction is the best among the five sections. These results are rather unexpected and contradict the model test results, and they might come partly from overestimation of polymer grid function and partly from insufficient compaction of the base owing to negative interlocking effect of the polymer grid. It may be early to make final conclusion at this stage. The field

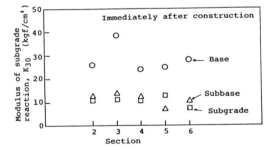

Fig.10 Comparison of modulus of subgrade reactions of test sections

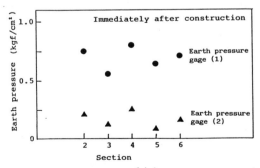

Fig.11 Comparison of vertical stresses of test sections

231

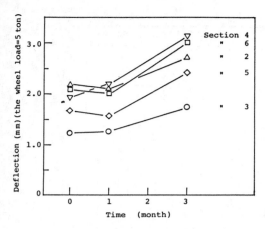

Fig.12 Benkelman beam test results

measurement will be continued for two years, and the final evaluation will be made.

By comparing the field and model test results, we find that polymer grid placed at the top of subgrade is more effective than at the top of subbase. Also the ss2 polymer grid is better than the ss3 if placed at the top of subgrade.

4 CONCLUSIONS

Model and field tests of reinforced pavement by polymer grids on the very soft ground were carried out and the following conclusions can be drawn.

1. Model and field test results showed that polymer grid placed at the interface of subbase and subgrade effectively suppresses the settlement under repeated loading.

2. The magnitude of surface settlement closely relates to the tensile force induced in the polymer grid.

3. FEM analysis, in which joint element was introduced, indicated that the polymer grid is not effective in suppressing the surface settlement of pavement. This contradiction suggests that an important function of polymer grid in the base may be the interlocking effect on which present study did not dealt with.

4. According to the field data obtained one and three months after the construction, the modulus of subgrade reaction, K of the reinforced pavements are smaller than that of conventional pavement. This might come partly from overestimation of polymer grid function and partly from insufficient compaction of the base owing to the action of the polymer grid.

5. To make polymer grid useful as a reinforcing material of the base in a pavement on the soft ground, compaction works should be performed carefully, not leaving any gap between polymer grids and underneath layers.

REFERENCES

Giroud, J.P.,Bonaparte, R. 1984. Design and unpaved roads and trafficed areas with geogrids, Proc. Symp.on Polymer Grid Reinforcement, London, 116-127.

Kutara, K. Gomado, M. Takeuchi,T. and Maeda Y. 1985. Use of geotextiles as a countermeasure for differential settlement in road embankments, JSSMFE, Vol.33, No.5 27-32.

Miura, N., Bergado, D.T., Sakai, A. and Nakamura, R. 1987. Improvements of soft marine clays by special admixtures using dry and wet jet mixing methods, 9th South east Geotechnical Conference.

Miura, N. Sakai, A. and Mouri, K. 1987 Model tests on reinforced pavement on soft clay ground, 1st Geotextile Symposium,1-6

Sakai, A., Miura, N. and Mouri, K. 1988 Model test and analysis on the reinforce base of pavement on soft ground, Report of the Faculty of Science and Engineering Saga Univ., Vol.16, No.2, 133-140.

International Geotechnical Symposium on Theory and Practice of Earth Reinforcement / Fukuoka Japan / 5-7 October 1988
© 1988 Balkema, Rotterdam. ISBN 90 6191 820 0

Design of geotextile and geogrid reinforced unpaved roads

G.P.Saha & M.H.Kabir
Bangladesh University of Engineering and Technology, Dhaka, Bangladesh

ABSTRACT: A new nonlinear approach to design of geotextile and geogrid reinforced unpaved roads on clay subgrades is presented in this paper. This considers the nonlinear behaviour generally exhibited by most geotextile and geogrid reinforcing materials as well as clayey subgrade soils. The design method is based on a closed form plane strain analysis of the deformation track produced under wheel loading, considering the compatibility of rutting between the reinforcing element and the clayey subgrade soil. Analysis under traffic loading is produced by using the so called "equivalent material" approach to take into consideration the degradation of behaviour of the constituting materials under repeated traffic loading which are modelled using newly developed appropriate constitutive relations.

1 INTRODUCTION

Although several researchers have reported results of model and prototype tests showing beneficial effects of reinforcing on unpaved roads, Giroud and Noiray (1981) should be credited for presenting an elaborate numerical treatment of the problem using soil, reinforcement, loading and the relevant geometrical parameters. However, like traditional AASHTO method (using CBR) in their analyses the elastic and ultimate limiting states of the subgrade clay soil is considered to be developed without considering the state of the accompanying strains and displacements. The only soil parameter which is used is the undrained shear strength (c_u). Moreover, their design is based on a single valued modulus of reinforcing geotextiles or geogrids which normally exhibit nonlinear load-strain behaviour. Again almost all soft clays also show nonlinear deviator stress-strain behaviour. As two materials (clay and reinforcement), with different characteristic properties and load transfer mechanisms, are involved here in sharing the pressure transmitted from the wheels, a sound design method should consider the amounts as well as the degree of mobilization of resistances in the materials under compatible deformation conditions. Therefore, the design method reported in this paper is based on two considerations. These are:

1. Nonlinear behaviour of both the subgrade soil and reinforcing material under repeatative loading conditions.

2. Compatibility of rutting between subgrade soil and reinforcing elements under all serviceability conditions.

Giroud and Noiray did not consider these, which in some cases may lead to unsafe design.

In developing the proposed design method, the deformation of clay subgrade under wheel loading is analysed by considering plane-strain situation. This is performed by using an analytical scheme, incorporating the true deviator stress-strain behaviour of subgrade clay, suggested by Prakash, Saran and Sharan (1984). Hyperbolic (Kondner 1963) stress-strain – number of load repeatations behaviour of subgrade clays as well as reinforcing materials are used. Almost all the clays and reinforcing geotextiles and geogrids were found to obey this law to a reasonable degree of accuracy.

A generalised approach of representation of the nonlinear mechanical behaviour of clayey subgrade soils as well as those for reinforcing elements under repeated loading condition are established and presented. Analysis of behaviour under repeated traffic loading condition is achieved by using the equivalent material approach. This assumes the subgrade soils and reinforcing elements to be new (equivalent) materials possessing degraded mechanical behaviour which is appropriate for the number of load repeatations under consideration.

Design charts showing the required thicknesses of aggregates for different degrees of rutting appropriate for both unreinforced and reinforced conditions are presented.

These are prepared for one type of clay and one type of geotextile reinforcement only. A comparison of the new approach of design with the conventional Giroud and Noiray method is also presented. An example is also produced to show the applicability of the design methodology.

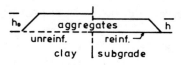

Fig. 1 Typical cross-section

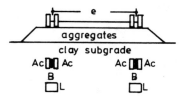

Fig. 2 Wheel contact area

2 DESIGN PARAMETERS AND THEIR CHARACTERIZATION

The geometric and material parameters, relevent to design of reinforced unpaved roads, and their characterization procedure are described briefly in the following sections. To keep uniformity, the notations used by Giroud and Noiray are also used here.

2.1 Geometry and loading condition

A typical cross-section of an unpaved road, with appropriate notations, is shown in Figure 1. These are (Figure 1), ho = thickness (m) of aggregate layer required under unreinforced condition; h = thickness (m) of aggregate layer required under reinforced condition.

The geometrical description of wheel contact area of American-British standard equivalent axle may be described as follows (Figure 2). For on-highway trucks: $L = B/\sqrt{2}$ and $B = \sqrt{P/pc}$ and for off-highway trucks $L = B/2$ and $B = \sqrt{P\sqrt{2}/pc}$. Where, P = axle load (= 80 kN) and pc = tyre inflation pressure (= 620 kPa). B and L are the equivalent dimensions of dual wheel contact areas and associated area between them.

2.2 Properties of aggregates and subgrade soils

The aggregates are assumed to be well graded, to provide sufficient interlock and effective load distribution. These should have CBR (California Bearing Ratio) larger than 80.

The design philosophy presented here considers different degrees of rutting. Therefore, a single limiting value description, using either undrained shear strength (c_u) or California Bearing Ratio (CBR) for the subgrade clay soils which normally exhibit nonlinear stress-strain behaviour is not sufficient. Moreover, a realistic representation of behaviour of clay soils under repeated wheel loading can only be modelled through properties and parameters obtained from repeated loading triaxial tests. Such tests on representative samples under representative confining condition are used here to model the behaviour of the clayey subgrade

soil. A new nonlinear constitutive relation for stress (σ) - strain(ε) - number of load repeatations (N) behaviour has been developed by the authors (Saha, 1988). Hyperbolic representation forwarded by Kondner (1963) was used to develop these relations. The $\sigma - \varepsilon - N$ relation for a silt clay, which is used to demonstrate the philosophy of the new design method is presented here as follows:

$$\sigma = \varepsilon (\sigma_1 - \log N/(c_1 + c_2 \log N))/((a_1 + c_3 \log N)\varepsilon$$

$$+ (1 - a_1 - c_3 \log N) \varepsilon) \qquad (1)$$

where σ_1 (=40 kPa), ε_1 (=0.2) and a_1 (=0.198) are limiting values and c_1 (=0.105 kPa-1), c_2 (=0.1098 kPa^{-1}) and c_3 (=0.02228) are calibration factors.

The $\sigma - \varepsilon - N$ relations, for the clay under consideration, obtained from tests as well as those from equation (1) are presented in Figure 3(a) showing reasonable degree of agreement between them.

2.3 Properties of reinforcing elements

The load strain behaviour of most geotextil and geogrid reinforcing materials exhibit nonlinearity under wide ranges of strain levels. Therefore a single modulus represen tation (2 percent strain level) may lead to considerable error in the analyses. Load (Pr) -strain (εr) - number of load repeata tions (N) relations similar to those sugges ted for clays were also developed using hyperbolic representation. Testing techniques for evaluation of load-strain behaviour of geotextiles and geogrids respectively, have been detailed by McGown, Andrawes and Kabir (1982) and McGown, et. al (1984). Similar appropriate techniques, with slight modifi-

ation for performing tests under repeated
loading was adopted for this study (Saha,
1988).

A woven tape geotextile constructed of
polypropylene tapes having weight per unit
area of 240 gsm, was used to demonstrate the
design philosophy. The $Pr - \varepsilon_r - N$ relation
of the geotextile, is presented as follows:

$$Pr = \varepsilon_r (Pr1 - \log N/(r1 + r2\log N))/((br + r3\log N)$$

$$\varepsilon_{r1} + (1 - br - r3\log N) \varepsilon_r) \qquad (2)$$

Where $Pr1 (=54.5$ kN/m$)$, $\varepsilon_{r1} (=0.1)$ and $br (=
0.55)$ are limiting values and $r1 (=0.3354$
m/kN$)$, $r2 (=-0.03357$ m/kN$)$ and $r3 (=0.0406)$
are calibration factors.

The $Pr - \varepsilon_r - N$ relations for the geotextile,
from tests as well as those from equation
(2) are presented in Figure 3(b), showing
reasonable degree of agreement between them.

3 ANALYSIS OF BEHAVIOUR OF ROAD STRUCTURE

Giroud and Noiray (1981) used two separate
analyses, one is for very light traffic and
the other is appropriate for heavy traffic.
For light traffic they used the so called
quasi-static analysis in which soil and reinforcement data from monotonic loading
tests are used. For analysis under heavy
traffic loading they used a modification of
the results obtained from quasi-static analysis by using empirical equation given by
Webster and Alford (1978), based on in-situ
test data produced by Hammit (1970). In contrast, the analyses poduced here uses unified
approach for any intensity of traffic loading. These are based on newly developed soil
and reinforcement constitutive relations
obtained from repeatative loading tests
which were described in section 2.

3.1 Load distribution through aggregate layer

A pyramidal distribution of load from the
wheels through aggregate layer is assumed.
The load distribution pattern and the geometrical representation of the problem for
cases with or without reinforcement are presented in Figure 4. The pressure at the base
of the aggregate layer, due to wheel loading
from off-highway trucks, may be represented
as follows (Figure 4). For cases without reinforcement, the pressure is:

$$po = P/2(B + 2ho \tan \alpha o)(B/2 + 2ho \tan \alpha o)$$

$$+ \gamma ho \qquad (3)$$

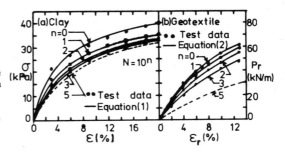

Fig. 3 Mechanical behaviour

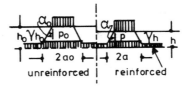

Fig. 4 Pressure distribution

For cases with reinforcement, the pressure
is:

$$p = P/2(B + 2h \tan \alpha)(B/2 + 2h \tan \alpha) + \gamma h \qquad (4)$$

In line with Giroud and Noiray recommendation
it is assumed that $\tan \alpha o = \tan \alpha = 0.6$. These
values are used in all subsequent calcula-
tions presented here.

3.2 Behaviour of clay subgrade

The clay subgrade is assumed to be semi-
infinite, homogeneous and isotropic. The load
deformation behaviour of clay subgrade under
loading from wheels transfered through the
aggregate layer is obtained by using the
calculation scheme suggested by Prakash,
Saran and Sharan (1984). The method utilizes
two parameter hyperbolic deviator stress-
strain relationship for clay soils. These are
obtained from the constitutive relations
$\sigma - \varepsilon - N$ described earlier. In this method
the clay layer beneath the loaded area is
divided into a number of horizontal slices.
The total deformation is then obtained as the
summation of deformations of each slice which
is denoted as 'S'. The calculations are repeated for different values of width of loaded area (2a) and prssure on loaded area
(po). The plot of pressure, po and the ratio
δ (=S/2a) is normally found to trace a unique
curve. Prakash, Saran and Sharan found very
good agreement amongst results obtained from,
this method, finite element analysis and test
data.

The $po - \delta$ curves for, different values of
load repeatation (N), for the clay,

described earlier, are presented in Figure 5(b). The hyperbolic representation of the po - δ relations are also presented in Figure 5(b).

3.3 Geometry of deformation

The wavy shape of the deformed unpaved road structure results from incompressibility of the saturated clay subgrade. Therefore, the volume of soil displaced downwards due to wheel loading must be equal to that displaced upward due to heaving. Schematic diagrams of unreinforced and reinforced unpaved road structures are presented in Figures 5(a) and 6(a) respectively. The wavy concave (AB) and convex (BB) parts were found to resemble arcs of parabolae by Webster and Watkins (1977), from full scale tests. The geometry of the three possible modes of deformation may be represented mathematically as follows (Figures 5(a) and 6(a)). Where, r = rut depth.

Mode1: $a' > a$, $S = ra'/(a+a')$ (5)

Mode2: $a' < a$, $S = 2ra^2/(2a^2 + 3aa' - a'^2)$ (6)

Mode3: $a' = a$, $S = r/2$ (7)

3.4 Unpaved road without reinforcement

The design of unpaved road without reinforcement involve determination of thickness of aggregate layer for different degrees of rutting. It has been discussed earlier that po-δ relation for clay subgrades may be represented by hyperbolic functions. Therefore, the resistance, po, of the clay subgrade may be represented as:

$$po = \delta/(s1 + s2\,\delta) + \gamma\,ho$$ (8)

where, s1 and s2 are hyperbolic parameters (Figure 5(b)) for clay subgrade appropriate for the number of load repeatation (N). Equating equations (3) and (8) and putting $\delta = S/2\,ao$.

$$P/(2(\sqrt{P\sqrt{2}/pc} + 1.2ho)(\sqrt{P/2pc\sqrt{2}} + 1.2ho))$$

$$= S/(2\,ao.s1 + s2S)$$ (9)

Values of 'S' as a function of 'r' may be obtained from equations (5), (6) and (7) depending on the mode of deformation. Design charts giving thicknesses (ho) of agreegate layer for different degrees of rutting may now be produced from equation (9) using appropriate soil parameters s1 and s2. Such chart for unpaved road on the clay subgrade

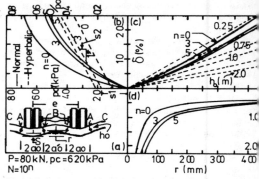

Fig. 5 Design chart for unreinforced unpaved road

for different intensities of traffic loading (N) conditions are produced in Figures 5(c) and (d).

3.5 Unpaved road with reinforcement

The use of reinforcement in an unpaved road structure may be described as producing two beneficial effects. These are:

1. Confinement of subgrade soils between and beyond the wheels, resulting in a system offering additional resistance against rutting.
2. Releaving the subgrade soil of some pressure transmitted from the wheels, thereby increasing the overall carrying capacity.

In this analysis, like that by Giroud and Noiray, the pressure releaving effect is only considered.

From equilibrium consideration, the pressure, p, transmitted from the wheels through the aggregates on part AB (Figure 6(a)) of the reinforced road should be equal to:

$$p = ps + pr$$ (10)

where, ps and pr are mobilised reactions (pressure) in the clay subgrade and the reinforcement respectively, appropriate for the intensity of traffic loading (N), under the same degree of rutting. Therefore, determination of ps and pr should be based on analysis of the kinematics of the deformation track produced due to wheel loading. The compatibility of rutting was not considered by Giroud and Noiray in developing their design charts.

In reinforced unpaved road, for a known degree of rutting, the pressure shared by the clay subgrade may be determined from the following equation:

$$ps = \delta/(s1 + s2\,\delta) + \gamma\,h$$ (11)

236

where $\delta = S/2a$ and the values of 'S' as a function of rut depth may be obtained from equations (5), (6) and (7). The force due to pressure, pr acting on zone AB (Figure 6(a)) is equivalent to the vertical component of the tension Pr in the reinforcing element acting at points A and B. Therefore:

$$a.pr = Pr. \cos\beta \qquad (12)$$

From property of parabola $\cos\beta = 4\sqrt{\delta^2/(16\delta^2+1)}$. Again from hyperbolic load strain representation of reinforcing elements:

$$Pr = \varepsilon r/(b1+b2\varepsilon r) \qquad (13)$$

The parameters b1 and b2, are dependent on the characteristics of reinforcing element and intensity of traffic loading (N). Values of b1 and b2 may readily be determined from equations (2), (12) and (13). Putting values of $\cos\beta$ and Pr in equation (12):

$$a.pr = (4\varepsilon r/(b1+b2\varepsilon r))(\sqrt{\delta^2/(16\delta^2+1)}) \qquad (14)$$

From the assumption of parabolic deformation the strain in the reinforcing element may be expressed as:

$$\varepsilon r = (1/2)(\sqrt{1+16\delta^2}+(1/4\delta)(\ln(4\delta+\sqrt{1+16\delta^2})-2) \qquad (15)$$

From equations (14) and (15) it can be found that pr.a may be expressed as a function of δ only for different values of N. pr.a – δ plots for the reinforcing element for different intensities of traffic loading (N), are presented in Figure 6(b). The pressure pr may be obtained from equation (14) as:

$$pr = (4\varepsilon r/a(b1+b2\varepsilon r))(\sqrt{\delta^2/(16\delta^2+1)}) \qquad (16)$$

Using equations (4),(11) and (16), equation (10) takes the form:

$$P/2(\sqrt{P\sqrt{2}/pc}+1.2h)(\sqrt{P/2pc\sqrt{2}}+1.2h))=\delta/(s1+s2\delta)$$
$$+(4\varepsilon r/a(b1+b2\varepsilon r))(\sqrt{\delta^2/(16\delta^2+1)}) \qquad (17)$$

The parameters εr and δ are functions of the degree of rutting. Design charts giving required thicknesses of aggregates, 'h' due to different degrees of rutting, may now be produced from solution of equation (17).

The design charts for the reinforcing element on the clay subgrade for different intensities of traffic loading (N) are presented in Figure 6(c) and (d).

3.6 Design example

To show the applicability of the suggested method, the following design example is produced. A reinforced unpaved road should be

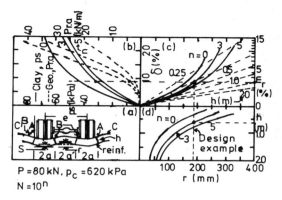

$P = 80\,kN,\ p_c = 620\,kPa$
$N = 10^n$

Fig. 6 Design chart for reinforced unpaved road

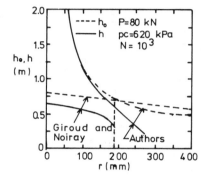

Fig. 7 Designs by authors' and Giroud and Noiray (1981) methods

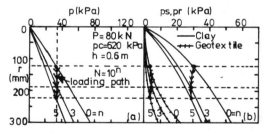

Fig. 8 Load sharing by clay and reinforcement

designed for American-British Standard, off-highway truck with, P = 80 kN, pc = 620 kPa, N = 1000 passages and r = 190 mm, on a clay subgrade and reinforcing element of the type described in this paper. Using Figure 6 or 7, for r = 190 mm and N = 1000; the required thickness of aggregate layer, h , are obtained as 0.31 m and 0.60 m by Giroud and Noiray method and the authors' method respectively. This show the Giroud and Noiray method to yield result on the unsafe side.

Table 1. Authors refinements over Giroud and Noiray (1981) method.

Area of refinement	Basic design parameters		Quasi-static (Q.S) analysis			Analysis under traffic loading	
Design method	Subgrade	Reinf.	Unreinf.	Reinf.		Unreinf.	Reinf.
				Subgrade	Reinf.		
Giroud & Noiray (1981)	c_u	K (modulus at $\varepsilon r=2\%$)	$P_o=\pi c_u$ $\neq f(r)$	ps= $(\pi+2)c_u$ $\neq f(r)$	pr= $f(K,r)$	po, ps, pr from Q.S.+ Webster and Alford (1978) equation.	
Authors	New $\sigma-\varepsilon-N$ relation	New Pr- $\varepsilon r-N$ relation	For clay:po,ps=$f(\sigma-\varepsilon -N,r)$ obtained from po or ps $-\delta$ - N relation using Prakash, Saran and Sharan (1984) method. For reinf.: pr = $f(Pr-\varepsilon r-N, r)$ from pr$-\delta-N$ relation.				

4 DISCUSSION AND CONCLUSIONS

Giroud (1982) in his closing remarks on discussion on the paper by Giroud and Noiray (1981) stated, "one of the goals of the authors of the original paper was to carefully document their approach so it can be used by researchers as a starting point for the development of improved methods of design". In line with Giroud's thought the method of analysis presented here may be considered as an improvement over that presented by Giroud and Noiray. The refinements introduced by the authors on their method are summarised in Table 1. The Giroud and Noiray method would produce results on the unsafe side, which is shown to calculate lower values of h (Figure 7), compared with those by the authors. To depict the load sharing mechanism of the subgrade soil and the reinforcement under repeated loading, the loading path appropriate for the design example is shown in Figure 8(a) and (b). These show the clay subgrade to loose load slightly and the reinforcing element to gain it, relative to their respective initial values.

From the studies presented in this paper the following conclusions could be made:

1. A single parameter representation of behaviour of both clay subgrade and reinforcing element is not adequate for design of reinforced and unreinforced unpaved roads.

2. In design of reinforced and unreinforced unpaved roads, the newly developed constitutive relations may be used to represent the behaviour of both clay subgrade and reinforcing element to a reasonable degree of accuracy.

3. Assumption of mobilization of elastic (in case of unreinforced road) and ultimate (in case of reinforced road) limit states irrespective of degrees of rutting may lead to unsafe design, especially at low values of rutting.

4. Overall, the design method proposed here is based on realistic representation of behaviour of both clay subgrade and reinforcing element under compatible degrees of rutting and therefore, may be adopted.

REFERENCES

Giroud, J-P & Noiray, L. 1981. Geotextile reinforced unpaved road design, PASCE, JGED. 107: GT9: 1233-1254.

Giroud, J-P. 1982. Discussion on geotextil reinforced unpaved road design. PASCE. JGED. 108: GT12: 1665-1670.

Hammit, G. 1970. Thickness requirements fo unsurfaced roads and airfield bare base support. Tech. Rep. S-70-5, U.S. Army En Waterways Exp. Stn. Vicks. Miss.

Kondner, R.L. 1963. Hyperbolic stress-stra response of cohesive soils. PASCE, JSMFD 89:SM3:115-143.

McGown, A., Andrawes, K.Z. & Kabir, M.H. 1 Load-extension testing of geotextiles confined in-soil. Proc. Sec. Int. Conf. on Geotex. Las Vegas:USA: 3: 793-798.

McGown, A. Andrawes, K.Z., Yeo, K.C. & DuBois, D. 1984. The load-strain-time be haviour of tensar geogrids. Symp on Poly Grid Reinf. in CE, I.C.E. London: 1-7.

Prakash, S.,Saran, S. & Sharan, U.N. 1984 Footings and constitutive laws. PASCE, JGED, 110: 1473-1487.

Saha, G.P. 1988. Nonlinear analysis and design of geotextile and geogrid reinfor unpaved roads. Forthcoming M.Sc.Engg. Thesis, CE Dept., Bangladesh Univ. of En & Tech., Dhaka: Bangladesh.

Webster, S.L. & Watkins, J.E. 1977. Invest gation of construction techniques for ta tical bridge approach roads across soft ground. Tech. Rep. S-77-1, U.S. Army Eng Waterways Exp. Stn. Vicks. Miss.

Webster, S.L. & Alford, S.J. 1978. Investi gation of construction concepts for pave ments across soft ground. Tech. Rep. S-7 U.S. Army Eng. Waterways Exp. Stn. Vicks Miss.

International Geotechnical Symposium on Theory and Practice of Earth Reinforcement / Fukuoka Japan / 5-7 October 1988
© 1988 Balkema, Rotterdam. ISBN 90 6191 820 0

Experimental studies to improve the surface land reclaimed from the sea

Hosei Uehara
University of the Ryukyus, Okinawa, Japan

Mitsuzo Yoshizawa & Seiryo Kohagura
Okinawa Gijustu Consultant Co., Okinawa, Japan

ABSTRACT: Silty and clayey soils dredged from bay bottom exhibit so extremery soft layers. Physical properties of those reclaimed soils and relationships between moisture content and density, shear strength after dredging have been investigated. Then the methods of treatment to improve those poor shallow layers were experimentally studied. Earth reinforcing techniques are, a) direct spreading, b) net spreading, c) sheet spreading, and d) lime stabilization. Observed results were compared with each other, and applicabilities of those methods to ultra soft layers were evaluated.

1 INTRODUCTION

1.1 Case study of Nakagusuku Bay Harbor, Okinawa

Nakagusuku Bay Harbor(New Harbor District) is located at the east coast of the middle-southern part on Okinawa Island. As shown in Fig.1, the area is surrounded by

Katuren Cape and Chinen Peninsula, consequently the bay may naturally be suitable for major harbor.

New harbor district was planned to be about 340ha area reclaimed from the sea by a quantity of about $10,500m^3$ of dredged soils which were sediments (inflow from

Fig.1 Location

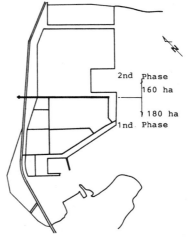

Fig.2 New Harbor District

Table.1 Physical Properties of Insitu soils

Depth (m)	Sand&Gravel (%)	Silt (%)	Clay (%)	Spesific Gravity	Moisture (%)	Wet Density (tf/m³)	Liquid Limit	Plasticity Index
-5.5	30.0	50.0	20.0	2.77	35.0	1.90	42.0	20.0
-7.5	20.0	40.0	40.0	2.77	35.0	1.76	60.0	36.0
-10.0	5.0	55.0	40.0	2.78	52.0	1.72	65.0	40.0
-13.0	2.0	38.0	60.0	2.78	55.0	1.70	72.0	47.0

surrounding land). For the present time this new district is divided into two blocks as shown in Fig.2 and about 180ha is now under construction since 1959.

1.2 Physical properties

Nakagusuku Bay appeared as a result of subsidence at the south east side of Okinawa Island for the reason of tectonic phenomenon, and so the bottom geological features consist of Shimajiri Formation (Tertiary mud rock; silty clay, and sandstone) overlaid by Holocene alluvium deposits. Most dredged soils contain much more fine grained soils, consequently the reclaimed area makes ultra-soft ground. Physical properties of dredged soils in depth are presented in Table 1 and typical grain size percentages in depth at some spots are shown in Fig.3. Immediately after dredging the surface lands exhibit potage-like aspects.

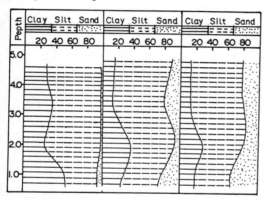

Fig.3 Texture of Dreged Soils

Photo.1 Surface Cracks

1.3 Examinations for reclaim, drainage, and surface layer stabilization methods

In order to keep trafficability on the newly reclaimed land, the examinations of suitable method of reclaim, drainage, and stabilization were made by the procedure shown in Fig.4. Considering that Okinawa Islands have subtropical climatic features, drying naturally these wet surface layers by the sun shine may be advisable (Photo 1). However the dry layers had only 2-3cm thickness, so it was considered that natural drying method would take much time to drain water from deeper layer. Then surface drainage by gravitation (slope, trench etc.) and subsurface drainage by inserting drain materials into the earth were examined. The test results relating to the relationships between depth and water content, vane shear strength, and wet density with lapse of time are shown Figs.5,6,7. It could been seen that it was applicable to adopt the procedure of some artificial methods with natural drying for the reclaimed land. Finally in order to improve the bearing capacity of the very soft layer, surface layer stabilization methods (over-layer method) were investigated.

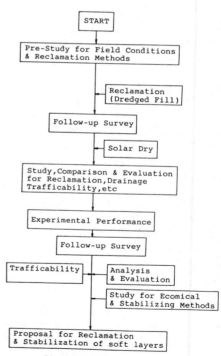

Fig.4 Flow Chart

240

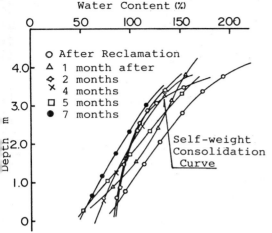

Fig.5 Relationship between Water content and Depth

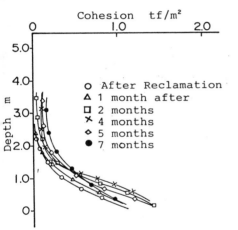

Fig.6 Relationship between Cohesion and Depth

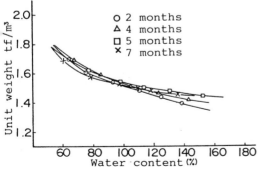

Fig.7 Relationship between Unit weight and Water content

2 INVESTIGATIVE STUDIES

2.1 Preparatory studies

There are some historical or geotechnical reviews of subsurface stabilization methods by T.Yamanouchi(1978), J.K. Mitchell(1981), and T.Okumura(1984) etc. In Figs.8 and 9, applicable earth reinforcement methods in relation to grain size ranges and water contents are shown. From those data the case of Nakagusuku Bay Harbor reclamation was examined relating with the physical properties of dredged soils. Also some trial examinations of natural (self weight) consolidation behavior was evaluated and by observation the settlement in 10-14 days after reclamation was nearly the same as laboratory tests. Since then it was observed that creep settlement (may be secondary consolidation) followed. Increments of strength were calculated by a normal procedure and those were compared with the results of sounding in the soft layer (by Vane shear test) for the lapse of time after reclamation had been performed.

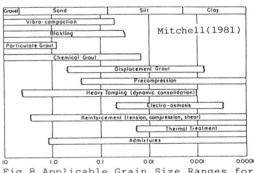

Fig.8 Applicable Grain Size Ranges for Different Stabilization Methods

2.2 Experimental studies

Experimental studies were performed to examine the surface layer for keeping trafficablity.The test program was carefully examined and the selections of some earth reinforcing methods which were applicable and suitable for the local soils in site were discussed. Experimental performances were then executed with special care being taken for the preparation, materials, the procedures, and the weather. The test spot was divided into several zones for each examination performance as shown in Fig.10.
To improve the trafficablity (bearing capacity) of soft ground, usually there are

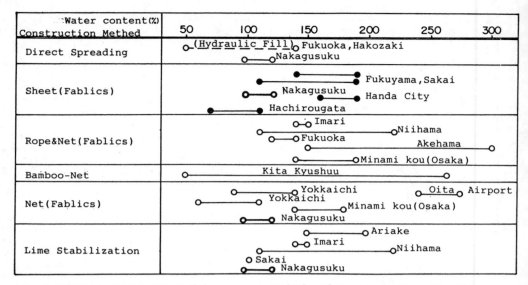

Fig.9 Relationship between improvement method and water content of soft
subsurface layers   O: Actual water content   ●:Estimated water content
---: Range of measurment

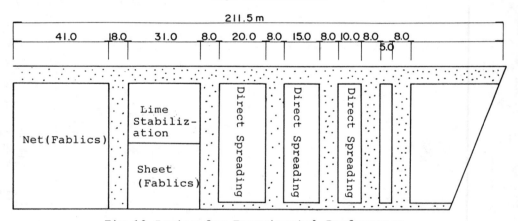

Fig.10 Zoning for Experimental Performance

two types of treatment of surface layers,
that is one using the competent soil materials and another using the materials
other than soils (reinforcement materials
or chemical materials). In this test program, the following methods were adopted.
a) Direct Fablics: Sandy,gravelly(corals)
soils were applied for inducing drainage.
b) Net Fablics: Nets A, B,and C were used
for supporting the point or concentrated
load (tension)
c) Sheet Fablics: Polyester sheet reinforced by making grill was used
d) Lime stabilization: Caustic (quick)
lime was used for the effect of chemical
reaction

2.3 Results and discussinos

As shown in Photo.1, thin surface layer
(2-3cm) can not stand a walk, here we
have to increase the strength of the layer
(thicken more than 10cm, lower water content less than 60%, and strength more than
$qu=0.6tf/m^2$) according to the data of site
investigations. Therefore we examined the
four kinds of reinforcement method by the
preparatory studies, and a few examples of
test results regarding variations of water
content, wet density, vane shear strength
and settlement etc. after the reclamation
with the lapse of time are presented in
Figs.11,12; Figs.13.14; and Figs.15,16.

242

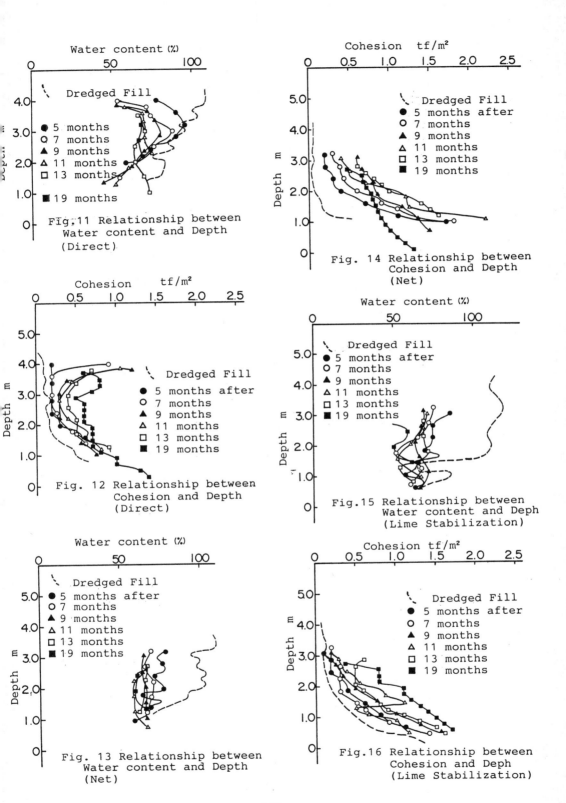

Water content (%)

Fig. 11 Relationship between
Water content and Depth
(Direct)

Cohesion   tf/m²

Dredged Fill
● 5 months after
○ 7 months
▲ 9 months
△ 11 months
□ 13 months
■ 19 months

Fig. 14 Relationship between
Cohesion and Depth
(Net)

Cohesion   tf/m²

Dredged Fill
● 5 months after
○ 7 months
▲ 9 months
△ 11 months
□ 13 months
■ 19 months

Fig. 12 Relationship between
Cohesion and Depth
(Direct)

Water content (%)

Dredged Fill
● 5 months after
○ 7 months
▲ 9 months
△ 11 months
□ 13 months
■ 19 months

Fig.15 Relationship between
Water content and Deph
(Lime Stabilization)

Water content (%)

Dredged Fill
● 5 months after
○ 7 months
▲ 9 months
△ 11 months
□ 13 months
■ 19 months

Fig. 13 Relationship between
Water content and Depth
(Net)

Cohesion tf/m²

Dredged Fill
● 5 months after
○ 7 months
▲ 9 months
△ 11 months
□ 13 months
■ 19 months

Fig.16 Relationship between
Cohesion and Deph
(Lime Stabilization)

a) Direct spreading; this method dries rather rapidly the surface layer and it is preferable for a wide reclaimed land to make the working roads. The working roads need the thickness of 1.0-1.5cm to stabilize, but sometimes there happens cracking where soft muds spout out. Width of spreading shoud be more than 6m, and it is desirable for this case to keep 7.0m. The depth of spreading was precisely checked by Swedish penetrometer and the average thickness was 3.0cm which was changeable with water content and vane shear strength of the reclaimed layers. It was found that the heaving and influence range of the spreading were respectively 30-60cm and 15-25m at the site.

b) Net spreading; For convenience sake to operate, 6m*15m net role was adopted, and the overlap had to be kept at least 30cm in length. Also joints of nets had to be tighten because of rupture after fill completion. Piling of the spreading materials or smaller bulldozer on the layer made surrounding nets often unstable (warping) and spreading depths were in the range of 0.8-1.2cm after completion. So there is a quite problem resulting from the lower strength of the layer in the first spreading works. It is advisable in this case to adopt the bulldozer of 0.11 -0.15kgf/cm^2 contact pressure type. The mechanism of bearing capacity of the net spreading may be similar to other cases, but it is still complicated for these soft layers of Nakagusuku Bay Harbor.

c) Sheet spreading; The most important difference between the net and sheet is that the net has rigidity of the materials, but the sheet has not, and the frictional force of the net is higher than that of the sheet. The fixation(anchoring) of the sheet ends were set into the temporary work roads and piling spots of sandbags. It was found sometimes that the sheet was torn up due to the full weight of spreadings (sand or gravelly soils) introducing tension forces

d) Lime stabilization; Some laborratory tests were performed to collect the data for lime stabilization of the dredged Shimajiri clayey soils. The content of the additive was fixed by the laboratory data to be 120kgf/m^3. Dehydration of the soils by mixing, with lime clearly improved the layer soils, and about two hours later after mixing, it was possible to walk slowly on the layer. It is advisable that the layer stabilized by lime should be covered by over-layer (sandy, coral gravelly soils) of at least 30cm thickness.

3 CONCLUSIVE REMARKS

Applying the method of direct spreading of sandy or gravelly soils with sola drying procedure is preferable, and may be reliable and economical if the layer i left for somewhat long duration having suitable thickness and width

Workability of gravelly (corals) soil applying the soft layer on the net spreading is preferable to the sand applying.

Stabilization effects of bearing capacity of the layer by net tensile strength and continuous drainage (consolidation) were recognized. Also stitches of the net may cooperate well with Shimajiri silty and clayey soils in generating frictional forces, consequently it is not necessary to fix the net ends.

Comparing with the net spreading, the sheet spreading may have difficulties to be repaired if it is damaged by coral gravels and shell fragments of dredged soils

The sheet spreading may not be suitable for impervious soils and the excess water can easily stay over the sheet.

The problem of lime stabilization is dependent on the weather condition (rainy or windy day should be avoided), and so lime mixing performances should be carefully controled.

Surrounding parts of lime stabilized soils are still very soft and drainage effect for those parts can not be expected.

Finally, even if the difference of the cohesive strength of the layer soils be so small, it may have an influence largely on the cost of the surface layer stabilization performance.

ACKNOWLEDGEMENT

Writers greatly appreciate Nakagusuku Bay Harbor Construction Branch Offices of Okinawa Prefeture and also Okinawa General Bureau for their helps and courtesies.

REFERENCES

Yamanouchi, T. 1978. A review and prospect for the surface layer stabilization methods.Jiban no Hyoso Anteisyori Koho:1-17(in Japanese)
Mitchll,J.k. 1981. Soil improvement-state-of-the-art-report. proc.10th ICSMFE Vol.4, Stokholm:509-565
Okumura,T. 1984. Earth improvement-the present and the past. Chishitsu to Chosa No. 2:51-54(in Japanese)

International Geotechnical Symposium on Theory and Practice of Earth Reinforcement / Fukuoka Japan / 5-7 October 1988
© 1988 Balkema, Rotterdam. ISBN 90 6191 820 0

Modelling for bearing capacity analysis of reinforced sand subgrades

B.P.Verma
Regional Institute of Technology, Jamshedpur, India

A.N.R.Char
Indian Institute of Technology, Kharagpur, India

ABSTRACT: Bearing capacity of sand subgrades is increased when reinforced with galvanised rods placed as vertical instrusions in the subgrade. The improvement is comparable with the results obtained by investigators using horizontal forms of reinforcements. The improvement is a function of the spacing, diameter, roughness and extent of the reinforcing element. The present investigation attempts a modelling for bearing capacity analysis of reinforced sand subgrades.

1 INTRODUCTION

Many investigators such as Akinmusuru and Akinbolade (1981), Binquet and Lee (1975), Fragaszy and Lawton 1984) and others have reported improvement in bearing capacity of sand subgrades under footing foundations when horizontal reinforcements are placed in the subgrade. However, the serious disadvantage with horizontal reinforcements is that it can not be used in in-situ conditions. Re-laying and compaction of the subgrade is essential after placement of the reinforcement. Bassett and Last (1978) investigated the possibility of using non-horizontal reinforcements. Installation of root piles for improving bearing capacity has been advocated by Lizzi (1979). If inclined or vertical reinforcements are established to be effective they can be installed more easily in new constructions and used for strengthening of existing foundations as well. With this objective a laboratory investigation was carried out by the authors to evaluate the efficiency of vertical reinforcing elements in improving sand subgrades, and the results were found to be encouraging and have been reported 1986).

2 TEST ARRANGEMENT

Two dimensional model tests were carried out in a wooden box of size 720 x 400 x 90 mm. A 7mm thick perspex sheet was used in the frontage for observing the failure surface. Special care was taken to make the box as rigid as possible. Model footings of 40 mm thickness were made out of well-seasoned teak wood and their bases were made rough to simulate the rough base of a prototype footing. The cohesionless test beds were prepared by pouring standard Ennore sand in layers through a funnel held at a constant height of 300 mm above the surface. The uniformity co-efficient and the effective size of the sand were 1.41 and 0.49 mm respectively. The dry density of the sand bed was found to be 1.58 mg/m^3 (R.D. = 71%) for all the tests performed. Galvanized iron rods of required length and size were pushed into the sand bed vertically at predetermined spacings (Fig.1). A single layer of sand particles were bonded onto the surface of rods with araldite to simulate a rough surface and were employed in a few tests. The footing was pushed into the sand bed at a constant speed of 1 mm per minute until failure. The applied load was recorded with the help of a

Table 1. Ultimate bearing capacity ratio for reinforced sand subgrades: Width of footing (B) = 100 mm

Diameter (mm)	Spacing (mm)	Spacing/Diameter	Ultimate Bearing Capacity Ratio					
			R=B		R=2B		R=3B	
			L=B	L=1.5B	L=B	L=1.5B	L=B	L=1.5B
1	2	3	4	5	6	7	8	9
1.7	18	10.59	1.48	1.49	1.67	1.69	1.71	1.76
1.7	15	8.82	-	1.50	-	1.69	-	1.91
1.7	13	7.65	1.70	1.89	1.79	2.31	1.91	2,51
1.7	10	5.88	1.94	2.31	2.08	3.08	2.66	3.20
1.7 (Rough)	13	7.65	2.51	-	2.79	-	3.91	-
2.51	22.5	8.96	-	1.69	-	1.79	-	2.11
2.51	18	7.17	1.79	1.91	2.11	2.31	2.51	2.50
2.51	15	5.98	2.11	2.31	2.51	2.51	2.74	2.69
2.51	13	5.18	-	2.91	-	3.31	-	3.91
2.51 (Rough)	18	7.17	2.50	2.66	2.89	3.91	3.11	-

calibrated proving ring. The settlements of the footing were recorded by two dial gauges fixed with adapters and resting on two extension plates fixed on either side of the footing.

3 TEST RESULTS

The variables of the investigation and bearing capacity ratios at failure are given in Table 1. Since a well defined failure point in the load-settlement curves was not present in most of the cases determination of experimental ultimate loads were done through the method suggested by De Beer (1970) and employed by Vesic (1973). Detailed test results are available elsewhere [Verma (1986)] .

4 MODELLING FOR BEARING CAPACITY ANALYSIS

The experimental results show improvement in bearing capacity of the sand subgrades due to vertical reinforcement of different lengths and of different extents as shown in Table 1.

The following alternative approaches for the evaluation of bearing capacity of reinforced subgrades are examined in this paper.
1. Reinforcements as piles: reinforcement rods are assumed to act as vertical piles and their contribution is added to the bearing capacity of the subgrade.

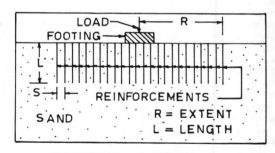

SECTIONAL ELEVATION

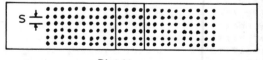

PLAN

FIG.1 REINFORCEMENT PATTERN

Table 2. Load carrying capacity of reinforcing rods acting as piles:
Diameter of rods = 1.7 mm, Length= 1.5B, Width of footing = 100 mm

Test No	Spacing of rods (mm)	Ext. of Reinforcement (R)	A_r/A_s x 10^{-3}	Total No.of rods used	Total pile load for all rods used in kN (Q_{PR})	Capacity of sand+pile load ($Q_{UR}+Q_{PR}$)	Expt. Load in kN (Q_R)
1	2	3	4	5	6	7	8
10	18	B	6.06	48	0.0215	4.5611	6.645
12	18	2B	6.06	96	0.0429	4.5825	7.543
13	18	3B	5.72	136	0.0608	4.6004	7.793
15	15	B	8.83	70	0.0313	4.5709	6.663
17	15	2B	8.83	140	0.0626	4.6022	7.506
18	15	3B	8.41	200	0.0894	4.6290	8.457
20	13	B	12.11	96	0.0429	4.5825	8.422
22	13	2B	12.11	192	0.0859	4.6255	10.235
23	13	3B	12.11	228	0.1020	4.6416	10.960
25	10	B	22.20	176	0.0787	4.6183	10.234
27	10	2B	21.19	336	0.1503	4.6898	13.679
28	10	3B	20.86	496	0.2218	4.7614	14.212

2. Equivalent surcharge depth: reinforcements are assumed to contribute an equivalent surcharge on the subgrade.

3. Apparent cohesion or shear strength increase: the reinforcements are assumed to impart an apparent cohesion or cause an increase in the frictional resistance or both.

4.1 Reinforcement as piles

Reinforcement rods may be considered as vertical piles whose load capacity can be computed through static formulae. An appreciation of their fundamental functions and a comparison with those of presently considered reinforcement system may clarify the functional difference of the two. Load carrying capacity of the unreinforced subgrade (Q_{UR}) along with the additional contribution of reinforcing rods as piles (Q_{PR}) for a few test results are tabulated in Table 2. Another set of values of all the reinforcing rods contribution as piles (Q_{PR}) are also given in the table. Comparison with the experimental loads (Q_R) on the reinforced sand subgrade for different conditions reveal that Q_R is always much more than ($Q_{UR} + Q_{PR}$) which suggests that the improvement in bearing capacity may not be due to the reinforcing rods acting simply as piles.

4.2 Equivalent surcharge depth

In this hypothesis, the increase in bearing capacity due to the introduction of reinforcements in the subgrade is assumed to be due to an increase in the effective surcharge pressure (D_{equiv}) at the footing base and consequent increase in the confining pressure [Denver et al (1983)]. Using the bearing capacity equation (1), D_{equiv} can be evaluated as all other terms in the equation are known.

$$q_{ult} = \left[D_{equiv} + (h_1 + h_2) \right] \gamma N_q + \tfrac{1}{2} \gamma BN_\gamma$$

$$\dots (1)$$

247

Table 3. Apparent cohesion and equivalent surcharge depth due to reinforcing rods. Dia of rods = 1.7 mm, Length = 1.5B, B = 100 mm

Test No	A_r/A_s x 10^{-3}	Surcharge depth in cm	Calculated UBC in kPa (q_c)	Expt. UBC in kPa (q_{expt})	$q_{expt}-q_c$ (C_F)	App. cohesion= C_F/N_c in kPa (C_a)	$D_{equiv}=$ $\frac{\Delta q \times 100}{N_q}$ (in cm)
1	2	3	4	5	6	7	8
10	6.06	3.30	517.7	755.1	237.4	1.10	6.57
12	6.o6	3.70	544.4	857.1	312.7	1.44	8.54
13	5.72	3.45	533.4	885.6	352.2	1.62	9.63
15	8.83	3.25	517.7	757.1	239.4	1.11	6.61
17	8.83	3.05	524.1	852.0	329.0	1.51	8.94
18	8.41	3.75	553.2	961.0	407.8	1.86	11.05
20	12.11	4.10	559.1	957.0	398.0	1.83	10.86
22	12.11	3.90	560.7	1163.1	602.6	2.75	16.29
23	12.11	4.00	551.7	1245.5	693.9	3.20	19.02
25	22.20	3.87	554.2	1163.0	608.8	2.79	16.55
27	21.19	3.90	584.3	1554.4	970.1	4.31	25.39
28	20.86	3.90	591.9	1615.0	1023.1	4.55	26.77

where

h_1 = height of soil which heaves above the original surface of the subgrade.

h_2 = height of soil above base of the footing.

In Table 3 values of D_{equiv} have been tabulated for a typical series of tests on reinforced sand subgrades. The obtained values are very large and appear to have no useful relation to the length or density of reinforcements. From the results it is seen that equivalent surcharge depth increases with A_r/A_s and is dependent on length and extent of reinforcement. (A_r represents area of reinforcement and A_s area of soil reinforced) .

4.3 Apparent Cohesion

Vidal (1969) and Schlosser and Long (1974) hypothesise that when reinforcement is introduced to a non-cohesive soil, the whole mass exhibits some cohesion arising from the friction of soil grains against the reinforcing elements. This concept has been supported by several investigators such as Gray (1978), Waldron (1977), and Verma and Char (1978). Based on this concept the value of apparent cohesion was evaluated from the experimental results. Since the unreinforced soil was frictional the bearing capacity q_c was calculated by the equation

$$q_c = (h_1+h_2)\gamma N_q + \tfrac{1}{2}\gamma BN\gamma \qquad ...(2)$$

where h_1 and h_2 are already defined through equation (1). The values of q_c have been tabulated in Table 3. It is noted that experimental ultimate bearing capacities (q_{expt}) of reinforced sand are always larger than values obtained by equation (2) . This difference ($q_{expt} - q_c$) is assumed to be the contribution due to apparent cohesion due to the introduction of reinforcements in the subgrade. This difference is termed as a cohesion factor (C_F) and the apparent cohesion is evaluated through

$$C_a = \frac{C_F}{N_c} \qquad \ldots (3)$$

where

C_a = apparent cohesion

and N_c = Terzaghi's bearing capacity factor for cohesion corresponding to ϕ_p .

The values of C_a for a few test results are tabulated in Table 3. The relation between A_r/A_s and C_a shows that the rate of increase of C_a with A_r/A_s is dependent on the length of the reinforcing bar and the extent of reinforcement and the variation is almost linear, the slope of the curve being a function of the extent of reinforcement.

5 CONCLUSIONS

The study shows the beneficial effect of using vertical reinforcing rods for sand subgrades. The greatest advantage of this method is that relaying of the subgrade is not required as in the case of horizontal reinforcements. The increase in bearing capacity of sand subgrade may be taken as due to apparent cohesion induced in the soil due to the presence of reinforcement. The apparent cohesion is dependent on the area ratio of reinforcement as well as the length and extent of reinforcement. Tests on bigger models or prototype testing may help in arriving at accurate analytical estimates.

REFERENCES

Akinmusuru, J.O. & Akinbolade, J.A. 1981. Stability of loaded footings on reinforced soil. Journal of Geotechnical Engineering, ASCE. lo7 : 819-827

Bassett, R.H. & Last, N.C. 1978. Reinforcing earth below footings and embankments. Proceedings of the symposium on Earth Reinforcement, Pittsburgh : 202 - 231.

Binquet, J. & Lee, K.L. 1975. Bearing Capacity tests on reinforced earth slabs. Journal of the Geotechnical Engineering Division, ASCE. 101 : 1241-1255.

De Beer, E.E. 1970. Experimental determination of the shape factors and bearing capacity factors of sand. Geotechnique 20 : 387-411.

Denver, H. et al. 1983. Reinforcement of cohesionless soil by PVC-grid. Proceedings of the 8th. European Conference on Soil Mechanics and Foundation Engineering, Vol. 2 : 481-489.

Fragaszy, R.J. & Lawton, E. 1984. Bearing capacity of reinforced sand subgrades. Journal of Geotechnical Engineering ASCE. 110 : 1500-1507.

Gray, D.H. 1978. Role of Woody Vegetation in reinforcing soils and stabilising slopes. Proceeding of the symposium on soil Reinforcing and stabilising Techniques, Sydney : 253-306.

Lizzi, F. 1979. Reticulated root pile structures for in-situ soil strengthening, theoretical aspects and model tests. Proceedings of the International Conference on Soil Reinforcement, Vol. 2, Paris: 317-324.

Schlosser, F. & Long N.T. 1974. Recent results in French research on reinforced earth. Journal of the construction Division, ASCE. 100 : 223-235.

Verma, B.P. 1986. Bearing Capacity of reinforced sand subgrades, Ph.D. thesis I.I. T. Kharagpur, India

Verma, B.P. & Char, A.N.R. 1978. Triaxial tests on reinforced sand. Proceedings of the Symposium on Soil Reinforcing and Stabilising Technique, Sydney : 29-39.

Verma, B.P. & Char, A.N.R. 1986. Bearing capacity tests on reinforced sand subgrades. Journal of Geotechnical Engineering, ASCE. 112 : 701-706.

Vesic, A.S. 1973. Analysis of ultimate loads of shallow foundations. Journal of the Soil Mechanics and Foundations Division, ASCE. 99 : 45-73.

Vidal, H. 1969. The principle of reinforced earth. Highway Research Record No. 282 : 1-16.

Waldron, L.J. 1977. Shear Resistance of root-permeated homogeneous and stratified soil. Soil Sci. Soc. of Amer. Jour. 41 : 843-849.

International Geotechnical Symposium on Theory and Practice of Earth Reinforcement / Fukuoka Japan / 5-7 October 1988
© 1988 Balkema, Rotterdam. ISBN 90 6191 820 0

Mattress foundation by geogrid on soft clay under repeated loading

Kazuya Yasuhara & Kazutoshi Hirao
Department of Civil Engineering, Nishinippon Institute of Technology, Fukuoka, Japan

Masayuki Hyodo
Department of Civil Engineering, Yamaguchi University, Ube, Japan

ABSTRACT: The small-scaled model tests were carried out at the laboratory to find a design method of unpaved roads on soft foundations reinforced with the geotextiles. The tests were mainly aimed at providing a basis for the selection of a compatible embankment geometry and reinforcement layout. According to the results from model tests for adaptation of the geomesh and geogrid, it was proved that application of a geogrid with cellular layout is most preferable for reducing the settlement of the fill on soft grounds under traffic-induced cyclic loading because of the high rigidity of geocell with the gravels contained inside and the eminent interlocking effect exerted between geogrid and gravels.

1 INTRODUCTION

Soft clay subgrades under traffic-induced cyclic loading sometimes exhibit the predominant deflection which may lead to the harmful settlements of pavements for motorways, railways and runway. Behaviour of clay subgrades subjected to cyclic loading is so complex that its mechanism has not been clarified yet. Adaptation of geotextiles is counted as one of the promising countermeasures against settlements of clay under cyclic loading (Yamanouchi et al., 1967: Yasuhara et al., 1986). However, reinforcement by the geomesh has not been successful for preventing it (Yamanouchi, 1967: Yasuhara et al., 1986).

Based on the laboratory small-scaled model footing tests, the present paper explores the best way how to use the geotextiles for controlling the settlements of unpaved roads founded on soft clay grounds. Three kinds of geotextiles so called geomesh, geomesh-pipe and geogrid were selected to investigate the behaviour of soft clay under cyclic loading with an inclusion of the geotextile as well as to determine the most suitable material on this subject. It is aimed in this study that laboratory model tests will provide a basis for selection of a compatible geometry and reinforcement layout.

2 EXPERIMENTAL PROGRAMME

To detect the selection of compatible geometry and reinforcement layout for reducing the settlements of soft clay under cyclic loading, several families of small-scaled model tests were carried out. Those were:
1) Without geotextiles,
2) With planar layout of geotextiles and
3) With cellular layout of geotextiles.

All the test series involved in the present paper are summarized in Table 1. The modeled ground with planar layout of geotextiles is reinforced by laying the geogrid between the upper soft clay layer and the lower sand layer. As shown in Table 1, in addition to this, the cellular layouts by the mattress-type of geogrid with gravels contained inside (Test no. G-6) and by the folding mattress-type geogrid (Test no. G-3) were also used for reinforcing the layered ground. The difference of those two layouts is schematically illustrated in Fig. 1.

The outline of the apparatus for model tests was shown in Fig. 2. The plane strain model test was carried out on the footing apparatus made of the steel frames with 2.0 m in width, 1.0 m in depth and 0.5 m in length. The clay layer with 20 cm and 40 cm in depth was covered by the sand layer with 17 cm in depth. The width of the loading plate was 15 cm for most of

251

Table 1 Details of model footing tests

Test no.	Material	Layout	Clay layer (cm)	Width of apparatus (cm)	Cyclic pressure (kPa)	Loading plate (cm)	No.of cycles
N-1	Non.	—	20	100	40	15	(static)
-2		—	"	"	10	30	18000
-3		—	"	"	"	"	130000
-4		—	"	"	40	15	"
-5		—	40	"	"	"	"
-6		—	20	"	"	"	"
-7		—	40	200	"	"	15
M-1	Mesh	Planar	20	100	10	30	18000
-2	(Z-31)	"	"	"	40	15	130000
MP-1	Mesh-	Hori.	"	"	10	30	"
-2	pipe	"	"	"	"	15	"
-3	(PDS-50)	"	"	"	40	"	"
-4		Ver.	"	"	"	"	"
G-1	Geogrid	Planar	"	"	10	30	"
-2	(SS-2)	"	"	"	40	15	"
-3		Cellular	"	"	"	"	95000
-4		Planar	"	"	"	"	130000
-5		"	"	"	"	"	10540
-6		Cellular	40	200	"	"	100000
Q-1			20		10	30	130000

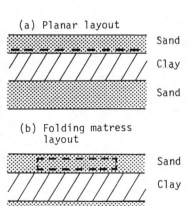

(a) Planar layout

Sand

Clay

Sand

(b) Folding matress layout

Sand

Clay

Sand

(c) Cellular mattress layout (bag-type)

Sand

Clay

Sand

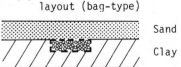

(— — — : geotextile)

Fig. 1 Layout of geogrid

tests and 30 cm for some tests, respectively.

The loading procedure of each model test to simulate the soft ground whose surface was covered by the thin sand layer is shown in Fig. 3. The preconsolidation pressure with $\sigma_{vo} = 0.1$ kgf/cm² was applied to a model ground until 100% primary consolidation of clay was attained. Then, an incremental cyclic vertical pressure of $\Delta \sigma_v = 0.4$ kgf/cm² or 0.1 kgf/cm² with the frequency of 0.1 Hz was applied to the surface of the ground through the oil-servo controlled bellofram cylinder. The number of load cycles was 130000 on the average. The strain gauge was attached to the geogrid and the earth pressure meters were instrumented among soil layers as was shown in Fig. 2.

The reconstituted Ariake clay with index properties of $G_s = 2.49$, $W_L = 108\%$, $I_P = 66$ was used to produce the clay layer with 20 cm and 40 cm in height. The properties of sand in the upper and lower layers are listed in Table 2.

Table 2 Index properties of sand

	G_s	ρ_d (t/m³)	D_r (%)
Upper sand	2.65	1.14	63
Lower sand	2.65	1.49	85

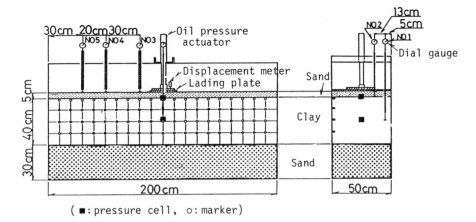

(∎: pressure cell, ○: marker)

Fig. 2 Model footing test apparatus

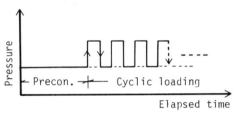

Fig. 3 Loading sequence

3 TEST RESULTS AND INTERPRETATION

Before starting discussion on the effects of geotextiles on stability and settlement of soft grounds, it is essential to illustrate the fundamental behaviour of clay grounds under cyclic loading without reinforcement and improvement. For this purpose, the first attempt was to run a model test on sand-clay ground without geotextiles.

Fig. 4 illustrates the settlement versus number of load cycles for this test. As can be seen from Fig. 4, the model ground (depth of clay layer is 40 cm) without reinforcement by geotextiles leaded to a complete failure at the number of load cycles of 15. Even in the case of a thin clay layer with 20 cm in depth, more eminent settlement was observed than in the case under static loading as also shown in Fig. 4 although this model ground was not completely failed. The settlement of the model ground under cyclic loading was approximately 5 times as the settlement under static loading as shown in Fig.

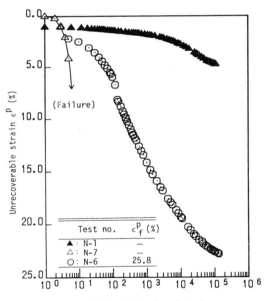

Fig. 4 Comparison of settlements under sustained and repeated loading

4. Those results from two tests without geotextile reinforcement indicate the necessity of reinforcement with geotextiles against settlement of soft grounds under cyclic loading, especially true is this on the ground whose clay layer is thicker compared with width of the loading plate in model tests.

3.1 Planar Layout of Geotextiles

It was indicated from the previous studies by Yamanouchi et al. (1967, 1970) and Yasuhara et al. (1986) that the use of geomesh was not effective to reduction of settlements of clay under cyclic loading though it was available for improvement of bearing capacity in soft grounds (Yamanouchi, 1967). This was made sure as well in the field test where the geomesh was placed by being combined with the quick-lime stabilisation (Yamanouchi et al., 1978).

To improve the defect of geomesh under this situation, the adaptation of geogrid has been attempted instead of geomesh by the laboratory model tests. Fig. 5

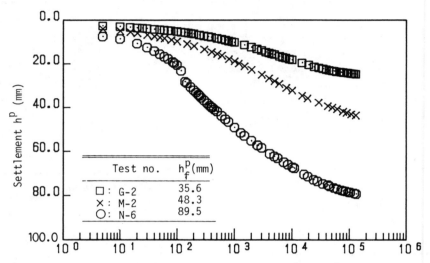

Fig. 5 Settlement versus number of load cycles relations with and without geotextiles

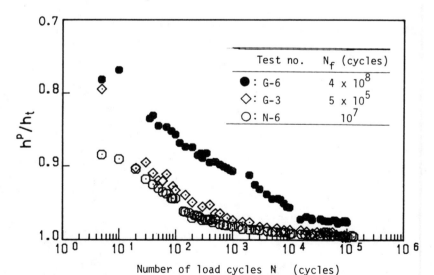

Fig. 6 Variations of settlement ratio with number of load cycles with and without geotextiles for extrapolation of final settlement under cyclic loading

illustrates the settlement versus time relations of model tests (20 cm clay layer and 17 cm sand layer) with non-geotextiles, geomesh and geogrid in order to compare the effect of geomesh and geogrid on reduction of clay settlement due to cyclic loading. Note that the converged displacement, h_r, under cyclic loading was determined by extrapolating the strain ratio, h^p/h_t (h^p, h_t: unrecoverable and total settlements, respectively) on each cycle versus number of load cycles relation as shown in Fig. 6. According to the final settlement determined by this procedure (Yasuhara, et al., 1983), it is pointed out from Fig. 5 that the geogrid is much useful for this purpose rather than the geomesh for settlement control.

As mentioned above, the difference of effectiveness between geomesh and geogrid may be derived from the fact that the larger pull-out resistance should be exerted between sand and geogrid than between sand and geomesh.

3.2 Cellular Layout of Geotextiles

The mattress foundation using the geogrid is expected to provide the effect of formation of the restraint layer among soils in addition to the effect of planar layout of geogrid for improvement of stability and compressibility of soft grounds. The reason for it is that the mattress foundation is possible to unify the reinforced area enclosed by the geogrid.

The characteristics of bearing capacity of modeled footing on sand reinforced by the cellular geogrid belonging to the mattress foundation were investigated by Fukuda et al. (1987). In the present study, in order to clarify the effect of mattress foundation on clay, the same type of test was carried out on the two layered model ground with folding mattress by geogrid which was placed among sand layer. This folding-mattress by geogrid was full of the sand inside. The results from the cyclic loading tests were illustrated in Fig. 7 to make sure of the difference in settlement versus time relations of model grounds reinforced with geogrid and folding geo-mattress. It is not obvious from Fig. 7 that the folding-mattress foundation should be the more desirable layout for controlling the settlement of clay-sand foundations under cyclic loading than the planar layout of a geogrid. A

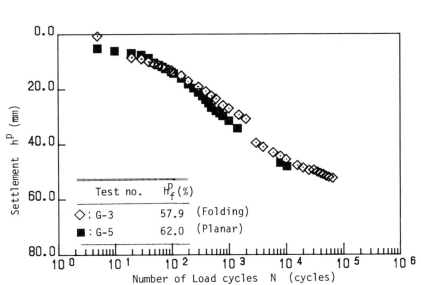

Fig. 7 Effect of planar and folding layout of geogrid on settlement versus number of load cycles relations

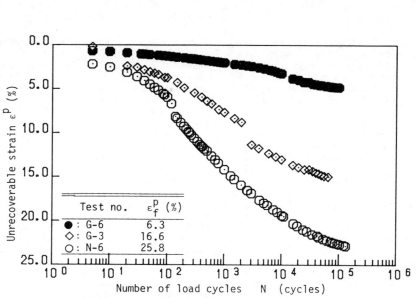

Fig. 8 Effect of mattress foundations on settlement versus number
of load cycles relations

test was, therefore, added to simulate the another type of geo-mattress foundation which belongs to the cellular mattress, but is called the bag-type mattress in the current paper.

In comparison of the effect between folding geo-mattress and bag-type geo-mattress, the settlement versus time relations were shown in Fig. 8. The figure indicates that the bag-type or cellular mattress by geogrid takes advantage in reducing the settlements than the folding geo-mattress by polymer grid. This advantageous feature must be due to interlocking effect exerted between gravels and geogrid.

CONCLUSION

1) Geogrid is more useful to reduction of settlement of clay under cyclic loading than geomesh.
2) The cellular layout of geogrid takes more advantage in reducing the settlement of soft clay under cyclic loading than the planar layout.
3) The bag-type geo-mattress by polymer grid containing the gravels inside is most desirable for reducing the settlements of soft grounds. This effect must be due to the interlocking effect exerted between gravels and geogrid in the form of bag-type mattress.

REFERENCES

Fukuda, N., et al. 1987. Foundation improvement by mattress foundation using the polymer grid, Proc. JSCE, Proc. 8th Asian Reg. Conf. SMFE, Kyoto, Japan

Yamanouchi, T. 1967. Structural effect of restraint layer of subgrade of low bearing capacity in flexible pavement, Proc. 2nd Intn'l Conf. Structural Design of Asphalt Pavement, Ann Arbor, USA. 1:381-389.

Yamanouchi, T. 1972. Experimental study on the improvement of the bearing capacity of soft ground by laying resinous net, Proc. Symp. "Foundations on Interbedded Sands", Perth, Australia, p.102-108.

Yamanouchi, T. and K. Yasuhara. 1975. Settlement of clay subgrades after opening to traffic, Proc. 2nd Australia and New Zealand Conf. Geomechanics, 1:115-120.

Yamanouchi, T., et al. 1978. A new technique of lime stabilisation of soft clay, Proc. Symp. "Soil Reinforcing and Stabilisation Techniques", Sydney, Australia, p.531-541.

Yasuhara, K., et al. 1983. Approximate prediction of soil deformation under drained-repeated loading. Soils and foundations, Vol.23, No.2, p.13-25.

Yasuhara, K., et al. 1986. The use of geotextile against settlement of soft clay under cyclic loading, Proc. 3rd Intern'l Conf. on Geotextiles, Vienna, Austria, 1:193-198.

International Geotechnical Symposium on Theory and Practice of Earth Reinforcement / Fukuoka Japan / 5-7 October 1988
© *1988 Balkema, Rotterdam. ISBN 90 6191 820 0*

Improvement of very soft ground

B.R.Ye
Research Institute of the 3rd Navigation Engineering Bureau, Shanghai, People's Republic of China

J.Zhang & X.Zhang
Research Institute of the 1st Navigation Engineering Bureau, Tianjin, People's Republic of China

ABSTRACT: The treatment of very soft ground consisting of hydraulic fill combined with vertical drainage at a jetty construction is presented. On the basis of the field study, a suitable method consisting of fascine sand cushion plus preloading by vacuum was suggested. This method was successful in treatment of 700,000 m^2 area at the jetty with cost saving, easy construction and proven performance. The function of geotextile and fascine sand cushion as well as preloading by vacuum in combination with packed sand drain or PVC drain is analysed and discussed.

1 INTRODUCTION

With development of national economy and opening of coastal cities to the outside world more wharves are in urgent need, some of which would be constructed on reclaimed land due to the lack of suitable sites. At present, it is economical and efficient to reclaim land from the surrounding sea with hydraulic mud filling. The mud having a high water content and almost zero strength is referred to as very soft ground. A jetty in China was constructed on such fill. The soft ground had to be improved immediately after hydraulically filling so that twelve berths scheduled to be completed at the jetty before 1992 will be put into service in time. Therefore, certain conditions must be met so that construction equipments can enter the site. In our case, this problem was dealt with using surface treatment and then using preloading by vacuum, sometimes in combination with PVC drains, which was the first application in China. This method was successful in treatment of 700,000 m^2 area of ground at the jetty.

2 TECHNIQUE FOR IMPROVEMENT OF VERY SOFT GROUND

In order to make the very soft ground suitable for construction as early as possible, the following techniques have

been adopted in China and abroad:

2.1 Natural evaporation

At the end of pumping the hydraulic fill is left for a certain period of time, allowing water to evaporate and to consolidate. According to the information available, after a duration of three months, the variation of water content with depth was close to that at complete consolidation under gravity. After a duration of two years, the water content at surface layer reduced considerably, resulting in the formation of soil layer where one could walk on it. After a duration of three years, the water content at depth of 0.5 to 2.0 m from the surface was about 80 %, resulting in the formation of a crust soil layer(Watari et al,1983). The method was used in Tianjin and Shanghai (Hu, 1982).

2.2 Drainage

Provision of a horizontal thin layer of permeable material between sedimented clays, such as sand, geotextile, and bagged sand will shorten drainage passage to accelerate drainage. A sand dredger was built in Japan for the purpose of construction

When the hydraulic fill is rather thick, sand drain, packed sand drain and PVC drains etc., are installed to form ver-

Table 1 Six proposed methods for surface treatment

Item	Description	Section	Remarks
1	A single layer of fascine mattress plus sand cushion: the subsoil was laid first with a layer of fascine mattress, then with a 30-cm-thick layer of sand	sand / fascine mud 30cm	overlap: 20 cm
2	Two layers of fascine mattress plus sand cushion: The subsoil was laid first with two layers of fascine mattress, then with a 30-cm-thick layer of sand.	sand / double layers of fascine mud 30cm	ditto
3	One and half layer of fascine mattress plus sand cushion: The subsoil was laid first with one and half layers of fascine mattress, then with a 30-cm-thick layer of sand.	sand / mud / one and half layer of fascine 30cm	ditto
4	Two layers of fascine mattress with sand between: The subsoil was laid first with one layer of fascine mattress, then with one 30-cm-thick layer of sand.	sand / fascine mud 30cm	ditto
5	Ditto	ditto	overlap: 40 cm
6	Geotextile and fascine sand cushion: The subsoil was laid first with a layer of geotextile, then with a layer of 30-cm-thick layer of sand, finally with a layer of fascine.	fascine / sand / geotextile mud 30cm	overlap: 20 cm

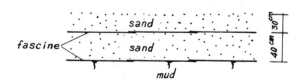

Fig.1 The final section adopted

tical drains. This method has been widely used because of economy and efficiency. For example, these methods have been already used in Lianyungang, Tianjin and Shanghai(Ye,1983; Gao, 1987) as well as in Osaka in Japan(Sasaki, 1982).

2.3 Surface treatment

Surface layer of the soft ground can be improved into the rigid plate by means of cement or lime stabilization. This method has been used in Tokanton, Japan (Nishio ,1981).

Geotextile or fascine sand mattress is placed on top of the subsoil in order to change stress distribution under loading. This method has been used to construct a highway at the port of Kanda in Japan(Yasuhara, 1982), a highway in the southeastern part in Mexico(Olivera,1982) and a breakwater in Tianjin, China(Li,1984).

3 SITE DESCRIPTION

The original ground at the jetty sloped down from west to east with an average elevation of +1.5 m, which was designed to construct to an elevation of +6.5 m by hydraulic filling. The field tests were carried out when the fill reached the elevation of +3.5 m. The test site consisted of 2 m hydraulically filled mud (W= 85-90 %, e = 2.2, γ =16 KN/m^3) underlying natural mud and silty loam (W=45-55 %, e = 1.1-1.4, γ = 16.5-17.5 KN/m^3).

4 SURFACE TREATMENT AND RETENTION EMBANKMENT CONSTRUCTION

With the principle of using locally available and inexpensive material, six surface treatment methods were proposed as shown in Table 1. They can be divided into two types. Type I, consisting of the first three methods, was to place sand on the fascine mattress. Type II comprising the remaining ones was to place sand between two fascine mattresses or between fabric and fascine mattress. Comparision of the results showed that Type II was superior to Type I because the working surface built using Type II had better rigidity. The 4th, 5th and 6th methods all had similiar effectiveness and could meet the requirements of the project. However, the geotextile suffered large deformation, and was difficult to construct and expensive

although its tensile strength was larger than that of the fascine. Finally, the 4th method was selected because it needed less fascine. In view of heavy construction equipment and requirements of preloading by vacuum an additional 30-cm-thick layer of sand was placed on the top of the upper fascine mattress and the sand cushion was increased from 30 to 40 cm. The final section adopted is shown in Fig.1. The practice showed that the working surface after treatment allowed construction equipment with wheel pressure of 30 KN and dumper truck with wheel pressure of 45 KN to perform normal operation on it.

5 FUNCTION OF GEOTEXTILE IN EMBANKMENT

During the last decade many researchers around the world have been studying the theory of the geotextile or fascine reinforced embankment but only a few have studied burried geotextile sand cushion. The researchers in China, based on laboratory model test in combination with case history, have analysized its behaviour(Zhang, 1987). A new fabric element was designed. A mathematical model of soil-geotextile interaction using finite deformation theory and relative movement method was derived as elaborated in a large computer program GNFT(Geotechnical Non-linear Finite Element Analysis Program). It was also found that the function of the reinforced soil layer is to reduce average stress, deviator stress and shear stress of the main bearing zone of the ground, and to distribute high strain along the narrow and deep zone just below the center line of the embankment, resulting in higher stability. It was also learned that, in order to keep the embankment and reinforced cushion as one unit, a suitable geotextile should be selected. Model and centrifuge tests indicated that, due to placement of geotextile, the slip circle moves a little deeper so the stability can be checked by using modified Miga Kawa method(Zhang, 1987) as shown in Fig.2. Let the tensile force of fabric be N = mτL, where L is length of slip circle ,and m is dimensionless coefficient. The following results were obtained.

When the soft soil is thick, the shear strength of the subsoil can be considered to increase linearly with depth, that is, $\tau = \tau_o + \lambda Z$ where Z is the depth as shown in Fig.2. λ is the slope of shear strength versus depth curve as shown in Fig.2. The factor of safety for slip

259

circle can be expressed by

$$K = \frac{\frac{2\lambda g^3}{\sin^2\theta}(1 - \frac{\theta\cos\theta}{\sin\theta}) + 2(1+m)\tau_o g^2 \frac{\theta}{\sin\theta}}{P(g-f)} \quad ---(1)$$

Where g and θ can be obtained by the following equations

$$g = \frac{(1+m)\tau_o}{\lambda} \cdot \frac{2\theta\, ctg\theta - 1}{\theta - 3ctg\theta + 3\theta ctg^2\theta} \quad -----(2)$$

$$\frac{g(2g-3f)}{(2f-g)} = \frac{(1+m)\tau_o}{\lambda} \cdot \frac{\theta}{1 - \theta ctg\theta} \quad ------(3)$$

The definition of notation is explained in Fig.2.

When m = 0 the above equations are applicable to embankment without geotextile. For homogenous subsoil, λ = 0, resulting in g = 2f and θ = 66°47'.

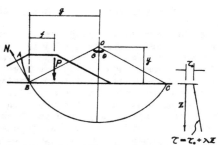

Fig.2 Schematic illustration of stability check

6 TREATMENT OF SOFT SOIL

The test site was divided into two zones: one was treated by preloading by vacuum in combination with packed sand drain; the other by that in combination with PVC drain in order to examine the effectiveness through comparisons.

Packed sand drains, 7 cm diameter, were installed to depth of 15 m in a 1.3 m square grid. PVC drain, 100 mm wide by 3.5 mm thick, were installed in the same pattern with the same spacing and depth. The latter is manufactured by Nanjing Plastic Development Factory which produces three products, that is, SPB-1, SPB-1B, SPB-1C. Their properties and dimensions are summarized in Table 2.

The two test sites, 42 m x 40 m each, 4 m apart, were evacuated using two vacuum systems. Around the zones drainage channels were provided. The results obtained are as follows:

6.1 Settlement of surface layer

Settlements occuring at different stages are shown in Table 3. From Table 3 it can be seen that the settlements of the two test zones were very close.

Table 4 shows consolidation at various time. From Table 4, it can be seen that the rate of consolidation at the two zones was almost the same, indicating that the two kinds of drainage material had the same effectiveness. Settlement during a period of 65 days after normal pumping was equal to 80 %.

6.2 Vane shear strength

Table 5 shows increase of vane shear strength. From Table 5 it can be seen that the strength of soil treated by preloading by vacuum increased considerably and the increase rate reduced gradually from the surface to bottom. The two zones had similiar increase.

Table 2 SPB series of PVC drain

Width	Thickness	Weight	Length	Vertical flow capacity	Transverse permeability	Tensile strength	Extension
mm	mm	g/m	m/roll	m³/s (1)	m/s	KN/10cm	% (3)
100±2	> 3.5	90–120	200	15–40x10⁻⁶	5x10⁻⁶	> 1.0((2)	< 10

Note:1) lateral pressure, 350 KN/m²; 2) at extension of 10 %; 3) tensile force, 1 KN/10 cm.

Table 3 Settlements at various stages(in centimeters)

	Installation of vertical drain	Preloading by vacuum		Total
		centre	mean	
	1		2	1 + 2
Packed sand drain	24.2	100.4	87.21	124.6
PVC drain	23.2	101.0	88.07	124.2

Table 4 Degree of consolidation, U, at various time after pumping

Zone	S_∞ cm centre	mean	U_{60} % centre	mean	U_{65} % centre	mean	U_{99} % centre	mean
Sand drain	143.0	127.03	79.86	78.88	80.10	80.45	87.10	88.70
PVC drain	141.1	127.04	78.95	78.09	81.08	80.05	88.02	88.37

Note: Subscript 60, 65, 99 refer to days after pumping

Table 5 Variation of vane shear strength with depth

Depth	Strength at sand drain zone before treatment KN/m^2	after treatment	increase %	Strength at PVC drain zone before treatment KN/m^2	after treatment	increase %	Strength at area between two zones before treatment KN/m^2	after treatment	increase %
0-2.0	0.2	18.4	9100	0.2	18.3	9050	0.2	11.2	5500
2.0-10.0	14.2	32.0	125	14.2	30.4	114	14.2	29.3	126
10-13.0	17.6	25.5	44.9	17.6	26.1	48.3	17.6	24.7	40.3

Table 6 Soil properties before and after treatment

	Water content %	Unit weight KN/m^3	Void ratio
Before treatment	90	16	2.2
After treatment	55	17	1.5

6.3 Physical properties of hydraulic fill

The physical properties of the hydraulic fill are shown in Table 6. From Table 6 the improvement is obvious.

7 CONCLUSION

Fascine sand cushion and geotextile sand cushion both were successfully used in surface treatment. The former was cheaper, more efficient, easy to construct and abundantly available. So it was adopted in the project.

Preloading by vacuum in combination with packed sand drain or PVC drain was successfully used in treatment of very soft ground. Which one is to be adopted depends on such factors as cost, efficiency, supply of materials and construction equipment. In this project PVC drains produced by Nanjing Plastic Development Factory were used in great quantities because of excellent properties, low price less weight and less hard labour.

REFERENCES

Gao,H.X. 1987. Field study of soft ground at district 14 of port Shanghai. Port Engineering Information, No.5 p.1-11
(in Chinese)

Hu,R.Q. & Li,L.B. 1982. Hydraulic fill foundation use and treatment. Selected Papers of Weak Soil Foundation Treatment in Shanghai, Shanghai Municipal Construction Committee, p.377-389
(in Chinese)

Nishio, T. & Hara,H. 1981. Soft ground improvement by surface layer mixing method. Construction Mechanization, No.377 p.9-15
(in Japanese)

Olivera,A. 1982. Use of non-woven geotextiles to construct a deep highway embankment over swamp soil. 2nd Int Geotextile Conf., Las Vegas, Vol.III p625

Sasaki,S. & Kiyama,M. 1982. Ground improvement technique for a vast reclaimed land by combination of vertical drains and dewatering. Osaka and its technology. Osaka Municpal No.1 p.28-37

Watari,Y.H. et al, 1983. Changing characteristics of reclaimed clayey soil and selection of the earth spreading method. Soil & Foundation, June, p.19-24
(in Japanese)

Yasuhara,K. & Tsukamoto,Y. 1982. A rapid

banking method using the resinous mesh
on a soft reclaimed land. 2nd Int Geotextile Conf.,Las Vegas, Vol.III,p625

Ye,B.R., Liu,S, Tan,Y. and Gao,Z.1983
Packed sand drain and atmospheric
pressure for strengthening soft foundation. Port Engineering, No.1 p26-30
                        (in Chinese)

Zhang, D. 1987. Study on geotextile re-
inforced road foundation. PhD thesis
Research Institute for Railway Science
(China)   (in Chinese)

Li, B. & Ye, B.R. 1984. Method for construction on the newly filled soil,
Port Engineering, No.3 P 5-8
                        (in Chinese)

3. Slopes and excavations

International Geotechnical Symposium on Theory and Practice of Earth Reinforcement / Fukuoka Japan / 5-7 October 1988
© 1988 Balkema, Rotterdam. ISBN 90 6191 820 0

Major slope stability problems in dam constructions in Thailand

N.Deeswasmongkol, P.Noppadol & C.Areepitak
EGAT, Nonthaburi, Thailand

ABSTRACT: During dam constructions in Thailand, several slope failures occurred and various expensive remedial measures for slope stabilizations and slope protections were employed. It is accepted that geology plays a leading and very important role in analyzing slope stability problems. Yet, a large sum of extra money were used in stabilizing slopes of dams which include slopes of powerhouse, slopes of switchyard, and slopes of spillways. This paper summarizes the causes and remedial measures of slope stability problems of four hydro dams. Geological information, in reality, can not be fully obtained. However, this paper shows that more attention should be paid to geological investigation budget. No matter how hard we try to obtain geological data for slope design, it is rather impossible to guarantee that one's design is perfect without extra cost of additional remedial measures of slope.

1 INTRODUCTION

After the completion of the first hydro dam, Bhumibol dam, in Thailand in 1963. Within 25 years, the Electricity Generating Authority of Thailand, EGAT, has constructed 12 more hydro dams, both large and small, here and there all over the country to cope with the ever-increasing demand of electric power of Thailand.

Nowadays, slope designs are more practical and less time consuming by the application of computers. In this study, the limiting equilibrium technique, together with microcomputer application, are applied to evaluate the stability condition of rock cut slopes of four major hydro dams in Thailand, namely, Chiew Larn Dam, Bang Lang Dam, Khao Laem Dam, and Srinagarind Dam. Their locations are presented in Fig.1.

2 SLOPE FAILURES DURING CONSTRUCTION OF FOUR MAJOR HYDRO DAMS

(1) Chiew Larn Dam
The slope stability problems occurred at so many places during construction. These can be described place by place as follows:
(a) Powerhouse slope or penstock slope. Rocks that have been cut to the gradient vary from 1:1 to 0.2:1 (horizontal distance to vertical distance) are mainly pebbly mudstone and some shale. There are reports of some plane failures occurred along the bedding plane of those rocks which generally daylight on the slope face, dipping 40-45 degrees. Consequently, two units of three multiple points extensometers have been installed to detect the movement of slope. After plane failure occurred, the slope was re-excavated to the gradient varying from 1:1 to 1:0.5. In addition, to stabilize the slope, 60 tons prestressed tendon, rock bolts, shotcrete, drain hole and chain link fabric were also installed.
(b) Flip bucket slope
Flip bucket locates on the downstream portion of the spillway structure. The rock type in this area comprises mainly of pebbly mudstone, shale, sandstone and some shallow intrusive jointed igneous rock. The fractures and minor faults have been discovered considerable, so the rock condition was fair to poor with some fractures favorable to the slope failure. From stability analysis, plan failure was likely to occur. Consequently, 60 tons prestressed tendon including, rock bolts have been installed to protect the slope.

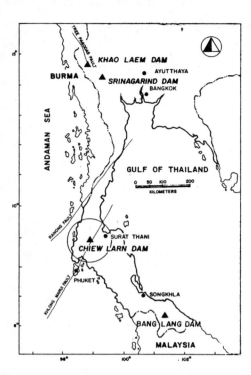

Fig. 1 Locations of hydro dams.

(c) Spillway slopes
The rocks in this area are mainly pebbly
mudstone and shale. The rocks that have
been excavated from the spillway channel
are also used as the rock fill material
for the main dam and the saddle dam. The
gradient of the spillway slope is of 0.5:
1 (h:v). During excavation, plane and
wedge failure occurred at many places.
Since the failed slopes do not influence
important structures, the slope stabilization is not necessary.

(2) Bang Lang Dam
Rocks dominantly comprise of sandstone,
shale, mudstone and some metamorphic
rocks which are subjected to high weathering. Slope failure occurred at many
places especially where highly weathered
rocks have been found. During construction period, circular failures generally
occurred after heavy rainfall at the residential area, some 1-1.5 km south of
the damsite.

(3) Khao Laem Dam
During construction, the rock cut slopes
of some appurtenance structures have been

reported of slope failure which can be
described separately as follows:

(a) Diversion channel (See Fig.2)
The diversion channel was excavated
across the Three Pagoda Fault which the
rocks comprise mainly of thin bedded
black shale, siltstone and some sandstone.
From design criteria, the slope gradient
varies from 0.5:1 to 1:1 (h:v) regardless
the geological conditions. Some small
wedge and plane failures occurred especially where the black shale exposed.
Consequently, the slope was re-excavated
which the gradient varied from 1.5:1 to
2.1 from the bottom to the top of slope.

(b) Left abutment (See Fig.3)
During excavation of left abutment slope,
single point extensometers revealed the
movement of slope along the clay seam 1
cm thick which daylighted at elevation
207.0 m (MSL). The result of direct
shear test revealed the cohesion and
friction angle of clay seam about 8-10
t/sq.m and 25-30° respectively. From
stability analysis after HOEK & BRAY
(1981), stabilization of slope was extremely necessary. Hence, 91 sets of 70
tons strand anchors have been installed
to stop the movement.

(c) Right abutment
The right abutment of Khao Laem Dam lies
on Ratburi Limestone or Khao Laem Massif
where the highly-steep fault scarp
exists. There have been reports on
small toppling and rock falls during
construction but no effect to the dam
structure. On December 16, 1981, 10000
tons of limestone fell down some 50 m.
upstream. Consequently, rock bolts and
shotcrete have been performed to stabilize the slope.

(4) Srinagarind Dam
Stability problem of rock cut slopes
occurred during the excavation of penstock slope and spillway slope. LEK
KANCHANAPOL (1978) stated that slope
failure occurred at the penstock slope
was due to the bed condition of shaly
limestone, quartzite which exposed in
between two adjacent faults. According
to the wide sheared zone in between those
faults, circular mode of failure was likely to occur. From circular failure analysis, the factor of safety of the slope
was 0.60. After stabilization by rock
bolt, strand anchor, the Factor of Safety
was 1.69 and considered to be safe for
the long period of time.

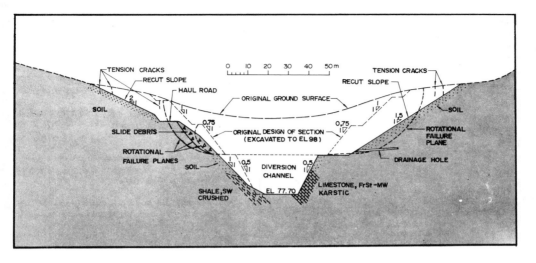

Fig. 2 A case study of slope design failure at the diversion channel of Khao Laem Dam, Western Thailand.

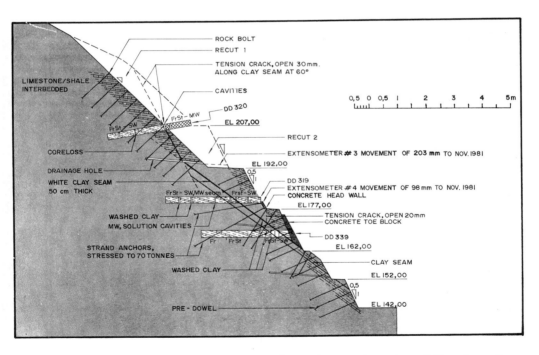

Fig. 3 A case study of spillway crest excavation and support system of Khao Laem Dam, Western Thailand.

Table 1. Results of Plane Failure Analysis (HOEK & BRAY Method and HENDRON Method)

DAM	SLOPE LOCATION	UNIT OF SLOPE	SLOPE HEIGHT (M.)	DIP OF SLOPE FACE	DIP OF FAILURE PLANE	UNIT WEIG. OF ROCK (T/CU.M.)	UNIT WEIG. OF WATER (T/CU.M.)	SEISMIC ACC. (g)	COHESION (C) (T/Sq.M.)	FRICTION ANGLE (DEGREE)	BOLTS&CABLES TENSION (T/M.)	BOLTS ANGLE (DEGREE)	F OF S AFTER HOEK &BRAY DRY NO BOLT	SATURATED NO BOLT	DRY BOLTED	SATURATED BOLTED	F OF S AFTER HENDRON DRY NO BOLT	SATURATED NO BOLT	DRY WITH BOLT	SATURATED WITH BOLT
1) CHIEW LARN	:SPILLWAY	1 BENCH SYSTEM SP1	15.0	63.4	44.0	2.63	1.00	0.00	0.0	45.0	-	-	1.04	0.53	-	-	-	-	-	-
		2 OVERALL SLOPE SP1	44.0	55.0	44.0	2.63	1.00	0.00	0.0	45.0	-	-	1.04	0.59	-	-	0.97	0.58	-	-
		3 BENCH SYSTEM SP7	20.0	63.4	52.0	2.63	1.00	0.00	0.0	45.0	-	-	0.78	0.31	-	-	-	-	-	-
		4 OVERALL SLOPE SP7	30.0	57.0	52.0	2.63	1.00	0.00	0.0	45.0	-	-	0.78	0.36	-	-	0.75	0.36	-	-
		5 BENCH SYSTEM SP8	15.0	63.4	38.0	2.63	1.00	0.00	0.0	45.0	-	-	1.28	0.72	-	-	-	-	-	-
		6 OVERALL SLOPE SP8	26.5	54.0	38.0	2.63	1.00	0.00	0.0	45.0	-	-	1.28	0.79	-	-	1.23	0.87	-	-
		7 BENCH SYSTEM SP10	15.0	63.4	56.0	2.63	1.00	0.00	0.0	45.0	-	-	0.67	0.21	-	-	-	-	-	-
		8 OVERALL SLOPE SP10	32.0	54.0	56.0	2.63	1.00	0.00	0.0	45.0	-	-	0.67	0.28	-	-	0.67	0.27	-	-
		9 FLIP BUCKET	14.0	63.4	53.0	2.63	1.00	0.15	3.0	35.0	13.7,11.5	45,6,11	1.20	0.68	2.05	1.60	0.54	0.27	1.26	0.96
	:POWERHOUSE	BERM NO. 1	11.0	65.0	40.0	2.63	1.00	0.00	8.3	37.0	12.225	50.0	2.45	1.88	3.06	2.31	0.96	0.63	1.34	0.93
		BERM NO. 2	10.0	65.0	36.0	2.63	1.00	0.00	8.3	37.0	4.95	50.0	2.62	1.98	2.88	2.16	1.15	0.63	1.34	1.07
		BERM NO. 3	13.7	45.0	42.0	2.63	1.00	0.00	8.3	37.0	-	-	3.00	2.50	-	-	0.81	0.48	-	-
		OVERALL SLOPE FACE	34.7	45.0	40.0	2.63	1.00	0.00	8.3	37.0	17.17	50.0	3.10	2.75	3.38	3.00	1.00	0.68	1.15	0.90
2) BANG LANG	:LEFT ABUTMENT	-	20.0	53.0	35.0	2.65	1.0	0.00	3.0	27.0	-	-	1.17	0.88	-	-	0.73	0.53	-	-
3) KHAO LAEM	:LEFT ABUTMENT	-	43.0	55.0	35.5	2.65	1.0	0.00	8.0	25.0	78,80	25,45	1.16	0.66	1.38	1.05	0.61	0.44	0.78	0.60

3 MAJOR SLOPE PROBLEMS AND MODES OF OCCURRENCES OF SOIL & ROCK SLOPES

In Thailand, several major slope problems occur on both natural and cut slopes in dam construction as following;

1 Chiew Larn Dam;
Plane failure of Penstock Slope

Table **2**. Summary of Stability Analysis of Slopes

Dam	Slope Location	Unit of Slope	Mode of Failure	Factor of Safety
1.Chiew, Larn	Spillway	sp1	Plane	1.04
		sp2	Wedge	1.32
		sp5	Wedge	1.15
		sp7	Plane	0.78
		sp8	Plane	1.28
		sp10	Plane	0.67
		Flip Bucket	Plane	2.05
	Powerhouse	Berm 1	Plane	3.06
		Berm 2	Plane	2.88
		Berm 3	Plane	3.00
		Overall Slope	Plane	3.38
2. Bang Lang	Powerhouse		Circular	1.17
	Left Abutment		Plane	1.17
	Right Abutment		Circular	1.40
3. Khao Laem	Left Abutment		Plane	1.38

2 Bang Lang Dam;
Circular slope failure at power house and residential area.

3 Khao Laem Dam;
Plane failure on the left abutment of rock slope.

4 Srinagarind Dam;
Plane failure on penstock slope.

In general, the common modes of slope failure is plane failure in rock excavation and the cause of failure is unanticipated geology conditions, construction technique of excavation (uncontrolled blasting). Plane failure usually occurs in both small and large scale failures while wedge failure normally occurs as small scale failure. Toppling failure is not common. Circular failure both small and large scale, is found in soil and highly weathered and heavily fractured rock slopes.

4 REMEDIAL MEASURES REQUIRE FOR CRITICAL SLOPES

Remedial measures required for critical slope depend on the factor of safety and the risk to life and/or structured recommended by Geotechnical Control Office, Public Work Department, HONG KONG (1979).

According to the analysis of existing slopes in terms of the factor of safety as shown in Table 1, summary of areas and their factors of safety can then be finalized as shown in Table 2.

Table **3**. Summary of Additional Cost of Stabilization Works due to Slope Problem of Chiew Larn Dam

Location	Description	Unit	Quantity	Thai Currency	Total Amount (baht)
Powerhouse Slope	- Rock bolts resin anchor Tendon 60 ton 20 m.long Shotcrete Grouting Recut slope Multipoints extensometers			Subtotal =	3625534.61
Flip Bucket Slope	- Tendon 60 ton	m.	308.00	1992.82	613788.56
	- Rock bolts	m.	750.00	273.39	205042.50
	- Anchor bars	m.	91.00	133.05	12107.55
	- Material	kg.	895.20	3.45	3088.49
				Subtotal =	834027.05
				Total =	4459561.66

Table **4**. Summary of Additional Cost of Stabilization Works

due to Slope Problem of Khao Laem Dam

Location	Description	Unit	Quantity	Total Thai Currency	Amount bath
Left	– Extensometer	LS	1.00	6000.00	6000.00
Abutment	– Dywidag rock anchor	m.	738.00	1200.00	885600.00
	– Rock bolts	m.	24873.00	906.00	22534938.00
	– Anchor bars	m.	21946.00	363.00	7966398.00
	– VSL rock anchors	m.	2156.00	3475.00	7492100.00
	– Anchor bolts	m.	86285.00	35.00	3019975.00
	– Dywidag washers	LS	1.00	49442.00	49442.00
	– Crane time to install rock bolts	hr.	299.00	780.00	233220.00
	– Materials	LS	1.00	382000.00	382000.00
				Total =	42569673.00

5 ADDITIONAL COST DUE TO CRITICAL SLOPES

Examples of addition costs due to critical slopes of Chiew Larn and Khao Laem dam are shown in Table 3 and Table 4 respectively. For penstock slope of Srinagarind dam, the total cost incurred for the protection of the rock sliding was totally US$ 1.14 million.

6 CONCLUSIONS

From data and tables obtained from several case studies of dam construction in Thailand as shown in this paper, it can be said that, in practical, dam constructions cannot avoid slope excavations especially rock slope stability problems.

Plane failures in dam constructions usually occur with a large rock mass and always very costly to treat while wedge failures occur here and there but of no vital hazards for dam construction. From several case studies during dam constructions, it has also been convinced that geological information obtained during investigation period were insufficient and caused delays which has many chain effect on construction schedule.

REFERENCES

Areepitak, C. 1980. Engineering Geology Engineering Geology and Foundation Treatment of Bang Lang Dam; Hydro-Electric Construction Department, EGAT.

EGAT, 1982. Geological Investigation of Chiew Larn Project, Geology and Soil Engineering Div., Hydropower Dept., Report No.842-2506.

Geotechnical Control Office, 1979. Geotechnical Manual for Slopes, Public Works Department, Hong Kong.

Hoek, E. & Bray, J. 1981. Rock Slope Engineering, The Institute of Mining and Metallurgy, London.

Kanchanaphol, L. 1978. Sliding of Penstock Foundation and Its Protection Srinagarind (Ban Chao Nen) Dam, Proc. Int. Symp. on Rocks Mechanics Related to Dam Foundations, ISRM, Brazil.

Nappadol, P. 1987. Stability Conditions of Rock Cut Slopes and Their Post Construction Evaluation of Some Jajor Hydro Dams in Thailand, AIT Thesis No.GT-86-32, Bangkok.

International Geotechnical Symposium on Theory and Practice of Earth Reinforcement / Fukuoka Japan / 5-7 October 1988
© 1988 Balkema, Rotterdam. ISBN 90 6191 820 0

Geotechnical approach to reinforcement system of underground openings in a gold mine

Tetsuro Esaki, Tsuyoshi Kimura & Tadashi Nishida
Kyushu University, Fukuoka, Japan

Nobuhiro Kameda
Kyushu Kyoritsu University, Kitakyushu, Japan

Ken-ichi Nakamura & Hisao Imada
Hishikari Mine, Sumitomo Metal Mining Co., Ltd, Hishikari, Japan

ABSTRACT: This paper describes rock mass classification techniques and monitoring of the behaviors of artificial roof supports which were performed to establish the stability of underground openings and the reasonable excavation method at a gold mine of vein type. The Q-system which is one of the classification techniques can provide optimum support systems, depending on the conditions of rock masses and openings at this mine. In addition it is confirmed from the monitoring results that the excavation design using the artificial roof supports, made of reinforced concrete and of H-beam with wire mesh, is valid for stability of the openings in the cut and fill stoping.

1 INTRODUCTION

One of the objects of rock mechanics practice is to assess the stability of underground structures, such as mines and tunnels. There are analytical methods, empirical methods and observational methods for the design of the underground structures (Bieniawski,1984).

The analytical methods include numerical procedures (finite element, finite difference, boundary element), analog simulations (electrical and photoelastic) and physical modeling, and utilize the analyses of stresses and deformations around openings. Since, unlike conventional structures such as buildings or bridges, the underground structures have an excessively large number of relevant parameters, they can not be easily treated by the analytical methods.

The empirical methods utilize statistical analyses of underground observations. The qualitative description of geotechnical materials by means of classification techniques has a long development history. While only one soil classification system, introduced by Casagrande, is widely used today for soils, many qualitative or quantitative rock mass classification systems have been proposed to describe rock mass conditions. From a practical point of view, the application of different classification systems to the same ground conditions may be desirable to benefit from individual advantages or to compensate for factors neglected by some (Kaiser et al.,1986).

The observational methods rely on actual monitoring of ground movement during excavation to detect measurable instability. These include the New Austrian Tunneling Method.

In this study, quantitative assessments of rock mass conditions were performed at the Hishikari gold mine to establish a more optimum support system, using two classification systems. In addition, the behaviors of two types of the artificial roof supports, one made of reinforced concrete and the other of H-beams with wire meshes, were monitored in order to reasonably design the cut and fill stoping method.

2 GEOLOGY, ORE DEPOSIT AND UNDERGROUND STRUCTURE

The Hishikari gold mine, which is now regarded as one of the highest grade gold mines in the world, is located in the Hokusatsu district, Kagoshima Prefecture, Japan. The deposit comprises an epithermal gold-silver bearing quartz-adularia vein (Abe et al.,1986). Several major veins and numerous veinlets have so far been encountered in an area some 800 m by 100 m. These veins strike 45° to 70°E and dip 70° to 90°N, and show extremely high gold contents in the region between 0 m and 100 m above sea level (250 m and 150 m below the portal, respectively; see Fig.1). Their widths vary from a few centimeters to 8 meters. The veins occur in both Pleistocene

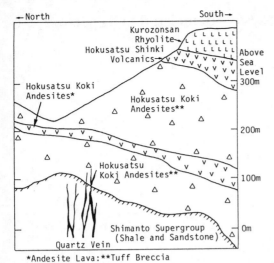

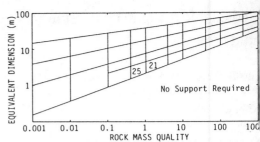

← North   South →

Kurozonsan
Rhyolite
Hokusatsu Shinki
Volcanics

Above
Sea
Level
300m

Hokusatsu Koki
Andesites*

Hokusatsu Koki
Andesites**

200m

Hokusatsu
Koki Andesites**

100m

Shimanto Supergroup
(Shale and Sandstone)

0m

Quartz Vein

*Andesite Lava:**Tuff Breccia

Fig.1 Schematic cross section of geology.

Fig.2 Support chart of Q-system.

andesitic volcanics and sedimentary rocks
of late Cretaceous to early Palaeogene age.
  This deposit was discovered in 1981
during a scout-drilling program carried out
by the Metal Mining Agency of Japan (MMAJ).
In December 1982, twin parallel inclined
shafts were developed to approach the orebody, and the first cross cut intersected
the orebody in July 1985. The shafts have
a grade of 17 % (10 degrees) and a size of
4.8 m width by 3.8 m height. Thereafter,
main levels were in turn driven at 100 m,
70 m, 40 m and 10 m above sea level.

3 ROCK MASS CLASSIFICATION

The excavation of the orebody is now being
advanced smoothly to a certain extent,
while it is necessary to establish a more
optimum support system. This is realized
by a rock mass classification. The present
study uses two classification systems: the
rock mass quality (Q-system) developed by
Barton et al.(1974) and the rock mass
rating (RMR) system by Bieniawski(1974).
  The rock mass quality number Q is
calculated by

$$Q=(RQD/J_n)\cdot(J_r/J_a)\cdot(J_w/SRF), \qquad (1)$$

where
RQD : the rock quality designation,
J_n : the joint set number,
J_r : the joint roughness number,
J_a : the joint alteration number,
J_w : the joint water reduction factor, and
SRF : the stress reduction factor.

The three terms on the right hand side of
Eq.(1) express rock block size, joint shea
strength and confining stress, respectively. The range in values of Q is from 0.00
for extremely poor rock to 1,000 for excel
lent rock. In addition Barton et al. defined the equivalent dimension (D_e) as

$$D_e=(span \text{ or height of opening})/(ESR), \quad (2)$$

where the term ESR is called the excavation support ratio and resembles the re-
ciprocal of a safety factor. The value of
1.6 is adopted in permanent mine openings
and 3 to 5 in temporary mine openings.
  The RMR is obtained by summing five
parameter values and adjusting this total
by taking into account the joint orientations. The parameters included in the
system are rock material strength, RQD,
joint spacing, joint roughness and separation, and ground water. The RMR value can
range between zero for extremely poor rock
and 100 for excellent rock, and provides a
relationship between the stand-up time and
the maximum unsupported span for a given
rock mass.
  The RMR system may give guidelines for
the selection of support, which depend on
such factors as the depth below the ground
surface (in-situ stress), opening size and
shape, and the method of excavation.
However, Bieniawski(1976) showed only the
support guideline for a typical opening
and excavation in his paper. On the
contrary, the Q-system can choose a
support system among 38 categories, based
on the relationship between the rock mass
quality and the equivalent dimension.
Figure 2 shows the support chart, where
the boundary of the maximum unsupported
span is drawn using the following
equation:

$$D_e=2\cdot Q^{0.4}. \qquad (3)$$

The Q-system considers not only the tunnel
sections in two dimensional plane strain

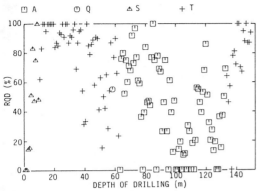

Fig.3 Distribution of RQD values along a drilling line. A;andesite, Q;quartz, S; shale and sandstone, and T;tuff.

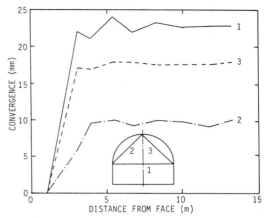

Fig.4 Convergences of the opening supported based on the Q-system.

condition but also crossings and bifurcations. Since, unlike tunnels in civil engineering, the Hishikari gold mine has a complicated underground structure such as inclined shafts, main levels, cross cuts, drifts etc. and their opening widths vary from point to point, the Q-system is preferable to the RMR system.

4 SUPPORT SYSTEM RECOMMENDED BY ROCK MASS CLASSIFICATION

For the reason described above, the support system recommended by the Q-system is mainly presented here. The parameters of RQD, J_n, J_r and J_a were evaluated for the typical and worst conditions of rock masses in this mine; the RQD values were not determined from the observation of rock walls but from that of drill cores which

Table 1 Evaluated Q values and recommended supports for two conditions of rocks without inflows.

Rock	Condition*	Q value	Support
Andesite	T	11	No support
	W	1.3	#21
Shale and sandstone	T	18	No support
	W	1.3	#21
Tuff	T	40	No support
	W	2.5	#21

* T:typical and W:worst.
#21:shotcrete of 5 cm thickness.

had been abundantly obtained during exploration and development. Figure 3 shows an example of the RQD distribution along a drilling line. Since in this line the RQD values considerably decrease at 100 to 110 meters in drilling depth, it is presumed that joint distribution is dense in this region and the rock mass is poor. The RQD values varying widely as shown in Fig.3 were handled with a statistical procedure to obtain their representative values for the typical and worst conditions.

The value of the joint water reduction factor was regarded as unity at the sections under dry condition, but as 0.5 at the inclined shafts for development of lower levels because of ground water inflows. The parameter of the stress reduction factor was also regarded as unity, considering that rock stress is medium in the Hishikari mine.

Table 1 shows the ratings for host rocks of andesite, tuff and Shimanto super-group (shale and sandstone) without inflows. Although the ratings for the quartz veins were additionally evaluated, the result does not appear in this table because the vein conditions varied considerably from point to point and it was difficult to arrange the ratings. For the drifts, therefore, the conventional support system was used: steel sets of H-beams with wood blocks. The recommended support systems is also shown in Table 1, using the equivalent dimension of 3 meters (the span of opening, 4.8 m and the excavation support ratio, 1.6). It can be seen from Table 1 that the typical rock masses have no need of support. Considering a loose part of rocks in the roofs, however, the support system for the worst rock masses was actually

273

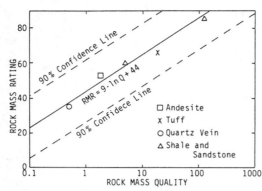

Fig.5 The Q values and the RMR evaluated
for rock masses in the Hishikari mine.

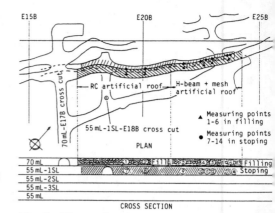

Fig.6 Plan and cross section of test site
of the downward cut and fill stoping.

chosen.

The recommended support system of the category 21 (shotcrete of 2.5 to 5 cm thickness) was tried in a site of tuff, and the convergences of the opening (4.0 m width by 3.457 m height) were measured there. The convergences, as shown in Fig.4, increased until the face stood about 3 m apart from the measuring point, but remained constant thereafter. It was judged from this result that the support of shotcrete is sufficient for stability of the opening, but it was also predicted that the shotcrete support would decrease the working efficiency. Therefore, instead of the shotcrete, the support system of wire mesh (ϕ5 x 100 x 100 mm) and rock bolt (ϕ25 x 2000 mm; 1 m spacing), corresponding to the ability of the shotcrete, was determined to use actually for all conditions.

For the support at the crossings without ground water inflows, the joint set number in Eq.(1) was multiplied by the value of 3.0 according to the suggestion by Barton et al.(1974); i.e. the Q value was decreased to one third and the support category 21 was changed into the category 25 consisting of wire mesh, rock bolt and shotcrete. The support category 25 was applied to the inclined shafts of the Shimanto super-group in the lower levels with inflows as well.

The RMR values were also evaluated to confirmed the following relationship which was developed from 117 case histories by Bieniawski(1976):

$$RMR = 9 \cdot \ln Q + 44. \qquad (4)$$

Figure 5 shows the Q values obtained for the worst condition and the RMR for the typical condition in the same rock masses at some places. Eq.(4) is valid when rock

masses are evaluated under such different standards in the Q-system and the RMR.

5 DOWNWARD CUT AND FILL STOPING METHOD

It is premised that all ores are excavated and no pillars remain, and that workers do not work below unsupported stoping areas for safety. Thus, the downward cut and fill stoping method with artificial roof supports was determined to be adopted.

The test stoping over a length of 71 m were performed directly below the artificially supported drift of 70 m level E17B which was filled over a length of 73 m. Figure 6 shows a plan and cross section view of the testing site of the downward cut and fill stoping. The two kinds of artificial roof, the reinforced concrete (RC) roof and the H-beam roof shown in Fig.7, were tested. The strains and deformations of the roofs were measured by mould gauges in two directions in RC and by strain gauges put on H-beam respectively during the advance of the face. The strain of the support of the lower stoping area was also measured by strain gauges on the middle of the supporting member simultaneously. In addition roof sags were measured by the deformeter connecting with anchored invar wire.

In the design of artificial roofs, the load of fill material only was considered and the load from side walls was neglected. The RC slab was 19 cm thick, and D19 steel bars were arranged with spacing of 14 cm and 30 cm at a right angle. Allowable stresses of reinforcement and concrete are 1,800 kgf/cm^2 and 112 kgf/cm^2 respectively. H-beams (H100 x 100 x 6 x 8 mm) were arranged with spacing of 50 cm and connected

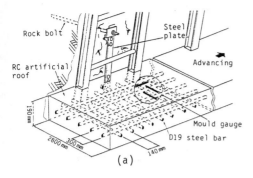

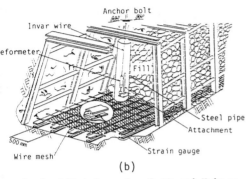

Fig.7 Artificial roofs of RC and H-beam with wire mesh, and their measuring systems. (a);RC and (b);H-beam with wire mesh.

at each end by splicing plates. The artificial roof of H-beams with two kinds of wire meshes were supported by steel sets with wood blocks.

The sags of the RC roof were -9.1 mm and -14.4 mm at 2 measuring points, respectively, just after the face passing. Since 96 % and 78 % of the deformations were recovered when the face had proceeded forward, the RC roof can be considered to be elastic. The H-beam roof caused a sag of 56 mm. The sag did not recover but maintained a

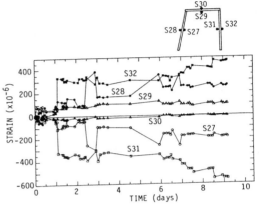

Fig.8 Strains of the steel sets at measuring point 8.

Table 2 Measured and calculated strains in steel sets at lower stoping area.

| Artificial roof | Measuring point | Condition* | Strain in steel sets ($\times 10^{-6}$) | | |
			Left vertical member	Horizontal member	Right vertical member
Reinforced concrete	7	A	350	80	-220
		B	230	40	-60
		C	360	430	-960
Reinforced concrete	8	A	130	50	-330
		B	140	60	-330
		C	380	330	-960
H-beam and mesh	9	A	150	40	-30
		B	80	0	-50
		C	370	490	-960
H-beam and mesh	10	A	300	100	-200
		B	700	900	0
		C	480	680	-1240

* A:value just after the face passing, B:final value and C:calculated value.

constant value. When the face passed through, the strains in the artificial roofs varied remarkably due to the change of moment. However, they became constant values with time and did not show excess values.

Figure 8 shows the strains of the steel set at the measuring point 8 in the lower stoping area, with advancing of the face. Table 2 shows the measured and calculated strains of the steel sets under different conditions at some measuring points. The estimated values were obtained from an assumption that all load of the fill material would apply to the steel sets through the wood blocks.

The RC roof is regarded as "self support roof" because the measured strains at the measuring points 7 and 8 are small. In the H-beam roof, the measured strains are relatively small compared with the calculated ones. Generally the load applied to the steel set is smaller than the total load from filling material. This reason is that the strength of fill material retained somewhat and the total weight did not apply to the support.

It is confirmed from these monitoring results that the artificial roofs can carry the applied load and the design is valid for the stability of the openings in the cut and fill stoping.

6 CONCLUSIONS

To obtain more optimum reinforcement systems of underground openings at a gold mine, two rock mass classification techniques were applied and investigated, and the behaviors of two types of artificial roof support were monitored. The results are as follows:

1. Considering that the gold mine has a complicated underground structure, The Q-system is preferable to the RMR.

2. A relationship is established between the Q value evaluated for the worst condition and the RMR for the typical condition in the same rock masses.

3. The Q-system provides a reasonable support system, made of rock bolt and wire mesh, for openings of host rocks under the conditions of two dimensional plane strain and no ground water inflow; tunnel crossings or inflows add the reinforcement of shotcrete to the support system.

4. The design of the artificial roof supports, made of reinforced concrete and of H-beam with wire mesh, are valid for the stability of the openings in cut and fill stoping.

REFERENCES

Abe,I., H.Suzuki, A.Isogami & T.Goto 1986. Geology and development of the Hishikari mine. Mining Geology 36: 117-130 (in Japanese).

Barton,N., R.Lien & J.Lunde 1974. Engineering classification of rock masses for th design of tunnel support. Rock Mechanics 6:189-236.

Bieniawski,Z.T. 1974. Geomechanics classification of rock masses and its application in tunneling. Proc. Third Congr. Rock Mechanics, Denver, Vol.2A: 21-32.

Bieniawski,Z.T. 1976. Rock mass classification in rock engineering. Proc. Symp. Exploration for Rock Engineering, Johannesburg, Vol.1:97-106.

Bieniawski,Z.T. 1984. Rock mechanics design in mining and tunneling. Rotterdom: Balkema.

Kaiser,P.K., C.MacKay & A.D.Gale 1986. Evaluation of rock classifications at B. C. Rail Tumbler Ridge Tunnels. Rock Mechanics and Rock Engineering 19:205-234.

International Geotechnical Symposium on Theory and Practice of Earth Reinforcement / Fukuoka Japan / 5-7 October 1988
© 1988 Balkema, Rotterdam. ISBN 90 6191 820 0

Field behaviour of two instrumented, reinforced soil slopes

R.J.Fannin & S.Hermann
Norwegian Geotechnical Institute, Taasen, Norway

ABSTRACT: The behaviour of reinforcement in soil forms the subject of this paper, in which the construction and monitoring of two field projects are reported. Fundamental behaviour of geogrid reinforcement in a cohesionless backfill is described from measurements of soil and reinforcement strain, and from measurements of force in the reinforcement. The results are used to compare field performance of the geogrid with the theoretical understanding that forms the basis of current design approaches.

1 INTRODUCTION

In 1985 a snow avalanche barrier was built of reinforced soil and instrumented to monitor field performance. Analysis of results from this working structure led to the design and construction in 1987 of a steep slope research structure. This research structure has been used to examine fundamental behaviour of the reinforcement in soil.

2 SNOW AVALANCHE BARRIER, ÅNDALSNES

2.1 Construction

A 6m high and 55m long reinforced soil embankment was constructed as a snow avalanche barrier at Åndalsnes, on the west coast of Norway. The embankment has steep (2:1) sides and rests on a competent foundation. A uniformly graded medium sand, one that is non-frost susceptible according to Norwegian guidelines, was selected as a backfill material.

The embankment is reinforced with four layers of Tensar SR2 geogrid, Figure 1(a), according to an arrangement and spacing based on the limit equilibrium design method of Jewell et al (1984). In addition to this main or primary reinforcement, layers of Tensar SS1 geogrid were placed as secondary reinforcement near the surface of the slope to take care of local stability, and a Terram 42A fine-grade mesh was used

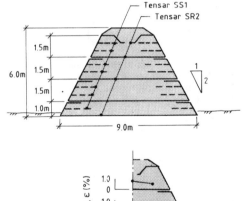

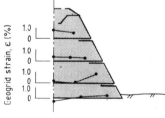

Fig. 1   Snow avalanche barrier

behind the Tensar on the slope face to prevent surface erosion of the backfill material.

The procedure adopted in construction was the "wrap-around" principle, in which the primary reinforcement is anchored higher in the slope to form discrete envelopes of enclosed soil layers, for which details of the fixing arrangement are reported by Hermann (1986). An external wooden frame was used to provide support to the sides of the embankment during placement and compaction of the backfill. While use of such an external scaffolding proved effective, it was nevertheless time-consuming and expensive to erect.

2.2 Field behaviour

Instrumentation was placed in a cross-section some 10m from the end of the embankment to enable field behaviour to be compared with that predicted by current design methods for steep reinforced slopes: soil strains and earth pressures were recorded in the backfill material, together with strains at several locations on the primary reinforcement. Full details of the arrangement of the instrumentation are given by Burd (1986).

Measurements of reinforcement strain have shown generally small strain magnitudes less than 0.8%. The distribution of strain nearly one year after completion of construction is shown in Figure 1(b). Yet a comparison of these measured strains with values of expected strains in the reinforcement (from the Jewell et al design method) has shown the limit equilibrium approach makes a significant over-prediction of the strains actually mobilized in the reinforcement, even allowing for differences between the infinite width assumed in the design models and the geometry of the barrier, Burd (1986). The nature of these observations in the avalanche barrier led to the construction of a fully instrumented field trial slope close to Oslo to enable an investigation of the fundamental behaviour of reinforcement in a steep slope.

3 STEEP SLOPE TRIAL, OSLO

3.1 Construction

A 4.8m high and 20m long reinforced slope was built on a competent, gravelly sand foundation. A uniformly graded medium-to-fine sand was used in construction. The steep (2:1) slope was reinforced with two different arrangements of geogrid reinforcement termed Section 'J' and Section 'N', see Figure 2, so that the influence of reinforcement length and spacing could be examined. In Section 'J' the Jewell et al method was used, while in Section 'N' the arrangement was based largely on the merits of a design method for reinforced soil block walls described

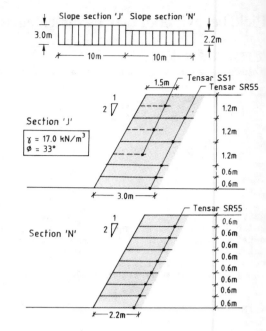

Fig. 2   Steep slope: arrangement of the reinforcement

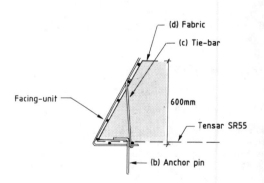

Fig. 3   Details of facing-units

by Knutson (1986). Tensar SR55 geogrid was used as primary reinforcement, and layers of Tensar SS1 included as secondary reinforcement when spacing of the primary reinforcement exceeded 0.9m.

Construction of the slope involved compacting the backfill material in a series of discrete layers behind

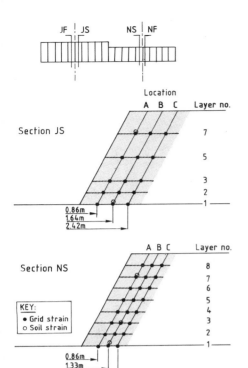

Plate 1.   Load cells: fixing details

Fig. 4   Steep slope: locations of strain
measurement

A layer of primary reinforcement comprises
ten strips of geogrid, and measurements of
force and strain are taken on the two
strips located either side of the centre-
line of the sections, that is sections JF
(or NF) and JS (or NS). Measurements of
force are taken close to the slope face at
location A, while values of strain in the
geogrid are calculated at locations A, B
and C, see Figure 4. Force in the
reinforcement is measured using an approach
similar to that described by Bassett
(1986), in which three load cells are
mounted in parallel across any one metre
wide strip of reinforcement using a special
arrangement of bearing clamps, see Plate 1.
Strain in the reinforcement is deduced from
the output of pairs of "Bison" inductance
coils fixed directly to the reinforcement.
The principle of operation is that of
elctromagnetic coupling and does not rely
on any mechanical connection. Glötzl
hyraulic pressure cells record earth
pressure and thermistors are used to
measure soil temperatures. A full

lightweight facing-units cut from a
commercially available steel mesh, see
Figure 3. The units obviate any need for
external scaffolding during construction
and provide support for a fabric which acts
to locally retain the soil behind the
apertures of the mesh. In spite of the
labour force being unfamiliar with
reinforced soil techniques, construction
proceeded as quickly as the backfill
material could be placed, Hermann and
Fannin (1988). The two design sections are
isolated from each other by a flexible
separator panel in the slope.

3.2 Field monitoring

Instruments were placed centrally in each
section of the slope during construction to
monitor behaviour of the reinforcement and
soil. Force and strain are measured at
selected locations on the primary
reinforcement, and soil strains and earth
pressures are measured in the reinforced
zone of the backfill.

Plate 2.   Cycle of surcharge load

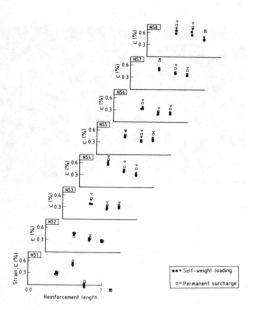

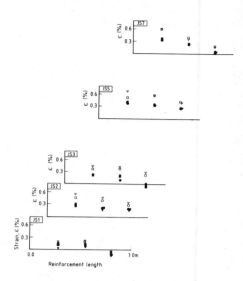

Fig. 5(a)   Section 'J':   reinforcement
strains

Fig. 5(b)   Section 'N':   reinforcement
strains

description of the design, construction and
instrumentation of the slope is given by
Fannin (1988).

The slope has been subjected to three
separate stages of loading. Following
completion of construction, behaviour of
the two sections was monitored for a period
of one month under self-weight loading
conditions. Thereafter the crest of the
slope was loaded so that greater forces
would be mobilized in the reinforcement. A
load/unload cycle was applied using large
tanks which were filled with water, see
Plate 2; a permanent surcharge load was
later applied by increasing the height of
the slope by 3m to 7.8m.

Typical profiles of strain in the
reinforcement during self-weight and
permanent surcharge loading are shown in
Figures 5(a) and 5(b) for Section 'J' and
Section 'N' respectively. The measurements
indicate a similar distribution in each of
the two slope sections. Larger strains are
measured close to the slope face, and
smaller strains are measured near the
embedded end of the reinforcement. The
response to surcharge loading is typified
by larger mobilized strains, and therefore
forces, with evidence of the increase being
greatest near the face of the slope. A very
limited response to the applied loading is

measured in the base layer of reinforcement
in each section.

3.3 Strain measurements in the geogrid and
soil

The mechanism by which a grid acts to
reinforce a granular soil is based on
interlock between the soil particles and
grid members. A suitable ratio of soil

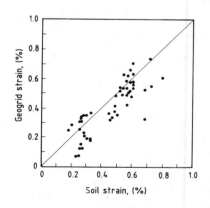

Fig. 6   Strain in the soil and geogrid

280

particle size to grid aperture size allows particles to interlock with the grid structure during placement and compaction of the soil. Any subsequent component of soil displacement in the plane of the reinforcement drives these particles to bear against the grid members, causing the reinforcement to be strained.

Strain in each layer of geogrid is measured at three locations along the longitudinal centre-line of the instrumented strip. Complementary measurements of strain in the backfill are taken at selected locations in the soil next to the geogrid. The relation between strain in the reinforcement and strain in the soil, Figure 6, shows similar magnitudes of strain. The data are for both slope sections and represent all three conditions of loading of the slope. The agreement in the measured values is clearly indicative of a good interlock between the geogrid and surrounding soil.

3.4 Force and strain measurements in the geogrid

In common with all polymeric materials, Tensar geogrids exhibit a viscoelastic behaviour, McGown et al (1984), which is to say the load-strain response under a constant rate of strain is significantly influenced by the test temperature and strain rate. In the trial slope, comparable measurements of force (T) and strain (ϵ) are taken at location A on the reinforcement. In total there are five such locations of measurement in Section 'J' and seven in Section 'N'.

Tensile stiffness of polymeric reinforcements is typically described by a value of tensile modulus K, established from laboratory tests on discrete samples of material. Values of tensile modulus for the geogrid reinforcement in the slope are calculated from the reported measurements of force and strain where,

$$K = \frac{T}{\epsilon}$$

and are summarised in Table 1 for all measurements taken during each of the stages of loading. The results indicate some variation in the deduced values of tensile modulus, particulary from pairs of data for which the strain is small, and show a mean value of 5.4kN/m/% throughout the period of study, for mean values of strain between 0.7 and 0.1%, and for temperatures in the range 0 to 20°C, Fannin (1988). Laboratory tests on samples

Table 1 Values of tensile modulus

Slope loading	Soil temp. °C	K (kN/m/%) Mean	S.D.
Self-weight	13-20	5.7	3.1
Cyclic of surcharge	10-15	5.2	2.3
Permanent surcharge	0-11	5.4	1.9

of Tensar SR55, Netlon (1987), show values of tensile modulus in the range 3.5 to 4.5kN/m/%, for strains between 1.0 and 0.5%, and for temperatures in the range 10 to 20°C.

4 DISCUSSION

The observations of field behaviour make some clear implications for the design and construction of steep slopes. The mechanism of reinforcement invoked by a geogrid, whereby the grid members interlock with the soil, is well understood. For the small strain magnitudes less than 1% measured in the Oslo field trial, strains in the soil and in the geogrid are nearly equal and indicate the presence of a good interlocking action. The distribution of observed strains shows small strain magnitudes close to the embedded end of the reinforcement, increasing to a more constant value near the front of the slope where the reinforcement is fixed to the facing-units.

McGown et al (1982) describe tensile tests made on unconfined ("in-air") and confined ("in-soil") samples of geotextiles. The results proved confinement of a non-woven geotextile to have a significant influence on the measured stiffness of the material, but little influence on that of woven geotextiles which are stiffer. The tensile modulus of Tensar geogrid for design is deduced from similar laboratory test data based on an approach using isochronous curves. The range of predicted moduli from these tests is similar to the range of values calculated from measurements of force and strain in the reinforcement in the slope: this agreement suggests that in the absence of any other test information, use of unconfined moduli to describe behaviour of the confined geogrid is quite reasonable.

281

5 CONCLUSION

1. The reinforcement is fixed to lightweight facing-units which are included as a permanent feature in the slope. While the arrangement proved very successful, selection of such facing-units in any kind of permanent structure must consider a requirement for protection against corrosion and a suitability against impact loading and vandalism. In this respect the use of concrete blocks, with provision for planting and vegetation growth in situations of enviromental concern, offers an attractive solution.

2. Interaction between the geogrid and backfill soil has been examined from instrumentation placed in the slope. A good interlock is observed, and the distribution of observed strains is in agreement with existing models of behaviour.

3. The difference between values of tensile stiffness of the geogrid deduced from field and laboratory measurements does not appear to be significant: it is therefore reasonable to use "unconfined" laboratory values of stiffness for design.

4. An important aspect of the behaviour of polymer reinforcement is the long-term behaviour under load. Polymeric materials are known to creep, with the creep strain that occurs being largely dependent on the polymer type and load magnitude (at constant temperature). Observations of force and strain in the geogrid are continuing so that the long-term behaviour of the reinforcement in the slope may be established.

ACKNOWLEDGEMENTS

The steep slope field trial is a co-operative project in association with the Norwegian Road Research Laboratory and Akershus Road Department, and the contribution of these organisations is gratefully acknowledged. Terram is a registered trademark of ICI Fibres. Tensar is a registered trademark of Netlon Ltd. Tensar geogrid reinforcement used in construction was supplied by Nor-Vest A/S; the lightweight facing-units used in the steep slope near Oslo were supplied by Christiania Spigerverk A/S. Use of similar facing-units, in combination with a non-woven vegetation fabric and geotextile reinforcement, is the subject of a Swiss patent application under the name of Textomur.

REFERENCES

Bassett, R.H. (1986). The instrumentation of the trial embankment and the Tensar SR2 grid. Reinforced Earth Embankment Symposium, King's College, London.

Burd, H.J. (1986). A snow avalanche barrier of reinforced earth at Åndalsnes Results and analysis of instrumentation data. Norwegian Geotechnical Institute, Internal Report 52757-4, Oslo.

Fannin, R.J. (1988). An instrumented fiel study of the analysis and design of geogrid reinforced slopes. Norwegian Geo technical Institute, Report 52757-10, Oslo.

Hermann, S. (1986). Sikringsvoll av armer jord, Åndalsnes vegstasjon. Oppfølging a byggearbeidene. Norwegian Geotechnical Institute, Internal Report 52757-2, Oslo

Hermann, S. and Fannin R.J. (1988). Con-struction experience in steep, reinforce soil slopes. Nordic Geotechnical Meeting Oslo, Vol. 1.

Jewell, R.A., Paine, N. and Woods, R.I. (1984). Design methods for steep reinforced embankments. Symp. on Polymer Grid Reinforcement, I.C.E. London.

Knutson, Å. (1986). Reinforced soil block walls. Norwegian Road Research Laoratory Internal Report No. 1292, Oslo.

McGown, A., Andrawes, K.Z. and Kabir, M.H. (1982). Load-extension testing of geotextiles confined in-soil. Second Int Conf. on Geotextiles, Las Vegas, Vol. 3, 793-798.

Netlon, (1987). Private communication, Netlon Ltd., Blackburn.

International Geotechnical Symposium on Theory and Practice of Earth Reinforcement / Fukuoka Japan / 5-7 October 1988
© 1988 Balkema, Rotterdam. ISBN 90 6191 820 0

Soil-nailing – Theoretical basis and practical design

G.Gässler
Forschungs- und Materialprüfungsanstalt Baden-Württemberg, Stuttgart, FR Germany

ABSTRACT: A failure mode for steep nailed walls and slopes is given. It is based on the kinematic failure mechanism of rigid bodies. Four possible failure modes are investigated. By means of variation of the slip planes the unsafest mechanisms, depending on soil properties and boundary conditions, are found. Partial safety factors for the relevant variables are presented. Two examples are given using graphic solutions and design charts.

1 INTRODUCTION

The method of soil nailing is based on the principle of reinforcing natural soils by means of tension carrying bars (so-called "nails"). Soil-nailing is being used at present to stabilize natural slopes, cuts or excavation walls in granular soils (with some capillary cohesion), stiff clays and soft rocks. In general, nailed walls or cuts are carried out in a sequence of three steps (Fig.1):

1 Excavation in layers of 1m – 2m depth
2 Covering the new surface by shotcrete
3 Nailing with steel bars and grouting

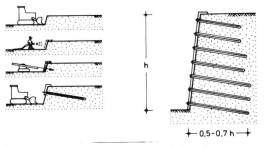

+— 0,5-0,7 h —+

Fig.1. Construction method of nailed walls

The thickness of the wall face should be 8cm up to 20cm, depending on the soil properties, the height of the wall and the expected service life (temporary or permanent nailing). In the most cases the nails consist of steel bars with diameters of 20mm up to 28mm or even 50mm. In general, the nailing density amounts 0.5 up to 1 nail per square meter. An extensive description of the construction of nailed walls is given by Gässler(1987).

The idea of stabilizing artifical cuts by means of steel bars was realized for the first time in France (Rabejac and Toudic (1974)). Within the scope of a German research and developing project that started in 1975, the new method of soil nailing (in German : "Bodenvernagelung") was theoretically studied and worked out for practical performance (Stocker(1976, Gässler(1977)). Using the literal translation from the German term "Bodenvernagelung", the method of soil nailing was then presented to the technical world (Stocker, Gässler, Gudehus(1979), Gässler and Gudehus(1981)). In the USA a simultaneous, but independent development of reinforcing natural soils took place in the years from 1976 to 1981, which was called "lateral earth support system(Shen, Bang, Romstad, Kulchin and De Natale(1981)). In France many practical projects of soil nailing (cloutage du sol) have been carried out since the end of the seventies Louis(1981). Today soil nailing is a world wide method in ground engineering. At present the best scope on its application is given by Bruce and Jewell(1986/87).

Although numerous practical projects have been carried out, it is necessary, to clear up the mechanical behaviour of walls or slopes in the limit state. In this paper a general theory on the failure model of nailed walls and steep slopes is presented. Until now, no generally accepted safety concept for designing has made its way. Hereat the paper contents a proposal.

2 REGULAR CROSS SECTION

The principle stability analyses of nailed walls were restricted to regular cross sections in homogeneous cohesive soils. Fig.2 shows a regular cross section that is determined by the nail length $0.5 \leq l/h \leq 0.7$ (l nail length at the foot of the wall, h height of the wall), the wall inclination $0 \leq \alpha \leq 20°$, the inclination of the slope above $0 \leq \beta \leq 20°$, and the nail inclination $\epsilon = 10°$. The inclination of the rear boundary of the nailed zone may amount $-10° \leq \rho \leq 20°$. The array of the nails is regular. The vertical distance a and the horizontal distance b have to be constant. Finally the ground level at the foot is horizontal.

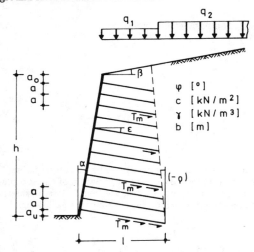

Fig. 2. Regular cross section of a nailed wall

In the regular cross section only axial nail forces are taken into account. The shear forces perpendicular to the axial forces are neglected, which is justified for granular and clayey soils by the results of several full scale tests (Gässler(1987)). The internal force of a nail at the slip surface is determined in the limit state by the mean shear force per unit nail length T_m and the section of the nail behind the slip plane. Apart from a maximum-depth of about 2.5m, the mean shear forces T_m are nearly constant and independent of the depth.

3 METHOD OF KINEMATICS OF RIGID BODIES

The kind of plastic limit state, which is attached by earth retaining structures after sufficiently large deformations, depends on the statical and kinematical boundary conditions. It was assumed for nailed walls and slopes that in the nailed zone as well as in the unnailed zone of the soil the plastic shear deformations are located in thin shear planes, whereas the greater part of the soil keeps rigid in the limit state. Thus, the failure model of nailed walls was based on the kinematics of rigid bodies that content two principles (Gudehus(1981)):
– Principle of the kinematic compatibility of the failure mechanism

This means the displacements of rigid earth bodies have to be correctly described by a hodograph.
– Principle of the minimum of safety by means of the variation of slip planes.

This means, one has to vary the inclinations (and radii) of the slip planes of a potential failure mechanism (e.g.: translation mechanism or rotation mechanism) until the most unstable configuration of the slip

planes is found.

As a provisional global safety definition

$$\eta_N := \frac{Z_a}{Z_g}$$

can be used. Herein Z_a denotes the available axial nail forces (determined from pull-out tests) and Z_g the axial nail forces in equilibrium or limit state. The principle of the minimum of safety now says, that failure will occur with that configuration of slip planes, which exactly fits the value 1 in the safety definition (1). This is an analogon to the principle of COULOMB's earth pressure theory (c.f. Gudehus(1981)).

Using the regular cross section of a nailed wall the safety of different potential failure mechanisms consisting of rigid bodies were investigated and compared for various boundary conditions. The procedure systematically developed in three steps :
1 Compiling of potential modes of failure mechanisms,
2 Variation of the slip planes to find the (relatively) unsafest configurations
3 Determination of the absolutely unsafest failure mechanism by means of comparing the relatively unsafest configurations of the different failure modes.

The investigated failure modes were as such as:
– translation of a rigid body (failure wedge) (TRA-I)
– translation of two rigid bodies (TRA-II)
– rotation of one rigid body (slip circle) (ROT-I)
– rotation of two rigid bodies (ROT-II).

4 PRINCIPLE STABILITY ANALYSES

4.1 Translation of one or two rigid bodies

The combined failure mechanism consisting of the rigid bodies (1) and (2), shown in Fig.3a, is determined by three slip planes. Their inclination angles $\vartheta_1 = 35°$, $\vartheta_2 = 55°$ and $\vartheta_{12} = 85°$ are assumed and do not represent the unsafest configuration. The data of the system are given in Fig.3a (horizontal nail distance b=1.25m). The soil may have an angle of internal friction $\varphi = 30°$ and a cohesion of 7.5kN/m^2. The available ultimate mean shear force per unit length T_m may amount to 30kN/m.

The external and internal forces acting on the sliding bodies are shown in Fig.3b. Herein P_1 and P_2 denote the resultant forces of the surcharge, W_1 and W_2 the dead weight of the earth bodies, and K_1, K_2, K_{12} the cohesion forces and Q_1, Q_2, Q_{12} the friction forces for the three slip planes. The velocities v_1, v_2 and v_{12} indicate the displacements that are determined by means of a hodograph (Fig.3c). (The scale is arbitrary.) The internal force N_i of a nail in one of the lower rows i, which is intersected by the slip plane with the inclination ϑ_1, is determined by the mean shear force T_m and the section of the nail behind the slip plane l_i :

$$N_i = T_m \cdot l_i.$$

The sum of the axial nail forces, referring to a unit width of the wall, can be expressed by:

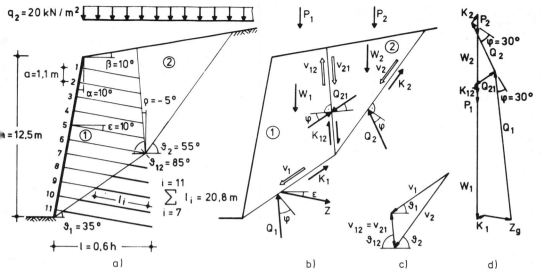

Fig.3 Stability calculation of a nailed wall: a) cross section with combined translation mechanism b) acting forces and displacements c) hodograph d) force polygon

$$Z = \frac{1}{b}\sum_{i=j}^{i=n} N_i = \frac{T_m}{b}\sum_{i=j}^{i=n} l_i \qquad (3)$$

Herein j denotes the upmost row, and n the lowest row that is intersected by the slip plane. With $T_m = 30$kN/m, $b = 1.25$m and $\sum_{i=7}^{i=11} l_i = 20.8$m the sum of the available nail forces amounts after Equ.(3) $Z_a = 500$kN/m. An equilibrium system of forces can be found graphically (Fig.3d). The sum of the nail forces in the limit state totals $Z_g = 260$kN/m. From that the factor of safety $\eta_N = 500/260 = 1.92$ is yielded.

Following the principle of the minimum of safety, one has now to vary ϑ_1 and ϑ_2, until the minimum of $\eta_N = Z_a(\vartheta_1)/Z_e(\vartheta_1,\vartheta_2)$ is found (it is not necessary to vary ϑ_{12}). The variation of ϑ_1 and ϑ_2 using graphic methods is very cumbersome. Hence a computer program was developped that finds the minimum of η_N, starting with roughly estimated values of ϑ_1 and ϑ_2. The case of the translation of one body, where only the inclination angle ϑ has to be varied, is covered by this program. For the example of Fig.3 the minimum $\eta_{N,min} = 1.75$ was found ($Z_a = 630$kN/m; $Z_g = 360$kN/m); $\vartheta_{1,a} = 43°$ and $\vartheta_{2,a} = 54°$). From $Z_g = 360$kN/m follows $T_{m,g} = 17.1$kN/m by means of equ.(3). The sum of the nail forces can also be expressed by the following formula (c.f. Gässler(1987)):

$$\overline{Z} = \left(\frac{l}{h}\right)^2 (\tan\theta_1 + \tan\epsilon)\cdot\frac{\cos\alpha}{\cos(\alpha - \epsilon)}\cdot\frac{T_m}{\gamma ab} \qquad (4)$$

with $\overline{Z} = Z/(\gamma\cdot h)$. The last term in this equation represents the specific nailing density

$$\mu := \frac{T_m}{\gamma\cdot a\cdot b} \qquad (5)$$

which is very useful for the following theoretical considerations and also for practical design.

In the example of Fig.3 the values μ_a and μ_g can be calculated using Equ.(5). With a=1.1m, b=1.25m;γ=20kN/m³ one yields $\mu_a=30/(1.1\cdot1.25\cdot20) = 1.09$ and $\mu_g = 17.1/(1.1\cdot 1.25\cdot20) = 0.622$.

Instead of the safety factor η_N (Equ.(1)) it is now proposed to use the equivalent safety definitions

$$\eta_T = \frac{T_{m,a}}{T_{m,e}} \quad \text{or} \quad \eta_\mu = \frac{\mu_a}{\mu_g}, \qquad (6)(7)$$

which can be obtained from Equ. (1),(3)and (5).

Consistently one yields the same minimal safety factors for η_N, η_T, and η_μ:

$$\eta_{T,min} = \frac{30}{17.1} = 1.75 \quad \text{and} \quad \eta_{\mu,min} = \frac{1.09}{0.622} = 1.75$$

Because of the fact, that both, $T_{m,a}$ and μ_a, are constant values (not so $Z_a = Z_a(\theta_1)$ in Equ.(1)), the unsafest slip plane configuration is immediately indicated by means of the maximum value of $T_{m,g}$ or μ_g (symbols : $T_{m,g,max}$, $\mu_{g,max}$).

4.2 Rotation of one body (slip circle)

In Fig.4a the same cross section is shown as in Fig.3a The assumed failure mechanism, however, is a slip circle, determined by the chord inclination angle $\vartheta = 50°$ and the radius r = 1.5h. This slip circle is arbitrary and not yet the unsafest configuration.

The external and internal forces acting on the sliding earth body are shown in Fig.4b. Z is located

285

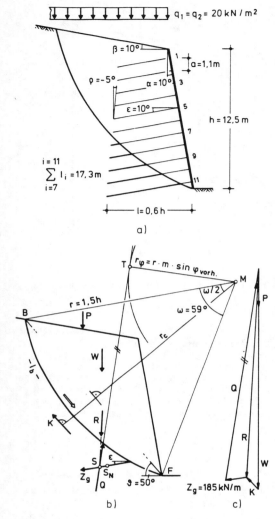

a)

b)

c)

Fig.4 Stability calculation of a nailed wall: a)cross section with slip circle b) acting forces c) force polygon

in the centre of gravity S_N of the nailed zone behind the slip circle. The forces R and Z are balanced in point S by the resultant slip circle force Q, the latter determined with the friction circle assumption after Krey(1932). Closing the force polygon in Fig.4c yields the sum of the axial nail forces in the limit state $Z_g = 185$kN/m. From Equ.(3) follows $T_{m,g} = 185 \cdot 1.25/17.3 = 13.4$kN/m and finally from Equ.(5) $\mu_g = 13.4/(20 \cdot 1.1 \cdot 1.25) = 0.49$. Similar to the TRA-II-mechanism one has to vary the parameters ϑ and r of the slip circle, until the maximum of $T_{m,g}$ or μ_g is found. By means of a computer program, the solution of the arithmetic variation is found: $\mu_{g,max} = 0.636$ with $\vartheta^* = 49.6°$ and $r^* = 3.4 \cdot h$.

4.3 Rotation of two rigid bodies

The rotation of two rigid bodies may occur in the case of a extremely high surface load in the rear. This failure mode has been observed in a model test. The graphical solution as well as the numerical analysis of this failure mode are more of theoretical than of practical interest. A description of the graphical solution is given by Gässler(1987).

5 RESULTS OF STABILITY ANALYSES

The third step is made, now, to find the unsafest failure mechanism. For various soil properties and boundary conditions the maximal specific nail densities μ of the four possible failure modes, presented above, have been calculated and plotted in diagrams. The instablest mechanism is the one that leads to the absolute maximum of μ. The diagrams are restricted to an exemplary regular cross section with l/h = 0.6, $\alpha = \epsilon = 10°$ and $\rho = 0$ or $\rho = 5°$. The cohesion $c[kN/m^2]$ and the surcharge q are transformed to the term $\bar{c} = c/\gamma \cdot h$ and $\bar{q} = q/(\gamma \cdot h)$.

In Fig.5 $\mu_{g,max}$ is plotted against φ for the slope inclination $\beta = \varphi/2$. The curve of the ROT-II-mechanism coincides with the one of the TRA-II-mechanism (the radii increase ad infinitum). The result is that the TRA-II- and the ROT-I-mechanism are nearly in like manner unsafe and that both, TRA-II and ROT-I, are (slightly) unsafer than TRA-I. Fig.5b presents the case of low surface loads on a slope ($\beta = 10°$) in a soil with medium cohesion ($\bar{c} = 0.1$). The result is here, that the ROT-I-mechanism is the unsafest mechanism with a significant distance to the TRA-II- and TRA-I-mechanism.

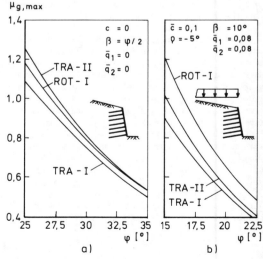

Fig.5 Maximal nailing densities in a) non cohesive soil b) cohesive soil

The results of numerous calculations performed by Gässler (1987) can be summarized as follows:

286

– In cohesionless soils the TRA-II-mechanism is the most critical mechanism for nearly vertical nailed walls ($\alpha \leq 10°$) with nails of constant length (or shorter nails in the upper rows), especially for high surface loads at the rear. For less steep walls ($\alpha > 10°$) and/or longer nails in the upper rows the ROT-I-mechanism (slip circle) is the unsafest mechanism. For $\phi > 35°$ the simple failure wedge TRA-I is nearly as unsafe as the TRA-II-mechanism.
– In soils with little cohesion, both, the TRA-II-mechanism and the slip circle are as far as safety is concerned nearly equivalent for nearly vertical walls (see examples of Fig.3 and Fig.4). In soils with medium or high cohesion (and for less steep walls) the slip circle distinctly is the unsafest mechanism.

6 PARTIAL SAFETY FACTORS

Instead of only one global factor η_μ it is proposed to take use of various partial safety factors. For the deduction of partial safety factors a limit state equation is necessary. For this the following limit state equation has been formulated based on the TRA-II-mechanism (Gässler(1987)):

$$\overline{Z}\cos(\vartheta_{1a} - \varphi + \varepsilon) - ((\overline{W} + \overline{P})\sin(\vartheta_{1a} - \varphi) +$$

$$\overline{E}_a\cos(\vartheta_{1a} - 2\varphi)) = 0. \qquad (8)$$

Herein $\overline{Z}$ is known from Eqn.(4). $\overline{W} = 1/h(2 - 1/h\tan\vartheta_{1a}) - \tan\alpha$ denotes the weight of body (1) (cf. Fig. 3b), $\overline{P} = 2q_1/(\gamma \cdot h)(1/h - \tan\alpha)$ the surcharge of body (1) and $\overline{E}_a = h'/h(h'/h + 2q_2/(\gamma \cdot h))K_a$ the earth pressure acting on the vertical intermediate slip surface of the length h' (K_a pressure coefficient). (Notice: $\overline{W} = W/(\gamma \cdot h)$, $\overline{P} = P/(\gamma \cdot h)$ and $\overline{E} = E/\gamma \cdot h)$). ϑ_{1a} is the inclination of the slip plane with the minimum of safety.

In Eqn.(8) the quantities φ, μ (or T_m), and q are scattering, i.e. they are basic variables in the sense of the new statistic - probabilitic safety theory following Eurocode 7. Concerning the statistical distribution of the basic variables assumptions had to be made. Nevertheless, partial safety factors can be proposed on the basis of numerous Level II approach calculations (Hasofer and Lind (1974)). The Level II approach applied on soil nailing is described by Gässler and Gudehus(1983) and Gässler(1987). Using the following partial safety factors a sufficiently safe dimensioning (safety index $\beta \approx 4,7$) of steep nailed walls and slopes can be achieved:

$$\gamma_\varphi = \frac{\varphi_k}{\varphi_d} = 1.1 \qquad (9)$$

with φ_k: characteristic value of φ (10% - fractile of a log normal distribution, truncated at 20°),

$$\gamma_\mu = \frac{\mu_k}{\mu_d} = 1.3 \quad \text{or} \quad \gamma_T = \frac{T_{m,k}}{T_{m,d}} = 1.3 \quad (10)(11)$$

with μ_k, $T_{m,k}$: characteristic value of m_μ, m_T (10 %

-fractile of a log normal distribution),

$$\gamma_q = \frac{q_d}{m_q} = 1.3$$

with m_g: mean value of q. The index d of the symbols φ_d, μ_d, q_d denotes the so-called design point, i.e. the assumed limit state with the most unfavourable combination of all basic variables. The safety factor γ_q is fixed in according to the Eurocode 7. For the case of friction soils with cohesion the partial safety factor $\gamma_c = 1.4$ referred to a suitable characteristic value is proposed for the present. The characteristic values φ_k or μ_k can be read from Fig.6, as depending on the mean values m_φ or m_μ and the coefficients of variation V_φ or V_μ.

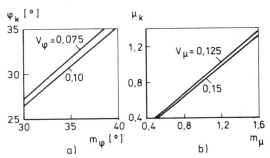

Fig.6 Charts for characteristic values φ_k(a) and μ_k(b)

7 DESIGN EXAMPLES

Nailed wall in cohesionless soil (cf. Fig.2):
geometry: h=10m, l=6m, $\alpha=\varepsilon=10°$, $\rho=0$
soil: $m_\varphi=36°$, $V_\varphi=0.075$, $\gamma=18kN/m^3$
surcharge: $m_q=20kN/m^2$
nails: a=1.1m; $T_m=30kN/m$, $V_T=0.15$
The horizontal nail distance, b, is to be determined so that $\beta = 4,7$ is kept.
One obtains the characteristic values $\varphi_k=32.7$ from Fig. 6a and $T_{m,k}=24kN/m$ via μ from Fig. 6b, and the design values $\varphi_d=32.7/1.1\approx 30°$ and $T_{m,d} = 24/1.3 = 18.5$ kN/m. The design value of the surcharge amounts to $q_d = 1.320 = 26kN/m^2$. Referring to chapter 5, the TRA-II-mechanism is the unsafest failure mode. The graphic solution may start with $\vartheta_1 = 40°$. The force Q_{12} interacting between body (1) and (2) (cf. Fig. 3b) can be substituted by the design earth pressure $E_{ad} = E_{ad}(\varphi_d, q_d) = 90kN/m$ (cf. Eqn.(8)). With $P_{1d} = 4.2q_d = 110kN/m$ and $W_1 = 655kN/m$ the graphic solution, similar to Fig. 3c, yields $Z_g = 240kN/m$. By means of graphic variation of ϑ_1 one finds the minimum of η_N (Eqn.(1)) for $\vartheta_{1a} \approx 40°$ (casually coinciding with the starting value of ϑ_1). From Eqn.(3) one obtains the mean shear force per unit length of the nail, referred to the unit width of the wall, $\overline{b} = 1.0m$, $\overline{T}_{m,d} = Z_g/\sum l_i = 240/15.8 = 15.2kN/m$. Finally, the wanted horizontal nail distance is given by

$b = T_{m,d}/\overline{T}_{m,d} = 18,5/15,2 \approx 1,20m$.
By means of a design chart, based on the TRA-II-mechanism, the solution can be found very easily. With $\varphi_d = 30°$ and $\overline{q}_d = 26/(18 \cdot 10) = 0.14$ one obtains from Fig.7 $\mu_d = 0.78$. b is given immediately by $T_{m,d}/(\mu_d \cdot \gamma \cdot a) = 18,5/(0.77 \cdot 18 \cdot 1.1) \approx 1.20m$.

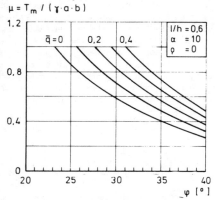

Fig.7 Design chart for cohesionless soil ($\beta=0$)

Nailed cut in cohesive soil (cf.Fig.2):
Geometry: h=11.7m, l=7m, $\alpha = 20°$, $\beta = 0$, $\varepsilon = \rho = 10°$
soil: design values: $\varphi_d = 23°$, $c_d = 11kN/m^2$, $\gamma = 19kN/m^3$. With $\varphi_d = 23°$ and $\overline{c}_d = 11/(19 \cdot 11 \cdot 7) = 0.05$ one obtains $\mu_d = 0.43$ from the design chart of Fig.8 based on the ROT-I-mechanism. For assumed distances a = 1.2m and b = 1.5m, e.g., $T_{m,d}$ gets 14.7 kN/m.

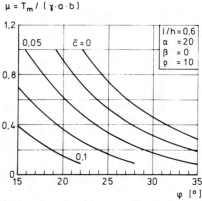

Fig.8 Design chart for cohesive soil

8 CONCLUSIONS

A comprehensiv failure model can be developed for nailed walls and slopes, based on the kinematic failure mechanism of rigid bodies. The instablest failure mode is to be found by variation of the slip planes. The results of the arithmetic variations coincide very well with model tests and field tests (Gässler, 1987).

The translation mechanism of one or two bodies and the simple rotation mechanism (slip circle) were foun to be the relevant failure modes to practical design. The design procedure, especially the application of design charts, gets very practicable by means of partial safety factors, either based on the new statistic - probalistic safety theory or obtained emirically.

REFERENCES

Bruce,D.A. and Jewell, R.A. 1986/87. : Soil nailing: Application and practice, part 1 and 2. Ground Engineering, Nov. 1986, Jan. 1987

Gässler, G. 1977. : Large scale dynamic test of in sitv reinforced earth. Proc. Dyn. Methods in Soil and Rock Mech., Karlsruhe, Vol.2, Balkema, Rotterdam: 333 - 342

Gässler, G. 1987. : Vernagelte Geländesprünge - Trag verhalten und Standsicherheit (Doctor thesis). Veröffentl. d. Inst. f. Bodenmechanik und Felsmechanik, University of Karlsruhe, FRG, Heft 108

Gässler, G. and Gudehus, G. 1981. : Soil nailing - some aspects of a new technique. Proc 10th Int. Conf. Soil. Mech. and Found. Eng., Stockholm, Vol. 3: 665 - 670

Gässler, G. and Gudehus, G. 1983. : Soil nailing - Statistical design. Proc. 8th Eur. Conf. Soil Mech and Found. Eng., Helsinki, Vol.2: 491 - 494

Gudehus, G. 1981. : Bodenmechanik. Enke Verlag, Stuttgart

Hasofer,A. M. and Lind, N. C. 1974. : Exact and invariant second - moment code format. Journ.Eng.Mech.Div.ASCE, Vol. 100, EM1: 111

Krey, H. 1932. : Erddruck, Erdwiderstand und Tragfähigkeit des Baugrundes. 4.Aufl., Verlag Wilhelm Ernst & Sohn, Berlin

Louis, C. 1981. : Nouvelle methode de soutenement des sols en deblais. Traveaux No. 553, March

Rabejac, S. and Toudic, P. 1974. : Construction d'un mur de soutenement entre Versailles - Chantier et Versailles - Matelots. Revue Generale des Chemins de Fer, 93 e Année, Avril: 232 - 237

Shen, C.K., Bang, S., Romstad, K.M. : Kulchin, L. and De Natale, J.S. 1981. Field measurement of ar earth support system. ASCE, J. of the Geot. Eng. Div., Vol. 107, No. GT 12: 1625 - 1642

Stocker, M. 1976. : Bodenvernagelung. Vorträge Baugrundtagung, Nürnberg, Dt. Ges. f. Erd- und Grundbau: 639 - 652

Stocker, M., Körber, G., Gässler, G. and Gudehus, G. 1979. : Soil nailing. Proc. Colloque Int. sur le Renforcement des sols, 2, Paris, 1979: 469 - 474

International Geotechnical Symposium on Theory and Practice of Earth Reinforcement / Fukuoka Japan / 5-7 October 1988
© 1988 Balkema, Rotterdam. ISBN 90 6191 820 0

Role of facing in reinforcing cohesionless soil slopes by means of metal strips

V.Gutierrez
Sumitomo Construction Co., Ltd, Japan
(Formerly: University of Tokyo, Japan)

F.Tatsuoka
Institute of Industrial Science, University of Tokyo, Japan

ABSTRACT: In order to develop a practical method of stability analysis for slopes reinforced by means of steel bars, a series of small model tests using sand were performed with measuring in detail the boundary pressures, tensile reinforcement forces and strain fields in model slopes. The results were analysed by means of the limit equilibrium method. A new method of stability analysis was developed in order to properly deal with inclined boundary forces and reinforcement forces.

1 EXPERIMENTAL METHOD

The sand model slopes (Fig.1) were made up by the air-pluviation method using air dried Toyoura sand and provisions were taken in order to avoid disturbance of the sand around the reinforcements during the placing of sand. The facing consisted of three 1.5 mm thick acrylic plates. The plates were divided vertically with each retaining one third width of the slope surface. After the sand slope together with the reinforcements were set up the facing plates were fixed to the heads of buried reinforcements by means of miniture screws and epoxy glue. A displacement control system by means of a bellofram cylinder was used for loading the footing. The displacement rate was kept at 0.1 mm/min throughout a test. In order to fail the slope model in such a manner as seen in a prototype slope the loading direction was fixed to 30 degrees from the vertical (see Figs.2~4). A lubricated footing was used in order to avoid the restraint to the lateral displacement of sand by the footing. The footing load was measured at its middle-third to avoid the effect of wall friction by means of eleven two-component small load cells, so that the distribution of normal and shear loads can be obtained. Pictures of the lateral surfaces of model slopes at appropiate displacements of footing were taken so that the displacement and strain fields in the models could be constructed. The coordinates of the targets drawn on the latex membrane used for the side wall lubrication were measured by means of a photogram metric method, to an accuracy of the order of 20μm.

2 EXPERIMENTAL RESULTS

2.1 Mechanism of failure

The mechanism of failure of the slopes was studied by constructing their strain fields. In the case of unreinforced slope (Fig.2) it was observed that the direction of minor principal strain ε_3 has two distinct tendencies: the one directed horizontally near the heel of footing and the other directed at an angle of approximately 45° to the horizontal in a deeper zone, directed towards the slope surface. Due to this fact it was considered that the best direction of the reinforcement should be such that its direction would be as much as possible in parallel with the tensile strains in deforming zones in the unreinforced slope.

Then, the strain field of reinforced slope without facing (Fig.3) shows that strains in the zone where strains were concentrated in the unreinforced slope were well restrained

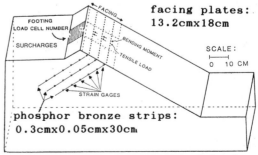

facing plates:
13.2cm x 18cm

SCALE:
0   10 CM

phosphor bronze strips:
0.3cm x 0.05cm x 30cm

Fig.1 Instrumentation of model slope

by reinforcing, whereas two other failure zones of strain concentration appeared. The first and shallower one was closer to the slope surface and penetrated the reinforced zone. The strength of this reinforced slope seems to have been controlled by this failure. A further development of the first failure was restrained by the reinforcements. A second and deeper failure zone developed and became a clear shear zone. This fact shows

that the strain field was altered significantly by the reinforcement.

When a facing structure was installed in the reinforced slope (Fig.4) the failure in the shallow zone near the slope surface which appeared in the reinforced slope without facing, was well restrained. At the same time, two zones of strain concentration may be seen; one starts from the heel of footing and passes through the center of the reinforcements and other starts from the lower end of the reinforcements.

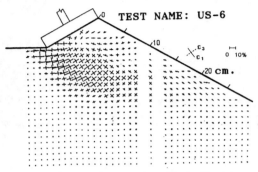

Fig.2 Principal strain distribution of unreinforced slope

2.2 Strength of slopes

The relationships between average axial stress and footing axial displacement for the three model slopes are shown in Fig.5. It may be noticed that despite the increase in the average void ratio of slopes, the strength of slopes increases significantly by means of three layers of metal strips and further by using a facing. After performing several tests with different arrangements of reinforcements and normalizing the results to the same void ratio e=0.69, the effects of facing could be better understood.

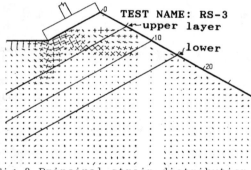

Fig.3 Principal strain distribution of reinforced slope

Table 1. Condition of reinforcing for each test and rate of increase in axial load.

Test Name	Condition of reinforcing (n=# of strips, l=# of layers)	Rn^*
US-6	Unreinforced	—
RS-3	Reinforced without facing (n=54, l=3)	3.2
RS-4	Reinforced with facing (n=54, l=3)	6.9
RS-5	Reinforced with facing (n=54, l=6)	7.5
RS-6	Reinforced with facing (n=27, l=3)	5.3
RS-7	Reinforced with facing (n=9, l=3)	3.6

*The rate of reinforcing Rn was defined for e=0.69 as follows:

$$Rn = \frac{\text{Max footing load of reinforced slope}}{\text{Max footing load of unreinforced slope}} - 1$$

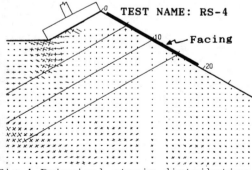

Fig.4 Principal strain distribution of reinforced slope + facing

Fig.6 shows the relationship between Rn and the plane density m. This shows a nonlinear pattern. It may be seen that the largest efficiency is obtained at a value of m of around 0.03, which means a light arrangement of reinforcements. The profiles (or distributions) of axial stresses along the length of footing base for the tests listed in Table 1 are shown in Fig.7.

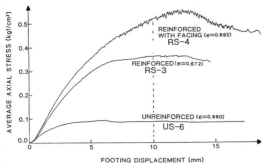

Fig.5 Comparative axial stress-displacement relationships

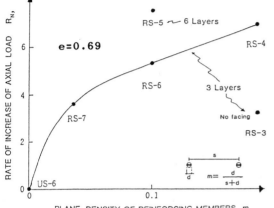

Fig.6 Rate of increase of axial load vs. the plane density

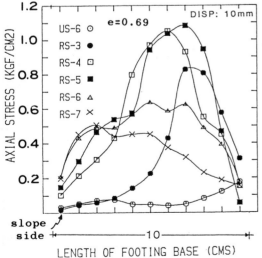

Fig.7 Axial stresses along the length of footing base

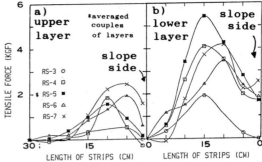

Fig.8 Tensile force along the length of the reinforcements

2.3 Tensile force of reinforcements

The measured forces along the length of strips for a footing displacement of 10mm are shown in Fig.8. This figure can explain the results shown in Fig.7 as follows.

The increase in axial stress and the shifting of its center to the slope surface in Test RS-4 with facing in comparison with Test RS-3 without facing is due to the large increase in tensile force near the slope surface in deep layers (Fig.8b). The enlargement (expansion) of the bell's shape of the axial stresses distribution in Fig.7 from test RS-4 to test RS-5, is due to the increase in tensile reinforcement force attained by a better arrangement of reinforcements (i.e. by increasing the reinforcement layers for the same total number). When the number of reinforcements decreases, the location of the center of axial stresses is shifted towards the slope surface. This corresponds to the increase in tensile force near the slope surface as the number of reinforcements decreases (compare the results by tests RS-4, RS-6, and RS-7 in Fig.8).

3 STABILITY ANALYSIS

3.1 Brief discussion on the Ordinary Method of Slices (Fellenius Method)

In this method, vertically divided slices are employed for all the system of forces acting on the slope regargless of the directions of forces. Thus, when the system of forces in this experimental work (Fig.9) is analysed by the Ordinary Method of Slices (the Fellenius Method), the total normal force ΔF_n on the base of each slice is obtained by resolving total forces normal to the base (Chowdhury, 1978), and then, the safety factor for this system is given as:

$$F = \frac{\sum (W\cos \alpha + P\sin(\beta - \alpha) + T\sin(\gamma + \alpha)) \tan\phi}{\sum W\sin\alpha + \sum P|_{P\cdot A} \pm \sum T\cos(\gamma + \alpha)} \quad (1)$$

where, W=weight of soil in the slice,
P=load of footing working on the top
  end of slice,
T=tensile force in reinforcement
  working on the base of slice,
$\Delta F \tan\phi$=total shear resistance of
  soil,
R is the radius of the trial circle,
  and lw, lp and lt are the arm
  lengths of those forces.

As can be seen in Eq.(1), the term defining
the component of the shear resistance due to
the footing load, $P\sin(\beta-\alpha)$, becomes zero
when $\beta=\alpha$, and further, when $\beta<\alpha$, it becomes
negative. This means that for the same in-
clination of footing load the shear resistance provided by this component decreases as
the angle of slice increases; thus, for the
footing pressure on the side of $\beta<\alpha$, for example when a deep failure surface is generated, as in the case where a large degree of
reinforcing is used, then the shear resistance decreases with the increase in α, as
illustrated in Fig.9b. Of course this situation is not realistic. For the experimental
conditions (an inclination angle of footing
load of $\beta=60°$ and an inclination angle of
reinforcement of $\gamma=30°$) and by using a constant angle of internal friction of sand
along the slip line ($\phi=45°$) the safety factor for the observed slip lines shown in
Fig.10 were obtained as listed in Table 2.

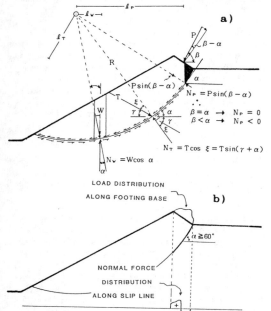

Fig.9 Analysis by Ordinary Method of
  Slices a) system of forces, b)
  normal forces along a segment
  of slip line

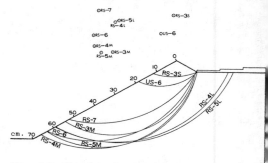

Fig.10 Observed slip lines

Table 2. Results by Ordinary Method of
Slices.

Test(slip line)	F
US-6	1.215
RS-3M	0.732
RS-4M	0.693
RS-5M	0.675
RS-6	0.591
RS-7	0.602

These results show that the values of safety factors are underestimated for the reinforced condition considerably. Consequently, in the following, a modification of th
direction of slicing is attempted.

3.2 Directional Slicing-Fellenius Method

It was assumed that each load on the surfac
of the slope acts on the slip line at the
point where the line parallel to the direction of the load intersects with the slip
line. This is only a generalization of the
assumption made for the vertical loads such
as the weight of soil or vertical surcharge
in the Ordinary Method of Slices. Thus, for
each of the forces W,P and T defined before
a different system of slices is produced
with keeping the force equilibrium on any
slice along the slip line. In each of these
systems the normal force on the base of eac
slice is obtained by resolving total forces
normal to the base of slice as in the Ordinary Method of Slices. Considering separately each system of forces working on the slope we have the following:
For the weight of soil, W (Fig.11): only
weight of soil is considered; this is the
same with the case of an unreinforced slope
under gravity forces without exterior loads
In this system the slices are divided vertically, as in the conventional method of slicing.

292

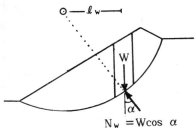

$N_W = W \cos \alpha$

Fig.11 Slicing for weight of soil

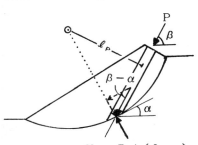

$N_P = P \sin(\beta - \alpha)$

Fig.12 Slicing for footing pressure

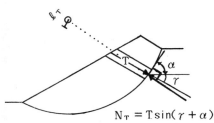

$N_T = T \sin(\gamma + \alpha)$

Fig.13 Slicing for tensile forces

For inclined footing load, P(Fig.12), and inclined tensile reinforcement forces, T (Fig.13): In this systems the slices are divided in the direction of either P or T. Thus, the safety factor for the overall system is obtained by summation of the resistant moments and the driving moments in these different systems. When the subscripts i, j,k are used to distinguish different components for different slices, we obtain the following equation:

$$F = \frac{\Sigma W_i \cos \alpha_i \tan \phi_i + \Sigma P_j \sin(\beta - \alpha_j) \tan \phi_j + \Sigma T_k \sin(\gamma + \alpha_k) \tan \phi_k}{\Sigma W_i \sin \alpha_i + \Sigma P_j l_P / R + \Sigma T_k \cos(\gamma + \alpha_k)}$$

$$(2)$$

This is a generalized expression for the directional slicing method by using the specific assumptions of the Fellenius Method. A typical result of the stability analyses performed for an observed slip line when the internal friction angle is considered constant

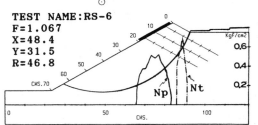

Fig.14 Typical distribution of normal stresses by directional slicing

$\phi = 45°$ is shown in Fig.14. In this figure, Np and Nt are the normal stresses on the slip line due to the footing pressures and the reinforcement forces respectively. This result shows that this proposed method is a more realistic approach as compared with that shown in Fig.9.

In the following analyses, the minimum values of F by Eq.(2) were obtained for the critical circles using the following three kinds of angle of internal friction; 1) a constant value of $\phi = 45°$, 2) $\phi(\delta)$ with taking into account the strength anisotropy of a model sand (air-pluviated Toyoura sand), i. e., ϕ is a function of δ (the angle of δ_1 direction relative to the bedding plane) with $\phi = 45°$ at $\delta = 90°$ with the assumption that the δ_1-direction is $45° - \phi/2$ from the direction of failure surface at each point; and 3) $\phi_{mob}(\delta)$ with taking into account further the progressive failure, i.e., the current mobilized angle of friction ϕ_{mob} at each point in the slope is estimated from the measured local values of the shear strain $\gamma_{max} = \varepsilon_1 - \varepsilon_3$. In this estimation, the relationships between ϕ_{mob} and γ_{max} in plane strain defined for 1cmx1cm element at e=0.69 as shown in Fig.15 were used. In these relations $\phi = 47.7°$ for $\delta = 90°$. Note that $\phi = 45°$ and $\phi(\delta = 90°)$ are underestimated for the test condition (i.e. e=0.69). A typical result of stability analysis is shown in Fig.16 for test RS-6 in which the contour lines for the shear strain also are shown. It can be observed that an accurate prediction of the critical slip line is obtained by using the actual ϕ_{mob} along the slip line. The summary of the analyses for the unreinforced slope and the slopes reinforced with three layers of reinforcements and facing is shown in Fig.17. In this case, the correct values for safety factors are unity. It may be seen that rather accurate prediction of the safety factor has been obtained.

A rate of the increase in stability by reinforcing may be defined as:

$$S_i = \frac{(FR - FU)}{FU} \times 100$$

$$(3)$$

293

where FR=the safety factor for the reinforced condition calculated for the
critical slip circle in the unreinforced slope, and
FU=minimum safety factor for unreinforced condition.

The rate of increase in stability Si is
plotted in Fig.18. An increase in stability
as the number of reinforcements increases
may be observed. This seems to be a very
reasonable result in accordance with the experimental results shown in Fig.6.

Conclusions: It was found that for the system where the inclined reinforcement tensile forces and the inclined footing pressure
are incorporated, reasonable values of safety factor and locations of critical slip
circles can be obtained only by modifying
the Ordinary Method of Slices (Fellenius

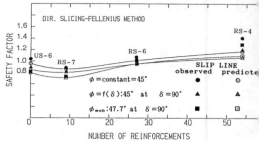

Fig.17 Safety factors vs.
number of reinforcements

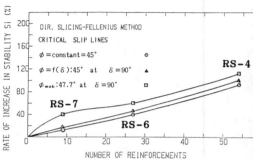

Fig.18 Rate of increase in stabilit
vs. number of reinforcements

Method) into a so called Directional Slicing-Fellenius Method. In order to obtain
accurate results using actual properties o
soil, the progressive failure should be
accounted for. This new method of stabilit
analysis proposed in this work is practica
simple, accurate and versatile.

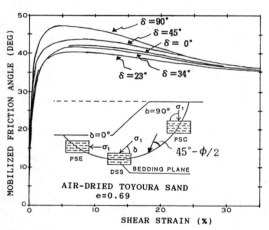

Fig.15 Mobilized friction angle as a
function of the shear strain

Condition of Critical Slip Line	safety Factor Fmin	
PC : constant ϕ	1.007	pc
Pd : $\phi(\delta)$	0.948	pm °o pd
Pm : $\phi_{mob}(\delta)$	0.951	

Test name: RS-6

Shear strain (%)

Scale
0   10
cm.

Fig.16 Critical slip lines by Dir.
Slicing-Fellenius Method
compared with shear strain

REFERENCES

Chowdhury, R.N. 1978.Slope analysis; Developments in Geotechnical Engineering,
vol.22.
Gutierrez, V. 1988.Sand slopes stabilized
with metal reinforcement and facing: mod
tests and stability analyses. Doctor
thesis, University of Tokyo.
Jewell, R.A. 1980.Some effects of reinforcement on the mechanical behavior of soi
Doctor thesis, Cambridge University.
Tatsuoka et all 1986.Strength and deformation characteristics of sand in plane
strain at extremely low pressures. Soil
and Foundations, 26-1: 65-84. Japan.

International Geotechnical Symposium on Theory and Practice of Earth Reinforcement / Fukuoka Japan / 5-7 October 1988
© *1988 Balkema, Rotterdam. ISBN 90 6191 820 0*

Study on the effect of the vertical RC bolting on the unstable slope

Motosuke Iwata
Japanese National Railways Settlement Corporation, Japan

Yoshio Mitarashi
Institute of Construction Technology, Kumagai Gumi Co., Ltd, Japan

Mitsuo Masuda
Tokyo Branch, Kumagai Gumi Co., Ltd, Japan

ABSTRACT: This paper describes the outcome of the research on effects of the vertical RC bolts on unstable slopes. Herein, the concept and case studies on vertical RC bolting are introduced, and its effectiveness is evaluated based on field survey results.

1. PREFACE

In the past, the tunnelling, using timber to support tunnels, was considered uneconomical in many uneasy works to be safely operated. The machines and materials to be used had not been sufficiently developed, so a great deal of time and labor was required to complete the works. Tunnel routes were chosen which would minimize the tunnel length, where the geological condition would be easy to excavate. However, tunnelling technology has made rapid progress in recent years, and it is now generally recognized that it is possible to construct tunnels with high safety and at low cost. Two particularly important breakthroughs which have deepened that impression were the introduction of the New Austrian Tunnelling Method (NATM), now the standard tunnelling technique in Japan, and the reflection of results of simulation performed beforehand by the numerical analysis etc. to the design and the execution of tunnels. In addition, the roles of tunnels, with the progress of urbanization and high density land use, have been increasingly growing. Faced with this social necessity, many tunnels nowadays have to be constructed in places with less favorable conditions. It is expected that this tendency will continue.

The key issue in the excavation of tunnels is the stability of the face.
The primary factors which determine this stability can be summarized as follows:
1. the strength of the natural ground
2. the shape and size of the tunnel
3. the initial stress condition of the natural ground

4. the excavation speed, etc.
Geological factors have a particularly important influence.

In poor ground, an auxiliary method must be found to strengthen the stability of the face. The aim of an auxiliary method is to reinforce the ground in front of the face to make it possible for tunnelling. The auxiliary method can be classified broadly into two categories, whether it can be performed by using the facilities and machines already existed in the field or not. The first, including the forepiling, facepiling etc., can be adopted easily and helpful to maintain the face stability until the support system has installed completely. The second, including well point drainage, deep well, grouting, artificial ground freezing, and the vertical RC bolting method etc., must use the new facilities and machines, and is applied to improve the ground condition.

It used to be common to consider the design of tunnels simply as a problem of earth pressure, Terzaghi's theory of loosening pressure being the typical example. Recently, however, more emphasis has been placed on deformation, applying NATM measuring or the numerical analysis method. Accordingly, the three-dimensional effects of tunnelling have come into focus, and the hitherto somewhat neglected behaviour of the natural ground beyond the face has been given more attention in the design and execution of tunnels.

The results of NATM measurement and three-dimensional FEM analysis have shown that looseness equivalent to at least one-third of the total displacement at the time of convergence has already occurred

in the natural ground in the neighborhood
of the face while the face is approaching.
These pre-displacements or looseness reduce the stability of the face. The "vertical reinforced concrete bolting method"
is an auxiliary technique which has been
developed to control this looseness and
raise the stability of the face or slope.

Briefly, the method is to drive fixed
pitch mortar reinforced concrete piles
into holes bored from the surface.

In this report, we present actual results detailing the effectiveness of the
vertical reinforced concrete bolting method, our view of this method and a case
study.

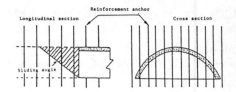

Fig.1 Reinforcement of Face

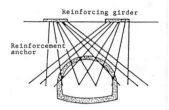

Fig.2 Reinforcing method by suspension
effect

2. THE VERTICAL REINFORCED CONCRETE BOLTING METHOD

The vertical reinforced concrete bolting
method is construction technique for maintaining the stability of the natural
ground, either temporarily or in the long
term, by reinforcing against the disloca-
tions caused by the approach of a tunnel.

This method is applied in places where
the geological conditions are extremely
unsuitable for tunnel construction, such
as talus layers in the neighborhood of
portals or soft ground.

The vertical reinforced concrete bolting
is normally executed by inserting bolts (ϕ
some 30mm) into boring holes (ϕ some 100-
120mm) filled with mortar.

The method:
1. improves the stability at the face,
2. has a suspension effect,
3. improves the stability of the slope,
and so on.

2.1 Rise in stability at the face

The most important point when excavating
tunnels is to stabilize the face. Fig.1
shows a face reinforced by vertical reinforced concrete bolting. As Fig.1(a)
shows, collapse of the face would usually
occur in almost all cases when slippage
happens toward the area in front of the
face.

By driving in mortar reinforced concrete
piles, we raise the shear strength of the
natural ground to prevent rupture of the
natural ground, and thus, the natural
ground in front of the face is stabilized.
This method also holds together discontinuous areas of natural ground composed of
joints and fissures etc., preventing the
rock mass from forming blocks, and the
weakened rock mass from segregating and
falling off.

2.2 Suspension effect

When overburden is small, the ground pressure can be reduced by suspending the
ground on bolts fixed to beams laid on the
surface. This supplements the natural
supporting ability of the ground.

2.3 Rise in stability of the slope

When we look at the topography and geology
in the neighborhood of the entrance to a
tunnel, we often find that there are several layers of weathering on the bedrock
and a talus layer covers the top. There
is a high possibility of slope collapse or
landslides when excavating tunnels under
these conditions.

We can raise the stability of a slope by
driving in bolts to hold together discontinuous layers and rock mass that are apt
to cause slope sliding.

It is supposed that none of these results are obtained individually, but that
there is interaction among them.

3. A CONCEPTION OF THE DESIGN FOR VERTICAL REINFORCED CONCRETE BOLTING

The basic conception of the design for
vertical reinforced concrete bolting is to

raise the inherent shear strength of the natural ground by driving mortar reinforced concrete piles into the ground.

The force needed to prevent deformation of the natural ground by the excavation is borne by the piles, utilizing the adhesive power of the mortar.

The mortar and bolts both serve an important function in restraining the deformation of the natural ground.

In sediment or weathering layers broken into flinders, the mortar permeates the soil or cracks to form cylindrical clods around the RC bolts, raising the cohesion or angle of internal friction of the natural ground. The shear and bending strength of the piles resists the sliding force which occurs at contact surfaces of sediment and rock or between rock layers.

The sizes and intervals of the borings and the thickness and length of the bolts are mostly calculated according to the physical properties of the natural ground. In this case, we include the following two considerations:

1. The axial intervals of the tunnel should roughly correspond to the intervals of the steel arched supports. Intervals at right angles to the face should be chosen by the size of the boring, and the dimensions of the bolts. The earth pressure at the face depends upon the quantity of earth between each bolt interval.

2. The choice of procedure must depend on the availability of machines and the suitability of the ground.

4. CASE STUDY OF VERTICAL REINFORCED CONCRETE BOLTING
(Example of the 2nd Shirasaka Tunnel at Yodogasawa)

The examples of the vertical RC bolting adopted in the past is shown in Table 1. Here, we describe a case study at Yodogasawa, the 2nd Shirasaka Tunnel of Shinonoi Line, where the tunnel was constructed under an unstable slope.

4.1 Outline

The 2nd Shirasaka Tunnel of the Shinonoi Line of Japan Railways is a 1765m double track tunnel ($A=65m^2$), situated almost in the center of Japan. About 800m from the Shiojiri-side portal, there is a marshland called Yodogasawa, and in one place where the minimum thickness of the ground above the tunnel would be only about 1m. Landslides have happened here several times in

Table 1 Examples of vertical RC Bolting

Execution	Pile diameter (mm)	Pile length (m)	Quantity
Rout 373 Shidosaka Tunnel	76.0 100.0	5.65 (average)	235
Misawagawa water branch tunnel	300.0	8.8-9.6	214
Minamihonjyuku manhole	116.0	12.5	195
Uchiumi Tunnel Rout 220, Miyazaki	101.0	4.1-9.6	208
Toi sewage tunnel	100.0	8.0-14.0	123
JR Second Shirasaka Tunnel	116.0	8.0-20.0	745
JR Kyobashi Tunnel	100.0	20.0-26.0	489

the past, and the topographic and geological conditions suggested that a landslide might be induced by tunnel excavation. Further, it appeared that it might be difficult to secure the stability of the tunnel. Therefore, it was decided to reinforce the stability of the natural ground by vertical reinforced concrete bolting. Execution of the tunnel was carried out under the NATM, with a conventional method adopted at the place where the covering was very thin.

The geology here consists of lower layers of Cenozoic Neocene Miocene Japanese red pine alternation with a high proportion of sandstone and sandy mudstone, mudstone alternation.

The spot where the tunnel was to cross the Yodogasawa Marshland is some 15m wide, at a point where both banks form oblique slopes of 35°. In the neighborhood of Yodogasawa, deposits of the present riverbed had accumulated to a thickness of 4 to 5 meters, and they were supposed to appear in the upper half of the tunnel's section.

4.2 Vertical reinforced concrete bolting

The vertical reinforced concrete bolting was executed over the area shown in Fig. 3.

The mean value of physical properties of the mudstone is shown in Table 2. The bolting zone was positioned in a sphere where a covering was thicker than the calculated Terzaghi's height of loosening bedrock; namely the earth covering thickness of 1.5D from the crown of the tunnel (D: diameter of tunnel).

The pitches of the reinforced concrete bolts were decided according to the equilibrium of the slope, a virtual sliding surface being imagined. Calculations on the assumption that the mortar reinforced

297

concrete bolts might provide shear-resistance against the sliding force showed that it was proper to drive in piles every 1.5m².

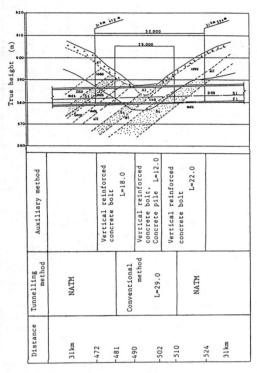

Fig.3 Outline of RC Bolting Zone

Table 2　Physical Properties Value of Mudstone(average)

Unit weight (kgf/cm²)	Unconfined compression strength(t/m³)	Angle of internal friction (°)	Cohesion (kg/cm²)
1.8	50	30	2

4.3 Measurement

Measurements were made to precisely ascertain the behavior of the natural ground to confirm the effects of the vertical reinforced concrete bolting, and, at the same time, to secure the safe execution of the tunnel excavation by rapidly feeding back the results of the measurements. Each measurement is given below for 31km 476m.
　1. Surface subsidence
Surface subsidence was measured by establishing five survey stations at 10m intervals in the direction of the tunnel's length and also five stations in the crosswise direction.

A diagram of surface change over time is shown in Fig.4. Movement of the surface began when the face came to the point at − 0.5D (D: diameter of tunnel) and, when passing the upper half of the face, it reached 50% of the total subsidence.

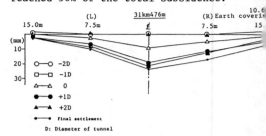

Fig.4 Diagram for Change of Ground Surface Subsidence by Time Lapse (Tunnel cross-section)

　2. Ground horizontal displacement
Measuring instruments were set up on either side of the tunnel, 9 meters away from the tunnel. Ground horizontal displacement volume and direction are shown in Fig.5. The left side instrument showed a movement in the axial direction of the tunnel, namely a movement along the slope with displacement especially notable at the talus part (GL-8m). By contrast on the right, the upper layer revealed prominent displacement in the axial direction,

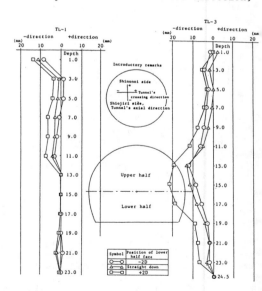

Fig.5 Ground Horizontal Displacement (Tunnel crosswise direction) Displacement by Lapse of time as lower half face advances

298

and the lower layer prominent displacement in the crosswise direction, these two layers being in the environs of the crown of the tunnel. These results showed that the surface talus layer was greatly influenced by the slant of the slope and, in the environs of the crown of the tunnel, a relatively greater side pressure was imposed from the right side as a result of the release of stress during the excavation.

3. Ground vertical displacement

Ground surface subsidence was more marked on the right side, centering on GL-14m. The relative degree of displacement shifted from the compressed side to loose side between GL8 and 10m before and after passing the upper half, with a loose layer apparently existing between them. A distribution diagram of strain caused by ground subsidence, classified by depth, is shown in Fig.6.

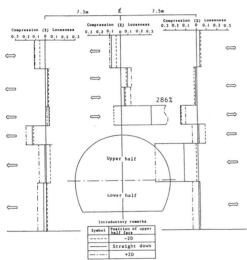

Fig.6 Strain Distribution Diagram by Ground Depth (Survey point 31km476m)

4.4 Behavior of the natural ground at Yodogasawa

1. Movement in the axial direction of the tunnel

When the upper half face arrived in front of the vicinity of the crown of the tunnel, the axial force of the bolt and ground displacement began to change. Ground surface subsidence showed little movement before the upper half face reached -0.5D. The stages at which each measurement point was first influenced by the

tunnel face are shown in Fig.7 . From this, it was found that the straight line distance at which the influence began to be felt was 10.5 to 11.0m. The influence of the tunnel excavation seemed to spread in concentric circles from the crown of the tunnel face. Pre-displacement, (as a ratio of the measured value at the time the face arrives to the final measurement) was some 40 to 55%.

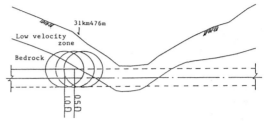

Fig.7 Sphere Affected by Excavation

Comparing the crown subsidence quantity at the time of the upper half excavation with that at the time of the lower half excavation, we found a ratio of subsidence quantity of almost 1:1. The quantity of the crown subsidence increased under the influence of the lower half excavation in the same way as during the upper half excavation. However, the subsidence of the ground surface showed only a slight increase, with little influence from the lower half excavation. In addition, when measuring the axial force on the reinforced concrete bolts driven into the central axis of the tunnel, we found little change caused by the approach of the lower half face. It was concluded that the reinforced concrete bolts seemed to restrict the displacement of the natural ground as predicted.

2. Movement in the crosswise direction

The inclinometer installed at equal distance from the tunnel center showed movement along the slope in the vicinities of the ground surface and the tunnel. There was a tendency for the side pressure to act from right to left.

The ground strain distribution diagram showed looseness some 7m away from the surface of the tunnel wall and compression on the upper parts. Regarding axial force distribution, a tensile force was felt some 7m away from the surface of the tunnel wall. Thus, the two results coincided exactly. The earth covering at this spot was some 1.0D, with the ground configuration of a slope. Judging from these conditions, it had been supposed that the

formation of a ground arch would be difficult during tunnel excavation. However,
the measurements showed that a ground arch
had been formed.

The reinforced concrete bolts can be
said to have formed a ground arch and
stabilized the ground surface in a geolo-
gically difficult site.

5. EFFECTS OF ThE VERTICAL REINFORCED CONCRETE BOLTING

It was clear that the vertical reinforced
concrete bolting successfully restrained
the displacement of the natural ground and
controlled the increase in displacement
quantity. The working effects can be
divided broadly into the following two
categories:

1. Controlling the quantity of subsidence at the upper part of the tunnel

2. Controlling the influence of the excavation on the front of the face

5.1 Effects of restraint on subsidence quantity at the upper part of a tunnel

In the case of the 2nd Shirasaka Tunnel,
the influence of the tunnel excavation
spread in concentric circles from the top
of the face. This had already reached to
the ground surface when the upper half
face arrived. Accordingly, the natural
ground on the upper part of the tunnel
began to subside, but this subsidence was
restrained in the vicinity of the reinforced concrete bolts. It is evident from
the axial force distribution on the bolts
that the restraint on subsidence was the
result of frictional force on the surface
of the bolts. It has been found that when
driving reinforcements, such as rock-bolts
or reinforced concrete bolts, into natural
ground, both the yield load of the natural
ground and the residual stress after
yielding increase. The absorption energy
also increases prior to the breaking of
the natural ground. There are two ways of
viewing these phenomena. One is to approach them from a macro-viewpoint, in
which the physical properties of the whole
natural ground, including those of the
reinforced concrete bolts, are supposed to
be improved, while the other is to view
them in terms of the so-called suspension
and reinforcement effects.

Restraining of subsidence by reinforced
concrete bolts is supposed to be achieved
by the reinforcement effect to make the
natural ground in the triaxial stress
state and by the suspension effect to

restrain the natural ground.

5.2 Restraint of the influence of excava-tion on the front of the face

It was made clear by the results of measurement and analysis that the influence on
the front of the face was large, especially at places like Yodogasawa where the
diagonal movement (tunnel axial direction)
is great. However, judging from the fact
that the influence of excavation beneath
this slope was in fact almost equal to
that of tunnel excavation beneath a flat
surface, the influence on the upper part
of the tunnel and on an oblique front
could be restrained by some working. This
was the restraint obtained by the vertical
reinforced concrete bolting method.

These effects are supposed to have been
produced by the facts that the bearing
force against exterior load increases with
the restraint imposed on the ground by the
reinforced concrete bolts, and stability
of the natural ground rises accordingly.
At the same time, the extent of the influence of the excavation is supposed to
become narrower, because the value of the
apparent modulus of elasticity rises and
the resultant strain becomes less.

6. POSTSCRIPT

The vertical reinforced concrete bolting
method, developed at the same time as NATM
was introduced into our country, has often
been used with perfect success and has
become commonly known as an excellent
auxiliary method for construction.

Hereafter, tunnels will be built in
increasingly difficult conditions. Under
these circumstances, the role the vertical
reinforced concrete bolting method will
be quite important as a means for execu-
ting economical tunnel construction. The
success of this method is achived after
first confirming the stability of the
natural ground, and then by application in
conformity with a precise estimate of the
behavior of the natural ground, using
cumulated data and careful construction
techniques.

BIBLIOGRAPHY

Y. Murakami, T. Yokota, March 1986,
Collection of Essays by Institution of
Civil Engineering, the 367 issue/VI-4,
Effects of the forepoling method with
RC-bolts on the ground stabilization

International Geotechnical Symposium on Theory and Practice of Earth Reinforcement / Fukuoka Japan / 5-7 October 1988
© 1988 Balkema, Rotterdam. ISBN 90 6191 820 0

Kinematical limit analysis approach for the design of nailed soil retaining structures

Ian Juran, Baudrand George & Farag Khalid
Louisiana State University, Baton Rouge, La., USA

Victor Elias
Elias and Associates, P.A., Bethesda, Md., USA

ABSTRACT: Soil nailing is an in-situ soil reinforcement technique which has been used during the last two decades mainly in France and Germany to retain excavations or stabilize slopes. Design of nailed soil systems has been traditionally done using slope stability analysis methods. These methods have been modified to incorporate the effect of the available tension and shear forces in the passive reinforcements on the slope stability. However, they provide only a global safety factor. This paper presents a limit analysis design approach which provides estimation of maximum tension and shear forces mobilized in each reinforcement. To verify the applicability of the method, the predicted forces are compared with those measured in both laboratory models and full scale structures.

INTRODUCTION

The fundamental concept of soil nailing consists of placing in the ground passive inclusions, closely spaced, to restrain its displacements and limit its decompression during and after excavation.

The currently used approach in the design of nailed soil structures is based on rather conventional slope stability analysis procedures that have been adapted to evaluate the safety factor of the nailed-soil mass and the surrounding ground with respect to failure along potential circular or wedge shaped sliding surfaces. When such a method is used for the design of a nailed-soil structure, the conventional slope stability analysis procedure is modified to account for the available limit shearing, tension, and pullout resistance of the inclusions crossing the failure surfaces.

Available design methods that are derived from this approach (Stocker, et al., 1979; Shen, et al., 1981; Schlosser, 1983) involve different assumptions with regard to the definition of the safety factors, the shape of the failure surface, the type of soil-reinforcement interaction and the resisting forces in the inclusions. Stocker and coworkers

proposed a force equilibrium method that assumes a bilinear sliding surface, whereas Shen, et al. proposed a similar design method with a parabolic sliding surface. Both methods consider only the tension capacity of the inclusions.

A more general solution, including the two fundamental mechanisms of soil-inclusion interaction (lateral friction and passive normal soil reaction), has been developed by Schlosser (1983). This solution involves a slices method (e.g., Bishop's modified method or Fellinius' method) with a multicriteria analysis procedure. This method takes into account both the pullout tension and the shear resistances of the nails as well as the effect of their bending stiffness.

These design methods can only be used to evaluate the global safety factor with respect to the shear strength characteristics of the soil and the soil-inclusion lateral friction. They do not allow for the evaluation of the local stability of the reinforced soil-mass at each reinforcement level. In nailed-soil retaining structures the local stability at the level of a sliding nail can be more critical than the global stability with respect to general sliding in the structure and/or its environment. Therefore, it appears necessary to develop design methods, which provide an estimation of nail forces under the expected working loads.

This paper presents a kinematical limit analysis design approach which allows for the evaluation of the effect of the main design parameters (i.e., structure geometry, inclination, spacings, and bending stiffness of the nails) on the tension and shear forces generated in the nails during construction. To verify the applicability of the method, the predicted forces are compared with those measured in both laboratory models and full scale structures.

2 KINEMATICAL LIMIT ANALYSIS DESIGN METHOD

This design method is based on a limit analysis solution associating a kinematically admissible displacement/failure mode, as observed on model walls, with a statically admissible limit equilibrium solution. The main design assumptions, shown in Figure 1, are that: (a) failure occurs by a quasi-rigid body rotation of the active zone which is limited by a log-spiral failure surface; (b) at failure, the locus of maximum tension and shear forces coincides with the failure surface developed in the soil; (c) the quasi-rigid active and resistant zones are separated by a thin layer of soil at a limit state of rigid plastic flow; (d) the shearing resistance of the soil, as defined by Coulomb's failure criterion is entirely mobilized all along the failure surface; (e) the horizontal components of the interslice forces acting on both sides of a slice comprising a nail (Figure 1) are equal; and (f) the effect of a slope (or horizontal surcharge) at the upper surface of the nailed-soil mass on the forces in the inclusions is linearly decreasing along the failure surface.

The effect of the bending stiffness is analyzed considering available elastic solutions for laterally loaded long (semi-infinite) piles, illustrated in Figure 1. This solution implies that at the failure surface, the moment (M_o) is zero whereas the tension and shear forces generated in the nails are maximum. It is assumed that the maximum shear stress in the nail (S) is mobilized in the direction (α) of the failure surface in the soil. Hence, the shear stress (τ_n) and the normal stress (σ_n)

acting on the normal plane of the nail are related by:

$$\tau_N = \frac{1}{2} \cot [2 (\alpha - \beta_{mod})] \cdot \sigma_N;$$

$$\beta_{mod} = \beta - d\beta \qquad (1)$$

$$d\beta = \frac{2}{(\frac{K_s \cdot \beta}{\gamma \cdot H}) \times (\frac{\ell_0^2}{S_v \cdot S_h})} \cdot TS; \qquad (2)$$

where, $d\beta$ is the maximum nail deformation attained at the failure surface, α is the inclination of the failure surface with respect to the vertical; β is the nail inclination.

$TS = T_c/(\gamma \cdot H \cdot S_v \cdot S_h)$ is the normalized maximum shear force; K_s is the lateral soil reaction modulus.

$\ell_0 = [4EI/K_s D]^{1/4}$ is the transfer length which characterizes the relative rigidity of the inclusion to the soil.

E and I are the elastic modulus and the moment of inertia, respectively, of the nail; D is the nail diameter; γ is the unit weight of the soil; H is the wall height; and S_h and S_v are the horizontal and vertical spacings, respectively, between two nails.

In this non-dimensional solution, the effect of nail bending stiffness depends on both the relative rigidity of the nail to the soil and the structure geometry. It can be defined by the rigidity parameter

$$N = [(K_s D/\gamma H)/(S_v S_h/\ell_0^2)]$$

This elastic solution is derived for the relatively flexible nails encountered in practice (i.e., length of the nail: L > 3 ℓ_0). For more rigid nails, a solution considering the limit case of perfectly rigid nails, has been derived.

A unique failure surface which verifies all of the equilibrium conditions of the active zone can be defined. In order to establish the geometry of this failure surface, it is necessary to determine the two geometrical parameters: A_0 - inclination of the failure surface at the upper ground surface, and A_f - inclination of the failure surface at the toe of the reinforced soil-mass. Observations on nailed-soil model walls suggest that for flexible nails the failure surface is practically vertical at the upper part of the structure ($A_0 = 0$). The angle A_0 depends upon the nail bending stiffness and the

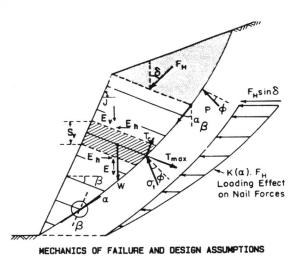

$T_{max} = \sigma_N \cdot A_S; \ T_C = \tau_N \cdot A_S$
A_S: SECTION AREA OF NAIL

STATE OF STRESS IN THE INCLUSION

$M = M_{max} = 0.32 \ Tc \ l_0$

$M = 0$

MECHANICS OF FAILURE AND DESIGN ASSUMPTIONS

THEORETICAL SOLUTION FOR INFINETELY LONG BAR ADOPTED FOR DESIGN PURPOSES

Figure 1. Kinematical limit analysis approach.

failure surface is practically perpendicular to perfectly rigid nails. The parameter A_f is determined from the moment equilibrium of the active zone.

The normal soil stress along the failure surface is determined using Kotter's equation. The maximum tension force in each reinforcement is calculated from the horizontal force equilibrium of the slice comprising the inclusion and the maximum shear force is calculated from Eq. 1. The normalized width of the active zone (S/H) and the values of the normalized maximum tension forces (TN) and shear forces (TS) are represented as non-dimensional parameters: $[TN = (T_{max})/(\gamma \cdot H \cdot S_v \cdot S_h)]$ and $[TS = (T_c)/(\gamma \cdot H \cdot S_v \cdot S_h)]$ at the relative depth (Z/H). Figure 2 illustrates the output variation of the normalized tension and shear force obtained for a typical nailed-soil wall with a vertical facing (J = 90°), a horizontal ground surface (V = 0°), soil strength characteristics of ϕ = 35° and (c/$\gamma \cdot$H) = 0.05, nail inclination β of 15° and different values of the rigidity parameter N (perfectly flexible; N = 1, 10; perfectly rigid).

3 DESIGN OF NAILED-SOIL CUT SLOPES

Design with the kinematical limit analysis is based on the evaluation of the local stability of each reinforcement with respect to failure by pull

out, breakage and excessive bending.

Detailed analysis with relevant parameters at each depth (Z/H) requires an appropriate computer code for design optimization. However, for structures in homogeneous soils with uniform nail lengths, simplified, yet conservative, design charts can be prepared considering the maximum values of S/H, TN and TS. Design charts for perfectly flexible nails and for #8 rebars which are frequently used in practice are shown in Figure 3. The design charts for #8 rebar are established for N = 0.33 corresponding to a wall height of 12 m in a silty sand with a K_s of 50,000 kN/m³. With these charts the following iterative design procedure can be used:

(1) Select the nail type (bending stiffness - EI, allowable tension stress - F_{all}, diameter - D, and spacing - S_v, S_H).

(2) For the specified soil properties (γ, K_s, c, ϕ), selected nail type (EI, F_{all}), nail inclination (β) and the structure height (H), determine the ratios S/H, TN and TS. The soil strength properties should be factored considering an acceptable safety factor.

(3) Verify that the selected reinforcement satisfies the breakage/excessive bending failure criteria.

(4) Select the limit interface lateral shear stress, f_ℓ, from field pullout data or available correlations with in-situ tests.

303

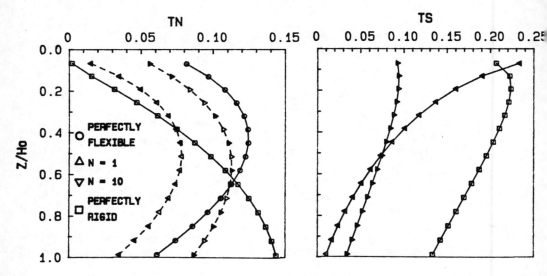

Figure 2.   Variation of TN and TS for different relative nail rigidities.

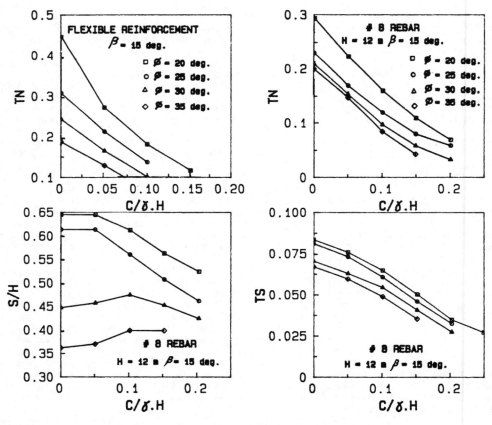

Figure 3.   Nail forces and geometry of the active zone (Elias and Juran, 1988)

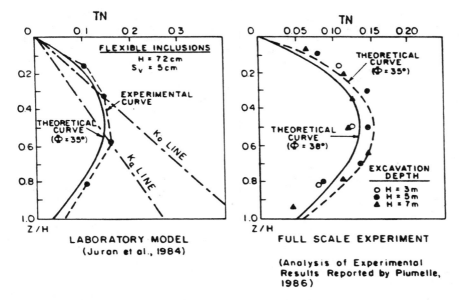

Figure 4. Comparison between predicted and measured nail forces.

(5) Establish structure geometry (L/H) to satisfy the pullout failure criteria with the required value of the safety factor, F_L.

4 THEORETICAL PREDICTIONS VS. EXPERIMENTAL RESULTS

In order to evaluate the proposed kinematical analysis design method, laboratory tests on nailed-soil model walls (Juran, et al., 1984) and full scale experiments on instrumented structures (Plumelle, 1986) have been analyzed and measured nail forces have been compared with predicted values. The use of this limit analysis method to predict at working stress structure behavior assumes that during construction, the soil resistance to shearing is entirely mobilized along the potential slip surface.

Figure 4 shows a comparison between predicted and measured values of maximum tension forces in a model wall and in a 7 m deep experimental nailed soil wall in granular ground (field data reported by Plumelle, 1986). The results obtained on this experimental wall are reported for successive excavation depths of 3, 5 and 7 m. They illustrate that the total excavation depth has only a negligible effect on the variation of the normalized tension force (TN) with

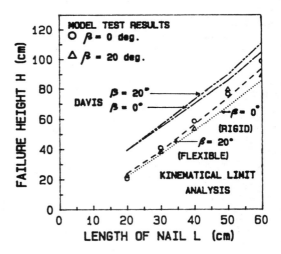

Figure 5. Predicted and measured failure heights of reduced scale model walls (Elias and Juran, 1988).

the relative depth (Z/H) and on the geometry (S/H) of the potential sliding surface. The comparison between predicted and measured values of the maximum tension forces indicates that the proposed design approach provides a reasonable estimate of nail forces.

305

Most failures of nailed-soil retaining structures that have been reported in the literature occurred as the result of nail pullout. It is therefore essential to evaluate the proposed design method through the analysis of observed pullout failures on laboratory model walls. For this purpose a series of laboratory model tests has been conducted (Elias and Juran, 1988). Pullout tests on the 6 mm diameter rigid steel nails, used in this study, provided the pullout resistance for the laboratory test conditions. Figure 5 shows the comparison between predicted and measured failure heights of the model walls. The heights are calculated using both the "Davis" slope stability design method, developed by Shen, et al. (1981), and the kinematical limit analysis. With the "Davis" method the failure height is attained when the safety factor reaches 1. For the kinematical limit analysis, the failure height is defined as the structure height that will generate sliding ($F_L = 1$) of the upper two nails. The failure heights were calculated considering perfectly rigid horizontal nails ($\beta = 0°$) and perfectly flexible inclined nails ($\beta = 20°$). Predicted failure heights correspond fairly well to the experimental results.

The "Davis" method was found to overestimate the critical height at pullout failure. The differences between the predicted and measured failure heights suggest that the definition of an overall "global" value of the factor of safety is not consistent with the observed progressive pullout failure due to sliding of the upper nails. The calculated global safety factor can be significantly higher than the local safety factor at the critical level of a sliding nail. Therefore, for safe engineering design local nail stability has to be investigated.

5 CONCLUSIONS

The kinematical limit analysis, presented in this paper provides a new engineering approach for the design of nailed-soil retaining structures. It allows for the evaluation of the effect of the main design parameters (inclination and bending stiffness of the nails, embankment slope, facing inclination, soil strength characteristics) on the magnitude and location of the maximum nail forces and on the structure stability. This design method enables the engineer to evaluate the local stability at the level of each reinforcement and therefore provides rational predictions of progressive pullout failure. The local factor of safety at the critical level of a sliding nail can be significantly smaller than the overall "global" safety factor predicted by the conventional slope stability design methods.

REFERENCES

Elias. V. & I. Juran 1988. Preliminary Draft - Manual of practice for soil nailing. Prepared for U.S. Dept. of Transportation, Federal Highway Administration, Contract DTFH-61-85-C-00142.

Juran, I., J. Beach & E. De Laure 1984. Experimental study of the behavior of nailed soil retraining structures on reduced scale models. Intl. Symp., In-Situ Soil and Rock Reinforcement, Paris.

Juran, I. & V. Elias 1987. Soil nailed retaining structures: analysis of case histories. Soil Improvement, ASCE Geotechnical Special Publication No. 12, 232-244.

Plumelle, C. 1986. Full scale experimental nailed soil retaining structures. Revue Francaise de Geotechnique, 40:45-50.

Schlosser, F. 1983. Analogies et differences dans le comportement et le calcul des ouvrages des soutennement en terre armee et par clouage du sol. Annales de L'Institut Technique du Batiment et des Travaux Publics, No. 418.

Shen, C.K., S. Bang & L.R. Herrmann 1981. Ground movement analysis of an earth support system. J. Geotech. Div., ASCE, 107(GT12).

Stocker, M.F., G. W. Korber, G. Gassler & G. Gudehus 1979. Soil nailing. Intl. Conf. on Soil Reinforcement, Paris, France, 2:469-474.

International Geotechnical Symposium on Theory and Practice of Earth Reinforcement / Fukuoka Japan / 5-7 October 1988
© 1988 Balkema, Rotterdam. ISBN 90 6191 820 0

Effect of rock bolts on volcanic sandy soil 'Shirasu' by floor lowering test

R.Kitamura
Kagoshima University, Kagoshima, Japan

M.Haruyama
Formerly Kagoshima University, Kagoshima, Japan

ABSTRACT: Floor lowering test was carried out to test the effect of reinforcement on volcanic sandy soil (Shirasu) which was sampled from Kagoshima City, Japan. The dimensions of the soil tank were 66.5 cm in width, 50 cm in height and 28 cm in length. Air dried Shirasu was poured into the soil tank in loose state. Cardboards which simulate rock bolts were inserted into the soil and the effect of reinforcement on Shirasu ground was investigated. It was found out that the reinforcement due to cardboards (rock bolts) works when the cardboards (rock bolts) cross the boundary between the region I, and the region II and III.

INTRODUCTION

In Kagoshima, which is located in the southern part of Kyushu Island in Japan, the volcanic soils and rocks are widely deposited. The main volcanic product is a non-welded part of pumice-tuff derived from the Pleistocene pyroclastic flow which occurred about 22,000 years ago. This material mainly composed of volcanic sandy soil and is called "Shirasu" in Japanese. Shirasu is classified as special soil and its mechanical and physico-chemical properties are different from the usual soil in Japan.

In this paper, the effect of reinforcement by rock bolts on Shirasu ground is investigated by carrying out the floor lowering test with cardboard reinforcement.

2 MATERIAL, APPARATUS AND PROCEDURE

Shirasu specimens were sampled from a telus by a cliff in Kagoshima City. The sampling material was seived and the particles larger than 9.52 mm were excluded. The specific gravity, the maximum and the minimum void ratios are listed in Table 1. The grain size distribution curve is shown in Fig.1. It is shown in Fig.1 that the used material, Shirasu, is sandy soil. The main mineral composition of Shirasu is silica and the shape of Shirasu particles is angular.

Figure 2 shows the testing apparatus.

Table 1. Physical properties of Shirasu

specific gravity	max. void ratio	min. void ratio
2.35	1.72	1.30

The dimensions of the soil tank were 100 cm in height, 100 cm in width and 28 cm in length. The front face was made of tempered glass so that the deformation of soil mass can be observed. The other faces were made of steel boards. The crossed straight lines whose interval was 5 cm were indicated on the surface of the glass. The width was controlled by inserting wooden boards. In these test series, the width was adjusted to be 66.5 cm as shown in Fig.2. The floor was 10 cm in width. Two load cells with individual capacity of 50 kgf and a jack were set under the lowering floor as shown in Fig.2. In order not to leak the material from the soil tank when the floor was lowered, wooden boards were also set on both sides of the lowering floor as shown in Fig.2.

Figure 3 shows cardboards which simulated rock bolts. The width of the cardboards was 1 cm. The length of cardboards varied from 7 cm to 21 cm in order to investigate the effect of rock bolts. Cardboards were put on the lowering floor as shown in Fig.3. After setting the cardboards, Shirasu

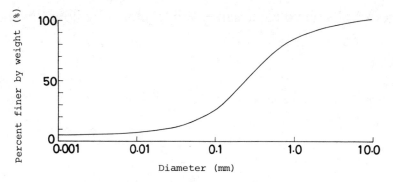

Fig.1 Grain size distribution curve of Shirasu

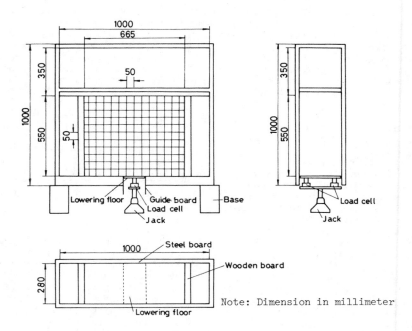

Note: Dimension in millimeter

Fig.2 Apparatus for floor lowering test

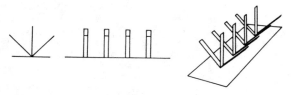

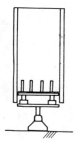

Fig.3 Cardboards which simulate rock bolts
and their setup in the soil tank

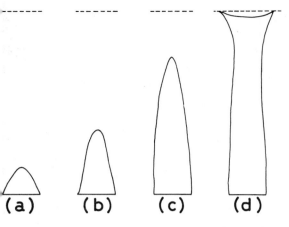

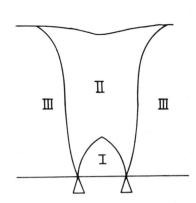

Fig.4 Development of loosened region for Shirasu

Fig.5 Three regions generated by floor lowering test for usual sandy soil (Murayama and Matsuoka, 1971)

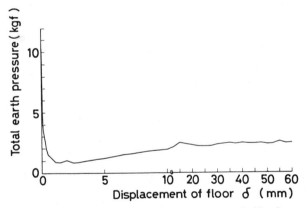

Fig.6 Relation between earth pressure and displacement of floor without cardboard

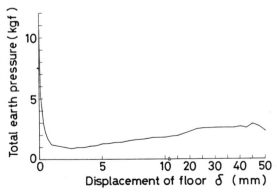

Fig.7 Relation between earth pressure and displacement of floor with cardboards of 7 cm

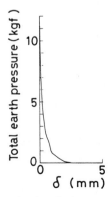

Fig.8 Relation between earth pressure and displacement of floor with cardboards of 21 cm

material was poured into the soil tank gently to produce the loose state. Along the crossed straight lines on the glassface, coloured soil (Shirasu) was installed to observe the deformation of the soil mass when the floor was lowering. After the height of sedimented soil reached to 50 cm, soil is compressed for 24 hours by self weight.

The floor was lowered by 0.05 mm interval until the total displacement of floor was 0.5 mm, by 0.1 mm interval until the total displacement was 1 mm, and by 0.5 mm interval after the total displacement is larger than 1 mm. At each stage, the settlement of ground surface, the deformation of coloured soil, and the earth pressure at the lowering floor were measured by using a dial gauge, photographs, and load cells, respectively.

3 TEST RESULTS

Figure 4 schematically shows the pattern of deformation of soil mass without rock bolts (cardboards), where δ denotes the displacement of lowering floor. The loosened region in Fig.4(a) was observed at about $\delta=5$ mm. The loosened regions in Fig.4(b), (c) and (d) was observed at about $\delta=10-15$ mm, 20-30 mm and larger than 30 mm, respectively. It is shown in Fig.4 that the loosened region was assumed to be parabolic until total displacement of less than 20 mm. Afterwards, the loosened region reached up to the ground surface. The width of the loosened region did not expand larger than the width of lowering floor. It may be considered that this mechanical behavior is due to the inter-locking of particles because the shape of Shirasu particle is angular. It is known that three regions were generated on the floor lowering test as shown in Fig.5 (Murayama and Matsuoka, 1971), region I where sandy soil moved with the lowering floor, while region II which was located in upper part of region I, and region III where the sandy soil did not move. In these series, region I was assumed by comparing the weight of soil which was calculated by parabolic loosened region with the earth pressure. The approximate relation between the width of lowering floor, B, and the height, H, of the loosened region I equals 0.8B without cardboards (rock bolts) reinforcement.

Figure 6 shows the relation between the displacement of floor and the earth pressure without rock bolts. At the beginning, the earth pressure decreased very fast between 0 and 1 mm of displacement δ. Then, the earth pressure gradually

increased until about $\delta=20$ mm and almost constant thereafter. In Fig.7, cardboards whose dimensions were 1 cm in width and 7 cm in length were inserted into the soil. Comparing Fig.7 with Fig.6, it was found that the relation between the earth pressure and the displacement was almost the same. It means that the rock bolts whose length is 7 cm cannot reinforce the soil mass. This fact relates to the loosened region I in Fig.5, i.e., the cardboards (rock bolt) may not reach to the boundary between the region I, and the regions II and III according to the relation H=0.8B. Figure 8 shows the same relation as Figs.6 and 7, but the cardboards (rock bolts) whose dimension is 21 cm were inserted into the soil in this case. The earth pressure decreased to zero when the displacement was only about 2.5 mm. It means that the reinforcement due to cardboards (rock bolts) works when the cardboards (rock bolts) cross the boundary between the region I, and the region II and III.

4 CONCLUSIONS

In this paper the experimental research was conducted although the quantitative and/or analytical research had not been conducted for the effect of cardboards (rock bolts). It may be summarized as follows.
1) Two kinds of cardboards which were different length were inserted into the soil to investigate the effect of reinforcement on Shirasu ground.
2) The reinforcement due to cardboards works when the cardboards cross the boundary between the region I, and the regions II and III.
3) The direction and length of rock bolt should be taken care for Shirasu ground because the mechanical behavior in the floor lowering test was not same as one of usual soil.

ACKNOWLEDGEMENTS

The authors wish to express their gratitude to Mr. Joumoto, the technical clerk of Kagoshima Univ., and Messrs. Toku Idemoto, Muranaga, Takahashi and Kawano, graduates of Kagoshima Univ. for their supports.

REFERENCE

Murayama, S. and Matsuoka, H. 1971. Earth pressure on tunnels in sandy ground. Proc. JSCE. No.187, pp.95-108.

International Geotechnical Symposium on Theory and Practice of Earth Reinforcement / Fukuoka Japan / 5-7 October 1988
© 1988 Balkema, Rotterdam. ISBN 90 6191 820 0

Model loading tests of reinforced slope with steel bars

Teruki Kitamura & Akira Nagao
Japan Highway Public Corporation, Tokyo, Japan

Seiji Uehara
Sumitomo Construction Co., Ltd, Tokyo, Japan

ABSTRACT: In order to study the effect of the steel bar reinforcement in a sandy slope, vertical loading tests for reinforced slope models were carried out. In these tests, only the placement angle of reinforcement bars was varied, and the surface deformation of the model slope and the stresses of the reinforcement bars were measured.
   The results show that 1) the effect of the reinforcement increases in the order horizontal, downward and upward placement of the reinforcement members, 2) the location where the largest increase in axial stress is observed at each loading step for reinforced and un-reinforced slopes, 3) bending and shearing resistance of reinforcing members contribute little to the reinforcing effect.

1. INTRODUCTION

In Japan, the land is being used for a wider range of applications. Therefore, various geotechnical methods have been developed in recent years, most of which are intended to be economical and simple. As one of these methods, reinforcing earth with steel bars is recognized as being effective in excavation and is becoming widely used. Basically in this method, the strength - deformation characteristic of the natural ground is improved by the insertion of steel bars.

However, though the mechanism of the reinforced earth with steel bars has been studied in some degree by the triaxial compression test and direct shear test (Jewell 1980, Tatsuoka et al. 1983), the loading tests on small or large in-situ model slopes have only been carried out a few times (Stocker et al. 1979, Hayashi et al. 1986, Kitamura and Nagao 1988).

Accordingly, in order to study the reinforcement effect, the authors carried out the vertical loading tests on reinforced model slopes with silty sand.

This paper describes the deformation of slope, mode of failure, axial and frictional stress of reinforcing members derived from the test results.

2. MODEL SLOPE AND LOADING TEST

The loading tests were performed on the model slopes, which had a width B of 900mm, a height H of 750mm and a length L of 2100mm shown in Figure-1. A general view of the loading test is shown in Figure-2. The model was made by compacting 15cm thick layers of sand in a steel frame, the reinforcing members were placed in locations before compaction. The frame was removed after compaction, and then it was cut to the profile shown in Figure-1.

The properties of the soil used for the test are as follows:

maximum particle size	: 29mm
gravel content	: 24%
fine particle content	: 13%
specific gravity of particles:	$Gs=2.65$
natural moisture content	: $W_n=11\%$

The properties of the model slopes are as follows:

wet density	: $\rho t=2.00g/cm^3$
void ratio	: $e=0.47$
cohesion	: $c=0.15kgf/cm^2$
internal friction angle:	$\phi=33°$

Reinforcing members made of aluminum plate, 450mm long, 25mm wide and 2mm thick, were used instead of steel bars normally used in practice. Its Young's modulous (E) was $7.03*10^5kgf/cm^2$ and the horizontal and vertical spacings were 225mm. Three placement angles were used in the test; sloping upwards at 20°, horizontal and sloping downwards at 20°. Sand was affixed to the surface of the reinforcing members with epoxy adhesive. Bond between the head of the reinforcing members and the slope surface was ensured by setting a 40mm *

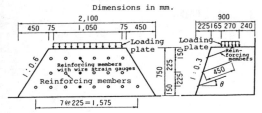

Dimensions in mm.

Figure-1 Reinforced slope model

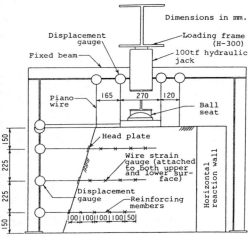

Figure-2   Instrumentation of the loading test

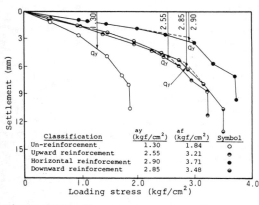

Classification	ay (kgf/cm^2)	af (kgf/cm^2)	Symbol
Un-reinforcement	1.30	1.84	○
Upward reinforcement	2.55	3.21	◐
Horizontal reinforcement	2.90	3.71	●
Downward reinforcement	2.85	3.48	◑

Figure-3 Loading stress vs. settlement of loading plate

40mm aluminum plate on the head. Wire strain gauges were set on the three reinforcing members on the centerline of the slope and the strain in these members was measured.

The model dimensions of slope height and front and side gradients of 750mm, 1:0.3 and 1:0.6 respectively, were decided upon to prevent failure of the side of the

specimen under loading.

The loading plate was 270mm wide (B) and 1050mm long (L). This is considered as a two dimensional strip footing because L/B is very nearly 4 (Muhs 1965).

The load was applied through a hydraulic jack system in steps of 0.5kgf/cm^2, and was maintained until the settlement rate was less than 0.5%/min. at each loading step. At each load increment the load was held for fifteen minutes. When the loading approached the yield condition, loading steps were reduced 0.25kgf/cm^2.

The deformation of loading plate and slope surface, and stress of the reinforcing members were measured.

3. DEFORMATION OF SLOPE SURFACE AND SLIP SURFACE

Figure-3 shows the relation between the settlement of the loading plate and loading stress. Yield stress q_y in the figure is equivalent to the first inflection point of settlement, and it is shown as the intersection of the broken lines. Failure stress q_f is equivalent to the stress at the time of a sudden increase of settlement. The yield stress is 2.9kgf/cm^2 in both horizontal and downward reinforcement the value, 2.9kgf/cm^2, is twice as large as that of a un-reinforced slope. The difference in the reinforcement effect between horizontal and downward placing appears in the difference between settlements at yield stress: the settlement of downward reinforcement is nearly twice as large as that of horizontal reinforcement.

The difference of reinforcement effect as shown in Figure-3 also has an influence upon the horizontal displacement of the slope surface. Figure-4, 5 and 6 show the relation between the loading stress and surface displacement of an un-reinforced slope, downward reinforcement and horizontal reinforcement respectively: the displacement increases in the order of horizontal, downward and un-reinforced.

The displacement for upward reinforcement shows a tendency similar to that of downward reinforcement, as shown on the curve of Figure-3. Further, it is observed that the horizontal displacements at points 2 and 3 are larger than at points 1 and 4 in all figures, and also that the horizontal displacements at points 2 and 3 are nearly equal. This means that the whole soil mass is pushed out uniformly above the measuring point 4. Thus, deformation of the slope surface is closely related to the slip surface.

Figure-7 shows observation results of the slip surface after the loading test. The slip surfaces for reinforced slopes are further back than the slip surface observed for un-reinforced slopes.

The degree in shift of slip surfaces is in the order of horizontal, downward and upward reinforcement. In particular, in the case of upward reinforcement the lower part of the slip surface lies along the lowest reinforcing members.

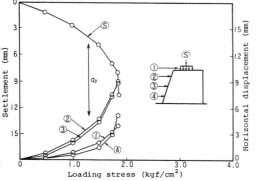

Figure-4 Loading stress vs. horizontal displacement of un-reinforced slope

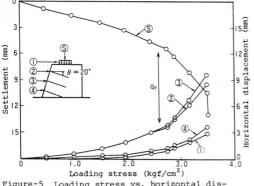

Figure-5 Loading stress vs. horizontal displacement of downward reinforced slope

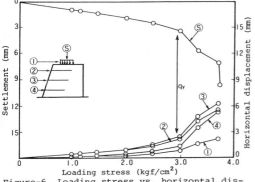

Figure-6 Loading stress vs. horizontal displacement of horizontal reinforced slope

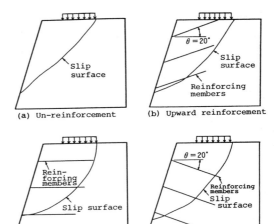

(a) Un-reinforcement (b) Upward reinforcement

(c) Horizontal reinforcement (d) Downward reinforcement

Figure-7 Observed slip surface

4. STRESS IN REINFORCING MEMBERS

It was observed from the deformation of the slope and the shape of slip surface that the upper part of the reinforced zone was pushed out with the slip surface passing between the middle and lower reinforcing members. Change of axial stress, which occurs in the reinforcing members with an increase of loading stress, is shown in Figures-8 to 12.

Axial stress of the middle reinforcing members increases in the order of upward, downward and horizontal reinforcing members. The stress is almost the same for horizontal and downward reinforcement, and that of upward reinforcement is particularly small. Also the axial stress increases beyond the value at the slope yield point q_y.

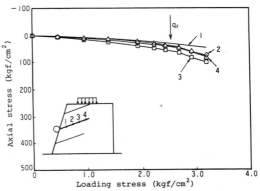

Figure-8 Loading stress vs. axial stress in upward reinforcement (middle member)

313

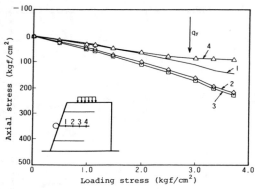

Figure-9 Loading stress vs. axial stress in horizontal reinforcement (middle member)

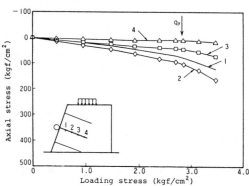

Figure-10 Loading stress vs. axial stress in downward reinforcement (middle member)

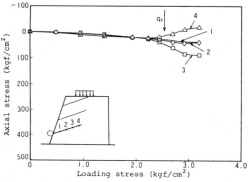

Figure-11 Loading stress vs. axial stress in upward reinforcement (lower member)

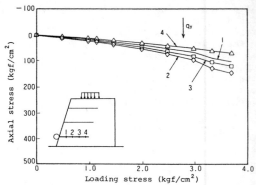

Figure-12 Loading stress vs. axial stress in horizontal reinforcement (lower member)

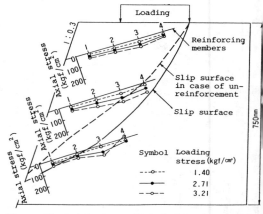

Figure-13 Axial stress distribution in upward reinforcement

Figures-13 to 15 show the axial stress distribution for each stage of the loading stress: the white circle dotted lines show the distribution before the yield stress, the black circle solid lines show that near the yield stress, and the broken lines show that at failure stress. From the relative position of the slip surface in the figures the maximum axial stress occurs near the slip surface. The loading increments are small in all cases. The maximum axial stress in the middle members lies near the point where the member cuts the slip surface. In the case of horizontal reinforcement, the increase of axial stress at measuring point 3 of the middle reinforcing members is larger than that of measuring point 2. Similarly, for the downward reinforcement the axial stress at point 2 increases more rapidly than at the other points. The points where the largest increase in axial stress occur lie between the slip surfaces for reinforced and unreinforced slopes. The reason for this is assumed to be that the increased slope deformation moves the slip surface further back from the face, thus increasing the effectiveness of the reinforcing members.

Figures-16 to 18 show the frictional stress occurring on the surface of the reinforcing members; the frictional stress

314

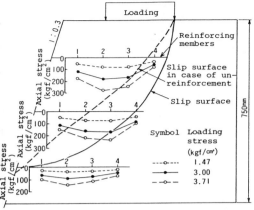

Figure-14 Axial stress distribution in horizontal reinforcement

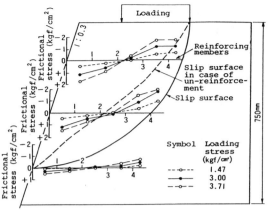

Figure-17 Frictional stress distribution in horizontal reinforcement

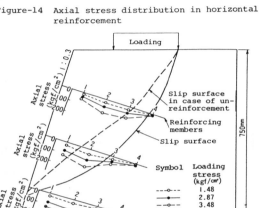

Figure-15 Axial stress distribution in downward reinforcement

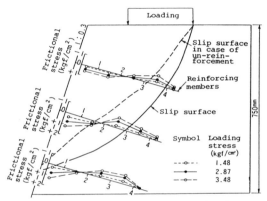

Figure-18 Frictional stress distribution in downward reinforcement

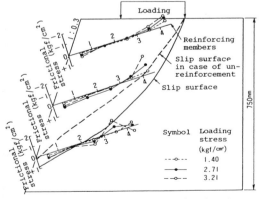

Figure-16 Frictional stress distribution in upward reinforcement

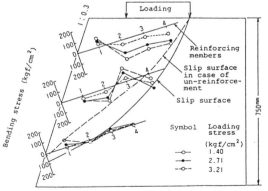

Figure-19 Bending stress distribution in upward reinforcement

was calculated by considering the frictional force on the face of the reinforcing members to be equal to the difference in adjacent axial forces. The frictional force is lower for the upward reinforcement than for the other cases. For the horizontal and downward cases the greatest increase in frictional stress is in the

315

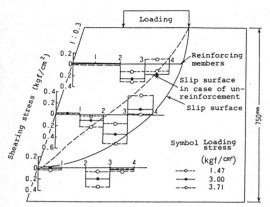

Figure-20 Shearing stress distribution in horizontal reinforcement

middle member, between the slip surfaces for reinforced and un-reinforced slopes, as for the axial stress.

Figure-19 shows the bending stress of reinforcing members shown typically for upward reinforcement. It shows that the distribution of bending stress bears little relationship to the location of the slip surfaces, the other cases show similar results. Further, when the curvature of reinforcing members is calculated, its radius is shown to be 2.3m even for the largest bending strains observed; the radius is extremely large compared to the model size.

Figure-20 shows the shearing stress occurring in the horizontal reinforcing members. The shearing stress was obtained by numerical differentiation of the measured bending strain. The value of the subgrade reaction calculated from the observed bending strain becomes extremely small.

Therefore, it is considered that reinforcing members do not contribute to the strength through bending and shearing resistance, because they behave identically with the surrounding ground.

5. CONCLUSIONS

In order to confirm the effect of reinforcement, the failure pattern of the slope and the stresses in the reinforcing members were measured by vertical loading tests for model slopes. The main conclusions obtained are as follows.

(1) The results of the vertical loading tests showed that the reinforcement effect increases in the order of horizontal, downward and upward reinforcement. The reason

is that the failure slip surfaces are shifted back from the slope in that order.

(2) The location where the largest increase of axial stress is observed at each loading step is between the slip surfaces of reinforced and un-reinforced slopes. The reason is that the increase of slope deformation shifts the location of the slip surface which increases the effectiveness of the reinforcing members.

(3) It was observed that bending and shearing resistance of reinforcing members contribute little to the reinforcement effect. Therefore, this suggests that only the axial force in reinforcing members should be considered for design in reinforced earth.

REFERENCES

Jewell, R.A.: Some effects of reinforcement on the mechanical behavior of soils, Ph. D. Thesis, Cambridge Univ., 1980.

Tatsuoka, F., Kondou, K., Miki, G., Haibara, O., Hamada, E. and Satoh, G.: Fundamental study on tensile-reinforcing of sand, TSUCHI-TO-KISO JSSMFE, Vol. 31, No. 9, pp. 11-19, 1983 (in Japanese).

Stocker, M.F., Korber, G.W., Gössler, G. and Gudehus, G.: Soil Nailing, Proc. ICSR, Paris, Vol. II, pp. 469-474, 1979.

Hayashi, S., Ochiai, H., Tayama, S. and Sakai, A.: Effect of top-plates on mechanism of soil-reinforcement of cut-off slope with steel bars, Proc. JSCE, No. 367, VI-4, pp.62-70, 1986 (in Japanese).

Kitamura, T. and Nagao, A.: Experimental study on the reinforcement effect of steel bars in the sandy soil slope model for vertical loads. Proc. JSCE, No. 391 VI-8, pp. 188-195, 1988.

Muhs, E.: On the phenomenon of progressive rupture in connection with failure of footing on sand, Proc. 6th I.C.S.M.F.E., Vol. III, pp. 419-421, 1965.

International Geotechnical Symposium on Theory and Practice of Earth Reinforcement / Fukuoka Japan / 5-7 October 1988
© *1988 Balkema, Rotterdam. ISBN 90 6191 820 0*

Finite element stability analysis method for reinforced slope cutting

T.Matsui & K.C.San
Osaka University, Osaka, Japan

ABSTRACT: An elastoplastic joint element was derived being based on the Coulomb yield criterion. A finite element stability analysis method for reinforced cut slope was proposed. Example problems were given to illustrate the applicability of the proposed method to the reinforced cut slope. The results of example problems showed that the proposed analysis method can be applied to obtain a more reasonable design solution for the reinforced slope cutting.

1 INTRODUCTION

Methods of stability analysis currently used for reinforced slope are mainly the limit equilibrium stability analysis. Failure mode of the reinforced slope is assumed either circular failure mode, e.g. Schlosser and Juran (1983), Catier and Gigon (1983), Guilloux, Notte and Gonin (1983) and Juran (1983), or plane failure mode, e.g. Gassler and Gudehus (1981). In the limit equilibrium stability analysis of reinforced slope, the tensile force developed in the reinforcements is considered either to reduce the force tending to cause movement or to increase the force resisting the movement. The evaluation of tensile force developed in the reinforcements is based on the mechanism of pull out test. Due to the uncertainty of the tensile force developed in the reinforcements, Gassler and Gudehus (1983) proposed a probabilistic stability analysis method for reinforced slope, in which the pull-out resistance of reinforcements is considered as stochastic variable.

In the analysis of reinforced slope cutting, the equilibrium of forces and the strain compatibility between the soil and the reinforcement should be satisfied. In the limit equilibrium analysis the strain compatibility condition is not satisfied. The limit equilibrium method could provide misleading safety factors due to its inability to represent the stress relief producing by excavation and the interaction between the soil and the reinforcement. Consequently, it would be desirable to apply the finite element method technique to the stability analysis for reinforced slope cutting.

In the finite element analysis of reinforced slope, the slippage between the soil and the reinforcement becomes prime concern. In this paper, first an elastoplactic joint element is derived being based on the Coulomb yield criterion. Then a finite element stability analysis method for reinforced slope is proposed. The average local safety factor approach is adopted, together with introducing a local safety factor surface. Shear strength reduction technique is employed to trace the failure slip surface. Finally, example problems are given to illustrate the applicability of the proposed finite element stability analysis method for the reinforced cut slope.

2 ELASTOPLASTIC JOINT ELEMENT

The elastic stress-strain matrix, D_e, of the joint element is

$$D_e = \begin{bmatrix} G & 0 \\ 0 & E \end{bmatrix}. \tag{1}$$

Coulomb yield function and its associated flow rule are used, such as

$$|\tau| = \sigma \tan\phi + c. \tag{2}$$

Eq.2 may be rewritten as

$$\tau^2 = (\sigma \tan\phi + c)^2 . \qquad (3)$$

The yield function, f, can be defined as

$$f = \tau^2 - (\sigma \tan\phi + c)^2 . \qquad (4)$$

Plastic normality flow rule states that the incremental plastic strain vectors, de_p, are orthogonal to a plastic potential function. If the associated flow rule is assumed, hence

$$de_p = \lambda \frac{\partial f}{\partial \sigma} . \qquad (5)$$

The quantity λ is a positive scalar parameter.

During plastic flow,

$$df = \frac{\partial f}{\partial \sigma} d\sigma = 0 . \qquad (6)$$

Substitution of Eq.4 into Eq.6 leads

$$df = \tau \, d\tau - (\sigma \tan\phi + c)\tan\phi \, d\sigma = 0 . \qquad (7)$$

The total strain increment, $d\varepsilon$, can be decomposed into elastic strain increment, $d\varepsilon_e$, and plastic strain increment, $d\varepsilon_p$, that is,

$$d\varepsilon = d\varepsilon_e + d\varepsilon_p . \qquad (8)$$

The elastic strain increment is defined as,

$$\begin{pmatrix} d\gamma_e \\ d\varepsilon_e \end{pmatrix} = \begin{bmatrix} 1/G & 0 \\ 0 & 1/E \end{bmatrix} \begin{pmatrix} d\tau \\ d\sigma \end{pmatrix} . \qquad (9)$$

Differentiation of Eq.4 leads

$$\frac{\partial f}{\partial \sigma} = [2\tau \quad -2(\sigma\tan\phi + c)\tan\phi]^T . \qquad (10)$$

Substitution of Eqs.5, 9, and 10 into Eq.8 leads

$$\begin{pmatrix} d\gamma \\ d\varepsilon \end{pmatrix} = \begin{bmatrix} 1/G & 0 \\ 0 & 1/E \end{bmatrix} \begin{pmatrix} d\tau \\ d\sigma \end{pmatrix} + \lambda \begin{pmatrix} 2\tau \\ -2(\sigma\tan\phi + c)\tan\phi \end{pmatrix} . \qquad (11)$$

The inverse relationship corresponding to Eq.11 can be obtained as,

$$\begin{pmatrix} d\tau \\ d\sigma \end{pmatrix} = \begin{bmatrix} G & 0 \\ 0 & E \end{bmatrix} \begin{pmatrix} d\gamma \\ d\varepsilon \end{pmatrix} - \lambda \begin{pmatrix} 2\tau \\ -2(\sigma\tan\phi + c)\tan\phi \end{pmatrix} . \qquad (12)$$

Substitution of Eq.12 into Eq.7 and its simplification lead

$$\lambda = \frac{\tau G d\gamma - E(\sigma\tan\phi + c)\tan\phi d\varepsilon}{2\tau^2 G + 2(\sigma\tan\phi + c)^2\tan^2\phi E} . \qquad (13)$$

Substituting Eq.13 into Eq.12, the stress-strain relationship can be written as,

$$D_{ep} = \begin{bmatrix} G & 0 \\ 0 & E \end{bmatrix} - \begin{bmatrix} D_{11} & D_{12} \\ D_{21} & D_{22} \end{bmatrix} , \qquad (14)$$

where $D_{11} = \dfrac{\tau^2 G^2}{(\sigma\tan\phi + c)^2\tan^2\phi E + \tau^2 G}$, $\qquad (15)$

$$D_{12} = D_{21} = \frac{-(\sigma\tan\phi + c)\tan\phi \tau E G}{(\sigma\tan\phi + c)^2\tan^2\phi E + \tau^2 G} , \qquad (16)$$

$$D_{22} = \frac{(\sigma\tan\phi + c)^2\tan^2\phi E^2}{(\sigma\tan\phi + c)^2\tan^2\phi E + \tau^2 G} . \qquad (17)$$

3 FINITE ELEMENT STABILITY ANALYSIS METHOD

The operation of the finite element analysis for reinforced slope cutting consists of two excavation processes. The original ground is assumed as horizontal. The first excavation is to form the natural slope to simulate the erosion process. The second excavation is to simulate the actual construction sequence of reinforced slope cutting.

The local safety factor, F_{SL}, of an element is defined as,

$$F_{SL} = \frac{2c \cos\phi + 2 \sigma_3 \sin\phi}{(\sigma_1 - \sigma_3) \cdot (1 - \sin\phi)} . \qquad (18)$$

The local safety factor surface, $F_{SL}(x,z)$, is then constructed being based on the local safety factor obtained from finite element stress analysis. If a trial slip surface, S, is defined, the local safety factor along the slip surface, $F_{SL}(S)$, can be calculated from the local safety factor surface, as shown in Fig.1.

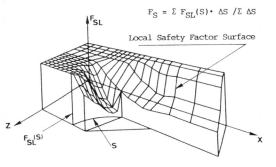

$$F_S = \Sigma F_{SL}(S) \cdot \Delta S / \Sigma \Delta S$$

Local Safety Factor Surface

Fig.1   The local safety factor surface

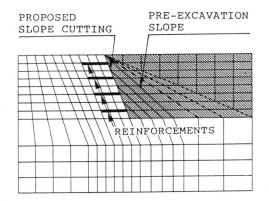

PROPOSED   PRE-EXCAVATION
SLOPE CUTTING   SLOPE

REINFORCEMENTS

Fig.2   The mesh used in the analysis

Table 1 Material properties of the ground

Elastic modulus E (tf/m²)	2.2×10^3
Unit weight γ (tf/m³)	1.8
Poisson's ratio ν	0.3
Friction angle ϕ (degree)	30
Cohesion c (tf/m²)	1.0
Coefficient of earth pressure at rest K_0	0.6
Hyperbolic constant K	210
Hyperbolic constant K_{ur}	420
Hyperbolic constant n	1.02
Failure ratio R_f	0.69

The average safety factor of a slip surface, F_S, is defined as,

$$F_S = \frac{\Sigma F_{SL}(S) \cdot \Delta S}{\Sigma \Delta S} . \qquad (19)$$

In order to trace the development of the failure slip surface of the slope, the shear strength reduction technique is employed. In this technique, the shear strength parameters, cohesion, c , and coefficient of friction, $\tan\phi$, where ϕ is the angle of internal friction, of the component materials are incrementally reduced by dividing them with a common shear strength reduction ratio, R.

The failure mechanism of cut slope is examined by using shear strain failure criterion. The shear failure of an element is defined as that the shear strain level exceeds the failure shear strain.

The failure of slope is defined when the shear failure slip zone developed from the toe to the crest of the slope. The corresponding shear strength reduction ratio is called the critical shear strength reduction ratio, R_c.

The availability of the presented analysis method for reinforced slope cutting will be examimed through the field test data (Matsui and San, 1987, Matsui, San, Amano and Otani, 1988) in our following paper.

4 EXAMPLE PROBLEMS

The mesh used in the analysis is shown in Fig.2. The soil is assumed as nonlinear elastic material with a hyperbolic modulus. The reinforcement is considered as one dimensional bar element. The nonlinear characteristic of slippage between the reinforcement and the soil is modeled by the elastoplastic joint element described in the previous chapter.   The material properties of the ground used in the present analysis are summarized in Table 1. The elastic modulus of the reinforcement is assumed as 2.1×10^7 tf/m², and the cross section area, 5.2×10^{-4} m².

In this chapter, first the failure mechanism of the cut slope is exmained. In the finite element analysis, the failure slip surface of the slope is not easy to be captured, when the failure criterion is based on the stress. Alteratively the strain failure criterion is used in this work.   It will show that the failure slip surface of the slope can be well

319

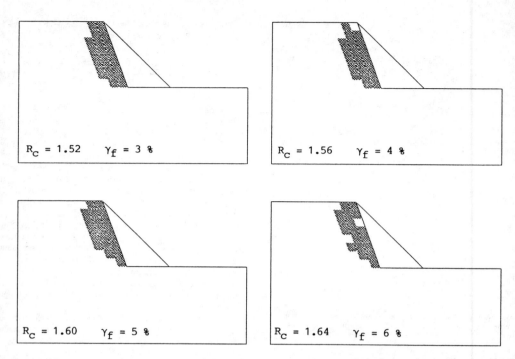

Fig.3   Failure mechanisms of cut slopes with different failure shear strains

captured by applying shear strength reduction technique to the non-linear hyperbolic finite element analysis. Then the influences of some pertinent design parameters, such as the K_o value of the ground, and the length, the total number and the inclination of the reinforcement, on the stability behavior of the slop cutting are examined.

4.1 Failure mechanism of cut slope

The pre-excavation slope of the slope cutting is assumed as 45° slope with the K_o value of 0.6. Fig.3 shows the failure mechanisms of cut slopes with different failure shear strains and the corresponding critical shear strength reduction ratios. From Fig.3, it can be seen that the critical shear strength reduction ratio R_c increases as increasing the failure shear strain γ_f. Although the choice of the value of failure shear strain affects the critical shear strength reduction ratio, the failure patterns of the slopes are almost similar.

Barata and Danziger (1981) used 5% shear strain as shear failure in the slope analysis. Fig.4 shows the comparison of the slip surface obtained by

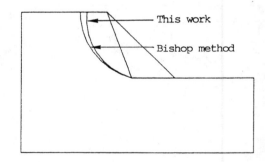

Fig.4   Comparison of the failure slip surfaces between this work and Bishop method

using shear strength reduction technique for the case of 5% shear strain failure criterion to that from the conventional Bishop method. The slip surface obtained by finite element method is close to that from Bishop method.

Once the failure pattern of the cut slope is obtained, the safety factor of the slope can be determined from the local safety factor surface , without any trial and error.

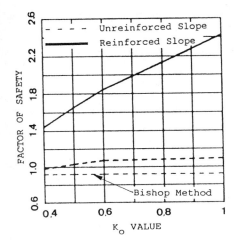

Fig.5 Safety factor versus K_o value

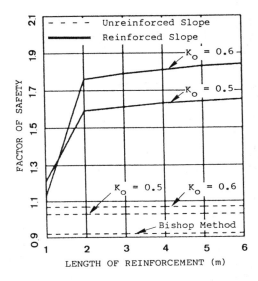

Fig.6 Safety factor versus reinforcement length

4.2 Effect of K_o value

Fig.5 shows the relationship between the safety factor and K_o value for both unreinforced and reinforced slopes. The safety factor of an unreinforced slope given by Bishop method is 0.92. The safety factors of the unreinforced slope obtained from finite element analysis method are little greater than that obtained by Bishop method. The effect of K_o value on the safety factor in the reinforced slope is much greater than that in the unreinforced one. It can be seen from this figure that the estimation of K_o value plays an important role in the stability analysis for a reinforced slope.

4.3 Effect of reinforcement length

Fig.6 shows the relationship between the safety factor and reinforcement length for K_o values of 0.5 and 0.6. For both cases, the safety factor increses as increasing the reinforcement length. The rate of increase in the safety factor decreases as the reinforcement length is greater than two meter. Vidal (1966) described the mode of action of the reinforcement as that the individual soil particles tied together under the effect of reinforcement, consquently a solid block of earth was formed. To achieve this function a minimum length of reinforcement is required, for this example it may be two meter. The performance of the stability of the reinforced slope cutting is not significantly improved when the

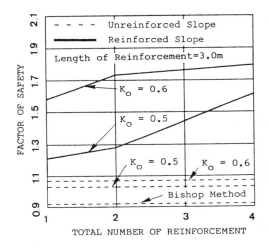

Fig.7 Safety factor versus number of reinforcement

reinforcement exceeds the minimum reinforcement length.

4.4 Effect of number of reinforcements

Fig.7 shows the relationship between the safety factor and number of reinforcements for cases of reinforcement of 3 meter with K_o values of 0.5 and 0.6. The safety

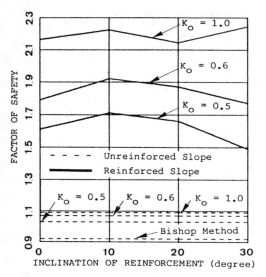

Fig.8 Safety factor versus inclination of reinforcement

factors of reinforced slopes increase as increasing the number of reinforcements. The effect of number of reinforcements on the safety factor of reinforced slope is affected by the K_O value.

4.5 Effect of inclination of reinforcement

Fig.8 shows the relationship between the safety factor and inclination of reinforcement for K_O values of 0.5, 0.6 and 1.0. The effect of the inclination of reinforcement on the safety factor of reinforced slope is affected by the K_O value. The optimum value of inclination of reinforcement depends on the K_O value.

5 CONCLUSIONS

A finite element stability analysis method for reinforced cut slope was presented, being based on the concept of local safety factor surface. From the results of the example problems, following conclusions can be made:

1. The failure slip surface of an unreinforced slope obtained by the proposed method is close to that from the conventional Bishop method.

2. The effect of K_O value on the safety factor in the reinforced slope is much greater than that in the unreinforced one. The K_O value plays an important role in the reinforced slope stability analysis.

3. There exists a minimum reinforcement length in the reinforced slope design, and the performance of the stability of the reinforced slope could not be significantly improved when the reinforcement exceeds the minimum reinforcement length.

4. The safety factors of reinforced slopes increse as increasing the number of reinforcements.

5. The inclination of reinforcement affects the safety factor of reinforced slopes. The optimum value of inclination of reinforcement depends on the K_O value.

6 REFERENCES

Barata, F. E. and Danziger, F. A. B. 1981. Design of slopes in residual soils by an allowable-strain method. Proc. of 10th Int. Conf. on SMFE, Stockholm: 347-351.

Catier, G. and Gigan, J. 1983. Experiments and observations on soil nailing structures. Proc. of the 8th Conf. of the ECSMFE., Helsiki: 473-476.

Gassler, G. and Gudehus, G. 1981. Soil nailing, some aspect of a new technique. Proc. of the 10th Int. Conf. on SMFE, Stockholm: 665-669.

Gassler, G. and Gudehus, G. 1983. Soil nailing - statistical design. Proc. of the 8th Conf. of the ECSMFE., Helsiki: 491-494.

Guilloux, A., Notte, G. and Gonin, H. 1983. Experiences on a retaining structure by nailing. Proc. of the 8th Conf. of the ECSMFE., Helsiki: 449-502.

Juran, I. 1983. Reinforced soil system - applications in retaining structures. Proc. of the 7th Asian Region Conf. of the SMFE: 96-114.

Matsui, T. and San, K. C. 1987. Reinforcement mechanism of cut slope with tensile inclusions. Proc. 8th Asian Reg. Conf. SMFE., Kyoto, Vol.1: 185-188.

Matsui, T., San, K. C., Amano, T. and Otani, Y. 1988. Field measurement on a slope cutting with tensile inclusions. Proc. 2nd Int. Conf. on Case Histories in Geotechnical Engineering, St. Louis. Vol.2 : 1099-1105.

Schlosser, F. and Juran, I. 1983. Behaviour of reinforced earth retaining walls for model studies. Developments in Soil Mechanics and Foundation Engineering - 1, Chap.6. Applied Science Publishers: 197-229.

Vidal, H. 1966. The principle of reinforced earth. Highway Research Record No.282.

International Geotechnical Symposium on Theory and Practice of Earth Reinforcement / Fukuoka Japan / 5-7 October 1988
© *1988 Balkema, Rotterdam. ISBN 90 6191 820 0*

Field monitoring procedure of cut slopes reinforced with steel bars

Sadamu Mino, Kunitomo Noritake & Shuuzou Innami
Sumitomo Construction Co., Ltd, Tokyo, Japan

ABSTRACT: The design of cut slopes reinforced with steel bars is not always reliable due to many unknown soil properties of the natural ground. Monitoring trial excavations is a suggested method to overcome this problem. If the performance of a trial reinforced cut slope is observed, collected data are useful for the safe construction of the reinforced cut slope.

The authors have constructed about 60 cut slopes reinforced with steel bars and have used monitored trials on some excavations.

As a result of these observations, it is confirmed that a monolithic reinforced zone was formed by inserting steel bars in the natural ground which then behaved like a concrete retaining wall.

1 INTRODUCTION

In recent years, many reinforced soil works have been constructed, though guidelines for the design have not been established. The mechanism of soil reinforcement have been studied in various laboratory tests (Hamada et al. 1984, Tatsuoka and Hamada 1984), field trials (Kitamura et al. 1987) and field monitoring (Noritake and Innami 1986). As a result, some design methods were proposed (Gässlar and Gudehus 1983). However, the actual behaviour of the reinforced cut slope may be significantly different from the behaviour predicted by these design methods, because of the non-uniformity of the natural ground. The authors considered that trial monitoring could be applied as effectively to reinforced cut slopes as to other earthworks e.g. embankments.

It is important to clarify the fundamental behaviour of a reinforced cut slope. This paper describes the monitored trial and the FEM prediction used for the construction of a cut slope reinforced with steel bars.

2 CONSTRUCTION PROCEDURE

The reinforced earth method is applied to a cut slope. This stabilizing technique is applied to the cut slope soon after each bench has been excavated so that the slope surface does not become loose due to the relaxation of stress. The construction sequence is as follows:

a) Excavation with bench height of 1.0 to 2.0 m
b) Shotcreting to protect the slope face
c) Boring holes, usually 45 mm diameter, in the ground at the designed spacing; inserting steel reinforcing bars into the bored holes and grouting.

This sequence is repeated for every bench until the designed cut slope is completed.

3 APPLICATION OF TRIAL MONITORING

3.1 Outline

A slope was excavated adjacent to a railroad track for the construction of a bridge pier.

Photograph 1 shows the excavation to which the reinforced earth technique was applied (Suda et al. 1984). Figure 1 shows a cross section of the reinforced cut slope.

1) Deformed bars (25 mm diameter, 3.0 to 7.0 m length) were inserted in the slope at a rate of $1/m^2$ of the slope surface.
2) The gradient of the cut slope was 1:0.5 and the height was 11.6 m.

Photo. 1 View of reinforced slope

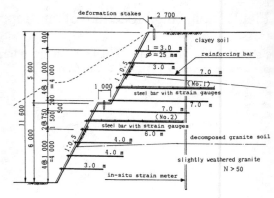

Fig. 1 View of reinforced slope and arrangement of field measuring equipment

3) The slope surface was protected with 7.0 cm thick wire-mesh reinforced shotcrete.
4) The slope was excavated by cutting benches.
5) The height of each bench was 1.0 m to 2.0 m.
6) The excavation took 20 days to complete.

The ground consists of three layers as shown in Figure 1. The first layer is clayey soil, the second layer is decomposed granite soil and the third layer is slightly weathered granite.

Table 1 shows the physical properties of the clayey soil and the decomposed granite soil obtained from laboratory and field tests.

Figure 1 shows the measuring apparatus used for monitoring the reinforced cut slope. One in-situ strain meter, nineteen deformation stakes and two reinforcing bars with strain gauges were used.

3.2 Results

Figure 2 shows the maximum value of the measured horizontal deformation in the ground with elapsed time from the ground surface to a depth of 7.0 m and the settlement of the ground surface. Progressive deformation occured during the excavation. The rate of deformation decreased after completion of the excavation. The slope was left for about one month until the construction of the pier was completed. The rate of deformation was very low after excavation.

Table-1 Physical properties of decomposed granite soil

		unit	clayey soil	decomposed granite soil	
unit weight	γ	tf/m^3	1.5～1.6	1.8～2.0	
water content	ω	%	——	1.5～21.0	
internal frictional angle	ϕ	degree	10～16	28～36	direct shear test
cohesion	C	tf/m^2	0.05～0.12	0.1～0.2	
coefficient of permeability	κ	cm/sec	——	6.3×10^{-5}	
modulus of elasticity	E	kgf/cm^2	32～59	140～180	plate loading test

These results show that the elastic deformation of the slope was caused predominantly by the release of the in-situ stress in the ground during excavation with more gradual deformation after completion of the excavation. The elastic deformation represented 85 to 90% of the total deformation and occurred during the period of excavation. It is necessary to observe the reinforced cut slope carefully to assess its stability. Measuring apparatus must be set up before the excavation, because the data from the beginning of the excavation are necessary to estimate and analyse the stability of the slope.

Figure 3 shows the measured surface deformation of the reinforced cut slope after excavation. Although the slope showed displacement, it remained flat and there was no cracking on the shotcrete surface. These results showed that the reinforced cut slope behaved as a rigid body. However, it was predicted from the results of an FEM analysis that a non-reinforced cut slope would not behave as a rigid body.

Figure 4 shows the measured displacement in the ground. From this figure, it was found that the displacement from the ground surface to a depth of 7.0 m was constant, and the first and second layers moved laterally as a rigid body. The conclusion drawn from this was that the area reinforced with steel bars behaved as a rigid body.

The displacements of the slope surface and the cracks in the shotcrete must be carefully observed. Elsewhere, irregular deformation of the slope surface and cracking of the shotcrete were observed at another temporary cut surface for which the reinforcement was insufficient.

Figure 5 shows the relationship between the axial forces in the reinforcing bars and elapsed time. Axial force worked in a complicated manner. A compressive force was observed in the section near the slope surface of both reinforcement bars and the compressive force increased gradually with time after completion of excavation. At the same time, the tensile force which occurred in the rest of the steel bar decreased slowly.

From these results it was confirmed that the area reinforced with steel bars behaved as a rigid body and showed the same behaviour as a concrete retaining wall.

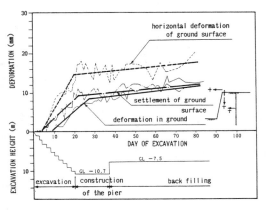

Fig. 2 Measured deformation and settlement with time

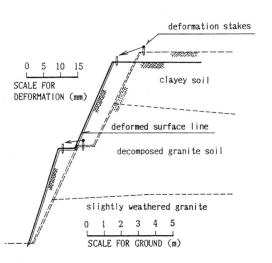

Fig. 3 Deformation of cut slope surface

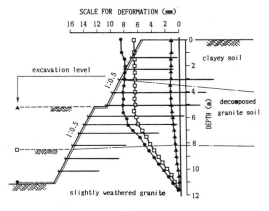

Fig. 4 Measured deformation in the ground

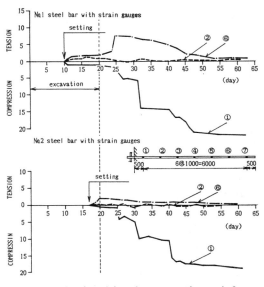

Fig. 5 Relationships between the axial force and elapsed time

325

4 TRIAL MONITORING

The trial monitoring takes the following process: design, trial construction, monitoring, analysis, judgement and full construction.

An FEM analysis (Kitamura et al. 1988) which considered the discontinuity between the reinforcement and the natural ground was used with the trial monitoring.

Figure 6 shows the calculated deformation of the reinforced cut slope using the FEM analysis and Table 2 shows the physical properties assumed for the analysis. The figure shows that the calculated slope remained a plane surface and was similar to the observed results.

Figure 7 shows the observed and the theoretical deformation at the position of the in-situ strain meter after completion of the excavation, and both deformations show the same tendencies and characteristics.

Figure 8 shows the measured and theoretical axial force distributions. The measured axial force was smaller than the theoretical one because the steel bars with strain gauges had to be placed after the excavation of each bench. However, the observed axial force distribution was similar to the theoretical one. The axial force in the lower steel bar with strain gauges was not so significant because the excavation was completed 4 days after the bar was set.

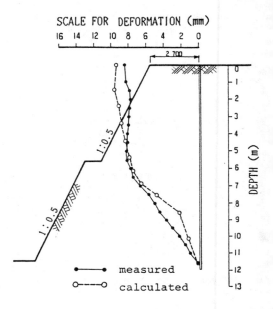

Fig. 7 Comparison of measured and analysed deformations

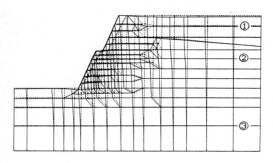

Fig. 6 Calculated deformation by FEM analysis

Table-2 Physical properties used for analysis

		unit	stratum①	stratum②	stratum③	steel bar
modulus of elasticity	E	tf/m²	400	1500	5000	104600
unit weight	γ	tf/m³	1.6	1.9	2.0	1.92
poissons ratio	ν		0.40	0.35	0.20	0.3
cohesion	C	tf/m²	1.0	1.5	10.0	——
internal frictional angle	φ	degree	15	30	3	——
c & φ between soil and bar		φ=30°	c =2.63 tf/m²			

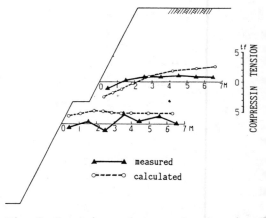

Fig. 8 Comparison of measured and analysed axial force distribution

Figure 9 shows the safety factor distribution calculated from the FEM analysis. This figure shows that the unstable elements are found only near the slope surface and it means that the effect of the reinforcement does not work satisfactorily near the slope surface. Therefore an appropriate protective work, such as shotcrete, fulfills an important function for stabilizing the slope.

It was confirmed that FEM analysis can be carried out as an effective part of the trial monitoring procedure.

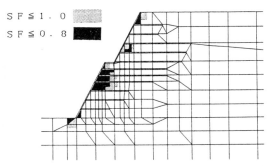

S F ≦ 1 . 0
S F ≦ 0 . 8

Fig. 9 Safety factor distribution by FEM analysis

REFERENCE

Gässlar, G. & Gudehus, G. 1983. Soil nailing-statistical design. Proc. 8th European Conf. on SMFE Helsinki Vol. 2: pp.491-494.
Hamada, E., Tatsuoka, F. & Morihira, K. 1984. Model test of reinforced sand slope with steel bars. Proc. 19th JNCSMFE: pp.1167-1170 (in Japanese).
Kitamura, T., Nagao, A., Okuhara, M. and Saitou, T. 1987. Some considerations on the application and the design of the steel bar reinforcement slope stabilization method. Proc. of JSCE No. 385, VI-7: pp.79-87 (in Japanese).
Kitamura, T., Nagao, A., Noritake, K. and Innami, S. 1988. Study on observational procedure for soil-reinforcement method with steel bars. Proc. of JSCE No. 391, VI-8: pp.151-160 (in Japanese).
Noritake, K. & Innami, S. 1986. Stabilizing method for cut slope by reinforcement bars. Proc. 13th ARRB-5th REAAA Combined Conference: pp.261-266.
Suda, T., Noritake, K., Segawa, T. and Sasaki, K. 1984. Slope stability method by reinforcement bars (No. 2). Proc. 19th JNCSMFE: pp.1319-1320 (in Japanese).
Tatsuoka, F. & Hamada, E. 1984. Laboratory study on reinforcing of sand slope with steel bars [I]-[XI]. Seisan Kenkyu, Tokyo Univ: 36(10)-37(9) (in Japanese).

5 CONCLUSIONS

The main conclusion obtained are as follows:

1) A reinforced cut slope shows the same characteristics and behaviour as a concrete retaining wall, because the area reinforced with steel bars behaves as a rigid body.
2) It is confirmed that the monitoring of elastic displacements during excavation is important for ensuring the stability of the reinforced cut slope. Thus, displacement of the slope and ground surface have to be measured, and deformation and cracking of the slope surface have to be observed.
3) An FEM analysis which considers the discontinuity between the reinforcing members and the ground is able to simulate the behaviour of the reinforced cut slope satisfactorily and it gives confidence in the trial monitoring procedure.

International Geotechnical Symposium on Theory and Practice of Earth Reinforcement / Fukuoka Japan / 5-7 October 1988
© 1988 Balkema, Rotterdam. ISBN 90 6191 820 0

Field experiment on reinforced earth and its evaluation using FEM analysis

Akira Nagao & Teruki Kitamura
Japan Highway Public Corporation, Tokyo, Japan

Jun Mizutani
Sumitomo Construction Co., Ltd, Tokyo, Japan

ABSTRACT: In order to explain the effect of steel bars as reinforcing members in natural ground and the mechanism of stability of reinforced earth, loading tests using large-scale specimens of reinforced earth were carried out. After that, the loading tests were simulated using two-dimensional and three-dimensional FEM analysis. As a result, numerical simulations were consistent with loading tests, and it was confirmed that the two-dimensional FEM analysis which considered the discontinuity between reinforcing members and natural ground was an effective analysis method for reinforced earth.

1 INTRODUCTION

In recent years, the soil reinforcement method using steel bars has come into widespread use as an effective earth retaining for excavation. The soil reinforcemernt is made by steel bars put into the natural ground.

As the characteristics of the natural ground and steel bars interact effectively, steel bars improve the strength of the soil mass. This method is mainly applied to the stabilization of excavation using relatively short steel bars from 1.0 m to 5.0 m long.

The fundamental characteristics of the soil reinforcement with steel bars have been explained by model tests (Gässler et al. 1983, Tatsuoka et al. 1984) and the stability analysis method (Kitamura et al. 1987) has been established. However, the actual behavior of a reinforced slope may be remarkably different from the theoretical behavior, because of the nonuniformity of the natural ground. In this case, the design has to be modified by a change of model or soil property and an analytical method is required to simulate the behavior of a reinforced slope.

This paper describes one useful method of analysis for observations of soil reinforcement compared with the results of a large scale field loading test and a numerical simulation of the loading test. Loading tests were simulated by two-dimensional FEM analysis considering the discontinuity between the natural ground and reinforcing members, and then the results

of analysis, were corrected by considering the three-dimensional stress dispersion of the loading test. In order to verify a three-dimensional stress dispersion of the loading test, a three-dimensional elastic FEM analysis was performed on a non-reinforced specimen.

2 FIELD LOADING TEST OF REINFORCED EARTH

Loading tests of reinforced earth were carried out at the Ishibe plant nursery of Japan Highway Public Corporation (Nagao et al. 1984, Nagao et al. 1985). The nature of the ground generally consists of homogeneous layers: the first sandy soil layer (ks1) and the second sandy soil layer(ks2).

The shape of the loading test specimen is 3.0 m-high, 7.0 m-wide and 3.0 m-deep as shown in Figure-1. Deformed bars (ϕ25 mm) were used as reinforcing members and were set at intervals of 1.0 m. Further, 3 insitu strain meters, 14 deflection gauges and 3 steel bars with strain gauges were located to measure the behavior of the specimen as shown in Figure-1. The large rigid bearing plate made of concrete was set as shown in the figure.

Although seven specimens were tested with varied length, angle and interval of steel bars, the following two specimens are described in this paper.

(1) length of reinforcing bars; L = 4.0 m, position; horizontal
(2) length of reinforcing bars; L = 2.0 m, position; horizontal

The specimens were loaded, using 12 jacks, until they were yielded. As a result, the specimen with 4.0 m reinforcing bars yielded at q = 47 tf/m^2, the specimen with 2.0 m reinforcing bars yielded at q = 37 tf/m^2 and non-reinforced specimen yielded at q = 29 tf/m^2. (See photo-1)

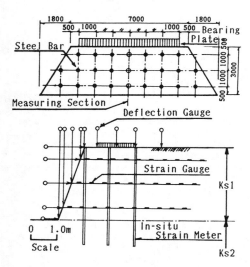

Fig.-1 Outline of Measuring Equipment

Photo-1 View of the Loading Test

3 TWO-DIMENSIONAL FEM ANALYSIS CONSIDERING DISCONTINUITY BETWEEN REINFORCING MEMBERS AND NATURAL GROUND

When the behavior of the specimen is simulated by FEM analysis, it is important to know how to evaluate the discontinuity between the reinforcing members and the natural ground.

The discontinuity is evaluated using the frictional resistance (ie. adhesion and friction force) between the reinforcing

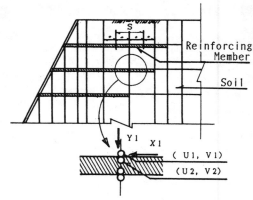

Fig.-2 Discontinuity Model between Reinforcing Member and Natural Ground

members and the natural ground (Iseda et al 1979). The discontinuous surface, as shown in Figure-2, is expressed by two different nodes which are located on the same coordinates and belong to different elements. The continuity is evaluated using the dimension comparison between the nodal force (X1) and the sliding resistance (Y1· tan∅* + C*·S). In Figure-2, node (U1, V1) belongs to the elements of natural ground, and node (U2, V2) belongs to the reinforcing materials.

(Criterion Formula)   (Continuous Condition between Nodes)

X1 $\leq$ (Y1·tan∅* + C*·S) ... U1=U2, V1=V2 Continuous Conditions

X1 $\geq$ (Y1·tan∅* + C*·S) ... U1≠U2, V1≠V2 Sliding Conditions

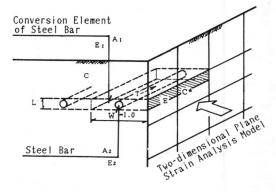

Fig.-3 Conversion Method for Two-dimensional Analysis Model of Reinforcing Member

The resistance force at the boundary surface (Y1·tanø* + C*·S) is expressed by the
friction angle (ø*) and the conversion
adhesion (C*) between the natural ground
and the reinforcing members.

In order to use two-dimensional plane
strain analysis, the reinforcing bars,
which were originally cylindrical, were replaced with thin flat reinforcing members
of uniform thickness, as shown in Figure-3.
The conversion of the physical properties
of the flat reinforcing members seemed to
be as follows:

$$E = \frac{A1 \cdot E1 + A2 \cdot E2}{A1 + A2}$$

$$C* = C + \frac{T}{2.0 \cdot W}$$

where, A1: sectional area of the flat
          reinforcing member
       A2: original sectional area of the
          cylindrical reinforcing bar
       E1: modulus of elasticity of the
          natural ground
       E2: modulus of elasticity of the
          reinforcing bar
       C : cohesion of the natural ground
       C*: adhesion between the natural
          ground and the reinforcing
          member
       T : pull-out resistance force of one
          reinforcing bar

However, the Poisson's ratio of flat reinforcing member was the same as the original
value.

The loading tests were simulated by
elastic analysis. Table-1 shows the
physical properties obtained by the soil
tests and used in the analysis. Figure-4
shows an example of two-dimensional FEM
analysis.

Table-1 Physical Properties Used for
Analysis

Description	Unit	Sand①	Sand②	Steel bar
Modulus of elasticity	tf/m²	3700	14800	206000
Unit weight	tf/m³	2.10	1.73	2.06
Poisson's ratio		0.35	0.25	0.3
C (soil)	tf/m²	4.5	4.0	——
φ (soil)	degree	35°	34 °	——
C* & φ* between soil and steel bar		C*= 6.26 tf/m² φ*= 35°		

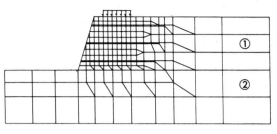

Fig.-4 Model of Two-dimensional FEM
Analysis

4 THREE-DIMENSIONAL FEM ANALYSIS

The behavior of the loading test specimen
was simulated by two-dimensional plane
strain FEM analysis considering the discontinuity between the reinforcing members
and the natural ground as described in the
previous paragraph. However, it is supposed that the specimen undergoes a three-
dimensional stress and strain distribution
rather than a perfect two-dimensional plane
strain, even in the central section of 7 m-
wide specimen. Therefore, three-dimensional FEM analysis of the non-reinforced
specimen was performed and compared with
the results calculated by the two-dimensional plane strain FEM analysis.

Figure-5 shows a three-dimensional model.
Physical properties were same as those for
the two-dimensional analysis.

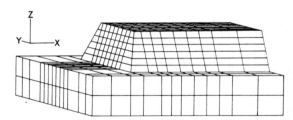

Fig.-5 Model of Three-dimensional Analysis

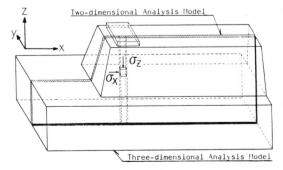

Fig.-6 Ground Stress (σx) and (σz)

Figure-7 and 8 show the stress-comparison of σx and σz at the center of the specimen and just under the bearing plate as shown in Figure-6. It was found that perfect two-dimensional plane strain condition was not indicated even in the central section of the specimen.

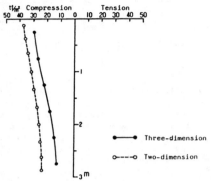

Fig.-7 Comparison of Ground Stress of Three-dimensional and Two-dimensional Analyses (σz)

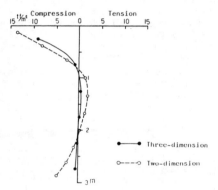

Fig.-8 Comparison of Ground Stress of Three-dimensional and Two-dimensional Analyses (σx)

The ground stress in the vertical direction σz (Figure-7) calculated by three-dimensional analysis was 0.75 ∿ 0.69 times as large as the stress calculated by two-dimensional analysis. Therefore, if the settlement of the specimen surface is simulated by two-dimensional analysis, it is necessary to be corrected by the follow-ing ratio (Nz).

Settlement $\delta = \int_0^h \sigma z \cdot dh$

$$Nz = \frac{\int_0^h \sigma z3 \cdot dh \text{(three-dimensional analysis)}}{\int_0^h \sigma z2 \cdot dh \text{(two-dimensional analysis)}} = 0.6$$

When the absolute value of the horizontal ground stress σx (Figure-8) is compared with the location of reinforcing bars, the result seems to be as follows.

$$Nx = \frac{|\sigma x3|}{|\sigma x2|}$$

Nx1 = 1/2 (upper reinforcing bar)
Nx2 = 1/3 (middle reinforcing bar)
Nx3 = 1/5 (lower reinforcing bar)

Therefore, when the axial force of rein-forcing bar is simulated by two-dimensional analysis, it is necessary to be corrected by the ratio (Nx).

If a perfect two-dimensional plane strain conditions are existed, it is not necessary for the two-dimensional simula-tion results to be corrected.

5 COMPARISON BETWEEN THE MEASURED AND CALCULATED VALUES OF LOADING TESTS

A comparison between the measured and cal-culated values of the specimen with 4.0 m reinforcing bars is shown in Figure-9 ∿ 12, and a comparison of the specimen with 2.0 m reinforcing bars is shown in Figure-13 ∿ 16. These figures show the settlement of ground surface, the deflection of slope surface and axial force of the reinforcing bars. According to these figures, each calculated value is similar to the measured value.

The original axial force distributions calculated by the two-dimensional plane strain analysis are shown in Figure-12 and 15. It is recognized that the non-corrected value is not similar to the measured value.

Safety factor distributions calculated by two-dimensional FEM analysis are shown in Figure-17. Although unstable elements are widespread in the non-reinforced specimen, many elements of the reinforced specimens maintain stability and unstable elements are distributed only near the slope surface. Further, the unstable elements of reinforced specimens are di-vided into thin layers by reinforcing bars and a failure zone caused by compression occurs near the slope surface of each layer. Consequently, this means that the slope can be protected by simple slope protection work, and the slope protection work for such a slope is very effective. The specimens were failed, except for the lower reinforcing bars, and the failure range calculated by two-dimensional FEM analysis corresponds closely with the test results. The specimens reinforced by 4.0 m steel bars and 2.0 m steel bars show the same tendency.

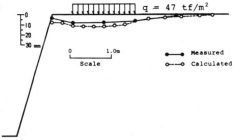

Fig.-9 Settlement of Ground Surface
(reinforcing members L = 4.0 m)

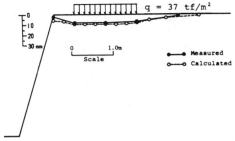

Fig.-13 Settlement of Ground Surface
(reinforcing members L = 2.0 m)

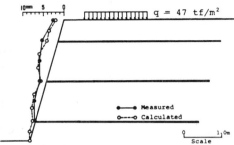

Fig.-10 Deflection of Slope Surface
(reinforcing members L = 4.0 m)

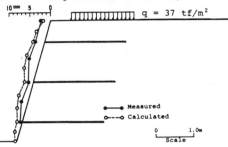

Fig.-14 Deflection of Slope Surface
(reinforcing members L = 2.0 m)

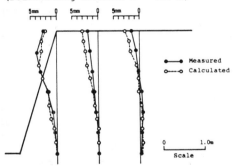

Fig.-11 Horizontal Deflection
(reinforcing members L = 4.0 m)

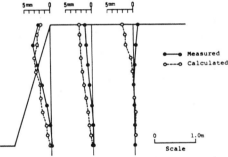

Fig.-15 Horizontal Deflection
(reinforcing members L = 2.0 m)

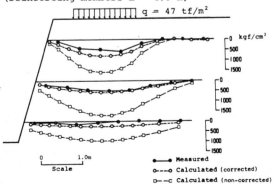

Fig.-12 Axial Force of Reinforcing Member
(reinforcing members L = 4.0 m)

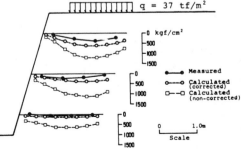

Fig.-16 Axial Force of Reinforcing Member
(reinforcing members L = 2.0 m)

333

SF ≦ 1. 0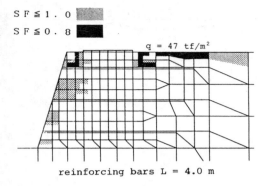

SF ≦ 0. 8

q = 47 tf/m²

reinforcing bars L = 4.0 m

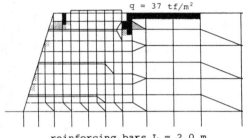

q = 37 tf/m²

reinforcing bars L = 2.0 m

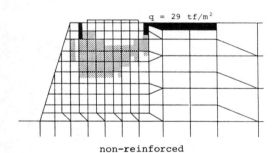

q = 29 tf/m²

non-reinforced

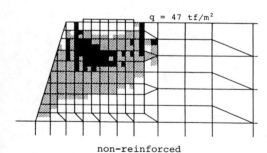

q = 47 tf/m²

non-reinforced

Fig.-17 Safety Factor Distribution of Specimen Simulated by Two-dimensional FEM Analysis

6 CONCLUSION

It was proven on the basis of the results that the behavior of an actual reinforced slope and the axial force of reinforcing bars can be estimated by two-dimensional plane strain FEM analysis considering the discontinuity between the reinforcing members and the natural ground. Consequently it was confirmed that the two-dimensional FEM analysis which considers the discontinuity between natural ground and reinforcing members is an effective analysis method for the observational procedure of reinforced earth.

REFERENCE

Gässler, G. & Gudehus, G. 1983. Soil nailing—statistical design. Proc. 8th European Conf. on SMFE Helsinki Vol. 2, pp.491-494

Hamada, E., Tatsuoka, F. & Morihira, K. 1984. Model test of sand slope reinforced with steel bars. Proc. 19th JNCSMFE, pp.1167-1170 (in Japanese)

Iseda, T., Tanahasi, Y. & Higuchi, T. 1979 FEM analysis of earth pressure considering friction of wall. Proc. 14th JNCSMFE, pp.989-992 (in Japanese)

Kitamura, T., Okuzono, S. & Nagao, A. 1985 Experimental study on soil reinforcement method with steel bars. Proc. 16th Japan Road Conference, pp.93-94 (in Japanese)

Kitamura, T., Nagao, A., Okuhara, M. & Saitou, T. 1987. Some considerations on the application and the design of the steel bar reinforcement slope stabilization method. Proc. JSCE No. 381 VI-7, pp.79-87 (in Japanese)

Nagao, A., Noritake, K. & Innami, S. 1984. Frictional registance between reinforcing material and ground on loading test of reinforced slope. Proc. 37th JSCE Conf, pp.383-384 (in Japanese)

Nagao, A., Kaneko, K., Uehara, S., Outa, M & Mikami, H. 1985. In site loading test of reinforced slope with steel bars -Stress and friction property of rein- forcing steel bars-. Proc. 20th JNCSMFE pp.1349-1352 (in Japanese)

Noritake, K. & Innami, S. 1986. Stabilizing method for cut slope by reinforcement bars. Proc. 13th ARRB-5th REAAA Combined Conference, pp.261-266

Tatsuoka, F. & Hamada, E. 1984. Laboratory study on reinforcing of sand slope with steel bars (I)-(X). Seisan Kenkyu, Univ of Tokyo 36(10)-37(9) (in Japanese)

International Geotechnical Symposium on Theory and Practice of Earth Reinforcement / Fukuoka Japan / 5-7 October 1988
© 1988 Balkema, Rotterdam. ISBN 90 6191 820 0

Theory and practice on reinforced slopes with steel bars

S. Okuzono
Japan Highway Public Corporation, Tokyo, Japan

N. Yamada & N. Sano
Laboratory of Japan Highway Public Corporation, Machida, Japan

ABSTRACT: This paper summarizes a part of a report entitled "Study on Reinforcing Slopes with Steel Bars", which has been carried out mainly by the Laboratory of the Japan Highway Public Corporation (Nihon Doro Kodan) during the last five years.

The relation between the theory and the practice of this work is studied through: (1) large scale (full-size) in-situ model test to confirm the effect of the steel bars, (2) survey of actual results of existing facilities and (3) follow-up survey after construction. The purpose of this study is to propose a design guide for reinforcing slopes with steel bars.

1 INTRODUCTION

The purpose of the method is to stabilize the whole natural ground (slope) by installing reinforcing materials into soil which is essentially weak in tension. With this method, slopes which do not need such protection works as anchor works or pile works, can be reinforced inexpensively, safely and easily. It has been applied in the cases shown in Figure 1.

i) Prevention of small scale failure of slopes

ii) Steeper slope with reinforcement

iii) Temporary reinforcement of an excavated slope

iv) Reinforcement of slope above tunnel entrance

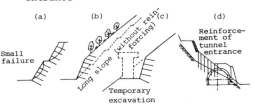

Figure 1. Simplified drawing of applications of reinforcing earthwork with steel bars

2 THEORETICAL REINFORCEMENT EFFECTS

The three kinds of effects shown in Figure 2 are considered.

i) Anchoring effect

ii) Shearing effect

iii) Effect like gravity retaining wall

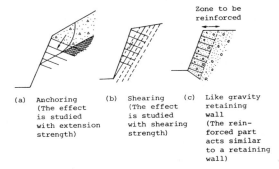

(a) Anchoring (The effect is studied with extension strength)   (b) Shearing (The effect is studied with shearing strength)   (c) Like gravity retaining wall (The reinforced part acts similar to a retaining wall)

Figure 2. Types of reinforcing effects

3 IN-SITU MODEL TEST

In this paper, we report on an in-situ model test, which was conducted with a model of full size on the diluvial terrace sand and gravel, and on the Kanto loam (cohesive soil like volcanic ash) formation plateau.

(1) Test equipment and methods

The outline of specimen is shown in Figure 3. In the case of a full size load test, the equipment consists of loading frames, twelve 50-ton jacks, a cast-insitu reinforced concrete loading plate 30cm thick and four earth anchors, both in front and behind the specimen as reaction points. The load was applied through a load-controlled system in steps of 3∿5 tf/m^2 depending on the reinforcement in the specimen until failure. At each load increment the load was usually held for fifteen minutes.

The ground where the test was performed consists of a homogeneous sandy soil mixed with silt; and a typical cohesionless soil. The other ground is a typical cohesive soil.

The basic experiment with a middle-size model, which was tested before the full size test, was performed in a manner similar to that of the full-size specimen; The specimen was scaled down to 1/4 size from 3m to 75cm, the drilling diameter was scaled down to 15mm; the aluminum pipe was scaled down to an external diameter of 8mm and to a wall thickness of 1mm; and the perforated plate was scaled down to 50mm x 50mm.

The amount of vertical settlement of both ground surface and loading plate, horizontal displacement of the slope, axial tension of the reinforcing materials, and the subgrade displacements were measured.

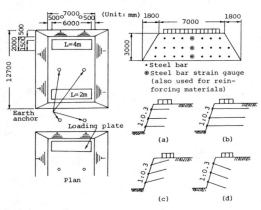

Figure 3. Schematic drawing of specimen

(2) Results of the experiment

The results of the loading experiment showed a tendency for the yield load to increase as the length of reinforcement was increased. In addition, from the curves of load and settlement after yield load, it was found that the longer the reinforcing materials were, the longer the time required for the slope to fail.

The factors affecting the reinforcement of natural ground include angle and density of reinforcement. In order to compare these factors, the following parameters were set up.

Length ratio of reinforcement

$$LR = \frac{\text{length of reinforcing steel bar L}}{\text{width of loading plate B}}$$

Density of reinforcement

$$D = \frac{\text{the number of reinforcing steel bars n}}{\text{area of slope A}}$$

Ratio of yield load

$$SR = \frac{\text{yield load of reinforcement ground q}}{\text{yield load of ground without reinforcement } q_0}$$

Figure 4 shows the relation between the length ratio of reinforcement and the reinforcing effect.

In figure 4, results are plotted for test specimen in which the reinforcement is horizontal and the spacing of reinforcement is kept constant (full size: 1m, scale model: 25cm). However, it is understood that the ratio of the yield load, between 1.0 and 3.0 of the length ratio of reinforcement, changes greatly, and, the yield load does not change much when the length ratio of reinforcement exceeds 2.0 (when the reinforcement length equals 1.5 times the distance from the top of slope to the end of the loading plate).

Figure 5 shows the relation between the angles of the reinforcement and the reinforcing effect. The results show that the angle of reinforcement in which the ratio of yield load is maximum depends upon the type of soil tested; it seems that the most effective angle of reinforcement is horizontal or downward ($\sigma = 60°$ or $90°$) for sand, and horizontal or upward ($\sigma = 90° \sim 110°$) for cohesive soil.

Figure 6 shows the relation between the density of reinforcement and the reinforcing effect. It is understood that there is an upper limit, even though the ratio of yield load increases when the density of the reinforcement of the reinforcing steel bars is increased.

4 PERFORMANCE OF THE TECHNIQUE IN REAL APPLICATIONS

The results of a survey conducted in 1985 of 73 applications of reinforced slope

336

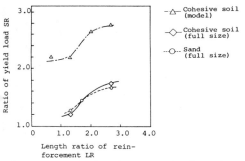

Figure 4. Relation between the length ratio of reinforcement and the ratio of yield load

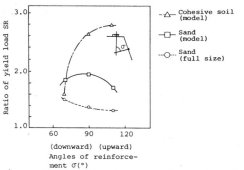

Figure 5. Relation between the angle of reinforcement and the ratio of yield load

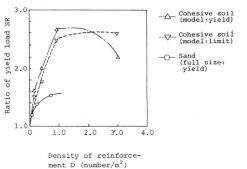

Figure 6. Relation between the density of reinforcement and the ratio of yield load

constructed by the Japan Highway Public Corporation and others are presented.

Figure 7 (a) shows the geological conditions of the reinforced natural ground. This shows that the method was used for equal numbers of soil and soft rock applications. Figure 7 (b) shows the frequency of steel bar length. This shows the frequency of use of bars over 2m but under 3.5m long is high. Figure 7 (c) shows the frequency of use of steel bars installed

at an angle. Most of them are installed perpendicularly to the side-slope or horizontally. Figure 7 (d) shows frequency of spacings between steel bars. The interval is concentrated between 1m and 1.75m.

5 THEORY AND PRACTICE

5.1 Theory and experimental results

By comparing the results of the full-size loading experiment with the value calculated by the simplified Bishop's method, the ultimate equilibrium stability was analyzed. The simplified Bishop's method can be applied to the stability analysis of reinforced slope with steel bars. The simplified Bishop's method, with tensile force T affecting the slip plane, was combined and expanded is as follows;

$$Fs = resistance/sliding\ force$$

$$= \frac{\Sigma \left\{ \dfrac{c \cdot li \cdot cos\alpha i + (Wi + Qi \cdot di) \cdot tan\phi + Ti \cdot sin\theta i \cdot tan\phi}{cos\alpha i + tan\phi \cdot sin\alpha i / Fs} \right\}}{\Sigma Wi \cdot sin\alpha i + \Sigma Qi \cdot sin\alpha i \cdot di - \Sigma Ti \cdot cos(\alpha i + \theta i)}$$

where,
Fs: safety factor
ϕ : internal friction angle of the soil
Wi: dead load of the slice
di: width of the slice
Qi: applied load
li: length of the sliding face of the slice
Ti: tensile force of the reinforcing materials affecting the base of the slice
αi: slope of the slip plane of the slice
c : cohesion of the soil
θi: angle of the reinforcement

The tensile force T used in the calculation is obtained from the distribution of the tension force (axial tension) in steel plate at the yield load during the full scale load test. Soil properties were derived from the results of the tests on unreinforced specimens. The value of cohesion c and the internal friction angle ϕ were selected so that the safety factor Fs_0 becomes 1.0 at yield, as in the triaxial compression test.

The load at yield (when safety factor Fs=1.0), was found from the tension force of the steel bars T, cohesion c and the internal friction angle ϕ of the soil. Figure 8 shows a comparison between the experimental value of the yield load and its calculated value. The alphabetic characters in Figure 8 show each of the reinforcing patterns (a) to (d) in Figure

337

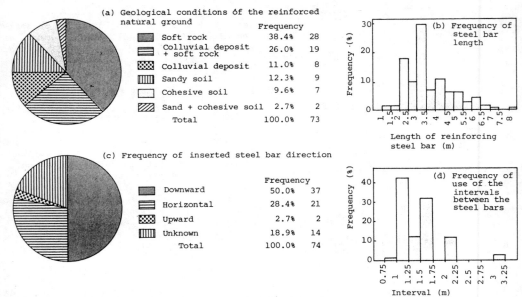

(a) Geological conditions of the reinforced natural ground

		Frequency	
	Soft rock	38.4%	28
	Colluvial deposit + soft rock	26.0%	19
	Colluvial deposit	11.0%	8
	Sandy soil	12.3%	9
	Cohesive soil	9.6%	7
	Sand + cohesive soil	2.7%	2
	Total	100.0%	73

(b) Frequency of steel bar length

Length of reinforcing steel bar (m)

(c) Frequency of inserted steel bar direction

		Frequency	
	Downward	50.0%	37
	Horizontal	28.4%	21
	Upward	2.7%	2
	Unknown	18.9%	14
	Total	100.0%	74

(d) Frequency of use of the intervals between the steel bars

Interval (m)

Figure 7. Actual conditions of execution at the job sites

3. Generally the experimental and calculated values of the yield load are the same; the experimental value tending to be slightly smaller than the calculated one.

5.2 Theory and results of in-situ survey

If conditions are constant, it is expected that, theoretically, more steel bars are used in soil than in soft rock.

In Figure 9, the density of reinforcement and the gradient of slope β are plotted on the two axes, vertical and horizontal, and the actual results are plotted by actual safety conditions. The density of reinforcement was found by dividing the total length of the steel bar used on the slope by the slope area. The hatched vertical lines in the two figures refer to average standard slopes (Sectional Committee of Slope Works and Slope Stability 1986). It can be seen that the slopes remain stable even though they were steeper than the standard one.

The solid and broken lines in Figure 9 (a) and (b) are lower limits at which the slopes remain stable. The difference between soil and soft rock is slight. It seems that reinforcement of soil is more effective than reinforcement of rock. We cannot conclude that the method has been over-designed when used for soft rock, because it may have been adopted to cope with other geological requirements (such as fissures).

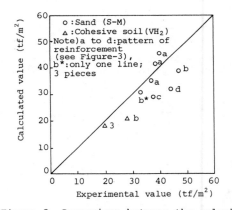

Figure 8. Comparison between the calculated value and the experimental value of the yield load

On soft rock, Mesozoic and palaeozoic strata (sand stone, slate, schist) have many fissures, such as bedding, joint and schistosity, and their fissures are the main cause of slope failure. In this case the number of fissures and the direction of fissures become a problem. It is thought that the number of fissures greatly affects the shearing strength, and that the direction of fissures affects the angle of slip surface, i.e. the sliding force.

Figure 10 and Figure 11 show actual conditions of slopes without reinforcement on mesozoic and palaeozoic strata in

338

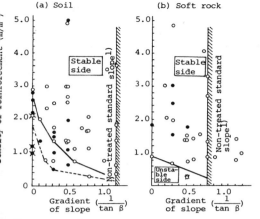

(a) Soil   (b) Soft rock

● : Temporary work
◐ : Permanent work
✕ : Deformation
— : Lower limit line of the execution result
- : Temporary limit line

Figure 9. Comparison of the reinforcing effect of soil and soft rock

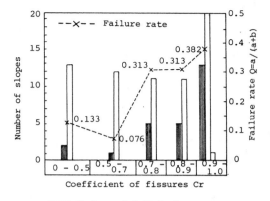

Number of failed slopes   a
Number of stable slopes   b

Figure 10. Slope failure rate by coefficient of fissures (Okuzono 1983)

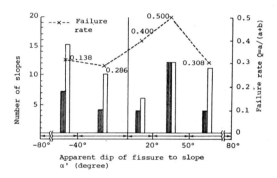

▦ : Number of failed slopes   a
▭ : Number of stable slopes   b

Figure 11. Slope failure rate by fissure dip (Okuzono 1983)

highways (Okuzono 1983). Figure 10 shows the coefficient of fissures Cr (see following formula) calculated from the elastic wave velocity (natural ground: V_{p1}, sample: V_{p0}) which was measured for the natural ground and samples. The figure shows the number of slope corresponding to the coefficient of fissures on bar graphs by failed and stable slope samples.

$$Cr = 1 - (V_{p1}/V_{p0})^2$$

The broken lines in the figure are a failure rate (Q). The values were obtained by dividing the number of failed slopes by the total number of slopes for that value of Cr. The value indicated an index of failure. The graph shows that Q increases as Cr increases.

The results are plotted against α' on Figure 11. α' is the apparent dip of fissure, which is the angle between the main fissure (bedding, schistosity) and a level surface on the cross section. Similarly, the failure rate for values of ' is shown by broken lines. The highest failure rate occurs particularly between 0° and 50° on the right side (dip slope) rather than the left side (receiving slope).

On the basis of the primary factors obtained and the failure rate, the evaluation marks of the two primary factors of each slope without reinforcement, which were found by the survey of actual condition, were added together. The added values are referred to as the evaluation

marks of each slope (total evaluation marks).

In Figure 12, horizontal and vertical axes show the total evaluation mark and the gradient of slope, respectively. The evaluations of the actual slope, stable or unstable conditions, are plotted. As shown, the upper right side of the broken lines shows slopes (mark:•) on which failure occurred, and the lower left side of the broken lines shows the stable slopes. Consequently, in the case where slopes without reinforcement were designed, the slopes steeper than the angles of the lines were unstable. The design of such slopes was unavoidable.

On the other hand, Figure 13 shows the actual research results of slopes reinforced by the same method. As shown, most of the plotted marks are over the broken line in Figure 12, the limit line of non-reinforcement. These slopes remained

339

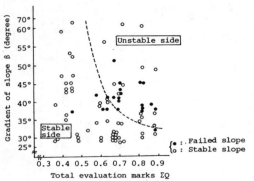

Figure 12. Total evaluation marks and limit gradient of slope of mesozoic and palaeozoic strata (non-reinforced) (Okuzono 1983)

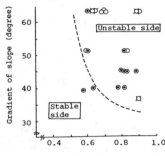

o : Advance measure
⊛ : Measure after failure
口 : Deformation
---- : Stable limit line at the time of non-reinforcement

Figure 13. Relation between total evaluated marks and gradient of slope (reinforced)

stable, even though the slopes were steep. However, some plotted marks, which extend far over the limit line, have deformed (mark: 口). A limit line may be drawn in the vicinity of the deformation.

6 SUGGESTIONS FOR PREPARING A GUIDE

The Japan Highway Public Corporation has been preparing a draft for a reinforced earthwork design guide. We intend to introduce a design procedure.

First, consideration is given to the size of failure that would occur on an unreinforced slope. The degree of failure is classified into three types: small (several tens of m^3), medium (several hundred m^3) and large (several thousand m^3). In case of a small failure, an empirical design is adopted. In case of a medium failure, a

Table 1. Specifications of empirical desi⟨

Parameter	Range
Drilling diameter	ϕ40mm (equivale⟨ to leg drill)
Diameter of steel bars	D19 ∿ D25
Length of steel bars	2m ∿ 3m
Density of reinforce-ment	one bar/2m²
Angle	from horizontal to right angle

slope is designed by stability analysis. In case of large failure, other methods, such as anchoring or recutting, is studie⟨ The empirical design in the case of a sma⟨ failure is generally according to Table 1 from the results of section 4. Stability analysis is calculated by failure types a⟨ shown in Figure 2.

7 ACKNOWLEDGEMENTS

The paper is a part of a "Research on reinforced slopes with steel bars" by the Japan Highway Public Corporation which started in 1982.
We are indebted to Dr. Gosaburo Miki, Professor Emeritus of Tokyo University, Dr. Fumio Tatsuoka, Associate Professor o⟨ Tokyo University, and Dr. Toyotoshi Yamanouchi, Professor of Kyushu Sangyo University. In addition, we appreciate t⟨ cooperation of Sumitomo Construction Co., and many other consultant companies in th⟨ experiment and research.

References

1) Sectional Committee of Slope Works and Slope Stability. 1986. Guide of Road Earthwork-Slope works and Slope Stability. Tokyo: Nippon Doro Kyokai.

2) Okuzono, S. 1983. From Design of Cutting Slope from Maintenance & Management. Tokyo: Kashima Shuppankai⟨

International Geotechnical Symposium on Theory and Practice of Earth Reinforcement / Fukuoka Japan / 5-7 October 1988
© 1988 Balkema, Rotterdam. ISBN 90 6191 820 0

Reinforcement of soil slopes by electrochemical methods

A.K.S.A.Perera
Open University of Sri Lanka, Nugegoda, Sri Lanka

ABSTRACT: This paper deals with a theory supported by practice and results of a method of stabilizing potentially unstable slopes of clayey soils by electrochemical methods. This investigation has established that the reinforcement of the soil massif lying on the pre-existing slip surface of ancient buried river bed by electrochemical methods combined with electrosilication have been found to be effective. As a result of this methodology of reinforcement, the strength of the the soil massif has increased by 1.5 - 2 times from the initial values. A reduction of humidity was also observed in and around anode and cathode locations by 7% - 4% respectively. Calculation carried out on the stability analysis of the slope after the electrochemical reinforcement show an increase of the factor of safety of about 1.25.

1 INTRODUCTION

Electrokinetic and electroosmosis phenomena can be observed in the higher dispersed soils under the action of direct electrical current. Passage of water through the interconnected pores between the solid particles is initiated not only by the hydrostatic or hydrodynamic pressures, but also due to the action of various physical and physico-chemical gradients which can be classified according to Sergev(1983) as:

1. Gradient of the field caused by direct electric current (electro osmosis).

2. Gradient due to the concentration of dissolved electrolytes (capillary osmosis)

3. Temperature gradient (Thermo osmosis)

During the passage of water between soil particles, displacement of liquid takes place along the surface of the soils under the influence of the surface forces. In clay soils with the particles of high specific surface, the flow of water takes place with velocities exceeding several times the velocity of the ionized liquid

cementing the loose soil particles. Simmultaneously the potential of the electric field created by the metallic electrodes (Fig.1) emits ions from the metallic surface. These cations easily bind with the molecules of soil particles which finally reinforce the soil. The main disadvantage of both electroosmosis and electrolysis processes is the emission of heat. (Reauter 1978).

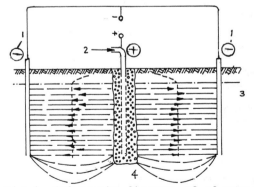

Fig.1 Schematic diagram of electrochemical reinforcement of soils. 1-Cathode, 2-Injection of solution 3-water table, 4-Anode filter, - - Direction of stabilization.

2 ELECTROCHEMICAL REINFORCEMENT

Artificial reinforcements of soils in the massif of potentially unstable slopes by electrochemical methods allow to control the over all stability of the slope. In order to improve the effectiveness of the electrochemical method of reinforcement of soils with different coefficients of permeability, the following methods were used in this work.

1. Reinforcement of soils with help of the emission of ions from the metallic surface of electrodes

2. Reinforcement with the help of solutions of electrolytes applied through the tube of the anode.

3. Reinforcement with the help of the quick hardening solutions applied on soils through the tube of anode or cathode.(liquid glass)

The investigation was carried out on the massifs of the potentially unstable slope of non uniform geological structures. In the area of slope failure,loamy clay of 7-10 M thickness was lying on Quartenary loams of 3-6 M thickness. It has also been found that the existance of buried ancient river beds under the slopes was mainly responsible for causing slope failures. In this pre-existing potential slip surface, a sudden increase of the humidity has been observed. The study area consisted of soils under the unfavourable geotechnical and hydrogeological conditions. The methodology developed for this study area is mainly dependent on the coefficient of permeability and the particle size distribution. It has been found that the soils of the slope consist of clay particles from 15-40% and the loamy clay particles of 22-70% with respective coefficient of permeability < 0.005 M/day and 0.5 to .005 M/day Under these conditions both electrochemical and electrosilication methodologies have been used to obtain maximum effectiveness.

3 MODEL STUDIES

In order to analyse the stress-strain state of the massif after the electrochemical reinforcement, model studies were carried out using model slopes prepared from optically sensitive materials (Illin 1985). Investigations have been carried out on 18 models of slopes corresponding to heights from 8 to 15 M. The main task of these investigations was to establish the most effective configuration of electrodes and also to establish the positioning of such configurations in the locations of the slope to obtain maximum effectiveness of the reinforcement. It has been found that the form of a hexagon with six anodes positioned at its corners with the cathode installed in centre was quite effective. Finally it has been found that the location of such a pattern is quite independent of the height of the slope. In these studies probable slip surfaces were drawn and the factor of safety (F) calculated. Fig.2 shows the most favourable positioning (in respect of the factor of safety) of electrodes for the slope of 10 M high. Fig.3 shows the changing of the factor of safety in respect of the distance of the configuration of electrodes positioning from the bottom of the slope of this model.

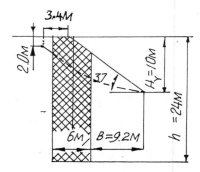

Fig.2 Positioning of electrodes for optimum effectiveness.

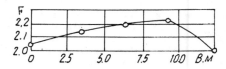

Fig.3 Plot of factor of safety (F) against the horizontal distance (B) from the toe of slope.

4 PROCESS OF REINFORCEMENT

During the reinforcement of the non operational benches of the open pit mine,a number of anode-cathode groups with the cathode positioned in the centre of the bore holes and the anodes in each of the corners of the hexagon were introduced as shown in Fig.4a. When the height of the slope exceeds 10 M,two rows of similar configuration were used (Fig.4b). Model studies conducted (Fig.2) for a slope with a height of 10 M showed that it is more effective to position electrodes partially under the inclined surface of the bench. This was necessary to reinforce the soil massif which is situated within the zone of active stresses.

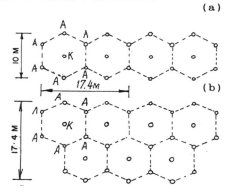

Fig.4 Plan of electrodes configuration in slope

The particular electrodes configuration and positioning adopted resulted in fully utilizing the electro potential energy and intensifying the electrochemical processes taking place within the space between the electrodes. Volume of soils thus reinforced was confined to the blocks of soil mass enclosed between the positive and negative electrodes (Perera 1983). The electrodes were made of metallic tubes provided with casing which pass through the full length of the bore holes. Diameters of anodes were between 42 - 50 mm and the cathode diameters ranged from 108 - 112 mm. Reinforcement of soils have been carried out in 67 blocks having an approximate volume of 80,000 cubic meters.

Because of the non uniform lithological profile and the existance of soils with different values of permeability, electrochemical and electrosilication methods of reinforcement of soils were also used. A 18% - 20% strength solution of calcium chloride and the liquid glass of density (1.05 to 1.10).10 Kg/M^3 were used for chemical injection. An electro potential of 320 volts was applied between the electrodes which maintained the potential gradient of 0.7 - 0.8 V/cm. This gradient of potential was maintained on each block of soil for a period of 360 hours.

As a result of this method of reinforcement the strength of the soil massif was observed to have increased by 50% to 75% of the initial value. A reduction of humidity was also observed in and around anode and cathode locations by 7%-4% respectively.Calculations carried out on the stability analysis of the potential slope after the electrochmical reinforcement of soils show an increase of the factor of safety by 25%. In addition, this methodology enabled construction of slopes with steeper angles. This helped to cut down the expenditure for additional stripping works. Further it was possible to increase the slope angle from 22^0 to 30^0 (Fig 5) and effected saving on expenditure by cutting down cost of extra stripping works by 51 cubic meters per meter length of slope (Perera 1986).

In order to evaluate the effect of the electrochemical reinforcement, experimental bore holes were drilled in the reinforced massif. This was carried out to cover areas between anodes and cathodes. Core samples were subjected to different laboratory tests for the determination of the strength properties. Calculation of the physico-mechanical characteristics of the massif before and after the reinforcement shows considerable increase of the strength properties of soils (Table 1).

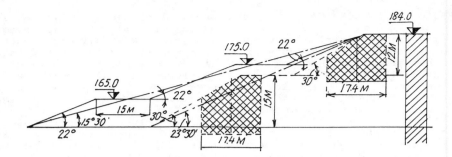

Fig.5   Diagram showing increase of slope angle obtained by electrochemical reinforcement of soil

Table 1.

Characteristics	Massif before Reinforcement	Massif After Reinforcement
Cohesion MPa	$3.0 \cdot 10^{-2}$	$6.0 \cdot 10^{-2}$
Angle of internal friction	$21°$	$30°$
Density Kg/M^3	$2.0 \cdot 10^{-3}$	$2.1 \cdot 10^{-3}$
Factor of safety	1.00	1.25
Humidity around A.	25%	18%
Humidity around C.	25%	21%

5 CONCLUSIONS

1. Basic theories of soil mechanics and physical chemistry are two important disciplines for the formulation and investigation of reinforcing potentially unstable slopes by electrochemical methods.

2. The concentration of tangential stresses is located in the depth range of 0.9 to 1.2 of the height of the slope as determined from model studies. The depth of the electrodes should therefore penetrate to a distance exceeding the height of the slope.

3. The reinforcement becomes more effective when the system of electrodes configurations is located in the regions covering the area where the soil slope intersects the horizontal top surface of the massif as shown in Fig.5.

4. Results obtained after electrochemical reinforcement was carried out, show that the soil strength increased by a range within 20% to 50% of the initial values. It was also observed that the humidity levels around the anode and cathode decreased by 7% and 4% respectively.

5. Electrochemical reinforcement methodology can be effectively used to control the stability of potentially unstable slopes situated on buried river beds where the presence of additional humidity can aggravate the stability criteria of the slip surfaces.

6. Steeper slopes can be formed using this particular methodology and extra expenditure for additional stripping works curtailed on of costs that would otherwise result in stripping, milder slopes in the open pit mining industry.

ACKNOWLEDGEMENT

The author wishes to express his gratitude to Prof. A. Thurairajah and Prof. K.B.E. Karunaratne for reading the draft manuscript and offering a number of suggestions for improvement. The author also wishes to thank Miss Rani Ponnamperuma for her assistance in preparing this paper.

REFERENCES

Galpirin A.M. and Perera A.K.S.A. 1983. Prognosis of dewatered rock masses in mining construction works. International symposium of the Association of Engineering Geologists, Lisbon, Portugal.

Illin A.I., Galpirin A.M., Streltsov V.I. 1985. Control over the long term stability of quarry slopes. Moscow, Nedra, USSR.

Perera A.K.S.A. 1986. Control over geomechanical processes in excavated slopes. Asian Regional symposium on geotechnical problems and practices in foundation engineering. Vol. 1, Colombo Sri Lanka : 91 - 99.

Perera A.K.S.A. 1986. Conrol over geomechanical processes during the formation of quarry slopes. Proceeding of the international symposium on geotechnical stability in surface mining. Vol. 1, Calgary, Canada: 305 - 312.

Reuter F., Klengel K., Pasek J. 1978. Ingenieur Geologie. VEB Deutcher Verlag Fur Grundstoffinndurie. Leipzig, G.D.R.

Sergev E.M. 1983. Gruntovedeniya. Moscow State University Press, Moscow, USSR.

International Geotechnical Symposium on Theory and Practice of Earth Reinforcement / Fukuoka Japan / 5-7 October 1988
© *1988 Balkema, Rotterdam. ISBN 90 6191 820 0*

Proposal for design and construction of reinforced cut slope with steel bars

N.Sano
Laboratory of Japan Highway Public Corporation, Machida, Japan

T.Kitamura & K.Matsuoka
Japan Highway Public Corporation, Tokyo, Japan

ABSTRACT: This paper describes the outline of a reinforced cut slope with steel bars in "The guide of design and execution of reinforced earthwork" compiled by the Japan Highway Public Corporation. For the reinforcement method on cut slopes, steel bars are mainly used as reinforcing materials. Applicable conditions, method of survey, design and construction are given in the guideline. The design method based on the past experiences or the one based on the calculation of stability is chosen depending on the scale of the expected failure. As for the construction method, a procedure is given and the main points are the control of construction work, and the field observation.

1 INTRODUCTION

Our concern in the report is to introduce an outline of reinforcement methods of natural ground in "The guide of design and execution of reinforced earthwork" by Japan Highway Public Corporation. Although some parts of mechanism and design technique in this method has still not been made clear, this method will be widely used for construction of roads, especially in mountainous districts. Here, when the method is applied for a design on the basis of present technical level, only a general idea for the design is suggested. Consequently, if the design is to be executed, it is necessary to study it well.

2 OUTLINE OF REINFORCEMENT METHOD

2.1 Features of reinforced earthwork

To strengthen natural ground with reinforcing materials, steel bars and similar materials are installed into the ground. These members provide tensile reinforcement and/or shearing reinforcement, and reinforced ground and the reinforcements act effectively with each other.

The features of the reinforcement method are as follows.

① Reinforced slopes excavated in stages can largely control the looseness of natural ground due to excavation.

② The method is very safe and economic because the length and intervals of reinforcing materials can be changed properly during excavation while conditions and deformation of the ground are observed.

③ For a delayed deformation after completion of the reinforced slope, counterplans and additional reinforcements are easily applied.

④ Because reinforcing materials are comparatively short and light, they are suitable for application even at sites which large machinery can not enter.

2.2 Selection of method and materials

The reinforced earthwork method has various features as described above, and is used for various applications as shown in Figure 1. It is necessary to select the proper method and materials on the basis of site conditions (including temporary or permanent use), resistance force necessary for reinforcing materials, reliability of design method, and durability and economical efficiency. General cautions on selection of reinforced earthwork are as follows.

① For permanent slopes, it is desirable to avoid designing steep slopes with the reinforced earthwork method. Eventually, it can be thought that application of the reinforced earthwork method is to complement the stability of a slope with the usual gradients.

② On long-term durability and deterioration of reinforcing materials, such as that caused by weathering of the ground,

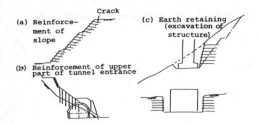

Fig.1 Application example of reinforcement method for natural ground

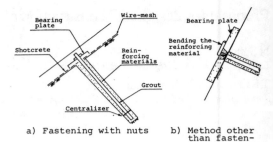

a) Fastening with nuts   b) Method other than fastening with nuts

Fig.2 Basic structure of reinforcement method for natural ground

some problems remain because the zone of reinforced ground is rather shallow. Consequently, it is necessary for repairing techniques or methods for that trouble to be considered in advance and its consideration is reflected in the design.

③ It is necessary to select appropriate materials for the reinforcement purpose, though there are many kinds of reinforcing materials.

2.3 Basic structure

As shown in Figure 2, the basic members of the method are divided into three sections: reinforcing materials, grout and bearing plates. As reinforcing materials, rock bolts, steel bars, steel pipes and others are used. Cement mortar, cement slurry and other materials are generally used for grout. By connecting the bearing plate to the reinforcing material, the reinforcing effect can be increased; however the quantitative evaluation of the effect has not been done yet. In a case of the cut slope, sometimes shotcrete is used for surface protection; however, various types of slope protections, such as concrete crib work, have been used.

2.4 Applicable slope types

In this guide, whether or not reinforced earthwork with steel bars can be adopted, depends upon the estimated extent of surface slope failure.

Further, it is important for the design method to be based on the idea of estimating the shape of failure zone of the slope.

Since slope failure is caused by a combination of natural ground conditions, slope shape and external force, their combination should be used in estimating the extent and shape of the failure zone.

3 SURVEY

3.1 Geological survey

Generally, cut slopes do not consist of homogeneous soil; they consist of heterogeneous soils. Most of them show complicated geological sections. Difference in the section changes the sliding force and the sliding shape, which are the basis upon which the length, interval and direction of reinforcing materials are decided. Therefore, in the case of designing reinforcement, it is necessary to find the soil properties, and to survey geological formations (section and shape of layer).

3.2 Pull-out test

In reinforced earthwork, it is important to estimate pull-out resistance, and this is also an indefinite factor. The tensile force of reinforcing materials used for design, is decided by means of a pull-out test of the reinforcing materials before construction. Since the pull-out resistance of reinforcing materials is affected by the construction methods, such as types of ground, qualities of grout, grouting method and types of drilling, the test shall be carried out on the same conditions of those of the construction. In this case, since it is supposed that the pull-out resistance is not proportional to the length of the reinforcing materials, it is necessary to test reinforcing materials of different lengths.

4 DESIGN

4.1 Design procedures

Figure 3 shows the basic flow of the design selection of reinforced earthwork with steel bars. From the slope conditions and from the standpoint of whether a small failure occurs or not, the design is decided; one is a design based on experience. The other is a design in which mechanical stability is confirmed by means of stability analysis. The small failure shown in Figure 3 is the case in which there is a possibility of failure of around 2 m depth.

Furthermore, since the application area of the empirical design overlaps the application area of concrete block retaining wall works or relatively simple retaining wall works, proper use of these methods should be considered carefully.

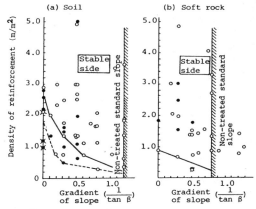

(a) Soil (b) Soft rock

●: Temporary work
o: Permanent work
×: Deformation
—: Lower limit line of the execution result
----: Temporary limit line

Fig.4 Comparison of the reinforcing effect of soil and soft rock

Table 1. Specifications of empirical design

Parameter	Range
Drilling diameter	ϕ40mm (equivalent to leg drill)
Diameter of steel bars	D19 $\sim$ D25
Length of steel bars	2m $\sim$ 3m
Density of reinforcement	one bar/2m^2
Angle	from horizontal to right angle

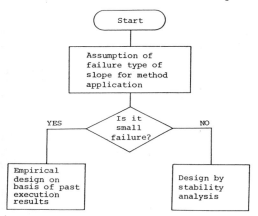

Fig.3 Flow of design procedures

4.2 Empirical design method

According to statistics of the relation between frequency of failure occurrence and failure depth of the steep slope, about 80% of the failures occur with a depth of 2 m or less. This means that relatively short reinforcing materials can restrain against most of the failure. For the ground which leg-drill with short rods can drill, this method is very effective because of its simplicity and economical efficiency.

Further, for the decision of lengths and intervals of reinforcing materials, Figure 4 is instructive. Reinforcement density in the figure is a value that is obtained by dividing the total length of inserted steel bar by the cut slope area. The limiting line of the execution result is the border between the stable side and the unstable side from the data of actual works.

Reinforced earthworks for these failures can be decided based on experience without special calculation because there are small differences, according to past experiences. Table 1 shows the specification of the empirical design method based on past executions.

4.3 Design method by stability analysis

4.3.1 Standard of design

The design method by stability analysis is based on the ultimate equilibrium method except in special cases. And, the results of this method have to satisfy at least the specification shown in Table 1.

(1) Safety factor

The safety factor of reinforced earthwork shall be considered, on the failure of reinforced slopes and reinforcing members.
① Planned safety factors for failure of a reinforced slope
Planned safety factors for failure of a

349

reinforced slope are treated in the same way as the case of a slope without reinforcement; it shall have an FS of 1.2 or more.

② Material safety factors of reinforcing members shall be considered as the material characteristics and pull-out resistance of steel bars, etc.

a. Material characteristics

Material characteristics shall be confirmed for tensile force and shearing force. The allowable stress shall be one of the values described in "Concrete Standard Specifications (Japan Civil Engineering Society)."

b. Pull-out resistance of reinforcing materials

In principle, the pull-out resistance should be determined after conducting the pull-out test. The safety factor of allowable pull-out force is 2.0 basically, however, in case of temporary works, the safety factor can be lowered to 1.5.

c. Others

On factors of shotcrete and members of reinforcing steel bar head, although safety factors are not clear, needed stability may be secured according to the construction method described later.

(2) Decision of reinforcing mechanism

This guide explains the design method based on tensile reinforcement or shearing reinforcement from among the several reinforcing effects. A detailed evaluation of each reinforcing effect is given in the following paragraphs.

a. Evaluation of tensile reinforcing materials

The tensile force of reinforcing materials which acts on slip surface is shown in Figure 5.

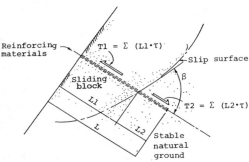

Fig.5 Evaluation of tensile reinforcing materials

Where, the length of sliding block is L1, the length of stable natural ground is L2, the allowable pull-out resistance of each soil layer is τ.

The maximum tensile force obtained from the sliding block is T1. The maximum tensile force obtained from the stable natural ground (pull-out resistance) is T2. The allowable tensile force is T3.

The above items can be considered, and the one with the smallest value is adopted.

b. Evaluation of shearing resistance of reinforcing materials

Although reinforcing materials basically shall be evaluated for tensile materials, reinforcing materials shall be evaluated as shearing resistant reinforcing materials in a case where slope ground consists of rock.

4.3.2 Stability analysis

(1) Study by circular slip failure

Study by circular slip shall be applied in a case where sliding surface is expected to be circular. On a slope which consists of uniform soil, and there is no external force to control the sliding shape, the sliding surface becomes a circular arc.

In the case of cohesive soil and sandy soil ground, the reinforcing materials are not sheared simply because the slope is not a simple sliding surface. Rather, it is thought that tensile force controls most of reinforcing effect, and tensile resistance reaches its limit before the shearing resistance acts effectively. Consequently, in a stability calculation, it is thought that the tensile force of reinforcing materials is acting on the base of sliding block of reinforces the ground.

(2) Study by wedge type slip failure

The study by wedge type slip shall be applied in the case where the sliding surface is considered to be a straight line. In case the cut slope consists of rock with many fissures, joints and bedding, the slope slips down, or is pushed out along these fissures. In that case, most sliding surfaces are regarded as straight lines. It is thought that not only the tensile force, but also the shearing force, acts on the reinforcing materials arranged throughout the sliding surface.

Consequently, the stability analysis is studied from the balance of the sliding force, shearing resistance of the soil on the sliding surface and the shearing resistance of reinforcing materials. Reinforcing materials shall be well arranged inside the sliding surface.

(3) Study by earth pressure

The study by earth pressure shall be applied to the place where gravity retaining wall or concrete block retaining wall works are to be constructed.

The study is based on a method in which reinforced natural ground can be replaced by a gravity retaining wall, which has some limitation. For the imaginary wall, the same stability analysis (check for sliding, overturning and bearing failure) that is used for gravity retaining wall, shall be performed. However, in this case, it must be confirmed that there is no external slip failure slide at the outside of the reinforced ground. In addition, a stability study for slide in the reinforced area is based on items (1) and (2) already mentioned.

Table 2. Relation between sliding shape and calculation technique

Calcu-lation tech-nique	① Study by circular slip failure	② Study by wedge type slip fail-ure	③ Study by earth-pressure
Model example	Circle center		Presumed breaking, reinforcing materials

5 CONSTRUCTION

5.1 Construction procedure and considerations

5.1.1 Construction procedure

First, an operation, which is drilling and insertion of reinforcing materials and grouting and fixing of head plate, shall be done after the top of the slope (the first stage from top of slope) is excavated. After that, the same operations are repeated with inserting the reinforcing materials in one or two lines. As shown in Figure 6, the excavation in each stage shall be the depth at which the slope is able to stand alone. The depth shall be decided by the test excavation or by the observation of earthwork during construction; generally its height should be 1.5 to 1.7 m on the basis of construction efficiency. When the excavation depth is great, even in the case of natural ground which is able to stand alone, scaffolds become necessary when reinforcing materials are inserted.

The reinforcing effect is provided by tensile force and shearing force of reinforcing members caused with release of stress or small displacement of ground at the time of excavation. Consequently,

steel bars shall be installed and shotcrete shall be applied as soon as possible after excavation to avoid loosening of the slope by release of stress or erosion and weathering.

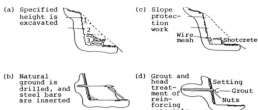

Fig.6 Details of reinforcing method

5.1.2 Drilling and inserting reinforcing materials

The cost and schedule of reinforcing earthwork are greatly influenced by drilling. Drilling efficiency is controlled by ground conditions and construction and scaffolding conditions. The type of drilling machine is limited by the above conditions. Since the properties of soil and geology of the natural ground are very complicated, drilling tests should be done in advance whenever possible.

After drilling, reinforcing materials with grout tube are inserted. After that, the holes shall be grouted as soon as possible before the hole-wall collapses (or after grouting, reinforcing materials can be inserted). Reinforcing materials shall be handled so that rust, oil and dirt, which reduce adhesion, do not adhere to them.

5.1.3 Slope protection work

As shown in Figure 6, slope protection shall be done as soon as possible after cutting the slope. In the case where the slope fails, sometimes small failure will precede large failure. Therefore, when slope protection is done, first of all, it is very important to protect against small failures. For this purpose, slope protection work must be done. Although there are many kinds of slope protection, shotcrete with concrete or mortar should be suitable because these methods can be executed immediately. The effects of the shotcrete method are to unite the earth with the reinforcing materials; heads of reinforcing materials are bound with earth. The reinforcing materials are also bound together. At this time, drain pipes or drainage ditches shall be provided to restrain water remaining in the slope when necessary.

351

5.1.4 Grouting and head treatment

Except where steel bars are driven into slopes directly, cement milk or cement mortar shall be grouted at low pressure before or after the steel bars are inserted. At that time, quality control of the grout materials and control of the grouting volume shall be done. Further, loosening of the natural ground by grouting pressure must be avoided, and an impermeable zone shall be made by grout leaked into the natural ground. Before grouting, any slime in the holes must be removed.

In a case where the head of reinforcing materials is fastened by nuts, plates shall be set at the head of steel bars, and they shall be fastened with nuts 24 hours after grouting. The torque shall be such that plates make contact with the slope tightly, but a large tensile force shall not occur in the steel bars.

5.2 Execution management

The items covered by the execution management are the shape, dimensions and quality (Japanese Industrial Standard) of the reinforcing materials, combination and strength of the grout materials, as well as the pull-out resistance of the reinforcing materials. The pull-out resistance of the reinforcing material must satisfy the design value within three days after construction.

5.3 Field observation

Reinforced earthwork with steel bars has unsolved problems, and it will be necessary to deal with the change of natural ground conditions. Consequently, an observational procedure like New Austrian Tunnelling Method of tunnels is recommended. In the case where an excavated slope is large and unstable without some protection works, or an influence on important structure in the vicinity of the slope is foreseen by excavation, survey by instruments is recommended. Further, when design is done by stability analysis, since there are still some unsolved problems, at least the surface displacement, the axial tension of reinforcing materials and the underground displacement shall be monitored. In the case of empirical design, the surface displacement shall be measured.

6 CONCLUSIONS

This report summaries research works per-
formed by the committees listed in the References. So far, though various research on reinforced earth with steel bars has been examined, some unanswered questions remain. However, since a favorable effect of the method for slope reinforcement is expected, it is desirable for the questions to be solved through the accumulation of future construction and test results.

REFERENCES

1) Research Committee of the Reinforced Earthwork Method. 1986. A Study of the Reinforced Earthwork Method. Tokyo: Express Highway Research Foundation.
2) Committee of the Reinforcing Method with Steel Bars for Slopes. 1986. A Study of the Reinforcing Method with Steel Bars for Slopes. Fuji: Japan Construction: Method and Machinery Research Institute.
3) Research Committee of Measures for Slope Works and Slope Stability on the Highway between Yatsushiro and Hitoyoshi in Kyushu. 1986. A Study of the Reinforcing Method with Steel Bars. Fukuoka: Expressway Technology Center.
4) Compilation Committee of the Reinforced Earth. 1986. The Reinforced Earth. Tokyo: The Japanese Society of Soil Mechanics and Foundation Engineering.

International Geotechnical Symposium on Theory and Practice of Earth Reinforcement / Fukuoka Japan / 5-7 October 1988
© *1988 Balkema, Rotterdam. ISBN 90 6191 820 0*

Analysis of a slope reinforced with rockbolts

Tatsuya Tsubouchi, Yuji Goto & Tetsu Nishioka
Tokyu Construction Co., Ltd, Kawasaki, Japan

ABSTRACT: The stress on a cut slope reinforced with rockbolts was analysed under the finite element method (FEM). In the stress field analysed, the opimization technique was employed in an attempt to find the minimum safety factor of a slip line. In comparison with a slip line examined on an unreinforced slope the slip line on the reinforced slope was deeper in the slope and had a higher safety factor.

1 INTRODUCTION

Many studies have been made on the use of rockbolts to reinforce slopes. However, the mechanics of this type of reinforcement are not totally clear. While there is still uncertainty on many points, the authors believe that the following explanation can be made.

Due to the release of stress when earth is excavated from the face of a slope the ground around the excavated surface deforms laterally. The lateral stress in the ground becomes less than it was before the excavation. Therefore, the resulting stress value on the new slope reaches Coulomb's failure criterion (Fig.1a).

When rockbolts are used to reinforce a newly excavated surface, the ground near the bolts is confined by them and does not easily deform. The result is that the decrease in lateral stress in the ground is less than it is without reinforcement and the stress value does not reach Coulomb's failure criterion (Fig.1b).

This is important in the design of rockbolt reinforced slopes. At present, however, it receives little attention during the design stage. FEM analysis can be used in predicting stress but the data obtained is incomplete. What is required is information concerning the safety factor of the slip line.

In this paper the optimization technique is applied to the stress data obtained through FEM analysis in order to find the slip line having the minimum safety factor.

2 PROCEDURE FOR SLIP LINE SEARCH

In this paper we use the same procedure as Yamagami(1984) to determine the slip lines in the ground under the footing. This is as follows.

The safety factor of a slip line in a given slope(Fig.2) is defined in the following equation using Coulomb's failure criterion.

$$F_s = \frac{\int_s (C + \sigma \cdot \tan\phi) \cdot d_s}{\int_s \tau \cdot d_s} \qquad (1)$$

Where

F_s : safety factor of the slip line
C : ground cohesion
ϕ : internal friction angle of the ground
σ : normal stress on the surface of the slip line
τ : shear stress on the surface of the slip line

and integration is executed along the slip line.

There are numerous slip lines to be considered. The line which has the minimum safety factor is determined by using the optimization technique. Here, Dynamic Programming(1973) is used. Dynamic Programming resolves multistage optimization problems.

In order to apply Dynamic Programming to this problem the appropriate number of stages in a given slope is established as shown schematically in Fig.3. At each stage the appropriate number of states is provided, which is indicated by the points in Fig.3.

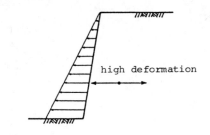

high deformation

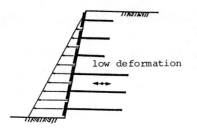

low deformation

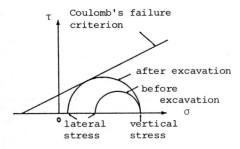

τ Coulomb's failure
criterion

after excavation

before
excavation

σ

0 lateral   vertical
stress   stress

a) unreinforced slope

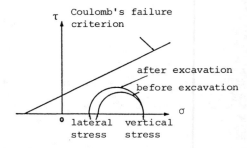

τ Coulomb's failure
criterion

after excavation

before excavation

σ

0 lateral   vertical
stress   stress

b) reinforced slope

Fig.1 Principle of earth reinforcement

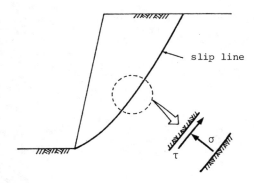

slip line

σ

τ

Fig.2 Definition of the safety factor

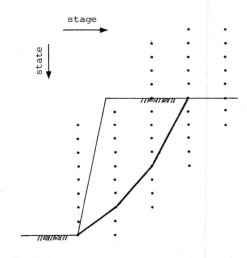

stage

state

Fig.3 Representation of stages and states

Now, we consider a slip line made by connecting points at two arbitrary successive stages as shown in Fig.3.
For this slip line, Eq.(1) is rewritten as

$$F_s = \frac{\Sigma R_i}{\Sigma T_i} \qquad (i=2, M) \qquad (2)$$

Where

$$R_i = \int_{s(j, k)} (c + \sigma \cdot \tan \phi) \cdot d_s \qquad (3)$$

$$T_i = \int_{s(j, k)} \tau \cdot d_s \qquad (4)$$

Here, S(j,k) denotes the line connecting point j at the stage i-1 and point k at the stage i(Fig.4). M is the total number of stages.
In the execution of equations (3) and (4), the stress factors σ and τ obtained from FEM analysis are used. Here, it is assumed that stress is constant in each FEM element

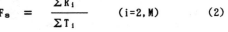

354

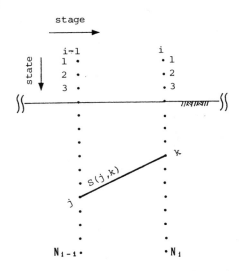

Fig.4 Application of dynamic programming

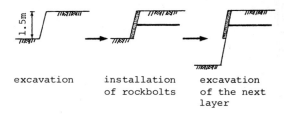

excavation	installation	excavation
	of rockbolts	of the next
		layer

Fig.5 Excavation procedure

Table 1. The material properties adopted for the analysis

Materials	Items	Values	
Ground	Young's modulus	850	(tf/m^2)
	Poison's ratio	0.33	
	Coefficient of lateral pressure	0.5	
	Unit weight	1.8	(tf/m^3)
	Cohesion	1.8	(tf/m^2)
	Internal friction angle	30°	
Bolt	Young's modulus	2.1×10^7	(tf/m^2)
	Diameter	2.5	(cm)
Plate	Young's modulus	2.0×10^6	(tf/m^2)
	Thickness	10.0	(cm)

Further, we define the new auxiliary function G as

$$G = \sum (R_i - F_s \cdot T_i) \qquad (i=2, M) \qquad (5)$$

It is known that minimizing the function F_s in Eq.(2) is equivalent to minimizing the new function G.

According to the "principle of optimality", which is the central concept in dynamic programming, the minimum value of G between the initial stage and point k, the function $H_i(k)$, is given by the sum of the minimum value of G between the initial stage and any state j at the previous stage i-1 and the change in G on passing between the two states j and k. This is expressed as

$$H_i(k) = \min_{1 \le j \le N_{i-1}} [H_{i-1}(j) + D_i(j,k)] \qquad (6)$$
$$(i=2, M)$$
$$(k=1, N_i)$$

where, N_i is the number of states at stage i. $D_i(j,k)$ is the change in G on passing from the point j to the point k and is expressed as

$$D_i(j,k) = R_i - F_s \cdot T_i \qquad (7)$$

After the calculation of Eq.(6) is the final stage, the minimum value of G is obtained in the following equation.

$$G_{min} = \min_{1 \le k \le N_N} H_N(k) \qquad (8)$$

The critical slip line is obtained by tracing back the path which gives G_{min}.

3 NUMERICAL RESULTS

The procedure described herein was applied in the stability analysis of two cut slopes. One slope was unreinforced. The other was reinforced by short rockbolts and large-sized bearing plates placed at 1.5m intervals on the excavation as shown in Fig.5.

The ground stress was analyzed by elastic FEM analysis. Rockbolts were treated as beam elements and plates were treated as plane strain elements. The material properties adopted here are shown in Table 1.

The area where stress exceeds Coulomb's failure criterion is shown in Fig.6. The danger area in the reinforced slope is smaller than that in the unreinforced slope.

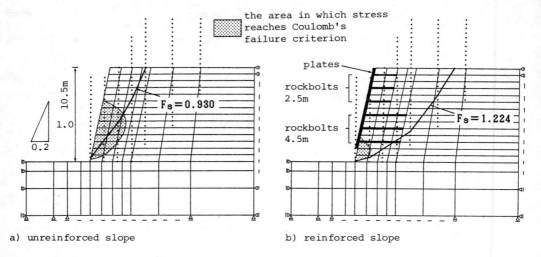

a) unreinforced slope   b) reinforced slope

Fig.6 Results of FEM analysis and search for the slip line

The slip line having the minimum safety factor was searched for in each slope as shown in Fig.6. Compared with the slip line of the unreinforced slope, that of the reinforced slope was deeper and its safety factor was higher.

4 CONCLUSIONS

The procedure of searching for the slip line having the minimum safety factor in the stress field obtained from FEM was adopted for the stability analysis of slopes reinforced with rockbolts.

FEM analysis is used in this procedure. As a result, the interaction between the ground and the rockbolts can be examined.

In the analysis performed herein, it was assumed that there was perfect cohesion between the rockbolts and the ground surrounding them. But in practice this may not be the case. Therefore, the safety factor obtained herein may be somewhat greater than that which would actually be achieved.

In the future, the results of field measurements or experimental studies on the interaction between the ground and rockbolts should be introduced into the analysis performed herein.

For example, field measurements or experimental studies might indicate that less rigid rockbolts could be used.

REFERENCES

Yamagami, T., Ueta, Y. & Koyama, M., 1984
Search for potential slip line by Dynamic Programming. Proceedings of the 39th Annual Conference of the Japan Society of Civil Engineers, 3: 157-158(In Japanese)
Ogata, K., 1973
Dynamic Programming, Baifukan Publishing Company(In Japanese)

International Geotechnical Symposium on Theory and Practice of Earth Reinforcement / Fukuoka Japan / 5-7 October 1988
© 1988 Balkema, Rotterdam. ISBN 90 6191 820 0

Rock anchoring of dosing tank for sewerage disposal

S. Valliappan & V. Murti
University of New South Wales, Sydney, Australia

B. Jayaram
Public Works Department, New South Wales, Sydney, Australia

Ming Bao Cao
Department of Geotechnical Engineering, Tongji University, People's Republic of China

ABSTRACT: An axisymmetric finite element analysis of a dosing tank for sewerage disposal which was carried out to investigate the necessity of rock anchor system to counter the uplift forces, is presented in this paper. The proposed tank site was found to be covered with a thin layer of sand overlying sandstone and conglomerate.

1 INTRODUCTION

This paper presents the results of an axisymmetric finite element analysis carried out for designing the rock anchor system proposed to counter the local uplift forces on the dosing tank for sewerage disposal. The concrete dosing tank is part of the inlet structure associated with the Norah Head outfall tunnel. The proposed tank is 27 meters in diameter and 7.1 meters in height, with a thickness of 600 mm for the floor and 500 mm for the wall. According to a preliminary analysis carried out by the Department, a system of 68 rock anchors was proposed to counter the uplift forces. Fig.1 shows a section of the tank. The anchor layout proposed originally is shown in Fig.2.

An engineering geological investigation by the Department indicated that the tank site was covered with a thin layer of sand upto 1.5 meters thick, overlying sandstone, pebbly sandstone and conglomerate. Due to the nature of weathering, the geotchnical report suggested that a free cable length of 6.5m above a bond length of at least 3 m be considered in the design of the anchor system. It was also proposed that anchors be placed at 3 meter centres providing a maximum working force of 500 kN based on assumed value of 55.6 kPa for the localized uplift pressure.

Thus, the original design of anchor system proposed was based on a spacing of 3m for a major portion of the foundation with 2.5 and 3.5 meters at certain areas near the edge of the tank. In order to determine an adequate and economic design of an anchor system which will counter the uplift forces, based on the interaction effects between

dosing tank, rock foundation and rock anchors, an axisymmetric finite element analysis was carried out.

2 FINITE ELEMENT ANALYSIS

It can be seen from Fig.1 and 2 that the geometrical configuration of the dosing tank and the proposed layout of the anchor system warrants a full three dimensional analysis for an accurate prediction of stability conditions. Due to the high cost and computational efforts involved in such a three dimensional analysis, it was decided to adopt an axisymmetric modelling which is likely to provide approximate but practically acceptable solution.

For the finite element discretization of the tank and the surrounding rock/soil, eight node isoparametric elements were used. The material properties for concrete and steel as well as rock and soil are shown in Fig.3. The proposed layout of the anchors and the loading conditions adopted for the analysis are shown in Fig.3. The proposed layout of the anchor system is similar to the original one but with a larger spacing of 4.33 meters (row-wise or column-wise) and a diagonal spacing of 6 meters. Fig.4 shows the anchor system comprising 32 anchors.

The assumption introduced in the axisymmetric analysis is that the anchors are in fact ring elements rather than individual anchor elements. Thus, for the initial analysis, four sets of ring elements were assumed and the total force required to counter the uplift pressure was determined. From this total force a modified layout of three sets of ring elements as shown in Fig.4 was considered. Fig.5 shows the finite element discretization of the tank, the surrounding soil and rock as well as the rock anchors.

For the modified layout of anchor system, two analyses were carried out - one with and the other without the self-weight of the tank. It should be realized that the critical condition for the design of anchor system is when the tank is empty and hence the weight of water in the tank was not included. However, the self weight of the tank should be included because this gives the final state of stress and deformation for the entire system.

3 DESIGN OF ANCHORS

From the initial finite element analysis with four sets of ring elements, the forces on the four sets were calculated from the stresses. The total force which has to be applied for counteracting the uplift pressure was determined as 18.74 MN.

In the modified layout of 32 anchors, the average force per anchor can be calculated as 585.6 kN. However, it should be noted that the forces should be calculated per set of ring elements rather than the average for a single anchor, taking the whole set in one unit. Thus, for the layout shown in Fig.4, the total force for each set of ring elements was determined and then the force per anchor within that set was calculated. This resulted in a force of 543.5kN per anchor within the inner ring, 550.9kN per anchor within the middle ring and 622.3 per anchor within the outer ring.

From the above calculation of average force per anchor, it was proposed that 6 strands of nominal 12.5 mm diameter be used. The breaking load for 65% in the case of 6 strands is 718kN which is more than the calculated value of 622.3kN per anchor at the outer ring.

4 RESULTS OF STRESS ANALYSIS

The plots of the vertical displacements at the floor of the tank, the horizontal displacements at the wall of the tank, the maximum and minimum principal stresses in the floor as well as the wall of the tank are given in Fig.6 to 9. For the floor, the values have been plotted along a radial line from the centre of the tank to the edge of the wall and for the wall, the values are plotted from base to top.

The curves have been plotted for the following two cases.
(1) uplift forces on the floor and soil and water pressures on the outside of the tank wall

(2) in addition to the above forces, self-weight of the tank and the anchor prestresses

It can be seen from Fig.6 that the maximum vertical displacement for the case (1) is 0.16mm upwards at the centre of the tank floor and the maximum horizontal displacement is 0.52mm inwards of the wall. For the case(2) Fig.7 shows the maximum vertical displacement of 0.10mm occurs at the floor near the wall and the maximum displacement for the wall is 0.55mm inwards. It should be pointed out that the values for the floor displacement is in fact small and wall displacement will be reduced when the tank is filled.

From Fig.8, it can be noted that the principal stress, σ_1 for the case (1) is tensile and the maximum value is 0.2 MPa in the floor and 0.4MPa near the wall. In the wall itself, the maximum is about 0.58 MPa at the junction. The principal stress, σ_2 is compressive with a maximum of 0.36 MPa in the floor and 0.57 MPa in the wall.

The principal stresses σ_1 and σ_2 for the case (2) are shown in Fig.9. It can be seen that the maximum tensile stress in the floor is 0.18 MPa and in the wall, 0.64 MPa. The maximum compressive stress is 0.46 MPa in the floor and 0.53 MPa in the wall.

It should be indicated that these maximum values of tensile and compressive stresses are within the allowable limits for reinforced concrete.

5 CONCLUSIONS

It can be concluded that the original design of the anchor system based on approximate analysis is not economical. An alternate layout of the anchor system proposed on the basis of the finite element method results in an economical system. The results of the finite element analysis of the alternate anchor system along with the tank and the surrounding rock indicate that the displacements and the principal stresses are within the allowable limits.

This investigation indicates that for certain situations such as this, it is essential to consider the structure-foundation interaction effects in order to obtain more realistic solutions.

6 ACKNOWLEDGEMENTS

The authors wish to acknowledge the support given by the Director of Public Works Department as well as the permission given by the Department to publish this work.

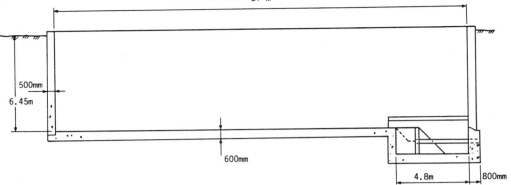

Figure 1. Section of Dosing Tank

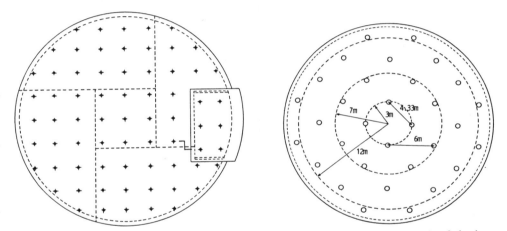

Figure 2. Original Layout of Anchors   Figure 4. Modified Layout of Anchors

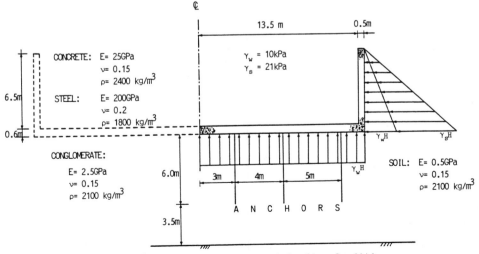

Figure 3. Dosing Tank - Material Properties and Loading Conditions

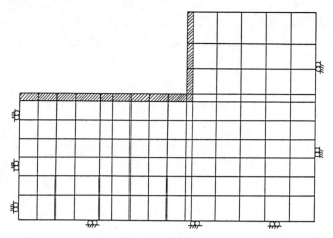

Figure 5. Finite Element Mesh

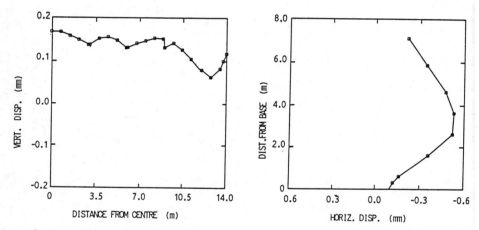

Figure 6. Vertical and Horizontal Displacements - Weight of tank not included

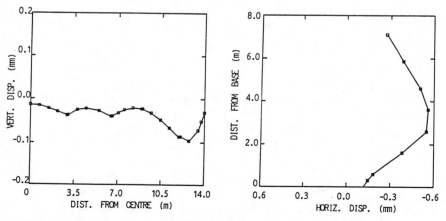

Figure 7. Vertical and Horizontal Displacements - Weight of tank included

360

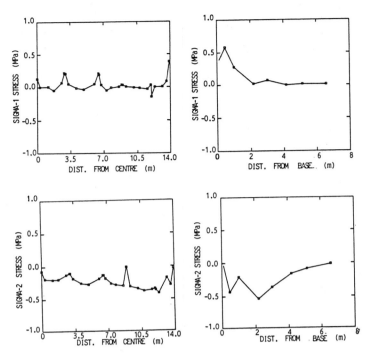

Figure 8. Principal Stress Distribution - Weight of tank not included

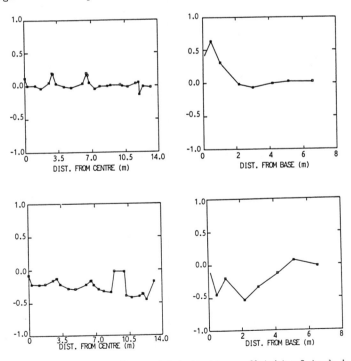

Figure 9. Principal Stress Distribution - Weight of tank included

International Geotechnical Symposium on Theory and Practice of Earth Reinforcement / Fukuoka Japan / 5-7 October 1988
© 1988 Balkema, Rotterdam. ISBN 90 6191 820 0

Insitu ground reinforcement for slope improvement in Hong Kong

K.P.Yim, A.T.Watkins & G.E.Powell
Geotechnical Control Office, Hong Kong

ABSTRACT: The Geotechnical Control Office of the Hong Kong Government is responsible for the Government's long term Landslip Preventive Measures (LPM) Programme. This programme is set up to reduce the risk to the public from landslide through the rectification of unsatisfactory slopes and retaining walls in the densely populated terrain of Hong Kong. Insitu ground reinforcement techniques using micro/minipiles and soil nails have been applied in the LPM Programme for stabilising slopes and retaining walls. The techniques are particularly useful for stabilizing existing slopes where more conventional regrading of the slope is not possible or where access or site congestion pose difficulties.

The paper describes typical design and construction practice for insitu ground reinforcement techniques using micro/minipiles and soil nails in Hong Kong. Some case histories are presented to illustrate the benefit gained by the application of the techniques, the range of the application, the design principles, and the construction methods and costs.

1 INTRODUCTION

Many cut slopes in Hong Kong formed before 1977 were designed empirically to an angle of 10 vertical to 6 horizontal without much regard for geological or hydrological characteristics of the slope (Lumb, 1975). Such slopes generally do not meet current safety standards, and, on investigation, are found to have unsatisfactory factors of safety; accordingly, remedial works are required to upgrade them. By far the simplest means of improving a cut slope is to cut it back to a flatter angle which, despite possibly exposing the slope to a greater risk of infiltration, provides a satisfactory and economic solution. Where space permits, this is the general approach adopted. However, many slopes have substantial buildings above the crest, as well as at the toe, rendering cut back solutions inappropriate. This means that structural buttressing or other support systems is required. More recently, slope reinforcement systems in the form of soil nails, passive tie-backs and pile stitching have become a preferred option. Indeed, in some cases, these methods have been shown to be more economic than a cut back solution.

Many fill slopes constructed prior to 1976 were formed by end tipping. The loose fill was liable to liquefaction following

heavy rainfall, and a number of disasters occurred (Vail 1984). The most common form of remedial work to these slopes has been to recompact the outer 3 m of the slope and provide surface and subsurface drainage. Site constraints do not always allow the recompaction approach and other methods, including reinforcement, have been used. Some of these older fill slopes now have well-established tree cover. Apart from the aesthetic value of the trees, the roots provide additional shear resistance near the surface. Engineering solutions which do not involve removal of the trees are therefore favoured in these circumstances.

This paper presents three case histories of stabilization works where reinforcement systems have been used. Site condition, design considerations, economic considerations and performance testing are discussed.

2 DESIGN PHILOSOPHY

In all of the cases outlined in this paper, additional shear capacity was provided to slopes which were considered marginally stable, but which had not yet failed. The conditions which would lead to failure on these slopes are the result of saturation of the soil and increased pore water pressures. Such conditions are likely to occur only very infrequently, as a result of rare intense rainstorms.

Investigation of the stability of marginally stable slopes in Hong Kong by means of conventional soil mechanics approach, as outlined in the Geotechnical Manual for slopes (Geotechnical Control Office, 1984), often leads to calculated factors of safety at or below 1.0 (Brand, 1985). However, because the majority of older slopes have been subjected to rainstorms of intensities greater than or equal to 1:10 year return period storms and have not failed, an increased degree of confidence can be held in designing remedial works. The Geotechnical Manual recognises this, and allows for remedial works on existing slopes to provide a 20% improvement in Factor of Safety for high risk to life situations, over the existing worst known condition for which a factor of safety of at least 1.0 may be assumed.

There are currently no codes or official guidance on standards to be applied to the design of insitu ground reinforcement when used for the improvement of existing slopes, although a recent publication by the Geotechnical Control Office (1988) deals with new reinforced fill structures.

The design approach has been to determine the force required on any given potential slip surface necessary to increase the factor of safety to the required standard (F > 1.2). The provision of this force is made through the reinforcement, either in shear in the case of minipiles, or through tension in the case of soil nails and passive tie-backs. The design methods for these are briefly outlined in each case history.

The actual margin of safety of the completed works is generally higher than that required to achieve an overall 20% improvement in the factor of safety, since additional allowance is made for possible long term corrosion of the reinforcement. A combination of low design stresses, generally 30% ultimate tensile stress (UTS), together with a sacrificial thickness of 2 mm on each surface is made. All the reinforcement is protected by a grout surround with a minimum thickness of 6 mm. More stringent corrosion protection is provided where adverse ground or water conditions are expected.

The reinforcement is not likely to be stressed to any significant level for the majority of its lifetime, for the reasons outlined above. It is therefore not considered productive to monitor the long term performance of such structures. Rather, a conservative approach is adopted for both design and corrosion allowance. Where

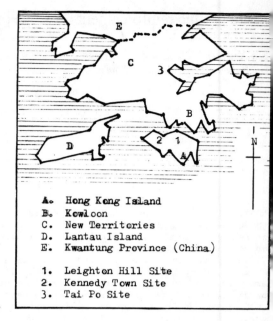

A. Hong Kong Island
B. Kowloon
C. New Territories
D. Lantau Island
E. Kwantung Province (China)

1. Leighton Hill Site
2. Kennedy Town Site
3. Tai Po Site

Figure 1. Location plan

possible, design assumptions, such as soil/grout bond and lateral resistance are confirmed on a site-specific basis.

3 CASE HISTORIES

Each of the following examples discusses a case history where steel inclusions grouted into drilled holes were used to improve the stability of an existing slope. The first case study deals with the reinforcement of a fill slope by the use of a minipile stitching system. The second case study deals with a case where micropile passive tie-backs were to introduce efficiency to a structural retaining and partial cut back system. The third case study deals with a soil-nailed slope. The location of the sites is shown in Figure 1.

3.1 Leighton Hill fill slope

At the crest of a 22 m high fill slope, the sole vehicular access to a number of residential buildings is supported on a mass concrete gravity wall. The slope overlooks two primary schools built very close to the toe of the slope, as shown in cross-section in Figure 2.

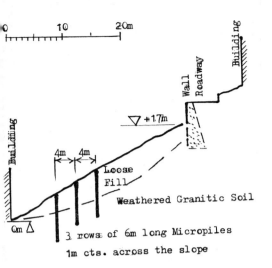

Figure 2. Cross-section of slope at
Leighton Hill

Site investigation showed that the fill
varied between 3 m to 7 m in depth and was
underlain by some 20 m of weathered granitic soil above granitic bedrock. The fill
material was decomposed granite in a loose
state, with an average insitu dry density
of only 1.4 t/m³. Based on consolidated
drained triaxial compression tests, the
shear strength of the fill was found to be
in the region of c' = 0, ∅' = 35°. Stability analysis, using the simplified method
of Janbu (1972), showed low factors of
safety for trial slips in the fill layer.
As the fill slope has a rather dense cover
of trees, a study was conducted to quantify
the tree root reinforcement effect on the
slope stability. This study followed the
procedures described by Greenway (1987).
However, the tree roots were found to be
only effective in increasing shear strength
in the top 2 m. The factor of safety for
deeper slips was still below the minimum
of 1.2 required for existing slopes which
pose a high risk to life. Preventive works
were therefore necessary.

Because of access constraints and the
wish to maintain the existing tree cover,
it was decided not to recompact the slope.
Instead, a design was adopted involving the
installation of minipiles vertically into
the slope face in three rows across the
slope at 1 m centres. Each pile was designed to be 6 m long and to comprise a 16
mm thick steel hollow circular section of
outer diameter of 190 mm encased in a 220
mm diameter grout column. A 2 mm sacrifi-

cal thickness allowance was made for
corrosion.

The design principle was to increase the
shear resistance of the slope by reinforcing the soil in the area of concern. It
was postulated that when the soil moved
over any potential surface, it would exert
pressure on the vertical reinforcement
which, in turn, would mobilize reaction
from the soil to stabilize the slope. The
interaction between the soil and the reinforcement is therefore very similar to that
of a pile subjected to lateral force. The
piles were analyzed as free head piles
using the method of Broms (1964). The
pile was also checked by the method derived
by Ito & Matsui (1975).

Because of concern that drilling might
cause large ground movement in the loose
fill air was used as a flushing medium for
drilling, and the drilling sequence was
staggered so that no drilling was permitted
within a distance of 10 m from any uncom-
pleted piles. The lateral resistance of the
piles was confirmed by jacking the piles
and measuring deformation.

The cost of the completed slope works
was $US50 per m².

3.2 Kennedy Town cut slope

The existing slope was 35 m high, 50° formed in completely weathered volcanic tuff,
below a 25° natural slope. A multi-storey
residential building was located 4 m to 5
m from the toe of the slope. Stability
analysis showed that the factor of safety
of the soil portion of the slope could be
marginal during heavy rainfall.

The conventional solution of cutting
back the slope was only found to be practical in improving that part of the slope
where bedrock outcropped at the toe of the
slope. However, at the central highest
part of the slope, where bedrock did not
outcrop and the groundwater table was high,
stabilization of potential deep-seated
slips extending to the toe of the slope
would have required cutting to a gradient
of 33°. This solution would have meant
substantial excavation into the uphill
natural slope, which would have been expensive, and would have presented problems
in the maintainance of a smooth transition
profile to adjacent parts of the slope.
For these reasons, and because of other
constraints such as access, environmental
disturbance, and the risk of disturbing
boulders on the natural slopes above, the
total cut-back solution was not selected.
The design option selected consisted of a
45° cut back of the upper portion of the

365

central part of the slope, together with provision of a support structure at the toe. This structure comprised a reclining concrete slab cast directly onto the surface of the slope, supported on rock socketed caissons at the toe and with a row of sub-horizontal micropiles acting as a passive anchorage system at roughly midspan of the slab.

Figure 3 provides a cross-section of the slope and illustrates these works.

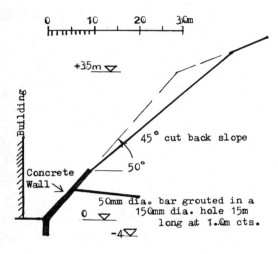

Figure 3. Cross-section of slope at Kennedy Town

The support structure was analysed conservatively as a propped cantilever. It was assumed that, when the soil mass moved over a potential failure surface, it would exert pressure on the reclining slab. The slab would then mobilise resistance from the caissons and the tie-back piles, to restrain the moved soil mass. The earth load on the reclining slab was assumed to act in the horizontal direction. The magnitude was determined by calculating the lateral earth force required to increase of factor of safety of any particular potential failure surface to the required standard. In the absence of knowledge as to how the earth pressure might be distributed, triangular and rectangular earth pressure distributions were both used.

The 1 m diameter rock-socketed caissons, spaced 3 m apart and connected by a capping beam, were designed as laterally loaded piles. Stiffness calculations indicated that the tie-back piles could, within acceptable deformation limits, be expected to resist about half of the earth load and to transmit the load beyond the region where the factor of safety was found less than the required value. The design of the tie-backs followed the anchor design principles outlined in Littlejohn & Bruce (1977) and DD81 of the British Standards Institution (1982). Calculations showed that a high yield steel bar of 50 mm diameter in a 150 mm diameter grouted borehole were required at 1.0 m centres, and that the required movement of the slab at the position of the pile was only some 2 mm to 6 mm in a sub-horizontal direction. This was considered to be an acceptably small movement.

Details of the tie-backs are shown in Figure 4. These are the same as bar anchors, except that the tie-backs were not stressed. Design stresses in the bar were limited to 30% UTS. A 2 mm sacrificial thickness was provided on the steel section and, as an additional protection against corrosion, plastic corrugated sheathing was specified. Testing was carried out to substantiate the pull-out capacity of the tie-back piles. Proof tests to 10% of the piles, whereby a load of up to 1.6 times the design load was applied over 5 cycles, as well, as acceptance tests to other piles where a single load of 1.6 times the design load were applied to the completed pile. Deformation limits were specified as a means of acceptance. The cyclic test sequence is illustrated in Figure 6.

Some thought was given to installing strain gauges to monitor the performance of the wall and tie-back piles in service. However, given that loads in the passive tie-backs would only develop when the factor of safety approached 1.0, it was felt that data useful for design substantiation would not result.

The cost of the slope works described was US$48 per m².

3.3 Retaining wall at Tai Po

The 10 m high stone masonry wall, 1 m thick, was built over 40 years ago against a steep slope cut into decomposed Granodiorite. Apart from the facing stones there was very little interlocking between the blocks which made up the wall as there is no mortar or other packing between the slightly rounded stones which comprised the bulk of the wall. The wall was in poor condition showing signs of lateral and

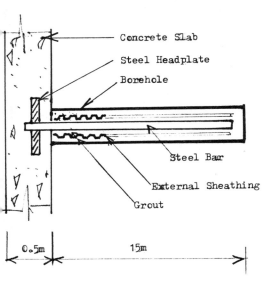

Figure 4. Details of the tie-back micro-pile

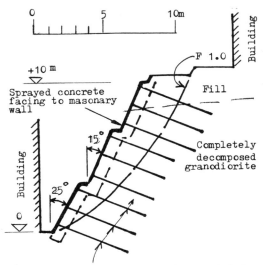

6m long 20mm dia. soil nails grouted in 50mm dia. holes at 1.5m cts. vertical and 2.0m cts. horizontal

Figure 5. Cross-section of slope at Tai Po

vertical displacement. There are 3 storey village houses near to the top and within 1 m from the toe of the wall. Only pedestrian access was available to the site and because of this, together with the very congested nature of the site, the soil nailing design option had a great deal of appeal. The site is illustrated in Figure 5.

The slope behind the wall was analyzed taking into account the effects of the wall as a surcharge, but ignoring any shear capacity it may provide. Site investigation and subsequent monitoring and testing allowed the development of a geological/hydrological model which revealed a factor of safety of only 1.0 when analysed using Janbu's rigorous method.

To design the soil nails a factor of safety of 2.0 was applied to the pull-out resistance between the soil nail and the ground, and the tensile stresses in the nail were limited to 30% UTS. An additional 2 mm was allowed on bar diameter as sacrificial thickness. The bars used were 6 m long, 20 mm dia. high-yield steel bars conforming to BS4449, spaced 1.5 m vertically and 2.0 m horizontally. A 100 mm thick mesh reinforced facing of sprayed concrete was applied to the outer surface of the wall, and head plates bearing against this surface were fixed to the soil nails.

Using the same analytical method, the horizontal force necessary to upgrade the minimum factor of safety to 1.2 was calculated, and a system of soil nailing was designed which could provide this resistance. In the design of the soil nails, a number of assumptions had to be made, all of which are considered to be conservative. These included :

(i) The required loading is distributed in either a rectangular or triangular distribution.

(ii) The boundary between the active and resistant zones is the slip surface with factor of safety = 1.0. This surface was similar to the Coulomb critical wedge surface.

(iii) All the load is applied to the nail through the 100 mm reinforced sprayed concrete facing and head plates on the nails.

(iv) The resistance provided by the nails follows the form :
Resistance = (c' + k Tan u) x Area
where k is a function of overburden pressure and u is a function of $\emptyset$'. Values for these coefficients were based on the measured results reported by Guilloux et al. (1979) and Cartier & Gigan (1983).

The cost of this work is US$65 per m² of slope face.

Tests were required to substantiate design assumptions, and pull-out tests were

367

carried out on at least one bar at every vertical level. These test bars did not form part of the permanent works. Figure 6 illustrates the testing sequence. Acceptability tests, whereby the nails were required to sustain a load of 1.5 times the design load for a period of 60 minutes, will be carried out on some working bars. The acceptability criterion was that deformation should not exceed 0.1% of the bar length.

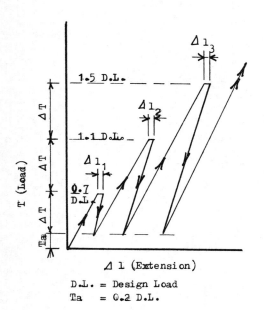

D.L. = Design Load
Ta = 0.2 D.L.

Figure 6. Test loading sequence for soil nails.

4 CONCLUSIONS

The use of soil reinforcement to increase the stability of marginally stable slopes in Hong Kong is relatively new. The method has clear advantages where conventional flattening a slope is not possible and where access difficulties limit the construction alternatives.

As experience is gained through the installation and performance testing of these types of structure in Hong Kong, increased confidence should allow review of the conservative aspects of the design currently being used, and should lead towards the development of Hong Kong Codes of Practice.

5 ACKNOWLEDGEMENTS

This paper is published with the permission of the Director of Civil Engineering Services of the Hong Kong Government.

6 REFERENCES

Brand E.W. (1985). Predicting the performance of residual soil slopes (Theme Lecture). Proceedings of the 11th International Conference on Soil Mechanics and Foundation Engineering, San Francisco, Vol. 5, in press.
British Standards Institution (1982). Recommendations for Ground Anchorages: Draft for Development. DD81:1982, 123 p.
Broms, B.B. (1964). Lateral resistance of piles in cohesive soils. Journal of the Soil Mechanics and Foundation Division, ASCE, Vol. 90, SM2:27-63 pp.
Cartier, G & Gigan, J.P. (1983). Experiment and observations on soil nailed structures. Proceedings of the 7th European Conference on Soil Mechanics and Foundation Engineering, Helsinki, Vol. 2, pp 473-466.
Geotechnical Control Office (1984). Geotechnical Manual for Slopes (Second Edition). Geotechnical Control Office, Hong Kong, 295 p.
Geotechnical Control Office (1988). Model Specification for Reinforced Fill Structures. Geotechnical Control Office, Hong Kong, in press.
Greenway, D.R. (1987). Vegetation and Slope Stability. Slope Stability : Geotechnical Engineering and Geomorphology, Edited by M.G. Anderson and K.S. Richards. John Wiley & Sons, U.K., pp 187-230.
Ito, T. & Matsui, T.(1975). Methods to estimate lateral force acting on stabilizing piles. Soils & Foundations, Vol. 15, pp 43-59.
Janbu, N. (1972). Slope stability computations. Embankment Dam Engineering : Casagrande Volume, edited by R.C. Hirschfield & S.J. Poulos. John Wiley & Sons, New York, pp 47-86.
Littlejohn, G.S. & Bruce, D.A. (1977). Rock Anchors, Stage of the Art. Foundation Publications Ltd., Brentwood, England, 50 p.
Vail, A.J. (1987). Two landslide disasters in Hong Kong. Proceedings of the 4th International Symposium on Landslides, Toronto, Vol. 1, pp 717-722.

4. Embankments

International Geotechnical Symposium on Theory and Practice of Earth Reinforcement / Fukuoka Japan / 5-7 October 1988
© 1988 Balkema, Rotterdam. ISBN 90 6191 820 0

The behaviour of a reinforced trial embankment on soft shallow foundations

R.H.Bassett & K.C.Yeo
King's College, University of London, UK

ABSTRACT: The paper describes the behaviour of a reinforced earth embankment constructed across soft alluvial deposits. The Project involved an instrumented trial embankment – which included monitoring the load in the reinforcement. The key data is presented, the findings derived from the data and the insight gained are discussed.

INTRODUCTION

Stanstead Abbotts lies 20 miles to the north of London. It is bisected by the River Lea running south and the A414 running east-west. The A414 is a primary route within Hertfordshire providing an important link between the nearby towns and an alternative circumferential route round north London. The development of these towns has resulted in a heavy rise of road traffic. The A414 has a poor alignment and a number of secondary and local roads join the route at junctions with poor visibility, further hazards included a gated rail crossing, two pedestrian crossings and a 1:10 hill. The number of accidents caused the council concern and local opinion was sought to choose an acceptable route for a Bypass. The route favoured contained a number of major civil engineering problems; the one which concerns this paper was a 1km length of embankment built across a 4.5m thickness of soft clay/peat/soft clay terminating at the river viaduct. The initial proposal had been for a viaduct crossing the complete valley but this was rejected on grounds of high cost, for similar reasons replacement of the soft materials beneath the embankment was also rejected. An acceptable economic solution required the embankment to be constructed on top of the marshland, thus stability and settlement together with timing became the critical issues. Calculations for settlement showed (i) that the magnitude of the primary settlement would be between 1.2m-1.5m and (ii) that a significant secondary settlement would need to be eliminated to avoid maintenance problems. Natural dissipation of pore water pressure would be too slow due to the sealing effect of the upper and lower clay layers and lateral migration in the peat would give rise to instability at the embankment toe. It was therefore decided that (i) wickdrains should penetrate the two clay layers to increase the vertical permeability and (ii) the fill material plus a surcharge load should be placed as quickly as possible to maximise the pore pressure gradient. These two features would enhance pore water pressure dissipation and induce rapid settlement.

The penalty was the short term instability of the side slopes and a layer of Tensar SR2 reinforcement was therefore included to safeguard against premature failure. As efficient action of these features was critical to both the economic success and the timing of the project, it was decided to construct a fully instrumented trial embankment to provide forewarning of any problems. Fifteen load cells and two pairs of Bison strain measuring devices were installed in the SR2 grid, with funding provided by Netlon and the SERC. The load cells were installed in sets of 3 (Fig.1a) at five locations (Fig.1b). The strain devices enabled direct correlation with the load measurements. The soils data and the instrumentation locations were used as the basis for an international symposium to which eleven predictions and the actual results were presented in September 1986 at King's College London.

The highlights of the measured data are summarised in the following sections.

2 SUMMARY OF THE MEASURED DATA

2.1 Pore Pressure

The main thrust of the field instrumentation was to measure the pore water pressure, its distribution and its decay together with the resulting settlement. Key instrumentation included inclinometers, hydraulic and pneumatic piezometers and a horizontal profile gauge. The data obtained from the piezometers (Fig.2a) showed that the build-up and the dissipation of the pore pressure during the 14 day loading period was very sensitive to the actual construction process (Fig.2b) marked falls being noted during weekends and overnight, this was attributed to the wickdrain system.

2.2 Settlement

At the end of construction the settlement was 600mm, thereafter the shape of the settlement pattern remained consistent, the maximum increasing to 1.2m over 18 months (Fig.3).

2.3 Inclinometers

The inclinometers were located at mid slope (see Fig.1b) and near to the crest of the embankment both showed similar deformation patterns, developing a maximum lateral displacement within the peat layer (Fig.4a). Detailed study of the slope changes indicated that the maximum shear strain occurred at the interface between the peat and the lower clay (Fig.4b). The shear gave an angle change of 11 degrees at this interface and 5 degrees at the embankment/upper clay interface.

2.4 Horizontal Profile Gauge

The horizontal profile gauge showed (Fig.5) that the displacement at mid-slope for the 1 on 1.5 side and 1 on 3 side were 290mm and 150mm respectively. Replotting this information as a direct strain showed zero strain positions towards the two toe areas. A compression zone occurred in the area between chainage 25 & 35m but no instrumentation was placed in the corresponding location beyond the steeper slope. A maximum tensile strain of 3% developed below the crest. The pattern of strain distribution remained consistent thereafter (Fig.6).

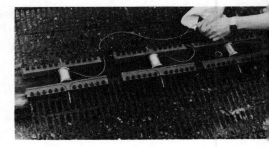

Fig.1(a)   Load Cell Installation

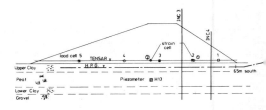

Fig.1(b)   Section of Trial Embankment

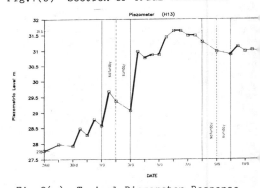

Fig.2(a)   Typical Piezometer Response

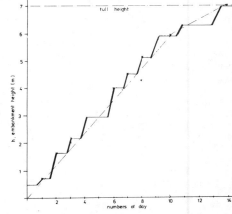

Fig.2(b)   Details of the Constructio Process

372

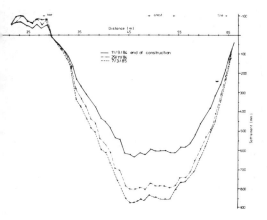

Fig.3   Settlement Profile

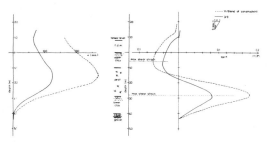

Fig.4(a)   Fig.4(b)

Inclinometer Data

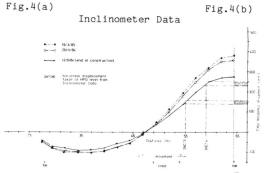

Fig.5   Horizontal Displacements

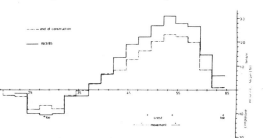

Fig.6   Horizontal Strains

2.5 Load and Strain in the Reinforcement

Data obtained from the reinforcement load cells is shown in Fig.7. The loads developed rapidly during construction but surprisingly continued to increase with time under the subsequent steady state condition. This disagreed with the currently held belief that the maximum should occur after construction and that thereafter stress relaxation would result in slowly falling load. Strain measurement Fig.8, increased from 2% to 3%, i.e. (50%) while the load in the Tensar at the same position increased by 10%.

3 COMPARISON OF LOAD AND STRAIN IN THE REINFORCEMENT AND IN THE SOIL

Correlation between the load and the strain data was assessed by the application of isochronous curves based on Yeo (1985) and Bassett & Yeo, (1988). Figure 9 shows the correlation between the directly measured strains and the strains interpreted from the load at the same point. The correspondence was good. A second comparison is the strain in the foundation soil and that developed by the

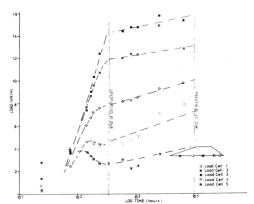

Fig.7   Loads Measured in the Tensar Reinforcement

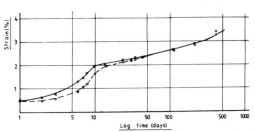

Fig.8   Strains Measured in the Tensar Reinforcement

373

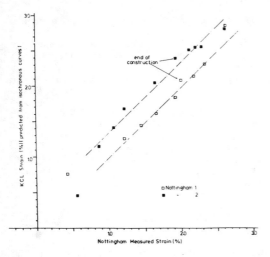

Fig.9 Comparison of Strains Derived from Loads and Directly Measured

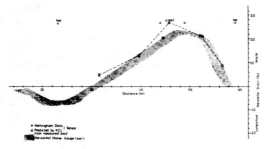

Fig.10 Comparison of Foundation Strain and Reinforcement Strain

Tensar layer. The Tensar strains and those from the horizontal profile gauge are compared in Fig.10. The close correlation implies little or no slip between the Tensar reinforced base layer and the underlying foundation soil.

4 ANALYSES

The analytical approaches adopted by predictors at the symposium divided into two categories (i) conventional, classical solutions and (ii) numerical models.

The conventional predictions divided into (a) those based on stress field solutions and (b) those based on variations of the slip circle analysis. For the particular geometry the stress field solutions generated acceptable answers but for the slip circle methods the shallow depth resulted in large radius circles which gave poor estimates of the forces and

poor estimates of the performance. It was concluded that the slip circle analysis was not satisfactory for the case of a wide load on a shallow foundation.

The various numerical predictions were based on finite element methods, some incorporating simple elastic formulations while others used highly developed elastic/plastic criteria. The predicted values showed (i) that the accuracy of the prediction depended on the interpretation of suitable parameters based on the soil data rather than on the sophistication of the computing model and (ii) that the least satisfactory aspect of the numerical methods was the representation of the drainage behaviour. It was shown that the fill properties have a marked effect on the behaviour. The conclusion reached was that conventional site investigation does not yield suitable parameters for incorporation in the rapidly developing field of numerical analysis.

5 POST CONSTRUCTION DATA

The trial section (Fig.1b) was of restricted width and was designed with one slope at 1 on 1.5 to test the reinforcement principles. In practice the rapid drainage resulted in foundation strengthening even during the construction period. This was an excellent result but as far as research was concerned it limited the significance considerably as the direct necessity for the reinforcement was removed. However, in view of the large deformations it is believed to have been beneficial in that the Tensar acted as a strain distributor and prevented the development of a single shear rupture as observed by Davies (1981) in centrifuge tests of unreinforced embankments on soft ground.

6 BEHAVIOUR OF THE MAIN EMBANKMENT

After 12 months consolidation the main embankment was completed to the north side of the trial, Fig.11. The newly placed material represents an embankment shaped load in its own right as shown by the shaded area. 800mm of vertical settlements occurred between construction in April '85 and May '87 when the surcharge was removed. Removal of surcharge was followed by 50mm of swell back. Fig.12 shows the horizontal tube gauge data for the same period. It can be seen that the whole trial embankment

Fig.11 Dimensions of Additional Main Embankment

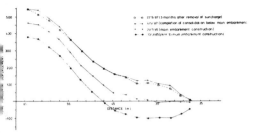

Fig.12 Additional Horizontal Movement

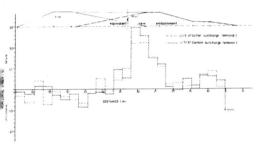

Fig.13 Horizontal Strain in the Foundations

from the 65m reference has been bodily displaced southward by some 150 to 170mm. Fig.13 shows the horizontal strain developed. As with the trial section the maximum tensile strain approached 3%, with small compression zones outside the toe areas. It was surprising that the strain fell to near zero between 12m to 22m, yet outside this zone 1% was developed. There is no obvious reason for this behaviour other than the fact that the 10m strip had earlier been used as a haul road. Fig.14 showed the loads measured in the Tensar SR2 reinforcement (see Fig.1b) over this same period. Cell groups 2, 3 and 4 showed falls in load when the main embankment was constructed, consistent with the compression observed in Fig.13. However, during the subsequent consolidation period they return to a steady rise with log time. Load cell group 5 (near the crest of the "new material") showed a marked rise in measured load. Removal of the surcharge seems to have produced a small fall in load throughout the five sets.

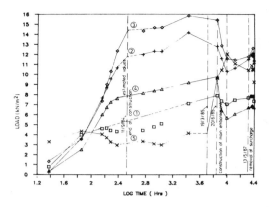

Fig.14 Loads in the Reinforcement During the Main Embankment Work

7 BEHAVIOUR OF THE STANDARD EMBANKMENT SECTION

Other sections of the main embankment, where the whole 60m wide section was constructed on a virgin foundation, were also instrumented for settlement and lateral spread but not for reinforcement loads. In Figs.15 and 16 the overall settlement profile at 26.1.87 and the horizontal strain distribution have been shown for chainage 1890. The settlement profile is nearly uniform at 1.1 to 1.2m between the crest points 24m and 44m respectively, the settlement below the slopes and the berms being pro-rata. The horizontal strain developed is uniform at 2.2% from the berm edge at 49m to the equivalent at 14m. A sharp drop occurred at the inner edge of the berm with a small rise to 1% below the berm shoulders. Assuming correspondence between strain in the horizontal profile gauge and in the Tensar SR2, this would indicate a near uniform load in the reinforcement between 49m and 14m of 14kN/m².

8 DISCUSSION

The most unexpected feature was the increase in the Tensar load during the consolidation periods from 14kN/m to 16kN/m while the strain increased from 2% to 3%. It is thought to be associated with the dissipation of suction within the cohesive fill, although at 3% strain recompacted London clay should be tension cracked and hence imposed no lateral forces on the system. This level of strain and loading was low compared with the recommended design values of 25kN/m to 30kN/m and 5% strain. The long term

375

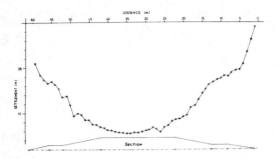

Fig.15   Settlement Under the Main
Embankment

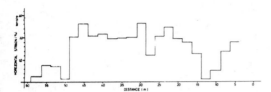

Fig.16   Horizontal Strain Measured Under
the Main Embankment

creep strains and the loads can be
correlated using laboratory isochronous
curves. Tensar SR2 was strong enough and
stiff enough to produce an ideally rough
footing situation and the geometry
allowed the fill material to interlock
within the mesh and thus provided a
non-slip relationship with the granular
material. The behaviour of reinforced
earth was a "self stressing" exercise and
the forces that developed were dominated
by two features (a) the properties of the
fill, in particular cohesion and (b) the
rate of dissipation of pore pressures in
the foundation materials. In the
prototype analysis of this project and in
all the prediction papers an engineering
judgement was applied to the
effectiveness of the wickdrains the
influence was invariably underestimated.
Similarly poor fill properties were
assumed.   Both are conservative
assumptions common to engineering
practice but both have significant
influence and cause considerable
discrepancies between the analytical
prediction and the actual performance.
This typifies the differences been
research for knowledge and the safe
design of engineering works. In theory
reinforcement was not required due to the
rapid consolidation achieved by the wick
drain system however the Tensar mesh
contributed in the manner of distribution
steel in that it maintained the integrity

of the embankment against serious
cracking during construction. Some base
friction must have developed as there was
a 5 degree inward rotation of the
principle strain direction. Rapid
drainage plus base reinforcement enabled
the fill to be placed quickly which gave
the shortest use of the expensive hired
plant and overall saved money.

9 CONCLUSIONS

The final construction cost for the
project using the reinforced earth
solution was £1.11m which compared
favourably with £9.9m for a viaduct
alternative. The project was an example
of how engineering design can provide an
acceptable economic solution which not
only provides the required facility but
minimises detrimental effects to the
overall environment.

10 ACKNOWLEDGEMENTS

The authors would like to express their
thanks to the Engineers of Hertfordshire
County Council for their help and support
throughout this work and to the Science
and Engineering Research Council and
Netlon Ltd., for their financial
assistance.

11 REFERENCES

BASSETT, R.H. & YEO, K.C. 1988. The
Behaviour of Tensar SR2 Geogrid on an
Embankment at Stanstead Abbotts. Report
to the TRRL, CROWTHORN, BERKSHIRE.

DAVIES, M.R. 1981. Centrifugal
Modelling of Embankments on Clay
Foundations. Ph.D Thesis, Cambridge
University.

YEO, K.C. 1985. The Behaviour of
Polymeric Grids Used for Soil
Reinforcement. Ph.D Thesis. Strathclyde
University, Glasgow.

International Geotechnical Symposium on Theory and Practice of Earth Reinforcement / Fukuoka Japan / 5-7 October 1988
© 1988 Balkema, Rotterdam. ISBN 90 6191 820 0

Design and construction of quay wall using geotextile

Didiek Djarwadi
P.T. Bangun Tjipta Sarana, Jakarta, Indonesia

ABSTRACT: The design and construction of quay wall using Geotextile as reinforcement in shore is described in this paper. Design method, fabric selection, backfill materials, method of construction are discussed. These include for example, required tensile strength, type of fabric, resistance against sea water, grain size range of backfill materials and compaction. Attention is given to the compaction of backfill materials below sea water level. The following advantages as fast in construction, lowest in cost comparing with sheet pile walls or piling were also considered.

1 INTRODUCTION

1.1 Description of the project

To the author's knowledge, this paper presents the first use of Geotextile for quay construction in Indonesia. The location of the project is at Batam Island of Indonesia, some 30 miles southeast from Singapore.

The main purpose construction of quay wall is to divert the loading and unloading of the interisland vessels which have 1,000 DWT to 2,500 DWT while Batuampar Port was being enlarged for 35,000 DWT vessels.

The project mainly consists of 100 m long reinforced quay wall which has 7.60 m in height above its foundation, with 30x30 sa-cm concrete piles at 5 m center to center spacing along the front of reinforced wall. Anchor rod with 40 tons capacity were installed in order to provide for the horizontal forces created by berthing impact of vessels. Wooden fender were installed at the concrete piles for absorbing the berthing impact, and sand bags barrier were placed along the reinforced wall face inside the Geotextile in order to prevent the erosion of material due to tidal fluctuation and wave actions.

The top elevation of reinforced wall is located at + 3.60 m, the same as the elevation of existing sheet piles quay walls which located at the right side of the reinforced quay wall. Prefabricated concrete slabs with 30 cm in thickness and 3.30 in length were installed for flooring at the top of quay wall, and mooring bolards were provided at

25 m spacing. Figures 1a and 1b show the typical cross section of quay wall and front view of the quay wall, respectively.

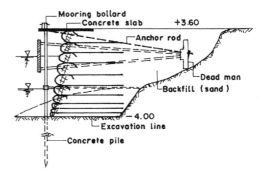

Fig. 1a Typical cross section of quay wall

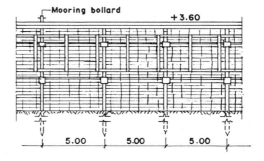

Fig.1b Front view of quay wall

1.2 Soil investigation

Two boreholes and three Dutch Cone Penetrometer Tests were carried out at the site (Sofoco,1986). Figure 2 shows the location of boreholes and cone tests.

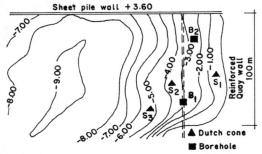

Fig.2 Location of boreholes and cone tests

Several laboratory tests such as specific gravity, grain size analyses, shear box and consolidation tests were undertaken from samples obtained from boreholes. Standard Penetration Tests were also done. Soil profiles shows that a homogenous dense sand layer is encountered from elevation -4.00 m (see figure.3).

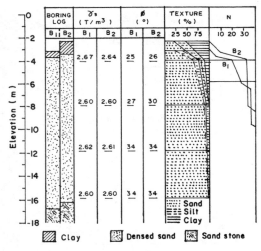

Fig.3 Soil profiles and properties

The N values indicated that the sand layer can be used as foundation of the reinforced wall system. Figure 4 shows the penetration resistantce of the soil (q_c).

Based on the data taken from the field investigation and laboratory tests, the foundation of reinforced quay wall was fixed at elevation -4.00 m, on the sand layer with angle of internal friction ($\emptyset$) of 25°.

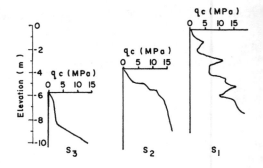

Fig.4 Penetration resistance of foundation

2 DESIGN CONSIDERATION

2.1 Fabric selection

Selection of fabric for reinforcement purposes was mainly influenced by 2 factors, namely: internal and external. Internal factors are considered as tensile strength of fabric, elongation and creep behaviour of fabric under prolonged loading condition, and resistant to the environment, while external factors are considered as height of the wall, working load, time of loading, a backfill material. Koerner and Hausmann (1987) tabulated the relationship betwee the required tensile strength of fabric an wall's height for reinforcement purposes where higher tensile strength are consider for higher walls.

The tensile strength of fabric is import to control the stability of a reinforced structure. The elongation of fabric should be low, since high elongation will result large deformation and failure of the structure. In Figure 5 the relationship between stress and strain of fabric according to t material formed fabric are represented (En bv, 1985).

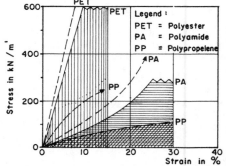

Fig.5 Stress strain relationship of fabric

The fabric tend to elongate when loaded. The elongation corresponds to the load applied to the fabric. The maximum tensile strength at break and simultaneous measured elongation at break are called the ultimate tensile strength and elongation at break, respectively. A 10% elongation at break is considered for reinforcement purposes (Enka bv,1985).

Creep is defined as the increase in length of fabric at constant loading. Creep behaviour of fabric is influenced by construction of fabric, material, working load, and time of loading (Den Hoedt,1986).

In woven fabric yarns should be positioned in straight and exactly to the direction of the load in order to control creep. In non woven fabric, it is not necessary to position the yarns to the direction of loading. In this case, creep is controlled by the bond strength of fabric.

The fabrics were mainly formed from polyamide (PA), polyethelene (PE), polypropelene (PP), and polyester (PET). They produce different properties to the fabrics, making it more and less suitable for reinforcement purposes (Enka bv,1985). Figure 6 shows the relationship between creep ratio (in percentage) and time of loading at 20% and 60% of ultimate load (Den Hoedt,1986).

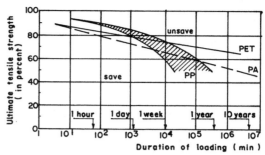

Fig.7 Time to fracture as function of creep load

and polyester (PET) is 40% (Enka bv,1985). The constituent materials of the fabric mainly influenced to the resistance against enviromental attack on fabric, since they have their own typical chemical reaction against acid, alcaline, sea water, and another special conditions. Figure 8 shows the resistance of fabrics against sea water with time. It shows that the residual tensile strength decreased with time (Risseeuw,1984).

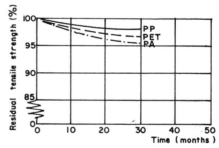

Fig.8 Resistance of fabrics against sea water attack

Based on the above reasons, a woven polyester (PET) with 200 kN/m in ultimate strength is selected for reinforcement in the quay wall.

2.2 Backfill Material

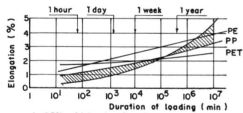

Creep at 20% ultimate load

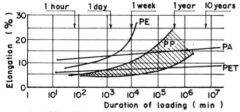

Creep at 60% ultimate load

Fig.6 Creep at 20% and 60% of ultimate load

When fabrics are used as reinforcement for prolonged periods of time, the load applied to the fabric should be related to the time to fracture of fabric. The relation between procentage of ultimate strength and time to fracture is represented in Figure 7 (van Zanten,1986).

In general, the acceptance of stress ratio for prolonged time of loading for polypropelene (PP) is 25%, polyethelene (PE) is 25%,

Since half of the reinforcement of the quay wall underwater, there was not recomended to use cohesive soil as backfill material. In addition, it is difficult to control field compaction to desired degree of compaction, since the water contents would exceeds from optimum moisture contents. Sand are recomended for backfill material. Since the fabric opening of woven fabric would permit fine sand to pass through, the grain size of backfill materials should be related to the opening size of the fabric. Figure 9 shows the recomended gradation of

379

sand backfill. Direct shear tests were undertaken in order to measure the angle of
internal friction of backfill materials.
Based on the tests results, the angle of internal friction was defined at 25° and a
sand-fabric friction coefficient at 67% was
adopted.

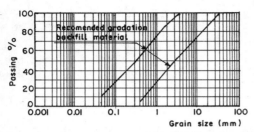

Fig.9 Grain size range of backfill materials

2.3 Design calculations

Due to time related factors in the design of
fabric for reinforcement, the following
should be considered in the design:
  1. Applied load and elongation during construction.
  2. Creep ratio during the lifetime of the
structure.
  3. Required safety factors against failure
due to internal, external forces and during
various stages of lifetime of structure (Van
Zanten,1986).
Broms (1977,1978) developed calculations
for reinforcement layer spacing and required
anchor length, while the overlapping length
was fixed at 2.50 m. Figure 10 shows the external design forces of reinforcement wall
system.

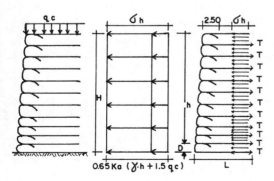

Fig.10 External design forces

Since the backfill materials was sand, the
lateral earth pressure is calculated as suggested by Terzaghi and Peck (1967) for
design of struts in open cuts in sand as

follows:

$$\sigma h = 0.65 \, Ka(\gamma.h + 1.5 \, q_s) \qquad (1)$$

where σh is the lateral earth pressure, Ka
is Rankine's coefficient for active earth
pressure, γ is bulk density of backfill, h
is the height of wall, and q_s is static ove
burden pressure, while 0.65 is a safety fa
tor to cover the variation of unit weight
and angle of internal friction of backfill
material.
Rankine's coefficient for active earth pressure can be calculated as follows:

$$Ka = \frac{1 - \sin \emptyset}{1 + \sin \emptyset} \qquad (2)$$

where $\emptyset$ is the angle of internal friction
of backfill material.
  The spacings of reinforcement layers was
determined by considering the distribution
of lateral earth pressure and the effectiv
load take up by fabrics. The layer spacing
can be calculated using equation 3:

$$D = \frac{Ta}{\sigma h} \qquad (3)$$

where D is the vertical spacing of reinfor
ment layers, Ta is the allowable permanent
load in fabric, and σh is the lateral eart
pressure.
  The required anchor length must be suffi
cient, so that the tension in the fabric c
be distributed, and safety factor should b
taken into the calculation to cover insufficiency in compaction and variation of
unit weight of backfill materials. The required anchor length can be calculated usi
the following formula:

$$L = \frac{Ta.FS}{\gamma.D.\tan \varphi} \qquad (4)$$

where L is the required anchor length, Ta
allowable permanent load in fabric, FS is
safety factor adopted for calculation, γ i
bulk density of backfill material, D is th
reinforcement layer spacing, and φ is the
friction between fabric and backfill mater
als.

2.4 Stability

In the design for reinforcement wall, the
modes of failure should be taken into cons
deration. The stability of structure shoul
be checked against the safety factor used
in the design. The modes of failure of rei
forcement system are internal and external
stability. Risseeuw and Voskamp (1984) c
veloped the internal stability calculation
using limit equilibrium method. External

stability consists of 4 modes namely: sliding
overturning, bearing capacity failure and
rotational.

A limit equilibrium analysis is usually
used to check whether the safety factor used
in the design is adequate. In this analysis,
the lowest possible safety factor was calculated using several configurations of sliding
planes from flat to circular surface by
varying point B and the angle of β and δ
(Enka bv,1985). Figure 11 shows the equilibrium of the reinforcement wedge.

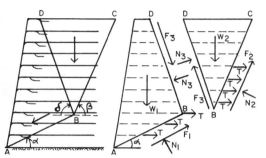

Fig.11 Equilibrium of two reinforced wedges

External stability of reinforced wall is verified using modes of failure as shown in
figure 12. These modes were due to sliding,
overturning, rotational and bearing capacity
failure.

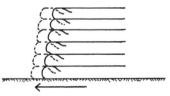

Fig.12a Failure due to sliding

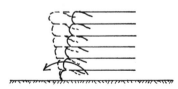

Fig.12b Failure due to overturning

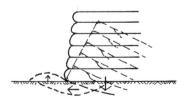

Fig.12c Failure due to bearing capacity
failure

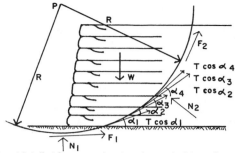

Fig.12d Failure mode due to rotational

Sliding failure occurs due to insufficient
friction resistance at the base between fabric and backfill material or the anchor
length less than required. The safety factor
against sliding can be defined as follows:

$$FS = \frac{\text{resisting forces}}{\text{sliding forces}} \quad (5)$$

Safety factor against overturning can be
defined as follows:

$$FS = \frac{\text{resisting moments}}{\text{disturbing moments}} \quad (6)$$

The overall stability is defined as stability against deep rotational sliding. Calculation of the stability is taken based on the
method defined by Bishop, modified by introducing of restoring moments contributed by
fabrics. The safety factor according to rotational can be defines as follows:

$$FS = \frac{M \text{ restoring} + \Sigma T \cos\alpha_x R}{M \text{ disturbing}} \quad (7)$$

Bearing capacity failure occurs due to insufficient strength of the foundation soil.
The Terzaghi bearing capacity of the foundation soil can be calculated using the following formula:

$$\sigma_{ult} = c.N_c + \gamma.D_f.N_q + \tfrac{1}{2}\gamma.B.N_\gamma \quad (8)$$

where σ_{ult} is the ultimate bearing capacity,
c is the cohesion, γ is bulk density, D_f is
the depth of foundation, while N_c, N_q, and
N_γ are bearing capacity factors obtained from
Meyerhoff's chart (Terzaghi and Peck,1967).
The safety factor against bearing capacity
failure can be defined using the following
formula:

$$FS = \frac{\sigma_{ult}}{\sigma_{act}} \quad (9)$$

where σ_{ult} is the ultimate bearing capacity
of foundation, σ_{act} is the maximum load to
the foundation. The safety factor in this
case should be taken greater than 3.

3 CONSTRUCTION

After the excavation reached the final elevation of foundation and completion of piling the reinforced quay wall was constructed. Compaction of backfill materials were carried out in two ways such as: compaction underwater and above water. Dynamic compaction method was adopted for underwater compaction from elevation -4.00 m up to elevation +1.00 meter. A weight of 2 tons with 1 sq meter plan dimensions hammer was dropped from a height of 3 meter at 1 meter spacing. Track crane was used with this method. The maximum thickness of compaction layer was not more than 50 cm. Compaction of backfill above water level started from elevation +1.00 m up to elevation +3.60 m at the top of quay wall. This portion was compacted using 11 tons vibratory compaction machine in 30 cm lift thickness in 8 passes at the speed not more than 10 km/hr. Sand replacement tests (sand cone tests) were utilized for checking the degree of compaction against the Standard Proctor Compaction Test determined by AASHTO T 99-74 method C from laboratory. Samples were taken by pressing a standard steel mold into the compacted fill, and then the degree of compaction of the sand inside steel mold was measured. The degree of compaction of backfill materials should not less than 90% Standard Proctor Compaction Test determined by AASHTO T 99-74 method C.

4 DISCUSSION

The project was completed within 6 months, and so far there were no special problems encountered during construction.

Compaction of backfill in underwater condition although there was no parameters given for checking the degree of compaction, the result was satisfactory enough. There was no deformation occurs when the compaction above water level was undertaken.

Monitoring of the structure were undertaken after completion the project. A simple horizontal movement measurement was carried out by measuring the change of distance the reinforcement quay wall face from the concrete piles, in order to detect the deformation of the structure. In the first year after completion of the job, the deformation of the structure was very small, under the predicted deformation calculated in the design.

5 CONCLUSION

From the presented case, the following may be pointed out:

1. The use of Geotextile in reinforcement purpose, especially in reinforced quay wall widened the area of application in Civil Engineering Construction.

2. Recent development in the design method of reinforcement wall gives a satisfactory result in construction, and open the various application of Geotextile.

3. Dynamic compaction method which adopted in the compaction for underwater compaction of sand backfill gives a satisfactory result although the degree of compaction could not be measured.

ACKNOWLEDGEMENT

The author sincerely thank Dr D.T. Bergado, Associate Professor, Geotechnical Division, Asian Institute of Technology, Bangkok, for his invaluable advice in preparing the final paper.

REFERENCES

AASHTO.1974. Standard Specifications for Transportation Materials and Methods of Sampling and Testing, Part II. Methods of Sampling and Testing. Washington D.C .

Broms,B.B. 1977. Polyester fabric as reinforcement in soil. Proc. International Conference on Use of Fabrics in Geotechnics, Paris:129-135

Broms,B.B. 1978. Design of fabric reinforced retaining structure. Proc. Symposium on Earth Reinforcement, Pittsburg:282-30

Den Hoedt. 1986. Creep and relaxation of g synthetics fabrics. Geotextiles and Geomembranes, Vol.4 No.2:83-92.

Enka bv System. 1985. Stabilenka, fabric mats for stabilizing, embankments and retaining structure, Enka bv, The Netherla

Koerner,R.M. and Hausmann,R.M.1987. Streng requirements of geosynthetics for soil r inforcement. Geotechnical Fabric Report, Vol.5, No.1:18-26.

Risseeuw,P.1984. Long term behavior of hea duty reinforcing mats/structural element in earth works. Proc. International Man-Made Fibre Congress, Austria:463-472.

Sofoco,P.T.1986. Report on Soil Investigation of Batuampar Quay wall (in Indonesi

Terzaghi,K. and Peck,R.B.1967. Soil Mechan in Engineering Practice, John Wiley & So New York.

Van Zanten,R.V.1986. Geotextiles and Geomebranes, A.T.Balkema Publ.The Netherlands

Voskamp,W. and Risseeuw,P.1984. Reinforcir fabric under embankments on soft soil, A calculation. Construction Industry Inter national,Vol.10, NO.3:74-76.

International Geotechnical Symposium on Theory and Practice of Earth Reinforcement / Fukuoka Japan / 5-7 October 1988
© 1988 Balkema, Rotterdam. ISBN 90 6191 820 0

Expanded polystyrene – A superlight fill material

T.E.Frydenlund & R.Aaboe
Norwegian Road Research Laboratory, Oslo, Norway

When road embankments are constructed across deposits of soft clay or peat, both bearing capacity and settlement problems may be solved using blocks of expanded polystyrene (EPS) as a fill material. Also when constructing roads on steep slopes, EPS may be used to solve stability problems. The unit density of polystyrene is about 100 times lighter than ordinary fill material, and the material properties match the load conditions in a road structure. In Norway more than 100 road projects involving the use of polystyrene blocks have been successfully completed with volumes varying from a few hundred to several thousand cubic metres of EPS.

1 SOFT GROUND PROBLEMS

In construction projects involving structures resting on soil, satisfying answers must be found to at least two main questions: is the soil strength sufficient to support design loads, and will deformations within the soil layers be of permissible magnitude? If the answer is no to either of these questions, special design solutions will have to be considered depending on the local site conditions and the availability and cost of suitable construction materials. The most obvious solution to load problems is, of course, to reduce the load if possible, but other methods like soil improvement, soil replacement, load transfer to firm ground etc. should also be considered technically and economically if applicable.

2 LIGHT FILL MATERIAL

Traditionally, wastes from the timber industry like sawdust and bark have been used as a light fill material in road construction in Norway. Wastes from the production of cellular concrete elements and building blocks with light expanded clay aggregate (Leca) as well as Leca-grains as such have also been employed.

The weight reduction obtained by using such materials is of the order of half the weight of ordinary fill material. When the idea of using blocks of EPS as a light

fill material emerged, the incentive was to cut weights more dramatically.

Since 1965, various sorts of insulation material have been tested for road frost protection purposes in Norway. An extensive programme was carried out on polystyrene materials, both expanded and extruded to investigate compressive strength under repeated loads, and water absorbtion properties in particular. Frost protected pavements were constructed using 5-10 cm thick polystyrene boards mainly of the extruded type, but also boards of expanded polystyrene were tested. No strength problems were detected with pavement covers of 30-50 cm thickness above the boards. Based on this experience, it was

Fig 1. EPS blocks in roadfill.

clear that from a technical point of view it would not matter if the thickness of polystyrene material was increased from 5 cm to 50 cm or even 500 cm.

In 1972, the use of polystyrene in greater thicknesses than the insulation boards was investigated at the Norwegian Road Research Laboratory for the first time. Excess settlements of road embankments adjoining a bridge were then successfully halted by replacing a one metre layer of ordinary fill material with EPS. Since then, the use of polystyrene blocks as a light fill material has become standard practice in road construction in Norway. Today more than 100 projects have so far been carried out with polystyrene fills totalling a volume in the range of 200.000 m^3.

3 ADVANTAGES

The major advantage of using EPS is, of course, the low unit density, 20 kg/m^3 when delivered from the producer. Although a design value of 100 kg/m^3 is applied for stability and settlement calculations to allow for some increase in water content over its service life, EPS is by far much lighter than any other light fill material commonly used in road construction.

The low unit density also makes EPS easy to handle on site. One EPS block of normal size (.5x1x3 m) can easily be handled by one man, since it only requires a lifting force of 300 N. Furthermore, cranes on site can handle whole truckloads in one unit, placing the blocks on the required spot. The material is also easily formed and adjustments to block shapes can be made by hand- or chainsaw or even with a knife if necessary to shape details when EPS fills join on to other structures like bridge abutments, drainage systems and so on.

With the low unit density, it is possible to haul large quantities in one truckload. In general, one truck with trailer can take up to 100 m^3 of EPS blocks, the limiting factor being freight volume rather than freight load. This, of course, keeps transport costs down.

4 DURABILITY

To prevent the polystyrene from being dissolved by petrol or other chemicals in case of a spill from an overturned tanker on the road, a 100-150 mm reinforced concrete slab is cast on top of the EPS blocks. The concrete slab also contributes

Fig. 2. EPS blocks are placed so that continuous joints are avoided.

to the strength of the pavement structure and reduces the total thickness of pavement material above the EPS blocks. The chances of a petrol tanker overturning just on an EPS fill are, of course, extremely small, and even if this should happen, only the outer part of the fill is likely to be affected. Necessary repair should also be easy to perform.

Otherwise, no decay of the EPS material is expected as polystyrene is a very stable compound chemically. Samples retrieved from existing fills show no sign of strength reduction. On the contrary, a slight increase in compressive strength is observed in some fills. From load cycling tests, it has been shown that the material will stand up to an unlimited number of load cycles as long as the repetitive loads are kept below 80 % of the compressive strength.

EPS is also resistant to biological destruction from bacteria and enzymes. Major attacks by animals are not likely as EPS does not represent a source of nourishment. The blocks could, however, easily be excavated to satisfy personal housing needs for smaller animals, but this is unlikely to affect the technical behaviour of the fill and has not been observed.

Ordinary EPS is not fire resistant and may be ignited by a fire when freely exposed to air. When covered by the concrete slab and soil on the embankment slopes, oxygen will not be available in sufficient quantities to allow a fire to develop. However, when EPS is placed in large quantities and/or will be exposed to open air for a long period during construction, and when nearby structures may be damaged by a fire, the selfextinguishing quality of EPS should be specified to a cost of 5-10 % above the standard quality.

5 MATERIAL REQUIREMENTS

For road construction purposes the following material requirements have been specified:

* The unconfined compressive strength measured on 5x5x5 cm cubes should have a mean value not less than 100 kN/m^2 and single measurements should not be below 80 kn/m^2.

These values are not selected to protect the EPS itself, but to limit pavement deflections.

* The EPS blocks should be cut with sides at right angles and with a thickness not less than 0.5 m unless otherwise specified. Deviations from specified dimensions should be within 1 %, and the evenness of the block surface measured with a 3 m straightedge should be within 5 mm.

As the material continues to expand for some time after production, these requirements will normally only be met if the producers cuts the blocks prior to delivery and not earlier than 24 hours after being taken out of the form. A new vacuum process is, however, now making it easier to meet dimension requirements.

6 PAVEMENT DESIGN

Normally, a 100-150 mm slab of lean reinforced concrete is cast on top of the EPS fill. The quality of the concrete should at least correspond to quality C 15 (15 N/mm^2 or 2100 psi, 28 days). The reinforcing net is placed in the middle of the slab.

For design purposes, EPS with a compressive strength of 100 kN/m^2 is considered as equivalent to a subgrade soil within bearing capacity class VI (similar to clay or silt). According to Norwegian specifications, the required pavement thicknesses are as listed in table 1.

Table 1. Minimum pavement thicknesses for 10-ton axle loads in Norway.

Annual average daily traffic	Pavement thickness, including concrete slab
0 to 1000	40 cm
1000 to 10.000	50 cm
over 10.000	60 cm

In some cases, at least in a country with a climate similar to that of Norway, the EPS may create unfavourable icing conditions. The pavement thickness is therefore sometimes increased depending on conditions at the adjacent road section. In general, road icing is reduced by using pavement materials with a high heat storage capacity. A gravel subbase will normally provide satisfactory conditions.

Fig 3. A layer of lean concrete is cast on top of the EPS fill.

7 ECONOMY

When considering the cost of EPS blocks, the present rates quoted in Norway are of the order of 30-35 US $/m^3$ from the producer. This is only half the price paid in 1972 when related to the present price level. With low transportation costs, even a long haul will not affect project cost too much.

When comparing the cost of different light fill materials, however, one should bear in mind that it is the cost per unit load reduction related to ordinary fill material that should be considered, since weight reduction is the aim. Assuming an average unit density of 2000 kg/m^3 for ordinary road fill materials, a cost comparison as shown is achieved for light fill materials in Norway with varying transport distances. Apart from material costs, volume reduction and cost of placing and compaction are also considered.

However, in cases where other materials compete favourably on price, they are not always readily available (e.g. sawdust and bark). In some cases, EPS represents the only choice if a certain weight reduction is to be achieved. In Norway, EPS will also compete favourably with soil improvement methods like salt wells and lime columns as well as vertical drains and piling. Depending on local conditions, soil replacement and counter weights will often provide a cheaper solution if applicable, but EPS may still be preferred for technical reasons.

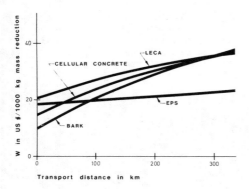

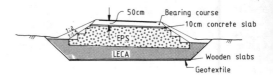

Fig 5. Typical cross section at Lysaker.

Fig 4. Cost comparison of various light fill materials.

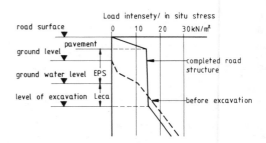

Fig 6. Load conditions at Lysaker.

8 PRESENT USE OF EPS FILLS

Typical projects where EPS is used today are in road embankments crossing clay deposits and peat bogs and for reconstructing slide areas. Special applications of EPS have been to reduce the horizontal pressure against bridge abutments. At the same time, the problem of differential settlements between the bridge and adjacent fills is reduced. Solutions involving the termination of the EPS fill in a vertical wall, covered with some protective material like steel sheets have also been adopted. Furthermore, designs utilizing the buoyancy effect of EPS when submerged have been successfully completed. In this case it is essential to control or predict the highest ground water level likely to occur in order to balance the buoyancy forces. Some examples from the present use of EPS are given below.

8.1 Road on very soft, sensitive clay

Most ramps in a major road intersection at Lysaker near the City of Oslo airport rest on bedrock or on foundations to firm ground. Two ramps 1 and 2 are founded on a very soft, sensitive normally consolidated clay layer of 15-25 m thickness. The ramps have an elevation of 1-1.5 m above the ground level and a marked cross slope since both are lying in a sharp curve.

To satisfy both bearing capacity and settlement conditions, it was decided to apply a design with EPS blocks and Leca grains in order to prevent stress increases in the ground. The design adopted is shown on the cross section in fig 5. The weight balance obtained from excavating

1,5-2 m of the top soil and replacing it with light fill material is shown in fig. 6. The total volume of EPS used is somewhat above 8000 m^3.

The ramps were completed during the first half of 1986. Observations show settlements of the order of 4-5 cm during the first year and decreasing with time. No adverse effects have been observed since the ramps were opened for traffic towards the end of 1987.

8.2 Bridge abutment

EPS was used as backfill material for a bridge abutment on road E 18 in the county of Vestfold, Norway. The bridge was completed in spring 1987. The EPS fill was placed in contact with the abutment wall. To monitor possible stress transfer from the EPS fill to the vertical concrete wall, 6 Gløtzl cells were installed at three different levels on the wall. Load tests were carried out during autumn 1987 using a lorry weighing 31 tons and containers applying a total force of 30 kN/m^2 over an area of 25 m^2. The resulting horisontal stress increase was of the order of 0.1-0.2 times the applied load intensity by the containers on the road surface.

On another bridge built in 1977, EPS blocks were used behind one of the abutments, but a small gap was left between the EPS and the concrete wall. Measurements performed in 1984 and 1987 show that the gap is maintained unaltered indicating

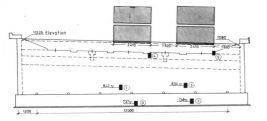

Fig 7. Location of load cells.

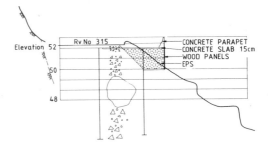

Fig 9. Vertical EPS wall on steep slope.

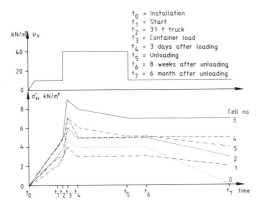

Fig 8. Results from loading test.

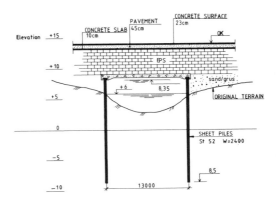

Fig 10. Reduced load on culvert.

that the EPS fill is stablle, and that
there is no force transfer between the
blocks and the concrete wall.

8.3 Vertical EPS wall

In the county of Vestfold close to the
City of Holmestrand, a widening of road
no. 315 was planned to allow for the construction of a separate pedestrian/cycle
path. The road passes through a rock scree
with a steep slope (~1:1). A conventional
retaining wall founded in the rock scree
could not be accomodated due to bearing
capacity problems. Two alternative solutions were considered, one involving a huge
retaining wall at the bottom of the scree,
the other consisting of an EPS fill
terminated in a vertical wall.

The latter solution was adopted both for
technical and economic reasons. Approximately 800 m^3 of EPS was used with a
vertical wall height up to 2.5 m. A 15 cm
reinforced concrete slab was cast on top
of the EPS blocks and continued 3 m horizontally into the existing road for
anchorage. A concrete parapet was erected

as an outside barrier and connected to the
concrete slab. To protect the EPS blocks
the verical wall was covered with wood
panels. The wall as shown in fig 9 was
completed in 1985. Prior to this case EPS
fills with vertical walls up to 5.5 m have
been constructed.

8.4 Simplified bridge design

Road E 18 trough the county of Vestfold is
presently being realigned to bypass urban
areas. At Sem the new road is passing over
a small erosion valley with a river at
the bottom. Site investigations indicate
more than 30 m of medium stiff clay of
shear strength 25-30 kN/m^2. Alternative
designs considered were a conventional
bridge, a concrete culvert with a concrete deck
and EPS blocks on top. The steel sheet
pile solution was selected and will be
completed during spring 1988. The EPS
layer will be 4 m high with a pavement of
78 cm on top. This will reduce the weight
to be carried by the sheet piles dramatically and also reduce the length of the

387

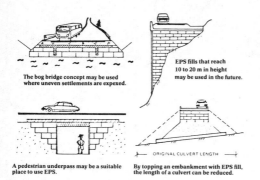

The bog bridge concept may be used where uneven settlements are expexed.

EPS fills that reach 10 to 20 m in height may be used in the future.

A pedestrian underpass may be a suitable place to use EPS.

By topping an embankment with EPS fill, the length of a culvert can be reduced.

ORIGINAL CULVERT LENGTH

Fig 11. Possible applications of EPS.

culvert since the EPS will be terminated in vertical walls along the road edges.

9 FUTURE POSSIBILITIES

The superlight fill material with the highest potentials seems still to be EPS both based on material quality and cost. A new superlight material with a honeycomb structure has recently been developed making it possible to accommodate changes in water level without introducing buoyancy forces. For certain uses this material may be superior, but future use will in the end depend on technical properties and price in relation to other materials available.

Superlight materials have mainly been used to reduce loads and settlements on soft ground. This approach allows fairly high fills to be constructed with an adequate factor of safety and tolerable settlements. The many uses for this purpose will probably develop further as more ingenious details are worked out.

It has been established that it is possible to reduce or prevent transfer of pressure on a retaining wall or bridge abutment by using EPS as a backfill. This fact must be taken advantage of when designing such wall in the future. The same approach may probably be persued for the design of sheet pile walls.

The application of EPS blocks as a structural element will certainly increase and probably lead to the use of higher vertical walls. Combined with the reinforced earth concept, new solutions based on the merits of both are likely to emerge. This will permit various types of structures on hillsides without creating stability problems.

A highway crossing deep ravines with normal soil embankments will result in long and expensive culverts. The use of an EPS fill with vertical walls or combined with a conventional fill in the lower part, may considerably reduce the culvert length and fill volume.

EPS used as building blocks could be applied for rapid construction of pedestrian underpasses and have certain advantages on soft ground. The future may also bring increased use of EPS for floating structures, whether it is on extremely soft ground or in water. Prefabrication of certain elements can insure quality and reduce construction time.

There are still room for more ideas when it comes to the use of superlight fill materials in road building and other construction activities. Many of the old soil mechanics problems may have new and surprising solutions. A satisfactory solution and cost advantage will depend on the solution of many small details of which we have only seen the beginning. The new materials required are available.

REFERENCES

Aaboe,R. 1987. Experience with expanded polystyrene as lightweight fill material in road embankments. Meddelelser nr 61 Norwegian Road Research Laboratory.

Myhre,Ø. 1987. EPS - material specifications. Meddelelser nr 61 Norwegian Road Reseach Laboratory.

Refsdal,G. 1987. EPS - design considerations. Meddelelser nr 61 Norwegian Road Research Laboratory.

Refsdal,G. 1987. Future trends for EPS use. Meddelelser nr 61 Norwegian Road Research Laboratory.

Hjorteset,A. 1987. Spenningsfordeling i EPS som konstruksjonsmateriale (stress distribution in EPS as a construction material). Thesis for B.Sc degree at the Norwegian Institute of Technology.

International Geotechnical Symposium on Theory and Practice of Earth Reinforcement / Fukuoka Japan / 5-7 October 1988
© 1988 Balkema, Rotterdam. ISBN 90 6191 820 0

Design and construction of steel bars with anchor plates applying to strengthen the high embankment on soft ground

Masami Fukuoka
Science University of Tokyo, Tokyo, Japan

Masaaki Goto
Kiso-Jiban Consultants, Chiba, Japan

ABSTRACT: A high embankment for interchange was to be constructed on soft ground of 10m in thickness without causing harmful displacement at the ground surface near the toe of the embankment. A new method i.e. the steel bar reinforcing method, was proposed to reduce deformation at the ground surface near the toe of the embankment and to strengthen the embankment. The precision of predicting displacement was improved by conducting soil surveys, building test embankments and analyzing divergencies of predicted values from values observed in the test embankments. Displacement was thus successfully minimized. For predicting the displacement of the ground surface near the toe and the proportion of tensile forces in each layer of the steel bars, finite element method (FEM) was found to be the best method available.

1 INTRODUCTION

A high embankment on soft ground was contemplated at Hayashima Interchange, at the north end of the expressway connecting Kojima (Honshu) and Sakaide (Shikoku). Though the area at the toe of the embankment was mainly used for rice fields, significant ground surface displacement at the toe could not be tolerated. For strengthen the embankment, the steel bar reinforcing method, i.e. placing steel bars with anchor plates at the bottom of embankment, was adopted. This method was used on a small scale at Muchiki about 20 km south of Hayashima Interchange. The results of the test were used in the construction work at Hayashima.

2 OUTLINE OF MUCHIKI EMBANKMENT

Figure 1 shows cross section of the embankments at Muchiki. Sand piles were driven into the soft ground and steel bars with anchor plates were placed at the bottom of the embankment.

Ac: ALLUVIAL SILTY CLAY
Ams: ALLUVIAL SILTY SAND
Dc: DILUVIAL SILTY CLAY
Ds: DILUVIAL SAND
N: N-VALUE OF S.P.T.
SD: SAND DRAIN
H: POINT FOR MEASURING HORIZONTAL DISPLACEMENT

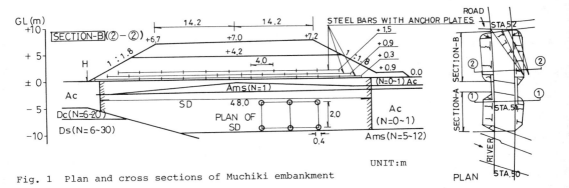

Fig. 1 Plan and cross sections of Muchiki embankment

Properties of the soils are described in Fig. 2. Triaxial compression tests and consolidation tests were also performed. The soil constants used for design were as follows.

For soft clays: unit weight γ is 17 kN/m^3, unconfined compressive strength qu is 20 kN/m^2 from the ground surface to the depth of 3m, and $\{20+6.6(z-3)\}$ kN/m^2 below the depth of 3m. Preconsolidation pressure σ_p' and effective overburden pressure σ_{vo}' increase with depth, as shown in Fig. 2. Coefficient of consolidation c_V is 3.3 x 10^{-7} m^2/s. For an embankment of weathered granite: unit weight γ is 20 kN/m^3, cohesion c is 10 kN/m^2, angle of internal friction ϕ' is 30°.

For FEM analysis, the following soil constants were used in addition to the above soil constants. Unit weight of submerged soil for As layer γ' is 10 kN/m^3. Poisson's ratio ν and coefficient of earth pressure K for As layer and embankment are 0.3 and 0.43, respectively. Poisson's ratio ν and coefficient of earth pressure K for Ac layer are 0.33 and 0.50, respectively. Coefficient of deformation E for clay is derived from the coefficient of volume compressibility m_V and the Poisson's ratio ν. Coefficient of defor-

mation E for sand is obtained by the results of pressuremeter test, and the formula E = 7N (where, N is the N-value of the SPT) was used for reference. Coefficient of deformation E for embankment is 10 MN/m^2.

3 DESIGN OF STEEL REINFORCING BARS WITH ANCHOR PLATES

In 1963 Fukuoka used steel reinforcing bars with anchor plates to strengthen an embankment of 40m in height on soft ground stabilized with sand compaction piles. Sand piles were used at Muchiki instead of sand compaction piles to avoid noise pollution during construction. Sand piles of 40cm in diameter were driven at intervals of 2m and to depths of 8 - 13m in pattern to square.

Anchor bars were inserted only at the bottom of Muchiki embankment as it was lower than the 1963 embankment. Deformed steel bars of 22mm in diameter having steel bearing plates 250 x 300 x 9mm in dimension were placed at intervals of 50cm in horizontal direction and 60cm in vertical direction. As it was very hard to predict tensile forces acting on the

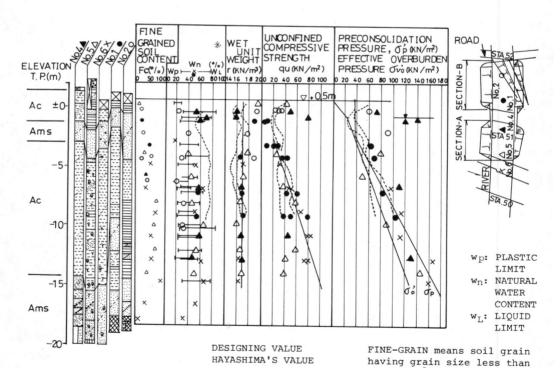

Fig. 2 Soil map of sections A and B at Muchiki

390

bars, three layers of steel bars were provided. Therefore, the number of bars per unit length of embankment was 6.

Pullout tests of bars were performed at the construction site. Pullout strength and pullout force at 1cm displacement were 120 - 140 and 70 kN/m, respectively.

The following methods were used for predicting maximum tensile forces on the bars.

(1) Circular arc analysis (Fig. 3(a)). Tensile force on the bars is obtained by the following formula.

$$T \cdot h + R \Sigma (Su \cdot \Delta \ell) \geq F \cdot W \cdot d \qquad (1)$$

$$T \geq \frac{1}{h} \{ F \cdot W \cdot d - R \Sigma (Su \cdot \Delta \ell) \} \qquad (1)'$$

where, T : total tensile force, h : vertical length between center of circular arc and the point where total tensile force act, R : radius of circular arc, Su : shear strength, $\Delta \ell$: length of element acting shear force, F : factor of safety, W : total weight of soil inside of circular arc, d : horizontal length between center of circular arc and the point where total weight of soil inside of circular arc act.

365 kN/m of tensile force was obtained.

(2) Method of using earth pressures on assumed vertical walls (Fig.3(b)). Tensile force on the bars T is calculated by the following formula.

$$S + T \geq (P_{A1} + P_{A2} - Pp) \cdot F \qquad (2)$$

$$T \geq (P_{A1} + P_{A2} - Pp) F - S \qquad (2)'$$

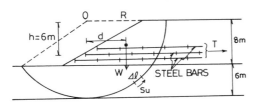

(a) Circular arc analysis

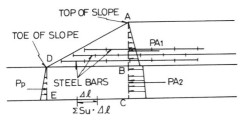

(b) Earth pressure method

Fig. 3 Method of estimating tensile force of reinforcing steel bars

Where, F is factor of safety, S is total shear stress along the bottom of the soft clay layer which is obtained by the following formula.

$$S = \Sigma (Su \cdot \Delta \ell) \qquad (3)$$

where, Su : unconfined compressive stress divided by 2, $\Delta \ell$: length of element between top and toe of slope. P_{A1} is horizontal earth pressure on the assumed vertical wall at the top of slope of the embankment which is obtained by the following formula.

$$P_{A1} = 1/2 \cdot K \cdot \gamma \cdot H^2 \qquad (4)$$

where, k : coefficient of earth pressure, γ : wet unit weight, H : embankments height. P_{A2} is the active earth pressure on the vertical wall at the same position in the soft foundation. Pp is the passive pressure on the vertical wall in the soft foundation at the toe of slope.

358 kN/m of tensile force was obtained.

(3) Estimation by observations at Ebetsu testing embankment.

The Japan Road Public Corporation tested an embankment of 8m in height over soft ground of 10m in thickness. Two layers of steel strips of 2.3mm in thickness and 140mm in width, were laid at the bottom of the embankment. Horizontal and vertical spacings were 0.5 and 1.0m, respectively. Maximum tensile stress on the strip of the lower layer at the center of the embankment was greater than the yielding stress of 3 MN/m². Analyzing this case, 386 kN/m of tensile force was obtained for the case of Muchiki.

(4) Model test.

A model test of 1/10 scale was performed. The tensile force on the steel bars was calculated as 292 kN/m for the case of Muchiki.

(5) FEM analysis (the first stage).

The soil constants used were given above. The steel bars were assumed to be bar elements resisting the deformation of the embankment by axial tensile force. We assumed that E_SA value of the bars was 25 MN/m (real E_SA value was 160 MN/m), in order that the tensile stress was nearly equal to the allowable stress. Where, E_S is Young's modulus and A is the cross sectional area of a steel bar. The tensile forces per unit length of the embankment were calculated as 112.1 kN/m at the lowest layer, 50.8 kN/m at the middle layer and -10.6 kN/m at the highest layer. Thus, the total tensile force was 163 kN/m. We knew later that this E_SA value was too small.

4 PREDICTED AND OBSERVED VALUES

The results of calculation using different methods were given in the previous paragraph for design. Based on engineering judgement, three layers of deformed steel bars of 22mm in diameter were provided with horizontal spacing of 50cm. Allowable tensile strength per bar was 69.7 kN, and for 6 bars, 418 kN/m. Values predicted by FEM and by Ebetsu test embankment were 163 kN/m and 386 kN/m, respectively.

(1) Observed tensile forces on steel bars. Figure 4 show the measured tensile stresses on steel bars. The total tensile force of steel bars at the center of the embankment was 426 kN/m, which is much larger than the predicted value by FEM (the first stage). Tensile forces of the bars in the lower layer were the largest among the tensile forces of bars in the three layers. The ratio of the tensile force on the lower layer to that on the middle layer was about 2. This ratio is similar to the result of FEM analysis. The tensile forces acting on the bars at the upper layer were negative when the height of the embankment was low, and positive when the height became high. Total tensile force can be obtained by conventional circular arc analysis, but this analysis cannot provide the proportion of tensile forces in each layer of the bars. It is necessary to use FEM for obtaining the tensile forces on bars in each layer.

(2) FEM analysis (the second stage). The $E_S A$-value of a steel bar was assumed to be 25 MN/m at the first stage of prediction ((5) FEM analysis), but this assumption was not very correct. Therefore, considering relative displacement between an anchor plate and surrounding soil, the $E_S A$-value was assumed to be 130 MN/m (real $E_S A$-value of steel bars was 160 MN/m).

The observed values and predicted values (the second stage by FEM) were given in Fig. 5 at the embankment heights of 4.2m and 7.0m. Calculated tensile force agreed with the measured value at the bank height 4.2m, but there was little difference between them at an embankment height of 7.0m. Observed tensile stress on lower bar was higher than the yield value (35 kN/m^2). The steel bars were elongated at the yield stress without breaking, and they were serving as strengthening bars for the soil. It was felt that this condition should be improved in the future.

(3) Pressure on each plate.
It was very difficult to predict bearing forces acting on the anchor plates at the

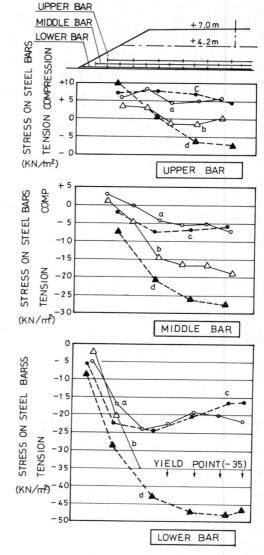

a: BANK HEIGHT 4.2m 122 DAYS AFTER
b: BANK HEIGHT 7.0m 193 DAYS AFTER
c: PREDICTED BY FEM (BANK HEIGHT 4.2m)
d: PREDICTED BY FEM (BANK HEIGHT 7.0m)

Fig. 4   Stresses on steel bars

beginning. The bearing forces acting on the plates were calculated from the tensile forces on the tensile bars, and they are given in Fig. 5. The ratio of the pressure between middle plate and lower plate was 50%. As the result of pullout test, maximum bearing capacity of

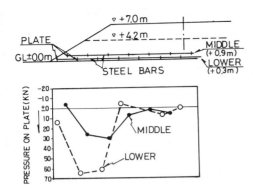

Fig. 5　Pressure on each plate

the anchor plate was 120 - 140 kN.
Therefore, bearing forces acting on the
plates were under allowable force.
(4) Settlement and displacement.
Observed settlement at the middle of the
embankment with sand drains and steel bars
reached 68cm at the end of construction
and 86cm 406 days after completion. The
observed value was 91% of the predicted
value of 95cm.
　　Observed horizontal displacement (3cm)
at the ground surface at the toe of the
embankment and the predicted value (2cm)
was nearly equal (Fig. 6).

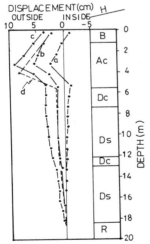

Toe of slope of the embankment

a: BANK HEIGHT 4.2m 122 DAYS AFTER
b: BANK HEIGHT 7.0m
c: BANK HEIGHT 7.0m 193 DAYS AFTER
d: PREDICTED BY FEM (BANK HEIGHT 7.0m)

Fig. 6　Horizontal displacement

5 EMBANKMENT AT HAYASHIMA INTERCHANGE

The difference between the observed and
predicted values of the Muchiki test
embankment was investigated fully before
constructing the embankment at the
Hayashima Interchange. Points of improvement for the methods of prediction and
design are stated as follows.
(1) Design.
The number of layers for the reinforcing
bars, three for the Muchiki embankment was
reduced to 2. Three layers were used at
Muchiki because the prediction was thought
to be not very accurate. Therefore, it
was thought that breakage of the lower
layer bars was quite possible. In that
case, the upper layer bars could work
instead of the lower layer bars. As a
result of the embankment test, it was
found that the lowest layer of bars
yielded, but did not break. It was
thought that the upper layer of bars could
have been omitted by improving the methods
of design. The ratio of the pressure
between upper plate and lower plate was
assumed to be 50%.
　　The dimension of the anchor plates was
225 x 225 x 9mm, which was smaller than
Muchiki's plates. The dimension was
decided based on an pullout test at the
real construction site.
　　The diameter of steel bars was 29mm,
which is larger than Muchiki's steel bars
of 22mm in diameter. The diameter was
decided based on allowable stress and
facility of construction work.
　　The length of steel bars was decided
based on circular arc analysis.
(2) Method of calculation by FEM.
Calculation by FEM was improved based on
the tests on the Muchiki embankment.
Another improvement was made at Hayashima.
Considering the relative displacement
between the anchor plate and the surrounding soil, a special joint element was
adopted for the reinforcing bars. The
constants for the joint element were
provided by an pullout test at the site.
(3) Method of construction.
Total area, total volume of embankment,
and maximum height of the Hayashima
Interchange were 200,000 m^2, 1,200,000 m^3
and 11m, respectively. Depth of soft clay
layer was about 10m. Vertical drains were
installed under the middle of the embankment. Steel reinforcing bars with gravel
compaction piles were placed under the
slope of the embankment to strengthen the
embankment and to prevent displacement of
the ground surface near the toe of the
embankment. Figure 7 shows the arrangement of reinforcing bars at Hayashima
Interchange.

Fig. 7 Arrangement of reinforcing bars

(4) Prediction and performance.
Figure 8 represent the observed values and the predicted values by FEM.

Observed settlement and recorded stress on the upper bar were nearly equal to the predicted value. But recorded stress on the lower bar under the slope did not agree with the predicted one. Observed horizontal displacement at the toe of embankment was about 25% of predicted value. Therefore, steel reinforcing bars with gravel compaction piles were more effective than predicted.

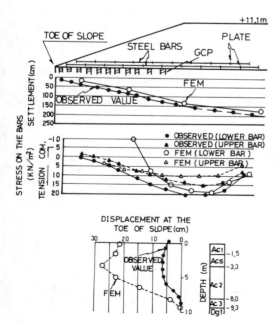

Fig. 8 Observed and predicted values by FEM at Hayashima Interchange

6 CONCLUSIONS

To strengthen the high embankment on soft ground of 10m in thickness, reinforcing steel bars with gravel compaction piles were used to prevent deformation at the toe of the embankment. This treatment was adopted on a small scale at Muchiki, and the predicted results were compared with the observed.

(1) FEM was used to analyze tensile forces acting on the reinforcing steel bars at the bottom of the embankment. Bar elements for the reinforcing bars were used.

(2) The EsA-value, which is the product of Young's modulus Es and cross sectional area A of the steel bars, was assumed to be 20 MN/m. When construction was half complete, this value was found to be too small, so it was raised to 130 MN/m. And calculated tensile force was nearly equal to the measured value.

(3) Number of layers, dimensions of anchor plates, and diameter of steel bars were changed based on the experiment at Muchiki.

(4) Joint elements were used in addition to bar elements for FEM analysis.

(5) The construction work proceeded safety without causing serious deformation at the ground surface near the toe of the embankment.

7 ACKNOWLEDGEMENT

This paper was prepared based on a report on the construction work by the Hayashima Interchange of the Honshu-Shikoku Bridge Authority. The authors express their sincere appreciation to the engineers of the Kurashiki Construction Office of the Authority and the committee members of Technology Center for National Land Development related to the construction work.

REFERENCES

H. OHTA, R. MOCHINAGA and KURIHARA. 1980. Investigation of soft foundations with surface reinforcement. Proc. 3rd Australia - New Zealand Geo-mechanics Conference.
M. FUKUOKA and M. GOTO. 1987. Embankment for interchange constructed on soft ground applying new methods of soil improvement. Prediction and performance in geotechnical engineering, Calgary, Canada.

394

International Geotechnical Symposium on Theory and Practice of Earth Reinforcement / Fukuoka Japan / 5-7 October 1988
© 1988 Balkema, Rotterdam. ISBN 90 6191 820 0

Application of RBSM analysis to earth reinforcement method

M.Hada, Y.Taguchi & K.Kagawa
Technical Research Division, Fujita Corp., Yokohama, Japan

ABSTRACT: The numerical analysis is carried out to many kinds of slope using the Rigid Body and Spring Model (hereafter called RBSM) proposed by Prof.Kawai (Kawai 1977). This paper reports on the application of the RBSM analysis to a model retaining wall test and a polymer-grid reinforced embankment assuming that RBSM is an effective analytical technique for earth reinforcing mechanisms.

1 INTRODUCTION

For designing the earth reinforcement system,it is desired that a numerical method, with which properties of the interface of soil and reinforcement can be assessed properly, be established.

In RBSM analysis, the interface of them will be modeled by the beam element and the plane element, as shown in figure 1, using stress (τ , σ) of the interface directly. Stress of the RBSM are transmitted by two springs (a shearing spring and a normal spring) distributed over the contact surface of two adjacent rigid elements. The deformation is defined by the relative displacement between two elements represented by displacement (u,v,θ) of the center of gravity of each elements, moving the elements themselves as the rigid body.

The discrete surface as illustrated in figure 2 can be easily assessed by properties of the RBSM and is being accepted as an analytical model useful for simulation of failure (Takeuchi 1981).

2 ANALYSIS OF MODEL TESTS

The model tests that were analyzed by the authors were conducted with the back filling as shown in figure 3, to which vinyl sheets were installed. The lower end of the movable wall on the left of the figure was connected by a pin. The wall was moved until horizontal displacement at the top reached 80mm (Ohkawa 1986a,1986b). Table 1 lists the conditions of sheets analyzed.

This paper studies the results of analyses of Case No.1,2,and 3. Table 2 outlines the material constants applied to the analyses.

2.1 Earth pressure

Figure 4,5,and 6 show the relation between the displacement of the movable wall and earth pressure, and figure 7 compares between the earth pressure distribution and the individual experimental values at the wall displacement of 30mm. Solid lines in the figures represent the experimental values and the broken lines the computed values.

The variation in earth pressure to the displacement of the movable wall is thought to be in good correspondence with experimental values, except for

Table 1. Condition of model tests

Case	Sheet condition
1	None
2	3 sheets
3	1 sheet
10	2 sheets
14	3 sheets(20cm away from the wall)

Table 2. Material constants applied to analyses

Materials	γ KN/m3	E MPa	ν	Φ degree
Onahama sand	14.4	0.92	0.28	38.0
Sheet	0	660	0.33	-
Movable wall	76.4	2100	0.33	-
Contact surface	-	0.92	0.28	19.0

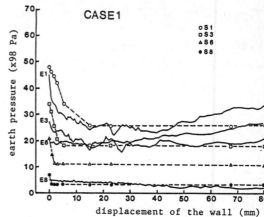

Figure 1. Model of the interface

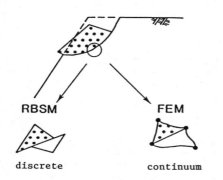

Figure 2. Difference between RBSM and FEM

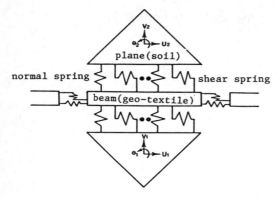

Figure 3. Diagram of the model test

the values of earth pressure gauge E6 in fig. 4 and 5, and E8 in fig. 6. As shown in fig.7,it has become clear that the earth pressure was reduced of depths where the reinforcing sheets were installed.

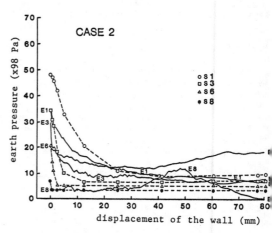

Figure 4. Variation of earth pressure

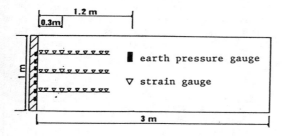

Figure 5. Variation of earth pressure

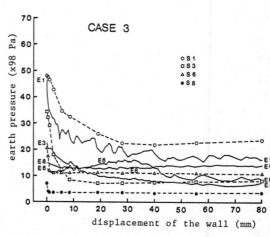

Figure 6. Variation of earth pressure

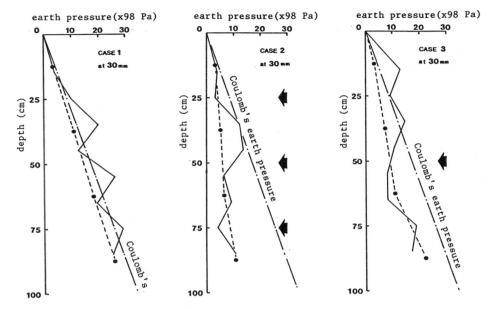

Figure 7. Earth pressure distribution

2.2 Slip line

Figure 8,9, and 10 respectively show the analysis results of slip lines inside the filling caused by shifting of the wall.

As shown in these figures, slip lines started commonly in the vicinity of a surface close to the wall, and they progressed deeper along Coulomb's active wedge. But the development of them differs according to intervals of sheet installation, which is thought to occur as follows :

Case 1 : Figure 8 shows that slip lines are formed along Coulomb's active wedge.

Case 2 : Figure 9 shows that developing of the slip lines is different from case 1 because of the tensile resistance of vinyl sheets. Horizontal lines appearing in figure (c) indicate the slip lines on sheet surfaces. The slipping caused the redistribution of stress, developing other slipping apart from these positions in the direction further from the wall. This tendency becomes clearer in figure (d) and (e).

Case 3 : Figure 10 shows that the slipping develops as if the filling is divided into the two parts because of the sheet installed in the middle. In other words, while the shifting of the wall is small, slip lines are formed in almost the same way as for filling without inner sheet installation. When the wall displacement reaches 5mm, slipping occurs above the sheet along Coulomb's active wedge. If the displacement progresses further, slipping occurs on this

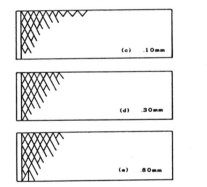

Figure 8. Slip line (case 1)

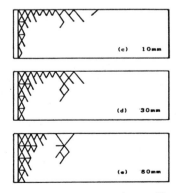

Figure 9. Slip line (case 2)

397

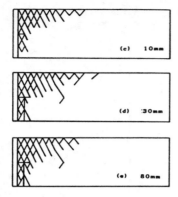

Figure 10. Slip line (case 3)

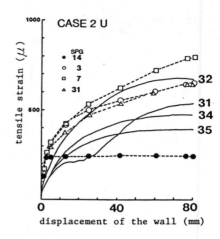

sheet. Finally, a slip area completed is similar to that of the filling having no sheet inside.

2.3 Tensile strain of sheet

Figure 11 and 13 show the variation of tensile strain caused to the sheet by the shift of the movable wall. Figure 12 and 14 show the distribution of the tensile strain in the longitudinal direction.

The solid lines in the figures indicate the experimental values and the broken lines the computed values. The numbers in rising order at the end of the solid lines in figure 11 and 13 are proportional to the distance from the wall. Markings ●, ○, □, and △ also correspond to the distance accordingly.

Case 2 : Figure 11 shows the computed values and experimental values correspond well to each other except for the lower sheet. The computed value ● of the upper sheet indicates that the slip line starting on the sheet surface at the wall displacement of about 3mm becomes constant thereafter and corresponds with the actually measured value up to the displacement of 30mm. The two computed values ● and ○ of the sheets in the middle correspond with the actually measured values. Computed values in the lower sheet show the occurrence of slipping on the sheet surface at the wall displacement of about 30mm, and the strain becomes greater than the actually measured values.

Case 3 : Figure 13 shows that the computed values correspond well with the experimental values (No.21). Furthermore, as shown in figure 10, it is thought that slipping occurs on the sheet surface from about 10mm of displacement. Figure 14 shows that strain distribution tends to be greater than the experimental values by two to three times.

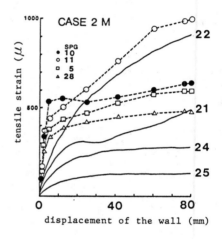

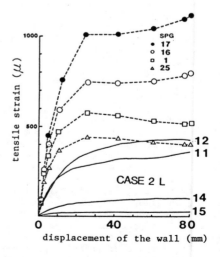

Figure 11. Variation of tensile strain

3 ANALYSIS OF REINFORCED EMBANKMENT

The mesh division diagram figure 15 shows a reinforced embankment of about 4 meters with slope grade of 1:0.2.

Strain gauge was attached to the reinforcement on the second and fourth layers from the bottom at five locations at intervals of 50cm. These measured strain during the construction and when the trucks drove on the embankment. Since the strain caused on the reinforcement by the trucks were so nominal, the following describes the computed values and actually measured values during the construction period.

Uniaxial oriented polymer-grid was used for the reinforcement at vertical intervals of 1m for the length of 3m.

The filling material used was a well-graded gravel of Japanese unified soil classification system with the unit weight of 20.1 KN/m3, water content of 6%, uniformity coefficient of 350, and 10% of fine particles of 74 μ or finer.

The slope is constructed with the many amount of sandbag packed with the above mentioned filling material and covered with the reinforcing material.

Table 3 outlines the material constants used for the analysis. The angle of friction of the contact area between the reinforcement and the filling soil was, same as used for the analysis of the aforementioned model test, one half of the internal friction angle of the filling soil.

Figure 16 shows the comparison between the computed values and actually measured values of reinforcement strain caused at the completion of the embankment. Both values show correspond well to each other with the maximum strain being about 0.65%. Since no slipping is observed by analysis, this embankment is thought to be stable.

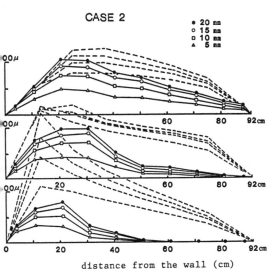

Figure 12. Distribution of tensile strain

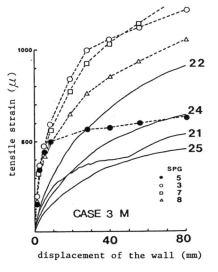

Figure 13. Variation of tensile strain

Table 3. Material constants applied to analysis

Materials	γ KN/m3	E MPa	ν	ϕ degree	c KPa
filling	20.1	0.49	0.4	35.0	10
polymer	0.1	220	–	–	–
sandbag	20.1	0.49	0.4	35.0	15
Contact	–	0.49	0.4	17.5	10

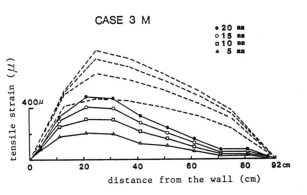

Figure 14. Distribution of tensile strain

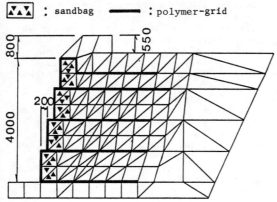

▨ : sandbag ▬▬ : polymer-grid

Figure 15. Mesh division of the embankment

REFERENCE

T.Kawai 1977. A new discrete model for analysis of solid mechanics problems. Seisan-Kenkyu. vol.29 No.4
N.Takeuchi,M.Hada,& T.Kawai 1981. Limit analysis of soil and rock foundations by means of new discrete models (7th report) Seisan-Kenkyu. vol.33 No.7 (in Japanese)
S.Ohkawa,M.Noda,K.Tanaka,M.Nakamura, & H.Ikemi 1986a. Effect of reinforcing sheets for the abutment in model test. Proc. of 41th the annual conference of JSCE (in Japanese)
S.Ohkawa,M.Noda,& K.Tanaka 1986b. Static and dynamic model test concerning the effect of reinforcing sheets laid in the back-filling to the abutment. Proc.of 41t the annual conference of JSCE(in Japanese

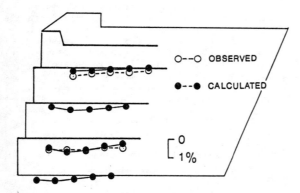

O--O OBSERVED

●--● CALCULATED

0
1%

Figure 16. Distribution of strain

4 CONCLUSION

As the results of the application of RBSM analysis to the reinforcement system, the following items have become clearer.
1) Reduced of earth pressure by rein-forcement
2) Peak strain to the reinforcement occurs to the boundary between plastic and elastic regions,and that RBSM can be effective means for analyzing earth reinforcement method.

ACKNOWLEDGMENT

The model test data were quoted from the "Report on Experimental Studies of Earthquake Resisting Reinforcement of Abutment" for 1985 by Japan Highway Public Corporation. The authors are grateful for the suggestion by persons concerned.

International Geotechnical Symposium on Theory and Practice of Earth Reinforcement / Fukuoka Japan / 5-7 October 1988
© 1988 Balkema, Rotterdam. ISBN 90 6191 820 0

Analytical method for geotextiles used for earth spreading work and applicability

Y. Higuchi & Y. Watari
Technical Research Institute, Penta Ocean Construction Co. Ltd, Tokyo, Japan

ABSTRACT: Experiment has confirmed that the geotextile used for any earth spreading work when a continuous load was applied on very soft clayey ground indicated a distribution of displacement that can be expressed by an ellipse. One experiment was done to clarify the upward pressure that was used to predict the geotextile displacement and the generating tension by numerical analysis using the pneumatic membrane theory. The result of numerical analysis for which measurements of the upward pressure in this experiment were used, agreed well with the result of the model test. Therefore, it is thought that the geotextile behavior analyzed by the pneumatic membrane theory has sufficient reliability for use.

1 INTRODUCTION

To clarify the interaction between the ground and the geotextile used for the earth spreading work on very soft clayey ground, an analytical method which regarded the geotextile as a membrane structure was proposed by the authors (Watari, et al, 1986) on the basis of results obtained from a model test. The following two assumptions were used for the analysis in this report:

(1) The displacement of the geotextile, when a continuous load is applied, can be approximated by an elliptic curve, and

(2) The upward load strength acting from the ground to the geotextile is proportional to the relative settlement from the maximum heaving point of the geotextile.

The assumptions were indicated by the model test and a back analysis of its results. Therefore, the validity had to be verified. This experimental model was first applied out in this research where the geotextile and ground materials were replaced by rubber-membrane and water models, respectively, to investigate the validity of the above assumptions. In this experiment, the deformed curve of the rubber membrane was measured three-dimensionally and, at the same time, the upward

load acting on the rubber membrane was measured with open piezometers. Subsequently, the loading experiment was carried out by using cohesive materials instead of water and the influence of the cohesion and unit weight on the deformation of geotextile was investigated.

2 OUTLINE OF EXPERIMENT

2.1 Experimental Method

The soil vessel was 105 cm wide, 100 cm long, and 50 cm deep. The soil vessel was filled with water, then a 0.6-mm-thick rubber membrane was spread on the water surface with its four edges fixed to the side walls of the soil vessel. A continuous load was applied by sticking lead shot on a 10-cm-wide tape and adjusting the loading strength per unit area to 5 gf/cm². Three loads (5, 10, and 15 gf/cm²) were applied by stacking these loads up to three steps.

The displacement of the rubber membrane due to load was determined by measuring the X, Y, and Z coordinates of lattice points prepared on the surface at 1.0-cm intervals.

Upward pressure acting on the rubber membrane was measured by five open piezometers mounted on the membrane. Measurement locations and loading locations are shown in Figure 1, and the experimental

scenes were as shown in Figure 2.
A watersoluble polymer (sodium carboxy-
methylcellulose, or CMC) and kaoline clay
were used in place of water in the experiment using clay as the ground material.
In this case, only the displacement of the
rubber membrane was measured.

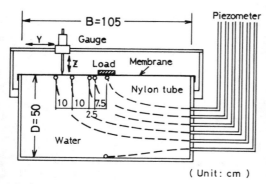

Fig. 1 Arrangements of model and measuring
instruments

3 RESULTS AND DISCUSSION

3.1 Displacement of the Membrane

Figure 3 compares the measured displacement of the rubber membrane near the loading points with an elliptic curve approxi-
mating the measurements. The distance
from the loading end to the point where
the displacement of membrane surface became nearly constant was used as the major
axis. The sum of respective maximum heaving and maximum settlement was used as
the minor axis. As can be seen in Figure
3, the displacement measurements agreed
well with the elliptic curve. So it can
be assumed valid to approximate the displacement of a membrane surface by an elliptic curve.

Fig. 2 View of loading test

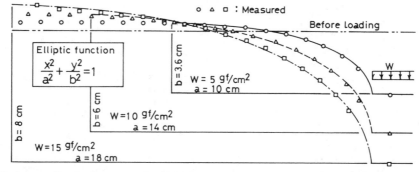

Fig. 3 Membrane displacement and elliptic curves under different load cases

.2 Upward Pressure Acting on the Membrane

Figures 4 and 5 show changes in the displacement of the membrane surface and
changes in the water level indicated on
the open piezometers, when 5, 10 and 15
gf/cm² loads were applied. It can be verified from this figure that the water level
indicated at each loading step agrees approximately

with the maximum heaving height of the
membrane. This result indicates that each
location on the membrane surface is acted
on by a pressure equal to the hydrostatic
pressure where the relative displacement
from the maximum heaving part of membrane
surface is regarded as the water depth.
Therefore, assumption (2) can be assumed
to be valid.

4 NUMERICAL ANALYSIS BY COMPUTER

Numerical analysis of the deformation of
the rubber membrane was done using measurements of hydrostatic pressure. The
tensional rigidity of the membrane material (which was one of input constants
used in the analysis) was determined by a
tension test of the rubber membrane. The
influence of the specimen's width on the
tensional rigidity was investigated by
changing the ratio of specimen's width to
length (B/L) in the tension test. Figure
6 shows the results of the tension test.
This figure shows that tensional rigidity
is not influenced by the specimen's width.
Further, a value of 1,000 gf/cm (corresponding to a measurement of 3% of tensile
strain for the membrane material when 5
gf/cm² load was applied) was utilized from
this figure as the value of tensional
rigidity to be used for the analysis.
Figure 7 compares the analyzed result with
actual measurements: the values agreed
very well over the entire membrane surface.
This agreement shows that the pneumatic
membrane theory can be applied to the deformation analysis of the geotextile.

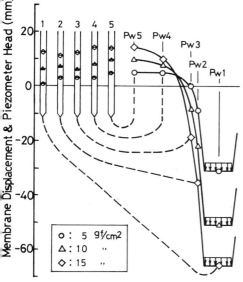

Fig. 4 Results of measurements

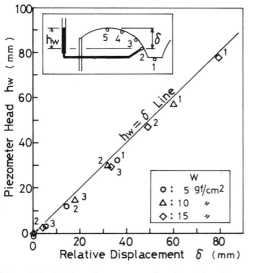

Fig. 5 Relation between relative displacement and piezometer head

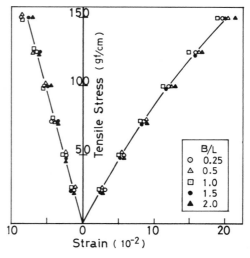

Fig. 6 Stress-strain curves of membrane

403

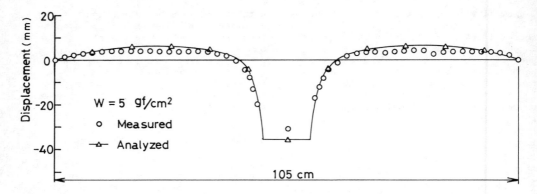

Fig. 7 Comparison of displacement between analysis result and measured values

5 LOADING EXPERIMENT ON COHESIVE GROUND

When water is used as the ground material, the ground cohesion, C_u, is zero. The geotextile behavior when the ground has some cohesion was investigated in the loading experiment using the materials shown in Table 1.

Table 1. Material properties

Material	Unit weight (gf/cm^3)	Cohesion (gf/cm^2)
Kaolin*	1.56	0.59
C.M.C.**	0.98	1.33

  * Moisture content 75%
** Concentration 2%

Features of the ground materials used in the model are as follows:
1) The CMC water solution has a unit weight nearly equal to water but has a larger cohesion than water, and
2) The kaoline clay has a higher unit weight and larger cohesion than water.
In this connection, these cohesions were determined by measurements with the vane shear test.
Figure 8 shows the loading test results, including also the results using water. These indicate that both settlement of the loaded part and maximum heaving at the membrane surface decreased with the cohesion of the kaoline clay (and with increasing unit weight). This phenomenon is found in all experiments of loading strength. It is clear that the soil unit weight greatly influences the displacement of

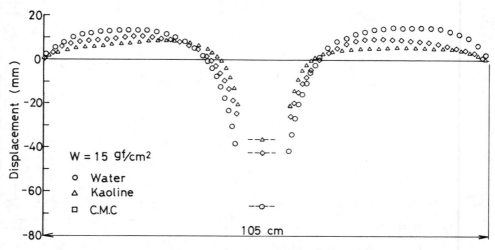

Fig. 8 Membrane displacement in the case of cohesive ground materials

geotextiles when the soil cohesion is small.

6 CONCLUSION

To analyze the behavior of a geotextile used for earth spreading work on very soft ground using the pneumatic membrane theory and to clarify the load acting on the geotextile, a series of experiments were carried out using a model. The conclusions derived from these results are as follows:

(1) The geotextile displacement near loading points agrees with an elliptic curve when a continuous load is applied.
(2) The upward pressure acting on the geotextile is proportional to the relative ground settlement from the maximum heaving part of the geotextile.
(3) The displacement of the geotextile is inversely proportional to the clay unit weight when the clay cohesion is small.

7 ACKNOWLEDGEMENT

Thanks are due to Professor Mamoru Kawaguchi of Hosei University for his many helpful suggestions and advise on analytical procedures given during the preparation of this paper.

REFERENCES

Watari, Y. & Higuchi, Y. 1986. Model test and analysis on geotextiles used for improvement of very soft ground under continuous loading. Soils and Foundations, Vol. 26, No. 4: 186-196 (in Japanese).

APPENDIX

ANALYTICAL METHOD FOR MEMBRANES

A deformation theory for a membrane surface can be systematized as a form of shell structures or suspension structures, with the basic formulas for a flat membrane surface being as follows.
The geometrical shape of an unstressed curved surface is defined by the following formula, for example an elliptical paraboloid shell, as shown in Figure 1:

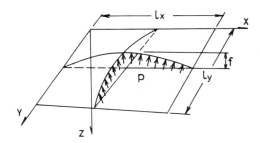

Fig. 1 Membrane surface expression

$$z = -16f(\xi - \xi^2)(\eta - \eta^2)$$

where
$$\xi = x/l_x, \eta = y/l_y$$

Expressing vertical displacement from the unstressed curved surface as when upward pressure is applied to the membrane surface, the following equations hold:

Equilibrium equation

$$H_x\frac{\partial^2}{\partial x^2}(z+w) + H_y\frac{\partial^2}{\partial y^2}(z+w) - p = 0$$

Elasticity law

$$H_x = \frac{E_x t_x}{2S_x - l_x}\left[-\int_o^{l_x}\frac{\partial^2 z}{\partial x^2}w\,dx + \frac{1}{2}\int_o^{l_x}\left(\frac{\partial w}{\partial x}\right)^2 dx\right]$$

$$H_y = \frac{E_y t_y}{2S_y - l_y}\left[-\int_o^{l_y}\frac{\partial^2 z}{\partial y^2}w\,dy + \frac{1}{2}\int_o^{l_y}\left(\frac{\partial w}{\partial y}\right)^2 dy\right]$$

By solving the above simultaneous equations, the displacement and the horizontal components H_x and H_y of membrane tension can be obtained. The symbols E, t and S represent Young's modulus of the membrane material, the membrane thickness and the initial length of the membrane, respectively. In addition, the analysis is based on the following conditions:

(1) As to the membrane material, the membrane surface has a large resistance against the tension in the fabric direction, with a linear relation between the tension and the strain, but a slight resistance against shearing force.

(2) Being nonlinear differential-integral of third degree, the fundamental equations necessitate numerical solution of nonlinear equations. However, judging from the slight influence due to neglecting the nonlinear terms in the case where the load is distributed on the overall membrane, the behavior may be taken as linear.

REFERENCES

Kawaguchi, M. & Chin, Y. 1968. On non-linearity of prestressed suspension roofs. Rep. Tech. Coll. Hosei Univ., No. 17: 1-26.

International Geotechnical Symposium on Theory and Practice of Earth Reinforcement/Fukuoka Japan/5-7 October 1988
© *1988 Balkema, Rotterdam. ISBN 90 6191 820 0*

Finite element analysis of grid reinforcement

K.Koga & G.Aramaki
Saga University, Saga, Japan

S.Valliappan
University of New South Wales, Sydney, Australia

ABSTRACT: Strip and grid reinforcements have often been used to strengthen various geotechnical structures such as retaining walls, embankments and foundations. Alternatively, geogrids and geocells can also be used for the same purpose. This paper describes the finite element analysis of soil reinforcement system which consists of geogrids used for two cases - one is an embankment on soft soil and the other is a footing foundation.

1 INTRODUCTION

Earth reinforcements are widely used at present for various structures such as embankments, foundations in soft soils, pavements and earth dams. For the purpose of reinforcing the soil, materials such as steel strip, geotextiles and polymer geogrids have been used. Even though the reinforcements in soil structures have become popular, the success or effective use of such reinforcements depend on the efficient design and construction of the reinforced soil structures. To achieve this aim, a proper analysis has to be carried out to determine the stresses in the materials used and the settlement or deformation of the structures. This in turn, requires the determination of material properties by proper testing procedures.

This paper describes the finite element analysis of soil-reinforcement system of geogrids with particular reference to an embankment on soft soil and a footing foundation.

In Japan, it has been found that the use of geogrids throughout the embankment permits higher compaction to be achieved. This allows a reduction in the width of the embankment, thus resulting in an economic construction. Embankments constructed on very soft subsoils may experience large settlements and geogrid mattress may improve the stiffness of the system and possibly reduce the amount of settlement. Similarly, in the case of foundations of structures on soft soils, the total settlement may be large or there may be differential settlement due to uneven loads. In such cases, the provision of geogrids may increase the bearing capacity and settlement characteristics of the subsoil.

2 FINITE ELEMENT ANALYSIS

For designing the soil reinforcements, empirical, experimental and numerical approaches have been used. The design of structures such as retaining walls, embankments and foundations on homogeneous soil situations can probably be done using conventional analytical methods. However, if the soil conditions are nonhomogeneous and the soil is of soft or weak type, then in order to arrive at efficient design, computer oriented analysis such as the finite element technique has to be adopted. The advantage of such an analysis is that displacement distribution and stress distribution can be obtained both in the subsoil as well as the soil reinforcement system. Nevertheless, it should be realized that the accuracy of the finite element results depends on the appropriate material properties used and the type of modeling adopted for the analysis. In the finite element analysis, the complete soil reinforcement system can be modelled using individual elements such as bar elements for reinforcements. continuum elements for soil and joint elements for interface behaviour or by using composite elements which comprise of the soil-reinforcement system as a whole. In the latter case, the properties of the composite element have to be evaluated either experimentally or by a separate numerical analysis.

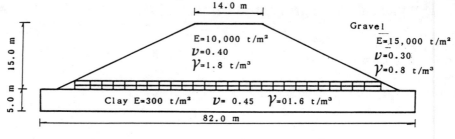

Bar element   E=0.18x10 t/m²   A=0.13x10⁻² m²

Joint element Ks=0.4x10³   Kn=0.4x10⁵ t/m²

Fig. 1   EMBANKMENT

In this paper, the former approach, that is, individual elements for soil, reinforcement and interface behaviour has been used.

3 NUMERICAL RESULTS

Fig.1 shows the geometrical configuration of the embankment analysed using the geogrid as the reinforcement. The assumed materials properties are shown in the figure. This embankment is similar to the one analysed by Jones(1985). Two variations of the grid design were considered

in the analysis - one with a single layer of 1m grid depth and the other two layers of each 1m depth. The settlement profile along a horizontal section in the subsoil just below the grid reinforcement, is shown in Fig.2(a) whereas Fig.2(b) shows the settlement profile along vertical section at the center line of the embankment. It can be noted that there is some minor advantage in using the geogrid regarding the reduction of settlement. However, there is practically no difference between the alternative designs. Fig.3 shows the stress distribution in the reinforcement for the two designs and the distribution is as would be expected.

Fig.4(a) shows the vertical stress (σ_y) distribution along the horizontal section in clay layer, just below the reinforcement. It can be observed that for all the three cases - no grid and two alternative grid designs - the distribution is almost the same. The maximum principal stress along the same horizontal section for the three cases considered has been plotted in Fig.4(b). As can be noted that using the geogrid reduces the maximum tensile stress by nearly 50 %.

Thus, it can be stated that in the case of embankment, providing the geogrid as

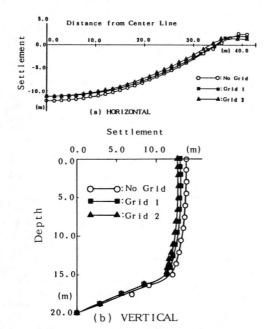

(a) HORIZONTAL

(b) VERTICAL

Fig. 2 SETTLEMENT PROFILE

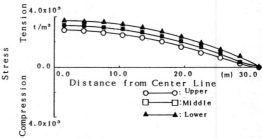

Fig. 3 STRESS DISTRIBUTION OF REINFORCEMENT

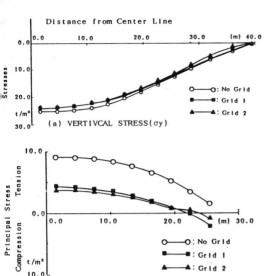

(a) VERTIVCAL STRESS(σy)

Distance from Center Line

○——○: No Grid
■——■: Grid 1
▲——▲: Grid 2

(b) PRINCIPAL STRESS

○——○: No Grid
■——■: Grid 1
▲——▲: Grid 2

Fig. 4 STRESS DISTRIBUTION

the reinforcement did not influence the settlement profile. This result is similar to the observation made by Symes(1985). For this case, the important thing is that the tensile stress in the soft layer has been reduced by nearly 50 % with the use of geogrid which would assist in preventing the failure of the subsoil.

Fig.5 shows the strip footing on cohesive soil containing horizontal layers of reinforcement, with two alternatives - one with strips and the other with geogrids. The material properties adopted for the analysis are also given in the figure. This problem is similar to the one analysed by Binquet and Lee(1975). Two cases of strip footings - surface footing and embedded footing were considered.

The vertical displacement(settlement) along the horizontal surface has been plotted in Fig.6(a) for the surface footing and in Fig.6(b) for the embedded footing. The maximum settlement of the surface footing for the case without any reinforcement is 8.44cm compared with the theoretical solution of 8.76cm. By providing only strip reinforcement the maximum deflection was reduced only very little (8.18cm) whereas the provision of grid reinforcement reduced the maximum deflection to 6.82cm. For the case of embedded footing the maximum displacement for the unreinforced soil medium is 4.28cm whereas for the strip reinforcement it is 4.11cm and for the grid reinforcement, it was reduced to 2.78cm.

Fig.7 shows the distribution of vertical stress(σy) for the case of surface footing, along horizontal sections between the four layers of reinforcement. As can be seen, the distribution for the three cases considered is similar. The stress distribution for the case of embedded footing is plotted in Fig.8. Again, it can be noted that practically there is no difference in results among the three cases.

It was observed that in the case of surface footing, the maximum tensile stress was not reduced by the use of reinforcement whereas in the case of embedded footing, it was reduced considerably, for example by about 30% with the use of grid reinforcement.

The maximum tie forces for both cases of reinforcements are given in Table 1. It can be noted that for both surface and embedded footing, the maximum tie force in the case of strip reinforcement is greater than that of grid reinforcement.

The bearing capacity ratio(BCR) calculated according to Binquet and Lee(1975) is 1.24 for surface footing and 1.54 for embedded footing. Binquet and Lee stated

$q=2.4MN/l in m$

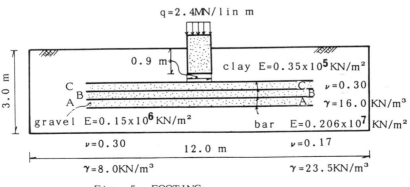

clay $E=0.35\times10^5 KN/m^2$
$\nu=0.30$
$\gamma=16.0 KN/m^3$
gravel $E=0.15\times10^6 KN/m^2$
bar $E=0.206\times10^7 KN/m^2$
$\nu=0.30$
$\nu=0.17$
$\gamma=8.0KN/m^3$
$\gamma=23.5KN/m^3$

0.9 m
3.0 m
12.0 m

Fig. 5 FOOTING

that the experimental data showed that the BCR values are between 1.5 and 4.0.

4 CONCLUSIONS

From the foregoing discussion of the numerical results related to the two examples - embankment and footing - it can be stated the use of geogrids as soil reinforcements
(1) Reduce the tensile stresses in the weak subsoil for the case of embankments.
(2) Does not influence the settlement characteristics of the embankment.
(3) Reduce the maximum settlement considerably both in the case of surface and embedded footings.
(4) Does not influence the tensile stress distribution in the case of surface footing
(5) Considerably reduces the maximum tensile stress for the case of embedded footing.
It was also noted that geogrids are better than strips as soil reinforcements.

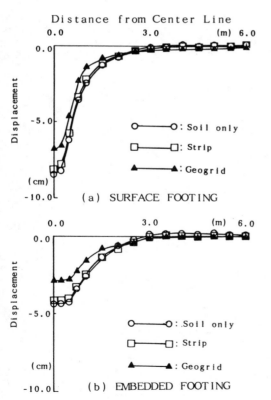

Fig. 6 VERTICAL DISPLACEMENT

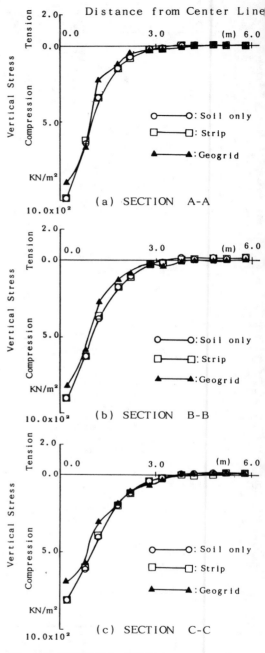

(a) SECTION A-A

(b) SECTION B-B

(c) SECTION C-C

Fig. 7 VERTICAL STRESS - SURFACE FOOTING

410

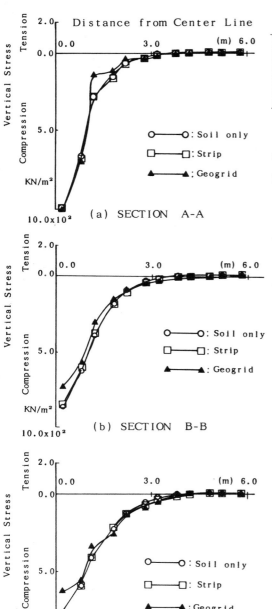

(a) SECTION A-A

(b) SECTION B-B

(c) SECTION C-C

Fig. 8 VERTICAL STRESS - EMBEDDED
    FOOTING

Table-1 Maximum Tensile Forces
(unit KN)

	strip	geogrid
surface footing	1580	936
embedded footing	1050	998

5 REFERENCES

Binquet,J. and Lee,K.L.(1975), Bearing
    Capacity Analysis of Reinforced Earth
    Slabs, Journal of Geotechnical Division,
    ASCE, Vol.101, GT12, pp. 1257-1276.

Jones,C.J.F.P.(1985), Earth Reinforcement
    and Soil Structures, Butterworths,
    London.

Symes,M.F.(1985), Written Discussion on
    Embankments, Polymer Grid Reinforcement,
    Thomas Telford Limited, London, pp. 113-
    114.

411

International Geotechnical Symposium on Theory and Practice of Earth Reinforcement / Fukuoka Japan / 5-7 October 1988
© 1988 Balkema, Rotterdam. ISBN 90 6191 820 0

Seismic resistance of reinforced embankment by model shaking table tests

Y.Koga, Y.Ito, S.Washida & T.Shimazu
Public Works Research Institute, Ministry of Construction, Japan

ABSTRACT: It was shown from 14 cases of model shaking table tests of embankments reinforced with nonwoven fabrics, plastic nets and steel bars, that the seismic resistance of an embankment is increased if a reinforcing element of high tensile stiffness is used to anchor the embankment to the bed slope. The local failure of the fill slope surface must also be prevented by an appropriate protection work. Moreover, the damage degree of embankments agreed well with calculation results by a circular slip surface method which considers the effect of the tensile force of reinforcing element.

1. INTRODUCTION

Reinforcing embankments with geotextile or steel plates, etc. has recently attracted wide interest. The evaluation of the effect of geotextile or other reinforcing elements on the seismic resistance of embankments is an important issue in such a country with high seismicity as Japan. This paper presents a seismic effect of reinforcement with geotextile and steel bar on the embankments constructed on an inclined ground by a series of model shaking table tests and their stability analyses (Koga et al., 1988).

2. MODEL SHAKING TABLE TEST

2.1 Models

Fig. 1 shows an example of cross section of test models. Each model consists of a bed slope and a fill part on it. The model was made in a steel box of 8 m long, 2 m high and 1 m wide. Both side slopes of one model were made in different conditions and two cases were simultaneously tested using one model. As summarized in Table 1, 7 models with 14 cases were so far tested.

Following conditions were varied in each case : slope surface gradient, existence of benchs on a bed slope, kind of reinforcing element, reinforcement spacing, slope protection, etc.

The bed slope was made of wet sandy silt (ML, w = 19 %) sufficiently compacted by a wooden block so as to be stable against vibration. The fill part was made of air dried sand (SP, w = 6-12 %) which was compacted to a relative density of about 50 % by human foot. Reinforcing elements were nonwoven fabrics, plastic nets and steel bars. These properties are described in Table 2. Nails, 15 cm long, were used to fix nonwoven fabrics and plastic nets to the bed

slopes. Steel bars were penetrated by 30 cm to the bed slope. The overlapped length of nonwoven fabrics and plastic nets were 30 cm irrespective of the spacing.

A following scaling law was adopted; the reinforcement ratio R defined by the following equation is same for prototype and model assuming the Poisson's ratio ν = 0 for reinforcing elements (Tatsuoka et al., 1985).

$$R = \frac{-\varepsilon_{3R} \cdot E \cdot t}{\sigma_{30} \cdot \Delta H} \qquad (1)$$

where ε_{3R} = average horizontal tensile strain of the reinforced soil ($\varepsilon_{3R} < 0$); E = Young's modulus of the reinforcing element; t = thickness of the reinforcing element; σ_{30} = horizontal confinig pressure; ΔH= spacing of the reinforcing element.

This reinforcement ratio represents the ratio of strength increase of a reinforced soil to an unreinforced one at a specified reference strain. The reinforcing element was chosen so that the model scale was to be 1/7 to the prototype.

2.2 Excitation conditions and measurement

Excition was conducted under sinusoidal wave loading of 4Hz and 10 sec. Excitation acceleration was increased step by step ranging from 100 to 800 gal. The shaking table was stopped for each excitation level and the damage degree was observed and horizontal and vertical displacements of the observation marks were measured. The acceleration, settlement and tensile force of nonwoven fabrics and steel bars were recorded during the excitation. Moreover the overall deformation of the model was observed by taking photos of drawn meshes with coloured sands in front surface of the model.

2.3 Test results

Fig.2 shows the overall deformation of the model 4

413

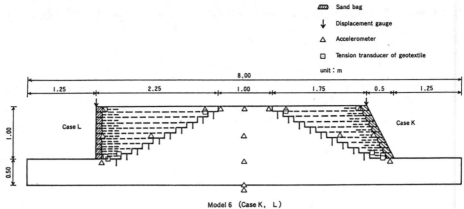

Model 6 (Case K, L)

Fig.1 Example of Test Models

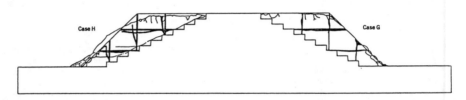

Model 4 (Case G, H) after 700gal excitation

Fig 2 Deformation after Excitation

Table 1 Test Model

Name of embankments (case)	Model		Kinds of reinforcing elements	Reinforcement spacing	Anchorage to bed slope	Overlapping	Slope gradient	Sand bag on slope surface	Water content (%)
A	1	R	None	——	——	——	1 : 1.5	——	11.5
B		L	Nonwoven fabrics	10cm	No	No	1 : 1.5	No	11.3
C	2	R	None	——	——	——	1 : 1.5	——	9.3
D		L	Nonwoven fabrics	10cm	Yes	No	1 : 1.5	No	10.1
E	3	R	None	——	——	——	1 : 1	——	8.1
F		L	Nonwoven fabrics	10cm	No	Yes	1 : 1	Yes	7.8
G	4	R	do	10cm	Yes	Yes	1 : 1	Yes	6.3
H		L	do	20cm	Yes	Yes	1 : 1	Yes	6.5
I	5	R	Reinforcing bar	Horizontally 10cm Vertically 20cm	Penetrated by 30cm	——	1 : 1	——	6.3
J		L	Plastic nets	20cm	Yes	Yes	1 : 1	Yes	6.5
K	6	R	Nonwoven fabrics	10cm	Yes	Yes	1 : 0.5	Yes	8.3
L		L	do	10cm	Yes	Yes	1 : 0	Yes	8.4
M	7	R	None	——	——	——	1 : 1	——	15.2
N		L	Nonwoven fabrics	10cm	Yes	Yes	1 : 1	Yes	15.4

Table 2 Properties of Reinforcing Elements

Kinds	Properties
Nonwoven fabrics	Nylon 70%, Polyesther 30%, Thickness 0.2mm
Plastic net	Polyethylene 100%, Grid 2.5×2.5mm
Steel bar	Pianowire, diameter 3.5mm

after the excitation of 700 gal as an example of test results. When amodel embankment was excited under such a large acceleration, it slided along a slip surface and its crest settled. Such a figure shows the damage mode and feature of each embankment model.

Fig.3. shows a relationship of a cumulative crest settlement and an excitation acceleration.
The damage features of reinforced embankments are summarized as follows.
1) Unreinforced embankments slipped near the boundary to the bed slope. Reinforced ones with slope surface gradients of 1:1.5 and 1:1.0 also deformed relatively largely near the above boundary, however, the deformation was smaller than that of unreinforced ones. The deformed region was in a shallower part for the reinforced one with steeper slope gradients.
2) The deformation of reinforced embankments were less than unreinforced ones for the same slope gradient (Fig.3(a)).
3) As the spacing of nonwoven fabrics becomes smaller, the crest settled less (Fig.3(a)).
4) An embankment reinforced with plastic nets, whose tensile stiffness is larger, settled less than one reinforced with nonwoven fabrics in the same spacing (Fig.3(a)).
5) Embankments settled and deformed less when the nonwoven fabrics were fixed to the bed slope (Fig.3 (b)).
6) Embankments deformed less when nonwoven fabrics or plastic nets were overlapped to their slope surface (Fig.3(a)),however, the overlapped geotextiles were pulled out during a large vibration at the upper layer where the overburden pressure was small.
7) As the slope got steeper,the deformation of the reinforced and unreinforced embankments became larger, however,the tendency of deformation increase of the reinforced ones was assumed rather smaller (Fig.3(c)).
8) Penetrated steel bars showed an effect to decrease the settlement of the embankment. However,its effect was not enough because the shallow part of the slope was removed when the slope protection was not effective. It suggests that steel bars must be connected to the slope protection works so that the embankment be effectively reinforced.

3. STABILITY ANALYSIS

3.1 Stability analysis method

A stability of a reinforced embankment must be investigated from various viewpoints because it can possibly fail in various modes. A stability of an unreinforced embankment is generally examined for its overall stability by a slip surface method. In the case of a reinforced embankment, not only the stability of an overall reinforced region, but such a local stability as a pulling out or a breakage of a reinforcing element and a bulging of soil of a slope surface must be examined.
The main failure mode of the embankment in these tests was that the fill slipped near the boundary to the bed slope and not that reinforcing elements were pulled out or broken apart. Therefore it was presumed that not a small strain occurred in the reinforcing element. Consequently it was considered important to

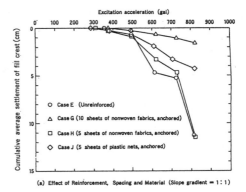

(a) Effect of Reinforcement, Spacing and Material (Slope gradient = 1 : 1)

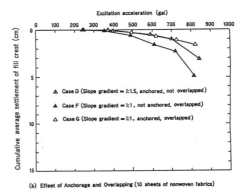

(b) Effect of Anchorage and Overlapping (10 sheets of nonwoven fabrics)

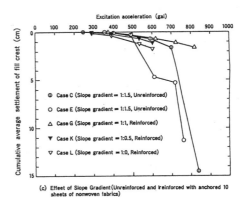

(c) Effect of Slope Gradient (Unreinforced and reinforced with anchored 10 sheets of nonwoven fabrics)

Fig.3 Relationship between Excitation Acceleration and Cumulative Settlement of Fill crest

examine the overall failure of the embankment which was observed in the tests.
There is not yet an established method as a seismic stability analysis method of a reinforced embankment. Since a circular slip surface method with a seismic coefficient is frequently adopted for seismic stability analyses, it was extended as follows to include the effect of reinforcing element.

415

$$Fs = \frac{\Sigma R[c\ell+\{W\cdot\cos\alpha-k_h W\cdot\sin\alpha+T_r\cdot\sin(\alpha+\theta)\}\cdot\tan\phi]}{\Sigma[R\cdot W\cdot\sin\alpha+y\cdot k_h W\cdot\cos\alpha-R\cdot T\cdot\cos(\alpha+\theta)]}$$

(2)

where R = radius of a slip circle; c = cohesion of soil; ϕ = friction angle of soil; W = self weight of a slice; ℓ = length of an arc of a slice; b = width of a slice; k_h = horizontal seismic coefficient; y = vertical distance between the center of a circular arc and the center of gravity of a slice; T_r = tensile force of reinforcing element when the reinforced earth reaches a failure state; T= tensile force of reinforcing element acting on the sliding earth mass in equilibrium; α, θ = angles shown in Fig.4.

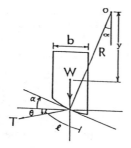

Fig.4 Slice for a Circular Slip Surface Analysis

The calculated factor of safety can be compared with that for unreinforced embankments, which enables to evaluate the stability increase by the reinforcement. Above equation is an extension of the Fellenius formula, which considers the effect of the tensile force of the reinforcing element to bind and to hold the sliding earth mass.

The characteristics of above equation is as follows.

The numerator of the equation to represent a resistant moment is obtained from a tensile force of the reinforcing element at failure, whereas the denominator of the equation to represent a sliding moment from a working tensile force in equilibrium. The tensile force of reinforcing element is determined from its tensile strain caused by the deformation of the surrounding soil, which was obtained so that the strain is compatible with shear strain of the soil along a failure surface (band) based on the following assumptions (see Fig.5).

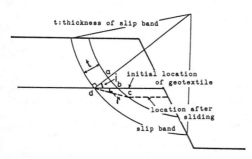

Fig.5 Calculation of Tension of Geotextile

1) A slip surface consists of a slip band with a constant thickness of a constant shear strain.

2) Reinforcing element deforms in a body with a soil therefore it cannot be separated from a soil.

3) The shear strain of the slip band generates a tensile strain of a reinforcing element intersecting the slip band.

4) The tensile strength of the reinforcing element is large enough to hold the slip of the earth mass.

As a consequence of these assumptions, the tensile strain of the reinforcing element ε_θ is given as below.

$$\varepsilon_\theta = \frac{c\,d-b\,d}{b\,d} = \frac{\cos i-\cos(i+\gamma')}{\cos(i+\gamma')}$$

(3)

where i = angle between the reinforcing element and normal line to a slip surface; γ' = corrected shear strain of a soil, generally not equal to that of the slip band.

The working tensile force T of the reinforcing element of the sliding mass in equilibrium is provisionally calculated from the following equation by assuming the linear relationship between all the mobilized forces on the mass and the deformation of the mass (see Fig.6).

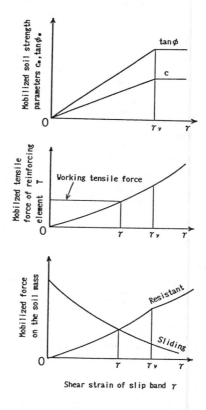

Fig.6 Relationship between Mobilized Force and Deformation of Soil Mass

416

$$F_s = 1$$

$$= \frac{\Sigma R \ [c_m \ell + \{W \cdot \cos\alpha - k_h W \cdot \sin\alpha + T \cdot \sin(\alpha + \theta)\} \cdot \tan\phi_m]}{\Sigma \ [R \cdot W \cdot \sin\alpha + y \cdot k_h W \cdot \cos\alpha - R \cdot T \cdot \cos(\alpha + \theta)]} \quad (4)$$

where c_m = mobilized cohesion of soil; ϕ_m = mobilized friction angle of soil.
The linear relationship implies

$$\frac{c_m}{c} = \frac{\tan\phi_m}{\tan\phi} = \frac{\gamma}{\gamma_y} \quad (5)$$

where γ = shear strain of slip band; γ_y = shear strain of soil at failure.

3.2 Stability analysis conditions

Stability analyses were conducted on some of the test models changing the excitation accelerations.
The analyzed models were unreinforced embankments and reinforced ones whose reinforcing elements were fixed to the bed slope.
The unit weight, cohesion and friction angle of the fill used for the calculation were 1.56 tf/m³, 0.3 tf/m² and 33° respectively. It was assumed that the shear strength of the bed slope was large enough and no slip surface to cut the bed slope was considered. While the friction angle of the fill was obtained from a triaxial test the cohesion of the fill was obtained from a critical height of a vertical cut of a ground made in the same manner as that of shaking tests.

The formula was as follows.

$$c = \frac{\gamma_t H_c}{2.67} \ / \ \tan\left(45° + \frac{\phi}{2}\right) \quad (6)$$

where γ_t: unit weight of ground; H_c: critical height.
The tensile stiffness of the reinforcing element to calculate a tensile force is shown in Table 2. The seismic coefficient was given based on the measured acceleration during shaking table tests.

3.3 Stability analysis results

Fig.7 shows a relationship between safety factors calculated from Eq.(2) and excitation acceleration.
The shear strain of 8% is taken as a failure criterion of an embankment on the basis of a triaxial compression test result, and the tensile force of the reinforcing element T_r was calculated to correspond with the above strain. Obviously no tensile force of the reinforcing element was considered for unreinforced embankments.
Fig.7(a) indicates the followings.
1) Among the embankments of a slope gradient of 1:1, the least factor of safety was obtained for an unreinforced embankment.
2) A little larger factors were obtained for the cases of 5 and 10 sheets of nonwoven fabrics.
3) The largest factor was obtained for that of 5 sheets of plastic nets.
4) The differences of safety factors among the unreinforced and 2 nonwoven fabrics-reinforced

embankments were slight, which indicates that the reinforcement effect by such a reinforcing element with a small tensile stiffness is small. Besides the plastic nets with a larger tensile stiffness gave fairly larger factors of safety.
Fig.7(b) shows that the safety factor of the unreinforced and reinforced embankments remarkably decreases as the slope gets steeper.

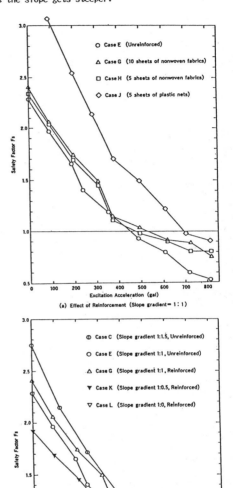

(a) Effect of Reinforcement (Slope gradient = 1:1)

(b) Effect of Slope Gradient (Unreinforced and reinforced with anchored 10 sheets of nonwoven fabrics)

Fig.7 Relationship between Excitation Acceleration and Safety Factor

417

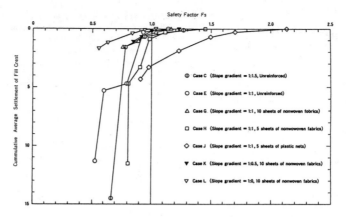

Fig.8 Relationship between Safty Factor and Measureed Settlement

Fig.8 shows a relationship between a calculated safety factor and a cumulative settlement of a fill crest. In general, a fairly unique relationship with little scatter exists in Fig.8 except the case of a slope gradient of 1:0.5 and 1:0 ,and that of plastic nets,irrespective of the existence of reinforcing elements and reinforcement spacing. Moreover the settlement rapidly increases when Fs is less than from 0.95 to 1.0 in the case of nonwoven fabrics and a slope gradient of 1:1 and 1:1.5, which suggests the calculatiuon method of Eq. (2) is useful for the assessment of a failure or deformation of a reinforced embankment whose slope surface is not so steep.

The reason that the calculated factor of safety is so small for the embankment of steep slope is because the safety factor is calculated by use of the Fellen-ius method which neglects the interslice force, which gives smaller safety factor than that with the above force, and also the Eq. (2) neglects the resistant effect of the overlapped geotextile slope facing and also the crest settlement of a steep slope becomes small against a certain strain of slip surface because the slip surface gets smaller as the slope gets steeper.

Followings are summarized from the above analysis results.

1) The calculation results by a circular slip surface method with a seismic coefficient method, which considers the binding and holding effect by a tensile force of reinforcing elements intersecting the slip surface, agreed well with the experimental results except the cases of steep slope.

2) It proved useful to calculate a tensile strain of a reinforcing element that is compatible with a shear deformation along a slip surface of an embankment to obtain a tensile force of a reinforcing element.

4. CONCLUSION

1) Model shaking table tests of reinforced embank-ments with nonwoven fabrics, plastic nets and steel bars showed that a reinforced embankment has a fairly large seismic resistance if a reinforcing element with a large tensile stiffness is effectively fixed to a

bed slope. Besides the slope protection to prevent a local slope surface failure is also critical.

2) The degree of deformation of the embankment corresponded well to the calculation results by a circular slip surface method which considers the effect of tensile force of reinforcing element inter-secting the slip surface except the cases of steep slope.

3) The selection of an appropriate slope protection works and the evaluation of its effect on the stabi-lity need a further research.

REFERENCES

Koga, Y., et al. : Experimental Study on Seismic Resistance of Reinforced Embankment (in Japanese), Technical Memorandum of PWRI, 1988.3.

Tatsuoka, F., et al. : Reinforcing of Cohesive-Soil-Embankment with Unwoven Fabric (in Japanese), Tsuchi-To-Kiso, JSSMFE, Vol.33, No.5, 1985.5.

International Geotechnical Symposium on Theory and Practice of Earth Reinforcement / Fukuoka Japan / 5-7 October 1988
© *1988 Balkema, Rotterdam. ISBN 90 6191 820 0*

Reinforcement against deep-seated failure: Stability analysis

Dov Leshchinsky & Deirdre S. Smith
University of Delaware, Newark, Del., USA

ABSTRACT: At the end of construction, embankments over soft clayey foundations have their minimal stability. At this stage, it may be necessary to bridge over a potentially short-term instability. This can be done by installing a reinforcing agent capable of carrying tensile force at the embankment-foundation interface. An analysis to determine the stability of such reinforced embankments against deep-seated rotational failure based upon limit-equilibrium is presented. The analysis accounts for a foundation with linearly changing undrained shear strength with depth. It also allows for the reinforcement's tensile force to act either horizontally, as installed, or tangential to the slip surface, contributing most to the resisting moment. The analysis results with the required distribution of the reinforcement's tensile resistance so that specified safety factors are exceeded through any cross-section of the embankment.

1 INTRODUCTION

Quite often embankments are built over soft clayey foundations. At the end of construction, the structure has its minimal stability against deep-seated failure. Since this stability increases with time it may be advantageous to design the embankment for lower initial safety while providing a means to bridge a potentially short-term instability. One possible, and often economical, solution is the use of a reinforcing matrix, such as a geotextile, geogrid, or steel mesh, which is capable of carrying tensile force. This reinforcement is usually placed at the embankment-foundation interface.

The objective of this paper is to extend the analysis presented by Leshchinsky (1987), so that the effects of a foundation with linearly changing undrained shear strength, as well as the effects of various orientations of the reinforcement tensile resistance, could be studied. The analysis determines the required distribution of the reinforcement tensile resistance so that adequate safety against deep-seated failure through any section of the structure is attained. It is done for two possible extreme inclinations of reinforcement force: horizontal (i.e., as installed)

and tangential to the potential slip surface.

It should be pointed out that the presented analysis is only one step in a design of such an embankment. Additional modes of failure, such as translational failure along the reinforcement or functional failure due to excessive deformations, as well as the selection of the appropriate reinforcing agent and verification of its capacity to withstand the required pullout, are other aspects of design which must be investigated.

2 FORMULATION

Fig. 1 presents a cross-section through a reinforced embankment. The embankment possesses a height H, a face inclination *i* and a free-draining cohesionless soil with an average unit weight γ and an internal friction angle ϕ. The soft foundation underlying the embankment is a saturated isotropic clay extending to depth d. The clay's undrained shear strength may remain constant or be linearly changing with depth as follows:

$$c_u(y) = c_{u_o} - s \cdot y \tag{1}$$

where $c_u(y)$ and c_{u_o} are the undrained shear strengths at any depth y ($d \geq y \geq 0$)

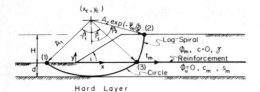

Fig. 1   Failure mechanism

and at the top of the foundation (i.e., at y=0), respectively; and s is the rate of shear strength increase.

A reinforcing sheet, which is capable of developing tensile resistance $t(x)$, is located at the embankment-foundation interface. Since the inclination of $t(x)$ is unknown, it is assumed to act horizontally, as installed, or tangential to the slip surface, contributing the most resistance to rotational failure.

The employed analysis is based upon limiting equilibrium. It is assumed therefore that the structure shown in Fig. 1 is on the verge of collapse. For this state to be realized, however, the strengths of all materials involved (i.e., ϕ, c_u and $t(x)$) must be mobilized simultaneously. As is always the case in design, the structure must possess margins of safety. In the present problem these margins are applied to the strength parameters of each material. Subsequently, each strength parameter is scaled to an appropriate value so that an artificial state of limit-equilibrium is produced:

$$\phi_m = \tan^{-1}(\psi_m) = \tan^{-1}\left[\frac{\tan\phi}{F_\phi}\right] \qquad (2)$$

$$c_m = \frac{c_{u_o}}{F_c} \quad \text{and} \quad s_m = \frac{s}{F_c} \qquad (3)$$

$$t_m(x) = \frac{t(x)}{F_r} \qquad (4)$$

where $\psi = \tan\phi$; the subscript m symbolizes mobilization; and F_ϕ, F_c and F_r are the factors of safety with respect to the friction, cohesion and reinforcement's tensile resistance, respectively. Clearly, failure occurs only when all strengths are fully mobilized, i.e., $F_\phi = F_c = F_r = 1$.

The reinforced-embankment problem can now be stated: *For a defined structure (i.e., given i, γ, H, ϕ, d, c_{u_o} and s), determine the required distribution of tensile resistance, t(x), of a reinforcing sheet so that the prescribed safety factors F_ϕ, F_c and F_r are exceeded everywhere.*

It should be pointed out that utilization of $t(x)$ inherently assumes that the reinforcement will not break before the soils' plastic strengths have already been reached.

To formulate the problem, first the failure mechanism must be postulated. Fig. 1 illustrates this mechanism: a log-spiral between points (2) and (3), and a circular arc between points (3) and (1). Both segments of the slip surface have their pole located at (x_c, y_c). This mechanism was implicitly suggested by Baker and Garber (1978) who arrived at it through variational extremization. Leshchinsky (1987) used it for reinforced embankment problems. For this mechanism, the moment equilibrium equation written about the pole (x_c, y_c) can be stated explicitly as

$$M = \int_{X_1}^{X_3} [c_m + s_m y][(y-y_c)-(x-x_c)y'] dx$$

$$+ \int_0^{X_3} \gamma y_c (x-x_c) dx + \int_0^{X_2} \gamma(\bar{y}-y_c)(x-x_c) dx$$

$$- \int_{X_3}^{X_2} \gamma(y-y_c)(x-x_c) dx - M_t = 0 \qquad (5)$$

where $y'=dy/dx$; $\bar{y}$=embankment surface function; and M_t=moment component due to the reinforcement's tensile resistance force. It is useful to introduce the following nondimensional notations:

$$N_m = \frac{1}{\gamma H} \frac{c_{u_o}}{F_c} \quad \text{and} \quad \alpha = \frac{1}{\gamma} \frac{s}{F_c} \qquad (6.1)$$

$$T_m = \frac{1}{\gamma H^2} \frac{t}{F_r} \qquad (6.2)$$

$$X = \frac{x}{H} \quad \text{and} \quad Y = \frac{y}{H} \qquad (6.3)$$

Eq. (5) can then be divided by γH^3 and the nondimensional notations be substituted to give

$$\bar{M} = \int_{X_1}^{X_3} [N_m + \alpha Y][(Y-Y_c)-(X-X_c)Y'] dX$$

$$+ \int_0^{X_3} Y_c(X-X_c) dX + \int_0^{X_2} (\bar{Y}-Y_c)(X-X_c) dX$$

$$- \int_{X_3}^{X_2} (Y-Y_c)(X-X_c) dX - \bar{M}_t = 0 \qquad (7)$$

420

where $\bar{M} = M/\gamma H^3$ and $\bar{M}_t = M_t/\gamma H^3$.

Introducing a polar coordinate system as defined in Fig. 1 into Eq. (7) and carrying out the integration gives:

$$\bar{M} - m_1 + m_2 + m_3 \, T_m = 0 \qquad (8)$$

where:

$$m_1 = A_2^3 \left\{ \frac{\cos^3\beta}{3} + \frac{\psi_m}{1+9\psi_m^2} \, (\sin\beta - 3\psi_m \, \cos\beta) \right.$$

$$\left. \times \exp(-3\psi_m\beta) \right\}_{\beta_1}^{\beta_2}$$

$$+ \frac{A_2^2 \, Y_c}{2} \exp(-2\psi_m\beta) \, \sin^2\beta \Big|_{\beta_1}^{\beta_2} + \frac{X_c^2}{2}$$

$$- \frac{A_2^2}{2} \exp(-2\psi_m\beta_2) \, \sin^2\beta_2$$

$$- \frac{\cot(i)}{6} \, [3X_c - \cot(i)] \qquad (9.1)$$

$$m_2 = 2A_1^2 \, [\beta_1 N_m + \alpha(A_1 \sin\beta_1 - Y_c\beta_1)] \qquad (9.2)$$

and

$$m_3 = A_1 \, \cos\beta_1 \qquad (9.3)$$

or

$$m_3 = A_1 \qquad (9.4)$$

where Eq. (9.3) is to be used when the reinforcement force inclination is assumed horizontal and Eq. (9.4) when it is assumed tangential; the constants A_1, A_2, β_1, β_2 are defined in Fig. 1.

Eq. (8) can be rewritten to fit the problem statement

$$T(X) = \frac{t(x)}{\gamma H^2} = -F_r \left\{ \frac{m_1 + m_2}{m_3} \right\} \qquad (10)$$

In Eq. (10) one seeks max(t) at a particular interface location x for specified values of F_r, F_ϕ and F_c. This is equivalent to searching min(F_r) for a problem where t(x) is given; however, Eq. (10) representation of the problem is more convenient in a design framework.

Notice that if $\alpha=0$ in Eq. (9.2), Eq. (8) degenerates then to the moment equation presented by Leshchinsky (1987). Examining Eqns. (8) through (10) it is clear that for a defined problem (i.e., given i, γ, H, ϕ, F_ϕ, c_{u_o}, s, d, F_c, F_r), there exist seven unknowns: x_c, y_c, A_1, A_2, β_1, β_2 and t(x). There is, however, only one equation available: Eq. (10).

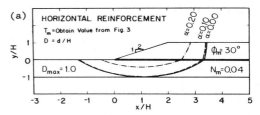

(a) HORIZONTAL REINFORCEMENT
T_m=Obtain Value from Fig. 3
$D = d/H$
$\alpha=0.20$ $\alpha=0.10$ $\alpha=0.00$
$\phi_m = 30°$
$D_{max} = 1.0$
$N_m = 0.04$

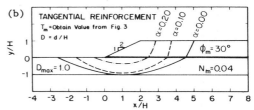

(b) TANGENTIAL REINFORCEMENT
T_m=Obtain Value from Fig. 3
$D = d/H$
$\alpha=0.20$ $\alpha=0.10$ $\alpha=0.00$
$\phi_m = 30°$
$D_{max} = 1.0$
$N_m = 0.04$

Fig. 2 Traces of critical slip surfaces:
(a) Horizontal inclination of reinforcement force
(b) Tangential inclination of reinforcement force

An additional five equations can be derived through geometrical relationships [see Leshchinsky (1987)]. Subsequently, a numerical procedure has been developed to find the required maximum tensile resistance, max(t), at a specified interface location, x, for all possible combinations of one of the unknown parameters. A similar procedure has been presented by Leshchinsky (1987). In the current problem, however, max(t) is determined for a foundation with linear change in strength as function of depth and d is taken just as a search limit.

3 RESULTS

Some special features of the formulation presented here as compared to Leshchinsky (1987) include the introduction of a foundation with linearly changing undrained shear strength and consideration of two extreme inclinations of reinforcement tensile resistance. To realize the effects of these features on the critical slip surface refer to Fig. 2. It can be seen that as α increases the extent of the critical slip surface decreases. For $\alpha=0$ the penetration depth of the slip surface equals the thickness of the foundation d. When α assumes a significant value, however, the penetration depth is less than d. It can also be observed that for small values of α, the critical location X, which defines the maximum required tensile resistance,

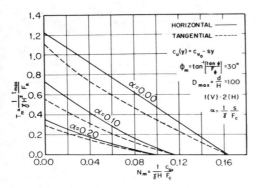

Fig. 3 Required maximum tensile resistance versus strength of foundation (D_{max}=1.00)

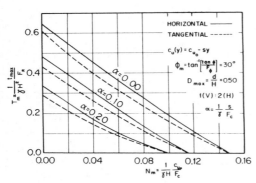

Fig. 4 Required maximum tensile resistance versus strength of foundation (D_{max}=0.50)

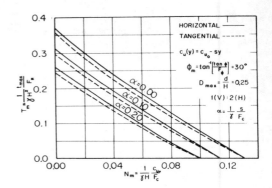

Fig. 5 Required maximum tensile resistance versus strength of foundation (D_{max}=0.25)

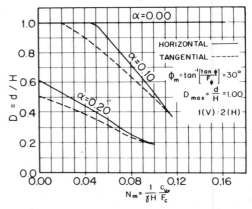

Fig. 6 Critical slip surface depth versus strength of foundation (D_{max}=1.00)

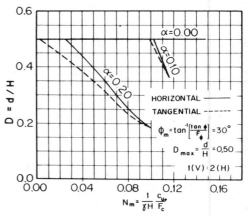

Fig. 7 Critical slip surface depth versus strength of foundation (D_{max}=0.50)

is larger for tangential inclination of the reinforcement force. This X and the difference of its values for horizontal and tangential inclinations are rapidly decreasing as α increases.

Figs. 3 through 5 are design charts illustrating the relationship between the undrained shear strength at the top of the foundation, N_m, and the required maximum tensile resistance , t_{max}, for both tangential and horizontal reinforcement orientations for some values of α and D=d/H, and for a 1(V):2(H) embankment with a typical value of ϕ_m=30°. For each figure, the maximum depth of the foundation layer is prescribed. This depth, however, is not necessarily the critical depth. In fact, the penetration depth, d, of the critical slip surfaces for which the results in Figs. 3-5 correspond, are presented in Fig. 6 through 8. These figures demonstrate that for larger values of α the effect of the foundation depth becomes irrelevant after exceeding

422

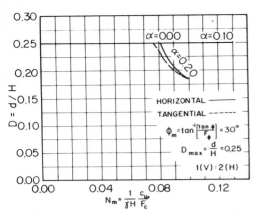

Fig. 8 Critical slip surface depth
versus strength of foundation
(D_{max}=0.25)

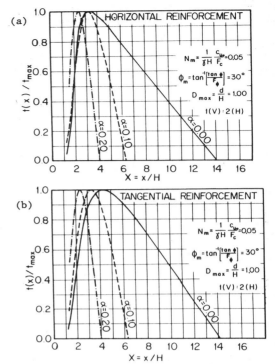

Fig. 9 Distribution of required tensile
resistance (D_{max}=1.00):
(a) Horizontal inclination of
reinforcement force
(b) Tangential inclination of
reinforcement force

a certain value. Furthermore, it shows
that the critical depth is always less
for the case of tangential reinforcement
than for the case of horizontal.

Figs. 9 and 10 complement some of the
design charts. These figures illustrate
the effect of α on the required tensile
resistance distribution for four cases.
Notice that as α increases the required
total reinforcement length, ensuring that
F_ϕ, F_c and F_r are exceeded everywhere,
decreases significantly. Furthermore,
the assumed orientation of the resistance
has very little effect on the distribution of the required normalized resistance.

When discussing tensile resistance
distribution, it is important to empha-
size that if the problem is symmetrical,
failure can occur at either end of the
embankment. Therefore, it is necessary
to superimpose a distribution with its
image mirrored about the centerline of
the embankment. For an embankment of a
short base length, this may result in the
overlapping of the required resistance
distribution curves. In such a case, the
greater of the values at each location x
should always be provided so that the
embankment is safe against failure at
either end. To illustrate this refer to
Fig. 11. This figure illustrates a
1(V):2(H) embankment having design parameters ϕ_m=30°, D=1.0 and N_m=0.05. The
mirrored slip surfaces shown are for the
cases of α=0.0 and α=0.2, and for horizontal reinforcement resistance. To
estimate the required tensile resistance,
Figs. 3 and 9a are utilized. Fig. 12a
shows the tensile resistance distribution
for α=0.0 and Fig. 12b for α=0.20, which

is required to insure adequate margins of
safety against deep-seated failure
towards either end of the embankment.

4 CONCLUSION

Presented is a method for analyzing the
short-term stability of reinforced granular embankment over soft foundation
having linearly varying strength. The
scope of the analysis is limited to deep-
seated rotational failures only (i.e.,
possible potential failures, such as
translational slide along the reinforcement interface, are not considered). The
analysis results can be condensed into a
convenient format of design charts.
However, because of the large number of
design variables involved in the problem,
many such charts are required. Subsequently, only a few of these charts,
which provide some insight, have been
presented. It seems more practical to

423

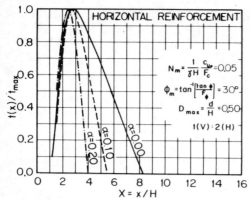

Fig. 10 Distribution of required tensile resistance ($D_{max} = 0.50$)

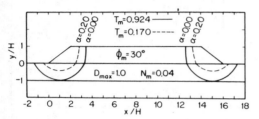

Fig. 11 Traces of critical slip surfaces signifying potential failures towards either side of a symmetrical embankment

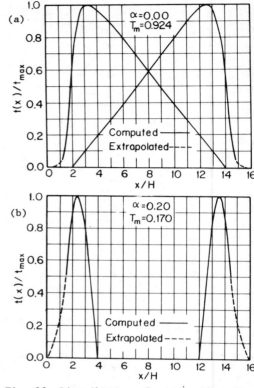

Fig. 12 Distribution of required tensile resistance mirrored about the embankment's centerline: (a) $\alpha = 0.0$, (b) $\alpha = 0.2$

use a computer program, such as the one used to develop the presented charts, for a specific problem.

Observing the trends exhibited by the analysis' results, the following conclusions may be stated:

1. The actual inclination of the reinforcement resistance probably lies between the two extreme assumptions of horizontal and tangential orientations. The required resistance for the horizontal is always greater than for the tangential inclination, and as feasible failure depth increases the difference between the two values, and thus the possible solution range, increases significantly.

2. When the undrained shear strength increases with depth, the prescribed foundation thickness is not necessarily the critical depth (i.e., does not define the failure surface penetration).

3. As the foundation shear strength increases with depth, the size of the sliding mass, the value of the required maximum tensile resistance and its location, and the required total length of reinforcement all decrease significantly.

4. The slope of the embankment face has little effect on the required distribution of the reinforcement tensile resistance. Therefore, the results presented here for a 1(V):2(H) slope can be approximately valid to slopes with similar inclination.

It should be pointed out that in two forthcoming papers, the authors demonstrate that their analysis compares well with a few case histories (reinforced and unreinforced). Furthermore, it compared well with predictions obtained from a rigorous slope stability analysis method.

REFERENCES

Baker, R. and Garber, M. 1978. Theoretical analysis of the stability of slopes. Geotechnique. 28(4):395-411.
Leshchinsky, D. 1987. Short-term stability of reinforced embankment over clayey foundation. Soils and Foundations. 27(3):43-57.

International Geotechnical Symposium on Theory and Practice of Earth Reinforcement / Fukuoka Japan / 5-7 October 1988
© 1988 Balkema, Rotterdam. ISBN 90 6191 820 0

Reinforcement of very soft ground using bamboo cross beam with polymer net

Sakuma Matsunaga
Penta-Ocean Construction Co., Ltd, Fukuoka, Japan

Yoshiharu Watari
Penta-Ocean Construction Co., Ltd, Tokyo, Japan

Sadao Isayama
Bureau of Ports and Harbours, Ministry of Transport, Japan

ABSTRACT: In construction of embankment on soft reclaimed ground, sheets and polymer nets are widely known as the suitable materials for reinforcement. This project site was reclaimed with a cutter suction dredger and is totally covered with very soft materials at present. Therefore, more rigid reinforcing materials should be used for the reinforcement of the embankment. The bamboo cross beams and polymer nets, as the rigid reinforcing materials, have been proposed to be incorporated into this project. However, the guidelines of the design and construction of the embankment using the above reinforcing materials have not been well-established due to the insufficient technical information resulting from scarcity of the preceeding works. To determine the design parameters and execution procedures, various field tests were conducted prior to commencement of the permanent work. As a result of these field tests, it is shown that the vertical load of the embankment is mainly supported by the bamboo cross beams.

1 FIELD TEST

1.1 Soil conditions

The field tests were conducted on the very soft ground reclaimed by a cutter suction dredger.

The properties of the soil on the site are shown in Fig. 1.

The soil profile at the site is mainly composed of clayey silt corresponding to C_H according to the Japanese Unified Soil Classification System. The natural water content (W_n), unit weight (γ_t) and cohesion (c) of the surface soil are W_n=150%, γ_t=1.3gf/cm³ and c=0.007kgf/cm², respectively. The surface soil is considered to be in an ultra soft state.

1.2 Outline of field test

Only few construction records are available for the embankment reinforced by bamboo cross beam and polymer net method therefore, there are still some uncertain parameters in the design and construction details. The field tests were conducted in advance of the permanent work to determine the design and construction parameters.

(1) Trial embankment

The location of a trial embankment was selected as shown in Fig. 2 taking into account the soil conditions of the existing ground at the project site.

The trial embankment as shown in Fig. 3 consisted of 60m in length, 60m in width and 0.5 to 1.5m in height and was constructed with the granulated slag. The properties of the granulated slag are shown in Table 1.

The trial embankment was constructed by the following procedure:

1) Assembly of the bamboo cross beams on the existing ground.
2) Installation of the polymer nets on the bamboo cross beams.
3) Spreading the granulated slag over the polymer nets in a slurry state in 2 layers (30cm thick each).
4) Levelling the embankment with a mini-backhoe and a hand-dozer.

(2) Laboratory tests and field measurements

The following laboratory tests and field measurements were conducted,

1) Laboratory tests
a. Tension test of bamboo
The tension test was carried out for the bamboo in accordance with Japanese Indus-

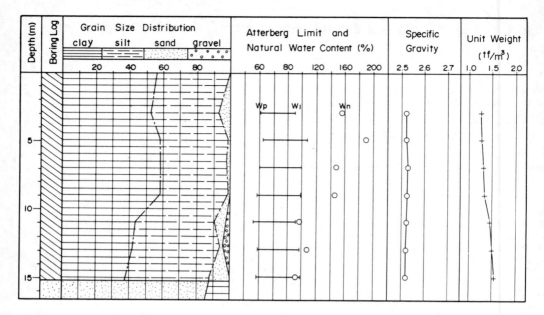

Figure 1. Soil profile and soil properties.

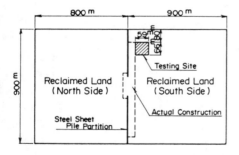

Figure 2. Location of field test.

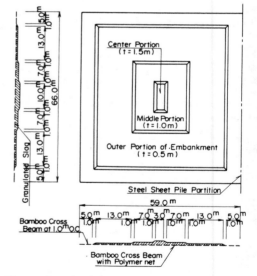

Figure 3. Dimensions of the trial embankment.

trial Standards (JIS) Z 2112 to determine its mechanical properties.

b. Tension test of bamboo joints
The tension test was carried out for the bamboo joints to determine the tensile strength.

2) Field measurements
a. Measurement of settlement
The settlement of the trial embankment was monitored to determine the height of the permanent embankment.

b. Measurement of vertical force
The vertical force acting on the bamboo cross beams was monitored to estimate the vertical force on the bamboo in the permanent embankment.

c. Measurement of tensile force
The tensile force acting on the bamboo cross beams was monitored to estimate the tensile force on the bamboo in the permanent embankment and to determine the spacing of the bamboo cross beams.

Table 1. Properties of granulated slag.

| Items | Granulated Slag | |
	Low Density	High Density
Specific Gravity	2.53~2.76	2.78~2.84
Unit Weight (tf/m^3)	1.3	1.3
Submerged Density	0.7 tf/m^3	0.9 tf/m^3
Internal Angle of friction	35°	35°
Coefficient of permeability (cm/sec)	$10^{-1}\sim10^{-2}$	$10^{1}\sim10^{2}$
California Bearing Ratio	20	20

Table 2. Laboratory test methods.

Test Items	Quantity	Methods of testing
Tensile strength test of bamboo	30 pieces	In accordance with JIS Z2112. Samples were taken from upper, middle, and bottom portions of bamboo
Tensile strength test of bamboo joint section	2 units	Prior to the construction of the bamboo cross beam, the joint section is subjected tensile test.

Japanese Industrial Standards (JIS)

Table 3. Methods of field measurements.

Measured Items	Measuring Device	Quan-tity	Method of Measurements
Settlement	Wooden Settlement Plate	19	Settlement of existing ground and embankment was measured using woo-den settlement plate.
	Pressure gauge	1	Settlement of existing ground was measured using a water pressure gauge.
Vertical Forces	Earth Pressure Gauge	6	The vertical forces were measured using an earth pressure gauge.
Tensile Forces	Load Cell	3	Tensile forces were measured from the stresses acting on the polymer net.
	Reinfocing Bar Stress Transducer	3	Tensile stresses were measured from the stresses acting on the bamboo cross beam.
	Strain Gauge	12	Tensile stresses acting on the bamboo cross beam and net were measured using a strain gauge.

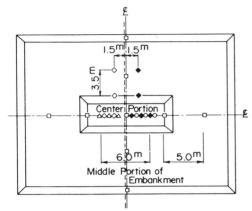

Legend:
○ Reinforcing Bar Stress Transducer
◆ Strain Gauge
◇ Load Cell
△ Load Cell
----- Plastic Pipeline (For settlement Measurement Using a Pressure Gauge)
□ Wooden Settlement Plate

Figure 4. Locations of measuring devices.

For methods of the laboratory tests and field measurements, refer to Tables 2 and 3. The location of the field measurements is shown in Fig. 4.

2 RESULTS OF LABORATORY TESTS AND FIELD MEASUREMENTS

2.1 Tension tests of bamboo

Test specimens were taken from upper, middle and lower portions of each piece of bamboo and subjected to the tension tests in accordance with JIS Z 2112.

The results obtained were 1,600kgf/cm² for the mean tensile strength and 87,819 kgf/cm² for the modulus of elasticity. And it was found that the tensile strength of

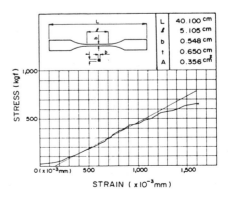

L	40.100 cm
ℓ	5.105 cm
b	0.548 cm
t	0.650 cm
A	0.356 cm²

Figure 5. Result of tensile strength test.

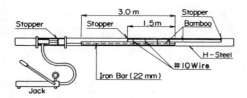

Figure 6. Hydraulic pressure jack, method.

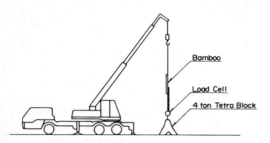

Figure 7. Pulling-up method employing crane.

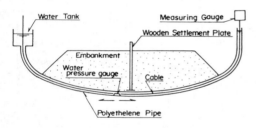

Figure 8. Pressure gauge method.

the bamboo depends on its age and not on the portion of the bamboo. Figure 5 shows the result of tests.

A modulus of elasticity of bamboo can be calculated in the following manner:

P = 650kgf (Breaking load)
σ = P/A = 650/0.356 = 1,826kgf/cm²
P_p= 440kgf (Yield load)
$$E = \frac{Pl}{PA} = \frac{(440 - 100) \times 5.105 \times 104}{(982 - 339) \times 0.356}$$
= 75,825kgf/cm² $\doteqdot$ 76,000kgf/cm²

2.2 Tension tests of bamboo joints

The tension tests were conducted with the following 2 testing methods,

(1) Method employing hydraulic pressure jack

As shown in Fig. 6, a test piece with a

Table 4. Test results of Hydraulic jack method.

Number of trials	Tensile strength
I st trial	2.0 tf/pieces
2 nd trial	2.1 tf/ ″
3 rd trial	2.0 tf/ ″
Mean value	2.0 tf/ ″

Table 5. Test results of pulling-up tests employing crane.

Number of trials	Tensile strength
I st trial	I.4 tf/pieces
2 nd trial	I.8 tf/ ″
3 rd trial	I.8 tf/ ″
Mean value	I.7 tf/ ″

1.5m joint section was pulled using a hydraulic pressure jack. Table 4 shows the results of the tests.

(2) Method employing crane

As shown in Fig. 7, a bamboo joint section was pulled up by a truck crane. The pulling force was measured with a load cell installed between the joint section and a tetra-block. The test results are shown in Table 5.

Comparing the above test results, the values obtained from the crane method appear to be smaller than those from the hydraulic jack method. This may be due to the difference in loading procedures. Based on the test results, the tensile strength of 2tf/m at the joint section was employed for the design.

2.3 Tensile strength of polymer net

The polymer net material is polyethylene with a tensile strength of 5tf/m.

The polymer net samples were tested and observed to yield the above tensile strength.

2.4 Measurement results of field test

(1) Settlement

The settlement was monitored by 2 measuring methods: the wooden settlement plate method and the pressure gauge method as shown in Fig. 8. Figure 9 shows the maximum settlement at the center of the embankment: this settlement was 2.1m.

(2) Tensile force of bamboo cross beam

The tensile force was monitored using the

reinforcing bar stress transducer and
strain gauge installed on the bamboo cross
beams, and maximum tensile force of 5tf/m
was observed. This value may be affected
by the bending tensile stress.

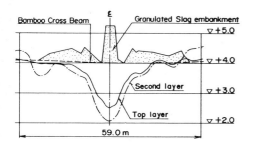

Figure 9. Measured settlement of
embankment.

(3) Tensile force of net

To determine the tensile force acting on
the polymer net, load cells and strain
gauges were used as measuring devices. The
results obtained from both instruments were
almost the same, that is from 10 to 100kgf/
m.

The difference in the tensile force between the bamboo cross beam and the net may
be related to the difference in the modulus
of elasticity between the materials.

(4) Measurement of vertical force

The vertical force was monitored using an
earth pressure gauge placed on the bamboo
cross beam. The values obtained using this
gauge are almost equal to the value evaluated by the unit weight of the granulated
slag and the thickness of the layers.

(5) Angle of inclination of materials and
    bulge radius of circumferential ground

For the cases where geotextiles are used,
the following bearing capacity equation
derived from Terzaghi's bearing capacity
theory was proposed. (Yamanouchi & Gotoh,
1979)

$$qu = \alpha \cdot C \cdot Nc + T(2\sin\phi/B + Na/R)$$
$$+ \gamma \cdot Df \cdot Na \quad \ldots \ldots (1)$$

where: qu = ultimate bearing capacity
       α = shape factor of foundation
       C = cohesion of soil
   Nc,Na = bearing capacity factors
       T = tensile force acting on
          geotextiles
       B = width of foundation

γ = wet density of soil
R = bulge radius of circumferen-
    tial ground covered by
    geotextiles
ϕ = angle of inclination of
    geotextiles

Figure 10 shows the measurement results of
ϕ and R on site: ϕ is 8° to 13° and R
11.0m. When only polymer net is used, R
should be a little smaller and ϕ a little
larger. The above difference may be due to
the difference in rigidity of the two
materials.

Right side ;

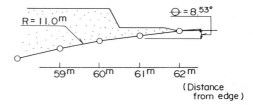

Left side ;

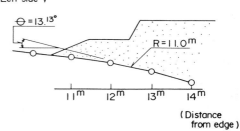

Fig. 10. Results of ϕ and R on site.

(6) Bearing capacity factor, Nc

The bearing capacity factor, Nc is obtained
by the back calculation using Formula (1).
Following values based on the laboratory
tests and field measurements are used for
the calculation, and as a result Nc=3.35
was obtained using the following parameters:

   C = 0.07tf/m²
   T = 3.5tf/m
   ϕ = 10°
   R = 11.0m
   B = 49.0m
   γ = 0.3tf/m³
  Df = 20cm

Terzaghi proposed to use Nc=5.71 in the

design of the embankment where the angle of internal friction of the soil is zero where the general shear failure is expected to take place, and Nc=3.81 where the ground is soft and the local shear failure is expected. In addition, (Watari et al, 1986) reported that Nc=3.19 was obtained from the tests for the similar soft ground.

Taking into account the rigidity of the bamboo cross beams, the calculated bearing capacity factor is considered reasonable as it lies between the values for the general shear failure and that for the local shear failure specified in Terzaghi's theory.

3 CONCLUSION

The results of the field tests on the trial embankment are summarized as follows:

(1) In the embankment constructed by the bamboo cross beam and polymer net method, the vertical load of the embankment is mainly supported by the bamboo cross beams. The polymer net protects the existing ground and embankment from being segregated and transmits the vertical load to the bamboo cross beams.

(2) The maximum tensile force acting on the bamboo cross beam was 5tf/m under the tested conditions.

(3) The angle of inclination of the embankment obtained in the tests was 8° to 13° and the bulge radius of circumferential ground was 11.0m.

(4) The bearing capacity factor, Nc of the existing ground was 3.53 when back calculated using Terzaghi's theory. The bearing capacity factor, if tested under similar ground conditions, might be considered to lie between the values for the general shear failure and that for the local shear failure specified in Terzaghi's theory.

4 ACKNOWLEDGEMENTS

The authors like to express sincere gratitude and thankfulness to the members of the 4th Division of the Bureau of Ports and Harbors, Ministry of Transport, for permission to use the necessary data and the full support extended by them in preparation and completion of this report.

REFERENCES

Yamanouchi,T. and Gotoh,K. 1979. A proposed practical formula of bearing capacity for earthwork method on soft clay ground using a resinous mesh, Technology Report of the Kyushu University, Vol.52, No.3, pp.201-207.
Watari,Y., Higuchi,Y. and Aboshi,H. 1986. Field studies of the behavior of geogrids and very soft ground, 3rd International Conference on Geotextiles, 2A/3, pp.187-191. Vienna, Austria.

International Geotechnical Symposium on Theory and Practice of Earth Reinforcement / Fukuoka Japan / 5-7 October 1988
© 1988 Balkema, Rotterdam. ISBN 90 6191 820 0

Experimental studies on the performance of polymer grid reinforced embankment

H.Miki & K.Kutara
Public Works Research Institute, Ministry of Construction, Ibaragi, Japan

T.Minami
Tokyu Construction Co., Ltd, Tokyo, Japan

J.Nishimura
Mitsui Petrochemical Industries Ltd, Tokyo, Japan

N.Fukuda
Fukken Co., Ltd, Yokohama, Japan

ABSTRACT: In order to establish an appropriate design and execution method for reinforcing the earth embankment with polymer grids (hereinafter grids) a series of experiments and analyses were conducted by constructing large-size test embankments. The dimensions of the experimental enbankment were: – height : 3 m, slope : 1:0.7, the length and spacing of grid laying in the embankment were varied and the reinforcing effects, when tested under a severe condition of 15 mm/hr artificial rain, were compared and evaluated. The integration effect of grid and earth was determined quantitatively through the analyses of slip-circle method and elasto-plastic FEM. The applicablity of FEM to the grid reinforced embankment analysis was discussed.

1 INTRODUCTION

The advantages of using the grids as reinforcements are various. The possibility of constructing steep embankments due to grid reinforcement enable us to effectively utilize the available construction space, which is one important factor in land scarce Japan, thus reducing the quantity of fill-material especially in road-construction works in mountainous areas. As a result of grid reinforcement, the embankment has got a resistance to torrential rain or earthquake. Morever, using the vegetative cover it was made possible to beautify the slope of embankment.

At present the design of grid-reinforced embankment is based for convenience on the concept of applying the tensile strength of reinforcing material to the slip-circle failure analysis (see Miki et al. (1986) and Kutara et al.

(1987)). Doubtless it could not necessarily explain the true nature of the mechanism of reinforcement.

Thus the present research is undertaken with a view to formulate a rational method of design and construction of grid reinforced embankment in corporating the mechanism of reinforcement with due consideration of the effect of grid laying pattern under a heavy rain.

2 PROCEDURE

2.1 Instrumentation and test conditions

The representative test embankments and salient features of reinforcement are as shown in Fig. 1 and Table 1. Three cases of grid laying length – 1 m, 2 m and 3 m and three cases of grid layers – 1-layer, 2-layer and 3-layer were investigated. The combination of grid laying length and grid layer is as shown in Table 1. Here the measurement of embankment's inner and surface deformations were conducted by using displacement gauges and inclinometers respectively. The strains on polymer grids were measured by foil strain gauges attached to grids. Ground water depths were measured by manometers inserted in the embankment. The moisture distribution inside the embankment were measured by radio-isotopes to determine the degree of saturation. The settings of those instruments were as shown in Fig. 1 and Fig. 2.

Table 1 Test cases.

Test case	Grid laying length L	Grid laying layer N
Case 0.0	0	0
Case 1.3	1	3
Case 2.3	2	3
Case 3.3	3	3
Case 3.1	3	1
Case 3.2	3	2
Case 3.3	3	3

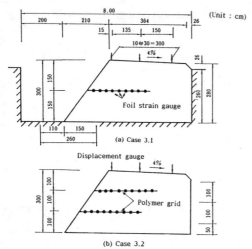

(a) Case 3.1

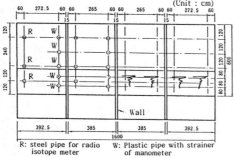

(b) Case 3.2

Fig. 1 Standard sections of reinforced embankment.

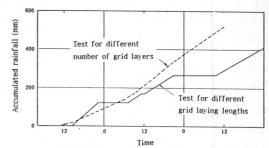

Fig. 2 Plan of test embankment.

R: steel pipe for radio isotope meter W: Plastic pipe with strainer of manometer

Fig. 3 Relationship between accumulated rainfall and elapsed time.

Table 2 Salient features of fill-material.

Natural water content	$\omega n(\%)$	22.4-24.3
Specific gravity	Gs	2.70
Gravel fraction	(%)	1-2
Sand fraction	(%)	70-74
Silt fraction	(%)	12-20
Clay fraction	(%)	9-12
Maximum grain size	(mm)	4.76
Uniformity coefficient	Uc	5.5-15.9
Compaction test method	(by JIS)	1.6.b
Optimum moisture content	$\omega opt(\%)$	17.0-18.6
Maximum dry density	$\rho_{dmax}(t/m^3)$	1.64-1.70
Permeability	k(cm/s)	$1.5-1.6\times10^{-4}$

2.2 Characteristics of fill-material and polymer grid

The choice of fill-material was done under the condition that the slope safety factor was greater than 1.0 during construction and less than 1.0 under heavy rain. Thus the mountain sand distributed in Ibaraki Prefecture was used as a fill-material for the embankment as it satisfied the above condition (see Table 2). The dry density of compacted soil at a degree of 80% compaction is 1.41 t/m^3 (average). The moisture content is 22% and degree of saturation is 70%. The shear strength of the soil under the above conditions is c_u=0.7 tf/m^2 and angle of internal friction ϕ_u=20.1°.

The bi-axial stretched polymer grids (SS-2) were used and their properties are : -weight : 320 gf/m^2, strength : 1.8 tf/m, peak strain: 8% in longitudinal direction.

3 TEST RESULTS

3.1 Test results for different grid laying lengths

The tests were conducted by constructing four cases of embankment-one case without reinforcement (Case 0.0), 3 cases of grid laying length 1m, 2m, 3m each having 3-layers of grids: Case 1.3, Case 2.3 and Case 3.3. The relationship between accumulated rainfall and the development pattern of deformation for each case are as shown in Fig. 3. When viewed from the accumulated rainfall, the failure of slope surface took place at 145 mm for Case 0.0 and when the accumulated rainfall reached 235 mm the sliding (collapse) occurred along the plane outside the reinforced zone in the Case 1.3; and also in the case of Case 2.3 only cracks occurred along the inner edge of the reinforced zone, but in the case of Case 3.3 the embankment was stable without any cracks even at the accumulated rainfall 450 mm but it was with only a minor erosion and settlement of several centimeters. The distribution of saturation degree of fill-material, distribution of grids strain and settlement of embankment surface are as shown in Fig. 4. These values are measured during the 15mm/hr artificial rain.

At the time of slope failure (collapse)

432

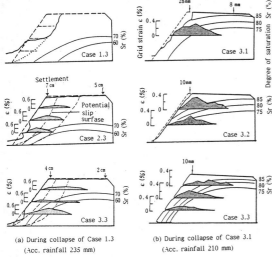

(a) During collapse of Case 1.3
(Acc. rainfall 235 mm)

(b) During collapse of Case 3.1
(Acc. rainfall 210 mm)

Fig. 4 Relationship between embankment deformation, grid strain distribution and saturation degree.

of Case 0.0, the peak strain ε max of Case 1.3 occurred at the inner side near estimated sliding plane of the embankment (Refer to Fig. 4). Similarly, in Case 2.3 and Case 3.3 the εmax occurred at the same position of the sliding plane estimated in Case 1.3.

At the time of slope failure (collapse) of Case 1.3 the range of ε max of Case 2.3 and Case 3.3 are increased to 0.16 – 0.33% and 0.22 – 0.41% respectively.

3.2 Test results for different number of grid layers

In this testing of embankment, the grids are laid horizontally with a length of 3m and three kinds of layers: 1-layer, 2-layer and 3-layer which are called Case 3.1, Case 3.2 and Case 3.3 respectively. The embankment deformations and the accumulated artificial rainfall are shown in Fig. 3. The embankment deformations, grid-strain distribution and the distribution pattern of saturation at the time of accumulated rainfalls 210 mm and 540 mm are as shown in Fig. 4.

When the accumulated rainfall reached 110 mm in the Case of Case 3.1, the deformation (horizontal displacement) of the top of slope increased rapidly and at the time of 210 mm rainfall, the slope was eroded for a depth of 50 cm. In the case of Case 3.2 and Case 3.3 there was only a surface erosion without any sliding until the accumulated rainfall

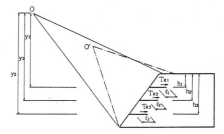

Fig. 5 Embankment section used in stability analysis.

reached 540 mm i.e the final rainfall of the test. At the initial stage of the test, the saturation degree of the inner fill was 65% to 70%. It increased gradually starting from the outer surface with the increase of accumulated rainfall. The sliding (collapse) area of the embankment of Case 3.1 was found to have a saturation degree of over 85%. As to the grid strain, it was less than 0.4% in all cases, which means that the grid strain was not considerably high during the rainfall test.

4 ANALYSIS OF REINFORCING EFFECT BY SLIP-CIRCLE METHOD

4.1 Basic equation for evaluating the internal stability of reinforced embankment

The basic equation that has been used for the usual practice of reinforced embankment design is:

$$Fs = \frac{Mr + \Delta Mr}{Md} = \frac{Mr + \Sigma (Tri \times yi)}{Md} \quad \ldots\ldots(1)$$

Refer to Fig. 5
Here Mr: the resisting moment of soil mass
    Md: the driving moment of soil mass
    ΔMr: the resisting moment due to grid reinforcements
    Tri: pull-out resisting force due to ith layer of grid reinforcement
    yi: the vertical distance of the ith layer of grid to the center of the slip-circle

This equation is meant for checking the internal stability of the reinforced zone of the embankment. Tri is determined by the smaller value of either the allowable tensile strength Ta (0.75 tf/m) of grid or the pull-out grid resistance Tpi which is given by the equation (2).

$$Tpi = 2 \cdot \sigma i \cdot \tan\phi \cdot li \quad \ldots\ldots(2)$$

Here σi = the vertical stress on the ith layer grid

li = the bonding length of ith layer grid

The constant 2 means that the frictional resestance is acting on both sides of grid.

4.2 An equation concerning the integration effect or increased apparent cohesion

The safety factor Fs* which is obtained by substituting the value of Tri in equation (1) with the tension T derived by multiplying grid strain εi with stiffness J (73 tf/m) is found to be smaller than the Fs obtained by the equation (1) with pull-out resistance force Tri. It is to be concluded that this is due to the fact that the grid and earth are integrated into a rigid body with a decreased deformation on account of the grid reinforcement. This integration effect will be quantitatively studied by conducting tests on embankments with different layered grids.

Here it is considered that the greater the difference of Fs and Fs* (ΔFs) is, the higher the integration effect will be. The increase of an apparent cohesion (Δc) which results in the ΔFs can be computed by the back-analysis using the equation (3).

$$\Delta Fs = Fs - Fs* = \frac{R \, \Delta c \, \Sigma 1}{Md} \quad \cdots (3)$$

where R: the radius of slip-circle

$\Sigma 1$; the length of slip-circle arc.

And also the rate of increase of apparent cohesion (Rc) can be determined by the following equation:

$$Rc = \frac{c + \Delta c}{c} \quad \cdots (4)$$

In that the increase of apparent shear strength due to laying of grids is considered as the increase of cohesion combined altogether.

4.3 Result of analysis

The analyses were done under the following conditions.

(a) At the time of rainfall (acc. rainfall 150 mm) when the sliding (collapse) of embankment took place in the case without reinforcement (Case 0.0).

(b) At the time of rainfall (acc. rainfall 265 mm) when the sliding (collapse) occurred in the case with 1-m long reinforcement (Case 1.3).

The strength parameters of the embankment (c,ϕ) can be checked by means of back-analysis for the Case 0.0 without reinforcement. In that the unit weight of the fill material will be increased from 1.70 to 1.85 tf/m^3 and the internal angle of friction is assumed to be 20°; that is taken constant throughout the testing and the cohesion only decreases with the increase of degree of saturation.

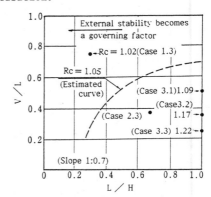

Fig. 6 Relationship between V/L, L/H and Rc.

Table 3 Analysis for the integration effect at collapse without reinforcement.

Test case	The drive moment Md (tf.m)	The resisting moment Mr (tf.m)	Arc length yi (m)	Measured strain ε m x10^{-6}	The resisting moment due to grid Δ Mr (tf/m)	Pullout resistance Tp (tf/m)	Allowable tensile strength Ta (tf/m)	Fs* based on measured strain	Fs by eq. (1)	Apparent cohesion C+ΔC (tf/m^2)	Rc $\frac{C+\Delta C}{C}$
3.1	15.48	13.48	5.00	9670	1.19	0.269	0.75	1.08	1.09	0.367	1.09
3.2			4.50	2600	0.85	0.364			1.08	1.19	0.393	1.17
			5.50	800	0.32	1.17	0.236					
3.3			4.25	1000	0.31	0.283			1.08	1.23	0.410	1.22
			4.25	1600	0.58	1.23	0.323					
			5.75	800	0.34	0.135						
2.3	14.23	14.23	4.25	1800	0.56	0.269			1.15	1.24	0.346	1.02
			5.00	2000	0.73	2.13	0.292					
			5.75	2000	0.84	0.136						

On the basis of the above conditions for Case 0.0 the cohesion is computed to be 0.337 tf/m^2. From the results of measurement of grid strains for the reinforced embankment the increase of apparent cohesion can be determined by using equation (3). And the rate of increase of apparent cohesion can be computed by equation (4).

Table 3 describes the results of computation for Fs*, Fs and Rc for different cases of reinforcement. Fig. 6 describes the relationship between the grid laying pattern and the rate of increase of apparent cohesion (Rc). In that the grid laying pattern is represented by the vertical spacing divided by grid laying length (V/L) and laying length divided by embankment height (L/H). From this graph it is learnt that the smaller V/L and the bigger L/H is the higher the value of Rc will be. That is to be considered that the integration effect is vividly increased.

5 NUMERICAL ANALYSIS BY FEM

5.1 Analysis model

The analysis model is as shown in Fig. 7. That is the elasto – plastic model of finite element method. Embankment fill-material is considered as a complete elasto-plastic body by Drucker-Prager's constitutive equation. And grids are

considered as elasto-plastic spring element. The analysis conditions are the same as those described in section 4.3.

The increase of unit weight and the decrease of shear strength (cohesion) due to increased saturation are considered in the analysis. The parameters of the material are as shown in Table 4. The analyses are done for the different cases as stated in Table 1.

5.2 Results of analysis

The horizontal displacement of embankment slope and the grid-strain distribution are as shown in Fig. 8. The comparison of reinforcing effects for the different number of grid layers can be seen in Fig. 8 (a) to (c). At the initial time of rainfall testing the difference of reinforcing effects is not clearly defined. But after the rainfall when the sliding (collapse) of the embankment without reinforcement occurs, the embankment deformation has turned out to be smaller with the increased number of grid layers. As to the grid strains, the results of the tests were found to be in close conformity with those obtained by FEM analyses. When the saturation get deeper inside the embankment, the peak strain on grid is found at a place nearer to the slope.

Thus this result in the increase of fill-material's unit weight and the decrease of cohesion at a comparatively high rate along the saturation line.

The local safety factor and the embankment deformations for Case 3.3 are changed with the increase of accumulated rainfalls. These results are shown in Fig. 9. These results are actually the standard results that are obtained from the analyses for different cases. And this analysis method explains sufficiently the reinforcing mechanism in the embankment.

6 CONCLUSION

In this report the large-size grid reinforced embankments were constructed and testings with analyses were done especially with artifical rainfalls and the effect of reinforcement is studied for the embankments with different layers of grids. The results of testing can be concluded as follows: -

(1) In the testings with the grids laying 0.75 m vertically spaced, for the Case 1.3 with the rates of laying length

Table 4   Properties of materials.

Earth fill (Drucker-Prager model)	Unit weigth (natural)	γ_t	(tf/m^3)	1.70
	Unit weight (saturated)	γ_{sat}	(tf/m^3)	1.85
	Elastic modulas	E	(tf/m^2)	30.0
	E after yield	E'	(tf/m^2)	0.0
	Cohesion (natural)	c	(tf/m^2)	0.7
	Cohesion (saturated)	c	(tf/m^2)	0.0175(*)
	Angle of internal friction	(o)		20.0
	Poisson's ratio			0.30
Polymer grid (Spring model)	Spring modulas	AE	(tf)	50.0
	AE after yield	AE'	(tf)	0.0
	Tensile strength	T	(tf/m)	1.8

(*) Determined by backanalysis at the collapse without reinforcement (Case 0.0, acc. rainfall : 150 mm)

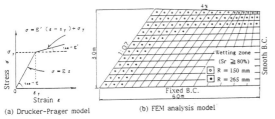

(a) Drucker-Prager model   (b) FEM analysis model

Fig. 7 Outline of analysis model.

to embankment height (L/H) ≤ 0.33 the external stability becomes a governing factor and for the case of L/H > 0.67 (Case 2.3), it is the internal stability that governs the overall stability of the embankment.

(2) When the grids are laid horizontally in several layers the deformation of the reinforced zone is so much decreased as a result of interaction between the grids and fill-material.

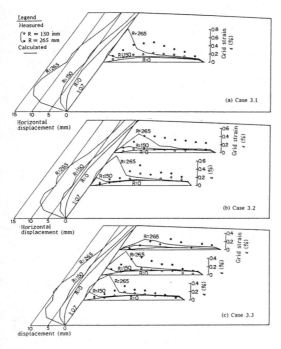

Fig. 8 Grid strain distribution and slope deformation.

Consequently the grids strain are considerably small i.e less than 0.4%. Thus it can be concluded that the reinforcing mechanism and its effect are quantitatively evaluated.

(3) As a quantitative estimate of integration effect of grid and fill-material the rate of increase of apparent cohesion Rc is defined and the relationship between Rc and grid laying pattern is analysed. As a result of it the smaller the value of V/L is, the bigger L/H and Rc will be. It means that the integration effect is quantitatively estimated.

(4) The FEM analysis is conducted assuming the embankment as an elasto-plastic body and the results are compared with those of the tests. The results are found to be in close conformity with each other. This means that this analysis method explains sufficiently the reinforcing mechanism in the embankment.

In addition to the present research, the testing of a practical size grid reinforced embankment (height. 6 m, slope 1:1.2) is in progress. And the authors hope that further important facts on reinforcing mechanism and construction method will be clarified in the near future.

REFERENCES

Miki,H. et al. 1986. Large scale experiments on the behavior of embankment reinforced with polymer grids. Proc. 1st Geotextile Symp., J.P.IGS, pp. 77-82. (in Japanese)

Kutara,k. et al. 1987. A numerical analysis of earth embankment reinforced by polymer grids. Proc. 2nd Geotextile Symp., J.P.IGS, pp.42-48. (in Japanese)

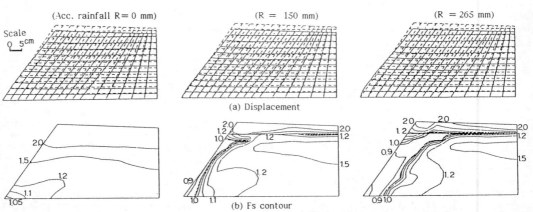

Fig. 9 Changes of deformation pattern and contour of localized safety factor(Case 3.3).

International Geotechnical Symposium on Theory and Practice of Earth Reinforcement / Fukuoka Japan / 5-7 October 1988
© 1988 Balkema, Rotterdam. ISBN 90 6191 820 0

Steel reinforced embankments on soft clay foundations

B.L.J.Mylleville & R.K.Rowe
The University of Western Ontario, London, Ontario, Canada

The results of finite element analyses are used to examine the behaviour
of steel strip reinforced embankments constructed on soft clay foundations. Some results are presented which illustrate that the mode of
failure and degree of improvement in embankment stability may vary
substantially depending on the amount of reinforcement used and the
properties of the foundation soil. The use of partial factors and definition of failure of a reinforced embankment system are discussed in the
context of using limit state design.

1 INTRODUCTION

The past several years have seen a
dramatic increase in the use of
reinforcement as a means of
ensuring the stability of embankments constructed on soft foundations. Common reinforcement
materials used include; geotextiles, geogrids, steel meshes and
steel strips (Rowe and Soderman,
1984; Lockett and Mattox, 1987;
Fowler et al., 1986; Elias and
Johnson, 1982 and others).

The objective of this paper is to
discuss the use of steel strips as
reinforcement for granular embankments constructed on soft clay
foundations which do not exhibit
strain softening. Results of finite
element analyses will be examined.
Details regarding the numerical
model which was used are discussed
by Rowe and Mylleville, 1988.

2 PARAMETERS CONSIDERED IN THE NUMERICAL MODEL

When constructing embankments on
soft clay foundations, the most
critical situation generally
corresponds to that at the end of
construction. In order to predict
the short term behaviour of
embankments on relatively soft
cohesive soils, the undrained

shear strength c_u ($\phi=0$) and
undrained modulus E_u (taking
Poisson's ratio to be 0.48) were
used. Figure 1 shows the general
arrangement which was adopted in
the finite element analyses. The
results examined herein are for the
case of a soft clay deposit with
strength and modulus which increase
linearly with depth from some
surface value. This type of
strength profile is commonly
encountered in soft, normally or
slightly overconsolidated clays.

Soft clay deposits of depth, D,
equal to 7.5 m and 15 m were
examined and assumed to be underlain by a rigid base. The ratio of
undrained modulus, E_u, to undrained
strength, c_u, considered was
$E_u/c_u=125$. The unit weight, γ, of
the clay was assumed to be 16.5
kN/m^3, and the coefficient of earth
pressure at rest was taken to be
$K_o'=0.60$.

The value of unit weight adopted
for the embankment fill was 20
kN/m^3 and the angle of internal
friction used was $\phi=36°$.

The steel reinforcement used was
in the form of Reinforced Earth (R)
(eg. Vidal 1968) ribbed steel
strips, of thickness, t=5 mm and
width, w=50 mm (see Figure 1). The
strips used have a yield strength
of 350 MPa and an ultimate tensile
strength of 490 MPa. Centre to

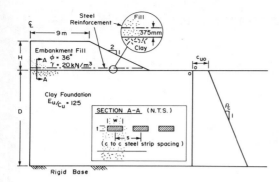

Figure 1. General Arrangement

centre spacings, S, of 375 mm and
125 mm were examined, with the
strips being placed transverse to
the alignment of the embankment
(see Figure 1). The strips were
located entirely within the
embankment fill at a distance of
375 mm above the clay/fill interface in order to develop frictional
resistance on both faces of the
reinforcement.

3 DEFINITION OF FAILURE AND THE USE OF PARTIAL FACTORS

For purposes of further discussion,
the practical definition of
failure of the reinforced embankment is the height at which the
increment in vertical displacement
is equal to or exceeds the increment in fill thickness added. The
addition of more fill will not
result in a net increase in
embankment height. For heavily
reinforced embankments, as is the
case when using steel strip
reinforcement, this practical
definition of failure height is in
fact very close to the plasticity
collapse height at which point
uncontained plastic flow occurs.

The cases which are going to be
discussed were analyzed using a
limit state type design philosophy,
where strength parameters are
factored down and loads are
factored up. The partial factors
used were taken from the Ontario
Highway Bridge Design Code, 1983.
For example, a factor of 0.65 was
applied to the clay foundation
strength parameters and factors of
0.8 and 1.25 were applied to the
tangent of the angle of internal

friction and unit weight of the
fill respectively. For the various
cases presented herein, the clay
foundation strengths quoted are
nominal (unfactored) values.

Collapse of a reinforced embankment on a soft clay foundation may
in many instances be preceded by
yield in the reinforcement. After
the onset of first yield, additional fill can be added prior to
collapse of the entire reinforced
embankment system.

In some instances, the amount of
fill that can be added after first
yield up to collapse is significant. Figures 2 and 3 show results
for a reinforced embankment with a
steel spacing of 125 mm, constructed on a clay foundation of
depth, D=15 m. The nominal foundation strength parameters are c_{uo}=30
kPa at the surface and a rate of
increase in strength with depth of
ρc=2.5 kPa/m.

The extent of plasticity at the
occurrence of first yield in the
reinforcement is shown in Figure 2,
where the cross-hatched portion in
the clay foundation and embankment
fill is that zone of soil which has
reached its shear strength. The
stiff reinforcement has limited the
extent of yield in both the fill
and foundation up to this point.
The failure in the foundation is
contained by a large region of
elastic soil and hence the embankment is still stable. Following
first yield of the reinforcement,
an additional 1.9 m of fill was
added before failure of the entire
embankment system occurred. This
represents an additional 30%
increase in fill thickness beyond
the point of first yield in the
reinforcement. Figure 3 shows the
plastic region for the embankment
at failure.

Cases such as the one just
discussed raise an important point
with regards to what one defines as
failure of the reinforced soil
system. In using limit state design
(ie. where partial factors are
applied to strengths and loads), it
is too conservative to associate
the embankment height required to
cause first yield with the failure
height of the embankment. It is
more realistic to define the
failure height as the height at
which the entire embankment system
fails.

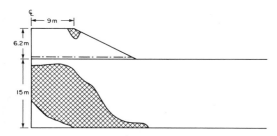

Figure 2. Plastic region at first yield of reinforcement for nominal parameters $c_{uo}=30$ kPa, $\rho_c=2.5$ kPa/m, S=125 mm and D=15 m

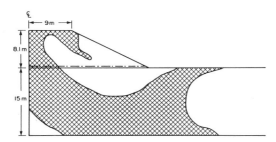

Figure 3. Plastic region at failure for nominal parameters $c_{uo}=30$ kPa, $\rho_c=2.5$ kPa/m, S=125 mm and D=15 m

4 BEHAVIOUR OF SOME STEEL REINFORCED EMBANKMENTS

The amount of steel strip reinforcement used in an embankment may have a definite effect on the failure mechanism, depending on the nature of the clay foundation.

Figure 4 shows the plastic region and velocity field for an 18 m crest width embankment constructed on a 7.5 m deep clay foundation with a nominal (unfactored) undrained shear strength at the surface of $c_{uo}=15$ kPa and a rate of increase with depth of $\rho_c=2.5$ kPa/m. For a steel strip spacing of S=375 mm, the calculated failure height was 4.4 m. Having adopted a limit state philosophy in the analysis, the failure height obtained using factored parameters corresponds to the allowable working height for the nominal (unfactored) conditions. The unreinforced failure height is 2.8 m, thus a steel strip spacing of 375 mm gives rise to a 57% increase in the allowable height to which

the embankment could be constructed on this deposit. The velocity field is also shown in Figure 4. The arrows indicate the direction and relative magnitude of movement in the soil at the onset of failure. Careful examination reveals an extensive lateral component to the deformation of both the fill and foundation soil. In this case, the failure mechanism is a function of yield in the reinforcement, some reinforcement pullout and general foundation failure.

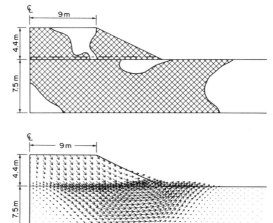

Figure 4. Plastic region and velocity field at failure for nominal parameters $c_{uo}=15$ kPa, $\rho_c=2.5$ kPa/m, S=375 mm and D=7.5 m

The effect of increasing the amount of reinforcement by a factor of three (ie. adopting a closer strip spacing) can be seen in Figure 5. The clay foundation has the same properties as the case just discussed. A steel strip spacing of S=125 mm increased the failure height to 5.0 m which corresponds to an 80% improvement over the unreinforced case. The increased amount of steel has also prevented yield of the reinforcement. In this case, the failure mechanism is a function of general foundation failure and some strip pullout.

A three-fold increase in the amount of reinforcement has improved the embankment performance significantly and as one might expect, resulted in an increase in the failure height. It should however be noted that the more heavily reinforced embankment (ie. S=125 mm) is approaching the maximum height which can be achieved for a perfectly reinforced embankment. Adopting a strip spacing closer than 125 mm is unlikely to achieve significant additional improvement in failure height for the foundation conditions examined.

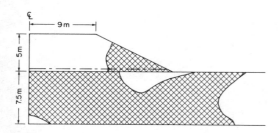

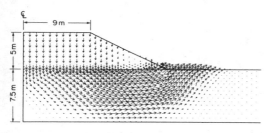

Figure 5. Plastic region and velocity field for nominal parameters c_{uo}=15 kPa, ρ_c=2.5 kPa/m, S=125 mm and D=7.5 m

There are some important differences in the failure mechanism for the more heavily reinforced case (ie. S=125 mm) compared to the embankment constructed with a strip spacing of 375 mm. Examination of the velocity field in Figure 5 reveals that there is essentially no lateral component of displacement in the embankment fill. Looking at the plastic region, there is no plasticity in the same area. This would suggest that the more heavily reinforced embankment moves downward as a rigid block over a width of approximately 30 m in this case.

The point about which rotation appears to be occurring in the fill (ie. the edge of an approxi-

mate rigid footing) corresponds to that point where the applied pressure γh equals the surface bearing capacity of $5.14c_{uo}$ when factored values of γ and c_{uo} are considered. This provides some additional evidence for an approximate method one might use to estimate collapse heights for heavily reinforced embankments proposed by Rowe and Soderman (1987). Based on the bearing capacity considerations of an idealized rigid footing of equivalent width, it provides one with a means of determining the maximum benefit that can be achieved by reinforcing an embankment. Keeping in mind that a reinforced embankment cannot be reinforced beyond being rigid, the simple approach proposed by Rowe and Soderman, provides the designer with an assessment as to whether a given height requirement can be satisfied.

The effect of ρ_c, the rate of increase in strength with depth in the clay foundation, on the failure mechanism in the underlying foundation can be seen if one compares the results shown in Figures 6 and 7. Figure 6 shows the velocity field at failure for an embankment with a steel strip spacing, S=125 mm, constructed on a clay foundation with a nominal undrained shear strength at the surface of c_{uo}=22.5 kPa and a rate of increase with depth of ρ_c=2.5 kPa/m. The failure mechanism in this case extends to a depth of approximately 10 m. Figure 7 is the velocity field at failure for the

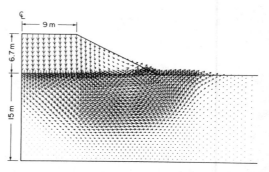

Figure 6. Velocity field at failure for nominal parameters c_{uo}=22.5 kPa, ρ_c=2.5 kPa/m, S=125 mm and D=15 m

440

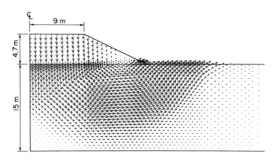

Figure 7. Velocity field at failure for nominal parameters $c_{uo}=22.5$ kPa, $\rho_c=1$ kPa/m, S=125 mm and D=15m

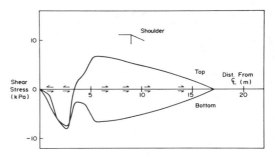

Figure 10. Distribution of shear stress along reinforcement at failure for nominal pararmeters $c_{uo}=15$ kPa, $\rho c=2.5$ kPa/m, S=375 mm and D=7.5 m

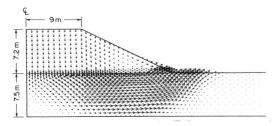

Figure 8. Velocity field at failure for nominal parameters $c_{uo}=22.5$ kPa, $\rho_c=2.5$ kPa/m, S=125 mm and D=7.5 m

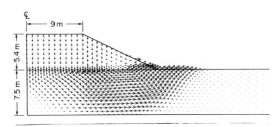

Figure 9. Velocity field at failure for nominal parameters $c_{uo}=22.5$ kPa, $\rho_c=1$ kPa/m, S=125 mm and D=7.5 m

same steel spacing and nominal parameters except that the rate of increase in strength in the clay foundation is lower, namely $\rho_c=1$ kPa/m. Comparing this velocity field with the previous case, one can observe that the failure mechanism extends deeper into the foundation, to a depth of about 12.5 m and encompasses a larger region of the foundation soil.

Having seen that the rate of strength increase with depth in

the clay foundation effects on the depth of the failure mechanism, one would expect that a rigid base at depth might also influence the overall embankment performance. The velocity field shown in Figure 8 is for an analysis performed using the same nominal parameters as for the case shown in Figure 6, except that now the depth of the deposit (or depth to a rigid base) is reduced to D=7.5 m. The result being an increase of about 7% above the failure height obtained for the case of D=15 m, if one considers a value of $\rho_c=2.5$ kPa/m in the clay foundation. Similarly, the velocity field in Figure 9 is for D=7.5 m and the same parameters used for the analysis reported in Figure 7. For a value of $\rho_c=1$ kPa/m, reducing the depth to a rigid base by half has resulted in a 15% increase in the failure height when compared to the failure height for D=15 m. In summary, the effect of depth to a rigid base is more significant for a lower ρ_c value, where the failure mechanism extends deeper into the underlying clay foundation.

The shear stress distribution above and below the reinforcing strips obtained from finite element analysis for the case of $c_{uo}=15$ kPa, $\rho_c=2.5$ kPa/m, S=375 mm and D=7.5 m is shown in Figure 10. The distribution is per metre width of embankment and corresponds to the point at which failure of the embankment occurs. The central 6 m of reinforcement is subjected to direct shear whereas the remainder of the reinforcement experiences "pullout" type shear. At failure,

441

pullout is only occurring along the outer 9 m of the reinforcement (ie. between the shoulder and toe of the embankment).

5 CONCLUSIONS

The results of finite element analyses have been used to illustrate the behaviour of steel reinforced embankments on soft clay deposits. Steel strip reinforcement can substantially improve the stability of granular embankments constructed on soft clay deposits where the undrained shear strength increases with depth.

The performance of these embankments is highly dependent on soil-structure interaction between the foundation, the embankment fill and steel strip reinforcement. It has been demonstrated that the failure mode for steel reinforced embankments on soft clay foundations may vary substantially depending on the amount of reinforcement used and the properties of the underlying foundation soil.

It has been shown that in some instances, yield of the reinforcement may occur well before collapse of the entire embankment system. It has been suggested that when using limit state design for reinforced soil systems, failure should be associated with failure of the entire system rather than a component of the system, such as the steel.

6 ACKNOWLEDGEMENT

The work presented in this paper is funded by the Natural Science and Engineering Research Council of Canada under grant A1007. Additional funding was provided by the Reinforced Earth Company (Australia). The authors gratefully acknowledge the value of discussions with Mr. M. Boyd of the Reinforced Earth Co. and with Professor H.G. Poulos, Professor J.R. Booker and Dr. J.C. Small of the University of Sydney.

REFERENCES

Elias, V. and Johnson, E.G. (1982). The use of a reinforced earth slab to reduce embankment loads at Auke Bay, Alaska. State of Alaska Department of Transportation and Public Facilities report.

Fowler, J., Peters, J. and Franks, L. (1986). Influence of reinforcement modulus on design and construction of Mohicanville Dike No. 2. Proceedings of IIIrd International Conference on Geotextiles, Vienna, pp. 267-272.

Lockett, L. and Mattox, R.M. (1987). Difficult soil problems on cochrane bridge finessed with geosynthetics. Proceedings of Geosynthetics '87 Conference, New Orleans, U.S.A., pp. 309-319.

Ministry of Transportation. Ontario Highway Bridge Design Code (and Commentary) 1983. Highway Engineering Division, Toronto, Ontario, Canada.

Rowe, R.K. and Mylleville, B.L.J. (1988). The analysis of steel reinforced embankments on soft clay foundations. Proceedings of ICONMIG '88 Conference, Innsbruck, Austria.

Rowe, R.K. and Soderman, K.L. (1984). Comparison of predicted and observed behaviour of two test embankments. International Journal of Geotextiles and Geomembranes, 1, pp. 143-160.

Rowe, R.K. and Soderman, K.L. (1987). Stabilization of very soft soils using high strength geosynthetics: the role of finite element analysis. 1st Geosynthetics Research Institute Seminar, Philadelphia, U.S.A., October 1987.

Vidal, H. (1969). The principle of reinforced earth. Highway Research Record No. 282, pp. 1-16.

International Geotechnical Symposium on Theory and Practice of Earth Reinforcement / Fukuoka Japan / 5-7 October 1988
© 1988 Balkema, Rotterdam. ISBN 90 6191 820 0

An experimental study on the performance of a prototype reinforced earth embankment

D.N.Naresh
National Thermal Power Corporation Ltd, New Delhi, India

M.Venkata Ratnam
Hyderabad, India

G.Subrahmanyam
Regional Engineering College, Warangal, AP, India

ABSTRACT: This paper presents a study on the performance behaviour of a prototype reinforced earth embankment during construction as well as when the structure is subjected to additional loading after its completion.

1 INTRODUCTION

Soils, as it exists insitu, if found suitable are used as foundation material to support various civil engineering structures. If net soil borrowing and soil replacement may become necessary which may not be economical, such situations stimulated new thoughts in civil engineerings, leading to development of a number of ground improvement techniques, which can be classified as:-

a. Ground improvement by insitu soil treatment.
b. Ground improvement by soil reinforcement.

The technique of inclusion of metallic or organic reinforcements into soil to improve its strength is named as reinforced earth or reinforced soil. This technique of soil reinforcement was practiced since ancient times in the crude form, until Nineteen sixties, when Henry Vidal(1966) proposed certain design methods for reinforcing soil, by the association of two materials (i.e.) the backfill material and, the reinforcements results in a new composite coherent material termed as Reinforced Earth. Consequently several model studies were initiated by several research workers (See Schlosser and Vidal (1964) in order to formulate some design procedures for the construction of reinforced earth structures. Several attempts were made to cover the problem of switching from model to actual structures (See Venkata-

Ratnam et al(1980)),but they were either not fully instrumented nor they were subjected to actual surface area loads.In order to study the performance behaviour both during construction and under applied surface area loads.A reinforced earth embankment of 4.5m hight, 10.5 wide and 75.0m long was constructed.This embankment was instrumented and tested under surface area loads. The results obtained are presented in this paper.

2 EXPERIMENTAL PROGRAMME

Model studies do provide cost effective and convenient research tool, but pose the problem of exact scaling of certain parameters of the actual structure, such as stress level, wall friction, gravity stresses, method of construction etc.(Al.Hussaini et al) (1978)).Hence in order to over come this practical difficulty and to study the performance of the reinforced earth wall, and compare it with its predicated behaviour when subjected to surface area loads, a proto-type reinforced earth embankment of 4.2m high with vertical faces and 10.5m wide and 60m long approach ramp with a slope of 1 in 15 on eastern side which joins the 7.0 meters wide B.T.Road, has been designed and constructed.Two types of skin elements were used for the construction of reinforced earth embankment.

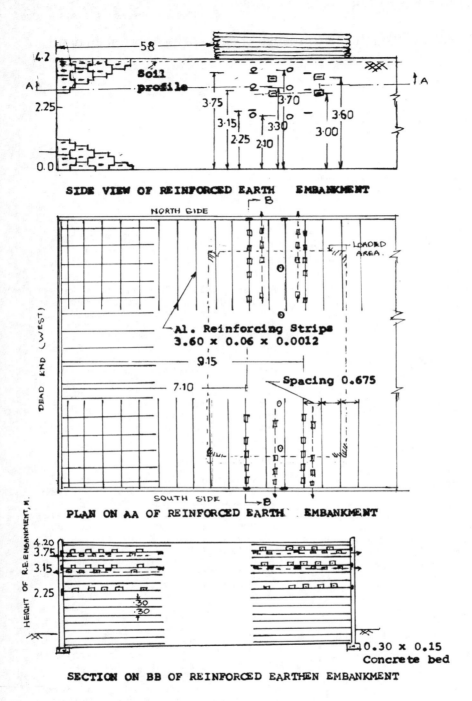

SIDE VIEW OF REINFORCED EARTH EMBANKMENT

PLAN ON AA OF REINFORCED EARTH EMBANKMENT

SECTION ON BB OF REINFORCED EARTHEN EMBANKMENT

Fig.1. Prototype Reinforced earth Embankment REC Warangal, India.

444

a. G.I.semi elliptical skin elements of size 240x25 and 0.10 cm thick
b. RCC cruciform panels with interlocking edges of overall size of 90x60 and 8.0cm thick and weighing about 85 kg.each.

The approach ramp, is provided with G.I. semi elliptical skin elements from 0 to 30m length and RCC cruciform panels, are used for the balance portion of ramp and level portion.

Locally available morum soil with a uniformity coefficient of 3.0 and an angle of internal friction of 35°, at a density of 1.89 gm/cc has been used as backfill. Considering various factors like, availability, of the material, low cost,and durability, two different materials, one of Aluminium of size 360.0x60x 0.12cm in the level portion, (in which the instrumented test zone is located) and another of G.I.(plain) of size 240x6.0x0.10cm in the approach. The tensile strength was determined from Houns-field tensometer and were found to be 3600 and 1450 kg/sq.cm. for GI and Aluminium strip, respectively. The value of coefficient of internal friction as determined by direct shear test(of size 30 x 30 x 15 cm) was 23° and 20° for aluminium and GI strip respectively.

2.1 Construction

The site of the embankment was cleared of bushes and other organic matter. The soil profile consists of Black cotton soil for the top 2.0m depth, underlaying it was poorly graded gravel. The earth was excavated upto a depth of about 0.65m below G.L.and a concrete bed(1:3:6 min) 30cm wide and 15cm thick was provided for the full length of the reinforced earth wall facing, Fig.1.C. Four different types of skin elements were used in the erection or reinforced earth wall with concrete panel facing. The concrete panels, which were castin moulds at site have been erected in position, and the first course of skin elements, which consisted of half and full panels (90x30cm) and (90x60cm) alternatively having a flat base have been erected in position on the top of the concrete bed. The soil was then filled up in thickness of 17 to 18cm

compacted to a finished thickness of 15.0cm. The compaction was done by 2.5 tonne roller operated manually. The compaction upto a distance of 0.30m from the inside face of panel was done by hand ramming. Necessary care was taken while compacting at the edges.The reinforcing strips were connected by means of two 6.3mm in dia. M.S. bolts & nuts to the projecting ends of the GI stub of 'C' shaped embedded in the concrete skin elements. The second course of panels comprising full panels, as placed on top of the first course, M.S. rods of 6.0 mm Ø were placed in the dowel hole provided in the elements to check plumb line. The soil was later filled upto a height of 33cm and compacted to a finished thickness of about 30cm, this procedure was repeated till the final height is reached. The top row consisted of full and half panels alternatively, having flat surface on the top.

In case of metallic skin elements, the G.I. sheet was cut to required size, and moulded using a specially prepared moulding device to obtain the semi-elliptical shape of required size. Similar procedure as mentioned earlier was adopted, except, that each vertical joints in successive layer of skin elements were kept staggering.

To obtain the performance data, during construction as well as under applied surcharge load, the following parameters were measured:-

a. The lateral deformations of the reinforced earth walls was measured from guide post, erected for this purpose. In the test zone,two guide posts on either size of the reinforced earth wall, were installed, these two were connected by GI wire to facilitate the measurement of lateral deformation at every strip location in a grid pattern.

b. The lateral earth pressure on panels were measured by means of 100mm Ø brass earth pressure cells (diaphragm type) placed on the inside face of the skin elements and kept flush with the panel surface. These are kept at distances of 7.10m and 8.45m

from the dead end and at an elevation of 2.10, 3.30 and 3.90m.
The location of pressure cells
are shown in fig.1.

c. To study the variation of
tensile stress along with
reinforcing strips, selected
reinforcing strip placed on both
sides of the embankment at
height of 2.25m, 3.15m and 2.75m
and at distances of were
instrumented, with five bakelite
electrical resistance type strain
gauges distributed over the
length of the strip. See fig.4
for further details.

The surface area load are applied
centrally on the top surface of
embankment. Edge distances of the
loaded area from the either side of
the panels was 1.5m(Ref.fig.1). This
surface area load was applied over
an area of 8.0m x 5.0m by means of
dead weights, in increment of about
8.5-9.0 ton each. Before applying the
load the initial readings of the
lateral measurements of the wall,
the tensile stress in the strips,
the stresses in the reinforced earth
mass and the lateral pressures on the
skin elements were recorded.

3 PRESENTATION AND DISCUSSION OF
RESULTS

3.1 Lateral Deformation

The bottom row of panels orginally
kept vertical, remained vertical at
the end of construction. However, the
panels did deform by about 23mm at
a height of about 2.25m and deformation increased with height
reaching a maximum average of about
11.00mm at the top of 4.2m height.
The loads were measured underneath
the loaded area in panel no.10,11,
12,13,14 and 15 on eitherside of the
embankment. For an applied load of
35.65 ton, the lateral deformation
at 4.05,3.75,3.45,3.15,2.85,2.55,
2.25,1.95,1.65,1.35 and 1.05 height
of the wall are, 0.6,0.7,0.8,0.5,0.
0.4,0.3,0.3,0.2,0.1,0.0 and 0.0mm
respectively on the southern side
of panel no.12. Similary the
values for the applied load of
76.78 and 104.62 tons are shown in
fig.2. The maximum lateral deformation of 12m and 15mm was observed
on the Southern and Northern side
respectively under the applied load
of 104.62 tonnes.

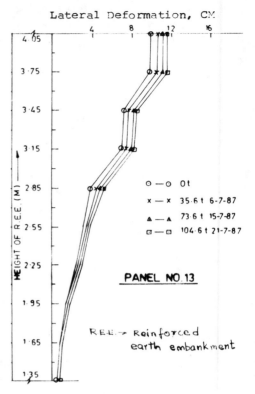

Lateral Deformation, CM

PANEL NO. 13

o — o 0 t
x — x 35.6 t 6-7-87
▲ — ▲ 73.6 t 15-7-87
□ — □ 104.6 t 21-7-87

R.E.E.→ Reinforced
earth embankment

Fig.2 Lateral deformation of R.E.E.

3.2 Lateral Pressure

The lateral pressures recorded at
2.10, 3.30 and 3.90m height at a
distance of 7.10 from the dead end,
in panel no.11 was 0.45, 0.15 and
0.035 kg/sq.cm.respectively, at the
end of construction on the southern
side. The measured values, were
about 1.2-1.5 times the computed
value. This may be due to some
times gap, during construction,
(which is reflected in the increased value) and also due to
compaction effect,close to the
vicinity of pressure cells. The
measured lateral pressure under
various applied surcharges are
plotted in fig.3 these pressures,
were converted into equivalent
height of embankment, and plotted
in fig.3.(panel no.13).

3.3 Tensile Stress Distribution

The tensile strain in the strips
were recorded from the digital

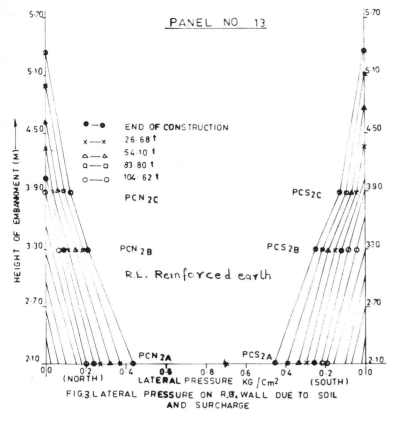

PANEL NO 13

END OF CONSTRUCTION
x——x   26·68 t
△——△   54·10 t
□——□   83·80 t
○——○   104·62 t

PCN 2C   PCS 2C

R.E. Reinforced earth

PCN 2B   PCS 2B

PCN 2A   PCS 2A

(NORTH)   LATERAL PRESSURE KG/Cm²   (SOUTH)

FIG.3.LATERAL PRESSURE ON R.E.WALL DUE TO SOIL
AND SURCHARGE

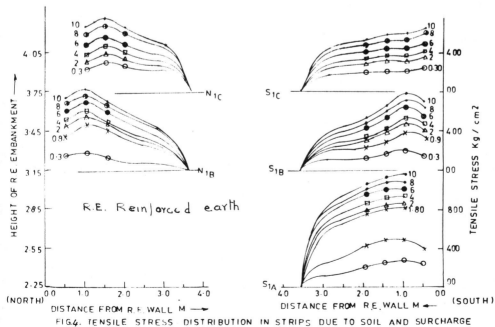

N 1C   S 1C

N 1B   S 1B

R.E. Reinforced earth   S 1A

(NORTH)   DISTANCE FROM R.E.WALL M →   DISTANCE FROM R.E.WALL M ←   (SOUTH)

HEIGHT OF R E EMBANKMENT →

TENSILE STRESS Kg/cm²

FIG.4. TENSILE STRESS DISTRIBUTION IN STRIPS DUE TO SOIL AND SURCHARGE

strain indicator, for each height of fill of the embankment. Knowing the Young's modulus of the strip material, and the strain recorded from the unit, the tensile stress was determined by using Hooke's Law. Adopting this technique, the stress in various strips both during construction and under applied loads were determined and are plotted in fig.4. for both sides of the embankment. From the figure it is observed that the tensile stress varied from the panel end to the free end of the strip. The maximum was observed to be at a distance of 1.0m from the panel end, and continuously decreased to zero, at the free end. The maximum tensile stress is found to be within middle third length of the strip, both during construction and under applied load.

4 CONCLUSIONS

During construction, the lateral displacement of wall was found to increase above a height of 2.25m. This was due to non provision of temporary supports. In practice, temporary supports may be used for maintaining vertical alignment.

The lateral moment at the end of construction was of 2.3% the height of the embankment. The increase in deformation under the maximum surcharge load was only 0.5% of the height of the embankment. The lateral deformation is found to be maximum between 0.8 and $1.0H$, where H is the height of embankment.

The lateral pressures on the skin elements, at the end of construction were about $1.2-1.5$ times the predicated value. However, under the applied loads, the recorded pressures were about $1.10-1.15$ times the theoretical value.

The tensile stress is varying between zero at the free end of the strip, and reached a maximum value at intermediate point(within middle third portion of the strip) and then tends to decrease towards the location of skin element.

The tensile stress distribution is somewhat parabolic both during construction and under applied loads.

5 ACKNOWLEDGEMENT

The authors acknowledge the assistance given by Central Board of Irrigation and Power, New Delhi, Regional Engineering College Warangal AP. and National Thermal Power Corporation Ltd., New Delhi.

REFERENCES

Al. Hussaini, M and Perry E.B. (1978) Field experiment of reinforced earth wall.Proc. symp Earth Reinforcement, ASCE, Pitts burgh, 127-157.

Schlosser, F. and Vidal,H.(1969) Reinforced Earth, Bulletin de liaison des.Laboratories routiers points etc. chaussees 41 paris.

Venkata Ratnam M Subrahmanyam G, and Balabhadra Rao M(1980)design, construction and performance study of reinforced earth embankment C.S.I.R. project Report REC Warangal.

Vidal H, (1966) Reinforced earth(A (A New Material for public works) Annales, de 1, institut Technique due Batiment et des Travans Pubbis, 1.9,224-233.

International Geotechnical Symposium on Theory and Practice of Earth Reinforcement / Fukuoka Japan / 5-7 October 1988
© 1988 Balkema, Rotterdam. ISBN 90 6191 820 0

A 14 metre high reinforced soil embankment as the abutment of a steel bridge in Tuen Mun, Hong Kong

H.Y.Ng & C.H.Mak
Maunsell Geotechnical Services Ltd, Hong Kong

ABSTRACT: A haul bridge to cross an existing stream was required to be built in Tuen Mun a satellite new town of Hong Kong, to serve the borrow area earth hauling activities. In order to minimize construction costs, two 14 metre high reinforced earth embankments were built as the bridge abutments making use of the earth materials locally available. A high density polyethylene grid (TENSAR SR2) was used to reinforce the abutment to maintain its stability. The design, construction and performance of this 14 metre high reinforced abutment are described in this paper. A 70° inclined facing was adopted as a compromise between minimising reinforcement required and limiting the bridge span. The applicability of the design requirements specified in the British Department of Transport Technical Memorandum (Bridges) BE3/78 for the design of the grid reinforced embankment is reviewed in this paper.

1 INTRODUCTION

Under the contract for the reclamation of the Tuen Mun New Town Area 14 and 16S, the Contractor was required to design and construct a haul bridge for the earthwork borrowing activities for his own contract and also for successive contracts for the development of Tuen Mun New Town. The site is located near Tan Kwai Tsuen between Tuen Mun and Yuen Long (see Fig. 1). Access to the bridge location was difficult, and extensive foundation works for an elevated bridge deck or abutments which would require heavy plant and equipment were undesirable. The Engineer's notional scheme involved a 3 span haul bridge with two piers and abutments as bank seats on the approach embankments. In order to maintain the required headroom for the existing stream as specified in the contract, if an elevated single span structure was to be built with abutment at the stream bank, it would be about 13 to 14 metres above the existing bank level. To avoid the access problems and to minimize construction costs, the abutments were decided to be built with reinforced soil technique at the river bank.

The reinforced soil abutment construction only required simple construction methods utilising mostly manual labour and light plant. It could also make use of the available fill as construction material. To minimize the length of the bridge span, the abutment locations were chosen right at the stream bank and an abutment about 14 metres high. The abutment wall face was slightly battered to 70° to reduce the amount of reinforcement required.

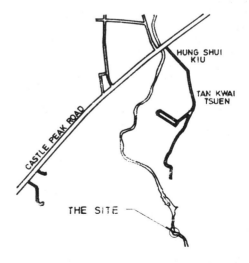

FIGURE 1: LOCATION PLAN

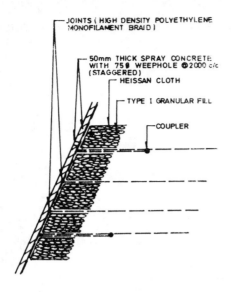

FIGURE 2 : TYPICAL DETAIL OF JOINTS

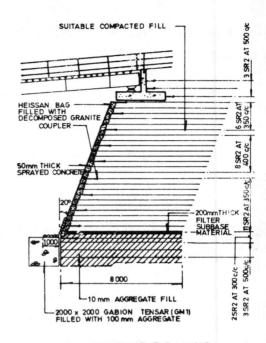

FIGURE 3 : REINFORCEMENT LAYOUT

There are two main types of soil reinforcement currently available in Hong Kong which could suit the design purpose :

1. Steel Strip (Reinforced Earth).
2. Grid (Tensar Geogrid).

Geogrid reinforcement was adopted because of its better anchorage properties and the flexibility of this material in various construction methods. No concrete facing panels were required, and the facing of the abutments was formed by the 'wrap around' construction method, (see Fig 2). This represented an additional cost saving. Gabion facing units were used at the toe of the slope to prevent scouring by the stream flow. After construction the whole of the steep slope surface was covered with shotcrete to prevent erosion of the slope surface, (see Fig. 3). This bridge abutment is the highest steep geogrid reinforced embankment built in Hong Kong.

2 CURRENT DESIGN METHODS

Reinforced soil structures have been studied extensively for the past thirty years. Most of the studies have been confined to vertical retaining walls, vertical reinforced abutments and reinforced slopes. Various design methods have been proposed and could generally be divided into the following five categories :

a. Finite element analysis (Rowe et al. 1984;).
b. Limit equilibrium method taking the strain compatibility into account (Rowe et al. 1985).
c. Conventional limit equilibrium method (Jewell et.al. 1984).
d. Tieback - Wedge analysis (DTp BE3/78 and Netlon published design guide 1986).
e. Coherent gravity analysis (Reinforced Earth Structure 1980 and Schlosser et al. 1983).

For routine design purposes, the last three methods are often used.

3 REINFORCEMENT PROPERTIES

The characteristic strength of TENSAR SR2 at 20°C for a performance limit strain of 10% for a design life of 120 years (McGown et al. 1984) is 29 KN/m and this value is usually used for design. The characteristic strength at other temperatures are also availiable from the manufacturer and are listed in Table 1.

For design purpose, a safety factor incorporating uncertainties in analysis, dimensions, site damage etc. is normally applied to obtain the permissible strength. The permissible strength for

Table 1 Geogrid Characteristic Strength at Different Temperatures	
Temperature (°C)	Characteristic Strength (kN/m)
10	32.2
20	29.0
30	26.1
40	24.7

Table 2 Permissible design loads (KN/m)

Fill type	Walls and Abutments
gravel	9
sand	10.5
	12
PFA	13

Notes:
(1) For particle sizes greater than 75mm, the TENSAR SR2 grid should be protected by a layer of finer material, the effect of which, on the design, must be considered.
(2) This table is for use with ϕ' peak, in accordance with DT_p BE 3/78.

TENSAR SR2 (for a design life of 120 years at 20°C) stated in the 'Roads and Bridges Agrement Certificate No. 86/27' is shown in Table 2.

4 DESIGN CONSIDERATIONS

4.1 Design criteria

In Hong Kong, reinforced soil structures are designed to BE3/78 standards in accordance with Government's recommendation. However, the code is only written for vertical reinforced soil structures. Strictly speaking, the code is not directly applicable to this case : abutments inclined at 70°. Nevertheless, in the absence of suitable design codes, the abutments were designed to BE3/78 with appropriate modifications.

4.2 Design strength vs temperature

The recommended design strength is generally based on the normal design temperature of 20°C. There is concern regarding the design strength to be adopted in Hong Kong which has a very hot summer of long duration. Soil temperatures generally tend to be lower than the ambient temperature away from the ground surface in view of the low conductivity of the soil mass. Ground temperatures from 0 to 3m below soil surface have been measured by the Hong Kong Royal Observatory in the past years. Plotting the average monthly temperatures for the past years, it can be observed that the yearly average for all depth is only about 25° (See Fig. 4). The characteristic strength differs very little between 20° to and 25°C and therefore the normal recommended design strength should be sufficient. During some months in the year, the temperature is higher than the annual average and a peak monthly temperature of 30° is recorded. Again, the characteristic strength at this temperature is lower than that for 20°. However, in view of the fact that the grid has sufficient strength reserve (i.e. the adopted design strength of 12 KN/m as compared with the grid charateristic strength of 29KN/m), it was considered to be capable to cater for higher design temperature situations.
Based on the above observations, design strength for the normal temperature of 20° was adopted.

4.3 Design Anchorage

As specified in BE3/78, clause 2.3.1.2, a minimum reinforcement length L shall be 5 metres or 0.8H for a vertical wall, whichever is the longer.
The reason for this requirement has not been stated in the code. The authors believe that the 0.8H and the 5 metre minimum criteria may be an 'empirical guide line' to ensure adequate anchorage length is provided to achieve the design strength of the reinforcement. To the best of the author's knowledge, this guideline has

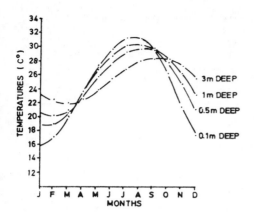

FIGURE 4 : MEAN MONTHLY SOIL
TEMPERATURES 1967-1985

been developed from case histories of
walls using steel strip reinforcement,
when the anchorage length is the
critical component for design.

The mechanism of stress transfer in
grid reinforcement and strip
reinforcement is different. In the
former case, the stress is transferred
by a combination of surface friction and
passive resistance, and hence the stress
transfer characteristic is much better
than that of the latter case, which is
solely relying on surface friction.
Large scale shear box tests using the
locally derived decomposed granitic
materials as backfill have been carried
out. The equivalent friction
coefficient is about 0.9. The required
anchorage length for the grid
reinforcement is very small, being
generally less than 1 metre and the
clause 2.3.1.2 requirement does not
appear to be necessary. Therefore the
above rule has not been used in this
design.

4.4 Analysis

The steep embankment can be regarded as
a 'vertical' wall using the Tieback
analysis (BE3/78 requirement) for the
local stability check. The coefficent
of earth pressure was modified for
sloping face and that as given in Caquot
& Kerisel, 1948 was adotped.

It is interesting to note that there
are still different approaches for the
distribution of vertical pressure
beneath the reinforcement due to the
self weight of backfill and uniform
surcharges. BE3/78 and Lee et al. 1973

suggest that the soil pressure is
uniform, while Bolton and Choudhury 1976
and the manufacturer (Netlon 1986)
suggest that the soil pressure is
trapezoidal. To err on the safe side,
the manufactuerer's recommendation of a
trapzodial distribution was adopted.

BE3/78 requires limit equilibrium
wedge analysis (see Fig. 5a) to be
carried out in order to check against
pull out failure. The authors
considered that more general failure
surfaces (see Fig. 5b) would be more
appropriate in this circumstance and the
Simplified Janbu Method was adopted
instead.

4.5 Design safety factor calculated from limit equilibrium method (F_1)

As specified in BE3/78 and the Agrement
Certificate No 86/27 for TENSAR SR2, a
gross safety factor (F_g) is applied to
the reinforcement while no safety factor
is required on the soil peak shear
strength. This gross safety factor has
taken into consideration all the
uncertainties in the method of analysis,
strain compatibility etc. Consequently,
trial wedge analysis specified in BE3/78
requires a minimum F_1 of unity at the
trial failure surface using the soil
peak shear strength (but a safety factor
of 2 is required on the anchorage).

In the authors' opinion, this approach
does not give a good indication of the
comparative degree of stability of the
various failure surfaces. As shown in
Fig. 5, a failure surface passing
through most of the reinforcement may
achieve a relatively smaller factor of
safety with a large reserve of strength
built in with the factor of safety on
the material via the permissible
strength of the grid. For a failure
surface which cuts across only some of
the reinforcement, the factor of safety
may be higher than the first one, this
however does not indicate the slope is
more stable since the reinforcement
strength reserve is less. A failure
surface with a factor of safety of just
unity could have little reserve on the
factor of safety. Consequently, limit
equilibrium analysis based on
permissible reinforcement strength would
lead to ambiguity in the minimum
acceptable safety factor for design.

It is suggested that when treating the
steep embankment as a slope using the
limit equilibrium method, partial safety
factors applied on both reinforcement
strength and soil peak shear strength

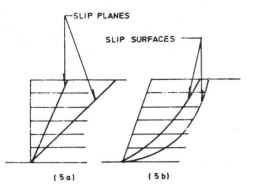

FIGURE 5 : SLIP PLANES AND SLIP
SURFACES IN LIMIT
EQUILIBRIUM ANALYSIS

are more logical and can eliminate some
of the ambiguity.

4.6 Limit equilibrium analysis with
    partial safety factors on soil peak
    shear strength and reinforcement
    capacity

The main procedures of analysis are as
follows :
  Reduce the soil peak shear strength
by a partial safety factor F_s

Design $c = \dfrac{c}{F_s}$

Design $\phi' = \tan^{-1} \dfrac{\tan\phi'_{peak}}{F_s}$

  Reduce reinforcement capacity by
partical safety factor F_t

Design capacity $= \dfrac{Ultimate\ capacity}{F_t}$

The above design parameters are then
used for the analysis. The layout of
reinforcement is modified until the
minimum F_1 evaluated from the limit
equilibrium analysis is greater than
unity.

  It should be noted, however, that the
slip surface with the lowest value of
F_1 is still not necessarily the most
critical one. Nevertheless, we can now
specify that a minimum F_1 of 1.0 is
required for all trial failure surfaces.

  The design approach can ensure
adequate reserve capacity has been
provided in the soil peak shear strength
and grid reinforcement.

5 CONSTRUCTION AND PERFORMANCE

The construction of the abutment
commenced by removing the loose alluvial
material at the base and backfilling
with aggregate fill. This allowed
compaction despite wet conditions. The
construction of the very steep battered
abutment is very simple and straight
forward. The grid reinforcement was
first laid flat on the ground with the
length for the 'wrap around' extended
beyond the slope face. The grids are
manufactured in a standard one metre
width, and they were placed normal to
the slope face with the main rib
perpendicular to the slope face. They
were laced together sideways with H.D.
polyethylene monofilament braid. The
reinforcement end at the slope face was
fixed in position by steel studs pinned
to the ground. The other end was pulled
taut and then pinned down to the
ground. Sand bags were placed at the
face position to retain fill and then
compaction proceeded. Subsequent grids
on top were placed similarly and were
tied to the face reinforcement with
polyethylene braid. The wrap around
face reinforcement was joined onto the
next grid with a HD polyethylene rod
coupler. Large pieces of rock and sharp
stones in the backfill were avoided to
prevent damage to the TENSAR grid.
Very light vibratory compaction plant
was used to undertake the earth
compaction.

  For fill slope formation in Hong Kong,
the slope is generally formed by
overfilling and retrimming to the
required final profile. This was
impossible in this situation in view of
the close vicinity of the existing
stream and also of the steep slope
face. Sand bags at the front slope face
as temporary formwork for each layer of
earth backfill were used to maintain the
70° slope angle. This turned out to
be a rather labour intensive activity.
Strict control on workmanship was
implemented to maintain the accuracy of
the slope face.

  During construction, difficulties were
also encountered in the orientations of
the reinforcement laying in the
transition zone between the side
embankment and reinforced abutment as
the slope face was curved. A large
over-lapping was used to keep the
curvature.

  Gabions were used as the face support
at the wall toe to avoid scouring of the
toe. The design flood level was based
on the catchment area and correlated

FIGURE 6: AS-BUILT ELEVATION (BEFORE SHOTCRETING)

with information provided by local villagers.

After construction, shotcrete was then applied to the formed slope face as a more permanent slope protection. The bridge seating was then constructed and the bridge deck was erected thereafter. All the remaining operations proceeded very smoothly.

The bridge has been opened for one year and the performance was found to be satisfactory.

Inclinometers have been installed in the two embankments to monitor ground movement. No measurable movements have been reported.

6 CONCLUSION

A case history of the two 14 metre high reinforced soil abutments with 70° facing in Tuen Mun is reported. It represents the first use of geogrids for construction of bridge abutments in Hong Kong. A combination of limit equilibrium analysis (simplified Janbu Method) and local stability analysis (with modified coefficient of earth pressure) was adopted in the design. Design aspects of the geogrid and the ambiguity in the definition of F_1 is discussed.

The construction procedure demonstrates the versatility and the convenience of the using geogrid to reinforce earth structures.

REFERENCES

Botton, M.D. and Choudhory, S.P. (1976). Reinforced Earth. U.M.I.S.T./T.R.R.L. Research Contract, first report.
Caquot, A & Kerisel, J. (1948). Tables for the calculation of Passive Pressure, Active Pressure and Bearing Capacity of Foundations. (Translated from the French by M.A. Bec, London) Gauthier - Villars, Paris, 120 p.
Janbu, N. (1974). Slope stability computations. Embankment-Dam Engineering, 1973, John Wiley and Sons, pp.47-86.
Jewell, R.A., Paine, N and Woods, R.I. (1984). Design Methods for Steep Reinforced Embankment. Symposium on Polymar Grid Reinforcement in Civil Engineering, ICE 1984, paper no. 3.1.
Lee, K.L. Adams, B.D. and Vagneron, J-M.J. (1973). Reinforced earth retaining walls. A.S.C.E. Soil Mech. and Found. Div., vol. 99, No. SM.10, pp.745-764.
Mcgown, A., Andrawes, K.Z. and Yeo, K.C. (1984[b]). The Load-Strain-Time behaviour of Tensar Geogrids. Symposium on Polymar Grid Reinforcement in Civil Engineering, ICE 1984, paper no. 1.2.
Netlon Ltd. (1986). Guidelines for the design & construction of reinforced soil retaining walls using TENSAR geogrids.
Reinforced Earth Structures (1980). Recommendations and Rules of the Art. French Ministry of Transport, August, Translation.
Rowe, R.K. Machaun M.D. and Soderman, K.L. (1984). Analysis of a geotextile - reinforced embankment constructed on part. Candaian Geotechnical Journal, vol. 21, pp. 563-576.
Rowe, R.K. and Soderman, K.L. (1985). An approximate method for estimating the stability of geotextile - reinforced embankments. Canadian Geotechnical Journal, Vol 22, pp. 392-398.
Schlosser, F. and Juran, I. (1983). Behaviour of Reinforced Earth Retaining walls from model studies. Development in Soil Mechanics and Foundation Engineering.
Technical Memorandum (Bridges) BE3/78. Reinforced Earth Retaining Walls and Bridge Abutments for Embankments. Department of Transport, U.K.

International Geotechnical Symposium on Theory and Practice of Earth Reinforcement / Fukuoka Japan / 5-7 October 1988
© 1988 Balkema, Rotterdam. ISBN 90 6191 820 0

Stability of reinforced embankment over soft cohesive foundation

H.Oikawa
Akita University, Akita, Japan

ABSTRACT: Stability analysis of a reinforced embankment over soft cohesive foundation has been carried out. A reinforcing sheet, capable of developing a horizontal tensile strength, is placed at the embankment-foundation interface. For simplicity of analysis, it is assumed that the failure surface is circular arc and a vertical tension crack develops in the embankment above the point at which the circular slip surface emerges from the foundation soil. It is also assumed that the undrained shear strength of foundation soil increases linearly with depth and $\phi_u=0$ analysis is applicable. Nondimensional stability charts for two different cases of ground condition are compiled from the results of analysis, i.e., one is the case where the thickness of cohesive soil stratum is infinite and the other is limited.

1 INTRODUCTION

Occasionally there is a need to construct an embankment over very soft cohesive foundation such as peaty ground. Adequate design of such an embankment has to ensure, among others, proper stability against failure throughout the embankment construction. One possible solution to such problems may be obtained by means of installation of a reinforcing agent such as a steel mesh, bamboo sheet, a geosynthetic, etc.

By the way, at the end of construction it is customary to assume that the foundation has not begun to consolidate (i.e., undrained condition), thus it has not yet gained any strength. Subsequently, the lowest margin of safety against failure is at the end-of-construction. Therefore, design related to stability of such an embankment is controlled by the undrained shear strength c_u of the foundation soil and in design it is customary to assume that the $\phi_u=0$ analysis is applicable.

The following is the stability analysis of a reinforced embankment over soft cohesive foundation where the $\phi_u=0$ analysis is applicable. It is based on the limit-equilibrium approach considering a rotational failure mechanism.

It should be noted here that the basic assumption of the present analysis is development of a vertical tension crack through the embankment. On the other hand, without the assumption of development of a vertical

tension crack in an embankment, Leshchinsky (1987) has presented a method to assess the short-term stability of a reinforced embankment over cohesive foundation. His method, however, is somewhat difficult for us to use in design of embankment construction.

2 STATEMENT OF THE PROBLEM

Let us consider an embankment with a height H and horizontal length of side slope L over soft cohesive foundation with a horizontal surface, as shown in Fig.1. The embankment has a unit weight of γ. A reinforcing sheet, capable of developing a horizontal tensile strength T, is placed at the embankment-

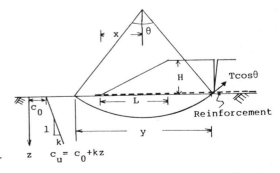

Fig.1 Selected failure mechanism and basic definitions

foundation-interface. The reinforcement sheet is assumed to remain horizontal. The foundation is a saturated isotropic and homogeneous soft cohesive soil and its undrained shear strength increases linearly with depth.

$$c_u = c_0 + kz \qquad (1)$$

where k is the rate of increase in strength with depth and is a constant for particular soil. z is the depth below the ground surface. The constant k has the same dimension as density. The failure surface is assumed to be circular arc. It is also assumed that a vertical tension crack develops in the embankment above the point at which the slip surface emerges from the foundation soil.

3 STABILITY ANALYSIS

3.1 Moment limiting-equilibrium equation

When failure takes place along a circular slip surface, the moment limiting-equilibrium equation is expressed as

$$M_r = M_d \qquad (2)$$

where M_r is the resisting moment and M_d is the driving moment.

Since it was assumed that a vertical tension crack develops in the embankment above the point at which the circular slip surface emerges from the foundation soil, the driving moment produced by a weight of the embankment within a failure surface in Fig.1 is written as follows, unit width of soil mass being considered

$$M_d = \frac{H\gamma}{2}(\frac{y^2}{4} - x^2 + xL - \frac{L^2}{3}) \qquad (3)$$

where x is the horizontal distance between the center position of the slip circle and the toe of the embankment. y is a chord length of the slip circle.

The resisting moment produced by both the shear resistance of the foundation soil along the circular slip surface and the tensile strength of reinforcement sheet at the embankment-foundation interface is written

$$M_r = \int_\theta (c_0 + kz)R^2 d\theta + TR\cos\theta$$

$$= \frac{ky}{4}\frac{\sin\theta-\theta\cos\theta}{\sin^3\theta} + \frac{y^2}{2}c_0\frac{\theta}{\sin^2\theta} + \frac{Ty}{2}\frac{\cos\theta}{\sin\theta} \qquad (4)$$

where R is the radius of the slip circle, θ is the sector angle and T is the tensile strength of reinforcement sheet per unit width.

Substituting equations (3) and (4) into equation (2) and rearranging, the following expression is obtained as the moment limiting-equilibrium equation for the reinforced embankment over cohesive foundation.

$$\frac{H\gamma}{c_0} = \frac{\frac{y}{2}(\frac{k}{c_0})\frac{\sin\theta-\theta\cos\theta}{\sin^3\theta} + y^2\frac{\theta}{\sin^2\theta}}{\frac{y^2}{4} - x^2 + xL - \frac{L^2}{3}}$$

$$+ \frac{y(\frac{T}{c_0})\frac{\cos\theta}{\sin\theta}}{\frac{y^2}{4} - x^2 + xL - \frac{L^2}{3}} \qquad (5)$$

The minimum value of $H\gamma/c_0$ in equation (5) corresponds to the stability factor of reinforced embankment over cohesive foundation

3.2 Stability factor and geometory of the critical circle

When the undrained shear strength of foundation soil, c_0 and k, the geometry of the embankment, H and L, and tensile strength of reinforcing sheet T are known, the value of $H\gamma/c_0$ in equation (5) is a function of x, y and θ. Therefore, minimum value of $H\gamma/c_0$, i.e., stability factor, may be obtained from the conditions $\partial(H\gamma/c_0)/\partial x=0$, $\partial(H\gamma/c_0)/\partial y=0$ and $\partial(H\gamma/c_0)/\partial\theta=0$.

The condition $\partial(H\gamma/c_0)/\partial x=0$ yields an expression

$$x = L/2 \qquad (6)$$

This equation indicates that the center of the critical slip circle lies on a vertical line passing through the middle of the side slope face. This is a well known matter in the analysis of the slope stability (e.g., Taylor, 1937; Baker 1981).

From the condition $\partial(H\gamma/c_0)/\partial y=0$, we get

$$3y (y^2 - L^2) (\frac{k}{c_0}) (\sin\theta - \theta\cos\theta)$$

$$= 4yL^2\theta\sin\theta+2(3y^2 + L^2) (\frac{T}{c_0}) \sin^2\theta\cos\theta$$

$$\qquad (7)$$

Equation (7) contains two unknown values, i.e., chord length y and sector angle θ of the critical circle. So, in order to obtain the values of y and θ of the critical circle, one more equation which expresses the relationship between the value of y and θ is needed. It may be obtained from the condition $\partial(H\gamma/c_0)/\partial\theta=0$, with the result

$$(\frac{k}{c_0})(\theta\sin^2\theta + 3\theta\cos^2\theta - 3\sin\theta\cos\theta)y^2$$

$$+ 2y(\sin\theta - 2\theta\cos\theta)\sin\theta - 2(\frac{T}{c_0})\sin^2\theta = 0 \tag{8}$$

This equation also contains two unknown values of y and θ. So, the values of y and θ which satisfy both equation, i.e., equations (7) and (8), may be obtained by eliminating one value from two equations mentioned above.

From equation (8), by considering that the value of y is larger than zero, we can get

$$y = \frac{B\sin\theta}{(\frac{k}{c_0})g} \tag{9}$$

where

$$B = \sqrt{f^2 + 2g(\frac{kT}{c_0^2})} - f \tag{10}$$

$$f = \sin\theta - 2\theta\cos\theta \tag{11}$$

$$g = \theta\sin^2\theta + 3\theta\cos^2\theta - 3\sin\theta\cos\theta \tag{12}$$

Introducing equation (9) into equation (7) results in

$$(\frac{kL}{c_0})^2 = \frac{3B^2\sin^2\theta[fB^2 - 2g^2(\frac{kT}{c_0^2})\cos\theta]}{g^2[2fB^2 + 4Bg\theta + 2g^2(\frac{kT}{c_0^2})\cos\theta]} \tag{13}$$

Equation (13) contains only one unknown value of sector angle θ. Therefore, we can get the value of sector angle θ by calculating equation (13) for various values of kT/c_0^2 and kL/c_0. The relationship between the sector angle θ and the parameter kT/c_0^2 is shown in Fig.2 for various values of kL/c_0. By the use of Fig.2, the sector angle of the critical slip circle for a given set of val-

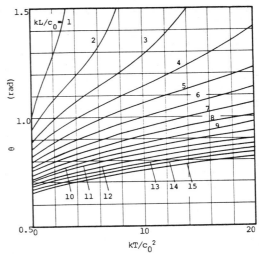

Fig.2 Sector angle of the critical circle

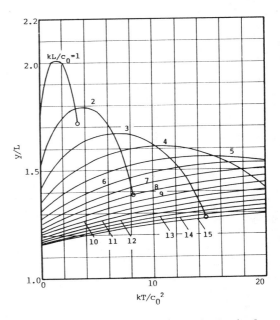

Fig.3 Chord length of the critical circle

ues of kT/c_0^2 and kL/c_0 is readily obtained.

After obtaining the value of sector angle θ by the use of Fig.2, chord length y of the critical circle can be obtained by means of following equation. That is, from equation (9) we get

457

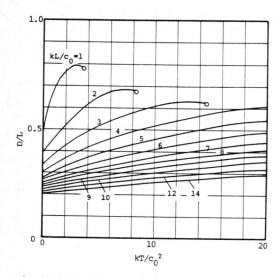

Fig.4 Depth of the critical circle

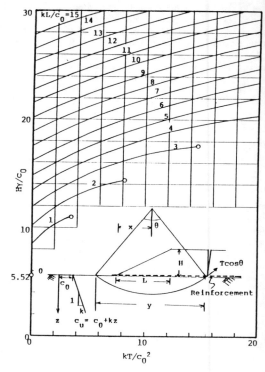

Fig.5 Stability chart

$$\frac{y}{L} = \frac{B\sin\theta}{(\frac{kL}{c_0})g} \qquad (14)$$

The right hand side of equation (14) contains the values of kT/c_0^2, kL/c_0 and sector angle θ. As shown in Fig.2, however, the sector angle is uniquely related to the parameters kT/c_0^2 and kL/c_0. Therefore, when the parameters kT/c_0^2 and kL/c_0 are given, chord length y can be obtained by substituting these values into equation (14) together with the value of sector angle which is obtained by the use of Fig.2. The relationship between chord length y of the critical circle and the parameter kT/c_0^2 is shown in Fig.3 for various values of kL/c_0. By the use of Fig.3, the chord length y of the critical circle for a given set of values of kT/c_0^2 and kL/c_0 is readily obtained.

The position of the critical circle may be determined from a set of values of x, y and θ mentioned above, i.e., from equation (6) and Figs.2 and 3. However, as for a geometry of the critical slip circle, its extent in depth may be most important problem. Because when soil stratum is shallow relative to the geometry of the embankment, we can not apply the design chart shown in Figs.2 and 3 for determining the position of the critical circle. Stability analysis for the case where the extent of thickness

of the soft cohesive foundation is limited will be examined in the next section.

In the case where the thickness of soft cohesive soil stratum is infinite, depth D of the critical slip circle is given as

$$D = L \frac{B(1 - \cos\theta)}{(\frac{kL}{c_0})g} \qquad (15)$$

The right hand side of equation (15) contains the values of kT/c_0^2, kL/c_0 and sector angle θ. As mentioned above, however, the sector angle is related to the parameters kT/c_0^2 and kL/c_0. Therefore, when the parameters kT/c_0^2 and kL/c_0 are given, depth of critical circle can be also obtained by substituting these values into equation (15) together with the value of sector angle which is obtained by the use of Fig.2. The relationship between depth of the critical circle and the parameter kT/c_0^2 is shown in Fig.4 for various values of kL/c_0. By the use of Fig.4, the depth of the critical circle for a given set of values of kT/c_0^2 and kL/c_0 is readily obtained.

458

.3 Stability chart

Substituting equations (6) and (9) into equation (5) results in

$$\frac{H\gamma}{c_0} = \frac{6B[B^2V + 2Bg\theta + 2g^2\left(\frac{kT}{c_0}\right)\cos\theta]}{g[3B^2\sin^2\theta - g^2\left(\frac{kT}{c_0}\right)^2]} \qquad (16)$$

where

$$V = \sin\theta - \theta\cos\theta \qquad (17)$$

When the values of kT/c_0^2 and kL/c_0 are known, the right hand side of equation (16) is a sole function of the sector angle θ. As mentioned above, however, the sector angle is related to the parameters kT/c_0^2 and kL/c_0. Consequently, when the values of kT/c_0^2 and kL/c_0 are known, stability factor can be obtained by substituting these values into equation (16) together with the value of sector angle θ which is obtained by the use of Fig.2. Stability chart is shown in Fig.5.

4 INFLUENCE OF THICKNESS OF COHESIVE SOIL STRATUM

In the preceding section, it has been assumed that the thickness of soft cohesive soil stratum is infinite. In the case where the soft cohesive soil stratum is underlain by a stiff layer such as sand gravel stratum or rock, it will be reasonable to consider that the extent of the slip surface is confined within the upper soft cohesive soil stratum, hence the possible maximum depth of slip surface is equal to the thickness of the cohesive soil stratum.

Fig.6 shows the situation considered in the section, where a cohesive soil stratum with thickness D is underlain by a stiff stratum. The condition where the depth of slip surface is equal to the thickness of the soft cohesive soil stratum gives the expression

$$y = D\frac{\sin\theta}{1 - \cos\theta} \qquad (18)$$

Substituting this relationship into equation (5) together with the equation (6), the moment limiting-equiliburium equation becomes

$$\frac{H\gamma}{c_0} = \frac{4\left(\frac{kD}{c_0}\right)V + 4\theta U + 2\left(\frac{T}{Dc_0}\right)U^2\cos\theta}{U[\sin^2\theta - \frac{1}{12}\left(\frac{L}{D}\right)^2U^2]} \qquad (19)$$

where

$$U = 1 - \cos\theta \qquad (20)$$

When the parameters kD/c_0, T/Dc_0 and L/D are known, the right hand side of equation (19) is a sole function of the sector angle θ. So, the minimum value of $H\gamma/c_0$ in equation (19), i.e., stability factor, is obtained from the condition $\partial(H\gamma/c_0)/\partial\theta=0$, with the result

$$\frac{kD}{c_0} = \frac{\frac{1}{2}\left(\frac{T}{Dc_0}\right)U\sin\theta[(J-I)\cos\theta - J\cos^2\theta + I]}{(\theta I - JV)\sin\theta}$$

$$+ \frac{(\theta J\sin\theta - I)U - \theta I\sin\theta}{(\theta I - JV)\sin\theta} \qquad (21)$$

where

$$I = \sin^2\theta - \frac{1}{12}\left(\frac{L}{D}\right)^2U^2 \qquad (22)$$

$$J = 1 + 3\cos\theta + \frac{1}{4}\left(\frac{L}{D}\right)^2U \qquad (23)$$

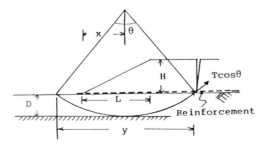

Fig.6 Slip circle within soil stratum of limited thickness

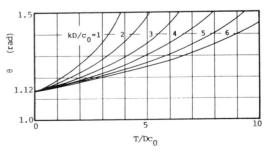

Fig.7 Sector angle of the critical circle for limited thickness (L/D=1)

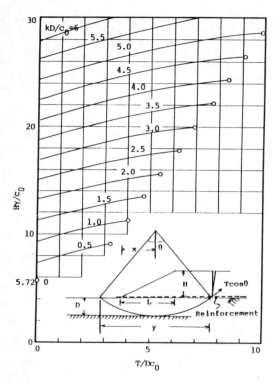

Fig.8 Stability chart for limited thickness (D/L=1)

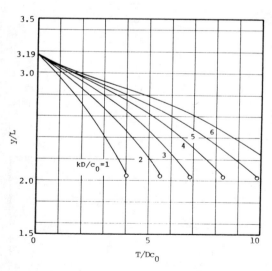

Fig.9 Chord length of the critical circle for limited thickness (L/D=1)

The sector angle θ is obtained by calculating equation (21) for a given set of parameters T/Dc_0, kD/c_0 and L/D. Consequently, for a given set of parameters T/Dc_0, kD/c_0 and L/D, the stability factor is obtained substituting these parameters into equation (19) together with the value of sector angle which is calculated by equation (21). Fig.7 shows an example of the relationship among T/Dc_0, kD/c_0 and sector angle θ for the case of L/D=1, and Fig.8 shows an example of design chart for the case of L/D=1.

The chord length y of the critical circle is obtained by substituting the value of θ which is calculated by equation (21) into equation (19). Fig.9 shows an example of the relationship among T/Dc_0, kD/c_0 and chord length y for the case of D/L=1.

5 CONCLUDING REMARKS

Stability analysis of reinforced embankment over soft cohesive foundation has been carried out. A reinforcing sheet, capable of developing a horizontal tensile strength is placed at the embankment – foundation interface. Nondimensional stability charts are compiled from the results of analysis. The basic assumption of the present analysis is development of a vertical tension crack in the embankment. Therefore, the present method of stability analysis of reinforced embankment should be applied to the problems in which the height of embankment is small compared with the length of slip surface within the foundation soil. As the height of embankment increases, the assumption of a vertical tension crack through the embankment tends to be unrealistic, and the shear resistance of the embankment material becomes appreciable. When applied to a relatively high embankment, the present method of analysis may be underestimate the factor of safety, because the shear resistance of the embankment material has been neglected.

REFERENCES

Baker, R. 1981. Tensile strength, tension cracks and stability of slopes. Soils and Foundations, Vol.21, No.2, pp.1-17.
Leshchinsky, D. 1987. Short-term stability of reinforced embankment over clayey foundation. Soils and Foundations, Vol.27, No.3, pp.43-57.
Taylor, D.W. 1937. Stability of earth slopes. Journal of the Boston Society of Civil Engineers, Vol.24, No.3, pp.197-246.

International Geotechnical Symposium on Theory and Practice of Earth Reinforcement / Fukuoka Japan / 5-7 October 1988
© 1988 Balkema, Rotterdam. ISBN 90 6191 820 0

Reinforced soil systems in embankments – Construction practices

I.Paul
Netlon Limited, Blackburn, UK

ABSTRACT: Extensive experience has shown that reinforced soil techniques are very effective in the construction of embankments over soft foundations. The paper uses case studies to examine the construction details of three separate systems which use the range of 'Tensar' polymer grids as reinforcement. Reinforced soil is also used in the repair of slip failures in embankments and cuttings. Case studies are again used to discuss construction methods and comparative costs.

1 NEW EMBANKMENTS

The problem of constructing embankments on soft foundations is common in many parts of the world. A number of solutions are available including piling, excavation of the soft layers and slow, staged construction.

Over the past ten years or so, reinforced soil techniques have been used very effectively to reduce the cost and time for construction of such embankments.

The reinforcement materials considered in this paper are the range of 'Tensar' polymer geogrids. The grids are manufactured by punching a regular array of holes in a sheet of polymer which is then heated and stretched at carefully controlled temperature and strain rates. The stretching process aligns the long chain molecules of the polymer thus increasing substantially the tensile strength and tensile modulus and correspondingly reducing the potential for creep. Stretching in one direction produces uniaxial grids which have enhanced properties in the direction of stretch. A second process of stretching in the transverse direction produces biaxial grids which have similar properties in each direction (Fig. 1).

Most embankment designs utilise uniaxial grids since this is a two dimensional problem with forces acting at right angles to the centre-line. However, when reinforcement stresses are low, the lightweight biaxial grids can often provide a cost-effective solution.

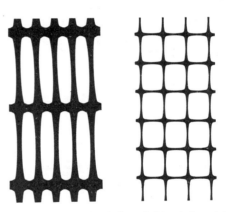

Figure 1. Uniaxial and Biaxial grids

1.1 Horizontal layers

The simplest application of reinforced soil techniques is to use the reinforcement to intercept potential failure surfaces passing through the embankment and into the soft foundation.

Such a system was adopted on the A12, Chelmsford By-Pass in Essex, England.

The road embankment is 8.5m high constructed of London Clay and has side slopes of 2.5:1. The foundation consists of weathered London Clay the upper 2.5m of which contains polished, pre-sheared surfaces aligned approximately parallel with the ground surface. The design parameters for this relatively weak surface layer are:- $c' = 1.5$ kN/m^2, $\emptyset' = 14°$ $r_u = 0.5$

The critical mechanism was a translational failure within the upper clay aligned with the direction of the weak shear planes. To avoid the cost of removing the weak layer a reinforced soil solution was adopted.

Unlike other grid-type products, 'Tensar' grids have integral, full strength junctions between the ribs and cross bars. They are therefore extremely efficient in transferring stresses from the soil into the reinforcement by abutment against the cross bars. For this reason the coefficient of interaction with the soil is very high and anchorage lengths are correspondingly short. It is quite possible therefore to place the reinforcement in the zone where it is required without the need to have layers continuous across the full embankment width. Since stress transfer is through abutment and not surface friction, the grids can be used with many different soil types including clays. The polymers used in their manufacture are very inert forms of high density polyethylene and polypropylene, the grids may be used in very acid or alkaline materials such as Fly Ash or other waste materials.

On the Chelmsford By-Pass four layers of 'Tensar' SR2 grid were used as shown in Figure 2.

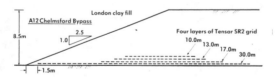

Figure 2. Cross section showing layout of reinforcement

The bottom layer was 30m long and extended to within 1.5m of the toe. This single layer of reinforcement was sufficient to prevent shallow toe slips. The three upper layers were 17m, 13m and 10m respectively which provided maximum reinforcement within the zone of highest stresses.

After preparation of the formation the bottom 30m length of grid was laid in position at right angles to the centre-line. 'Tensar' SR2 grids are uniaxial products manufactured at 1m wide. Since the embankment is long and is built up evenly no forces are induced parallel to the centre-line and there is no need to overlap adjacent rolls. The grids were therefore laid side by side to form a continuous strip 1.5m to 31.5m from the toe. Embankment fill was then spread and

compacted on top using standard construction equipment and standard techniques, the compaction specification being exactly the same for reinforced and unreinforced areas. When the embankment level was raised by 250mm the second layer of reinforcement was placed as before. At 500mm and 750mm the third and fourth layers were placed. Figure 3 shows the second layer of grid ready to receive the next layer of fill.

Figure 3. Placing fill on reinforcement grid

By using grid reinforcement the designer was able to minimise costs by placing the reinforcement only in areas where the calculated factor of safety was insufficient. The techniques for construction were very simple and required no alteration to the standard earthworks specification for fill placement or compaction.

1.2 Reinforcement mattress

In areas of deep soft deposits there are two main problems, short term stability and differential settlement. Such was the case on the A47, Great Yarmouth Western By-Pass in Norfolk, England. Details of the design have previously been reported by Williams and Sanders (1985). Essentially the embankment rises to a height of 8m and is founded on very soft organic alluvial deposits 22m deep.

The design solution adopted was to encapsulate the basal drainage blanket with the biaxial grid 'Tensar' SS2. In this way the toes of the embankment were prevented from moving outwards by the two layers of reinforcement, preserving the integrity of the high shear strength granular drainage blanket. Instrumentation was monitored during fill placement to avoid excess pore pressures and over-stressing the grid.

'Tensar' SS2 has quality control tensile strengths of 31.5 kN/m across the 4m width of the roll and 17 kN/m along the 50m length. For maximum economy the higher strength was required at right angles to the centre-line. That meant that the rolls had to be laid parallel to the centre-line but that the high tensile strength should be continuous across the full base width.

To provide this full strength joint between adjacent rolls a high density polyethylene braid was used to stitch the rolls together. Several configurations of stitching were tested and two were found to provide the required strength. In practice one was easier to form on site and was therefore adopted.

The construction procedure involved rolling out the grids parallel to the centre-line and overlapping adjacent rolls by one aperture (approximately 40mm). The high density polyethylene braid was then interwoven through every aperture of the two geogrids using a crescent shaped needle.

While it appears to be a labour intensive operation it is in fact fast and efficient. Output rates of 50 linear metres of joint per man hour were easily achieved and the cost of the joint is only equivalent to the cost of approximately 200mm of overlap. In all, 250,000m² of grid were used on the' project with all edges stitched as described.

A team of two men collecting and unrolling the 4m wide grid and forming 80 linear metres of joint per hour is equivalent to an all-in laying and fixing rate of 160m² per man hour for the fully stitched mattress.

On one particularly soft area a tracked excavator had sunk into the soft clay. However, when the first layer of reinforcement and the first layer of drainage material had been laid all traffic was able to cross the area without damage. This is an indication of the effectiveness of the grid in interlocking with the fill particles and spreading load. Limitation of differential settlement is achieved by the stiff composite action of the reinforcement mattress and on the Great Yarmouth By-Pass the embankment, which was constructed in 1983, is performing extremely well. In contrast, an adjacent section which used only a geotextile as a separation membrane is suffering from significant differential settlement.

1.3 Geocell mattress

The third reinforced soil system is one which can only be constructed using 'Tensar' polymer grids:- the Geocell mattress.

The Geocell is particularly useful where an embankment is to be constructed on a relatively thin layer of soft soil (less than 12m in depth). A method of analysis and design has been developed based on plasticity theories and is reported elsewhere (Jenner Bassett and Bush, 1988).

It is also used on deep layers of soft material where the stresses within the mattress would be higher than could be carried by the system using biaxial grids as described above. Once again the main purpose is to increase short term stability and to regulate long term differential settlement.

The Geocell mattress is a honeycomb structure formed by constructing a series of interlocking cells using the 1m wide uniaxial grids in a vertical orientation. The mattress is therefore 1m thick and is connected to a biaxial grid base. When filled with a granular material it forms a very rigid base to the embankment.

Site erection begins by unrolling a biaxial grid onto the soft foundation parallel to the centre-line of the embankment with 300mm overlap between adjacent rolls. The overlap is used rather than stitching to increase speed of installation and to balance erection and filling times. A uniaxial grid is then laid transversely across the embankment and one edge stitched to the base. A second grid is laid abutting the first and again one edge is stitched. This procedure continues, advancing out over the soft area. When a number of transverse grids have been stitched in place they are rotated about the stitched edge into a vertical position and temporarily tensioned (Fig. 4).

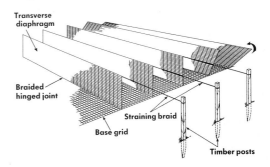

Figure 4. Construction of geocell mattress

463

The cell structure is formed by unrolling a roll of uniaxial grid between two transverse diaphragms and connecting it to the diaphragm using a hooked steel bar (Figs. 5 and 6)

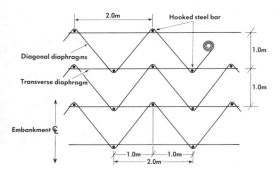

Figure 5. Geocell mattress - plan view

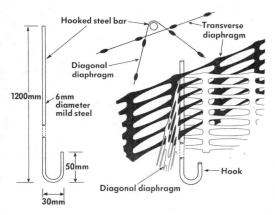

Figure 6. Connection detail

During the winter of 1985/86 a Geocell mattress was used to carry a 4.5m high embankment over very soft foundation soils at Auchenhowie, near Glasgow, Scotland. The foundation comprised:
1. Peaty topsoil - 0.4m thick
2. Loose, brown, organic, clayey silt and peat - 1.5m thick
3. Very soft brown, laminated silty clay - 2.5m thick
4. Weathered, friable, grey mudstone

Design was based on a soft layer depth of 4m having an average undrained shear strength of 15 kN/m².

Use of the Geocell mattress increased the factor of safety against foundation failure from 0.9 to 1.5.

Four alternative methods of construction were examined:
a. Partial excavation of the soft material and replacing by displacement of the remainder with rock fill. Lack of control over costs and performance made this option unacceptable.
b. Installation of drainage followed by staged construction. The need for rapid construction of the embankment to provide access to other sections of the contract ruled out this option.
c. Complete excavation of the soft layer and replacement with rock fill. Estimated cost - £152,000. In addition, this option would have caused a substantial increase in heavy site traffic on adjacent minor roads in this predominantly residential area.
d. Construction of a high strength geogrid mattress to provide a stiff base to the embankment eliminating stability problems and providing an effective drainage layer. Estimated cost - £104,000 (saving 31%).

The Geocell mattress option was therefore adopted.

Construction of the mattress at Auchenhowie involved working out over an old lake bed.

The construction team comprised three men laying, stitching and tensioning the grids plus an hydraulic excavator and driver to fill the Geocell with imported granular fill.

Figure 7. Filling geocell mattress

As can be seen from Figure 7 the excavator sits on previously filled sections of the mattress. Stone delivery wagons also travel on the mattress and dump adjacent to the excavator - thus no mechanical plant needs to run on the very soft foundation, the Geocell providing a stiff, 1m thick construction platform.

The filling procedure is to fill two rows of cells to half height before filling the first to full height. This system is continued, always ensuring that no cell is filled to full height before adjacent cells are at least half filled. In this way, potential distortion of the cell structure is avoided.

It is impractical to compact the fill in the cells therefore 150mm of overfilling is normally specified to cater for the slight compaction settlements which will be induced by construction traffic. It also protects the top edges of the diaphragms.

In this fairly typical project, erection and filling rates for the mattress averaged 350m² per day for the 3 men plus the excavator team.

Construction of the embankment to its full height of 4.5m was completed in about 7 days following completion of the mattress.

Settlement pins were installed within the embankment on top of the mattress. Approximately 90mm of initial settlement was recorded at the end of the embankment construction with no significant differential settlement noted either longitudinally or transversely.

Major advantages of this system are that construction can take place rapidly over very soft foundations without the normal problems of access for mechanical plant and that construction can continue in all weather conditions.

2 REPAIR OF SLIP FAILURES

A study carried out in England by TRRL showed that of five systems used to repair a slip in a 7m high road embankment, the reinforced soil technique using 'Tensar' geogrids was the cheapest (Johnson, 1985).

The five methods and associated costs for a 20m length are shown in Table 1.

Table 1. Repair options

Reinstatement techniques	Time taken (days)	Total cost (£)
Gabion wall	18	8360
Granular replacement	5	5020
Anchored tyre wall	8	4760
Lime stabilisation	7	4730
Geogrid reinforcement	6	3430

Only the lime stabilisation and geogrid reinforcement solutions re-used the soft clay. All other systems involved excavation and disposal off site along with importation of substitute granular fill.

Of the five systems investigated the reinforced soil repair using 'Tensar' geogrids was one of the fastest solutions and substantially less expensive than all of the others.

The A406 forms the North Circular Road close to the centre of London, England. Several years after its construction in 1968, slip failures began to occur in a cutting near Waterworks Corner where the A406 runs through Epping Forest. The 500m long cutting is up to 8m in depth through London clay and was constructed with side slopes of 1:2. Although the cutting was stable for the first 7 years after construction, slips began to occur with increasing regularity causing damage to fence lines and spillage onto the carriageway.

Several remedial measures were considered, including granular replacement and toe retention with crib walling or sheet piling combined with flatter slopes. These options were prohibitively expensive. Lime stabilisation was ruled out mainly due to doubts on the viability of achieving uniform mixing on site and uncertainty about long term performance.

A reinforced soil solution was selected on the basis of cost, ease of construction and confidence in the design, which involved reinstatement of the failed London clay, reinforced with horizontal layers of 'Tensar' geogrids. Drainage was also provided.

To achieve the specified factor of safety, horizontal layers of 'Tensar' SR2 grids were installed at 1.5m vertical spacing. To prevent sloughing at the surface, short horizontal layers of the lightweight biaxial grid 'Tensar' SS1 were installed at 0.5m vertically between the primary reinforcement layers (Fig. 8).

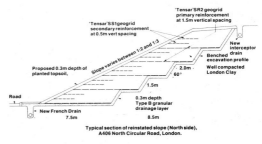

Figure 8. A406 - cross section showing reinforcement layout

465

Work was begun in September 1985 using a standard earthworks team of one Cat 215 tracked excavator, one Volvo four wheel drive dumptruck and one Cat 951 dozer with a four-in-one bucket and towed vibrating roller. Manual operations were carried out by two labourers.

Use of the reinforced soil system reduced traffic management measures to a minimum since there was a minimal removal of spoil and no importation of granular fill which would have involved large numbers of trucks reversing on to and off the carriageway.

Main earthworks began with the excavation and removal from site of a 35m long strip of slipped soil. Excavation extended beyond the failure plane with benched steps cut into the undisturbed clay. The general sequence then adopted was to reinstate the first strip using fill excavated from an adjacent second strip, to minimise double handling. The second strip was then reinstated using fill excavated from a third strip and so on.

Fill was tipped from the Volvo, placed using the Cat 951, and compacted to a maximum layer depth of 200mm using the vibrating roller towed by the Cat 951. The 2m widths of 'Tensar' SS1 secondary reinforcement were obtained by cutting the standard 4m wide rolls in half with a disc cutter.

'Tensar' SR2 rolls were cut to the required lengths and laid perpendicular to the slope face. Adjacent rolls were butt jointed (Fig. 9). The slope face was over filled and trimmed in the conventional manner.

Along part of the slope where additional land was available the gradient reduced to 1:3.

Approximate quantities involved in the 500m long repair were:

Excavation	23,000m³
Refilling	12,800m³
Gravel drainage layer	5,400m³
'Tensar' SR2 primary reinforcement	17,000m²
'Tensar' SS1 secondary reinforcement	8,000m²

In spring 1986 approximately 5000 cubic metres of topsoil were placed on the reconstructed slope for subsequent planting.

In the South East of England the more traditional repair involving removal of the slipped material and replacement with granular fill typically costs around £25/m³. The reinforced soil technique using 'Tensar' geogrids has been shown to cost between £7 and £12/m³ i.e. savings of between 50% and 70% (1985 prices).

3 CONCLUSIONS

The application of reinforced soil techniques using polymer geogrid reinforcement has in all cases shown substantial cost savings over more traditional construction systems.

The 'Tensar' Geocell mattress is a unique structure which allows rapid construction of embankments on soft foundations and avoids costly and time consuming excavation and replacement of the foundation soils. In a number of cases the system has been shown to produce substantial cost savings by reducing land-take due to the elimination of berms required for unreinforced constructions.

In all of the other systems discussed, the incorporation of 'Tensar' geogrids required no specialist equipment or techniques and required no amendment to standard earthwork specificaitons.

REFERENCES

Jenner, C.G., Bassett, R.H. and Bush, D.I. 1988. The use of slip line fields to assess the improvement in bearing capacity of soft ground given by a cellular foundation mattress installed at the base of a embankment. I.S. Kyushu.

Johnson. P.E., 1985. Maintenance and repair of highway embankments: studies of seven methods of treatment. TRRL Research Report 30.

Williams, D. and Sanders, R.L. 1985. Design of reinforced embankments for Great Yarmouth By-pass. Proc. XI ICSMFE, San Francisco.

Figure 9. Laying reinforcement grids

International Geotechnical Symposium on Theory and Practice of Earth Reinforcement / Fukuoka Japan / 5-7 October 1988
© *1988 Balkema, Rotterdam. ISBN 90 6191 820 0*

Analysis of slopes reinforced with bamboo dowels

H.B.Poorooshasb
Concordia University, Montreal, Canada

R.Azevedo & K.Ghavami
Catholic University of Rio de Janeiro, Brazil

ABSTRACT: Many slopes in the vicinity of the city of Rio de Janeiro and other localities in Brazil develop large land slides during the rainy seasons, many of them with catastrophic results in terms of human and property losses. A possible mechanism of failure is the loss of suction in the unsaturated residual soils which form the surficial layer of such slopes. To avert these failures the use of bamboo dowels have been proposed and indeed the paper presented is a first step in the envisaged investigation.
It is shown that the equation governing the mode of deformation of a dowel is in the form;

$$\frac{d}{dz}\left[\beta(z)\frac{d^4u}{dz^4} + u\right] + \lambda(z)\frac{d^3u}{dz^3} = \lambda(z)\sin\alpha$$

where $\beta(z)$ and $\lambda(z)$ are functions of the rigidity of the dowels, the shear modulus of the wet and dry soil, the spacing between dowels and their diameter. The parameter $\gamma(z)$ is a function of the unit weight of the dry and wet soil and the respective shear moduli. The angle α represents the inclination of the slope.
The governing equation is solved numerically using a microcomputer. The numerical scheme used in the analysis is described in detail and the results are presented in the form of graphs showing the distribution of the bending moments and the shearing forces in the dowels.

1 ANALYSIS

As mentioned in the abstract the failure of the slopes is contributed to a loss of suction in the active zone of the ground. Let the depth of this zone be denoted by z_a and for convenience measure the z axis from the surface of the slope as shown in Fig. (1). During the dry seasons the surficial layer for which $z < z_a$ is hard with a unit weight of γ_d and shear modulus of G_d both of which are likely to be constants but in general are functions of depth z. After prolonged rainfall the unit weight of the soil changes to $\gamma_r = \gamma_r(z)$ and its shear modulus to $G_r(z)$. It is fairly obvious that $\gamma_r > \gamma_d$ and that $G_r < G_d$ caused by saturation of the soil and a loss of rigidity due to the reduction in the suction in the soil. These changes bring about a downhill movement of the surficial layer which, if not checked, may lead to catastrophic results. A possible preventive method is, of course, the use of dowels installed to depth well below the active zone.

Let L represent the length of a dowel with a flexural rigidity equal to EI and an effective diameter d. Then the equilibrium of the dowel requires that

$$EI\frac{d^4u_b}{dz^4} = p(z).d \qquad (1)$$

where p(z) is the interaction pressure given by;

$$p(z) = (u_s - u_b)k(z) \qquad (2)$$

k(z) being the modulus of lateral reaction. Combining equation (1) and (2) results in;

$$\frac{EI}{kd}.\frac{d^4u_b}{dz^4} = u_s - u_b \qquad (3)$$

where u_s is the lateral displacement of the soil and u_b that of the bamboo dowel.

Next consider the equilibrium of a block of soil of depth z supported by a typical dowel and let the area of this block be denoted by a. Note that if the dowels are arranged in a square pattern then $a = b^2$ where b is the distance between the dowels. If the dowels are arranged in a triangular pattern then $a = b \cdot 3^{1/2}/2$. Assuming the average shear at depth z to be τ_r then it is clear that;

$$\gamma_r \, a \, z \, \sin \alpha - d \cdot \int_0^z p \, dz = \tau_r a$$

or

$$\tau_r = \gamma_r \, z \, \sin \alpha - \frac{d}{a} \int_0^z k \, (u_s - u_b) \tag{4}$$

During dry periods the average shear stress is given by;

$$\tau_d = \gamma_d \, z \, \sin \alpha \tag{5}$$

where the average shear stress under dry conditions is given by τ_d. Therefore the shearing strain caused by the reduction of soil suction is given by;

$$\frac{du_s}{dz} = \frac{\gamma_r \, z \, \sin \alpha}{G_r(z)} - \frac{d}{G_r(z) a} \int_0^z k \cdot$$

$$\cdot (u_s - u_b) \, dz - \frac{\gamma_d \, z \, \sin \alpha}{G_d(z)} \tag{6}$$

which upon substitution for $k \, (u_s - u_b) = (EI/d) \, (d^4 u_b/dz^4)$ reduces to

$$\frac{du_s}{dz} = \left[\frac{\gamma_r}{G_r(z)} - \frac{\gamma_d}{G_d(z)} \right] z \, \sin \alpha -$$

$$- \frac{EI}{G_r(z) a} \int_0^z \frac{d^4 u_b}{dz^4} \, dz \tag{7}$$

Integration of the second term on the right hand side of Eq. (7) yields;

$$\int_0^z \frac{d^4 u_b}{dz^4} \, dz = \frac{d^3 u_b}{dz^3} - \frac{d^3 u_b(0)}{dz^3} = \frac{d^3 u_b}{dz^3} \tag{8}$$

since $d \, u_b(0)/dz^3 = 0$ there being no shearing force at dowel at $z=0$; the surface of the slope. Thus Eq. (7) may be rewritten as

$$\frac{du_s}{dz} = \gamma z \sin \alpha - \frac{EI}{a G_r(z)} \frac{d^3 u_b}{dz^3} \tag{9}$$

where $\gamma = \gamma_r/G_r(z) - \gamma_d/G_d(z) = \gamma(z)$

Let the factor EI/kd be denoted by $\beta(z)$ then from Eq. (2) it may be concluded that;

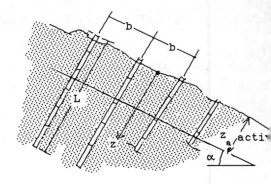

Fig. (1) – Reinforced Slope.

$$\frac{du_s}{dz} = \frac{d}{dz} \left[\beta \frac{d^4 u_b}{dz^4} + u_b \right] \tag{10}$$

Combining Eqs. (9) and (10) results in the governing equation of the problem viz:

$$\frac{d}{dz} \left[\beta(z) \frac{d^4 u}{dz^4} + u \right] + \lambda(z) \frac{d^3 u}{dz^3} =$$

$$= \gamma(z) \, z \, \sin \alpha \tag{11}$$

where, for convenience the subscript b has been dropped and the factor EI/aG_r has been represented by the variable $\lambda(z)$. For numerical evaluations it is easier to use a new spacial variable ζ where $\zeta = L-z$, L being the length of the dowel as shown in Fig. (1). With this new variable Eq. (11) reads;

$$\frac{d}{d\zeta} \left[\beta(\zeta) \frac{d^4 u}{d\zeta^4} + u \right] + \lambda(\zeta) \frac{d^3 u}{d\zeta^3} =$$

$$-\gamma(\zeta) \, (L-\zeta) \sin \alpha \tag{12}$$

The order of Eq. (12) may be reduced by one if both sides are integrated with respect to ζ and results in;

$$\beta(\zeta) \frac{d^4 u}{d\zeta^4} + \lambda(\zeta) \frac{d^2 u}{d\zeta^2} + u =$$

$$= C + R(\zeta) + S(\zeta) \tag{13}$$

In Eq. (13) the constant C is given by the expression;

$$C = \beta(0) \, d^4 u(0)/d\zeta^4 + u(0)$$

and the variables $R(\zeta)$ and $S(\zeta)$ by the definite integrals

$$R(\zeta) = \int_0^\zeta \frac{d^2 u}{d\xi^2} \frac{d\lambda(\xi)}{d\xi} \, d\xi$$

468

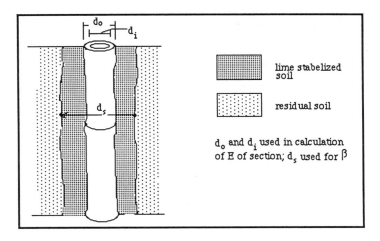

Fig.(2) – Enchased Bamboo Dowel.

and,

$$S(\zeta) = \int_0^\zeta \gamma(\xi)(\xi - L)\sin\alpha\, d\xi$$

Eq. (13) must be solved subject to the boundary conditions that the bending moments (BM) and the shearing forces (SF) at the extremeties of the dowel must be zero i.e.

$$BM_{\zeta=0} = 0 \qquad\qquad (14,a)$$

$$SF_{\zeta=0} = 0 \qquad\qquad (14,b)$$

$$BM_{\zeta=L} = 0 \qquad\qquad (14,c)$$

$$SF_{\zeta=L} = 0 \qquad\qquad (14,d)$$

In the numerical procedure used by the authors the governing differential equation is solved in terms of $U=u-c$. Since the value of c is constant the form of the governing equation does not change, neither would the boundary conditions which, when expressed in terms of U, would read as follows:

$$\partial^3 U/\partial\zeta^3 = \partial^2 U/\partial\zeta^2 = 0; \text{ at } \zeta = 0 \quad (15,a)$$

$$\partial^3 U/\partial\zeta^3 = \partial^2 U/\partial\zeta^2 = 0; \text{ at } \zeta = L \quad (15,b)$$

After completion of the numerical (finite difference scheme) procedure the value of constant c is evaluated noting that the total forces acting on the dowel must balance out. The nodal values of u are then calculated using the equation $u = U + c$.

Before closing this section it is worth pointing out that the matrix of the coefficients of unknown parameter u is not banded. The definite integral $R(\zeta)$ contains

the unknown value of U and hence its contribution to the matrix must be taken into account in the numerical procedure. Thus if a finite difference scheme is used and the section is divided to ns sections each of length $T(=L/ns)$ a typical subroutine for calculation of the coefficients matrix would read:

```
FOR i = 1 TO ns
FOR j = 1 TO i
if j=i then a(i,j)=(del(j-1)-2*del(j))/T:
 :GOTO 1
a(i,j)=(del(j-1)-2*del(j)+del(j+1))/T
IF j=i THEN a(i,j+1)=del(j+1)/T
NEXT j
a(i,i-2)=a(i,i-2)+beta(i)/T^4
a(i,i-1)=a(i,i-1)-4*beta(i)/T^4+lamda(i)
 /T^2
a(i,i)=a(i,i)+6*beta(i)/T^4-2*lamda(i)
 /T^2+1
a(i,i+1)=a(i,i+1)-4*beta(i)/T^4+lamda(i)
 /T^4
a(i,i+2)=beta(i)/T^4
NEXT i
```

From the above subroutine in which del(i) is the value of $d\lambda/d\zeta$ evaluated at node (i) it may be seen that the matrix of the coefficients of the unknown U(i) is non-symmetric.

2 NUMERICAL EVALUATIONS

Before proceeding with the presentation of the numerical evaluations it should be pointed out that it is necessary to encase the individual dowels in a protective sleeve, see Fig. (2). The material used for protection

max deflexion = 1.93812 Cm., ns= 41
max S.F.=-82.37607 Kg., dowel space= 2
max BM= 7150.338 Kg. Cm
max normal stress= 123.3614 Kg/Cm^2

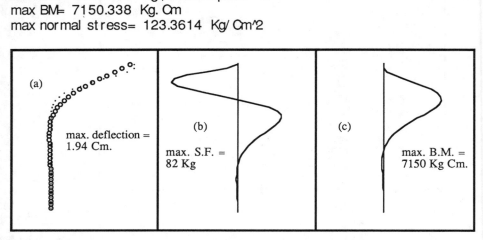

Fig. (3) - Performance of a set of bamboo dowels spaced at 2 meters intervals. (a) Deformed shape of a typical dowel, (b) Shearing force and (c) Bending moment diagrams for the dowel. Angle of slope 30 degrees.

max deflexion = 1.786847 Cm., ns= 41
max S.F.=-63.72635 Kg., dowel space= 1
max BM= 5649.52 Kg. Cm
max normal stress= 97.46849 Kg/Cm^2

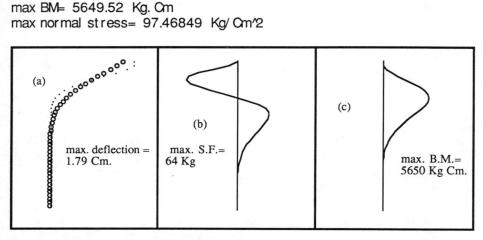

Fig.(4) - Performance of a set of bamboo dowels spaced at 1 meter intervals. (a) Deformed shape of a typical dowel, (b) Shearing force and (c) Bending moment diagrams for the dowel. Angle of slope 30 degrees.

max deflexion = 1.461587 Cm., ns= 41
max S.F.=-32.7416 Kg., dowel space= .5
max BM= 3028.413 Kg. Cm
max normal stress= 52.24777 Kg/Cm^2

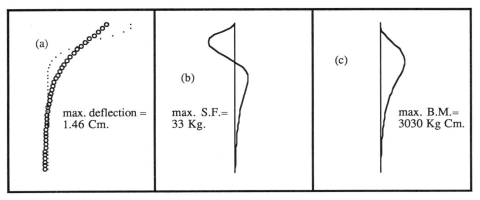

Fig. (5) - Performance of a set of bamboo dowels spaced at .5 meters intervals. (a) Deformed shape of a typical dowel, (b) Shearing force and (c) Bending moment diagrams for the dowel. Angle of slope 30 degrees.

of the dowels could be lime stabilized or portland cement stabilized soil. The former is preferable, however, in view of this low cost. The purpose of the sleeve is two fold. First it fills the space between the dowel and the soil surrounding it and thus prevents the dowel from acting as an unsupported beam, the resistance of which to bending would, of course, be very small. Second it increases the strength of the bamboo dowel in compression by prevention of local buckling. It has two undesirable side effects however. First, and this is particularly important for the lime sleeve, a deterioration of the bamboo may take place due to what is generally referred to as exothermic reaction. Second the presence of sleeve will add to the loads that each individual dowel must carry without effectively adding to its strength. Stated otherwise, since the strength of lime in tension especially is very low it is good practice to neglect the contribution of the sleeve to the second moment of area of the dowel but use its diameter (i.e. the outer diameter of the sleeve) to evaluate the value of $\beta = EI/kd$, see Fig.(2). Finally it should be noted that the magnitude of k, the modulus of lateral reaction is dependent, amongst other factors, on the spacing of the dowels. If the dowels are placed too closely to each other then a reduction of k value would result in that must be taken into account in the evaluations. In this paper this reduction is accounted for by the relationship.

$$k(z)_{group} = k(z) * (b - ds)/b \qquad (16)$$

where $k(z)$ is the modulus of lateral reaction corrected for depth and size (total diameter of the dowel d_s, see Fig.(2)) and b is the spacing of the dowels as defined previously. For a set of dowels spaced at 1 meter intervals the $k(z)$ value of the group is %80 that of a single dowel with an effective diameter of 20Cm.. For the spacings of 2 and 5 meters the corresponding values would be %90 and %60 respectively. Other more elaborate relations may be used in place of Eq.(16) the results, however, are unlikely to be of consequential effects.

The results of some typical evaluations are shown in Figs.(3) to (5) inclusive. The slope to be protected is assumed to be quite steep having an angle of 30 degrees. The bamboo dowels are assumed to have an external diameter of 10 centimeters with a wall thickness of 1 centimeter with a 5Cm. cover -of lime stabilized soil. The value of Young's modulus is taken equal to that of concrete 1.6×10^6 Tons/m² (Ghavami and Zielinski, 1988) and the modulus of lateral reaction (for a single dowel) is assumed to vary linearly from 1000T/m³ at the surface increasing by 100T/m³ for every meter of depth. The depth of the active zone is taken as 2 meters and other mechanical parameters (such as dry and wet unit weights of the soil and the corresponding shear moduli) are assumed such that they would produce a total downhill creep of about 2 centimeter per year.

When the dowels are spaced at 2 meter intervals the maximum bending moment experienced by the dowel is 7150 Kg Cm. causing a normal stress of 124 Kg/Cm^2 in the fibers of the dowel. The corresponding values for 1 meter spacing and 0.5 meters spacing 5650 Kg Cm. and 3030 Kg Cm. respectively producing, in turn, maximum fiber stresses of 97 Kg/Cm^2 and 52 Kg/Cm^2. All these stress levels are well within the limits of the allowable tensile strength (about 800 Kg/Cm^2) and compressive strength (about 300 Kg/Cm^2) for the bamboo dowel (Ghavami and Zielinski, 1988 and Ghavami and Hombeeck, 1981).

The deformation values are, however, quite large. As may be noted from Fig. (5) even for closely spaced dowels (.5 meters in this case) the presence of dowels has reduced the amount of downhill creep by only about 5 milimeters. This is, of course, because of the relatively low values of E of bamboo and the low value of the second moment of area of the section used in the reinforcement. The downhill creep is, nevertheless, confined to the first rainy season after the instalment of the dowels. The drying process which takes place in the summer would relax the forces taken by the dowels and infact they would tend to move the surficial layer uphill. Subsequent wetting would only restore the stresses in the bamboo to their original values i.e. the values registered at the end of the first rainy season.

3 CONCLUDING REMARKS

From the above evaluations it is concluded that the use of bamboos is indeed a feasible method of preventing downhill movement of slopes and their eventual failure.

The analysis, however, assumes that the slope is undergoing a form of mountain creep caused by a reduction of suction pressure in the soil skeleton during the wet seasons. A second mechanism postulating the existence of a moisture sensitive layer parallel to the slope surface is treated in a companion paper.

4 ACKNOWLEDEMENT

The authors wish to thank the Brazilian Council for Research and Development (CNPq) and the Natural Science and Engineering Research Council of Canada (NSERC) for their financial support.

REFERENCES

Ghavami, K. and Zielinski, Z.A. 1988. Permanent Shutter Bamboo Reinforced Concrete Slab. Civil Engineering Dept., Concordia University, Report BRCS 1, Montreal, Feb., pp. 1-34.

Ghavami, K. and Hombeeck, R.V. 1981. Aplication of Bamboo as a Construction Material, Part I, Mechanical Properties and Water-Reppellent Treatment of Bamboo. Latin-American Symposium on National Civil Construction. IPT-São Paulo, Oct., pp. 40-66.

International Geotechnical Symposium on Theory and Practice of Earth Reinforcement / Fukuoka Japan / 5-7 October 1988
© 1988 Balkema, Rotterdam. ISBN 90 6191 820 0

Collapse of dispersive soil and its prevention using geotextiles

Shou-Tian Sheng, You-Wei Hong & An-Jiang Wang
Heilongjiang Provincial Research Institute of Water Conservancy, Harbin, People's Republic of China

ABSTRACT: A large number of hydraulic works had been constructed and were being constructed in the western part of Song-Nen Plain of Heilongjiang Province, China. Most of them were damaged and eroded. This paper has examined primary reasons for an erosive damage due to the behavior of dispersive clayey soil. These are: soil dispersion; a dynamic water flow of erosive damage; a by-pass of water flow; etc. This paper introduces the model test method and the test process to protect erosion using geotextiles. Based on the test results in 1985, we found that the geotextiles used to prevent the soil from erosive damage were very effective and durable.

1 INTRODUCE

Song-Nen Plain is $1*10^5$ km^2 in area and locates in the western part of Heilongjiang Province, China. High natrium content soil is distributed in the area. The so-called "alkali gullies".

In the Plain some earth structures were built from high natrium content soil, when they are being constructed or have been built, we found more than hundreds erosive pockets due to collapse and damage by rainwater, as shown in Photo 1 and 2.

From a large number of tests, it has been proved: some earth structures have been damaged because of soil dispersion. Based on the damage mechanism of dispersion and the preventive function of geotextiles, a model pinhole test had been undertaken in the laboratory. We are successful to explore the possibility of using geotextiles to prevent damage and

Photo.2

bring to civil engineering practice, some good effects were obtained.

2 IDENTIFICATION OF THE DISPERSIVE SOIL

From the field investigation, we had secured samples to determine the engineering properties. Sampling location was shown in Fig. 1.

The dispersive soil is identified by the following method:

1. Clastic test (wetting and drying);
2. Two-hydrometer test;
3. Chemical analysis of pore water ;
4. Pinhole test.

The results of the tests were given in Table 1. The results of four types of tests are consistent, so the soil is dispersive. We conclude some earth structures are damaged because of dispersive behavior of soil.

Photo. 1

Table 1 The summary results of various test methods

NO.	drying clastic test	wetting clastic test	two-hydrometer test	pinhole test	proe water analysis
S-1	◎	◎	△	◎	◎
S-2	◎	◎	△	◎	◎
S-3	○	○	○	○	○
S-4	◎	◎	◎	◎	◎
S-5	◎	◎	◎	◎	◎
S-6	◎	◎	△	◎	◎
S-7	◎	◎	◎	◎	◎
S-8	◎	◎	◎	◎	◎
S-9	◎	◎	◎	△	◎
S-10	◎	◎	◎	◎	◎
S-11	◎	◎	◎	◎	◎
S-12	○	○	△	○	◎
S-13	◎	◎	○	◎	◎
S-14	◎	◎	◎	◎	◎
S-15	◎			◎	
S-16	◎			◎	
S-17	◎			◎	
S-18				◎	
N-1	○	○	△	○	○
N-2	○	○	△	○	○
N-3	◎	◎	◎	△	○
N-4	◎	△	◎	◎	○
N-5			○	○	○
F-1	◎	◎	○	◎	◎
F-2	◎	△	○	○	◎
F-3			△	○	△

◎ dispersive   ○ non-dispersive   △ middle-dispersive

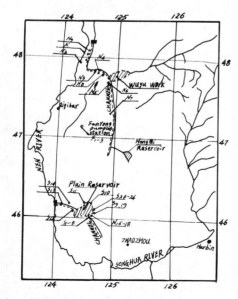

Fig. 1 Sampling location

* right is non-dispersive sample
** left is dispersive sample

Photo. 3 Wetting clastic test

3 THE EROSIVE CONDITION OF DISPERSIVE SOIL STRUCTURE

The dispersive clayey soil contains very high natrium ions around the soil particles because of the large thickness of the soil hydro-membranes existing in the double electronic layer. The attractive force between the soil particles, therefore, is very small. As the soil wetted, or immerssed in water, particles on soil surface began to separate one by one. No critical velocity exists in erosive damage, in the still water, the clayey soil particles were in suspention, as shown in Photo 3.

But the test data of dispersive clayey soil in the western part of Heilongjiang Province indicate that the coefficient permeability of soil is very low, for example, sample S-2: $k_{10}=3.3*10^{-9}$ cm/s. When seepage gradient is not high, and the coefficient is very small, the seepage cannot occured in soil. As to S-2, the soil sample is 4 cm long, the seepage gradient is 70, after 15 days there were only a few pollution area in the sample end. So we can infer, if the thickness of clay is 20 m, and a gradient is 70, it will take twenty years to saturate the clay. Therefore, it seems that some damage conditions of dispersive earth structure are: the soil is dispersive; the water is pure and there exists a by-pass to allow the water flow.

4 TEST RESULT OF PREVENTION FROM EROSION ON GEOTEXTILES

Under purified water flow, the dispersive clayey soil is a kind of non-cohesive soil, therefore, we undertake conventional treatment by a sandfilter, it is very successful to control erosive damage if the sandfilter is made carefully. According to the filter function of the geotextile it can be used instead of sandfilter.

4.1 The selection and property of the geotextiles

Based on engineering property of dispersive soil; the erosive damage condition and products available in China, we selected geotextiles, manufactured by Jiamusi No-woven Textile Mill of Heilongjiang Province. They are needle-knitted DS450-Ⅱ and polyester combined with polypropylene 1:1. The thickness is 0.35 cm; the weight is 450 g/m^2; the elongation is 57.0 %; the tensile strength is 37.8 kg/$_5$cm; the coefficient of permeability $K_{10}=4.83*10^{-3}$ cm/s; the effective diameter $O_{90}=0.114$ mm.

4.2 The selection of protected soil

The problem only concern with dispersive soil, so we believe that non-cohesive soil had a property of non-cohesive soil in pure water, then a typical damage soil (higher clay content) had been selected, used for testing.

The properties of dispersive soil sample S-4 collected from the field were given as followed:
Dispersion degree: D=49 %;
Natrium percent: 95.7 %.
Pinhole test: white fog appeared after several seconds, grain scattered and suspended; other details are shown in Fig. 2. From Fig.2 we found: the sample S-4 is

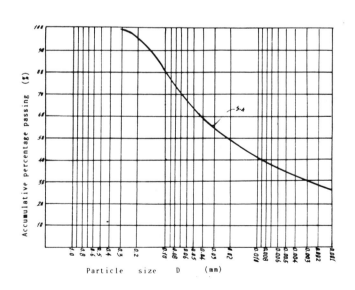

Fig. 2 Sample S-4 grain composition

excellent composited soil. Let us consider from the conventional method of sandfilter design, compare it with a poor composited soil, it is easier to treat. The key to the question is that the dispersive soil has both physical and chemical effects of damage, it seems that there is not a critical volocity in pure water, in general the critical volocity of non-dispersive soil is less than 10 cm/s. If the volocity is more than 100 cm/s, the dispesive soil becomes poor erosion-resistance.

4.3 Test result from laboratory

Because of the damage condition, we found, the key to the problem that was the earth structure cracked from upstream to downstream in the dikebody. In order to obtain this erosive condition, we undertake with pinhole test, designed by J.L.Sherard (U.S.A.), the property of sample S-4 is given as following: $r_d = 1.6$ g/cm^3; W=15.3 %;the sample is 4 cm long; a hole sized 1 mm or 2 mm at the center of the sample. As shown in Fig.3, the sample is in the pinhole unit. Geotextile DS450-II is used as filter. And also for safety, distilled water is used in testing. The water flows horizontally. The load applied are water heads of 5, 15, 45, 135 cm respectively. From the relations between the varity of water flow and the collapse of dispersive soil, we can evaluate the filter effects. See Table 2.

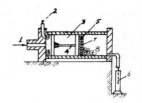

1--constant water head
2--upstream piezometer
3--dispersive clay sample
4--pinhole
5--wire mesh
6--graduated cylinder
7--geotextile
8--gravel

Fig. 3 Geotextile filter test equipment

4.4 Analysis of test results

Six sets of test were carried out, NO.01-04 gave good effects, infiltration water turning clear quickly, no soil particle in water, the pihole is not enlarged obviously, and very small area is polluted only around the pinhole; NO 05-06 has large pollution area, and pinhole enlarged, after water flowed through the hole, some of them escaped and flowed along the edges of the container.

According to these data mentioned above, which gives the common feature, the water flow turned clear gradually and the erosive damage was stopped.

Table 2 Goetextile as filters in pinhole test

NO	r_d (mg/m^3)	WATER CONTENT (%)	LAYER OF GEO-TEXTILE	LAYER OF WIRE MESH	HOLE SIZE mm BEFORE TEST	HOLE SIZE mm AFTER TEST	WATER HEAD (cm)	FLOWING WATER COLOR	NOTE	
01	1.60	15.3	1	no	1.0	1.5	135	milkwhite clean	diameter of the pinhole	pollution
02	1.60	15.3	1	no	1.0	1.5	135	as above	as above	pollution
03	1.60	15.3	1	no	1.0	1.5	5.0	as above	as above	pollution
04	1.60	15.3	1	no	2.0	2.0	5.0	as above	as above	pollution
05	1.60	15.3	1	2	2.0	5.0	45	little muddy	pollution at the below part	pollution
06	1.60	15.3	1	2	1.0	4.0	5.0	as above	as above	pollution

1 up right is distilling water only;
2 up left is high dispersive sample after 10 min

Photo4 Wetting clast test

5 CONCLUSION

5.1 The geotextile is the effective mean to control the dispersive soil from erosive damage.

5.2 When designing geotextile instead of sandfilter, the geotextile and soil must be quite fit well in the engineering practice, as samples NO.05-06, otherwise, the earth structures will be escaped partially due to flow along the edges of the by-pass.

5.3 The damage mechanism of geotextile used to prevent the dispersive soil from erosion was difference from ordinary clayey soil or non-clayer soil. It is not a catalyst function based on geotextile, which makes a bridge structure in net and accompanies with filter region. But the geotextile filter effect in dispersive soil works is, the soil is dispersed in water, a part of latex dispersive flocculation materils occured and a part of fine soil particles flow through the by-pass or geotextile. Second, the particles on the soil surface bagan to separate one by one, large particles and non-dispersive particles made a discharge curtain (pollution area) by clogging or blocking in the geotextile filter body, with forming the process of this curtain, (forming the process of discharge system). The permeability of geotextile began to reduce due to clogging and blcking. The particles which moved further distance arranged one by one and made the filter region enlarged, in the end a steady hydraulic system was formed and a discharge steady condition was satisfied (the water was clear, pinhole not enlarged, the current was steady).

5.4 For the ordinary soil (clayer soil and loose soil), the clogging problem of geotextile usually is the important basis of design. For the dispersive soil, because of the damage condition, which exists in water by-pass, particles began to erode, so that a part of flow was controlled, the object will be obtained. Because the fine particles were clogged and more fine particles were blocked, the geotextile became the materials of forming

discharge curtain, there was no problem when the geotextile was used for discharge design in dispersive soil.

5.5 For the geotextile the seepage standard of discharge design on dispersive soil must have a higher pemeability (due to a pipe water flow), also it has a homogenous and a discharge pore (formed a steady discharge curtain).

6 THE EARTH STRUCTURES DESIGN AND PERFORMANCE

The Southern Division Work in Heilongjiang was located in the distributed region, a filling reservoir was made of 33 dikes, NO. 16 dike was built from dispersive soil, the total length of NO.16 dike is 430m long, the height of dike is 3m, the upstream slop ratio is 1:3, the downstream is 1:2.5, r_d=1.5 g/cm^3.

The properties of soil are:

S.P.	WL (%)	WP (%)	IP	0,002 (%)	Activity degree	
2.67	21.2	18.4	2.8	18	0.156	ML

The pinhole test result is:

No	water head (mm)	Q (1/ min)	water con- tent	r_d g/cm^3	dia- (mm)	color	classify
1	5	350	13	1.64	5	muddy	dis- persive

The results of dispersion degree are:

without dispagent 0.005(%)	dispersive agent 0.005(%)	dispersive degree (%)	classify
13		61.9	dispersiv

The geotextile tested is SD450-Ⅱ, other indices are as above.

Beside the design indices mention above, we must think about the protected range of prevention from water-flow erosion: to prevent dam crest from raining, because the dam was located in subdry land, always appeared rainstorm, for more dry dispersive soil before raining, it is very strong,but it can be eroded by pure water easily; to prevent dam slop from erosion, for the dispersive soil which will be stood a non-stready wave action, also the erosive damage is very serious; to prevent the downstream dam body from the seepage water, the water flow by-pass which made the soil grain moved it will be important part of protection. Particular attention must be paid to the erosive damage for dispersive soil which must have a by-pass of seepage, so the location of design is dependent upon the water level of reservoir.

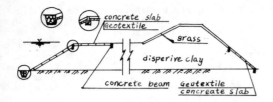

Fig.4 16# Dam at Southern Diversion Works

Fig. 4 demonstreated the structural design, The NO. 16 dike was constructed in 1978 and completed in 1987. It performs very well.

There are various methods to treat the erosion of dispersive soil, but using geotextile instead of sandfilter, is both durable and cheap.

There is no design criteria yet about using geotextile filter, people often prefer to adopt the design from the field and the data from the laboratory which, we believe, is reliable and feasible.

REFERENCES

1.   Heilongjiang Provincial Research Institute of Water Conservancy "Research of Dispersive Clay in Westen Area of Heilongjiang Province, China"
2.   Sheng Shou-Tian & Hong You-Wei "Geotextile Application in Hydraulic Works in Heilongjiang Province, China" Third International Conference on Geotextlie, 1986, Vienna, Austria,
3.   Sheng Shou-Tian & Wang An-Jiang "The Filter Test Research of Dispersive Soil", Chinese Journal of Reservior Engineering Manage" Vol. 4.1985.
4.   "Engineering and Frozen Soil" Vol. 18. No. 5. 1987.

International Geotechnical Symposium on Theory and Practice of Earth Reinforcement / Fukuoka Japan / 5-7 October 1988
© *1988 Balkema, Rotterdam. ISBN 90 6191 820 0*

Full scale testing and numerical analysis for polymer grid reinforced embankment

M.Taki & S.Renge
Fukken Co., Ltd, Hiroshima, Japan

T.Hachiman, Y.Morioka & H.Matsuda
Okayama Prefectural Government, Okayama, Japan

J.Natori
Dia Consultants Co., Ltd, Osaka, Japan

ABSTRACT: Using the polymer grid (hereinafter grid) as a reinforcing material for earth embankment is a new way of steep and high embankment construction. The present research is undertaken with a view to verify the design procedur and construction technique that are to be adopted in constructing grid reinforced embankments.

The works include the construction of a full scale test embankment, an observation of dynamic behaviour of the embankment, and a series of FEM analysis. The parameters of the embankment are : height-14 m, base width-55 m, slope 1:0.5 up to a height of 6 m and 1:0.8 from 6 m to 14 m height, a case of embankment without grid reinforcement, 4 cases with reinforcement. Comparative studies of test embankments were conducted through the back analyses of the observed dynamic data. The reinforcing mechanism is established and the reinforcing effects evaluated. From the results of the studies it is confirmed that there exist as a result of grid reinforcement an increased strength and a decreased deformation in the grid reinforced zone of the embankment. Moreover the laying of reinforcing grids in a checkered pattern is found to be structurally effective.

1 INTRODUCTION

Recently the use of grid as a reinforcing material in the earth embankment is becoming very popular in Japan, and the authors are also engaged in the laboratory testing of grids (Fukuda et al. 1987).

The reinforcing material and reinforcing methods are various. The application of grids to actual construction works is more advanced than ever. But the design considerations and reinforcing mechanism are not clarified as yet.

In that the present research in this paper is undertaken by constructing a full scale grid-reinforced embankment model and testing it through destruction in line with the embankment works for the site preparation of a housing complex in Okayama prefecture, Japan. The observed data of the test were brought under analysis by using FEM simulation and the reinforcing effects evaluated. The field measurement of grid stresses and strains, and embankment deformation are conducted with a view to confirm the dynamic behaviour of the embankment.

On the other hand the elasto-plastic analysis by the finite element method which is the back analysis using the observed data is carried out with an aim to evaluate quantitatively the effects and mechanism of grid reinforced embankment design. It is to be concluded that the breaking pattern obtained from in-situ destruction tests reveals the reinforcing mechanism and effects quantitatively. The elastic modulus (E) and the cohesion (C) of the reinforced zone are increased in the same pattern with the stress-strain relation.

2 FIELD TESTING

2.1 Test Cases

The test embankment is as shown in Fig. 1. In each case the embankment has a width; 10 m (15 m without reinforcement), height : 14 m. After filling the whole embankment with a side slope of 1:1.2 (initial slope), the slope is again cut to a finished slope of 1:0.5 for the lower 6 m and 1:0.8 for the upper 8 m (No.1 cut slope). The section reinforced by grid is 6 m high and 8 m wide. The upper 8 m fill section is constructed as a surcharge.

The reinforcing pattern is as shown in Fig. 2 with the Case 1 reinforced by one grid layer and Case 3 by 2 grid layers and Case 4 by 4 grid layers in a continuous pattern, and Case 2 by 5 grid layers in a checkered pattern, and Case zero is without reinforcement. Thus the quantities of grids laid in the embankment from Case zero to Case 4 are in the ratio of 0:1:2: 2:4.

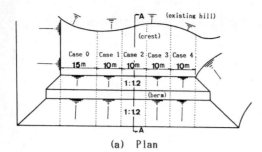

(a) Plan

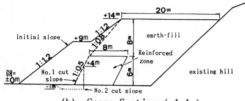

(b) Cross Section (A-A)

Fig. 1 Plan and Section of Test Embankment

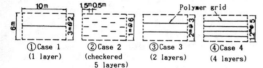

① Case 1 (1 layer) ② Case 2 (checkered 5 layers) ③ Case 3 (2 layers) ④ Case 4 (4 layers)

(a) Laying Patterns of Polymer Grids

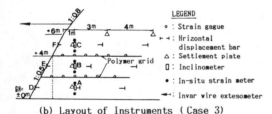

LEGEND

○ : Strain gague
⊢⊣ : Hrizontal displacement bar
△ : Settlement plate
▯ : Inclinometer
● : In-situ strain meter
◀⟶ : Invar wire extesometer

(b) Layout of Instruments (Case 3)

Fig. 2 Grids Laying Pattern and Instrumentation Layout

2.2 The Features of the Test Embankment

(a) Embankment Material: It is a highly weathered rock of greenschist and pelitic schist. Its properties are as shown in Table 1.
(b) Foundation of the Embankment: The foundation is the same kind of soil as the embankment material or filled up by a better quality soil.
(c) Reinforcing Material : Bi-axial stretched polymer grids are used. Tensile strength is 1.8 tf/m at a strain rate of 8 %.

2.3 Compaction

The fill material is spread out by bulldozers forming a layer of 50 cm thickness. Each 2-layer is compacted by 1~2 runs of vibrating rollers. The dry density of soil after compaction is 1.59 t/m³ at 90 % degree of compaction.

Table 1   Index Properties of Soil

Natural water content	Wn	(%)	21.30
Specific gravity	Gs		2.71
Gravel fraction		(%)	4
Sand fraction		(%)	51
Silt fraction		(%)	32
Clay fraction		(%)	13
Compaction test method		(by JIS)	2.5.b
Optimum moisture content	Wopt	(%)	15.00
Maximum dry density	ρ dmax	(t/m³)	1.76

2.4 Field Measurement

The following parameters of soil are measured using the measuring apparatus shown in Fig. 2:- ① the grid strain by foil strain gauge, ② relative horizontal displacement of the fill material from the slope by displacement bar, ③ the settlement of embankment by settlement plate, ④ horizontal displacement of embankment by inclinometer, ⑤ fill-material strain by strain gauge attached to a vinyl pipe, ⑥ horizontal displacement of slope by invar wire wire extensometer.

Measurement for ①, ④ and ⑤ are done by an automatic measuring and recording system and ⑥ by an automatic recording device; and measurement for ② and ③ by manual recording.

3 RESULTS OF MEASUREMENT

3.1 Deformation of the No.1 Cut Slope

The No.1 cut slope is completed by cutting the initial slope but no slope failure takes place.

In so doing, for Case zero the crest area around 8 m from the top of the slope was covered with several cracks with an opening of 2~3 mm and width 10 m. The horizontal displacement of the embankment after cutting of the initial slope is 1~3 cm as shown in Fig. 3. The deformation is decreased and the strength gain sharply noticed with an increase of reinforcing grids

The grid strain distribution is as shown in Fig. 4. It is in such a trend that the grid strain is higher as it approaches the slope. When grids are laid in several rows the upper ones have higher strain values than the lowers. These results are approximately the same as those obtained from laboratory testing (Kutara et al. 1987).

3.2 Destruction by Vertical Cutting

The vertical cutting of the No.1 cut slope (Fig. 1) does not lead to sliding although the horizontal displacement of the embankment is 1~4 mm/day within a period of 5 days.

But the slope failure occurs in each case
with Case zero after 20 minutes and all other
cases after 6~9 hours as shown in Fig. 5 and
Photo. 1. The larger the reinforcement quantity
of grids is, the more the sliding will be deep
and severe, and grids break round about the
sliding plane. Just before sliding, the horizontal displacement has gone up to 10 cm and
grids's maximum strain is 2~3 % which is the
limit of strain measurement.

The strain distribution is the same as that
of the No.1 cut slope. The inner strain at

1.5 m from the inner edge of the grid is found
to be as small as 0.1~0.3 % and hence the laying
length of grids is considered to be sufficient.

Moreover, in Case 2 the reinforced section
only stands firm but it is displaced forward
as affected by the sliding force of the adjoining surcharge fill. This means that the embankment with the reinforcing grids laid evenly
distributed just like a checkered pattern is
found to be effective from the structural point
of view.

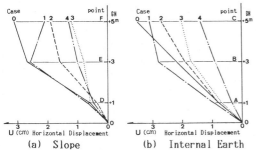

(a) Slope (b) Internal Earth

Fig. 3 Measured Horizontal Displacements of
Slope and Internal Earth (No.1 Cut
Slope)

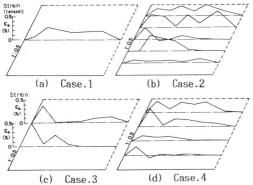

(a) Case.1 (b) Case.2

(c) Case.3 (d) Case.4

Fig. 4 Distribution of Grid-Strains
(No.1 Cut Slope)

(a) No.1 Cut Slope

(b) Final Filure State (for No.2 Cut Slope)

Photo. 1 The Scenes of Test Embankment

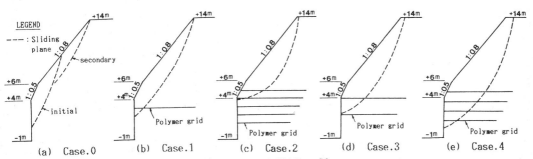

(a) Case.0 (b) Case.1 (c) Case.2 (d) Case.3 (e) Case.4

Fig. 5 Predicted Sliding Planes

4 METHOD OF NUMERICAL ANALYSIS

The analysis is done by elasto-plastic model of finite element method assuming that the integrated body of grids and earth forms a complete elasto-plastic body. And the constitutive equation is the relationship by Drucker-Prager and the grids laid horizontally in layers in the fill body is considered as a spring element of the elasto-plastic type.

Stress-strain relationship of the earth and local safety factors are as shown in Fig. 6. Here the local safety factor Fs is determined by using the Mohr-Coulomb's failure criterion for a stress condition of each element.

The analysis model and the properties of the material used in the model are shown in Fig. 7 and Table 2 respectively. The properties of

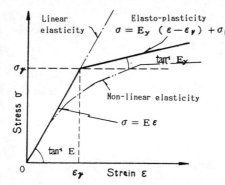

(a) Stress-strain Curve

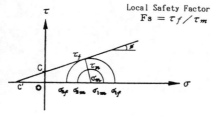

(b) Local Safety Factor

Fig. 6 Explanation of Stress-Strain Curve and Local Safety Factor

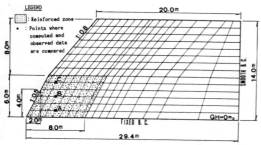

Fig. 7 Finite Element Mesh and Boundary Conditions

the fill material in the Table 2 are obtained from the back analysis of the observed data of the embankment, without reinforcement (Case zero). And the elastic modulus and cohesion of the reinforced zone are determined numerically by the back analysis using the observed data from Case 1~4.

Table 2 Properties of Materials

Earth fill (Drucker-Prager model)	Unit weight (wet)	γt (t/m³)	1.90
	Elastic modulas	E (tf/m²)	1400
	E after yield	Ey (tf/m²)	5
	Poisson's ratio	ν	0.30
	Cohesion	c (tf/m²)	0.50
	Angle of shear resistance	ϕ (°)	35.00
	Apparent tensile	C (tf/m²)	0.50
Polymer grid (Spring model)	Spring modulas	AE (tf)	50 (20)
	AE after yield	AEy (tf)	1
	Tensile strength	T (tf)	1.8 (0.7)
	T after yield	Ty (tf)	0

Note : Numbers in ( ) are for Case 2.

5 RESULTS OF BACK ANALYSIS (SIMULATION)

Fig. 8 describes the typical results obtained from FEM analysis:- (a) the deformation pattern (b) the vector diagram of principal stresses (c) the contour diagram of local safety factors The changes of elasto-plastic zone and horizontal displacement of fill body are described in Fig. 9. The results obtained from the analyses in all cases are in close conformity with those observed values of the fill body displacement. Thus the analyses are found to be precise and reliable. The horizontal displacement is reduced as the grid reinforcement is increased.

The effects of grid reinforcement can be confirmed by the occurance of the decreased deformation. And from Fig. 9 (b) it is obvious that in the case of vertical cutting of the No.1 cut slope i.e No.2 cut slope the stresses are not converged to an equilibrium state for both cases: Case zero and Case 1 and the failure of the embankment is expected. On the other hand in the cases, Case 2, 3, 4, the developments of plastic zone are in progress. But the embankments are structurally safe as latters are limited. The plastic zone has developed rapidly in the case of the surcharge fill as compared to the reinfoced zone of the embankment; thus it is to be surmised that the effect of reinforcement is obvious since an increase of strength is observed in the reinforced zone of embankment.

On the other hand the grid strain distribution is as shown in Fig. 10. The conformity of the computed grid strains with observed ones is of a lesser degree but they are considered to be satisfactory. Although the grid reinforcement quantity is the same in Case 2 and Case 3, Case 2 has relatively bigger strains. And with respect to Case 3 the No.2 cut slope

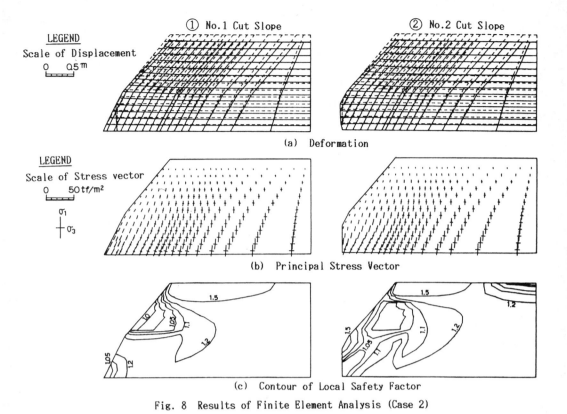

① No.1 Cut Slope   ② No.2 Cut Slope

LEGEND
Scale of Displacement
0   0.5 m

(a) Deformation

LEGEND
Scale of Stress vector
0   50 tf/m²

σ_1
σ_3

(b) Principal Stress Vector

(c) Contour of Local Safety Factor

Fig. 8 Results of Finite Element Analysis (Case 2)

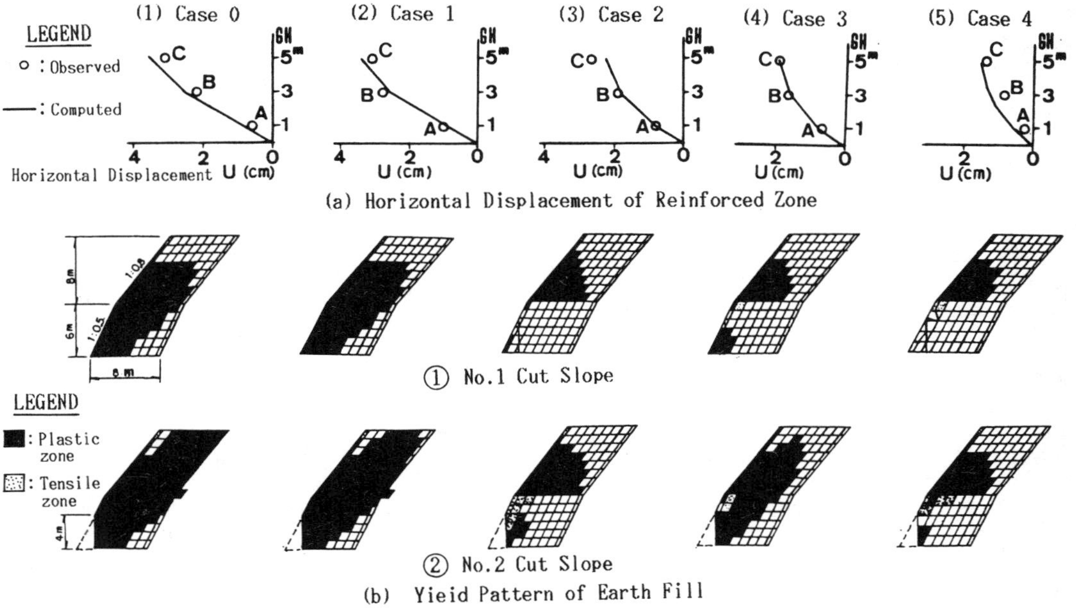

LEGEND
o : Observed
— : Computed

(1) Case 0   (2) Case 1   (3) Case 2   (4) Case 3   (5) Case 4

Horizontal Displacement U (cm)   U (cm)   U (cm)   U (cm)   U (cm)

(a) Horizontal Displacement of Reinforced Zone

① No.1 Cut Slope

LEGEND
■ : Plastic zone
▨ : Tensile zone

② No.2 Cut Slope
(b) Yield Pattern of Earth Fill

Fig. 9 Horizontal Displacement and Yield Pattern

483

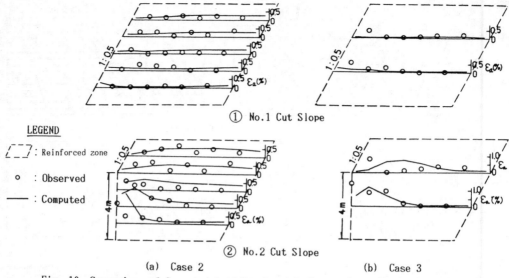

① No.1 Cut Slope

LEGEND

◻┐ : Reinforced zone

○ : Observed

—— : Computed

② No.2 Cut Slope

(a) Case 2   (b) Case 3

Fig. 10 Comparison of Computed Grid-Strain with Observed Data (Case 2 and Case 3)

has a peak strain which is distinguished from the others. These phenomena are considered as an indication of the grid and earth integration effect due to the reinforcements the checkered pattern of which is especially effective uniformly bearing the stresses on the embankment.

The summary of the results obtained from the FEM analyses of the above cases is shown in Table 3. As obvious from this table the effects of reinforcement which are the increased strength of the earth-fill and the decreased deformation of the embankment are pronounced with an increase of the reinforcement quantity in the embankment.

Table 3 Results of Computed Reinforcing Effect by FEM

Parameters		Case 0	Case 1	Case 2	Case 3	Case 4
Elastic modulus E (tf/m²)		1400	1400	1700	2000	2500
Cohesion c (tf/m²)		0.5	0.5	3.0	2.0	4.0
Horiz. displacement U (cm)	point C(GH=5m)	3.53	3.42	2.21	1.95	1.47
	point B(GH=3m)	2.55	2.50	1.81	1.63	1.43
	point A(GH=1m)	0.86	0.86	0.77	0.69	0.68
Maximum strain of polymer grid ε max (%)		--	0.48	0.33	0.20	0.09

Note : Locations of points A,B,C are shown in Fig.2 and Fig.7.

6 CONCLUSION

The major works that are done in this research include the construction of a full scale grid reinforced embankment, the destruction test with a change of slope and elasto-plastic model analysis by finite element method which verify the reinforcement effects and mechanism. The computed results by FEM are found to be in close conformity with the observed data both dynamically and quantitatively. It can be concluded that this analysis is a useful tool to solve the reinforcing mechanism of grids

laid in the embankment. It is clarified that there exists an increase of strength accompanied by a decrease of embankment defomation as a result of the grid reinforcement. And the method of reinforcement like the checkered pattern is quite effective as it results in the grid-earth integration accompanied by the uniform bearing of the stresses on the embankment.

ACKNOWLEDGEMENT

This project was carried out as a research work which is also a part of the safety measures for the construction of the high embankment required for the animal husbandry site-preparation in Okayama prefecture, Japan. The authors herewith would like to acknowledge those concerned with the embankment construction. And the authors express sincere thanks to Dr. K. Kutara of Public Works Research Institute, Ministry of Contruction and Dr. N. Fukuda of Fukken Co., Ltd. for their instructive discussion and support. Also thanks are due to Mr. Aung Swe of Fukken Co., Ltd. for his assistance in preparing this report.

REFERENCES

Fukuda, N., M. Taki and Y. Sutoh (1987) : Foundation Improvement by Polymer Grid Reinforcement, Proc. 8th Asian Regional CSMFE, Vol.1, pp. 365 ~368.

Kutara, K. et al. (1987) : A Numerical Analysis of the Earth Embankment Reinforced by Polymer Grids, Proc. 2nd Geotextile Symp., pp.42~48. (in Japanese).

Drucker, D. C. and W. Prager (1952) : Soil Mechanics Plastic Analysis of Limit Design, Quart. Appl. Math., Vol.10, pp.157~165.

International Geotechnical Symposium on Theory and Practice of Earth Reinforcement / Fukuoka Japan / 5-7 October 1988
© 1988 Balkema, Rotterdam. ISBN 90 6191 820 0

A study on stability analyses of reinforced embankments based on centrifugal model tests

E.Taniguchi & Y.Koga
Public Works Research Institute, Tsukuba, Japan

S.Yasuda
Kyushu Institute of Technology, Kitakyushu, Japan

I.Morimoto
Kiso-jiban Consultants Co., Ltd, Tokyo, Japan

ABSTRACT: Centrifugal model tests were performed for embankments reinforced with non-woven fabric by tilting the models and applying vertical load to them. The test results showed that the fabric reinforcement can sufficiently strengthen embankments to prevent their collapse. Stability analyses assuming a circular slip surface showed good agreement with the centrifugal model tests.

1 INTRODUCTION

Centrifugal model tests were performed on embankments reinforced by non-woven fabric. The purposes of model tests were 1) to grasp the effects of the length of fabric reinforcement on the stability of embankments against surcharge and horizontal force, and 2) to comprehend the validity of a simplified stability analysis which takes into account fabric reinforcement.

2 CENTRIFUGAL TESTING APPARATUS

A Centrifugal testing apparatus of the Public Works Research Institute, was used in the tests. The effective radius of the apparatus is 1.15m, and maximum centrifugal acceleration is 300G.

3 MATERIAL AND SCALING LAW

A model ground and embankment was constructed using Toyoura sand dried in the air. Static triaxial tests (CD) on Toyoura sand of the same density as that in the model showed that cohesion C was 0 and the angle of internal friction ϕ was 43.4 degrees.

Table 1 shows the physical properties of the non-woven fabric used in the tests. These data were obtained from uniaxial tension tests using a cylindrical fabric (diameter = 50mm).

The following scaling law was adopted in this series of tests; the reinforcement ratio R (non-dimension, Tatsuoka et al. (1985)) defined by the following equation coincides in the model and the prototype. Poison's ratio ν of the non-woven

fabric is assumed to be zero.

$$R = \frac{(-\overline{\varepsilon}_{3R}) E \cdot t}{\sigma_{30} \cdot \Delta H} \qquad (1)$$

where $\overline{\varepsilon}_{3R}$: average tensile strain of reinforced soil in the horizontal direction ($\overline{\varepsilon}_{3R} < 0$), E : Young's modulus of reinforcing material, t : thickness of reinforcing material, σ_{30}: horizontal confining pressure acting on reinforced soil, ΔH : spacing of reinforcing material.

A centrifugal model test provides the condition, $(\overline{\varepsilon}_{3R})m/(\overline{\varepsilon}_{3R})p = 1$ and $(\sigma_{30})m/(\sigma_{30})p = 1$ (m: model, p: prototype). Therefore in order to take (R)m equal to (R)p, the following equation should hold.

$$\frac{(E)m(t)m}{(E)p(t)p} = \frac{(\Delta H)m}{(\Delta H)p} = \frac{1}{n} \qquad (2)$$

where $\frac{1}{n}$: model scale

If the height of the actual embankment is assumed to be 5 m, the 1/n becomes 1/50. And if two kinds of prototype non-woven fabric, whose young's modulus and thickness are shown in Table 2, are selected in the tests, the value of (E)m(t)m/(E)p(t)p is 1/50.4 and 1/55.1, respectively. Therefore the use of this non-woven fabric essentially satisfies the scaling law described above. Actually 50G centrifugal acceleration is applied to the 1/50 scaled model.

4 TEST PROCEDURES

Eleven tests were performed using the models shown in Fig. 1. Table 3 shows the test conditions. Centrifugal accelerations of 50G and 100G were applied to the models. Air dried Toyoura sand was air pluviated to the container and model ground, and an embankment with the relative density of about 90% was constructed.

The embankments have an inclined slope and a vertical slope in series I and II, respectively. In latter case, the vertical slope was protected by four sandbags, 2.5cm in diameter. These non-woven fabric sandbags contained Toyoura sand. The sandbag and reinforcing fabric were sewn together.

To reduce the friction between the soil model and the side walls of the container,

grease (Dow Corning, for high vacuum) was applied to the side walls to a thickness of 100 μm. One side wall consisted of glass, and the other, of steel. A latex membrane, 0.2mm thick, was placed between the side wall and the soil model. As a mesh was drawn on the membrane at 5mm intervals, the deformation of the soil model could be measured from the deformation of the mesh.

To apply horizontal seismic force to the models, the models were tilted up to 16 degrees for Tests No. 6 to No. 8, and photographs were taken at the tilting angles of 2, 4, 6, 8, 10, 12 and 16 degrees. In the other tests a steel plate, 50mm in length and 100mm in width, 15mm in height, 210gf in weight, was placed on the top of the embankment, and a static load was vertically applied to the plate by a hydraulic piston. Strain controlled loading was adopted, and the loading speed was 0.012 mm/sec. The applied load and the settlement of embankment were measured using a load cell and displacement gauge which were installed on the piston.

Table 1 Properties of non-woven fabric

material	nylon 70%, polyester 30%
thickness (mm)	0.217
tensile strength at peak σ_f (kgf/cm)	13.2 – 13.6
strain at peak ε_f (%)	38 – 29
young's modulus at ε =15% E_{15} (kgf/cm)	47.3 – 57.3

Table 2 Data for non-woven fabric of model and prototype

model		prototype		
Young's modulus E (kgf/cm)	thickness t (mm)	Young's modulus E (kgf/cm)	thickness t (mm)	$\dfrac{(E \cdot t)_m}{(E \cdot t)_p}$
56.3	0.217	200	3.08	$\dfrac{1}{50.4}$
		150	4.49	$\dfrac{1}{55.1}$

Table 3   Test conditions

series No.	test No.	model No.	centrifugal acceleration (G)	relative density of model ground and embankment (%)	reinforcement length and number of fabrics	method of loading
I	1	I-A	100	89.8	0 sheet	
	2	I-B		90.0	307mm×4 sheet	applying vertical loads
	3	I-B		87.2		
	4	I-A	50	91.7	0 sheet	
	5	I-C		84.3	307mm×6 sheet	
II	6	II-A	50.1	87.3	60mm×4 sheet	tilting models
	7	II-B	50.8	87.3	100mm×4 sheet	
	8	II-C	49.7	87.1	150mm×4 sheet	
	9	II-D	50.2	90.7	60mm×4 sheet	applying vertical loads
	10	II-E	49.6	87.4	100mm×4 sheet	
	11	II-F	52.4	87.4	242mm×4 sheet	

Reinforcement lengths of fabrics in series I show average value among all sheets.

5 TEST RESULTS

Fig. 2 shows the relationship between the applied load and the settlement of the embankment with an inclined slope. Tests No. 2 and No. 5 were stopped before the load reached a peak value, because of the limitation of the capacity of the piston. Fig. 2 (b) indicates that the peak value of the load in the reinforced embankment (Test No. 3) was 19% larger than that in the un-reinforced one (Test No. 4), and the settlement when the load was at the peak. In the region where the settlement was larger than 4mm (=4% of the height of

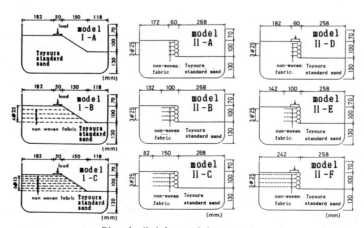

Fig. 1 Models used in tests

486

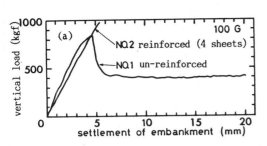

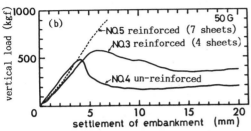

Fig. 2 Vertical load and settlement
of top of embankment

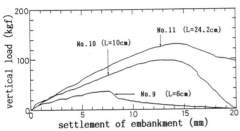

L:reinforcement length of fabric

Fig. 3 Vertical load and settlement
of top of embankment

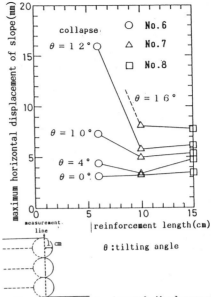

Fig. 4 Maximum horizontal displacement
of slope and placement length of
non-woven fabric

embankment), the load in the reinforced
embankment was larger than that in the
un-reinforced one for the same settlement,
and this was caused by the fabric reinforcement. However in the region where the
settlement was less than 4mm, the load in
the reinforced embankment was slightly less
than that in the un-reinforced one.
Therefore it can be concluded that the
effect of reinforcement by an extensible
material, such as non-woven fabric, appears
after the settlement of an embankment exceeds 4% of the height of the embankment.

According to the observation of model
after loading tests in the case of Test
No. 3, the first and second non-woven fabric
from the top of embankment were torn at two
lines and the third one was torn at one
line and the forth one was not damaged.

Fig. 3 shows the relationship between the
load and the settlement of the top of an
embankment with a vertical slope when a

vertical load was applied to the model embankment with a hydraulic piston. This
figure indicates that the peak vertical load
and the settlement of the top of an embankment at peak increase with the increase of
fabric reinforcement length L.

For example the peak vertical load of Test
No. 11 (L = 24.2cm) was about 3.5 times
larger than that of Test No. 9 (L = 6cm).

Therefore the reinforcing effects of
increasing the reinforcement length of
fabric are clearly observed in the figure.

Fig. 4 denotes the relationship between
the maximum horizontal displacement of a
slope and the reinforcement length of non-
woven fabric. The maximum displacement of
the slope was measured using photographs.

The maximum displacement of slope in Test
No. 6 with the reinforcement length of 6cm
was about 40% larger than that of Test No. 7
and No. 8, with the reinforcement length of
10cm and 15 cm, respectively, when the
tilting angle was 10 degrees. It was almost
the same in Tests No. 6, No. 7 and No. 8 if
the tilting angle was less than 4 degrees.
The embankment in Test No. 6 collapsed when
the tilting angle increased to 12 degrees,
but it did not collapse in Tests No. 7 and
No. 8 even if the tilting angle was 16
degrees.

These test results indicate that the reinforcement length of fabric considerably
affects the stability of reinforced embankment subjected to horizontal force.

487

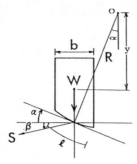

Fig. 5 Slice for stability analysis

6 STABILITY ANALYSES

Stability analyses were performed assuming a circular slip surface. The safety factor for stability of a slope with reinforcing fabric is expressed by the following equation.

$$F_s = \frac{\sum R[cl+\{(W-ub)\cos\alpha -k_H W\sin\alpha +S\sin(\alpha+\beta)\}\cdot\tan\phi]}{\sum[RW\sin\alpha +yk_H W-RS\ \sin(\alpha+\beta)]} \quad (3)$$

where R: radius of slip circle, c: cohesion, l: length of circular arc of a slice, W: weight of a slice, u: pore water pressure, b: width of a slice, α: angle between a vertical line and a line which connects the center of the slip circle to the center of the circular arc of a slice (see Fig. 5), β: angle between a horizontal line and reinforcing material (see Fig. 5), S: tension which acts in reinforcing material, ϕ: angle of internal friction, k_H: horizontal seismic coefficient.

The tension S which acts in the reinforcing fabric is taken as the smaller of the pull-out strength S_1 which is determined from Eq.(4) and the tensile strength S_2.

$$S_1 = 2\times \mu \times \sigma'_v \times B \times L' \quad (4)$$

where μ: friction coefficient between soil and reinforcing fabric, σ'_v: effective overburden pressure, B: width (= 1m), L': reinforcement length from the intersection of slip surface and reinforcing fabric.

The friction coefficient μ was 1.0 for the fabric and the sand used in this series of tests, which was determined from a pull-out test of fabric in sand. It was assumed that the angle of internal friction ϕ = 0 for the half of sandbags opposite to the backfill. For Tests No. 6, No. 7, and No. 8 the minimum safety factor of stability was calculated by changing slip surface, because it was quite difficult to determine a particular silp surface from the distribution of principal strains for Tests No.7 and No.8.

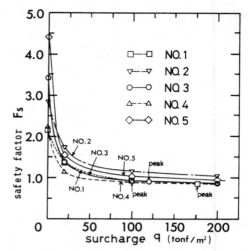

Fig. 6 Safety factor and surcharge (Test No. 1 ~ No.5)

For other tests, the safety factor of stability was calculated assuming a circular slip surface for each model from the distribution of principal strain.

In the tests, a slip surface was compulsively developed by vertical loading, and therefore the safety factor was calculated against a slip circle which was most similar to the observed slip surface. The results of calculation are shown in Fig. 6. In the calculation the maximum tensile strength at peak of non-woven fabric was substituted for S in Eq. (3). The arrows in Fig. 6 show the point where the load reached a peak value, and the corresponding safety factor was about 0.9 for Tests No. 1, No. 3, and No. 4. It can be estimated that the sliding soil mass began to create a large slide when the load reached a peak value. Fig. 6 shows that the safety factor calculated by Eq. (3) for such a condition was about 0.9, therefore the evaluation of stability by Eq. (3) gives reasonable results. The comparison among the results of Tests No. 3, No. 4 and No. 5 in Fig. 6 indicates that the safety factor increases with an increase in the number of non-woven fabrics placed in an embankment.

Fig. 7 shows the relationship of the minimum safety factor vs. the model tilt angle θ. This figure indicates that the minimum safety factor for Test No. 6 becomes almost 1.0 when the tilting angle θ is 10 degrees and becomes smaller than 1.0 when θ is 12 degrees. This calculation provides a reasonable explanation of why Test No.6 collapsed when the tilting angle was 12 degrees during the test. The safety factor for Test No.7 and Test No.8 is larger than 1.0

488

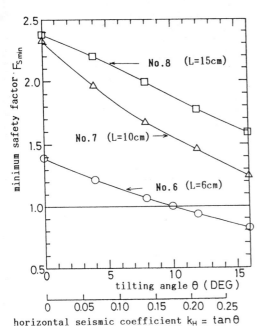

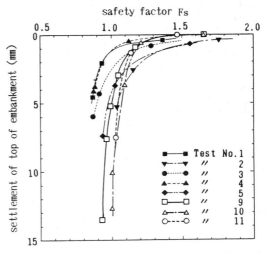

Fig. 9 safety factor vs. settlement of embankment

Fig. 7 Minimum safety factor vs. tilting angle
(Test No. 6 ~ No.8)

horizontal seismic coefficient $k_H = \tan\theta$

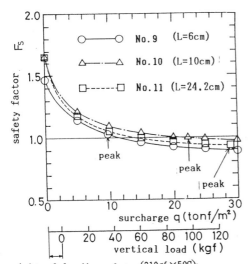

weight of loading plate (210gf ×50G)

Fig. 8 safety factor vs. surcharge
(Test No. 9 ~ No.10)

for the tilting angle θ = 0 - 16 degrees,
which well agree with the test results that
Tests No. 7 and No. 8 did not collapse when
the tilting angle became 16 degrees. It can
be noted from Fig. 7 that the minimum safety
factor is larger for models with longer
fabric reinforcement lengths.

Fig. 8 shows the relationship between the
safety factor for the observed slip surface
and the surcharge for Tests No. 9, No. 10,
and No. 11. In this figure the safety
factor is in the range of 0.95 - 1.05 when
the applied vertical load is at peak. The
change of safety factor due to the
change of reinforcement length of fabric
is as small as about 0.1.

Fig. 9 shows the relationship between the
safety factor and the settlement of the top
of the embankment, shown in Fig. 2 and Fig.
3, for the eight cases in which vertical
loads were applied to the top of the embankment. In this figure most data show that
the settlement of the top of the embankment
is very small if the safety factor is more
than about 1.0, but the settlement increases
extremely if the safety factor becomes less
than about 1.0. Similar tendency appeared
also in shaking table tests conducted by one
of the authors et al. (Koga et al. 1988)

7 A COMMENT ON STABILITY ANALYSES

Eq. (3) to calculate a safety factor of a
reinforced embankment was derived by extending the Fellenius method to incorporate
the effect of tension of non-woven fabric to
intersect a slip surface. The tension S has
different meaning in a strict sense in the
numerator and denominator in the equation.
While the former, resistant moment, should
reflect the value at the failure occurrence,
the latter, driving moment, should be that
to balance the slice.

The same value of tension was used in the
above stability analyses for their simplicity. However, it must be noticed that this

489

equation becomes irrational if the tension S is great, because the denominator becomes negative.

Further study is needed to select the most appropriate equation for stability analysis of reinforced embankment.

In the proper equation, it is desired that a safety factor of embankment against sliding should have the following continuity under several conditions.

① The safety factor without reinforcement is not so different from that with little reinforcement.

② The safety factor under no seismic force is not so different from that under a small seismic force.

8 CONCLUSIONS

(1) The reinforcement by fabric can be strengthened by increasing the reinforcement length, or the number of sheets of fabric in both cases of an inclined slope and a vertical slope. In the case of tilting tests, the horizontal displacement of the slope of reinforced embankment became smaller for the longer fabric reinforcement length.

In the case of applying vertical loads to the top of an embankment, both the peak value of load and the settlement of the top of the embankment when the load was at peak increased with the increase of the reinforcement length or the number of sheets of fabric.

(2) An nearly circular arc slip surface was observed when a fabric reinforced embankment collapsed or had large deformation due to tilting or applying vertical loads. Calculations on the slope stability assuming a circular slip surface well agreed with the centrifugal test results.

ACKNOWLEDGEMENTS

The authors would like to express their appreciation to Mr. S. Washida of P.W.R.I. for his help during the model tests and analyses.

REFERENCES

1) Tatsuoka,F. et al. (1985); Reinforcing of Cohesive-Soil-Embankment with Un-Woven Fabric (in Japanese), Tsuchi-To-Kiso, Vol.33, No.5, pp.15-20.
2) Taniguchi,E. et al. (1986); Centrifugal Model Tests on Geotextile Reinforced Embankments, Proc. of 8th Asian Regional Conf., ISSMFE, Kyoto, Vol.1, pp.499-502.
3) Taniguchi,E., Koga,Y., Morimoto,I. and Yasuda,S. (1988); Centrifugal Model Tests on Reinforced Embankments by Non-Woven Fabric, Proc. of the Int. Conf. on Geotechnical Centrifuge Modelling, Paris, pp.253-258.
4) Koga,Y., Ito,Y., Washida,S. and Shimazu, T. (1988); Seismic Registance of Reinforced Embankment by Model Shaking Table Tests, Int. Geotechnical Symposium on Theory and Practice of Earth Reinforcement, Fukuoka, submitting.

International Geotechnical Symposium on Theory and Practice of Earth Reinforcement / Fukuoka Japan / 5-7 October 1988
© *1988 Balkema, Rotterdam. ISBN 90 6191 820 0*

An application of non woven fabrics to embankment of cohesive soil

Yuji Yunoki
Nippon Koei Co., Ltd, Tokyo, Japan

Akira Nagao
Japan Highway Public Corporation, Kanagawa, Japan

ABSTRACT: For recent years, geotextiles have been used to increase the stability of embankments and other earth works. The authors tried to apply non-woven fabrics to a fill slope of cohesive soil which is about 20 m in height and comfirmed that the following effects should be included in the design of relevant earth works; ① to accelerate consolidation so as to increase shearing strength of the embankment during construction. ② to reinforce the fill slope with tensile strength of non-woven fabrics. A conventional slip circular method was employed taking into account both effects as above mentioned in the stability analysis of the fill slope. Tensile strength of non-woven fabric used for the slope stability analysis could be obtained from a series of tensile tests which were devised to reduce influence of necking for test specimens.

1. INTRODUCTION

This paper deals with a fill slope for the improvement of Ashigara parking area of Tomei expressway as shown in Fig. 1. It is located about 80 km west of Tokyo. Fine-grained volcanic ash soil of which origin is Mt.Fuji is distributed widely in this area. This soil is identified as "loam" in Japan and has high water content Fig. 2 shows a typical cross section of the site of Ashigara parking area. There is a soft layer of volcanic ash soil "Flm"

on the stiff layer of mud-flow "Fmf" and the embankment was planned to be filled by using excavated very soft loam material about 20 m height on the "Flm" layer. As shown in Fig. 2, the embanking site was limited in a narrow area due to the existing railway at the toe of this slope, and the minimum inclination of this slope should have been kept at 1 : 1.8. Therefore, this fill slope was considered to be less stable without any soil stabilization. In order to increase the slope stability, the authors employed non-woven fabrics to be placed horizontally in the embankment for an effective countermeasure.

There are not many examples of fabrics used like this. It may be worth describing an approach to use them for the reinforcement of inclined slope.

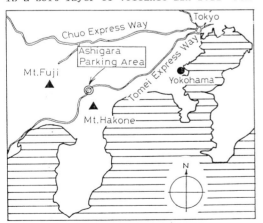

Fig. 1 Location of the Embankment

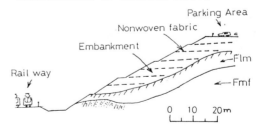

Fig. 2 A Typical Cross Section
of the Embankment

2.CONCEPT OF REINFORCEMENT OF FILL SLOPE

Since water content and degree of saturation of "loam" are considerably high, self weight consolidation is occurred during and after the embankment. Therefore, the safety factor of slope can be smallest immediately after the embankment. It gradually becomes larger with the progress of consolidation and the incerase of shearing strength. In general horizontal drainage is provided in the embankment. The role of drainage is, among others, to dissipate excess pore water pressure of the embankment during construction and to drain water fed by the rain or ground after construction. For recent years, non-woven fabrics have been used for drainage in many cases (Yamanouchi, 1985). Comparing with other geotextiles such as polymer grid,non-woven fabrics generally have high permeability, low tensile strength and large strain at rupture. Therefore, the strength of non-woven fabrics has not been included in the design of fill slope in Japan. However, the recent study with model tests of reinforced embankment using non-woven fabrics manifested that the effect of reinforcement is very large(Tatsuoka et al., 1985). Non-woven fabrics may be used for reinforcement of a fill slope. Up to now, there has been no acceptable method to evaluate the strength of non-woven fabrics for the design of fill slope of cohesive soil.

3.STABILITY ANALYSIS OF FILL SLOPE

For this project, the authors employed a conventional slip ciecle method. It can include the effects of consolidation and reinforcement, i.e. tensile strength by non-woven fabrics for analyzing the stability of fill slope. The tensile strength of non-woven fabrics used for the analysis is given in the next section.
First, increase of strength by consolidation is considered. Initial and final effective stresses in the embankment may be defined as follows;
P_0 -- effective stress immediately after compaction (which can be obtained from e~logp relationship)
P_f -- effective stress after the completion of embanking (which can be obtained from overburden pressure)
The result of Cu-test of compacted loam is obtained as shown in Fig. 3. Accordingly, the shearing strength on the sliding surface of the fill slope at its completion is given by the following

equation;
$$\tau = C_0 + \{P_0 + U (P_f - P_0)\}(C/P) -- (1)$$
Where U is a degree of consolidation and (C/P) is a rate of strength increase. It can be assumed that P_0 is equal zero because P_0 is generally very small. Then, equation (1) can be reduced as follows
$$\tau = C_0 + U\ P_f\ (C/P) ------------- (2)$$
Where consolidation pressure is assumed to be effective overburden pressure of slice for simplification of calculation. Therefore strength increase of banked material due to consolidation can be taken into account in the stability analysis of fill slope. The shearing strength on slip circle is calculated from applied effective overburden pressure by using equation (2). It is possible to control the strength increase by changing the pitch of holizontal drains.

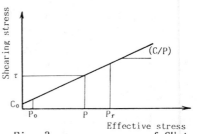

Fig. 3 $\sigma \sim \tau$ curve of CU-test

Next, the effect of reinforcement of non-woven fabrics is discussed hereunder. The effect of reinforcement can be modeled as shown in fig. 4 (a)~(c). Safety factor by slip circle method is given as below (Nagao et al., 1987).

$$Fs = \frac{Mr + \varDelta Mr}{Md}$$
$$= \frac{\sum R\ l\ \tau}{\sum R\ W\ \sin\phi} +$$
$$\frac{\sum R\ (S\ \cos(\theta - \alpha) + S\ \sin(\theta - \alpha)\ U\ (C/P))}{\sum R\ W\ \sin\phi}$$
$$-------- (3)$$

Where the second term represents the increase of safety factor by reinforcement of geofabrics. Notations in the equation are as follows.
τ : shearing strength
U : degree of consolidation
Fs : safety factor
Mr : resisting moment
$\varDelta Mr$: added resisting moment by geofabrics
Md : driving moment
R : radius of slip circle
l : width of a slice
W : weight of a slice

θ : angle between horizontal plane and tangential plane of slip circle

α : angle between holizontal plane and plane of geotextile

S : mobilized tensile strength of geofabrics
$$S \leqq Min[T, T_A, T_R]$$
where
T : tensile strength of geofabrics (tf/m)

T_A : pull-out regivtance of geofabrics (tf/m)

T_R : earth retaining ability (tf/m)

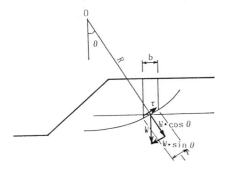

(a) Acting Stresses at a Slice

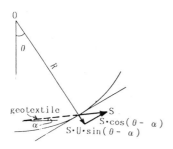

(b) Mobilized Tensile Strength of Geotextile on Slip Surface

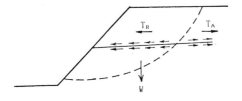

(c) Tensile Stress Induced on Geotextile in Embankment

Fig.4 Geotextile Reinforced Embankment

4. TENSILE TEST OF NONWOVEN FABRICS

There are many kinds of non-woven fabrics and they are classified to several types by materials and manufacturing processes. It is generally recognized that permeability and tensile strength for each kind of fabrics are very different. A series of extention tests that were devised to decrease the effect of necking were carried out to obtain strengths of geotextils for design and to compare ones of various non-woven fabrics under the same test condition.

TEST METHODS
Tensile tests on non-woven fabrics have been, in many cases, carried out using relatively narrow strip. In such tensile tests, necking develops entirely (Veldhuijzen van Zanten et al., 1986). Using non-woven fabrics, the necking has been hardly occurred in reinforced soil layers. Some testing methods have been devised recently which can reduce the necking to a large extent. Among these, the authors carried out the cylindrical tensile test proposed by Tatsuoka et al.(1985) and the wide width tensile test.

a) The cylindrical tensile test
A specimen 15cm × 35cm in size was stitched cylindrically by using cotton thread and was set up in the apparatus. Then specimen was fixed on the pedestals by wire and metal strips. The actual tensile length of specimen between the upper and the lower pedestals was 4 cm. See Photo. 1.

b) The wide width tensile test
Since the cylindrical tensile test described above could not be applied at high level of loading, the wide width tensile test also was employed which could reduce the necking to a large extent using relatively wide strip(10cm × 30cm). The apparatus is shown in Photo. 2.

TEST PROGRAMA
A series of tensile tests were carried out using several kinds of non-woven fabrics. Materials and manufacturing processes are different. Table 1 shows the test program and summary of the test results. Though geofabrics are generally wet in the soil layers, tensile tests were carried out not only on dry specimens but also on wet ones which were kept in the water for 24 hours for comparison. In Table 1, materials A to G were tested with the cylindrical apparatus, F to I with the wide width apparatus.

Photo. 1 Cylindrical Tensile Test

Photo. 2 Wide Width Tensile Test

TEST RESULTS

Fig. 5 shows load-elongation curves of the cylindrical tensile tests, and results of the wide width tests are given in Fig. 6. Since a kind of adhesive agent was used for bondage of fabrics in manufacturing process of specimen A, C, E, test results of the dry specimens were different from those of the wet specimens in strength and strain. The test results of other specimens showed similar values in the dry and the wet conditions except specimen A, C and E. Since specimen H and I had reinforcing members inside, their stiffness coefficients were initially very large. Nevertheless, after the reinforcing members were broken, their strengths were decreased a lot. Comparing test results of specimen G and I of which fabric materials were the same, both load-elongation curves became close after the reinforcing members

had been broken. Fig. 7 shows the comparison results of cylindrical tests and wide width tests. In the cylindrical tests, peak strengths of F and G were not obtained though load-elongation curves were similar to the results of wide width tests. Therefore it can be concluded that the wide width tests which have enough large width/length ratio can give the same accuracy as the cylindrical tests. Although the range of the tensile strength of non-woven fabrics was very wide at this time, it is found that the strain at ruptuer is very large as common characteristic. Because failure strain of soil is not so large comparing with that of non-woven fabrics, peak strength of geofabrics can not be mobilized in soil layers. Consequently, the strength of geofabrics at failure strain of soil is recommended as the design value.

Table-1 Test Program and Summary of Test Results

	specimen	manufacturing process	thickness (mm)	weight (kgf/m)	tensile strength (kgf/cm)			
					cylindrical		wide width	
					dry	wet	dry	wet
	A Polyester	Chemically bonded	3	0.26	5.47	3.23		
	B Polyolefin	Spun bonded	10	0.70	13.49	12.57		
	C Palm fiber	Chemically bonded	10	0.55	5.51	2.67		
	D Polyester	Spun bonded	10	1.00	3.03	3.63		
	E Polyester	Chemically bonded	10	0.41	4.47	3.99		
	F Polyester	Needle punched	3.5	0.40	*2 —	—	40.00	38.27
	G Polypropylene	Spun bonded	4	0.42	—	—	24.39	26.24
*1	H Polypropylene	Spun bonded	3	0.26			20.19	20.93
	I Polypropylene	Spun bonded	4	0.40			24.40	23.59

*1 : H & I have reinforcing member inside
*2 : The peak strength of F & G could not be reached by cylindrical tensile test

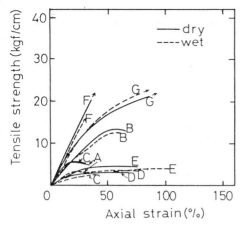

Fig. 5 Test Result of Cylindrical Tensile Test

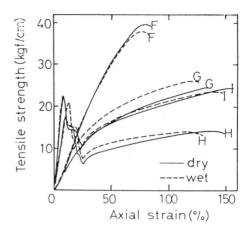

Fig. 6 Test Result of Wide Width Tensile Test

5. APPLICATION TO AN ACTUAL DESIGN

A stability analysis was carried out to examine the effect of reinforcement of non-woven fabrics using equation (3) described in the foregoing section. Fig. 8 shows an example of cross section of the embankment of Ashigara parking area. This embankment will be banked up in two years. The cross section and the banking material cannot be changed because of the restrictions as mentioned previously. The problem for this embankment, therefore, is how to keep the safety factor under the given local conditions; the soil is rather soft and slope inclination is fixed. Two kinds of countermeasures are considered. One is to replace soft soil "loam" with gravel in the toe area of the fill slope. Another is to increase the safety factor by using non-woven fabrics. The embankment material is local soil, "loam". Table-2 shows the properties of the soils used in

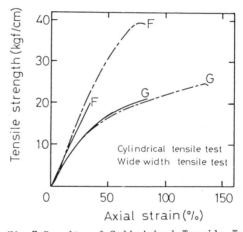

Fig.7 Results of Cylindrical Tensile Test and Wide Width Tensile Test

Table-2 Properties of the soils

Material	Unit Weight t/m³	Strength Parameters	
Embankment	1.42	Co=1.2 t/m²	C/P=0.3
Flm	1.60	Co=1.6	C/P=0.4
Gravel	1.90	Cd=0.5	$\phi_d = 35°$

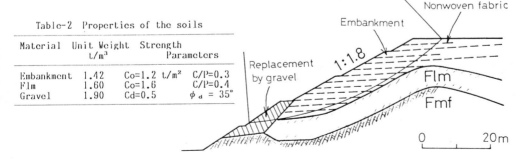

Fig. 8 An Example of Cross Section

495

the stability analysis. The strength of non-woven fabrics T was determined to be 0.6 t/m from Fig. 6, which is the strength of specimen H, I at 15% strain. (The failure strain of compacted "loam" was assumed to be 15%.) For this case, the effect of reinforcement of non-woven fabrics for safety factor was about 0.05. The embankment is under construction at present settlement, deformation, and pore water pressure are being measured. The authors will report on comparison of measuring results and estimated ones made in this report elsewhere.

6. CONCLUSIONS

① The authers used non-woven fabrics, 1.67m long, to a fill slope of 1 : 1.8 inclination and 20 m in height, then reinforcement effect of non-woven fabrics for slope stability was estimated to increase safety factor about 0.05.

② Since it was found that the effect of reinforcement by non-woven fabrics had significant influence to slope stability, it is supposed that the effect of reinforcement should be included in the design of cohesive-soil-embankments employing non-woven fabrics.

③ The strength of non-woven fabrics used in the design has to be determined by the tensile tests that can reduce the necking such as cylindrical tensile test or wide width tensile test.

ACKNOWLEDGEMENT

For this study, Dr.F.Tatsuoka, Associate Professor of Institute of Industrial Sience, University of Tokyo, gave us many useful advices, especially on estimation of strength of nonwoven fabrics in stability analysis, and conviniences to use facilities of his laboratory. The authors wish to acknowledge him with sincere gratitude.

REFERENCES

Nagao, A., Tamemitu, Y. and Yunoki, Y. (1987) The Use of Geotextiles for Cohesive-soil-embankment, Proc. 22th Annual Meeting of JSSMFE (in Japanese)

Tatsuoka, F., Ando, H., Iwasaki, K. and Nakamura, K. (1985) Reinforcing of Cohesive-Soil-Embankment with Non-Woven Fabric, Tuchi-to-Kiso, Vol.33, No.5 (in Japanese)

Veldhuijzen van Zanten et al. (1986) Geotextiles and Geomembranes in Civil Engineering, A.A.Balkema

Yamanouchi.T (1985) Current state and future development of geotextile engineering, Tuchi-to-Kiso Vol.33, No.5 (in Japanese)

5. Wall structures

International Geotechnical Symposium on Theory and Practice of Earth Reinforcement / Fukuoka Japan / 5-7 October 1988
© 1988 Balkema, Rotterdam. ISBN 90 6191 820 0

Reinforced earth bridge abutments

M.S.Boyd
Reinforced Earth Pty, Limited, Sydney, Australia

ABSTRACT: Reinforced Earth abutments combine retaining and load bearing functions. A key factor in the observed behaviour of Reinforced Earth abutments is the compatibility of steel reinforcement and dense granular earth, which allows for the superposition of the effects of applied loads on to the well established retaining function behaviour at design stress levels. Design methods which have evolved from the experience of model tests, full scale instrumented structures and finite element analyses have allowed the confident application to bridge structures in a wide range of configurations and environments.

1 INTRODUCTION

Retaining walls may have both retaining and load bearing functions, however, many retaining wall systems have limited ability to support large loads. At the beginning, Henri Vidal recognised that the effect of increased load on a Reinforced Earth structure was to mobilise more effectively the internal resistance of the structure itself. The use of Reinforced Earth as a bridge abutment was seen, therefore, as a logical application of the technique and one which had many practical and technical benefits for bridge construction including simplicity, economy and flexibility.

Reinforced Earth is a composite material formed by the association of High Adherence steel reinforcement and compacted granular earth. The compatibility of these materials promotes an elastic behaviour at design stress levels which allows for the confident superposition of stress fields resulting from both retaining and load bearing functions.

Almost twenty years of development, using models, instrumented full sized structures and finite element analyses, has shown how the retaining and load bearing stresses are mobilised and how they may be confidently predicted in the design of bridge abutments over a wide range of configurations and loading conditions.

2 DEVELOPMENT

In 1970, at the Port of Dunkirk in France, a double sided wall, 15 m high, 550 m long and 18 m wide was constructed to support a travelling gantry crane applying loads of 280 and 380 kN/m, 0.8 m and 2.7 m respectively back from the face. This project was extensively instrumented and provided a unique opportunity to investigate the effect of concentrated loads on a Reinforced Earth structure.

The first major abutment structure was built at Thionville over the Moselle River in 1972 to support the 38 m end span of a continuous concrete bridge structure. Subsequently in 1973–74 the French Road Research Laboratory conducted an extensive series of tests on a highway bridge abutment to Lille to determine the evolution of tensile stresses in the reinforcement strips and of the state of stress within the soil. Further structures were monitored at Triel in 1975 and Angers in 1977.

The present design method is defined in the French Ministry of Transports' "Recommendations and Rules of the Art" (1979). The method, based on the data available at the time, is generally considered to be conservative.

The evolution of Reinforced Earth technology has led to the conception of structures with trapezoidal and/or narrow sections (width/height < 0.7), at greater extremes of applied load and a wider range of geometry (skew, span etc.). Recent research has extended into the application of finite element analyses backed by the instrumentation of control structures to improve our understanding of the behaviour of abutments over a wider range of geometrical configurations. To this end, bridge abutments have been instrumented at Fremersdorf (Germany) in 1981 and Amersfoot (Netherlands) in 1984. Surcharged narrow and trapezoidal block sections have been tested at Milleville (USA) in 1983.

In 1982, TAI (Terre Armee Internationale) and CERMES (Centre d'Etude et de Recherches de Mecanique des Sols), undertook a research programme on three dimensional models with applied loading configurations. Despite their limitations (especially their ability to model friction behaviour and reinforcement characteristics) the models give valuable qualitive information on parameters relating to the behaviour and modes of failure in Reinforced Earth structures.

3 CONFIGURATIONS

The true Reinforced Earth abutment directly supports the bridge loads by way of a sill beam seated on the Reinforced Earth block immediately behind the facing panels (Fig. 1). Both horizontal and vertical bridge loads are transmitted directly to the Reinforced Earth block by the sill beam.

A Reinforced Earth structure may be applied to support an embankment either in front of, or behind, piles to form a "mixed" abutment (Fig. 2). On firm foundations, the piles may even be incorporated with the Reinforced Earth facing to form a "pier" abutment, to minimise the bridge span (Fig. 3). Mixed abutments allow the embankment to be separated from the bridge superstructure which may be useful where the bridge structure is sensitive to movements, however, the interaction of the embankment and the piles needs to be carefully evaluated. Pier abutments can only be considered, however, where there is no potential for post construction settlement as the flexibility of the facing is no longer available once construction is complete.

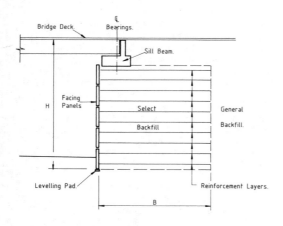

1. Limit of R.E. block shown thus ___ __ __ __
2. B/H> 0.7 for H > 20m
3. B > 0.6H + 2 for H< 20m
4. B min = 7m

Fig.1 Reinforced Earth abutment

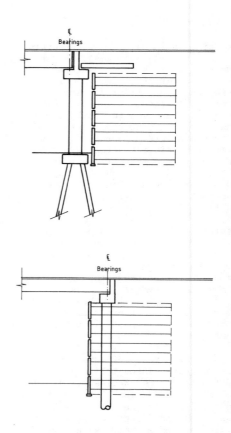

Fig.2 Mixed abutments

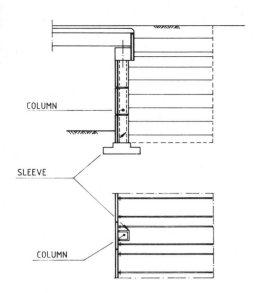

COLUMN

SLEEVE

COLUMN

Fig.3 Pier abutment

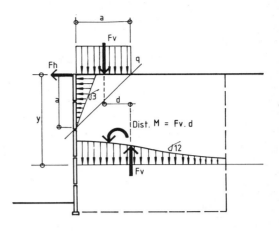

Fig.4 Load bearing stresses

4 DESIGN

The proportions of the sill beam are
determined to satisfy local stability
criteria for sliding, overturning and
bearing. Factors of safety appropriate to
bridge design standards are selected for
sliding and overturning while for bearing,
a limiting pressure of 200 kPa (150 kPa
under sustained loading conditions) is
recommended based on the need to limit
differential movements and local stress
concentrations immediately beneath the
sill beam. These limits reflect the
normal criteria for the selection and
placement of the granular material used in
the Reinforced Earth block.

As a load bearing wall, the stresses
distributed from the applied loads need to
be assessed for each layer. The vertical
pressures (from the sill beam) is diffused
following the Boussinesq analysis or, more
simply, at 2:1 through the block. The
horizontal force (applied by the sill
beam) is distributed linearly over a depth
at the face equal to the loaded width
(Fig. 4).

As a retaining wall, the calculation of
vertical stress at each layer needs to
consider the overturning effects of
embankment and overburden loads and

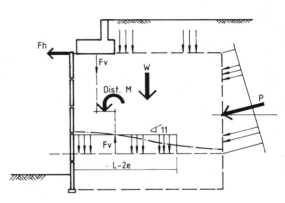

Fig. 5 Retaining stresses

surcharges as well as the overturning
effects resulting from the diffusion of
applied vertical loads through the
Reinforced Earth block and applied
horizontal loads from the sill beam (Fig.
5).

The tensile forces in the reinforcement
are then calculated by superimposing all
of the above effects, according to the
formula:-

$$T = A (K \sigma_{11} + K \sigma_{12} + \sigma_3)$$

where:

501

A	=	tributary area for each reinforcement
K	=	earth pressure coefficient
σ_{11}	=	vertical pressure (retaining)
σ_{12}	=	vertical pressure (load bearing)
σ_3	=	horizontal pressure (applied load)

Internally, a "potential failure line" is identified which equates to the locus of the points of maximum tension in each reinforcement layer.

The "potential failure line" defined by the retaining function is determined from the overall structure geometry and usually varies from a maximum of 0.3 H behind the face at the surface (where H is the functional height of the structure) to the face itself at the toe. This may be modified by the sill beam where its width exceeds the above limits.

The "potential failure line" defined by the load bearing function, passes through the centre of the load and approaches the facing at a depth defined by the "critical wedge" of the sill beam – that is, the Coulomb wedge defined by the rear of the sill beam. (Fig. 6)

For a structure to provide a satisfactory safety level, the tensile resistance of the reinforcement must be sufficient along each "potential failure line" as well as at the facing, while the friction resistance mobilised in the resistant zones must be sufficient for the tensile forces calculated at each "potential failure line".

The compatibility of dense granular earth and high modulus steel reinforcement is an important factor in the internal behaviour of load bearing Reinforced Earth structures. For typical reinforcement densities ranging from 2 to 4 strips/m^2,60 mm X 5 mm in section, and steel modulus of 200 MPa, the effective reinforcement modulus is 100 to 200 kPa, which equates to the modulus of dense granular earth. Such compatibility is not able to be achieved with present generation polymer reinforcement materials such as geotextiles, geogrids or strips, whose effective modulus may be less than 1/10 that of steel. Furthermore, the time dependent behaviour of such materials will preclude it from application in bridge abutments.

5 APPLICATION

Reinforced Earth abutments have been designed to support bridge loads in excess of 1000 kN/m (width). Live load to dead

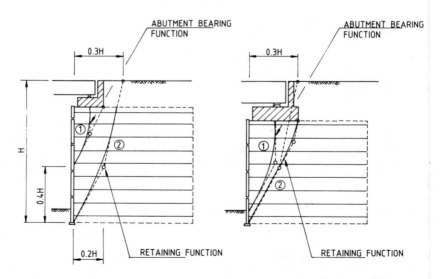

Fig.6 Potential failure lines

load ratios vary from less than 0.5 (highway structures) to more than 3 (mining structures). Abutment wall heights range from a few metres to over 20 m while bridge skew angles up to 75 degrees have been accommodated (square bridge skew angle equals zero).

The relative influence of the embankment (retaining) function and the abutment (load bearing) function depends on the magnitude of the applied loads and the size of the abutment. This will effect the internal and external design of the structure and its interaction with the foundation,both during and after construction. A primary advantage of the Reinforced Earth bridge abutment is its ability to be constructed on poor foundations. The Reinforced Earth block spreads the load more evenly on the foundation and its flexibility allows it to accommodate considerable settlements due to consolidation of the foundation. In some cases, pre-loading of the completed abutment can be applied to limit post construction settlements within acceptable structural and clearance tolerances.

In Spain, an abutment on the Bilbao – Behobie highway settled 1050 mm without distress. In Oregon, USA, a 55 m single span bridge was constructed on Reinforced Earth abutments founded on compressible soils to accommodate an anticipated 300 mm of settlement without loss of structural integrity. In Australia, an unusual bridge structure incorporating one Reinforced Earth abutment, 22 m high, was constructed by the Highways Department of South Australia over the Field River. Here the foundation was extremely variable including an ancient river bed and existing poorly compacted embankment material, however, the structure movements are within the limits required for the bridge and substantiated the use of the system to avoid the expensive foundation treatment required by conventional abutment structures.

The ability to design for dynamic loads is an important consideration for railway and mining bridge structures. In South Australia, a railway bridge of 42 m span over the River Torrens, is directly supported by 7.7 m high Reinforced Earth abutments where poor foundation and existing structures precluded the use of piles. In Queensland, Australia, a 16.3 m span steel bridge supported on 14.5 m high Reinforced Earth abutments is designed to support 300 tonne coal dump trucks

unloading at the rate of 2000 tonnes per hour.

In many areas Reinforced Earth abutments are being designed to accommodate the extreme loading conditions required for urban and industrial bridge structures. The performance of these structures in a wide range of applications has confirmed the Reinforced Earth technology as a sound, appropriate and economic bridge abutment construction technique. Worldwide, over 1000 bridge structures incorporate Reinforced Earth abutments.

6 REFERENCES

Juran, I., Schlosser, F., Long, N. T. and Legeay, G. "Full Scale Experiment on a Reinforced Earth Abutment in Lille", ASCE Symposium on Earth Reinforcement, Pittsburgh, 1978.

Verge, G. C. "Reinforced Earth Applications for Bridge Abutments in Australia" C. R. Coll. Int. Reinforcement des Sols, Paris, 1979.

Boyd, M. S. and Thomas A. I., "Reinforced Earth Bridge Abutments in South Australia" IE Aust., Annual Conference, Adelaide, 1980.

Guilloux, A. and Schlosser, F. "Reinforced Earth Uses on Soft Soils" Symposium on Recent Developments in Ground Improvement Techniques, Bangkok, 1982.

LCPC/SETRA "Ouvrages en Terre Armee; Guide Specifications et Regles de l'Art" Ministere des Transports, France, 1979.

Bastick, M. J., "Behaviour of Reinforced Earth Abutments" Internal Report R37, Terre Armee Internationale, 1985.

International Geotechnical Symposium on Theory and Practice of Earth Reinforcement / Fukuoka Japan / 5-7 October 1988
© 1988 Balkema, Rotterdam. ISBN 90 6191 820 0

Laboratory model test and research of the reinforced-earth wall

Mingde Cheng & Zhonglun Shi
Fourth Survey and Design Institute, Ministry of Railways, Wuhan, People's Republic of China

ABSTRACT: According to a number of model tests and data analyses this paper shows the working stress, deformation and stability property of wall mass for the reinforced-earth wall.

1. PURPOSE

Reinforced – earth wall has been widely applied in practical engineering, but its working stress, deformation are still not understood completely, the design and the calculation used for the structure are not consistent and perfect. The laboratory model tests upon the reinforced-earth wall have been made in order to make further the mechanism clearer, to find out the magnitude of tensile force at the reinforcing elements of soil mass, its deformation behaviour and stability, and to verify and improve the current design and calculation methods. The model test was divided into two parts: the actual working behaviour and the stability of reinforced-earth wall.

2. TEST EQUIPMENTS AND METHOD

2.1. The model test for the actual working behaviour of reinforced-earth wall

The test was carried out in the experimental tank with 1500x1700x1300, the wall areas were 1500x800mm. The front plates and the strips were made of the plexiglass, fill material was a machine-processed quartz sand, its mechanic overall index are $\phi = 38°$, $\gamma = 15t/m^3$. Under the condition of supporting by the special angle steel, the model wall

was built up alternatively by filling with sand, installing front plates, laying strips and placing instruments. When all the above work was finished, we demolished the supports and started to measure the bearing capacity of wall mass untill the measured values tend to be constent.

2.2. The model test for the stability of reinforced-earth wall

This test was similar to the above one. The area of model wall was 1500x800mm or 1500x600mm. The model wall was filled up after the measuring plate with 50x50mm grid graph drawn by toothpaste was sticked at the both sides of test tank. After the model wall was built up according to the design, loading units (loading beam, synchronous jack) were placed. Then, the supports were demolished and the loading was performing untill the wall was broken.

3. DATA PROCESSING AND ANALYSIS

3.1. The model test for the actual working behaviour of reinforced-earth wall

3.1.1. Lateral pressure applied to the front plates

The overall table of lateral pressure (average value) could be deduced from a number of actual measured data, and then the relative distribution curves were drawn out (Fig 1).

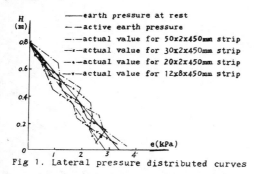

Fig 1. Lateral pressure distributed curves

As shown in Fig 1,the lateral pressure applied on the wall face presented the curve distribution along the height of the wall. Because of a significant difference between the model wall and the in-situ reinforced-earth wall (mainly the front plates of model wall were built up by supports and the wall face was laid on rigid foundation), the wall produced the overall displacement after demolishing the supports. The lateral pressure was similar to that which happened before demolishing it, and the distribution pattern of lateral pressure of model wall was different from that of the full-scale reinforced-earth wall.

Also Fig 1. shows that the lateral pressure applied on the front plate bears a relationship to the cross section of strips.For the isopachous strips, the wider the strip was, the higher the lateral pressure would be. For the model wall built up with 12 x 8 x 450mm strips , the pressure applied on the front plates was the minimum. This was due to the fact that the interface of both the faces of strip and the fill material was small, and the vertical stiffness of strip was relatively high, it caused the partial bottom face of strip not to be close up to the fill material, which has influenced the monolithity of the reinforced-earth wall. This means that the better the monolithity of reinforced-earth wall is , the higher the lateral pressure applied to the front plate will be.

The above analysis indicated the optimal design of how to select the cross section of strip and its laying density in the design of the rein-

forced-earth wall is to be further explored.

3.1.2. Soil column stress in the reinforced-earth wall mass

The overall table of soil column stress (average value)was collected and its distribution curves were drawn out in Fig 2.

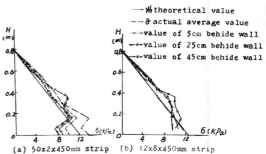

(a) 50x2x450mm strip   (b) 12x8x450mm strip

Fig 2. Soil column stress distribution curves

As shown in Fig 2, the actual stress of soil column mainly presents a linear distribution along the heigh of wall and is about 20% higher than the theoretical value, but the actual value measured in the bottom layer of the front plates obviously decreased.During the built-up of the model wall, the above distribution pattern was observed when the height of wall was about 40cm. How to explain this kind of distribution has to be waited.

Fig 2. also shows that, for the wall built up with 50x2x450mm strips, the actual pressure of soil column near the front plate was higher,which was determined by the construction method; for the model wall filled with sand by the free-falling method, the area near the front plate needs to be tamped because of hollow hole due to the wider strips, and this leads the area to be denser, so after having demolished the supports the displacement of the front plates was small and it had no evident influence on the density. That determined the magnitude of stress distribution. In the model wall built up with 12x8x450mm strips,the stress of soil column near the centre of strip was high, this was due to the fact that a great displacement occured on the wall face after demolishing the supports, the density of the filling sand near the front plate decreased and led the strip to act as a dis-

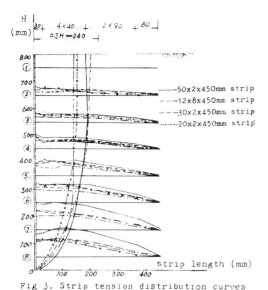

Fig 3. Strip tension distribution curves

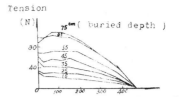

Distance to the front plate (mm)

Fig 4. Comparison of the forces acting on the different layers of strip (50x2x450mm)

charging beam, it made the stress of soil column in this area reduced and that outside the centre of strip increased, so that the above mentioned distribution pattern was formed.

3.1.3. Stress applied to strip

The tension distribution curves of strips could be drawn out according to the actual results measured (Fig 3).

As shown in Fig 3 , when the strip was shallow-buried , the distribution of tension within a certain length near the front plate was horizontal, which indicates that the horizontal distribution force is not fully mobilized and the tension at the end of strip is reduced gradually to zero; when the strip is deeply-buried, the tension along the length of strip primarily was a parabolical distribution . Linking smoothly the sections of minimum tension, the so-called " potential failure surface " is obtained, which approaches the normally called "0.3H failure surface ".

It is also seen from the figure, the magnitude of the tensile stress of strips bears a relation to the section, The wider the section is, the higher the tensile stress will be. This means the value of stress horizontally distributed on the strip is referred to the interface between the strips and fill material,and its

relationship is in direct proportion. In addition, as the height of fill material increased at the top of strip, the strip tension become relatively high. As shown in Fig 4,the magnitude of horizontal component applied on the strip bears a relation to the overburden pressure.

The above mentioned analyses show the horizontal component applied on the strip is approximately propertional to the height of fill material and the width of strip under the condition of the loose fill material. It indicates that the above situation is mainly caused by the friction between the soil and the strips.

Moreover, as the joint between the strip and the front plate is undesired, the partial strip near the front plate has a phenomenon of stress concentration, which results in the measured value being relatively high.

3.1.4. Wall mass displacement

The actual displacement curve of the wall face is shown in Fig 5. The model wall under this spacing of strips has primarily formed a composite massif. Its upper deformation is mainly the horizontal displacement, while the lower deformation is considered as a "rotation".

As shown in the Fig5,the displacement of the wall was affected by the number of strips,the more the number of strips are,the less the displacement will be. It also indicates the more the number of the strips are, the better the monolithity of the mall mass will be obtained.

3.1.5.Relationship between the strip tension and the lateral pressure acting on the front plate

Theoretically, the tension at the joint of strip and front plate is

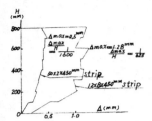

Fig 5. Actual displacement data of wall
after demolishing the wall

Tab 1. Comparison of actual lateral pressure and tension
on the model wall with different strips

Layers	12×8×450mm strip					20×2×450mm strip					30×2×450mm strip					50×2×450mm strip				
	E	T_o	T_o'	$\frac{T_o'}{E}$	$\frac{T_{max}}{E}$	E	T_o	T_o'	$\frac{T_o'}{E}$	$\frac{T_{max}}{E}$	E	T_o	T_o	$\frac{T_o}{E}$	$\frac{T_{max}}{E}$	E	T_o	T_o'	$\frac{T_o'}{E}$	$\frac{T_{max}}{E}$
2	10.4	15.0	12	1.54	1.19	15.2	7.8	18.9	1.5	22.9	1.92	2.94	11.2	22.4	12	1.54	1.07	1.39	20.6	32.4	20	2.61	0.97	1.17
3	14.7	25.7	18	20.9	1.22	1.42	21.3	33.1	0.21	24.9	0.99	1.17	15.5	32.6	28	3.26	1.81	2.10	23.3	33.9	27	3.07	1.16	1.32
4	17.7	36.5	25	29.5	1.41	1.67	24.4	42.7	21	32.3	0.88	1.32	34.2	49.4	31	3.6	0.91	1.06	35.5	52.9	40	4.22	1.04	1.10
5	32.0	39.6	32	43.6	1.00	1.36	37.8	47.5	43	55.4	1.44	1.47	45.8	46.5	43	5.53	0.94	1.21	34.3	61.4	45	5.61	1.31	1.65
6	56.8	50.5	45	47.5	1.22	1.29	43.8	71.8	63	70.3	1.44	1.61	44.1	57.4	51	5.74	1.08	1.21	52.0	68.0	63	7.02	1.21	1.44	1.35
7	47.4	50.5	40	54.2	1.01	1.14	57.2	72.7	62	69.6	1.08	1.22	45.1	73.8	55	69.5	1.22	1.54	55.4	95.4	80	1.21	1.44	1.85
8	65.8	61.7	61	63.0	23.0	0.97	50.4	66.9	59	64.1	1.17	1.29	51.5	61.5	60	69.3	1.17	1.35	69.3	86.4	85	10.6	1.25	1.53
Σ	284.5	277.3	241	274.9			242.7	366.5	284.5	343.1			259.7	346.0	280	336.1			293.4	432.6	360	451.1		
				1.07	1.22				1.17	1.40				1.12	1.34				1.23	1.47	

Not. E-- Actual lateral pressure acting on the front plate(N)

T_{max} --Actual maximum tension of strip(N)

T_o -- Actual tension at the joint of strip and front plate

T_o' -- T_o Analyzing value

equal to the relative lateral pressure acting on the front plate. In the test there was an influence of stress concentration was deducted according to the stress state , revising the actual results, and then the comparative table of the actual lateral pressure and the strip tension on the adopted strips could be obtained (Tab.1).
As shown in Tab.1, the sum of actual tension of the strips near the front plates was about that of the actual lateral pressure of the relative front plate, and the difference was only 10-20%. The sum of actual max tension measured on various strips was about 1.2-1.5 times greater than that of actual lateral pressure of the relative front plate. To a single strip and its corresponding front plate, there was a certain difference between the tensile force T and the lateral pressure E, some even a great difference. This was probably caused by the interference among the front plates. It should point out that the model wall on some tests was calculated according to the current design method and its stability factor was less than 1. This

shows that the current design theory is rather conservative.

3.2. Model test of stability for the reinforced-earth wall

According to the test results, the failure of the model wall was divided into three types.

3.2.1. Failure formed due to the large displacement of wall mass

This type of failure mainly appears in the case of the model wall with the narrow strips or with loading at the end of wider strip, or with a loose fill material. The characteristics of failure were that, under the condition of load, the wall face was in convex, the high settlement of fill material occured at the loading beam, which obviously displaced with the fill material. There was a obvious slickenside on the measuring plate but no apparent boundary line there, the overall slip surface was primarily a logarithmic spiral which crossed over the rear edge of the load beam and the spring of retaining wall. The failure was similar to that of subbase, there was only a difference that the side of wall face was an open-to-space and there were some strips in the reinforced-earth wall mass.

3.2.2. Failure due to the collapse of wall mass

It occured in the range of length of strip subjected to load and in the model wall with the great density of fill material. Under the condition of the load, the retaining wall was suddenly collapsed, but the overall wall only had a little deformation before the collapse. The characteristics of failure was extremely similar to that of the wall with the brittle material. To simulate a reinforced-earth wall in 9m high, the loading test was performed on the model wall built up with the strip which was equal to point six five times of the height of the wall. When the actual distributed spacing of strips was 1/8 defined by the design, and the fill is sand of medium compactness (Dr=57.6%), the limit bearing capacity of retaining wall was three times greater than the relative design load. If the wall was built up according to the current design theory and the construction

508

requirments,its bearing capacity will increase by ten times.This means that the current design theory is conservative.

3.2.3.Failure due to the lack of the strength of elements

It appeared in the case that the section of strip was rather wide, the fill material loose and the load was applied on the top of retaining wall. The failure characteristics were that, under the condition of the load, the settlement of loading beam was relatively large, and the displacement of wall face was small.The front plates at the top were broken out. The front plate have come out where the fill material ran off seriously, and the bearing capacity of wall was relatively high, which normally was monitored by the strength of elements For the reinforced-earth retaining wall which produces this failure,its pull strength is greater than the strength of elements, so that the elements broken first.

4. CONCLUSION

4.1. The soil column stress in the wall mass is approximate to the
γh value, it could be calculated by the current calculation method.

4.2. The linking line of the maximum stress acting on the strip, which is the so-called " failure surface ", presents a logarithmic spiral , the recommendation is given for using " 0.3H method " to calculate the effective length of strip.

4.3. As the actual maximum tension of strip is about one or two times greater than the lateral pressure applied to the relative front plate, the strip section and the joint elements should be designed according to two times larger than the lateral pressure acting on the relative front plate.

4.4. The wall stability is considerably affected by the density of the fill material, it should attach a great importance to the quality of tamping for the fill material.

4.5. The more perfect the monoli-

thity of reinforced-earth wall is formed, the better the overall stability will be, and the higher the stress acting on the element materials are used, therefore, the design of reinforced-earth retaining wall still needs to be optimized.

4.6. The current design method tends to be conservative and needs to be improved.

International Geotechnical Symposium on Theory and Practice of Earth Reinforcement / Fukuoka Japan / 5-7 October 1988
© 1988 Balkema, Rotterdam. ISBN 90 6191 820 0

Geotextile-reinforced retaining structures: A few instrumented examples

Ph.Delmas
Laboratoire Central des Ponts et Chaussées, Paris, France

J.P.Gourc
IRIGM, University of Grenoble, France

J.C.Blivet
Laboratoire des Ponts et Chaussées, Rouen, France

Y.Matichard
Laboratoire des Ponts et Chaussées, Nancy, France

ABSTRACT: Based on a few examples of instrumented real structures, this paper emphasizes the influence, on the behaviour of polymeric reinforced structures, of various parameters such as :
  - the method of placement, of both the formwork and the facing ;
  - the possible use of clayey soil ;

The conclusion to be drawn from these few instrumented examples is that the tensile forces, and in particular, their distribution in the structure, depend to a large extent on the conditions of placement, but also on the way structure is stressed.
In addition, on attempt should be made to optimize the mobilization of the forces by choising a type of form that allows pretensioning of the geotextile and holds deferred deformations of the structure to a minimum.

1 INTRODUCTION

The principle of retaining structures consisting of a reinforced soil mass has led to routine applications primarily through the work of la Terre Armée (TA). From this viewpoint, geotextile-reinforced structures may be compared to TA reinforced-earth structures. These two processes do have a number of points in common. But an analysis of their internal behaviour reveals differences.

The stiffness of the reinforcements, again calculated for one layer and a width of one metre, is between 25 and 100 times smaller in the case of geotextiles than in TA. This results in differences in behaviour, characterized by larger strains within the geotextile-reinforced structure for similar forces.

These findings explain why, in observations of geotextile-reinforced structures, an effort has been made to measure the strains of the reinforcements and of the soil. In the case of real structures, it is important to be able to make sure of a level of strain that is acceptable for the structure, whether locally, in the soil (to remain below the peak shear strength values), in the geotextile or at the contact surface, or more generally with respect to the superstructures or the facing.

It should also be noted that the design of all of the structures examined below assumes a maximum mobilizable strength in the tiers of geotextile that is less than 10 % of the ultimate tensile strength. This affords some protection against creep and means that there is no risk from damage to the geotextile during compaction.

In the course of specific tests carried out on various products placed between two layers of flint-bearing clay dumped from a height of 1.5 m and compacted to 95 % of the normal Proctor optimum in layers 30 cm thick, Perrier (1986) found that the compaction could decrease the maximum tensile strength of the geotextile by as much as 30 % , and alter its stiffness, either increasing or decreasing it.

Based on a few examples of instrumented real structures, we shall attempt in what follows to determine the influence on the behaviour of the structures of :

- the method of placement of the formwork and of the facing ;
- the use of clayey soil .

In practice, the stability of the structures with respect to the equilibrium of forces is ensured by the tensioning of the reinforcement sheets inside the soil mass. However, the overall behaviour of the structure will depend primarily on how the sheets of geotextile are tensioned and how the strains are induced. Indeed, for a given structure, overall static equilibrium can be ensured for different stress conditions in the geotextiles depending on how the strains occur in the soil mass.

Accordingly, placement will undeniably be a key stage in the generation of the forces in the reinforcement sheets. The few instrumented structures described here show, in effect, that in certain cases the final forces can be reached as soon as the construction of the structure is completed, independently of the of the subsequent loading conditions.

This is explained by the stiffness of the geotextiles, which means that rather large strains of the soil mass are required to generate the tensile forces that ensure stability. It follows that correct placement of the sheet is critical: no pleats, if possible manual tensioning before compaction of the fill, etc. However, the type of compaction and the way in which the fill is cased will in general be the two factors that determine the final strains in the soil mass. In addition, the facing, and its relative stiffness, will also be a factor to be taken into account in understanding the limit-state strains of the structure.

2 LA HOUPETTE EMBANKMENT

To carry out the widening of a road on an embankment 4 m high and 300 m long, it was decided to build a geotextile-reinforced structure

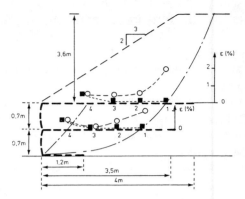

Figure 1: La Houpette embankement. Deformations of geotextile during construction: h = 1.4 m, h = 4 m.

consisting of a vertical part 1.4 m high reinforced by two layers of UCO 44614 topped by a bank having a batter of 3 in 2 (figure 1). For practical reasons, the contractor chose to build the lower part in forms placed all at once.

The fill is on aggregate having a continuous 0/250-mm grading and the following geotechnical properties :
γ_d = 22 kN/m^3 c' = 0 kPa
φ 35° to 40°.

This embankment was instrumented with cable-type strain measuring devices. Each measuring point consists of two cables attached to the geotextile 20 cm apart (in the direction of the stress), protected from the friction of the fill material by a sheath, that extend out through the front of the soil mass.

The deformations measured immediately after the removal of the forms from the two layers, before the upper embankment was built, and at the end of its construction, are given in figure 1.

It will be noted that, as regards the strains of the sheets, the form removal stage corresponds with a mobilization along a surface of maximum tension located near the facing, while the stage of embankment construction mobilizes the reinforcements farther back in the soil mass, so justifying the anchorage lengths determined in the preliminary design stage.

512

We point out that the final distribution of strains in the geotextiles in fact corresponds to the sum of the strains resulting from each of the two stages of constructions.

3 STRUCTURE ON A7 MOTORWAY (geotextile as reinforcement and geogrid on slope)

For the widening of the A7 motorway between Saint Rambert and La Galaure, an approach combining geotextile reinforcement with a prefabricated wall was chosen for the embankment portions. The structure consists of a soil mass reinforced by six layers of woven polypropylene geotextile (UCO44615) together with a polypropylene-ethylene grid placed on the facing to retain the topsoil on the embankment slope, 50° from the horizontal. This structure, which has a mean height of allow piping to pass. The fill is a pea gravel having good mechanical properties : c' = 0 kPa, $\wp$' = 41°, for a placement density γ_d = 22.5 kN/m^3.

The tensile characteristics. (NF 38014) of the geotextiles used are: for the woven polypropylene (tPP) α_f = 72.1 kN/m and $\mathcal{E}_f$ = 13 % for the geogrid polypropylene-polyethylene (G PP PET), α_f = 9.1 kN/m and $\mathcal{E}_f$ = 8 %.
Figure 2 shows the strains measured on the sheets and the settlements of the sheets at the end of construction and six months later. It will be noted that the lack of ties between the geotextiles and the facing grid, together with the low coefficient of friction of the two products, results in a relative slippage of about 2 cm. The local decompaction at the facing leads to local settlements, together with a small local increase in the tensile forces near the end of the reinforcing sheet and a slight rotation of the upper retaining structure.

It may be concluded from this particular example that proper compaction of the structure, in particular of the facing, is the key to the proper behaviour of the structure, and that the "reinforcing" and "facing" geotextiles must be tied together to confine the soil and pretension the sheet.

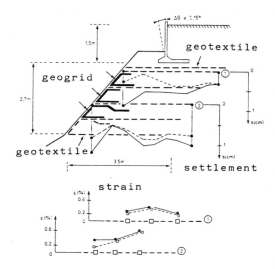

Figure 2: A 7 Motorway. Settlements (s) and strains of geotextile ($\mathcal{E}$) measured (O end of construction 04-86; ● 6 months later 10-86).

New, we shall accordingly restrict ourselves to analyzing the deformations engendered by the process in which the form for each layer is placed directly above a fixed point and supported by a fixed reference, derived from the international patent held by the Laboratoires des Ponts et Chaussées (1985) and worked by the MUR EBAL company (FRANCE).

In this connection, we may mention the Langres structure.

4 LANGRES STRUCTURE

Where it goes round the ramparts of the town of Langres, national highway 19 is on the uphill side of a retaining wall 4 m high and includes a sharp bend with its convex side outward. To correct the disorders of the existing wall, it was decided to build, under the cover of this structure, a geotextile-reinforced structure having its facing set back about 20 cm from the uphill facing of the existing structure, designed to take out thrust (Delmas et al., 1984).

The instrumentation implanted at the site included glued strain gauges 10 mm long capable of mea-

513

suring strains up to 10 %. The gauges were bonded to the geotextile on a rubber cement that ensured a suitable surface condition. A laboratory calibration was carried out because we had little experience with gauges of this type and, in particular, this way of bonding them. This revealed, notably, that while in the short term the measured strain values were reliable up to 2 %, in the longer term the creep of the cement made adequate precision impossible.

In addition, the deformations of the facing were measured using inclinometer tubes set in PVC tubes placed on the outside of the facing and attached by straps anchored in the structure.

Figure 3 shows the strains measured in the sheets when the road was reopened. It can be seen that the measured strains do not exhibit the distribution normally expected in reinforcing structures, and in particular exhibit no maximum. Moreover, the measurements made immediately after the removal of the form from the layer corresponding to the sheet measured show that, at this stage, from 70 to 95 % of the final strains have been reached, with the balance appearing when the next layer is placed.

The structure was built using one form per layer, supported by the existing structure. Given the permanent character of this structure and the state of knowledge when it was built, the fill material chosen

was a crushed limestone aggregate having a continuous 0/31.5-mm grading, placed at 95 % of the normal Proctor optimum, or w = 5 %; γ_d = 20 kN/m^3. The shear strength was estimated to be : φ' = 45°, c' = 0 kPa.

It would seem, in this particular case, that the placement of the fill soil, and especially its compaction, account for a large share of the final strains measured in the structure. This compaction resulted in a lateral deformation of the structure towards the wall in the layers from which the forms had already been removed. This deformation resulted in tensioning of the geotextile, which uniform strain along the sheets of reinforcement.

5 LUCHON STRUCTURE (Inflatable Formwork)

To restore access to the Hospice de France (the existing road was destroyed by a landslide), it was decided to build a new road on the opposite slope of the valley. For this work, a retaining structure 60 m long and 5 m high was built using a geotextile-reinforced soil mass. The patented construction process developed by the LPC, in conjunction with the facing of the EBAL construction company, was used to avoid disfiguring the site. The formwork bears against the facing during the placement of each layer, and can be removed thanks to a suitable inflation system, after the layer has been compacted (Delmas et al., 1986).

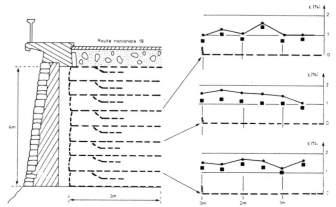

Figure 3: Langres structure. Strains of geotextile: ■ end of compaction of the soil layer just above the measured layer; ● end of construction.

514

The fill consists of materials from the site - shales having a particle size range from 0 to 200 mm, with 70 % smaller than 20 mm. The geotechnical properties of the 0/5-mm fraction are as follows :

$$c' = 10 \text{ kPa}$$
$$\varphi' = 35°$$

The compaction characteristics determined on the 0/20-mm fraction are :

$$\gamma_{opn} = 18.7 \text{ kN/m}^3$$

$$w_{opn} = 12.5 \%$$

The actual placement conditions were 7 % < w < 16 %, because of heavy rainfall at the site during construction ; compaction was carried out using a 50 kN vibratory roller. The structure was reinforced with a woven polyester multifilament (UCO 84464) having a tensile strength $\alpha_f = 217$ kN/m and a rigidity J = 800 kN/m. Soil-geotextile friction, as measured in the laboratory, was $\text{tg } \varphi \text{ geo/tg } \varphi' = 0.86$. The instrumentation included inductive and cable transducers to measure deformation of the geotextiles, horizontal inclinometers to measure settlement of the tiers, and levels to measure rotations of the formwork support facing. In addition, the total deformation of the geotextile facing with respect to the concrete facing was measured.

The main results of the deformation measurements are shown in figure 4. The most important point is the advantage of the type of formwork used to prestress the tiers of geotextile. 80 % of the final deformation is reached during compaction of the corresponding layer.

6 ROUEN EXPERIMENTAL RETAINING STRUCTURE (use of clayey soil)

In so far as it is not necessary to use dilatant soils in geotextile-reinforced structures, soils having a large fraction of fine materials can be used. If the soil used meets the usual specifications for fills, its use in a reinforced structure does not, a priori, pose any special problems if its mechanical properties are properly taken into account in the design and if the water content of the soil at placement is not likely to result in pore overpressures during subsequent loading.

The example of the experimental embankment at Rouen provides some additional information about the actual behaviour of reinforced fine soils having a high water content (Blivet et al., 1986). This experimental structure, 5.6 m high, was built with a silt having a water content of $w_{npo} + 5$ % and was reinforced with various types of geotextiles. Here we shall consider only two of them, on which pore pressure measurements were made : an Enka woven polyester and a needle-bonded nonwoven polyester in conjuction with a Rhône Poulenc grid.

The pressure sensor inside the embankment, outside the structures,

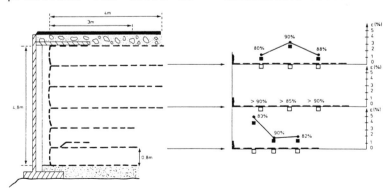

Figure 4: Luchon structure. Strains of geotextile: ■ end of compaction of the soil layer just above the measured layer (ε_1); • end of construction (ε_r) $(80\% = \varepsilon_1/\varepsilon_r)$.

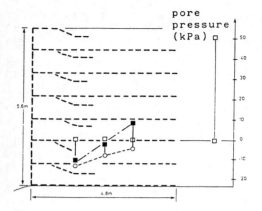

pore pressure (kPa)

Figure 5: Rouen experimental wall. Pore pressure mesured 120 days after the end of construction: o in the needle punched nonwoven with grid, ■ in the woven.

reveals placement overpressures of as much as 50 kPa.

On the woven sheet, the pore pressures are positive at the back of the structure, then disappear and finally become negative near the facing.

One the composite geotextile, on the other hand, the pressures are negative over the whole length of the reinforcement (figure 5).

This difference in local behaviour can lead to large changes in overall stability : on a nearby test section reinforced with a woven polyester with its surface treated to be non-wetting, the soil mass turned over because of anchorage failure. An after-the-fact calculation revealed an effective angle of friction of 5°, as against a soil-geotextile angle of friction of 21° in a drained condition.

7 CONCLUSION

The conclusion to be drawn from these few instrumented examples is that the tensile forces, and in particular their distribution in the structure, depend to a large extent on the conditions of placement, but also on the way the structure is stressed.

In addition, an attempt should be made to optimize the mobilization of the forces by choosing a type of form that allows pretensioning of the geotextile and holds deferred deformations of the structure to a minimum.

This must be taken into account in designing these structures, and for this reason preference should be given to design methods in which the actual conditions of placement can be simulated and taken into account. In this connection, we may note the interesting approach made possible by the "displacements method" (Gourc et al., 1986) .

REFERENCES

Blivet J.C., Jouve P., Maillot R. (1986) Numerical modelization of earth reinforcement by geotextile hydraulic function. C.R. IIIe Cong. Int. Geotextiles, Vienne, Avril, IV, pp 1061-1066.

Delmas P., Favre J.M., Matichard Y. Lehmann M., Prudon R., Rebut P. (1984) Renforcement par géotextile d'un mur de soutènement sur la RN 19 à Langres. Revue Générale des Routes et Aérodromes 609, Juin, pp 61-66.

Delmas P., Puig J., Schaeffner M. (1986) Mise en oeuvre et parement des massifs de soutènement renforcés par des nappes.Eléments de coût. Bulletin de Liaison des Laboratoires des P. et C., Paris, n°143, pp.65-78.

Gourc J.P., Ratel A., Delmas P. (1986) Design of fabric retaining walls : the "displacement method" C.R. IIIe Cong. Int. Geotextiles, Vienne, Avril, IV, pp 1067-1072.

Perrier H., Lozach D. (1986) Essai de poinçonnement sur geotextile - Influence des sollicitations de compactage. Compte rendu de mesures. Rapp. Labor. P.et Ch., 45 pp.

International Geotechnical Symposium on Theory and Practice of Earth Reinforcement / Fukuoka Japan / 5-7 October 1988
© *1988 Balkema, Rotterdam. ISBN 90 6191 820 0*

Design of geosynthetic retaining walls: 'Displacement method' and 'Two blocks method'
Comparison and charts

J.P.Gourc & Ph.Gotteland
IRIGM, University of Grenoble, France

Ph.Delmas
Laboratoire Central des Ponts et Chaussées, Paris, France

Abstract: The most widely used method of dimensioning geosynthetic-reinforced retaining walls is the "two blocks method". The authors of the article presented the "displacement method" in 1986, the salient feature of wich is that it allows for additional parameters, namely the stiffness J of the geosynthetics (and possibly k_s, the stiffness of the soil) and the soil-geosynthetic interface limit slip u_p. Comparative results and a chart model are presented.

1 INTRODUCTION

To simplify the expose, we consider the general case: the facing of the reinforced retaining wall is vertical and the section of the reinforced wall is trapezoidal (angle $\lambda > 0$) or rectangular ($\lambda = 0$). The soil of reinforced wall is identical to that of adjacent backfill (density γ (kN/m^3), angle of friction φ, cohesion c = 0).
The width of the base of the reinforced retaining wall is L and its heigt H. The gap between the geosynthetic layers is considered constant at Δ H (the charts envisaged in chapter 5 do however consider that certain layers can be eliminated. A double spacing 2.Δ H is then obtained).
The dimensions of such a reinforced retaining wall structure require:
- study of the external stability (chapter 2).
- study of the internal stability: by the "two blocks method" (chapter 3) and by the "displacement method" (chapter 4).
Chapter 5 will be devoted to an example of a chart obtained from the "CARTAGE" software of LCPC, based on the "displacement method".

2 EXTERNAL EQUILIBRIUM

The principle of triple calculation of the slip, punching and over-turning stability is well known,
since it is commonly used for monilithic retaining walls.
The authors considered it interesting to make a number of modifications to the method so as on the one hand to optimize the geometry of the reinforced retaining wall and on the other to improve the coherence between the study of the external stability and internal stability.

2.1 Slip stability.

Let us consider the case of a rectangular retaining wall ($\lambda = 0$) limited at its base by a geosynthetic layer with a length L. Let φ_g be the soil-geosynthetic friction angle such that
$tg \ \varphi_g = f \ . \ tg \ \varphi$.

Several procedures are used:

- procedure (a): It is considered that the vertical rear face of the wall undergoes horizontal Rankine thrust (normal to the face of the wall). A margin of security F_f = 1.5 with regard to the slip on the geosynthetic layer at the base of the wall gives the minimum with L:

(1) $(\frac{1}{2} K_a \gamma H^2 + K_a qH)/(\gamma H + q)L$
$< tg \ \varphi_g \ / \ 1.5$

where $K_a = tg^2 \ (\pi/4 - \varphi/2)$

- procedure (b): procedure (a) is

517

not coherent with internal stability (chapters 3 and 4) which assumed an identical coefficient of
safety along the slip surface. We
therefore consider a new Rankine
status behind the reinforced retaining wall obtained for a value
γ_c = arctg (γ) / 1.5
(i.e. F_s = 1.5).
Formula (1) is still valid, but
with Ka = tg_2 (π /4 - γ_c/ 2).

- procedure (c): the internal equilibrium considered in the
"displacement method" uses an exact
slices method (so called method of
"Perturbations", Raulin et al.,
1974), used in slope stability
calculations. In this way, the continuity between the internal and
external equilibrium is ensured,
using this slices method in association with a mixed slip line, a
circular line in the soil behind
the wall connected to a slip plan
along the base of the reinforced
block. The values γ_c and γ_{gc} are
identical to those of the previous
procedures.
The minimum values of L in terms of
the height H are shown in figure 1a
for a specific case. In what follows, we propose to adopt the procedure (c). However, in practice
continuous variation of the with L
with H is incompatible with a simple and rational application and
the stepped envelope curve will be
adopted (d).
Accordingly, the "rectangular"
block with a height H = 8 m will
have a width L = 3.5 m to comply
with slip condition, but the "trapezoidal" block of figure 1b also
verifies the slip stability at any
height H ≤ 8 m since its dimensions
correspond to the envelope (d) of
figure 1a. The mean slope of its
rear face is λ = 11°.
The overturning stability is not
shown here, but follows the same
procedure.

2.2 Punching stability

Conventional study of the stability
enables the resultant of the torser
of the forces applied on the base,
to be obtained on the base AD of
the block (figure 2). This force,
inclined and offcentered is compared against the limit load of the
soil of foundation. In practice,
the limit load is often obtained

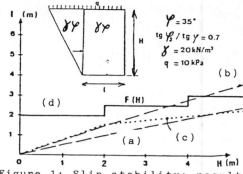

Figure 1: Slip stability: results
of procedure (a), (b) and (c).

from the limit pressure P_l determined with a pressuremeter.

3 INTERNAL EQUILIBRIUM BY THE "TWO BLOCK METHOD"

3.1 General principle

Internal rupture of the reinforced
retaining wall is assumed, i.e. a
slip line intersecting the layer j
of geosynthetic at the maximum
tension point (α_j). The active
sliding zone consist of two blocks
(figure 2) with a vertical interface (Gourc, 1987).

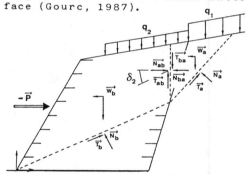

Figure 2: Internal equilibrium by
the "two blocks method".

By analogy with the behaviour of a
conventional retaining wall that
must stabilize the soil behind the
wall by balancing the soil thrusts,
the reinforced block must internally balance the thrust of the soil P
resulting from the slip of the
active double block.
Conventional application of this
method consist in checking the 4
equations of equilibrium of the
forces, though not the equilibrium
of the moments. From mechanical
standpoint, this method is hence

not "exact" or statically acceptable.
Just as in the case of external
equilibrium, we set the safety
factor F_s on the maximum shear
strength of the soil. The value of
the inclinaison of the force between the two blocks (ξ) is the one
corresponding to the least favourable case.
Solution of the 4 equations of
equilibrium yields 4 unknowns N_a,
N_b, N_{ab}, and P (the force to be
balanced by the geosynthetics).

3.2 Overall equilibrium design.

For given geosynthetic (failure
tensile strength α_f), we will
define an acceptable tensile
strength (or intrinsic tensile
strengh) $\alpha_i = \alpha_f / F_T$, where the
coefficient F_T enables account to
be taken of the creep and of possible loss of strength due to compaction.
In addition, the slip line divides
each geosynthetic layer j into an
active zone length L^a_{aj} and a pas-
sive zone length L^p_{aj} ($L = L^a_{aj} + L^p_{aj}$
for a rectangular block). We will
define the tension that can be
mobilized as a passive anchor:

(2) $\alpha_{aj} / F_f = 2 \int_0^{L^p_{aj}} \sigma_z (tg \varphi_g) / F_f \, dL$

σ_z, vertical stress applied on the
geosynthetic layer.
$\sigma_z \neq (\gamma \cdot H_j + q)$ owing to the
overturning moment of the block.
(H − H_j) represent the level of the
geosynthetic layer j measured from
the base.
The tensile strength that can actually be brought into play at each
level is
(3) $\alpha_j = min(\alpha_f/F_T , \alpha_{aj}/F_f)$

$\quad\quad = min(\alpha_i , \alpha_{aj}/F_f)$
if the geosynthetics are identical
for the entire reinforced block.

The overall equilibrium makes for:
(4) $\vec{P} + \sum_k \vec{\alpha_j} = \vec{0}$
(k layers intersected by the slip
line).

For a block where H, q, γ, φ, F_s,
α_j, F_f are fixed, the critical slip
line will be that giving the ma-
ximum values for α_i with a fixed
value of L, or the maximum of L
with a fixed value of α_i. One
should note that the critical slip

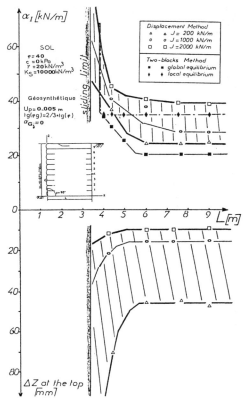

Figure 3: Compared results of "two
blocks method" (global and local
equilibrium) and of "displacement
method".

line is not systematically the same
as the maximum line of thrust P.

Figure 3 shows the acceptable pairs
(α_i, L) obtained for a rectangular
block and allowing the external
stability equilibrium.

3.3 Local equilibrium design

The overall condition (4) is replaced by a more restrictive condition: the thrust P is distributed
trapezoidally from top to bottom of
the block and the layers must balance this thrust locally (Juran et
al., 1978).
(5) $\vec{\sigma_x} \Delta H + \vec{\alpha_j} = \vec{0}$

In figure 3, we give the new acceptable pairs (α_i, L) ensuring
local equilibrium. In figure 4, we
consider the case L = 7.5 m of figure 3: the distribution of the

tensions α_j to the overall equilibrium on the one hand and the local equilibrium on the other is represented.

4 INTERNAL EQUILIBRIUM BY "DISPLACEMENT METHOD".

4.1 General principle

In the "two blocks method", the distribution of the tensile forces α_j obtained is independent of the type of geosynthetics used. The "displacement method" enables a number of additional conditions to be taken into account (Gourc et al.,1986; Delmas et al., 1986).
- **condition 1**: the stiffnes of the geosynthetics J deduced from a simple tensile test ($\alpha = J.\mathcal{E}$) is the elongation in the pseudo-static domain.
- **condition 2**: the relative displacement between soil and layer (u_p) needed for complete mobilization of friction γ_g, deduced from a friction test.
In the examples given here, we choose a value of u_p that is indepedent of Δ_z, though we proposed elsewhere (Ratel, 1987):

$$u_p = u_{p0} + u_{p1} \, \sigma_z \, / \, (1 + \xi_1 \, \sigma_z)$$

with u_{p0}, u_{p1}, ξ_1 constant.

- **condition 3**: the conditions of liaison of the layer at the vertical facing at point G. For example $\alpha_{jG} = 0$ for a layer not linked to the surface and $u_{jG} = 0$ for a layer linked to the facing.
- **condition 4**: the kinetics of failure. For a circular slip surface, for instance, (figure 3), and a rotation at the end of the block, a field of displacement can be set in the vicinity of the slip line $\Delta \theta_j = g (\Delta \theta)$.
Each geosynthetic layer intersected by the slip line undergoes distortions compatible with the failure kinematics: bending-shear in the vicinity of the slip line and elongation-slip in the anchorage zone (L^a_{aj}, L^p_{aj}).

- **condition 5**: the conditions of bending-shear in the case of the general "displacement method" requires knowledge of stiffness k_s of soil so that the local distortion in the membrane can be deduced and

also knowledge of the geosynthetics (Gourc et al., 1986). The tensile forces thus obtained in the vicinity of the slip line are inclined at an angle of β_j with the horizontal initial position.
In the case of CARTAGE dimensioning software used in chapter 5, the influence of the stiffness k_s has been neglected and the shape of the layer in the vicinity of the slip line has been simplified.

4.2 local equilibrium of layer j.

The law of behaviour of the anchored active and passive zones of the layer j can be deduced from the following conditions: (1) "linear behaviuor under tensile forces", (2) "elastoplastic behaviour at the interface" and (3) "limit conditions at the facing". The displacements at the anchorage head are:

(6a) $u^a_{aj} = v(\alpha_j, \, L^a_{aj}, \, u_{Gj} \text{ or } \alpha_{Gj})$

(6b) $u^p_{aj} = v(\alpha_j, \, L^p_{aj})$

The membrane behaviour in the shear zone is obtained in the general method on the basis of the assumption of a bi-circular convexity deformation depending on the soil stiffness k_s. The relation obtained provides the link between the anchored zones and the membrane zone.

(7a) $\Delta z_j = f(k_s, \alpha_j, \beta_j, \, u^a_{aj} + u^p_{aj})$

In the CARTAGE design programme (chapter 5), the equation (7a) is replaced by the geometrical compatibility relations (figure 4).

(7b) $\Delta z_j = (u^a_{aj} + u^p_{aj}) \cdot \cos \omega_j$
$\beta_j = 0.$

4.3 Overall equilibrium

The limit equilibrium of the active zone is derived from a slices method statically acceptable (method of "perturbations"; Raulin et al., 1974), also allowing the inclined tensile forces to be taken into account. The normal procedure is as follows:
- set the safety coefficient on the shear strength of the soil, F_s (here $F_s = 1.5$) and the mobilization of anchoring tensile forces F_f.
- choose a slip line and a displa-

cement field (here $\Delta\theta_j = \Delta\theta$ for a slip circle and $\Delta z_j = \Delta z$ for a straight line).
- increase $\Delta\theta$ (or Δz) by small increments. The verification of the local equilibrium of each anchored layer (6 and 7), enables increasing values of α_j and β_j with $\Delta\theta$ (or Δz) to be obtained.
- end the computation when the values of α_j and β_j are sufficiently large to satisfy the limit equilibrium.
- reiterate the computation for a new slip line. For a given geometry, we hence obtain the maximum tensile force brought into play $\alpha_i = \max_{\text{slip line}}(\max_j \alpha_j)$ and a maximum displacement $\Delta\theta$ (or Δz).

For the case already dealt with in chapter 3 by the "two blocks method", figure 3 shows the values obtained by the "displacement method" (with the membrane effect). One can see the considerable influence of the stiffness J of the geosynthetics. In addition, this method also enables a twofold design criterion to be used: limitation of the tensile forces α_i and limitation of displacements Δz.
In figure 4, we compare the distributions of the tensile forces obtained by different methods for the case where L = 7.5 m.

5 CARTAGE PROGRAMME AND DESIGN CHARTS

The chart model presented corresponds to the following specific conditions:
- soil: $\gamma = 20$ kN/m^3, c = 0, $\varphi = 35°$, $F_s = 1.5$

- geosynthetics: tg $\gamma_g = 0.7$ tg , $u_p = 15$ mm, layers related to the facing: $u_{Gj} = 0$
- reinforced block: H = 7.00 m, H = 0.40 m (or 0.80 m locally)
- mobilisation of the maximum anchoring force is assumed to be acceptable for each layer considered independently ($F_f = 1$), but will check that the sum of the anchoring forces brought into play is less than 2/3 of the sum of the maximum anchoring forces.
- the head displacement $\Delta z \leq 50$ mm.

In this way, the standard chart (figure 5) is obtained giving the relation between the intrinsic tensile force α_i and the number N of layers need for a rectangular or trapezoidal retaining wall. The profile of the trapezoidal retaining wall is obtained from the chart shown in figure 6.

The "trapeze" configuration gives $\alpha_i = 29$ kN/m and the "rectangle" one $\alpha_i = 24.5$ kN/m for the same number of layers (though for a greater layer length for the rectangle). The minimum value of the stiffness J verifying the criterion $\Delta z \leq 50$ mm appears in an additional chart (J ≥ 200 kN/m for the trapezoidal retaining wall).
On the basis of the observation that the overall cost of the geosynthetics is proportional to the length of the layers and the cost of the square meter of geosynthetic (itself practically proportional to its intrinsic tensile strengh), we added the value of the excess cost p% depending on the pair (α_i, N) selected, to the standard charts of figure 5.

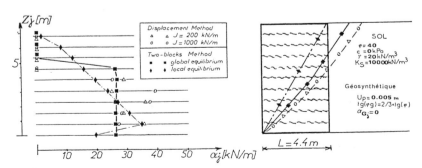

Figure 4: Distribution of tensions given by overall and local equilibrium.

521

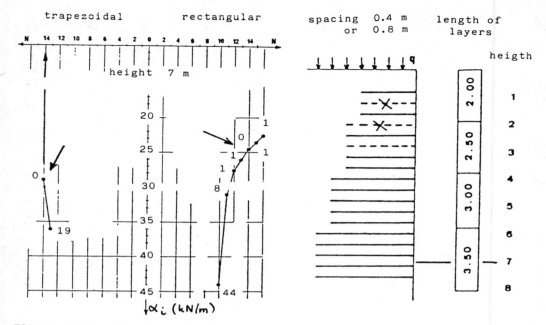

Figure 5: Standard chart using the "displacement method" (tensile strength).

Figure 6: Standard chart using the "displacement method" (length of the layers).

6 CONCLUSION

The range of geosynthetics avaible is increasingly wide. The "Two blocks method" provides a choice of the geosynthetic based exclusively on the value of α_i (for a fixed value of γ_g) namely the acceptable tensile strength (or intrinsic tension). The "displacement method" also enables the stiffness J of the layers to be selected in terms of the deformations that can be tolerated for the structure and the construction layout (geosynthetic-facing link,...). This method is easy to apply (installed on a microcomputer) an represents an advance in the problem of design.

REFERENCES

Delmas P., Berche J.C., Gourc J.P., 1986. Le dimensionnement des ouvrages renforcés par géotextile: programme CARTAGE, Bull. de Liaison des LPC, n° 142.

Gourc J.P., Ratel A., Delmas P., 1986. Design of fabric retaining walls : the "displacement method", IIIe Inter. Conf.on geotextiles and Geomembranes, Vienne

Gourc J.P., Ratel A., Gotteland P., 1987. Design of reinforced soil retaining walls. Analysis and comparison of existing methods and proposal for a new approach. Nato Grant. The use of polymers to reinforce soil. Kingston.

Juran I., Schlosser F., 1978. Theoretical analysis of failure in reinforced earth structures. Convention ASCE Pittsburg.

Ratel A., 1987. Modélisation d'un sol renforcé par géosynthétique. Application de la "méthode en déplacement. Thesis Dr Ing. Université of Grenoble, IRIGM.

Raulin P., Rouqués G., Toubol A., 1974. Calcul de la stabilité des pentes en rupture non circulaire Rapport de Recherche des LPC, n° 36.

International Geotechnical Symposium on Theory and Practice of Earth Reinforcement / Fukuoka Japan / 5-7 October 1988
© *1988 Balkema, Rotterdam. ISBN 90 6191 820 0*

Long term performance requirements for polymeric soil reinforcement in the United Kingdom

T.S.Ingold
St. Albans, UK

ABSTRACT: Broad guidelines have been set down by the Department of Transport for the long term performance requirements for polymeric soil reinforcement and these are reviewed. A broad approach to design is then presented and this suggests minimum values of partial factors of safety which might be employed to ensure an adequate margin of safety against long term tensile rupture of polymeric soil reinforcement.

1 DEPARTMENT OF TRANSPORT REQUIREMENTS

The principal document dealing with requirements for reinforced soil walls and abutments is Technical Memorandum BE3/78 (DTp 1987). Although this only strictly applies to reinforced soil structures under the jurisdiction of the Department of the Environment, its recommendations are widely accepted for permanent structures in the UK. The Memorandum was originally issued in 1978 and primarily applied to metallic strip reinforcement. Proprietary reinforcing materials falling outside this spectrum were permitted provided they were approved and duly issued with a current Roads and Bridges Certificate. Among other things this certificate quantifies reinforcement design loads consistent with the 120 years design life specified in Technical Memorandum BE3/78.

2 REVISED TECHNICAL MEMORANDUM

Technical Memorandum BE3/78 was revised in 1987. Among other things this revision sets out the principles for the assessment of the tensile strength of non-metallic reinforcement such as polymeric strips and grids. The principles are applied in assessing the permissible tensile strength of materials which exhibit significant long term creep behaviour.

2.1 Principles for assessment of tensile strength

The Memorandum requires the basic permissible axial tensile strength to be derived on the basis of the following two principles:

i) At the end of the design life of the structure strains in the reinforcement shall not exceed a prescribed value. This is 0.5% for abutments and 1% for walls after completion of construction.

ii) During the life of the structure the reinforcement must not fail in tension, for example by brittle failure or though ductile instability.

The permissible axial tensile strength is taken as the lesser of either the permissible average axial tensile load based on long term creep considerations, or the permissible peak tensile stress based on reinforcement failure at a temperature of 10°C. The permissible values incorporate factors of safety but the Memorandum does not define numerical values for these.

2.2 Approval of new materials

Approval is based on the issue of a Roads and Bridges Certificate by the British Board of Agrement. The Memorandum gives no advice on how design strengths are to be determined, however, it does present a check list of factors which should be considered in assessing the mechanical properties and durability of reinforcement. In assessing mechanical properties the Memorandum makes reference to short and long term data relating to load-strain characteristics, creep,

ductility and fatigue. Similarly
in assessing durability consideration
must be given to agencies such as
site induced damage, chemicals commonly
transported on highways, water, ultra-
violet and infra-red light attack,
bacteriological attack, fire and vandalism.
With the exception of the last two
categories these agencies can be divided
into the two broad categories of construction
induced damage and environmental attack.

3   SERVICEABILITY COMPLIANCE

Compliance with reasonable serviceability
requirements does not appear to be
problematical for selected polymer
reinforcement. For example Carroll
and Richardson (1986) report measured
short term geogrid strains generally
less than 0.6% and state that the
tied back wedge analysis, as presented
in BE3/78, significantly underestimates
reinforcement strains. Since observed
reinforcement strains are generally
small it is reasonable to analyse
these assuming the peak angle of shearing
resistance of the fill to be mobilised.
Using the techniques prescribed by
Andrawes et al (1986) an allowance
may be made for the effects of creep
on calculated strains. However, any
such calculations should make due
allowance for the effects of temperature
variations as described in later sections.

4   DETERMINATION OF DESIGN STRENGTH

The design strength of the reinforcement
governed by tensile rupture must at
all times during the design life of
the structure be greater than or at
least equal to the worst expected
design force exerted by the fill and
any superimposed loading. It is necessary
to assess how the tensile rupture
strength will decrease with time.
This can be achieved by loading different
samples of the reinforcement at different
load intensities so that the times
to failure fall inside a predetermined
range of time at a standard test temperature
of, say 20°C. Where the duration
of the longest test is less than the
required design life extrapolation
of results can be facilitated by testing
at higher temperatures, subject to
certain limitations, to accelerate
time.

This is the essence of the technique
used by the British Gas Corporation
to determine the 50 year design strength

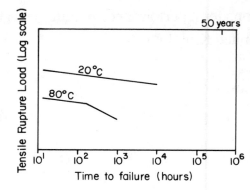

Figure 1.   Raw test data for gas pipe

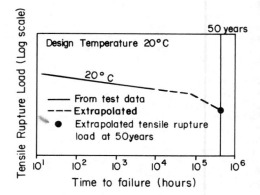

Figure 2.   Extrapolation of test data
for gas pipe.

of certain polymer gas pipes. Tests
are run at temperatures of 20°C and
80°C with times to failure up to 10^4
hours (Greig 1976). Typical results
are shown in Figure 1. The tests at
the design temperature of 20°C show
a log-log linear relationship between
rupture load and time. The perils
of extrapolating the 20°C test data
to 50 years are reflected in the knee
in the 80°C results which define a
transition between ductile and brittle
failure. By combining the test data
at 20°C and 80°C it is possible to
calculate the time, beyond the maximum
test duration of 10^4 hours, at which
any knee might develop.

As shown in Figure 2 these combined
data are extrapolated to define a tensile
rupture load at the end of the required
50 year design life.

This involves extrapolating over 1.64
log-cycles of time from 10^4 hours
(1.14 years) to 50 years. Since this
is in excess of the maximm extrapolation
of one log-cycle of time prescribed
in BS.4618:1970, a factor of safety
of four is applied to the extrapolated
rupture load to give the 50 year design
load (Greig 1981).

Similar techniques may be applied
to determine the long term tensile
rupture strength of soil reinforcement,
however, a very clear distinction
must be drawn between undamaged control
samples tested in a benign medium,
such as air at constant temperature,
and operational samples which will
be damaged during construction and
be subject to environmental attack
through agencies such as water and
chemicals or bacteria in the fill.
For a product subject to strict quality
control there should be little variation
in the extrapolated 120 year characeristic
strength from batch to batch. However,
the degree of mechanical damage and
aggressiveness of the fill will vary
from fill to fill as might the operational
temperature. It is vital that tensile
rupture tests are carried out on operational samples to determine how mechanical
damage and environment will reduce
the long term rupture strength. This
will allow the determination of various
partial factors to be applied to the
characteristic control strength to
reduce it according to the nature
of the particular fill to be employed.
Ideally laboratory tests on operational
samples, which have been pre-damaged,
should be carried out in an aggressive
environment since the combined effects
of environmental and mechanical damage
may be synergistic. This means that
the combined effects of environmental
and mechanical damage may be greater
than the sum of the effects of testing
damaged samples in a benign environment
and undamaged samples in an aggressive
environment.

To determine the 120 year design strength
the following minimum partial factors
are suggested. Those relating to
the effects of mechanical damage and
environment should be determined by
exhaustive testing along the lines
described above.

4.1 Material factor: γ_m

This relates to the probability that
the control strength of the soil reinforcement may occasionally fall below the
specified characteristic strength.
The suggested value of γ_m is 1.2.

4.2 Test data extrapolation factor: γ_t

This relates to the decreasing degree
of confidence in extrapolated data
as extrapolation is made over increasing
time intervals. No test data should
be extrapolated at the design temperature
without the aid of accelerated testing
at appropriate higher temperatures.
Ideally extrapolation should not exceed
one cycle of common log-time, that
is logarithmic time to the base 10.
Extrapolation should never exceed two
log-cycles of time. For n log-cycles
of extrapolation where $1<n<2$ the suggested
value of γ_t is 1.1n. Laboratory tests
should be conducted at the design temperature
which should equal the maximum operational
temperature in the soil. For temperate
climates a standard test temperature
of $20°C$ should be adequate.

4.3 Construction induced damage factor: γ_c

This relates to the long term effect
of mechanical damage suffered by the
reinforcement during installation.
Among other things it will be a function
of fill type, layer thickness and type
of compaction plant. The effects of
mechanical damage should be assessed
using long term tensile rupture tests
such as those employed to assess the
long term tensile rupture strength
of intact and undamaged control samples.
Short term constant rate of strain
tests have indicated reduction factors
in the range 1.1 to 1.6 for geogrids
(Mitchell and Villet 1987). The minimum
suggested value of γ_c is 1.2.

4.4 Environmental attack factor: γ_e

This relates to the long term affect
of the fill environment on tensile
rupture strength. Both chemical and
bacteriological attack must be considered
and their effects quantified by carrying
out long term tensile rupture tests
in an appropriate aggressive environment
at the design temperature or higher

temperatures as appropriate. The minimum suggested value of γ_e is 1.1.

4.5 Overall factor of safety: γ_r

The partial factors γ_m, γ_t, γ_c and γ_e are applied to the long term characteristic tensile rupture strength to reduce this to the basic design value. Where the ramifications of attaining the ultimate limit state of tensile rupture of the reinforcement are more serious the basic design design strength may be reduced by applying an overall factor of safety γ_r. The suggested minimum values of γ_r are in the range 1.0 to 1.2.

4.6 Design Strength

The design strength of the reinforcement for permanent structures is the 120 year characteristic tensile rupture strength, determined for intact control samples in a benign environment at the design temperature, divided by the partial factors γ_m, γ_t, γ_c, γ_e and γ_r. Minimum values of these factors have been suggested. Actual values of γ_c and γ_e due to construction damage and environmental effects are product and fill specific and must be determined directly by long term testing. As tensile rupture test data are gathered over longer test periods, the uncertainty of extrapolation decreases and therefore γ_t may be decreased as longer term data become available. Depending on the value of γ_r, the compounded minimum values suggested for the above partial factors varies from the range 1.7 to 2.1 for extrapolation through one log-cycle of time to the range 3.5 to 4.2 for extrapolation over two log cycles of time. Although on first sight the latter range of factors may appear severe they do relate to extrapolation of test data of 1.2 years duration by a factor of one hundred to a 120 years design strength. In comparison for extrapolation over 1.64 log-cycles of time the suggested partial factors result in a compounded factor in the range 2.9 to 3.4 which is significantly lower than the factor of 4 used by British Gas. It should be remembered that the above factors are suggested minimum values and these are likely to increase significantly for reinforcement susceptible to environmental attack or mechanical damage induced by the construction process.

5   OTHER FACTORS AFFECTING SAFETY

To obviate tensile rupture during the design life of the strucure the design strength of the reinforcement must never be less than the design force generated under the worst expected loading conditions. The design force will be dramatically affected by the angle of internal shearing resistance mobilised in the fill. The design strength of the reinforcement will be radically affected by operational temperatres in the fill and how these relate to laboratory test temperatures used to predict long term reinforcement design strength. For simple earth retaining structures the majority of the design force will be derived from the active thrust generated by the fill. This is the basis of design in Technical Memorandum BE3/78 where the coefficient of lateral earth pressure is calculated on the assumption that the peak angle of shearing resistance of the fill prevails.

5.1   Fill-reinforcement strain compatibility

The peak angle of shearing resistance is mobilised at small lateral strains which under plane-strain loading would be of the order of 1%. Such small lateral strains may be as much as an order of magnitude smaller than the axial tensile strains required to generate the long term rupture strength of the reinforcement. Consequently there is incompatibility between fill and reinforcement strain. To overcome this the design force for extensible polymeric reinforcement should be based on the constant volume angle of shearing resistance of the fill, $\emptyset_{cv}$. This is smaller than the peak value, however, it is a value which can be relied on even at large strain. For a particular frictional fill with a maximum particle size of 40mm Brady (1987) measured a peak angle of shearing resistance of 61° compared to a constant volume value of 41°. Based on active earth pressure analysis the thrust developed at $\emptyset_{cv}$ is in excess of three times greater than that developed at the peak angle of $\emptyset$ shearing resistance. This large difference endorses the need to design on the basis of $\emptyset_{cv}$ when guarding against tensile failure of the reinforcement.

5.2 Operational and test temperature compatibility

The performance of thermo-plastic polymers will be radically affected by temperature. For this reason the laboratory test temperature should equal or exceed the maximum operational temperature in the reinforced soil structure. To aid extrapolation of labortory tests data, tests may be carried out at elevated temperatures provided these do not change the basic mechanisms controlling the behaviour of the reinforcement. The concept of extrapolation using time-temperature transposition has been reported by Andrawes et al (1986) who suggest an acceleration in time by a factor of 10 by raising the test temperature from 10°C to 20°C. This behaviour would be governed by an Arrhenian relationship of the form given in Equation 1.

$$\ln (\mu) = \alpha(1/\theta_2 - 1/\theta_1) \qquad ...1.$$

where:-

μ = a time multiplier

α = a constant (polymer and product specific)

θ = temperature in degrees Kelvin.

Taking a time multiplier of 10 for a temperature shift between 10°C and 20°C allows evaluation of the constant α. Equation (1) can then be used to assess the relationship between operational temperature θ_1, laboratory test temperature θ_2, and the time multiplier μ. If the operational soil temperature is constant and equal to the laboratory test temperature the time multiplier is unity.

For permanent highway structures the creep properties of polymeric soil reinforcement are required to be determined at 10°C (DTp 1987). Recent work on the measurement of operational temperatures in a reinforced soil wall confirms 10°C as a reasonable mean soil temperature but shows that the seasonal variation, especially near the face of the wall is approximatel ±10°C (Murray and Farrar 1988). The seasonal variation was approximately sinusoidal with a maximum temperature of some 20°C minimum temperature close to 0°C.

Taking this sinusoidal variation Equation (1) can be used to determine the seasonal variation and mean value of the time multiplier μ. The results are shown in Figure 3. Due to the asymmetrical distribution of the seasonal time multiplier about the mean operational temperature the mean value is 2.85. The implication of this is that predictions made from laboratory tests run at the mean operational temperature of 10°C will underestimate performance since "effective" time under operational conditions is running 2.85 times faster than "real" time. This implies that a 120 year service life under the operational temperatures shown in Figure 3 is equivalent to 340 years, i.e., 2.85 x 120 years, at a constant temperature of 10°C. Consequently an extrapolation of test data at 10°C from say 1.2 years to 120 years must be extended to 340 years. This may increase extrapolation from 2 log-cycles of time to 2.5 log-cycles of time in which case γ_t would increase from 2.2 to 2.8. Conversely if the laborator test temperature was 20°C, i.e., equal to the maximum operational temperature in Figure 3, then a degree of conservatism is introduced into the extrapolation.

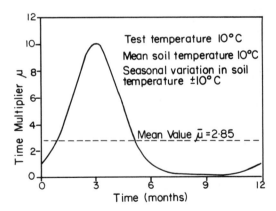

Figure 3. Seasonal variation of time multiplier

Care should be taken to ensure that laboratory test temperatures and operational temperatures are compatible. Consideration should extend to the variations of diurnal temperatures in the close vicinity of preformed facing units and any spontaneous heating in fill containing industrial waste. For example West and O'Reilly (1986) comment on heating in unburnt colliery shale and relate reductions in strengths of plastic reinforcing elements of 30% for temperature increase of 10°C above a 20°C ambient temperature. Risk of fire should also be considered.

6   CONCLUSIONS

The Department of Transport Technical Memorandum BE3/78 sets down broad requirements for the long term performance of polymeric reinforcement. Assessments of proprietary reinforcing materials are made by the British Board of Agrement who issue product specific Roads and Bridges Certificates and so confer compliance with the requirements of the Memorandum. A broad approach to design has been presented and this suggests minimum values of partial factors which might be employed to ensure an adequate margin of safety against tensile rupture of the reinforcement.

REFERENCES

Andrawes, K.Z., McGown, A. and Murray, R.T. 1986. "The load-strain-time-temperature behaviour of geotextiles and geogrids". Proc.III Int.Conf. on Geotextiles, Vienna, Vol.3.

Brady, K.C. 1977. "Performance of a reinforced earth bridge abutment at Carmarthen". Transport and Road Research Laboratory Research Report 111.

BS.4618 : 1970. "The presentation of plastics design data : Subsection 1.1.1 Creep in uniaxial tension or compression". British Standards Institution.

Carroll, R.G. and Richardson, G.N. 1986. "Geosynthetic reinforcd retaining walls". Proc.III Int.Conf. on Geotextiles, Vienna, Vol.2.

Department of Transport 1987. "Reinforced earth retaining walls and bridge abutments for embankments". Technical Memorandum (Bridges) BE3/78, (Revised 1987).

Greig, J.M. 1981. "Specification and testing of polyethylene gas distribution systems for a minimum 50 year operational life". Plastics and Polymer Pocessing and Applications, Vol 1, No.1.

Greig, J.M. 1976. "Fracture and its prevention in plastic gas distribution systems". Gas Engineering and Management, Vol.16, No.2.

Mitchell, J.K. and Villet, W.C.B. 1987. "Reinforcement of earth slopes and embankments". National Co-operative Highway Research Program Report 290. Transportation Research Board, Washington D.C.

Murray, R.T. and Farrar, D.M. 1988. "Temperature distributions in reinforced soil retaining walls". Int.Journal of Geotextiles and Geomembranes, Vol.6.

West, G. and O'Reilly, M.P. 1986. "An evaluation of unburnt colliery shale as fill for reinforced earth structures". Transport and Road Research Laboratory, Research Report 97.

International Geotechnical Symposium on Theory and Practice of Earth Reinforcement / Fukuoka Japan / 5-7 October 1988
© *1988 Balkema, Rotterdam. ISBN 90 6191 820 0*

The analysis of vertical loads applied to the top of reinforced earth structures

P.H.Jackson
HETS, West Yorkshire, UK

C.J.F.P.Jones
University of Newcastle upon Tyne, UK

ABSTRACT: Reinforced soil structures are frequently used as bridge abutments or as walls supporting industrial structures or other facilities including roads. In both conditions, concentrated line roads are applied to the top of the structure and positioned close to the face. The problem of calculating the distribution of concentrated loads has been considered in a number of design codes. However, a number of deficiencies and inconsistencies in the various methods are apparent. The paper introduces the Incremental Mirror Method (IMM) for determining the relative pressure distribution under a load and demonstrates how this can be used in the analysis of structures. Comparison is made between the IMM method and the United Kingdom design methods using parametric studies based upon a specially written computer program. This illustrates the strengths and weaknesses of the various techniques over a range of structural geometries. It is suggested that the IMM method offers the most realistic approach to this problem.

1 LOAD DISTRIBUTION UNDER A STRIP-LOAD NEAR A BOUNDARY

Several methods may be used to determine the increased pressure on an element or layer of soil at some depth in the strata below a foundation member. The simplest method is to use a stress zone defined by some angle (30° – 45°) with the vertical, this is frequently reduced to the use of a defined slope (2-1 or 1.5-1).

1.1 Elastic methods

An alternative method to the simple pressure distribution is to use elastic methods such as those developed by Bouss-inesq and Westergaard. The Boussinesq expression for the pressure at a point depth (z) below and displace laterally from the centre of a circular area acted upon by intensity of pressure (qv) is given as; Figure 1

$$q_v = \frac{3 \, Q \, z^3}{2 \pi R^5} \qquad 1.$$

$$R^2 = r^2 + z^2$$

$$Q = \text{Surface Load}$$

This expression can be adapted to cater for strip loads and is frequently used in bearing pressure calculations. The implicit assumptions of the Boussinesq equations include:
  i. The soil is weightless.
  ii. The soil is elastic, homogenous, semi-infinite and isotropic.
  iii. The soil is unstressed before the application of the load.
  iv. Stress distribution is symmetrical with respect to the vertical (Z axis).

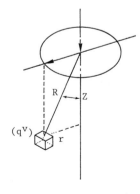

Figure 1. Intensity of pressure based upon Boussinesq approach.

In the case of reinforced earth structures supporting a strip loading applied close to the edge of the reinforced earth structure these assumptions are not fulfilled.

A better idealisation is provided by the Westergaard (1938) model

$$q_v = Q \frac{1}{Z^Z} \frac{1}{[1+2 \ (r/z)^2]^{3/2}} \qquad 2.$$

q_v = intensity of stress at a point in the soil due to surface loading Q

Z, r = the same as Boussinesq equation

This method was derived from a model with alternating thin layers of an elastic material sandwiched between layers of an inelastic material which permits vertical deformation, but which prevents lateral spread. The model is recognised as the first theoretical description of reinforced soil. Comparison of the Boussinesq and Westergaard methods shows that the Boussinesq equations give a larger intensity of stress beneath the footing, but that the Westergaard influence penetrates deeper into the soil.

As with Boussinesq model, the Westergaard method assumes a semi-infinite continuum, and whilst being a theoretical expression of reinforced soil, it is not truly applicable to the case of a strip load positioned close to a boundary.

1.2 Empirical methods

The presence of a strip load near to the facing of a reinforced soil structure presents a complex analytical problem in which the distribution of vertical stress is influenced by a number of variables, including the breadth of the footing (B), the distance of the footing from the boundary (X), and the eccentricity (e) of the load (S), Figure 2.

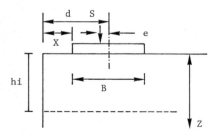

Figure 2. Strip load near to facing.

Three empirical methods have been considered in the United Kingdom design standard for reinforced earth to determine the pressure (q) beneath a footing similar to that shown in Figure 2. These expressions cater for the effects of eccentricity and the distance of the load from the boundary, Department of Transport (1978).

$$\text{Method I} \quad q = \frac{S}{Di} \left[1 + \frac{6e}{B} \right] \qquad 3.$$

where $Di = (hi + B)$ if $hi \leqslant 2d - B$

and $Di = d + \left[\frac{hi + B}{2} \right]$ if $hi > 2d - B$

The 6e term is to be ignored where $hi > \frac{2B}{B}$

$$\text{Method II} \quad q = \frac{S}{Di} \left[1 + \frac{6e}{Di} \right] \qquad 4.$$

where $Di = (hi + B)$ but $Di \not> 2d$

$$\text{Method III} \quad q = \frac{S}{Di} \left[1 + \frac{6e}{Di} \right] \qquad 5.$$

where $Di = (hi + B)$ and $hi \leqslant 2 \ (d - B/2)$

$$\text{and} \quad q = \frac{S}{Di} \left[4 + \frac{6e}{Di} - \frac{6d}{Di} \right] \qquad 6.$$

where $Di = d + \left[\frac{hi + B}{2} \right]$ and $hi > 2 \ (d-B/2)$

Each equation determines the maximum pressure applied to the soil by means of a simple trapezoidal distribution within an envelope determined by lines sloping at 2 in 1 from the edge of the footing or the facing if this lies within the 2 in 1 envelope. Di is the width of the envelope at level hi.

In the equations 3-6 it can be seen that the pressure at depth hi is determined by:-

Method I; reducing the maximum contact pressure in the proportion B/Di.

Method II; using the original load eccentricity as a constant as Di increases. At levels lower than that at which the 1 in 2 distribution line intersects the facing the pressure remains constant at its value at the intersection level.

Method III; regarding the line of action of the applied vertical force to be fixed in position so that the eccentricity of the applied load increases with increasing depth. This is the current method.

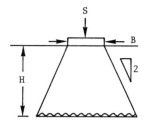

Figure 3.   Pressure at depth $H = \dfrac{S}{(B + H)}$ per unit length of footing.

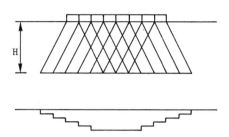

Figure 4.   Pressure distribution under a load considered as a series of elements.

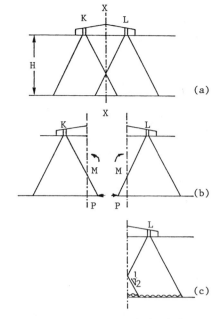

Figure 5.   Distribution of load from elements in a strip footing.

1.3   Incremental Mirror Method

A fourth method to determine the pressure beneath the footing has been developed by Jackson (1985). As stated above, the simplest methods of determining the pressure below a strip load is to consider the load to be uniformly distributed within an envelope limited by lines drawn at fixed gradients from the edges of the load, Figure 3. It can be shown that the use of a 1 in 2 gradient for load distribution gives a conservative approximation to an elastic analysis, although the pressures are underestimated at depths upto twice the foundation width and over estimated at greater depths, Table 1. By dividing the footing into a series of narrow elements (typically 100) the pressure distribution at depth can be determined using the principle of superposition, Figure 4. Comparison with an elastic (Boussinesq) distribution shows that the method is conservative at all depths, Table 1.   The method of combination has the added advantage that it can be used in cases where the load in each element is different or where an eccentric load is applied, Figure 5. Figure 5 shows a strip foot situated on the surface of a semi-infinite soil layer, symmetric about an axis X-X. If the footing and soil mass is split along the plane

Table 1.   Pressure coefficients by approximate and elastic methods.

Strip Loads Vertical Pressure Coefficients

H	1IN 1	1 IN 1.5	1 IN 2	IMM	Elastic
0	1.000	1.000	1.000	1.000	1.000
1	0.500	0.600	0.667	1.000	0.818
2	0.333	0.429	0.500	0.952	0.550
3	0.250	0.333	0.400	0.645	0.396
4	0.200	0.273	0.333	0.488	0.306
5	0.167	0.231	0.286	0.392	0.248
6	0.143	0.200	0.250	0.328	0.208
7	0.125	0.176	0.222	0.282	0.179
8	0.111	0.158	0.200	0.247	0.158
9	0.100	0.143	0.182	0.220	0.140
10	0.091	0.130	0.167	0.198	0.126
11	0.083	0.120	0.154	0.180	0.115
12	0.077	0.111	0.143	0.165	0.106
13	0.071	0.103	0.133	0.153	0.098
14	0.067	0.097	0.125	0.142	0.091
15	0.063	0.091	0.118	0.132	0.085

X-X into two blocks, it can be assumed that there are no vertical forces acting on the plane of the interface. There will, however, be horizontal forces and it would

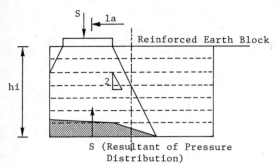

Figure 6. Pressure distribution at depth hi determined by I.M.M. method.

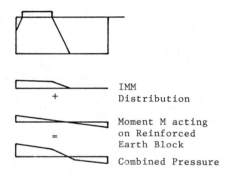

IMM Distribution

+

Moment M acting on Reinforced Earth Block

=

Combined Pressure

Figure 7. Combined pressure distribution.

be possible to separate the two blocks and maintain unchanged the vertical load distribution provided that a horizontal load of the correct magnitude and distribution were applied to the interface. This load is represented by the horizontal force P acting in conjunction with a moment M. In Figure 5 (a), the envelope of the two load elements K and L overlap and cross on the axis X – X. In Figure 5 (b) the blocks are split and the elements cannot distribute load across the interface and yet the vertical distribution of load in each block is unchanged. Element K must apply to its own block, a load which would have previously been applied by element L, and vice versa, resulting in the overlapping distribution shown in Figure 5 (c). In a sense the interface X-X acts as a mirror reflecting load back into the block. Using this technique, referred to as the Incremental mirror method (I.M.M.) it is possible to calculate the variation of vertical pressure beneath a footing comprising a series of strips situated close to a boundary, Figure 6. The I.M.M. considers the width of the

footing, the eccentricity of the load and the location of the footing relative to the boundary and structure.

The repositioning of a load by the mirror method has the effect of pushing the line of action of the reactant force away from the interface so that a moment (M) is developed by the vertical forces

$$M = S \times 1a \qquad\qquad 7.$$

This moment always acts as a positive overturning moment and is in no way related to the centre line of the reinforced earth block. If the envelope at level hi is not the facing then the lever arm (1a) and the moment are zero. If the moments are considered to act on a totally rigid reinforced earth block then the combined pressure distribution is as that shown in Figure 7.

2 COMPARISON OF EMPIRICAL METHODS

It is possible to compare the empirical methods by considering the relative pressures at depth developed by the four models. The Department of Transport Methods II and III calculate the global pressures and compare these with the local pressures, the greater value being used to determine reinforcement forces. Method I does not consider global pressures but it is possible to calculate the appropriate moment (M) using a similar technique to the I.M.M. method and thereby permits comparison. It should be noted that the moment from equation 7 may change direction depending upon the position of the load relative to the facing and the eccentricity of the applied load.

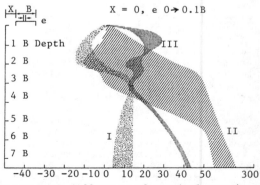

Percentage difference of vertical pressure at depth (B) relative to IMM method.

Figure 8. Comparison of methods used to determine vertical pressure, (x = 0).

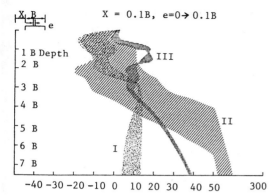

X = 0.1B, e=0→0.1B

Percentage of difference of vertical
pressure at depth (B) relative to IMM
method.

Figure 9. Comparison of methods used
to determine vertical pressure,
(x = 1.0B).

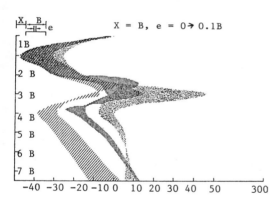

X = B, e = 0→0.1B

Relative difference of vertical pressure
at depth (B) relative to IMM method.

Figure 10. Comparison of methods used
to determine vertical pressure, (x = B).

The comparison of the four methods is
shown in Figures 8, 9 and 10 in which the
percentage in applied vertical pressures
developed by the Department of Transport
Methods I, II and III are compared with
the I.M.M. method. In the comparison the
distance (X) of the strip width (B) from
the edge of the boundary of the structure
was varied from (X = 0, 0.1B, B), whilst the
eccentricity of the applied load relative
to the centre line of the strip was varied
from (e = 0-0.1B).
The comparison suggests that Methods I,
II and III are very sensitive. Method II
is particularly sensitive to the eccentricity of the applied load when (X = 0),

whilst Method I is very sensitive at depths
between (1B - 3B) where (X = B). At depths
greater than (4B) Method I seems to produce
results consistent with the I.M.M. method,
whilst Method II can be shown to be unsatisfactory in almost any condition.

3 CONCLUSIONS

The determination of vertical pressures
beneath the strip footing adjacent to the
edge of a reinforced soil structure has
been a matter of uncertainty to designers
for some time.
A number of empirical methods have been
suggested to cater for the problem, all of
which are flawed to varying degrees. The
introduction of the incremental mirror
method (I.M.M.) is an improvement,
particularly suited to computerised
analytical techniques.
In order to improve the I.M.M. method
it would be necessary to compare the
forecast forces against experimentally
determined pressures, and then to adjust
the existing 2 in 1 gradient of the
element envelope to obtain a closer fit.
Of the three United Kingdom design methods considered for the calculation of the
pressure coefficients, beneath strip loads,
the modified Method I generally produces
the most realistic values. Method II is
unsatisfactory and whilst Method III is an
improvement on Method II it is not as
effective as Method I and is based on a
statically inaccurate model. Method I
can be improved by introducing controls
on the location of the reactant which
would prevent the negative over turning
moments from occurring. It would also
be of benefit if Method I could be altered
so that higher coefficients were calculated
in a zone from the surface to a depth of
approximately twice the footing width.

REFERENCES

Department of Transport 1978. Reinforced
and anchored earth retaining walls and
bridge abutments for embankments.
Technical Memorandum (Bridges) BE3/78,
Revised 1987.
Jackson, P.H. 1985. The design of
reinforced earth retaining walls and
abutments to BE3/78. MSc Dissertation,
University of Bradford.
Westergaard, H.M. 1938. A problem of
elasticity suggested by a problem in
Soil Mechanics: soft material reinforced
by numerous strong horizontal sheets in
contributions to the mechanics of solids.
Stephen Timoshenko 60th Anniversary,
Macmillan, New York.

International Geotechnical Symposium on Theory and Practice of Earth Reinforcement / Fukuoka Japan / 5-7 October 1988
© *1988 Balkema, Rotterdam. ISBN 90 6191 820 0*

Predicting the behaviour of reinforced soil structures

C.J.F.P.Jones
University of Newcastle upon Tyne, UK

ABSTRACT: The paper considers the factors which influence the behaviour of reinforced soil structures and gives details of the computer prediction which won a NATO competition held in Canada in 1987. Information on parameter selection and results not previously published are provided. In addition, the advantages of the use of finite element models in the design and analysis of reinforced soil is discussed, particularly, with regard to their ability to take into account the influence of different construction techniques and the presence of weak subsoils, both of which are recognised as being of major significance in determining the behaviour of this form of structure.

1 INTRODUCTION

The analysis of reinforced soil structures is conventionally divided into consideration of the external or global stability of the structure and the internal stability. Analysis of the external stability follows normal practice used with gravity retaining walls.

The usual analytical methods used for internal analysis are semi-empirical. Two methods are predominant; the tie-back method which is used by the Department of Transport (1978) in the United Kingdom Design Manual, and the coherent gravity method developed in France by LCPC (1979). Both methods are robust design aids, they make no pretence to model accurately the behaviour of field structures either with respect to strains developed during or after construction or resulting from site conditions. In the case of site conditions the presence of weak subsoils is recognised as being of importance in determining the true behaviour of structures particularly in respect of global movements, Jones and Edwards (1980). The nature and scale of some of these movements is such that it is unrealistic to ignore their effects on the distribution of stress within the reinforced soil mass, Brady (1987). In addition, the construction technique used, particularly that associated with the facing/reinforcement connections or the presence of yielding back-fill, has been

shown to have a dominant effect on reinforcement stresses, Naylor (1978), Jones (1979) and McGowan et al (1987).

The factors which influence the behaviour of any reinforced earth structure are identified in Table 1. The information sought by a designer is typically; strain – both of the soil and of individual reinforcing elements within the structure occurring both during construction and post construction; tension – along the reinforcement and at the connections; bearing pressure distribution and the global stability criteria; and the effects of surcharge loading.

The system capable of providing this range of information whilst at the same time taking into account the significant factors which influence the behaviour of the structure is the finite element method.

2 FINITE ELEMENT MODELS

A range of finite element models and structural idealisations may be used. The structure can be modelled in two dimensions (2D) or three dimensions (3D), the latter is expensive, but may be necessary with unusual structural geometries.

The simplest finite element analysis typically idealises reinforced earth construction as in Figure 1. The structure is assumed to be built in one

Table 1

Reinforcement	Soil	Construction
Composition	Particle Size	Construction System
Durability	Grading	Compaction
Form	Index Properties	
Surface Properties	Mineral Content	Facing
Dimensions	Durability	
Strength	Availability	
Stiffness		

Reinforcement Distribution	Soil State	Structures
Location	Density	Geometry
Spacing	Confinement	End Use
Orientation	State of Stress	Aesthetics
	Degree of Saturation	
	Drainage	

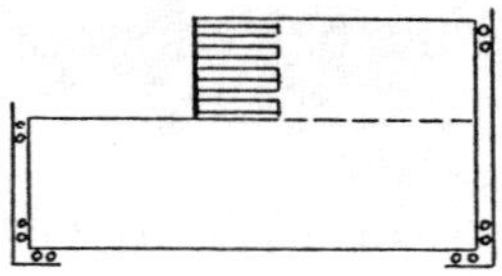

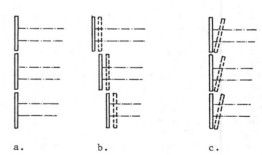

Figure 1.

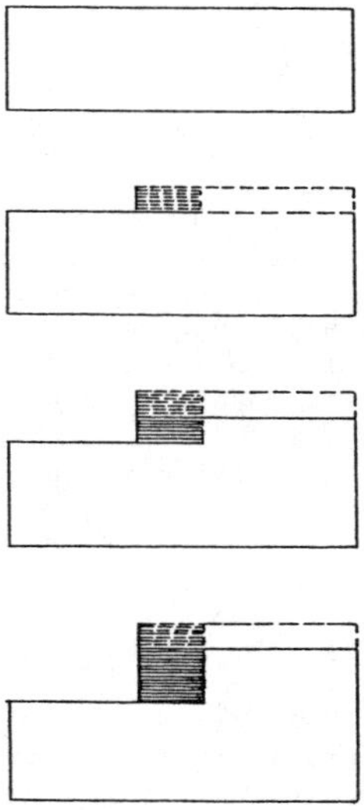

a.

b.

c.

d.

Figure 2.

step and loading is applied by 'switching on' gravity. A closer idealisation is to model the construction in stages, Figure 2 (a-d). This has the advantage of providing a more realistic loading for the subsoil and accurate modelling of non linear soils using this technique may be achieved. However, true idealisation can only be achieved if the construction of the reinforced soil block is mimicked by the finite element model as different construction techniques impose their own in-built stresses and strains. A comparison between the construction of a vertical reinforced soil structure built using individual facing panels and that built using a full height facing propped during construction illustrates the nature of these differences, Figure 3 and Figure 4.

Using a discrete panel system the elevation of the facing will follow the movements shown in Figure 3 (b or c), although the static condition shown in Figure 3 (a) is frequently believed to occur. At stage 1 of the construction, the facing panel is typically rotated backwards towards the fill; during fill placing and compaction, the facing

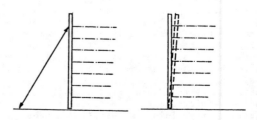

a.   b.   c.

Figure 3.

Figure 4.

536

Table 2

VARIABLE	COMPOSITE MATERIAL (Unit Cell)	SPECIAL ELEMENTS	'FULL' MODEL
Foundation	✓	✓	✓
Stage Construction	✓	✓	✓
Fill Properties	×	(✓)	✓
Reinforcement Properties	×	(✓)	✓*
Reinforcement Prestress	×	×	✓
Stress Distribution in Reinforcement	×	✓	✓
Facing/Reinforcement Connection Stresses	×	✓	✓
Compaction Stresses in Fill	×	×	✓
Construction Technique (Incremental – Full Height)	×	×	✓
Compressible Backfill Layers	×	✓	✓

(✓) Assumes part composite material
✗ 2D analysis sheet or grid reinforcement 3D analysis strip reinforcement

straightens and moves forward an element (δ 1) dependent upon the nature of the reinforcement and its interaction with the soil. The small movement (δ1) will be greater with extensible reinforcement and may be negligible with very stiff reinforcement. In the case of a propped full height facing the rotation and movement is primarily dependent upon the stiffness of the reinforcement and the effects of the fill are of secondary importance, WYMCC (1983).

Idealisation of the reinforced soil block can be made in a number of ways which can be conventionally grouped into three forms:

A. Unit cell approach, in which reinforced earth is modelled as a pseudoelastic material.
B. Use of special elements.
C. Full idealisation with the fill and reinforcement being considered separately.

Method A is convenient when global effects predominate but suffers from the disadvantage of being a simple model. The use of a special element, such as that developed by Naylor (1978), which permits reinforcements to slip (slip strip), reflects the influence of sub soil conditions on the structure but usually does not provide a true idealisation of the construction process. A full idealisation of the fill and the reinforcement using standard finite elements is the preferred approach but one which few have attempted because of the complexities introduced by the number of individual reinforcing elements and the subtleties of the construction techniques. Some forms of reinforcement, typically strips, demand a 3D analysis if a full idealisation is to be attempted. When the reinforcement is in the form of a sheet or grid producing a complete layer of reinforcement the analysis can be reduced to a 2D plane strain condition, the geometry of the overall structure permitting.

The ability of the different finite element models to accommodate the criteria which influence the behaviour of reinforced earth structures is detailed in Table 2.

3 REINFORCED EARTH PREDICTION USING FINITE ELEMENT METHODS

The best test for any analytical system is its ability to produce accurate (Class A) predictions of the behaviour and performance of a structure.

A Class A prediction competition was held as part of a NATO Advanced Research Workshop on the applications of polymeric reinforcement in soil reinforced structures, held at the Royal Military College (RMC) Canada in June 1987. Two vertical reinforced earth walls were constructed by RMC prior to the symposium. The walls were formed using an incremental facing and a propped full height facing. Participants at the symposium were asked to provide Class A predictions of the behaviour of both structures at elapsed times (100 and 1000 hours) after construction and following the addition of a range of surcharge loadings. The required information included: deformation /time response of the facing; strain

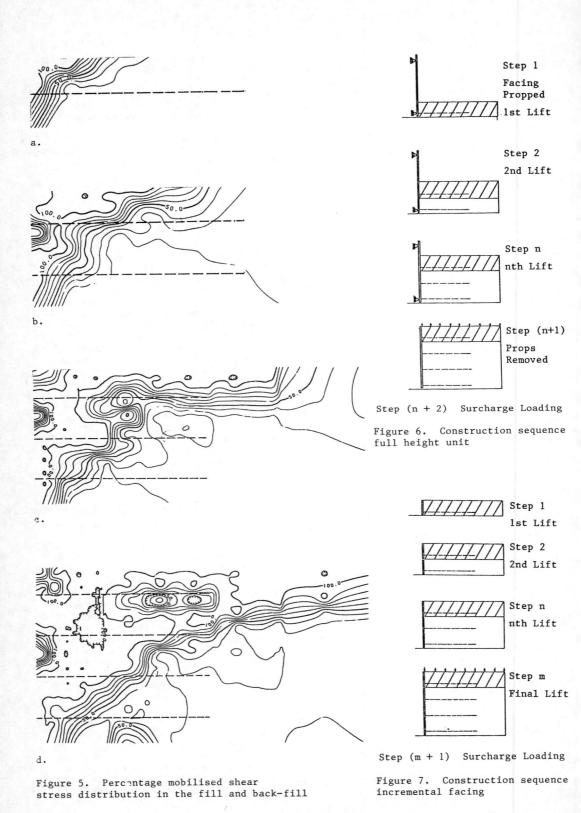

a.

b.

c.

d.

Figure 5. Percentage mobilised shear
stress distribution in the fill and back-fill

Step 1
Facing
Propped
1st Lift

Step 2
2nd Lift

Step n
nth Lift

Step (n+1)
Props
Removed

Step (n + 2) Surcharge Loading

Figure 6. Construction sequence
full height unit

Step 1
1st Lift

Step 2
2nd Lift

Step n
nth Lift

Step m
Final Lift

Step (m + 1) Surcharge Loading

Figure 7. Construction sequence
incremental facing

distribution within the reinforcement including the tensions at the connections; vertical stresses at the base of the reinforced soil block. Concise details of the materials and construction techniques used, including the loading, were provided by RMC.

One of the predictions at the symposium used a 'full' finite element model, it was the only method able to provide an answer to the 282 questions set by the prediction competition. In addition, the results from the analyses were consistently in accord with the measured values.

Complete details of the Class A prediction exercise for the NATO reinforced earth walls are given by Bathurst et al (1987).

Details of the model used in the prediction are given below. Also included are details of the mobilised shear stress distribution within and behind the reinforced soil mass, Figure 5. This information was not requested as part of the prediction exercise, however the results further illustrate the power of the finite element technique and support the use of the tie-back method for the conventional analysis particularly when extensible reinforcement is used.

The prediction was based upon a mathematical simulation of the construction and loading procedures used to construct the two models. The model used was based upon the hyperbolic equation proposed by Kondner et al (1963) and described by Duncan and Chang (1970).

$$(\sigma_1 - \sigma_3) = \frac{\varepsilon}{a + b} \qquad 1.$$

in which σ_1 and σ_3 = major and minor principle stresses.

ε = axial strain
a + b = constants
The asymptotic value of the soil at failure $(\sigma_1 - \sigma_3)$ is greater than the ultimate compressive strength of the soil $(\sigma_1 - \sigma_3)$ ult. but can be related to the ultimate value by the failure ratio R_f.

$$(\sigma_1 - \sigma_3) = R_f (\sigma_1 - \sigma_3) \text{ ult.}$$

By expressing the parameters a and b in terms of the initial tangulant modulus value (E_i) and the compressive strength, equation 1 may be written as

$$(\sigma_1 - \sigma_3) = \frac{\varepsilon}{\dfrac{1}{Ei} + \dfrac{ER_f}{(\sigma_1 - \sigma_3)_f}} \qquad 2.$$

Equation 2 may be used to model the non linearity of soil stress-strain behaviour.

4 SELECTION OF PARAMETERS, CONSTRUCTION AND LOADING PROCEDURES

An incremental non linear elastic model was used with a constant Poisson's ratio $\nu = 0.35$. No unloading-reloading was assumed. The parameters used in the model were generated from the information provided and included compaction stresses within the fill developed during stage construction, prestressing of the reinforcement, and the use of reinforcement whose stiffness varied with time. The facing elements were assumed to be elastic as was the compressible material used between the incremental facing units. Modelling of the construction process and the loading procedures are shown in Figures 6 and 7. The model covered every variable detailed in Table 2 with the exception of a compressible back-fill; the latter was not used in the construction of the trial walls.

5 CONCLUSIONS

The finite element technique can be used to model the subsoil conditions, the construction method, the loading and the reinforcement and soil used in a reinforced soil structure. If the correct idealisation is made and the selection of parameters is accurate, very close agreement can be achieved between the performance of the model and of the real structure. In addition, a study of the development of shear stress within the soil, the strains in the reinforcement and the movements of the facing, provide an insight into the behaviour mechanism of reinforced soil.
Although a finite element analysis represents a very large number of mathematical computations, the increase in computing power available to design offices, coupled with the reduction in computing costs, suggest that this approach to the analysis of critical or complex structures is possible and complimentary to the semi-empirical methods in general use.

REFERENCES

Bathurst, R.J., Wawrychuk, W.F. and Jarrett, P.M. 1987. NATO advanced research workshop. Application of polymeric reinforcement in soil retaining structures. Royal Military College, Canada.

Brady, K.C. 1987. Performance of a
reinforced earth bridge abutment at
Carmarthen. Research report III, TRRL,
Berkshire.

Department of Transport 1978. Reinforced
and anchored earth retaining walls and
bridge abutments for embankments.
Technical Memorandum BE3/78 (Revised
1987), HMSO.

Duncan, J.M. and Chang, C.Y. 1970. Nonlinear analysis of stress and strain in
soils. Jour. Soil Mech. and Foundation
Eng. ASCE, SM5, pp 1629-1653.

Jones, C.J.F.P. 1979. Lateral earth
pressures acting on the facing units of
reinforced earth structures. C.R. Coll.
Int. Reinforcement des Sols, Paris.

Jones, C.J.F.P. and Edwards, L.W. 1980.
Reinforced earth structures situated
on soft foundations. Geotechnique,
June, pp 207-211.

Laboratoire Central des Ponts et Chausees
1979. Reinforced earth structures -
Recommendations - Rules of the art,
Paris.

Kondner, R.L. 1963. Hyperbolic stress-
strain response: cohesive soils, Jour.
Soil Mech. and Foundation Div. ASCE,
Vol. 85, pp 115-143.

Kondner, R.L. and Zelasko, J.S. 1963.
A hyperbolic stress-strain foundation
for sands. Proc. 2nd Pan-Amer. Conf.
Soil Mechanics and Foundation Eng.
Brazil, Vol. 1, pp 289-324.

McGown, A., Murray, R.T. and Andraws, K.Z.
1987. The influence of boundary yielding
on the lateral stresses exerted by backfills. Proc. NATO advanced research workshop, Application of polymeric reinforcement in soil retaining structures, Canada.

Naylor, D.J. 1978. A study of reinforced
earth walls allowing slip strip. Proc.
ASCE Conv., Pittsburgh, pp 618-644.

West Yorkshire Metropolitan County Council,
1983. First report on trial reinforced
earth retaining walls, Dewsbury ring road,
Wakefield, August.

International Geotechnical Symposium on Theory and Practice of Earth Reinforcement / Fukuoka Japan / 5-7 October 1988
© 1988 Balkema, Rotterdam. ISBN 90 6191 820 0

TEXSOL: Already more than 50 successful applications

E.Leflaive
Laboratoire Central des Ponts et Chaussées, Paris, France

Abstract. The TEXSOL method of soil reinforcement presented at the Las Vegas and Vienna conferences on geotextiles is now developing on many job sites in France. The paper presents some significant jobs and indicates the results of tests on retaining structures and on repeated loading behaviour of the material.

1. Introduction.

The TEXSOL technique is a three-dimensional development of the geotextile technology. Its principle has been described in papers presented at the Las Vegas (1982) and Vienna (1986) conferences on geotextiles (1), (2) and at the 11th Soil Mechanics Conference in San Francisco (1985) (3). It is the mixing of continuous polymer yarn and soil to form a composite in which the yarn brings its tensile resistance. In the present practice, the soils used to produce TEXSOL are essentially natural sands ; this is for several reasons : the mixing of soil and yarn is easier with granular materials than with cohesive materials ; mechanical performance of sands is very strongly improved by the tensile strength of yarn ; in many areas, local natural sands are cheap materials. However the principle of the TEXSOL method is also applicable to other types of soils ; such applications are experimented and will be developed both from the point of view of production equipment and new uses of the TEXSOL material.

The TEXSOL technology has been developed by the Research Network of the French Ministry of Publics Works which owns the original patent. For contracting and site application, the TEXSOL Company has been established ; the TEXSOL Company is performing jobs in France and developing the method abroad through Joint Ventures and Licensees. The first license contract has been for Japan in 1987.

TEXSOL is a new material with unusual performances. The effect of yarn is to create a cohesion in the granular material ; however, in contrast with materials bound with cement, bitumen or other binders, the TEXSOL material has a fairly high deformability : simple compression tests lead to failure for 6 to 10 % of axial deformation and, secondly, TEXSOL is as permeable as the material without yarn.

According to the method and equipment used for production, the geometrical arrangement of the yarn within the material may vary ; in present practice, this arrangement is preferably horizontal and it has been shown that the resulting cohesion is anisotropic.

Anisotropy of cohesion is taken into account in the design of TEXSOL structures.

The TEXSOL material has a wide variety of potential applications : retaining structures, foundation layers under railroads and roads, shear resistant drains and filters, foundation blankets on soft soils, antiseismic structures, protection against erosion, shock resistant structures, etc... Presently, the application which has been essentially developed are retaining structures. Some of the corresponding jobs are described below, along with more special applications already performed or at the design stage. After these examples of application, tests results are presented, dealing with failure loading of full size retaining structures and with the behaviour under dynamic stresses and repeated loading conditions, in view of seismic applications and railroad construction.

2. Retaining structures in France.

So far, in June 1988, TEXSOL has been used on 50 job sites in France ; half of these have

TEXSOL, *a new material with unusual performances.*

Mixing of soil and yarn by special machines.

TEXSOL retaining wall on A7 Motorway (after turfing).

been for retaining structures for cuts in natural ground, fifteen for retaining fills and ten for more particular uses.

The sites where the largest amount of TEXSOL has been used are located on the A7 Motorway,

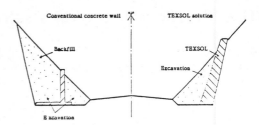

Fig.1 Comparison of concrete wall and TEXSOL solutions for retaining structures.

south of Lyon, towards the Mediterranean sea. The problem was to add one lane to the motorway in each direction without enlarging the right of way ; therefore the slope of the cut sections had to be steepened. For long term stability these sections needed a retaining structure.

As shown on figure 1, the TEXSOL solution is a trapezoïdal inclined structure. As compared with conventional retaining walls, this solution has two main advantages : cost and appearance.

Savings are mainly due to the reduction of earthmoving quantities and to the use of local natural sand for the production of TEXSOL.

Environmental performance is excellent due to grass growing on the TEXSOL surface, as shown on figure 2. In fact, the appearance of TEXSOL after one or two years is such that the motorist does not even notice that there is any retaining structure along the motorway, which is just the best result that can be expected. Experience has also shown that maintenance of the vegetative cover is performed without any problem, which is not the case when the retaining structure is made of concrete elements intermixed wih soil and vegetation. The total amount of TEXSOL produced on the three sites of A7 motorway is 44.000 m3. On these sites, the outer slope of TEXSOL is 60°, the height being between 2 and 9 m.

Another interesting site is the A12 motorway in the Paris area. The situation was similar to that of A7. The quantity of TEXSOL has been 3.500 m3 on this site ; the soil being retained is calcareous clay on one part of the site and clean sand on the other part. The outer slope of TEXSOL is 65° and the maximum height is 12 m.

In such situations, the calculation of the structure is performed according to conventional methods, where the parameters used for calculating the strength of the TEXSOL material for internal stability evaluation are the only specific elements. TEXSOL is characterized by angle of friction and cohesion. As mentioned earlier in this paper, the present method of production leads to an anisotropic cohesion. This has been introduced in the calculation program. For practical purposes, charts have been established for the most usual situations. For design, usual practice is to use values of angle of friction and cohesion that are known to be normally obtained with natural sands and usual proportion of yarn (between 0.15 and 0.2 percent), and then, to check and adjust the type and quantity of yarn with laboratory tests on the sand to be used on site.

Another interesting example of retaining structure is given by the work performed at Deauville. The job was to create a 4 m deep cut in a water-bearing low-plasticity clay slope which was already at the limit equilibrium state before the cut was made. Because of this situation even short term stability was not ensured and the cut had to be made by short lengths, the TEXSOL retaining body being constructed at the same rate to follow closely the excavation. An interesting feature in that case is that TEXSOL being made of sand could perform the filtering and the draining functions that were necessary for such a water-bearing finegrained soil. Therefore, The TEXSOL body is at the same time the retaining structure, the filter and the drain. It is then obvious that large savings may be obtained with this technique, due to the extreme simplification of design and execution.

The last example of retaining structure presented here deals with a case where TEXSOL is used as a fill retaining material. At Longefoy in the Alps region, a mountain road had suffered a slip failure of the natural slope under the road. Reconstructing the road platform is almost impossible with an ordinary fill because of the steep slope of the mountain and is very expensive using piles and concrete structures due to lack of space and because of required time and manpower. TEXSOL allowed a fast repair of the slip failure with a 60° slope angle and a maximum height of 9 m.

3. Other types of applications.

In a few instances TEXSOL has been used to improve the stability and the erosion behaviour of earth or sand levees built for protection of industrial or military areas. In one case sand levees were displaced by wind action ; a fairly thin cover of TEXSOL avoided sand to be blown away and solved the problem. In another instance earth levees were progressively washed down by rain action ; TEXSOL allowed to restore their initial slope,

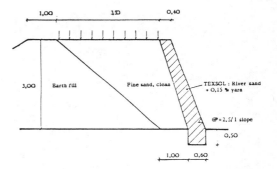

Fig. 2 Diagram of full size tests of retaining structures.

in some places with a steeper slope, while keeping the shock absorbing ability of the levees, which was required on this site where explosions could occur.

Other applications are presently planned where the shock absorbing ability of TEXSOL will be put to use. One is for protection of roads or constructions against falling rocks in mountain areas. To stop very large rocks heavy structures are the cheapest means but large embankments often cannot be built because of the slope angle of the natural ground. Thus a TEXSOL wall with a steep slope retaining a mass of ordinary fill is a good solution. In addition to static stability, TEXSOL is a deformable material which can absorb considerable shock energy through large deformations.

Another shock absorbing application presently at the testing stage is for road safety.

Vibration absorption and seismic performance are commented upon later in this paper, as well as performance under repeated loading.

4. Full-size testing of retaining structures.

Loading tests of 3 m high retaining walls of TEXSOL have been performed in Rouen in 1987. The purpose of these tests was to observe the type of failure of a TEXSOL wall retaining a sand fill and to see if the failure conditions were to be different according to the method of laying the TEXSOL, either in horizontal layers or inclined inwards or outwards with respect to the retained fill.

The testing layout is shown on figure 2. The materials were as follows :

TEXSOL : clean 0-5 mm river sand with 0,15 % polyester yarn 330 dtex 36 CN/tex.

TEXSOL retaining wall on A7 Motorway (before turfing).

Fill : clean fine sand 0-0,3 mm.

The loading was obtained by a fill constructed above the fill with vertical walls (either concrete wall elements or geotextile reinforcement), with the addition of iron blocks on top of the fill, to reach the failure load.

Three tests have been performed, two on six meters long TEXSOL walls and the last on an 18 m long TEXSOL wall.

The results did not show significant differences between the tests, in which the angle of deposit of TEXSOL during production was different. This is in some manners logical since the angle of deposit has an effect on the shear strength of the material (according to the value of the angle between the deposit plane and the shear plane) and that failures in all three tests were not due to shear failure of the TEXSOL material but occured as an overall overturning rotation of the TEXSOL wall.

This behaviour has been the most interesting result of these tests ; displacement measurements during failure showed a rotational movement around the base of the wall. The failure load above the fill has been approximatively 75 kN/m2 for the three tests.

5. Laboratory testing of fatigue and dynamic behaviour.

a) Fatigue tests. In view of the use of TEXSOL under traffic (either under railroads or road pavements) the behaviour of the material under repeated loading has to be tested. Two types of tests have been performed : repeated loading triaxial tests performed in Rouen and simulation testing in a railroad testing facility in the Paris area.

Triaxial tests used a uniform fine natural sand 0-0.3 mm, without fines, with a 50 dtex polyester yarn at a proportion of 0,3 % in weight. Static triaxial tests had given a value of $\sigma_1 - \sigma_3$ at failure, for $\sigma_3 = 50$ kPa, of 1,8 MPa. Repeated loading tests were performed with $\sigma_3 = 50$ kPa and a value of $\sigma_1 - \sigma_3$ oscillating between 0 and 0,7 MPa (i.e. 40 % of the failure load).

Total axial deformation was 1 mm after 10 cycles, 2 mm after 100 cycles, 2,6 mm after 10^3 cycles and 3 mm after 10^6 cycles. The compressive failure strength after 10^6 cycles was equal to that obtained without repeated loading. These results show a good behaviour under repeated loading, with no degradation of the yarn under the repeated stress applications.

The fatigue simulation tests used a very aggressive sand obtained from a railroad ballast quarry ; it was a 0-8 mm material with flat particles showing many sharp edges. The yarn was a 330 dtex polyester at the proportion of 0,21 %.

The testing machine is a system developed by the French Railways to simulate the effect of vibration loading on railway materials. A mechanical vibrator with eccentric wheels is attached to a beam connected with the two rails. These rails (a few meters long) are statically loaded at their ends. The static and dynamic loads applied on the rails are transmitted to one sleeper placed on the ballast and sublayer material to be tested.

This device has been used for many years and empirical correlations have been established between the effect of vibrations resulting from the machine and the actual behaviour of materials under railway traffic. A normal test is 100 hours long, corresponding to 18 million cycles (considered to be equivalent, as far as material behaviour is concerned, to 10 years of very heavy traffic).

For TEXSOL testing, a 21 cm thick layer of TEXSOL was made on the firm foundation clay soil, 3 m long x 1 m wide. A 25 cm thick layer of ballast was put above the TEXSOL, then the sleeper, rails and vibration system.

Measurements made dealt with density, settlements, stresses and acceleration.

For TEXSOL testing the total vibration time has been 143 hours, corresponding to more than 25 million cycles.

Measured accelerations in the TEXSOL layer varied between 10 and 20 g, with amplitudes between 1,5 and 2 mm.

During the test the density of TEXSOL increased from 1,85 to 2,06, with a corresponding reduction in layer thickness of 2,4 cm, from 21 cm to 18,6 cm. The total settlement of the sleeper has been 6,3 cm.

Directly under the sleeper the TEXSOL was disturbed on a few centimeters, but at a 20 cm horizontal distance the interface between the ballast and the TEXSOL was not damaged. The interface between the TEXSOL and the subgrade was not disturbed.

Fatigue damage on the yarn could be observed, specially at the upper part of the TEXSOL layer ; these damages were essentially local transverse compression of individual fibers pinched between grains.

The overall result is considered as quite excellent in view of the very severe testing conditions, since total deformations remained very limited and that no interpenetration occured neither between ballast and TEXSOL nor between TEXSOL and subgrade.

Two test sections are presently being built on an actual railway line, with thicknesses of 20 and 25 cm of TEXSOL under ballast ; measurements of deformation, stresses and acceleration will be made regularly.

b) Laboratory tests for evaluaton of dynamic behaviour.

Tests have been performed by Mr LUONG at Ecole Polytechnique near Paris in view of analyzing the mechanical behaviour of TEXSOL as compared with that of simple sand. These tests have explored cyclic loading behaviour, liquefaction and dynamic response as a function of frequency. The results have been presented in (4) and show that TEXSOL ductility and energy absorption capability are of great interest when vibrations are involved.

6. Conclusion.

TEXSOL, new construction material, has already proved its ability to solve in very economical and ecological conditions most of the problems of retaining walls, slope savings, anti-seismic and anti-vibration courses or foundations, protection against noises and shocks. New applications and other research are already moving and the Owners and Engineers are looking for its development with high interest all around the World.

7. Acknowledgements.

This paper describes works performed by the "Société d'Application du TEXSOL" (SACLAY, Boîte Postale n° 62, 91401 ORSAY Cédex, France) and tests conducted by C.E.T.E. Rouen (Boîte Postale 245Bis-247, 76120 GRAND QUEVILLY, France) - (part of the Research Network of the French Ministry of Public Works) and by Ecole Polytechnique ; their cooperation is gratefully acknowledged.

8. References.

(1) LEFLAIVE E., "The reinforcement of granular materials with continuous fibers", 2nd International Conference on Geotextiles, Las Vegas, U.S.A., 1982, pp. 721-726.

(2) LEFLAIVE E., LIAUSU Ph., "The reinforcement of soils by continuous threads" 3rd International Conference on Geotextiles, Vienna, Austria, 1986, pp. 523-528.

(3) LEFLAIVE E., "Soil reinforced with continuous yarns : the TEXSOL", 11th International Conference on Soil Mechanics and Foundation Engineering, San Francisco, U.S.A., 1985, pp. 1787-1790.

(4) LUONG M.P., LEFLAIVE E., KHAY M., "Propriétés parasismiques du TEXSOL", 1er Colloque National de Génie Parasismique, Saint-Rémy-lès-Chevreuse, 1986, pp. 3/49 to 3/58.

International Geotechnical Symposium on Theory and Practice of Earth Reinforcement / Fukuoka Japan / 5-7 October 1988
© *1988 Balkema, Rotterdam. ISBN 90 6191 820 0*

Optimised computer programs for design of reinforced soil structures

P.F.McCombie & D.I.Bush
Netlon Limited, Blackburn, UK

ABSTRACT: The overall process of engineering design can be considered as a sequence of problem definition, data acquisition, conceptual design and detailed design. The detailed design stage is usually most time consuming, while the initial stages require most use of engineering judgement. Detailed design is time consuming when it requires repetition of straightforward calculations and presentation of the design to a high standard. In both these areas the computer can offer considerable benefits. The authors set out to examine the overall design process for design of reinforced soil walls in the light of considerable experience of engineering design. A computer program was then written which embodied the entire detailed design process from raw data to presentation of the design. The program is highly versatile and interactive.

1. INTRODUCTION

The aim of engineering design is to solve problems. The first stage of engineering design is therefore to define the problem (Fig. 1). This definition may be quite complex; for example, to support a road with a given loading at a certain profile in space, which may be above, at or below existing ground level. Allowing for passage of such things as other roads, paths, pipes, etc., within defined areas. Arriving at a compromise between cost and land take, while satisfying environmental requirements with regard to appearance and noise during construction, in use and during decommissioning. The next stage is to collect the initial data about the problem required for the conceptual design. Omissions in the collection of data could lead to failure to fully solve the problem. In the conceptual design, the engineer decides how to solve the problem in general terms. In the example given above, we might now have cuttings, embankments, bridges, culverts and retaining walls proposed as a means to carry the road along the defined profile. This stage requires considerable engineering judgement and knowledge of what is possible, where, and at what approximate cost. The next stage is to carry out the detailed design; this is likely to require collection of further data to provide parameters for the design methods used. The more precise

understanding of a problem which emerges at this stage might well lead to revision of the conceptual design, which may lead back to a requirement for more basic data, and perhaps even to revise the problem to be solved in the light of what is possible. This final stage often requires far less engineering judgement than the initial stages of a design, but can take far longer, due to the tedious nature of many of the calculations involved. The aim of using computers in the design process is to free the engineer's time from carrying out routine calculation. This has two benefits. It allows the distribution in the time spent between the various stages to more closely reflect their importance producing pressure to rush the initial stages. It also allows the conceptual design to be revised with a less severe penalty in terms of time required to repeat the detailed design.

Historically, computer programs have tended to concentrate on one particular aspect of the detailed design calculations, processing data from input files to output files, with little or no user interaction. More recently, attempts have been made to provide "user-friendly" interfaces which provide interactive "front ends" to the programs, and perhaps some sort of graphical output in addition to the data files. The problem usually remains that the method of data entry is cumbersome and difficult to check or change to represent a

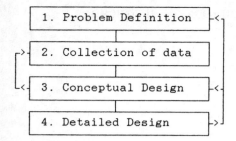

Figure 1 Flow diagram for design process

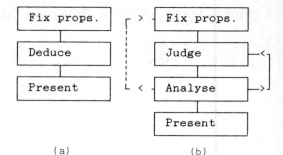

(a)   (b)

Figure 2 Example flow diagrams

different situation, and that the output requires considerable interpretation. Such programs can offer considerable savings in time over hand calculation, but are still far from providing the full potential benefits.

The authors have developed a computer program for the design of reinforced soil retaining walls. The overall design process has been embodied in the program, maximising the benefits of using the computer. The program is written for use on standard micro-computers, which allow the engineer easy and cheap access to powerful facilities for computing, including screen graphics and both graphical and printed paper output. The scope of the program was determined by considering the use which would be made of each feature compared with the time taken to implement it and the time it would save in use.

2. THE PROCESS OF DETAILED DESIGN

In the conceptual design stage the engineer has decided upon how to solve the problem, and will have decided factors such as the required appearance of the solution. In detailed design the engineer decides exactly what must be done, in terms of detailed specification of the nature and location of the solution. The engineer must also communicate this information to whoever must act on it.

There are five different types of process in detailed design:
  1. Fix properties of materials
Characteristics of materials affecting the solution of the problem are fixed. This includes deciding upon the particular material to be used out of a range of materials determined in the conceptual design.
  2. Deduction
Given one factor, then a parameter in the solution follows directly.

  3. Engineering Judgement
Given a set of factors, the engineer estimates parameters in the solution which will be satisfactory - these parameters may be wrong and should be checked by analysis.
  4. Analysis
Data relating to the problem, and possibly to a parameter, or parameters in the proposed solution, is processed to give further data. This further data may be a final indication that one aspect of the solution is satisfactory, or may be data needed in a further stage of the design.
  5. Presentation
The solution must be presented. This is likely to take the form of drawings and written information giving the assumptions made, the results of the analyses which demonstrate the solution to be satisfactory, and the details of the solution.

Two simple flow diagrams are shown in Figure 2.

Figure 2a shows a design process which is simple deduction - the correct design follows automatically given the problem. A retaining wall design might be reduced to this simple process; however, it is more likely that some engineering judgement is needed to ensure that the design is practical in a given situation. Figure 2(b) shows a design process in which no simple deduction can be made, and the engineer has to propose a complete solution, analyse it, and then make changes to the proposal until the results of the analysis indicate it to be satisfactory. An example might be the design of a slope with a fixed geometry and fixed materials. If the geometry and materials may be varied, then the route shown dotted is added.

Computer programs have commonly been used for the processes of analysis and deduction. Programs for deduction often have the disadvantage that what is deduced does not lead to a practical design; it is often better to leave the decision-making to the engineer. Programs for analysis commonly require the preparation of data files, which may require a lot of work to change if the results are not satisfactory. Typically, the program has to be left, the data file edited, and the program re-run at each stage of re-analysis. If a single analysis needs several runs of the program, for example to find the critical circle in a slope stability analysis, the procedure can become extremely tedious.

With this understanding of the design process, the aim is to make maximum use of the computer in ensuring that the engineer's time is not wasted in carrying out operations which do not need judgement.

3. THE 'TENSAR' REINFORCED SOIL RETAINING WALL DESIGN PROGRAM

The detailed design process for reinforced soil retaining wall design is illustrated in Figure 3. This entire process has been embodied in the Walls program. The engineer begins with the basic requirements and the conceptual design, and ends with all the written information and drawings that are needed. For a straightforward design, the entire process can take as little as five minutes. This would compare with about one hour by hand for an experienced engineer. For a complex design, the process could take fifteen minutes, compared with many hours by hand. As well as the speed of operation, an important feature is the wide range of problems which can be tackled. It was in this area in particular that a balance had to be struck between the time taken to implement a feature in the program and the potential savings in time which would result. At various stages, additional features have been added as it has become apparent that they would get sufficient use to justify the time spent in developing them. These additional features are noted in the description of the program given below.

It was decided to allow as much freedom as possible to the user in choices of factors of safety and other parameters, while following the tie back-wedge or coherent gravity methods.

The first feature of the retaining wall design process to note is that if the right decision is made every time then each stage is gone through only once. The process is therefore straightforward to embody in a sequential program; the complexity of the order of execution is limited to going back to previous stages in the main sequence. It is simple to impose an additional path immediately before presentation allowing flow to go back to any previous stage. This

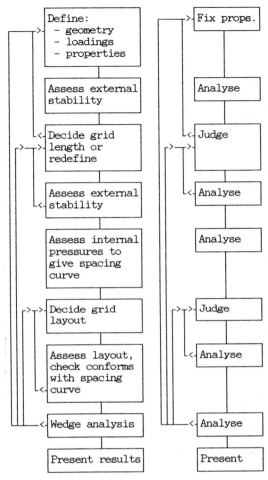

Figure 3. Flow diagram for wall design

allows changes to be made in any part of the design. Also, the presentation is given only as an option. This facilitates the use of the program as a tool for preliminary examination of a problem. The main structure of the program is now defined.

The first stage of the program is the definition of the problem. It is in this stage that the scope of the program must be decided. The scope was defined on the basis of the minimum requirements of a reinforced soil retaining wall design program, then by looking at the usefulness of each possible additional feature compared with the time required to implement that feature in the program.

The minimum requirement for the scope of a reinforced soil retaining wall design program is:

1. Vertical wall with identical purely frictional fill in and behind the wall, and no external loads.

The following additional features were considered to be required sufficiently often to justify the time required, which was considerable for item (6):

2. Different properties for the fill behind the reinforced soil.

3. Cohesion in both fills.

4. Surcharge on the top of the wall, starting at the face of the wall, and extending over the backfill.

5. A strip footing at the top of the wall, of any dimensions and at any position on the reinforced soil or the backfill, carrying vertical and horizontal loads, either of which can be zero, the vertical load being at a defined eccentricity.

6. A backfill slope of a defined height at a defined angle, the bottom of the slope being at the top of the wall, and the loadings in (4) and (5) above being on the level ground at the top of the slope.

7. The ability to use more than one reinforcement type in the same wall.

8. Design to the Tie Back-Wedge method, the Coherent Gravity method (French Ministry of Transport 1979), and to the requirements of the U.K. Department of Transport (1987) which incorporates the British Board of Agrément certificate (1986) for Tensar SR2 Geogrids.

At a later date the following additional features were found to be worth the time needed to implement them:

9. Build up of design strength of the reinforcement from characteristic strength and factors of safety.

10. Design to allow horizontal gaps to be left between grids.

Where appropriate, the program was written so that all the features could be combined in a single problem.

A large number of users, or a changing pattern of requirements, may lead to other features being added at a future date.

Another factor to be considered is that the program is needed quickly. It is therefore desirable to plan the writing of

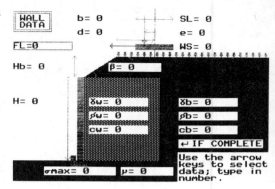

Figure 4. Main data input

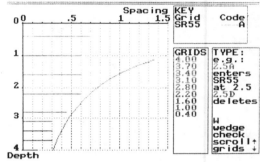

Figure 5. Spacing curve and placing of grids

a program so that it will begin to give benefits as early as possible in its development.

Much of the data is entered into the program on a schematic diagram of a retaining wall on the computer screen. This is illustrated in Figure 4. This clarifies the data being requested, and the sign conventions.

A feature of the flow diagrams in Figure 3 is the absence of any deductive step. Instead, analysis is carried out and the results presented to the engineer for the engineer to make all decisions. The two points where this occurs are in the choice of grid length and in the placing of the grids.

The program analyses the external stability of the wall for a range of grid lengths, and presents the results on the computer screen in the form of graphs of factor of safety versus grid length. The

550

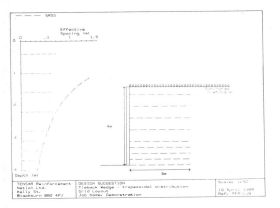

Figure 6.   Example plot

4.   CONCLUSION

By considering the design process as a whole and looking for every way in which the computer can be used to save the engineer's time and enhance his performance, a computer program has been produced which is optimised for its application.

The resulting design program may be compared with the previous generation of analytical programs which sought simply to take over the repetitive calculations involved in engineering design.   The availability of a design program allows the engineer to examine a problem in far more detail in a fraction of the time which used to be needed.

Consideration of the potential benefits of individual features in the program led to a program which began to pay for itself before it was even completed, and in which the time required to write features was carefully balanced against the time which might be saved by having those features available.

The particular program described in this paper has allowed the design department which produced it to handle an unprecedented number of enquiries in the same period in which it was written.   Further development of the program continues alongside its use.

REFERENCES

French Ministry of Transport/LCPC, 1979. Reinforced Earth Structures, Recommendations - Rules of the Art.

Roads and Bridges Agrément Certificate No 86/27 (1986).   'Tensar' SR2 Polymer Grid for Reinforced Soil Walls.   British Board of Agrément, Hertfordshire.

Technical Memorandum (Bridges) BE3/78 (Revised 1987).   Reinforced Earth Retaining Walls and Bridge Abutments for Embankments.   Dept. of Transport, UK.

engineer can then decide upon a grid length which satisfies the requirements of the particular code to which the design must comply.   The program then gives the bearing pressure at the toe corresponding to that grid length, and exact values of the factors of safety against sliding and overturning. The engineer can then use that grid length or enter a new grid length or even go back and change the data.

Next the program analyses the internal stability of the wall using the chosen grid length and presents the results as a spacing curve.   An example of the appearance of the computer screen at this stage of the program is given in Figure 5.   This gives the engineer all the information needed to choose a practical grid layout;   this layout will have the grids placed at levels compatible with the facing being used, while minimising the number of grids without the load in any grid exceeding its safe design load.   The layout is then checked by analysing the stability of a number of wedges.

A part of the design process which is often very time consuming is the presentation of the results.   As all the relevant information is in the computer at the end of the design, it is possible for the computer to complete the design process by presenting all the input data, the results of the analyses and the final design.   This is done using a printer and a plotter.   The final design is shown in cross section on the plot, and the grid length and layout is given in the printout, together with the quantities for a unit length of wall.   An example of the plot is shown in Figure 6.

International Geotechnical Symposium on Theory and Practice of Earth Reinforcement / Fukuoka Japan / 5-7 October 1988
© 1988 Balkema, Rotterdam. ISBN 90 6191 820 0

Roles of facings in reinforcing steep clay slopes with a non-woven geotextile

K.Nakamura & Y.Tamura
Tokyu Construction Co., Ltd, Tokyo, Japan

F.Tatsuoka
Institute of Industrial Science, University of Tokyo, Tokyo, Japan

K.Iwasaki
Mitsui Petrochemical Industries Ltd, Tokyo, Japan

H.Yamauchi
Penta Ocean Construction Co., Ltd, Tokyo, Japan

ABSTRACT: Various roles of facing structures for stabilizing steep reinforced slopes of clay embankments are classified and defined. These roles are examined based on the behaviors of four full scale clay test embankments reinforced with relatively short non-woven geotextile sheets. The behavior clearly shows that a slope covered with a rigid facing structure can be very stable, whereas a slope covered with a flexible facing exhibits a relatively large deformation.

1 INTRODUCTION

Metal strips as used in the Reinforced Earth retaining walls are inadequate as a reinforcement for clays, because they lack the drainage function. For cohesive soils, planar geotextile sheets having a function of drainage such as non-woven geotextile sheets are adequate, despite their relatively lower tensile strength. Further, planar geotextile sheets have another advantage of a larger contact area with soil. Consequently, theirs ratio of the pull-out resistance to the tensile strength becomes much larger than that of metal strips. This results in a much smaller anchoring length needed for geotextile sheets than metal strips.

So far, four full-scale clay test embankments reinforced with a non-woven geotextile have been constructed by the authors (Tatsuoka and Yamauchi 1986, Tatsuoka et al. 1987, Yamauchi et al. 1987). Fig.1 illustrates three of them. Non-woven geotextile sheets were used as (1) a drainage material, (2) a tensile reinforcement, and (3) a material to facilitate better compaction. The length of non-woven geotextile sheets was made so short that some sheets placed at higher levels did not extend beyond the potential failure plane. It has been found from the long-term behavior and the behavior during artificial heavy rain-fall tests that steep clay slopes can be made very stable by using both relatively short non-woven geotextile sheets and an adequate facing structure.

2 ROLES OF VARIOUS KINDS OF FACING RIGIDITY

When a short reinforcement is used for a steep slope, the following different kinds of facing rigidity, as illustrated in Fig.2, contribute to the stability in various ways (Tatsuoka et al. 1987).

When the facing is flexible without the local rigidity in the sense that large earth pressures are not activated on the back face of facing, the local compressional failure in soil near the facing tends to occur, in particular near the toe, as illustrated in Fig.3. This type of local failure can induce the three failure modes illustrated in Fig.4. Such type of failure has been observed for Test Embankment I as reported in details by Tatsuoka and Yamauchi (1986). In particular with a shorter reinforcement, the tendency of rotation of the reinforced zone about the toe can be larger.

The two slopes of Embankment II as shown in Fig.5 had partially the local rigidity (Type B-2 in Fig.2). When compared to the facing type B-2, the type B-1 has a worse degree of the local rigidity, whereas the type B-3 has a better degree. However, these facing types B-1, B-2 and B-3 lack the overall axial rigidity in the sense that a sufficient amount of the weight of back fill is not transmitted to the base ground through the facing. The facing type C has this function. However, the facing type C lacks the overall bending rigidity, thus such failure modes as illustrated in Fig.6 are not

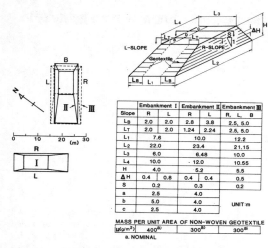

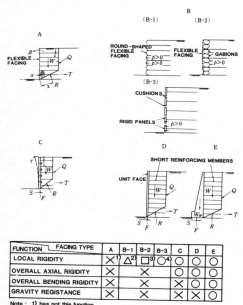

MASS PER UNIT AREA OF NON-WOVEN GEOTEXTILE

μ(g/m²)	400[a]	300[a]	300[a]

a. NOMINAL

Slope	Embankment I		Embankment II		Embankment III
	R	L	R	L	R, L, B
L_B	2.0	2.0	2.8	3.8	2.5, 5.0
L_T	2.0	2.0	1.24	2.24	2.5, 5.0
L_1	7.6		10.0		12.2
L_2	22.0		23.4		21.15
L_3	6.0		6.48		10.0
L_4	10.0		12.0		10.55
H	4.0		5.2		5.5
ΔH	0.4	0.8	0.4	0.4	0.5
S			0.2		0.3	0.2
a	2.5		4.0		
b	5.0		4.0		
c	2.5		4.0		UNIT:m

Fig.1. Test clay embankments constructed at Chiba Experiment Station, Institute of Industrial Science, Univ. of Tokyo.

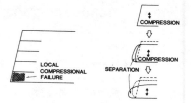

Fig.3. Local compressional failure of soil by using a flexible facing (Type A).

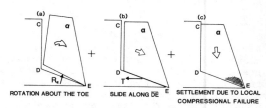

ROTATION ABOUT THE TOE   SLIDE ALONG DE   SETTLEMENT DUE TO LOCAL COMPRESSIONAL FAILURE

Fig.4. Three major failure modes of slope reinforced with a short reinforcement.

FUNCTION \ FACING TYPE	A	B-1	B-2	B-3	C	D	E
LOCAL RIGIDITY	×[1]	△[2]	□[3]	○[4]	○	○	○
OVERALL AXIAL RIGIDITY	×		×		○	○	○
OVERALL BENDING RIGIDITY	×		×		○	○	○
GRAVITY REGISTANCE	×		×		×	×	○

Note : 1) has not this function.
2) has this function only to a limited extent.
3) has this function to a large extent.
4) has this function sufficiently.

Fig.2. Illustration of various facing structures and their functions.

effectively restrained. The facing type D has this kind of rigidity.

When a long reinforcement is used, as illustrated in Fig.7, the reinforcement in each soil layer is designed so

as to resist againt the earth pressure acting to each soil layer. However, in the reinforcing method using a short reinforcement and a type D facing, the reinforced zone together with the facing are expected to behave more or less as a monolith. The degree of resembling a monolith depends on the degree of the rigidity of facing and the degree of reinforcing.

The facing type E shown in Fig.2 has further a gravity resistance. This type of facing may be very practical. In order to avoid the damage to the reinforcement at the connection to the rigid facing during the filling of embankment, the following method is effective; i.e., a slope having a type B-2 facing is first constructed. After an initial settlement has sufficiently occurred, a rigid facing structure is placed on the existing slope surface. In fact, this method was adopted for a test slope of Test Embankmnet III.

3 CONSTRUCTION OF TEST EMBANKMENT III

Based on both the above consideration and the experiences with Test Embankments I and II and the other similar one constructed at another place (Kami-Onda Embankment, Tatsuoka et al., 1977), Test Embankment III having three test slopes with different facing structures was constructed in October

1986 (see Fig.8). A volcanic ash clay called Kanto loam was used. At filling, the average values of the water content w, the degree of saturation S_r, and the dry density γ_d were 110%, 85% and 0.6gf/cm³, respectively. A spun-bond 100% polypropylene non-woven geotextile was used as reinforcing sheets as for Embankment II. The force per unit width at 15% elongation $\alpha_{0.15}$ and those at peak α_f are a function of normal pressure σ_0(tf/m²):i.e., $\alpha_{0.15}$(tf/m)=0.453+0.00675σ_0 and α_f(tf/m)=1.5+0.017σ_0 (Tatsuoka et al., 1987). The lowest two sheets were made longer as 5m in order to drain better the soil at the lower levels in the embankment. The top two sheets were also made longer as 5m, in order to prevent the possible development of cracks from the crest and also to increase the resistance against the overturning of the reinforced zone. A mass of crushed gravel was placed near the toe in order to collect water effectively from the interior of embankment and also to increase the resistance against overturning by increasing the strength of the toe of the slope. The three test slopes have three different facings as follows:
(1) Precast-concrete panel facing (as Type C in Fig.2). A facing structure, consisting of precast concrete panels, as shown in Fig.9, was placed on a foundation which had been made to avoid its settlement. This type of facing has been used also for Kami-Onda Embankment. Each panel has a dimension of 50cm×50cm×5cm and a weight of 34kgf for easier manual handling. Further, projections and grooves were provided at its four edge surfaces for easier connection to other panels. A 10cm-wide strip of non-woven geotextile reinforced with a polypropylene sheet having a high tensile stiffness was connected to each panel by means of a round steel bar and a sliding connection embedded in the panel. The reinforced strip geotextile has a yielding load of 1.18 tonf/m at a tensile strain of 4.7%. The sliding connections allow the settlement of the strip relative to the face. One end of the strip was heat-bonded to another part of the strip over 15cm to make a hoop to hook the steel bar. The strip was placed between two planar non-woven geotextile sheets so that eventually the geotextile sheets were to be connected to the concrete panels. Strains in geotextile sheets were measured only in this slope.

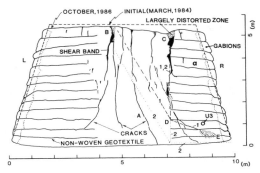

Fig.5. Cross-section of Embankment II exposed one year after relatively large deformation by an artificial heavy rainfall test; Lines denoted by 1 and 2 are failure surfaces by the analyses of the limit equilibrium method with and without water pressures in cracks (Yamauchi et al., 1987)

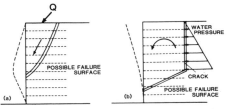

Fig.6. Possible failure mechanisms due to lack of overall rigidity in facing.

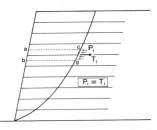

Fig.7. Assumed horizontal force equilibrium for a long reinforcement.

(2) Shotcrete skin facing (as Type D in Fig.2). After the slope had been completed by using gabions at the face as the slopes of Embankment II (see Fig.5), a skin of shotcrete reinforced with wire mesh, having a thickness of about 8cm, was placed on the existing slope face as shown in Fig.8a. The shotcrete skin was anchored with 10cm-wide reinforced non-geotextile strips as mentioned above to the main body of the embankment. Drain holes were placed as needed. It was expected for

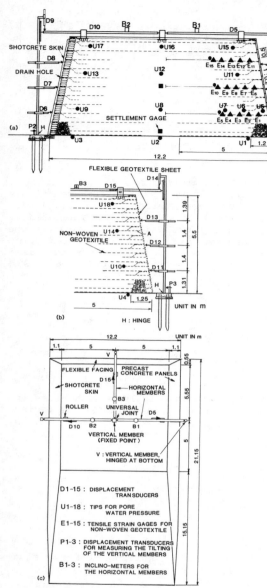

SHOTCRETE SKIN

DRAIN HOLE

SETTLEMENT GAGE

UNIT IN m

(a)

FLEXIBLE GEOTEXTILE SHEET

NON-WOVEN
GEOTEXITILE

UNIT IN m

(b)

H : HINGE

12.2
UNIT IN m

FLEXIBLE FACING | PRECAST
CONCRETE PANELS

SHOTCRETE
SKIN

HORIZONTAL
MEMBERS

ROLLER

UNIVERSAL
JOINT

VERTICAL MEMBER
(FIXED POINT)

V : VERTICAL MEMBER,
HINGED AT BOTTOM

D1-15 : DISPLACEMENT
TRANSDUCERS

U1-18 : TIPS FOR PORE
WATER PRESSURE

E1-15 : TENSILE STRAIN GAGES FOR
NON-WOVEN GEOTEXTILE

P1-3 : DISPLACEMENT TRANSDUCERS
FOR MEASURING THE TILTING
OF THE VERTICAL MEMBERS

B1-3 : INCLINO-METERS FOR
THE HORIZONTAL MEMBERS

(c)

Fig.8. Cross-sections and plan of
Embankment Ⅲ.

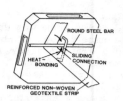

ROUND STEEL BAR

SLIDING
CONNECTION

HEAT-
BONDING

REINFORCED NON-WOVEN
GEOTEXTILE STRIP

Fig.9. Precast concrete panel.

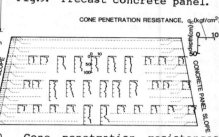

CONE PENETRATION RESISTANCE, q_c(kgf/cm²)

SHOTCRETE SLOPE

CONCRETE PANEL SLOPE

Fig.10. Cone penetration resistances
of soils as compacted (Embankment Ⅲ).

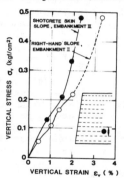

Fig.11. Compression of soil near slope
face during filling.

the gabions to work as buffers for
possible further relative settlements
between the shotcrete skin and the
embankment.

(3) <u>Flexible facing (as Type A in Fig.2).</u>
The flat slope faces of soil were made
and they were wrapped around with
geotextile sheets (Fig.8b). No structural measures was used. A thin
non-woven geotextile sheet having a

nominal thickness of 1.5mm (which is
half that of the non-woven geotextile
used as the reinforcing material) and a
length of 1m, as denoted by the letter
A in Fig.8b, was placed horizontal at
the intermediate height of each soil
layer for better compaction of the soil
near the face. This slope was expected
to exhibit the worst performance among
the three slopes, providing a good
evidence for the importance of facing
structural measures for such steep clay
slopes.

Each soil layer was compacted by a
heavy compaction plant with a weight
of 12tf. A light compaction plant with
a weight of 90kgf was used to compact
the soil within about 80cm from the
slope face. By this mechanical
compaction, the soil near the face was
well compacted, as seen from the result
shown in Fig.10: i.e., the cone penetration resistances near the face were

almost the same as those for the main body of the embankment. Fig.11 shows the relationship between the overburden and the vertical compressional strain at the slope face in each soil layer observed during filling in two slopes constructed using gabions, as indicated in the inset figure. The shape of the curve resembles the one under the one-dimensional compression of soil in the sense that the rate of axial compression decreases as the overburden increases. This implies that the geotextile restrained the horizontal tensile strains in the soil. It may also be seen that the vertical compression of soil near the face is smaller for this slope than that for Embankment Ⅱ, where manual compaction was employed near the slope face.

Fig.12 shows the tensile strains in the non-woven geotextile sheets for the concrete panel slope which occured during further filling over each soil layer. It may be seen that the geotextile functioned as a tensile reinforcement during filling. These data also suggest that earth pressures were activated at the back face of the facing, indicating that the precast concrete panels confined the back fill.

4 POST CONSTRUCTION BEHAVIOR OF EMBANKMENT Ⅲ

Fig.13 shows the time histories of the displacements at some representative points for a period from about two months after the construction. Fig.14 compares the displacements of the three slopes for about a year after the construction. Fig.15 shows the deformation of the flexible facing slope. It may be clearly seen from

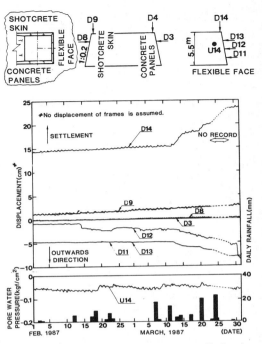

Fig.13. Behavior of Embankment Ⅲ during 1 Feb. ~ 31 March, 1987.

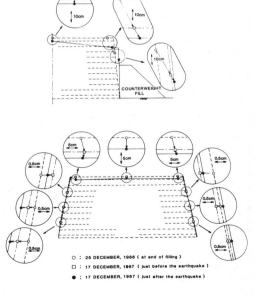

Fig.14. Total displacements at representative places at crest and slope faces, Embankment Ⅲ.

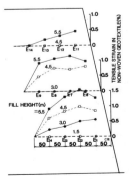

Fig.12. Tensile strains in geotextile sheets in concrete panel slope during filling (Embankment Ⅲ).

557

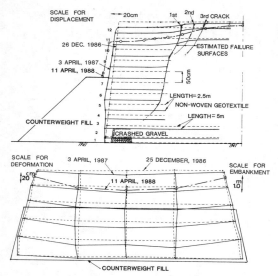

Fig.15. Deformation of flexible facing slope, Embankment Ⅲ (note that the scale for displacement is 2.5 times the scale for embankment).

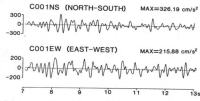

Fig.16. Earthquake motions at 1m-underground on 17 Dec. 1987, recorded at a distance of 210m from Embankment Ⅲ.

these results that the displacements of the flexible facing slope were much larger than those of the other two slopes, the one having gabions and a shotcrete skin and the other one having precast concrete panels. It may also be seen from Fig.13 that the displacements at the flexible facing slope were induced by rainfall. To avoid too large deformation of the flexible facing slope, a counterweight fill as shown in Fig.14 and Fig.15b was constructed in May, 1987.

The total rainfall for thirteen months after the construction to the end of 1987 is 1400mm. For this amount of rainfall, the displacements at the two rigid facing slopes were very small. In particular, it is to be noted that the horizontal outwards displacements near the crest (D3 and D8), which

are the good indicators of the instability of the steep slopes, were remarkedly small.

The test embankment experienced a relatively large earthquake motion on 17 Dec. 1987 (Fig.15). The deformation by the earthquake was the largest also for the flexible facing slope despite the counterweight fill (see Fig.14), and the deformations of the other two slopes were very small. This behavior suggests that also for increasing the resistance against seismic effects, the use of facing structures having various kinds of rigidity is effective.

From the field observation, a clear difference was not found in the behavior of the two slopes with facings having some degrees of rigidity. However, a clear difference has been observed between these two facing types C and D in the laboratory model tests in which the model slopes were brought to failure by loading them at their crests. The results will be reported in the future.

5 CONCLUSIONS

The behavior of full-scale test embankments of clay reinforced with a non-woven geotextile during natural rainfall, an artificial heavy rainfall test and a relatively strong earthquake motion shows that the facing structure having various kinds of rigidities should be used for increasing the stability of the steep slopes. These various kinds of rigidity can be classified into the local rigidity, the overall axial rigidity and the overall bending rigidity.

REFERENCES

Tatsuoka, F. & H. Yamauchi 1986. A reinforcing method for steep clay slopes using a non-woven geotextile. Geotextiles and Geomembranes, 4, 241-268.

Tatsuoka, F., K. Nakamura, K. Iwasaki, Y. Tamura & H. Yamauchi 1987. Behavior of steep clay embankments reinforced with a non-woven geotextile having various face structures. Proc. of the Post Vienna Conf. on Geotextiles, Singapore, 387-403.

Yamauchi, H., F. Tatsuoka, K. Nakamura, Y. Tamura & K. Iwasaki 1987. Stability of steep clay embankments reinforced with a non-woven geotextile. ditto, 370-386.

International Geotechnical Symposium on Theory and Practice of Earth Reinforcement / Fukuoka Japan / 5-7 October 1988
© *1988 Balkema, Rotterdam. ISBN 90 6191 820 0*

FEM analysis of polymer grid reinforced-soil retaining walls and its application to the design method

Eiji Ogisako
Shimizu Corporation, Fukuoka, Japan

Hidetoshi Ochiai & Shigenori Hayashi
Kyushu University, Fukuoka, Japan

Akira Sakai
Saga University, Saga, Japan

ABSTRACT: Finite element analyses for the polymer grid reinforced-soil retaining walls are performed using the method which is capable of taking account of the displacement dependence property of the pull-out resistance of the polymer grid in soils. The reduction effect of the earth pressure acting on the wall and the wall deformation by the reinforcement is discussed. Furthermore, the application of the analytical results to the design method is proposed.

1 INTRODUCTION

In recent years, much interest has been taken in the method of soil reinforcement with the polymer grids. Most of the design methods for reinforced soil structures of this kind currently being employed are based on the theory of rigid-plasticity which takes no account of displacement and deformation of reinforcing material in soil (Netlon Ltd. 1984). However, for some types of earth reinforcements such as polymer grid reinforced soil structures where the soil on upper and lower sides of the polymer grid is partially continuous, a pull-out resistance of the polymer grid in soil depends on the displacement of the grid junction (Hayashi et al 1985).

In this paper, finite element analyses for the polymer grid reinforced-soil retaining walls are performed using the method which is capable of taking account of the mechanism of pull-out resistance of

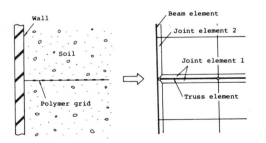

Fig.1 Modeling of polymer grid reinforced-soil retainig wall

polymer grid in soil obtained from the pull-out tests on polymer grid (Ochiai & Sakai 1987). Analyses are performed for some various height of the wall, and spacing and length of the polymer grid. The relation between these parameters and the reduction effect of the earth pressure acting on the wall and the wall deformation is discussed. Finally, the application of the analytical results to the design method is proposed.

2 ANALYSIS OF POLYMER GRID REINFORCED-SOIL RETAINING WALL

2.1 Modeling of polymer grid reinforced-soil retaining wall

For the case of deformation analysis of a reinforced soil structure with polymer grid, it is necessary to use an analytical method which is capable of expressing a behavior of discontinuous plane between polymer grid and soil with a peculiar friction. A modeling of the polymer grid reinforced-soil presented herein is a combination of the joint element expressing the property of discontinuous plane with the truss element transmitting the axial force only. This truss element whose ends are connected by the pin joint is used for modeling of polymer grid. The mechanism of pull-out resistance of the polymer grid in soil, especially the mobilizing process of the coeficents of pull-out resistance with shear displacement may be evaluated by introducing the dependence of shear

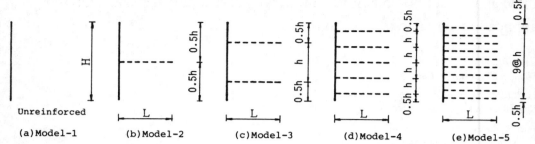

Fig.2 Analytical cases

| Unreinforced | | | | |
| (a)Model-1 | (b)Model-2 | (c)Model-3 | (d)Model-4 | (e)Model-5 |

displacement into a shear stiffness ks in the joint element.

The wall is modeled by the beam element which has a stiffness for bending moment and joint elements are also used on the boundary of soil-wall where the discontinuous plane may be caused. The finite element model for polymer grid reinforced-soil retaining wall is shown in Fig.1.

2.2 Analytical model and procedure

Fig.2 shows the cases to be studied, in which Model 1 is unreinforced one. The height of retaining wall, H, are 4.0 and 8.0m. The spacing, h, of the polymer grid laid in the backfill varies in h/H which are 1.0, 0.5, 0.2, and 0.1, and the length, L, of them varies in L/H which are 0.2, 0.4, 0.6 and 0.8. The end of the polymer grids are anchored to the wall.

The region of analyses is the foundation ground as well as the backfill because the retaining walls are not always built on the firm ground. Fig.3 shows an example of analytical models for the case of h/H=0.1 and L/H=0.8. As shown in Fig.3, the width of ground, B, equals to 4H, and the depth of it, D, equals to H, and the length of wall underneath the ground, Df, equals to 0.15H.

The soil backfill and the polymer grid used in analyses are a beach sand and an uniaxially orientated grid, SR-2, respectively. The relation between them resulted from the laboratory pull-out tests is used in the analyses in order to take account of the mechanism of pull-out resistance of the polymer grid in soil. In the analyses, Duncan-Chang model is used as a soil model. The parameters of soil backfill are determined by the results from laboratory tests and for those of foundation ground the past reference (Mitchell & Gardner 1971) are

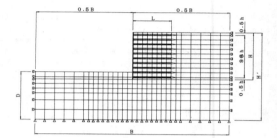

Fig.3 Finite element mesh

Table 1 Soil parameters

	γ' (kN/m³)	c (kN/m²)	ϕ (°)	K	n	R_f	ν
Backfill	17.7	0	38	1000	1.0	0.9	0.333
Foundation	7.8	0	35	300	0.5	0.7	0.333

Table 2 Material parameters

	E (kN/m²)	A (m²)	I (m⁴)
Polymer grid	1.62X10⁶ *	0.0012	—
Wall	2.45X10⁷	0.18	4.86X10⁻⁴

* Initial value

quoted. The material of the wall is a concrete of 18cm in thickness. Soil parameters of the backfill and the foundation ground are shown in Table 1 and material properties of the polymer grid and the wall in Table 2. Constant values of normal stiffness in the joint element are used in compression and tension sides, respectively. The values of compression side, kn1, and tension side, kn2, are 10^6kN/m³ and 10^{-2}kN/m³. The shear stiffness in the joint element used on the soil-wall boundary is constant value of 10^3kN/m³. The deformation analyses by self weight of soil backfill are performed.

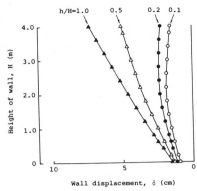

Fig.4 Distribution of wall displacement
(H=4.0m, L/H=0.6)

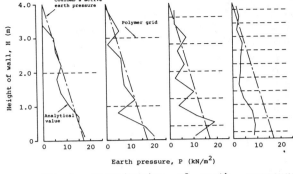

Fig.5 Distribution of earth pressure
acting on the wall (H=4.0m, L/H=0.6)

2.3 Analytical results

(1) Property of the wall deformation and
distribution of the earth pressure
acting on the wall

Fig.4 shows the lateral displacement of
the wall in the case of H=4.0m and
L/H=0.6. The lateral displacement of the
wall decreases as the spacing of the
polymer grid becomes smaller. As for the
distribution of the displacement, when the
spacing is large of h/H=1.0, the
displacement increases almost linearly
from bottom toward top. On the other hand,
in the case of h/H=0.5 the increment of
the displacement in the upper part of the
wall becomes smaller than the lower part
and the shape of the deformation shows a
lower bowed curve, the reason of which may
be thought that the polymer grid laid in
the upper part prevents the wall in the
upper part from displacing. And moreover
as the spacing becomes smaller of h/H=0.2
and 0.1, the inflection point of the
displacement appears near the middle of
the wall and the displacement has the
maximum value at this point and decreases
gradually at the upper part. In this
manner, the shape of lateral displacement
of the wall changes from a linear one to a
bowed one, as the spacing of the polymer
grid becomes smaller.

Fig.5 shows the distribution of the earth
pressure acting on the wall in the case of
H=4.0m and L/H=0.6. In this figure the
solid lines and dot-solid lines indicate
the analytical values and Coulomb's active
earth pressure respectively, and the
broken lines show the positions of the
polymer grid laid in soil. The earth
pressures of the part in which the polymer
grids are laid are decreased by the
reinforcement and the amount of earth
pressures reduced becomes larger as the

spacing of them becomes smaller. When the
spacing is large, there is a difference of
earth pressure between the parts in which
the earth pressures are reduced and the
parts not reduced, so that the
distribution of the earth pressure is
zigzag as a whole. On the other hand, in
the case that the spacing is the smallest
of h/H=0.1, there is less difference of
earth pressure between the parts reduced
and not reduced and the distribution of
earth pressure becomes smooth. The
distribution of earth pressure in this
case is different from a triangular shape,
and it is larger in the upper part of the
wall and smaller in the lower part than
Coulomb's active earth pressure. The
distribution of earth pressure like this
is associated with the condition of the
wall deformation, of which shape is bowed,
shown in Fig.4 and it may be thought that
the soil in the upper part approaches the
condition of the earth pressure at rest
than the soil in the lower part.

(2) Relation between the spacing of the
polymer grid and the reinforcing
effects

Reductions of earth pressures resulted
from the reinforcement in the case of
L/H=0.6 are plotted against the spacing of
the polymer grid in the backfill in Fig.6.
Here, the reduction ratio of the earth
pressures, Rp, is expressed by the
following equation;

$$Rp = 1 - (Pg / Po) \qquad (1)$$

in which Po is total force of Coulomb's
active earth pressure acting on the wall
in the unreinforced case and Pg is that of
earth pressure resulted from the analysis
with reinforced soil backfill. And the
spacing, h, is normalized by the wall
height, H, and the ratio, h/H, is used as
an abscissa in Fig.6. The relation between

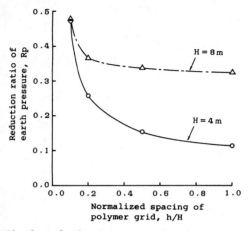

Fig.6 Relation between reduction ratio of earth pressure and spacing of polymer grid (L/H=0.6)

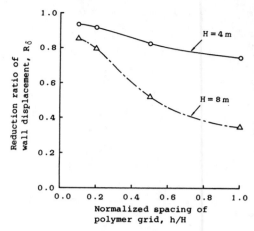

Fig.7 Relation between reduction ratio of wall displacement and spacing of polymer grid (L/H=0.6)

the reduction ratio of earth pressure and the normalized spacing of the polymer grid is like a hyperbola and the reduction ratio of earth pressure decreases rapidly as the spacing becomes smaller. And although the values of reduction ratio are different due to the difference of the wall height, these differences become smaller as the spacings become smaller and reduction ratios in both height become almost equal in the case of h/H=0.1.

Fig.7 shows the relation between the reduction ratio of lateral displacement of the wall, R_δ , and the normalized spacing of the polymer grid, h/H, in the case of L/H=0.6. Here, the reduction ratio, R_δ , is expressed by the following equation;

$$R_\delta = 1 - (\delta g / \delta o) \qquad (2)$$

where δg and δo are maximum values of the lateral displacements of the walls with and without reinforced soil backfill, respectively. The reduction of the wall displacement by the reinforcement of backfill becomes larger as the spacing becomes smaller. And although the values of reduction ratio are different due to the difference of the wall height, these differences become smaller as the spacings become smaller as well as the case of the earth pressure.

(3) Relation between the length of the polymer grid and the reinforcing effects
Reductions of earth pressures resulted from the reinforcement in the case of h/H=0.1 are plotted against the normalized length of the polymer grid in the backfill in Fig.8. The reduction ratios of the earth pressures have the almost same

values independently of the wall height, and they show the salient distributions, which have a peak at L/H=0.6, against the normalized length of the polymer grid. Accordingly, from the results of Figs.6 and 8, it may be said that the earth pressure acting on the wall can be reduced about 50 percent at a maximum against Coulomb's active earth pressure.

Fig.9 shows the relation between the reduction ratio of the wall displacement and the normalized length of the polymer grid in the case of h/H=0.1. Although the values of reduction ratio are different due to the difference of the wall height, both of them increase as the length of the polymer grid becomes longer.

By the way, as shown in Fig.10, the earth pressure acting on the wall is changed according to the condition of the wall deformation and it shifts from the condition of the active earth pressure to the condition of the earth pressure at rest as the wall displacement becomes smaller. On the other hand, the earth pressure is reduced by the tensile force acting on the polymer grid. And so it may be supposed that finally the earth pressure acting on the wall becomes such a distribution as shown by the solid line in Fig.10. Whereas it seems that the amount of the earth pressure reduced, which is shown by the shade in Fig.10, becomes larger as the length of polymer grid becomes longer, at the same time the distribution of the earth pressure shown by the broken line in Fig.10 approaches rapidly to the distribution of the earth pressure at rest according to the reduction of the wall displacement. And

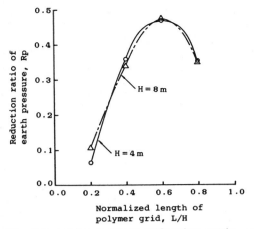

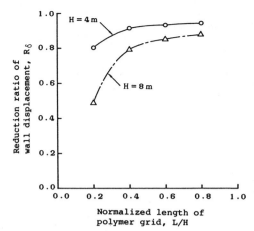

Fig.8 Relation between reduction ratio of earth pressure and length of polymer grid (h/H=0.1)

Fig.9 Relation between reduction ratio of wall displacement and length of polymer grid (h/H=0.1)

then the final earth pressure acting on the wall seems to be determined by the quantitative relation between them. It can be supposed that the reason why the reduction ratio of the earth pressure based on Coulomb's active earth pressure without reinforced soil backfill decreases between L/H=0.6 and L/H=0.8 is that the increment of the earth pressure due to the reduction of the wall displacement is larger than that of the earth pressure reduced by the reinforcement. Accordingly when the length of the polymer grid becomes longer than some value, the rear part of the polymer grid plays the part of the anchor and reduces the wall displacement, and at the same time the distribution of the earth pressure approaches the condition of the earth pressure at rest. And so it can be said that there exists the length of the polymer grid which minimizes the amount of the earth pressure acting on the wall.

3 APPLICATION OF THE ANALYTICAL RESULTS TO THE DESIGN METHOD

Up to this time, in the design of the retaining wall with polymer grid reinforced soil backfill, the stability for sliding, overturning and bearing failure etc. has been investigated by assuming that a reinforced soil wall behaves as a rigid gravity structure. It is the investigation of stability based on what is called the rigid-plasticity theory and does not mention anything about the displacement and deformation of the wall and reinforcing material caused by the

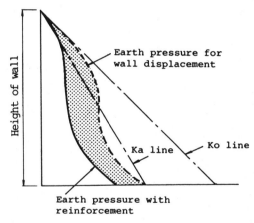

Fig.10 Distribution of earth pressure for wall deformation and reinforcement

self weight of soil backfill and surcharge. However, in the practical constructions the deformation of the structure is not seldom an important design factor because of the recent works with adjacencies in the city and importance of a fine view based on the environmental assessment. And then from now on, it will become a problem in the design how large the wall deformation is. In other words, the check of allowable displacement of the wall will be required.

The numerical methods for which the finite element method stands are one of the powerful methods to grasp the deformation of the structure. However, there is a limitation of time and cost to

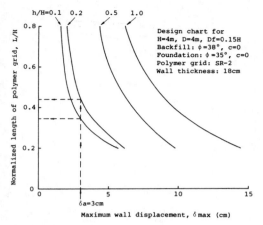

Fig.11 Design chart for checking wall displacement

analyse by such numerical methods on the design condition in each case, so that it seems to be very significant that the procedure to investigate the amount of deformation easily and quickly for the comparatively simple design condition is proposed. The procedure to check the amount of deformation of the reinforced soil wall using the design chart obtained from the analytical results is presented as the following. This is useful to investigate the wall deformation in the stage of the preliminary design.

At first define the design condition of the problem. That is, define geometry such as the height of wall, H, the length of wall underneath the ground, Df, and the depth of ground, D, and the material properties such as parameters of soil backfill and foundation ground and those of polymer grid and wall. To illustrate the design procedure, consideration will be given to the condition in the analyses mentioned above, as shown in Fig.11.

Next, determine the spacing and the length of the polymer grid for allowable displacement of the wall. The relations between the maximum lateral displacement of the wall and the length and spacing of the polymer grid obtained from the analytical results for given design conditions are shown in Fig.11. The design chart is entered for a known value of allowable displacement, moving vertically until the normalized spacing, h/H, is reached, and then sidewards; the required normalized length of the polymer grid, L/H, may then be read off. When there are more than one value of L/H for a given allowable displacement, the case in which the total length of the polymer grid per unit height, 1m, of the wall, Lt/H, is the smallest should be selected. In the case of an example shown in Fig.11, the required values of (L/H,h/H) for a given allowable displacement, δa=3cm, are (0.44,0.2) and (0.33,0.1). Therefore the value of Lt/H are 2.2 and 3.3 m/m, respectively, and then L/H=0.44 and h/H=0.2 are adopted finally.

4 CONCLUSIONS

The leading conclusions are summarized as the following.

(1) The earth pressure acting on the wall decreases and the distribution of it becomes smooth, as the spacing of the polymer grid becomes smaller.

(2) When the spacing of the polymer grid is small of h/H=0.1, the earth pressure acting on the wall is controlled by the length of the polymer grid independently of its spacing and is minimized in the case of L/H=0.6 and its value is about 50 percent of Coulomb's active earth pressure.

(3) The wall displacement decreases, as the spacing of the polymer grid becomes smaller and the length of it becomes longer. And the shape of the wall deformation changes from a linear one to a bowed one, as the spacing of the polymer grid becomes smaller.

(4) The procedure to investigate approximately the amount of wall deformation in the stage of preliminary design is presented using the design chart in which the relation resulted from analyses between the maximum displacement of the wall and the length and spacing of the polymer grid is shown.

REFERENCES

Hayashi, S. et al 1985. Mechanism of pull-out resistance of polymer grids in soils, TSUCHI-TO-KISO, JSSMFE, Vol.33, No.5: 21-26.
Mitchell, J.K. & Gardner, W.S. 1971. Analysis of load-bearing fills over soft subsoils, J. Soil Mech. and Found. Div., ASCE, SM11: 1549-1571.
Netlon Ltd. 1984. Guidelines for the design and construction of reinforced soil retaining walls using 'Tensar' geogrids.
Ochiai, H. & Sakai, A. 1987. Anatytical methods for geogrid-reinforced soil structures, Proc. 8th Asian Regional Conf. on SMFE: 483-486.

International Geotechnical Symposium on Theory and Practice of Earth Reinforcement / Fukuoka Japan / 5-7 October 1988
© 1988 Balkema, Rotterdam. ISBN 90 6191 820 0

Theory and practice of 'Reinforcing' steep slopes with nonwoven geotextiles

S. Resl, H. Schneider & G. Werner
Polyfelt GmbH, Linz, Austria

ABSTRACT: A new design theory for geotextile reinforced earth structures is presented combining horizontal reinforcement and the effect of earth filled bags as gravity retaining wall. Beside the principle and a design example resulting in a much smaller required tensile strength of the reinforcing geotextile, two important topics are dealt with: the improvement of shear strength of the fill material by compaction and drainage in the geotextile plane, and the long-term in-soil stress-strain characteristic of nonwoven needlepunched geotextiles. Finally, some practical examples of constructed walls using a nonwoven needlepunched PP-endlessfibre geotextile are reported.

1. INTRODUCTION

Reinforcing steep slopes or earth walls by the installation of tensile resistant components is a very old construction method: Thousands of years ago, reed or willow branches have been used for this purpose, e.g. in some parts of the Great Chinese Wall. However, this system has become technically impeccable when rot-resistant materials have been used, especially various kinds of "geosynthetics".

Geogrids, woven and nonwoven geotextiles are used to "reinforce" steep slopes. The word "reinforce" is set between quotation marks, as the relatively high elongation at break of these materials, especially nonwoven geotextiles, makes it difficult to calculate these systems by conventional design procedures, and as the actual "reinforcement"-mechanism is not yet clearified.

2. DESIGN

2.1. Conventional Design Procedures

Various design methods have been developped, which are all generally based on introducing tensile forces into the calculations. The two basic methods are:

a) Introducing horizontal forces into a slip circle or block sliding analysis (see Fig. 1a)
b) Taking up the horizontal earth pressure by tensile forces (see Fig. 1 b)

These methods have been modified slightly by various authors, trying to approach the design to the actual failure mechanisms as close as possible. Without any respect to technical accuracy, these methods allow a quick and safe approximate design of geotextile reinforced walls, being a highly economical alternative to other retaining structures, even when highly extensible nonwoven geotextiles with a relatively low tensile strength are used, as shown in Fig. 2 according to Studer/Meier (1986) and Chemie Linz/Polyfelt (1986).

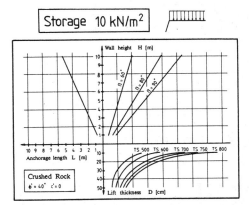

Fig. 2: Example for design charts based on conventional stability analysis acc. Chemie Linz/Polyfelt (1986)

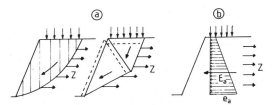

Fig. 1: Basic design procedures for reinforced slopes

2.2. Reflections on a New Design Theory

This theory is based on a combined functional mechanism of horizontal reinforcement and a gravity retaining structure. Without gravity retaining wall, the resulting tensile forces Z must be so high that the resulting force R from active pressure E and tensile forces Z is transfered into the basement (see Fig. 3).

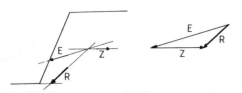

Fig. 3: Forces without gravity retaining wall

When a gravity retaining wall with the weight G is placed in front of the slope, the required tensile forces Z to transfer the resulting force R into the soil can be reduced significantly (see Fig. 4).

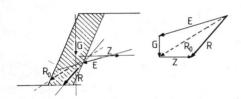

Fig. 4: Forces with gravity retaining wall

Gravity retaining walls are usually concrete walls or gabion walls. However, also flexible structures can act as a gravity retaining wall, when its internal stability is guaranteed. When looking at the construction procedure proposed for example by Chemie Linz/Polyfelt (1986) where completely closed earth filled bags are installed at the edge of the wall (see Fig. 5), it can be assumed that these "earth bags" act as a gravity retaining wall.

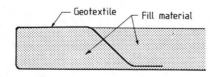

Fig. 5: "Earth bag" at the edge of geotextile reinforced wall

For checking the internal stability of the earth bag wall the following types of failure have to be considered (see Fig. 6):
a) Overturning around the edge point in every level.
b) Horizontal sliding in every level; for this type,

the friction angle between soil and nonwoven needlepunched geotextiles can be assumed as $\delta = 0,9 \div 1,0. \varphi$, whereas for heatbonded nonwovens and wovens with their smoother surface this value lies within $\delta = 0,6 \div 0,9. \varphi$, as proven by various authors, e.g. Richards/Scott (1985).
c) Internal stability of each bag.

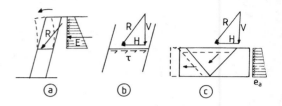

Fig. 6: Possible types of internal failure of the earth bag wall

The most critical failure type seems to be c), the internal stability of each bag. Fig. 7 shows possible configurations of the active failure zone and the involved forces.

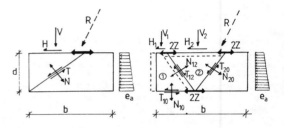

Fig. 7: Possible configurations of the active failure zone and the involved forces

Assuming the relationship between N and T as

$$T = 1 . c + N . tg \, \varphi$$

where
1 ... length of slip line
c ... cohesion
φ ... friction angle

the required tensile forces can be calculated. However, there are still some open questions:
a) How are the vertical and horizontal forces from the resulting force R distributed over the width b?
b) What friction angle φ and cohesion c can be assumed?
c) What tensile strength can be introduced in the calculations?

Calculations have shown that question a), the stress distribution over the width b, is of utmost importance for the stability. In concrete retaining walls, the stresses from the resulting force R is distributed triangularly, as shown in Fig. 8a. In the case of earth filled bags, the stiffness modulus of the bags is equal to that of the surrounding soil, therefore a part of the resulting force R is taken up by the soil (see Fig. 8b).

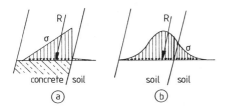

Fig. 8: Distribution of the resulting force R

The following sections 2.3. and 2.4. deal with questions b) and c).

As an example, the retaining wall illustrated in Fig. 9 has been designed. According Schulze/Simmer (1977) the active earth pressure coefficient λ_a can be calculated as

$$\lambda_a = \frac{\cos^2(\varphi + \alpha)}{\cos^2 \alpha \cdot \cos(\delta - \alpha)\left[1 + \sqrt{\dfrac{\sin(\varphi + \delta) \cdot \sin(\varphi - \beta)}{\cos(\alpha - \delta) \cdot \cos(\alpha + \beta)}}\right]^2}$$

with $\alpha = 10°$, $\beta = 0°$ and $\delta = \varphi = 30°$ this leads to $\lambda_a = 0{,}227$, resulting in the earth pressure distribution illustrated in Fig. 9.

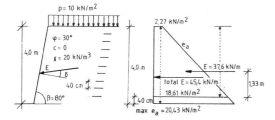

Fig. 9: Design example: given data and earth pressure

According to the different design methods, the following calculatory tensile strengths Z are required:

a) slip circle analysis acc. Chemie Linz/Polyfelt (1986) by using the design charts and taking into account the used factors of safety (1,3 for soil parameters, 3,0 for geotextile tensile strength) --> Z = 7,0 kN/m
b) by taking up total earth pressure:
   Z = total E/n = 45,4/9 = 5,0 kN/m
c) by taking up maximum earth pressure:
   Z = max e_a . d = 20,43 . 0,40 = 8,2 kN/m
d) acc. "earth bag wall theory" described in this paper --> Z = 3,0 kN/m

Fig. 10 shows the principles and assumptions of the

design acc. d). The single rupture of the lowest bag is assumed to be the most critical type of failure.

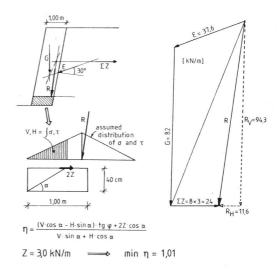

$$\eta = \frac{(V \cdot \cos \alpha - H \cdot \sin \alpha) \cdot \text{tg } \varphi + 2Z \cdot \cos \alpha}{V \cdot \sin \alpha + H \cdot \cos \alpha}$$

Z = 3,0 kN/m $\implies$ min η = 1,01

Fig. 10: Principles, assumptions and results of the design-example

Based on the simplifying assumptions illustrated in Fig. 10, a computer program has been developped for designing reinforced earth walls according to this new method.

2.3. Shear Strength Characteristics of the Geotextile Wrapped Soil

The actual shear strength situation inside the earth bags can be stated as higher than in the surrounding soil. The reasons for this can be found in various factors:

a) A better compactability:
   As shown by Tatsuoka et al (1986) and Werner/Resl (1986) the friction between soil and nonwoven geotextile reduce the lateral movement of the soil grains, resulting in a better compaction and thus in an increase of shear strength.
b) The introduction of a 3-dimensional state of stresses, leading to much higher allowable shear stresses, acc. Werner/Resl (1986)
c) The drainage function of the geotextile:
   Especially with cohesive fill material the drainge function is of great importance, in order to drain off pore water during compaction and consolidation as well as seepage water caused by rainfall or by groundwater flow, see Tatsuoka et al (1986).
   The positive effect of the drainage function has also been demonstrated by Fabian/Fourie (1987) in traxial tests, where various types of geotextiles have been installed horizontally in the middle of the soil sample. The tests have shown that "high

permeability geotextiles" (needlepunched nonwovens) show a higher increase in shear strength as "low permeability geotextiles" (heatbonded nonwovens, wovens), which show in some cases even a decrease! Therefore the lower tensile strength of needlepunched nonwovens seems to be overcompensated by their transmissivity.

2.4. In-Soil Tensile Strength of the Geotextiles

The question which tensile strength should be inserted into the stability calculations is influenced by 3 factors:
a) Factor of safety
b) Stress-strain-curve in soil confinement
c) Long-term behaviour

ad a)
When needlepunched nonwovens are used as reinforcing elements, a factor of safety FS = 3,0 is recommended, e.g. by Studer/Meier (1986) and Chemie Linz (1986), as long as a more detailed analysis of the functional mechanisms is not possible.

ad b)
The stress strain characteristic of needlepunched nonwovens is characterized by slippage and straightening of the fibres and fibre obliquity, acc. Hearle (1972). Due to the interlocking effect between soil and geotextile (see Fig. 11) the fibre slippage is reduced and therefore higher strength and lower elongation are yielded compared with the standard tensile test where the geotextile is examined without soil confinement, as proven by Fock/McGown (1987), and illustrated in Fig. 12.

Additionally, the stiffness of the geotextile is increased by preloading during compaction, acc. Studer/Meier (1986).

ad c)
As the load is sustained over a long period of time, the long-term behaviour is of importance for the stability of the structure. Many authors are concerned about the different long-term behaviours (i.e. creep) of the different raw materials, namely polyester and polypropylene, which is undenieable testing the geotextile without soil confinement. Fock/McGown (1987) however have shown that embedded in soil creep is no relevant factor even when polypropylene fabrics are used (see Fig. 13).

3. PRACTICAL EXPERIENCE

Numerous projects have been carried out using a nonwoven needlepunched PP-endless-fibre geotextile with the brand-name "Polyfelt TS". The construction was done according the recommendations by Chemie Linz/ Polyfelt (1986) and as illustrated in Fig. 14.

The design was in all cases based on conventional design methods, as basically described in section

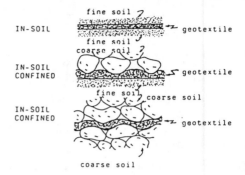

Fig. 11: Influence of in-soil confinement

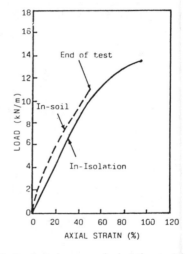

Fig. 12: Load strain curves in isolation and in-soil

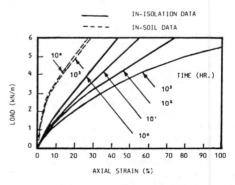

Fig. 13: Isochronous load-strain curves for a needlepunched PP-nonwoven, acc. Fock/McGown (1987)

568

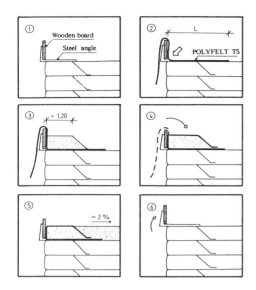

Fig. 14: Construction procedure

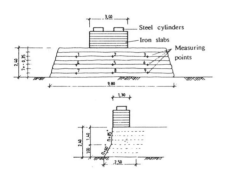

Fig. 15: Test embankment acc. Werner/Resl (1986)

Fig.16: Temporary retaining wall, Queensland/Australia

Fig. 17: UV-protection by planting, Steyr/Austria

2.1. This may lead to an overdesign, as shown in section 2.2., and Werner/Resl (1986), where a test embankment showed a factor of safety FS = 0,67, calculated with conventional methods, and no sign of failure, high deformations or creep have been observed (see Fig. 15).

Nevertheless, the economical benefits have been undenieable:
- low material costs
- low transportation costs
- the in-situ material can be used as fill material
- easy installation with unskilled workers and no heavy equipment.

When the retaining wall has to fulfill its function permanently and not only temporary (Fig. 16), a UV-

Fig. 18: UV-protection by shotcreting; storage yard, Linz/Austria

protection has to be provided. Possible methods are planting (Fig. 17), shotcreting (Fig. 18), non-constructive brick walls etc.

4. CONCLUSION

Even geotextiles with low modulus offer an economical method of "reinforcing" steep slopes in spite of their relatively low tensile strength. The described stability mechanism combining horizontal reinforcement and "earth bags" as a gravity retaining wall tries to give a more detailed approach to the actual stress situation. An example showed that the required tensile strength is up to 65 % lower than calculated with conventional design methods.

Additional mechanisms, especially the increase of shear strength of the geotextile wrapped soil due to compaction and drainage in the plane of the fabric have to be more closely analyzed in future research.

REFERENCES

Chemie Linz/Polyfelt (1986): "Polyfelt TS - Design and Practice", issued by Chemie Linz AG, Linz/ Austria, chapter 12.

Fabian K./Fourie A. (1987): "Current Research in Clay-Geotextile Interaction", Australian Geomechanics, pp. 14-21.

Fock G./McGown A. (1987): "The Creep Behaviour of a Needlepunched Polypropylene Geotextile", 9th Southeast Asian Geotechnical Conference, Bangkok, pp. 8-101 to 8-112.

Hearle J.W.S. (1972): "A Theory of the Mechanics of Needle Punched Fabrics", in Needle Felted Fabrics, P. Lennox-Kerr, Manchester/U.K., pp 51-64

Richards E.A./Scott J.D. (1985): "Soil Geotextile Frictional Properties", 2nd Canadian Symposium on Geotextiles and Geomembranes, Edmonton, pp. 13-24.

Schulze W.E./Simmer K (1977): "Grundbau", B.G. Teubner, Stuttgart

Studer J.A./Meier P. (1986): "Earth Reinforcement with Non-Woven Fabrics: Problems and Computational Possibilities", 3rd Int. Conference on Geotextiles, Vienna, pp. 361-365

Tatsuoka F./Ando H./Iwasaki K./Nakamura K. (1986): "Performances of Clay Test Embankment Reinforced with a Nonwoven Geotextile", 3rd Int. Conference on Geotextiles, Vienna, pp. 355-360.

Werner G./Resl S.(1986): "Stability Mechanisms in Geotextile Reinforced Earth-Structures", 3rd Int. Conference on Geotextiles, Vienna, pp. 1131-1135

International Geotechnical Symposium on Theory and Practice of Earth Reinforcement / Fukuoka Japan / 5-7 October 1988
© 1988 Balkema, Rotterdam. ISBN 90 6191 820 0

A review of field measurements of the behaviour of geogrid reinforced slopes and walls

Pietro Rimoldi
RDB Plastotecnica, Vigano, Italy

ABSTRACT: The paper present a review of papers dealing with field measurements for geogrid reinforced walls and slopes. Some comparisons and hypothesis are presented, in order to explain the field tests results. A proposal for the future research ends the paper.

1 INTRODUCTION

For some products used for earth reinforcement there are a good number of field and laboratory data currently available, particularly for steel bars and geotextiles. Instead, geogrids are currently well characterized by means of many laboratory tests which make the design engineers confident with the product, but both engineers and researchers are not confident with the mechanism of interaction between geogrids and soil structures, mainly because there are few field data from instrumented sections which can support design hypothesis. With the published papers on field measurements for geogrid reinforced soil, it is not possible to draw final conclusions but only to do some considerations to help the ongoing research. This is the aim of this paper.

2 DESIGN METHODS

The two design methods currently more widely used for geogrid reinforced soil projects are: the two-part wedge method for steep slopes (Jewell et al 1984), and the tie-back wedge method for vertical walls (Jones 1985). The method of Jewell for steep slopes is based on design charts which allow to obtain the coefficient of earth lateral pressure K* and the reinforcement length L, as a function of the slope angle β and the angle of internal friction Φ', for a given value of the pore pressure ratio Ru. The charts were obtained from an extensive computer calculation of limit equilibrium based on two-part wedge failure mechanism. It assumes that the foundation is stable and the cohesion of the fill soil is equal to

zero. The tie-back wedge method assumes a kinematic mechanism for wall movement of outward rotation around the toe; consequently it assumes that the active earth pressure is mobilized along the Rankine theoretical failure surface. The earth pressure is assumed to be resisted by the tensile forces developed in the geogrid layers, neglecting the resistance given by the wall face. In both methods the direct sliding on geogrids and the pull-out resistance are considered for the design of the length of reinforcement, based on a vertical pressure distribution assumed to be uniform of Meyerhof type.

3 CHECKING THE DESIGN

In order to check the currently available design methods or in order to develope new and more accurate design methods, the following data are needed:
- the distribution of σ_v and σ_h in the reinforced soil mass: they are essential for the proper calculation of settlements, of forces and displacements induced by direct sliding and pull-out mechanisms, and of the thrust to be resisted by the reinforcement;
- mechanism of soil reinforcement: can the soil develop the active pressure? how is the relative influence of interlocking and tensile resistance of geogrids?
- kinematic mechanism: in which cases it can be assumed to be a rotation around the toe or a two-part wedge rigid movement;
- distribution of tensile force along each geogrid layer: it will allow a correct evaluation of geogrid elongation and the comparison with the allowable tensile strength;
- diagram of geogrid strain versus time: these data will allow to know the in-soil

Table 1. List of the considered field tests.

No.	Location	Reference	Type	Description
1	Kagoshima (Japan)	Yamanouchi et al 1986	Wrap-around slope	78° slope, 6 m high, well graded volcanic sand, geogrid wrapped around sand bags at face;
2	Tucson (Arizona)	Berg et al 1986	full height panels concrete face wall	vertical wall, 1 ÷ 6 m high, sand soil, full height precast concrete facing panels;
3	Lithonia (Georgia)	Berg et al 1986	concrete face incremental wall	vertical wall, 6 m high, gravel soil, cruciform precast concrete facing panels;
4	Stanstead Ab-botts (U.K.)	Bassett et al 1986	embankment on weak soil	embankment on peat soil, with one geogrid at base to distribute the load;
5	Gaspe Penin-sula (Canada)	Berg et al 1987	concrete face in-cremental seawall	vertical wall for wave protection, 5.3 m high, sand-gravel soil, con-crete panels of different shapes;
6	Cascade Dam (Michigan)	Christopher 1987	wrap-around vertical wall	retaining wall supporting a heavy weight crane, 3 m high, sand-gra-vel soil, wrap-around geogrids;
7	Kingston (Canada)	Bathurst et al 1987	full height panels timber face wall	large scale laboratory test, 3 m high, sand soil, variable surchar-ge, full height timber panels;
8	Kingston (Canada)	Bathurst et al 1987	timber face incre-mental wall	large scale laboratory test, 3 m high, sand soil, variable surchar-ge, square timber panels;
9	Modena (Italy)	Cazzuffi et al 1988	steep sides embankment	embankment with 45° ÷ 60° side slopes, clay soil, wrap-around geogrids.

Table 2. Instruments used in the field tests

No. Location	Horizontal movement of face	Internal horizontal movements	Foundation settlements	Total stresses σ_h	Total stresses σ_v	Pore pres-sure	Strains in geogrids	Forces in geogrids
1 Kagoshima	Survey reference points	–	–	–	–	–	strain gauges	–
2 Tucson	Survey reference points	–	–	load cells at face	load cells at base	–	strain gauges, inductance coils	–
3 Lithonia	Survey reference points	–	–	load cells at face	load cells at base	–	inductance coils	–
4 Stanstead	–	Horizontal profile gauge at base	Horizontal profile gauge, in-clinometers	–	total pres-sure cells	p. p. gauges, pneum. piezom.	strain gauges, inductance coils	clamp load cells
5 Gaspe	Survey reference points	Multiple positions extensome-ter	–	total stress cells	–	pneum. piezom.	strain gauges	clamp load cells
6 Cascade	Survey reference points	wire extensome-ter	–	load cells at face	–	–	inductance and resi-stance strain g.	–
7 Kingston	electrical displacem. measuring device	–	–	–	earth press. cells at base	–	strain g. and tensio-ned steel wires	load cells at face
8 Kingston	electrical displacem. measuring device	–	–	–	earth press. cells at base	–	strain g. and tensio-ned steel wires	load cells at face
9 Modena	–	–	settlements plates	–	–	–	strain gauges	clamp load cells

creep and consequently to fix a realistic limit to the tensile strength.
These data are needed for different types of load (uniform, linear, concentrated), for their practical range of values (for uniform surcharge it may be $0 \div 100$ kPa).

4 FIELD TESTS AND INSTRUMENTATION USED

The field tests with published results are reported in Tab. 1: as it can be seen, two tests refer to steep slopes or embankments, one refers to an embankment on weak soil and six to vertical walls. The instruments used for each type of measurement in the considered field tests are described in Tab. 2. As it can be seen, in any case all the parameters were measured. Only in three cases both strains and tensile forces were recorded along some geogrid layers. The horizontal pressure was measured only at the face of the wall, but not in the soil mass and particularly not in the zone of maximum geogrid tensile forces. In each field test there were some interesting findings but, due to the little number of cases, they cannot be correlated in a general rule.

5 RESULTS

Tab. 3 shows a comparison of the results obtained in the considered field test; Stanstead Abbotts case was omitted in this list, since it is not really a slope reinforcement project but a base reinforcement one.
The values reported in Tab. 3 refer to three different conditions:
– Design condition (D): design values of parameters, used also to calculate strains ( ) and forces (α) in the geogrids;
– Calculation condition (C): real-situation-values of parameters, used also to calculate strains ( ) and forces (α) in the geogrids;
– Measured condition (M): the actual values. In Tab. 3 the subscript D, C, M are used to indicate the above situations; symbols and formulas used are reported at the bottom.
In order to compare the results of the different field tests, some ratios, which can be regarded as partial Factors of Safety, were used; just from the analysis of Tab. 3, the following considerations can be drawn:
– the calculation with real-situation-values of parameters can give a gross estimate of actual forces and strains in the geogrids (FS1 = $1.0 \div 2.0$), but if the soil is cohesive (Modena) they can be largely overestimated: this fact suggests that the mechanism of reinforcement for cohesive soils is slightly different from non-cohesive ones;
– a value of FSg of $1.1 \div 1.7$ can lead to anticipate in the design far greater tensile

forces than the actual ones (FS3 = $2.5 \div 10$), and to a high ratio between allowable tensile strength and actual tensile forces (FS5 = $2.4 \div 13$): this fact suggests that the actual design methods are conservative and can be used with a certain confidence;
– the previous consideration is not true if high line loads occur: in this case it's better to have more geogrid layers near the top than required from current design methods (Christopher 1987).
With a more accurate analysis of the published papers, all the cases can be reduced to three categories, based on the type of surcharge and of diagram of tensile forces:
Category 1) – slopes or embankment sides: as shown in Fig. 1, the tensile force has a smooth distribution along the geogrids; the maximum value is not at the base but in one of medium-low layers. If the slope angle β increases, the ratio $(x/H)^\wedge$ decreases, in fact: $(x/H)^\wedge \approx 0.30$ if $\beta = 45°$ (Modena); $(x/H)^\wedge \approx 0.15$ if $\beta = 78°$ (Kagoshima). These results seem to be in agreement with the theory based on two-part wedge failure mechanism, but with cohesive soils the forces in the geogrids are lower than the calculated ones. If important settlements occur, then the tensile force in the base geogrid can arrive to relatively high values like $15 \div 20$ kN/m (Bassett et al 1986).
Category 2) – vertical walls with uniformly distributed surcharge: as shown in Fig. 2 and Fig. 3, the tensile force diagram is different if the ratio $(q/\gamma H)$ is lower or greater of 0.8. In the former case there is a smooth pattern, with tensile forces increasing from top to bottom geogrid layer; the locus of maximum tensile force is a surface with an inclination $\theta = (45° + \Phi'/2)$, coinciding with Rankine failure surface (Berg et al 1987). In the latter case, the tensile force diagram has a pronounced peak and the upper layers are more in tension than the lower ones; the locus of maximum tensile force is a surface close to the vertical ($\theta = 0° \div 20°$), with $(x/H)^\wedge \approx 0.15$. So, if the base can be considered not subjected to lateral movements and $(q/\gamma H) < 0.8$, then the failure mechanism is a rotation around the toe and the results seem in agreement with the tie-back wedge method. For higher values of the surcharge, the real mechanism needs to be further investigated.
Category 3) – vertical walls with high line loads: as shown in Fig. 4, the failure surface is approximately a logarithmic spiral, starting from the rear edge of the line load (Christopher 1987). The tensile forces are greater for the upper geogrid layers than for the lower ones, and for the upper layers forces can be greater than calculated with real-situation-values of parameters. With reference to the lateral pressure deve-

Table 3. Results of field tests

D = Design conditions; C = calculation conditions; M = measured conditions

No. Location	1 Kagoshima			2 Tucson			3 Lithonia			5 Gaspe			6 Cascade			7 Kingston			8 Kingston			9 Modena		
Geogrid used	Tensar SR2 (HDPE)			Tensar SR2 (HDPE)			Tensar SR2 (HDPE)			Tensar SR2 (HDPE)			Signode TNX250 (Polyester)			Tensar SR2 (HDPE)			Tensar SR2 (HDPE)			Tenax TT1 (HDPE)		
	D	C	M	D	C	M	D	C	M	D	C	M	D	C	M	D	C	M	D	C	M	D	C	M
H (m)	6.0	7.0	7.0	4.65	4.65	4.65	6.0	6.0	6.0	5.3	5.3	5.3	3.0	3.0	3.0	3.0	3.0	3.0	3.0	3.0	3.0	5.0	5.0	5.0
β (deg)	78	78	78	90	90	90	90	90	90	90	90	90	90	85	85	90	90	90	90	90	90	45	45	45
q (kPa)	9.8	0	0	12	12	?	12.9	12.9	?	37	37	37	80	80	80	12 / 50	12 / 50	12 / 50	12 / 50	12 / 50	12 / 50	10	10	0
γ (kN/m³)	17.7	14.6	14.6	19.6	19.6	19.6	21.2	21.2	21.2	21.0	21.0	21.0	19.6	19.6	19.6	17.6	17.6	17.6	17.6	17.6	17.6	20.0	17.2	17.2
FS_g (-)	1.70	1.0	1.0	1.11	1.0	1.0	1.16	1.0	1.0	1.16	1.0	1.0	1.30	1.0	1.0	?	?	?	?	?	?	1.10	1.0	1.0
Φ' (deg)	30	45	45	40	43	43	34	38	38	34	38	38	30	37	37	?	43	43	?	43	43	23	25	25
c' (kPa)	0	0	2.45	0	0	0	0	0	0	0	0	0	0	0	0	0	0	0	0	0	0	0	0.25	0
Ru (-)	0	0	0	0	0	0	0	0	0	0	0	0	0	0	0	0	0	0	0	0	0	0	0	0
α_f (kN/m)	78.5	-	-	80.0	-	-	80.0	-	-	80.0	-	-	50.0	-	-	80.0	-	-	80.0	-	-	66.0	-	-
α_ω (kN/m)	31.4	-	-	29.0	-	-	29.0	-	-	21.5	-	-	(40)	-	-	29.0	-	-	29.0	-	-	30.0	-	-
N (-)	6	6	6	10	10	10	13	13	13	13	13	13	10	10	10	4	4	4	4	4	4	5	5	5
K* (-)	0.27	0.11	?	0.28	0.24	?	0.22	0.19	?	0.28	0.23	?	0.34	0.25	?	?	0.18	?	?	0.18	?	0.38	0.15	?
δ (cm)	-	7.1	≈18	0.5	-	6.5	12.5	-	6.5	-	-	4.8	-	-	(25)	-	0.3; 1.5	2.3; 3.8	-	0.3; 1.5	2.3; 3.8	-	-	?
ε (%)	≈2.0	≈0.3	≈0.3	≈0.4	≈0.3	0.2÷0.6	≈0.4	≈0.3	?	≈1.2	≈1.0	0.75	0.85	0.75	1.0	?	0.6; 2.0	0.1; 0.9	?	0.6; 2.0	0.1; 0.9	3.7	1.2	0.2
α (kN/m)	17.5	6.6	2.9÷6.9	7.6	6.5	2.5÷7.0	7.8	6.7	?	11.3	9.3	3.8÷8.8	14.0	12.3	16.5	?	5.4; 13.5	2.5; 7.0	?	5.4; 13.5	2.5; 7.0	23.0	6.4	2.2
FS1	0.9 ÷ 2.3			0.9 ÷ 3.2			?			1.05 ÷ 2.4			0.75			2.16; 1.93			2.16; 1.93			2.9		
FS2	2.7			1.2			1.16			1.21			1.14			?			?			3.6		
FS3	2.5 ÷ 6.1			1.2			?			1.28 ÷ 2.9			0.85			?			?			10.9		
FS4	11.3 ÷ 27.0			11.5 ÷ 32.0			?			9.10 ÷ 21.0			3.0			32.0; 11.4			32.0; 11.4			27.2		
FS5	4.5 ÷ 10.8			4.2 ÷ 11.6			?			2.40 ÷ 5.6			(2.4)			11.6; 4.1			11.6; 4.1			13.6		

Reported data: height of slope or wall H; slope angle β; uniform (or equivalent) surcharge q; unit weight of fill soil γ; global Factor of Safety FSg; angle of internal friction Φ' and cohesion c' of fill soil; pore pressure parameter Ru; peak and allowable tensile strength of geogrids used, α_f and α_ω; number of geogrid layers N; coefficient of lateral soil pressure K*; outward movement of top of slope or wall face δ; maximum tensile strain and force along geogrids, ε and α.

Formulas used for the conditions D and C:
- $\Phi' = \Phi'_m / FS_g$
- $K^* = f(\Phi', \beta, Ru)$ from design charts for slopes (Jewell et al, 1984)
- $K^* = \tan^2(45 - \Phi'/2)$ for walls
- required total tensile force:
  $T = \tfrac{1}{2} \cdot K^* \cdot \gamma \cdot (H + q/\gamma)^2$
- $\alpha = T/N$
- $\varepsilon = f(\alpha)$ measured from stress strain curve of geogrid.

- ε_M = from measurements
- α_M = measured or from stress-strain curve of the geogrid as a function of ε_M.

Partial Factors of Safety:
- FS1 = α_c / α_M
- FS2 = α_D / α_c
- FS3 = α_D / α_M
- FS4 = α_f / α_M
- FS5 = α_ω / α_M

loped against the face, an hypothesis can be done based on the results of Lithonia wall measurements (Berg et al 1986): as shown in Fig. 5, it seems that the actual lateral pressure diagram is parallel to the Rankine theoretical line; this fact suggests that the lateral constriction due to interlocking, typical of geogrids, makes the soil to have a fictitious cohesion $c^\wedge$, which shifts the lateral pressure diagram downward. So the thrustes against the wall, and therefore the forces in the geogrids, are lower than predicted with usual Rankine theory.

With refererence to the mechanism of movement of the wall face, a hypothesis can be done, based on the results of some field tests where an outward rotation around the toe was noted (Berg et al 1986; Bathurst et al 1987). Then if we suppose that the toe can also have a translation, we have the situation shown in Fig. 6. If the geogrids are equally spaced, we can write:

$F_4 = K \cdot \gamma \cdot h_4 \cdot \delta h$ ... $F_4 = K \cdot \gamma \cdot h_4 \cdot \delta h$

and: $\varepsilon = f(K \cdot \gamma \cdot h \cdot \delta h) = f(\gamma h)$.

Called δx the total movement of the face, then δx is the sum of the elongation of the geogrids due to tensile forces, $\varepsilon \cdot L$, and of the pull-out movement S. We can write that S is a function of the difference between the tensile force and the pull-out resistance: $S = f(K \cdot \gamma \cdot h \cdot \delta h - 2 \cdot f_p \cdot \tan \Phi' \cdot \gamma \cdot h) = g(\gamma h)$ where f_p is the pull-out coefficient ($f_p \approx 0.7 \div 1.0$). Then we can suppose that the diagrams $S = g(\gamma h)$ and $\varepsilon = f(\gamma h)$ have the patterns shown in Fig. 7. Therefore the two components $\varepsilon \cdot L$ and S combine to form δx as shown in Fig. 8a: this is the mechanism of movement of a wall which toe can translate. If, instead, the toe cannot translate, then we have to add a third element W to obtain the diagram of δx, as shown in Fig. 8b. W is the variation of movement of the wall due to bending stiffness of the face: it gives a negative contribution at toe and a positive contribution at top, the latter only if the vicinity of surcharge can locally increment the curvature. This mechanism says also that near the toe the wall face supports a part of the thrust: so if the surcharge is high, the lower geogrid layers have less tension than the upper ones.

With reference to the σ_v distribution, we can compare the results of Tucson and Kingston walls: as shown in Fig. 9, there were different results from similar walls; so the matter has to be furthermore investigated.

6 CONCLUSIONS AND PROPOSALS

Some hypotheses were done in this paper to explain the field results but, since the comparable data are too few to draw a definitive evaluation or to have a statistic

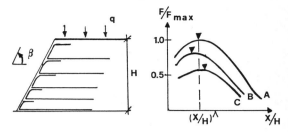

Fig. 1 – Tensile forces in a slope. A = medium-low layers; B = lower layers; C = upper layers.

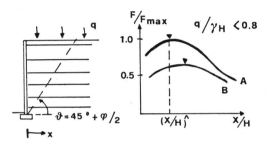

Fig. 2 – Tensile forces in a wall. A = lower layers; B = upper layers.

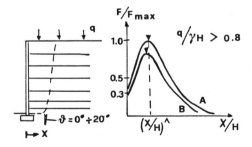

Fig. 3 – Tensile forces in a wall. A = upper layers; B = lower layers.

confirmation, it would be useful to run firstly many small scale tests in order to investigate each of the unclear matters; then other full scale tests will give confirmation of the obtained results, allowing to control that no scale effect has occurred in the small scale tests. Only at this stage, FEM models will allow to investigate in details the distribution of each parameter throughout the reinforced soil mass. The knowledge so gained will finally allow to issue new and more accurate design methods, specific for each category.

575

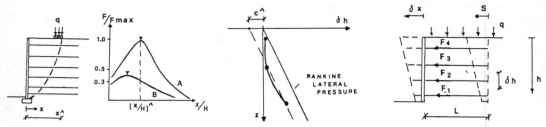

Fig. 4 - Tensile forces in a wall. A = upper Fig. 5 - Lateral earth pressure and Fig. 6 - Movements of wall face.
layers; B = lower layers.   fictitious cohesion

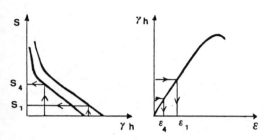

Fig. 7 - $S = g(\gamma h)$ and $\varepsilon = f(\gamma h)$.

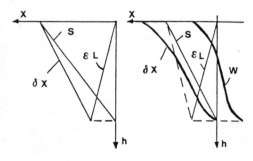

Fig. 8 - Superposition of movements with and
without toe translation.

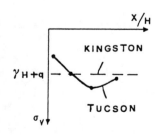

Fig. 9 - Vertical pressures at base.

REFERENCES

Bassett, R.A. et al 1986. Proceeding of
    Reinforced Embankment Prediction Sympo-
    sium. King's College. London.
Bathurst, R.J., Wawrychuk, W.F, Jarret, P.M.
    1987. Laboratory investigation of two
    large scale geogrid reinforced earth wall.
    Proc. NATO A.R.W. on Application of Poly-
    meric Reinforcement in Soil retaining
    Structures. RMC. Kingston.
Berg, R.R., Bonaparte, R., Anderson, R.P.,
    chouery, V.E., 1986. Design, construction
    and performance of two geogrid reinforced
    soil retaining walls. Proc. III Int. Conf.
    on Geotextiles. Vienna.
Berg, R.R., La Rochelle, P., Bonaparte, R.,
    Tanguay, L., 1987. Gaspe Peninsula reinfo-
    rced soil seawall case history. Proc. ASCE
    Symp. on Soil Improvement. Atlantic City.
Carroll, R.G., Richardson G.N. 1986. Geosyn-
    thetic reinforced retaining walls. Proc.
    III Int. Conf. on Geotextiles. Vienna.
Christopher, B.R. 1987. Geogrid reinforced
    soil retaining wall to widen an earth dam
    and support high live loads. Proc. Geosyn-
    thetics'87 conference. New orleans.
Cazzuffi, D., Pagotto, A., Rimoldi, P. 1988.
    Behaviour of a geogrid reinforced emban-
    kment over waste material. Proc. II Int.
    Conf. on Case Histories in Geotechnical
    Engineering. St. Louis.
Jewell, R.Q., Paine, N., Woods, R.I. 1984.
    Design methods for steep reinforced emban-
    kments. Proc. Symp. on Polymer Grid Rein-
    forcement in Civil Engineering. London.
Jones, C.F.J.P. 1985. Earth reinforcement
    and soil structures. London: Butterworths.
Yamanuchi, T., Fukuda, N., Ikegami, M. 1986.
    Design and techniques of steep reinforced
    embankment without edge supporting. Proc.
    III Int. Conf. on Geotextiles. Vienna.

International Geotechnical Symposium on Theory and Practice of Earth Reinforcement / Fukuoka Japan / 5-7 October 1988
© 1988 Balkema, Rotterdam. ISBN 90 6191 820 0

Seismic design of Reinforced Earth[r] retaining walls – The contribution of finite elements analysis

Pierre Segrestin & Michel J.Bastick
Sté Terre Armee Internationale, Puteaux, France

ABSTRACT : After experiments on reduced models and actual structures, dynamic finite elements analysis has recently been introduced into the continuing study of the response to seismic loads of Reinforced Earth[R] retaining walls. An overview is given of the results obtained at different levels of seismic excitation in a variety of foundation-soil conditions, and of their practical design implications.

1- INTRODUCTION

Being both heavy and flexible, Reinforced Earth[R] structures stand up well to vibrations and seismic loads. This has been demonstrated in reduced models, and in experimental structures subjected to strong vibration ; above all, though, it is confirmed by the excellent performance of the many structures – in Japan, Italy, New Zealand, USA, Mexico and even in Belgium – with in-service earthquake experience.

The usual approach at this stage was to carry out tests on models built on shaking tables, or on full size walls to which strong vibrations were applied from above. In more recent years, however, in order to validate and further refine the calculation methods derived from such investigations, Terre Armée Internationale has been making increasing use of finite elements models.

2- EXPERIMENTS

2.1 Tests using laboratory models.

Tests on reduced scale models provide a good insight into the type of phenomena which occur ; quantitatively, however, they are much less informative, due to the problems of similitude.

30 cm high models, used for example by Richardson and Lee (1974), were built of sand stabilized with magnetic tapes 6 mm wide and 25 μm thick. The facings were made of aluminium elements 0,4 mm thick

and 25 mm high. Vibrating table tests revealed two possible failure modes, similar to those found in static conditions : either failure occured fairly suddenly, as a result of strips breaking, or it was very gradual and due to insufficient adherence ; in the latter case, there was considerable associated deformation, but the wall recovered its equilibrium when vibration ceased.

2.2 Tests using a half size model.

Professor Chida's model (1980) was big enough to avoid the problem of similitude. As in actual applications, it had concrete facing panels, and steel strips. Its vibrating table allowed tests to be carried out at frequencies of between 2 and 7 Hz, for horizontal accelerations of between approximately 0,1 and 0,4g (Fig 1).

During the tests, measurements were made of dynamic increments in tension along the strips (Fig 2) ; these proved to be relatively evenly distributed, hardly changing the location of the point of maximum total tension (which governs the pull out resistance).

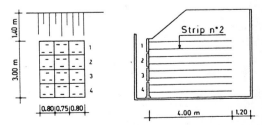

Fig.1 - Prof. Chida's half size model.

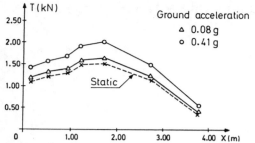

Fig.2 - Chida's test wall. Tension along strip n° 2.

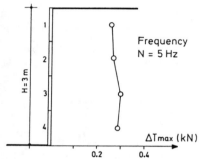

Fig.3 - Chida's test wall. Dynamic increments.

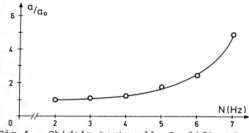

Fig 4 - Chida's test wall. Amplification of acceleration between base and top versus frequency.

At the same time, such dynamic increments seemed to vary very little with depth (Fig 3).

Professor Chida also measured amplification of horizontal acceleration as between the base and the top of the model. Relatively high values were found at higher frequencies ($a/a_o = 2$ at 5 Hz), though this might well be largely due to the model's rigid frame (Fig 4).

2.3 Tests in actual structures

The tests undertaken by Terre Armée (France) in a 6 m high structure at Triel (1976) were in fact specifically aimed at evaluating transmission of vibration caused by a roller on top of the wall.

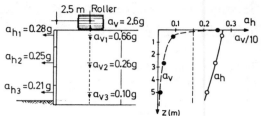

Fig.5 - Triel wall. Transmission of vibration.

Here, horizontal acceleration was found to vary little as a function of depth (Fig 5).

Meanwhile, in a general theoretical study commissioned by Terre Armée Internationale, Professors H.B. Seed and J.K. Mitchell estimated average amplification coefficients likely to be found in routine Reinforced Earth walls, as a result of their particular vibratory modes. A ratio of between 1,20 and 0,85 was suggested, for maximum accelerations at the wall base of 0,1 g to 0,5 g.

3- RESEARCH BASED ON FINITE ELEMENTS

3.1 Programme

Finite elements analyses had already been undertaken, on a considerable number of models, for a thorough investigation of the static behaviour of R.E. structures : Terre Armée Internationale's research in this direction (1982-84) relied on the Rosalie programme by LCPC (France). The dynamic study, on the other hand, was carried out using the SUPERFLUSH programme, a recent development, by Dr T. Udaka (Earthquake Engineering Technology, California), of LUSH, the programme created in 1974 by Lysmer, Udaka, Seed and Hwang.

Instead of rendering an accelerogram discrete in time steps, introducing instantaneous acceleration into the model for each step, Superflush uses a less costly approach, analysing the accelerogram in Fourier series per frequency steps. The elastoplastic behaviour of the soil is simulated by varying the modulus of elasticity as a function of observed deformation, this process being repeated until moduli and deformations are compatible.

The first Terre Armée model was gradually refined by, first of all, adjusting its results to fit those of Professor Chida's test wall. The close agreement between the experimental and the calculated results demonstrated the reliability of both the programme and the model (Fig 6). The representation of conditions encountered in actual structures was then improved, by introducing fictitious "half spaces" simulating the foundation soil and the rear backfill.

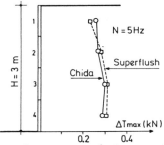

Fig. 6 - Agreement between Chida's measurements and Superflush results.

3.2 Models

The Reinforced Earth retaining walls represented in the last phase of the study (1987) are 6 m and 10,50 m high respectively, and similar (particularly as regards the length and number of their strips) to those most comprehensively analysed in the static finite element study (Fig. 7). The strip density in the 6 m wall is uniform throughout (4 40 X 5 strips per panel), whereas in the lowest third of the 10,50 m wall the strips are 1,5 times larger (4 60 X 5 strips per panel). The dimensions of the strips being those one would find in an actual structure, there is no risk of any soil / strip slippage.

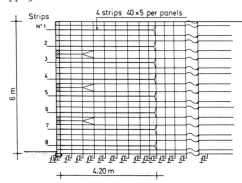

Fig.7 - F.E. model of the 6 m high wall.

These strips are in mild steel : obviously, the results of this research do not apply for other reinforcement materials having totally different moduli. The values assigned to the principal properties of the materials included in the models are summarized in the table below.

		Backfill	Steel	Concrete
Unit weight	pcf	125	460	150
	kN/m³	19,65	72,28	23,57
Poisson's ratio		0,20	0,30	1/3
Shear modulus	ksf	$70\sqrt{\sigma m}$	$1,661 \times 10^6$	$2,050 \times 10^5$
	MPa		$79,56 \times 10^3$	$9,82 \times 10^3$
Damping ratio		$\sqrt{G/\rho}$	3%	7%

3.3 Environment and motions

The structures were placed on three types of foundation soil in order to analyse the influence exercised by local conditions. These soils are homogeneous and are distinguished by their shear wave velocity V :

- hard rock soil where V = 1000 m/sec
- medium stiff soil where V = 400 m/sec
- loose soil where V = 250 m/sec (due account being taken, in selecting this value, of the presence of the wall).

In addition, two different typical accelerograms were applied to the structures : first, that of the so-called Golden Gate S80E earthquake, which occurred in 1957 at San Francisco, and then that of the 1971 San Fernando earthquake, known as Castaic N21E. In the former there is a fairly marked predominating frequency around 8 Hz (Fig 8).

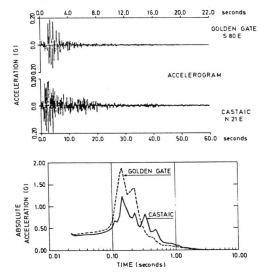

Fig.8 - Accelerograms and response spectra (top level).

And finally, each of these accelerograms was adjusted to fit several different maximum accelerations, chosen from amongst 0,1 g, 0,2 g and 0,4 g - the last of which in fact corresponds to a particularly severe earthquake.

3.4 Results

3.4.1 The influence of the environment

The results presented in graph form below are limited to those of particular comparative interest as regards maximum dynamic forces affecting the strips (which must,

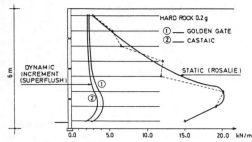

Fig. 9- Variation of maximum tension with depth. Influence of the type of accelerogram.

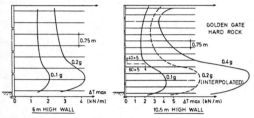

Fig. 10 - Influence of the maximum acceleration.

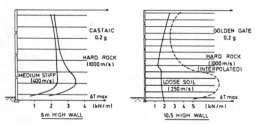

Fig.11 - Influence of the foundation soil stiffness.

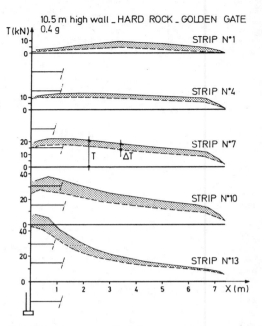

Fig.12 - Superflush. Tensions along strips.

of course, be added to forces of static origin).

On the same foundation soil (rocky) and at the same maximum acceleration (0,2g), the Golden Gate earthquake is a little more "aggressive" than Castaic ; this is not surprising, given their respective spectra, though the difference is really extremely small (Fig 9).

The structures respond much more sensitively to the maximum acceleration value of a single accelerogram, in respect of one and the same soil (Golden Gate / rock) (Fig 10)

There is also a very marked reduction in dynamic forces as a function of foundation soil stiffness, for a given seismic motion (Fig 11).

3.4.2 The distribution of dynamic forces

The finite element results confirm those obtained by Professor Chida : the distribution of dynamic tensile forces along the

strips (or, to be more accurate, the envelope of dynamic tensile increments) is fairly uniform and, as can be seen in the graphs below, does not give rise to any significant change in the position of the points of maximum tension, even for strong accelerations (Fig 12).

In line, once again, with the Chida model findings, the maximum dynamic increment varies little with depth in structures reinforced in a uniform manner. There is, however, a tendency for lower strips to be relatively more affected, i.e. those strips having the longest resistant length (beyond the point of maximum tension). This result is a striking confirmation of what was forecast in 1981 by Professors H.B. Seed and J.K. Mitchell.

The tendency for lower strips to be more affected is particularly pronounced when they are wider or more numerous, as is the case at the bottom of the 10,50 m wall, i.e. when these strips have what might be termed a higher "bonding-stiffness".

3.4.3 Acceleration

The amplification of acceleration as between the base and the top of the model is by a factor of about 2, at least in the unfavourable case of structures sited on a hard rock foundation and subjected to the Golden Gate accelerogram. This ratio is quite comparable to those found experimentally by Chida. However, if an average value is calculated for the curves depicting variations in maximum acceleration over the height of the structure, and this average is set against the reference maximum acceleration at ground level, the increase is then found to be only by a factor of about 1,35 for 0,1g, and 0,95 for 0,4g (Fig 13). Such values, albeit somewhat higher, are nonetheless comparable to those suggested by Seed and Mitchell. Possibly the way in which boundary conditions are simulated in the finite elements model is still a little unfavourable ; on the other hand, if design calculations are indeed based on these values, any bias will be on the safe side.

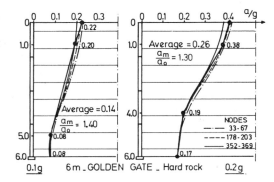

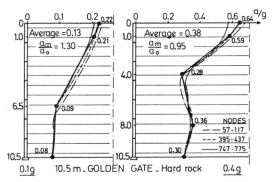

Fig.13 - Superflush. Amplification of acceleration.

4- PRACTICAL DESIGN METHOD

4.1 Internal stability

Analysis of these various results has led to the definition of a practical calculation method based on estimating a global inertia force E_I for the mass of the structure's active zone where the increments in tension occurring in the strips originate.

Given the envelope form adopted for the line of maximum tension in usual structures, we have (Fig 14) :

$$E_I = 0,20 \; \frac{am}{g} \; \gamma_1 \; H^2$$

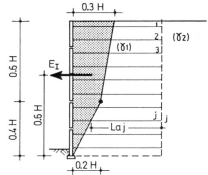

Fig.14 - Internal stability. Global inertia force E_I.

The value assigned to a_m is an estimation of the average maximum horizontal acceleration in the structure. In practice, for a reference acceleration a_o of between 0,05 and 0,5 g, it is assumed that :

$$\frac{am}{g} = (1,45 - \frac{ao}{g}) \; \frac{ao}{g}$$

for structures on rock foundations.

The load E_I is then distributed among the different layers of strips pro rata to their "resistant area" Nj bj Laj.

The graphs in figure 15 show that there is excellent agreement between the results achieved using these simple formulae and those calculated using finite elements.

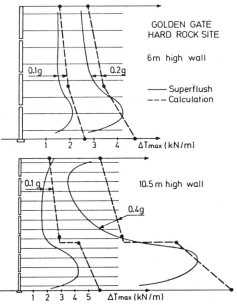

Fig.15 - Dynamic increments. Comparison between practical calculations and Superflush results.

These increments in tension are additional to static tensile loads resulting from the own weight of the structure as increased by the overturning effect due to the earth pressure of the backfill behind. This approach also includes a proportion of the supplementary dynamic (or pseudo-static) earthpressure E_{AE} calculated according to Mononobé Okabé (Fig. 16) (Taking only some 60% of dynamic pressure into account is justified by the fact that E_I and E_{AE} are unlikely to peak simultaneously).

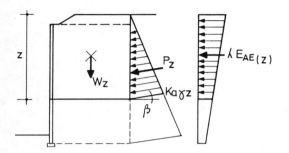

Fig.16 - Internal stability. External loads.

The strips are then checked in the usual way, applying the safety coefficients appropriate to accidental events such as earthquakes. When checking adherence, however, the apparent friction coefficient f* is limited to 80% of the value used in the static calculation.

4.2 External stability

When checking overall stability, the elements to be added to the structure's own weight and the static earthpressure of the soil to the rear are, this time, pseudo-static pressure E_{AE}, preponderant here, and a fraction ($\approx$ 60%) of the structure's inertia load E_I.

5- CONCLUSION

Thus further improved, practical earthquake design of Reinforced Earth structures is probably still non-perfect and conservative. Nevertheless, it is based on a varied range of experimental and analytical results (very possibly unparalleled even in the field of conventional retaining structures) which are largely in agreement.

In conjunction with Reinforced Earth's inherent adaptability, then, this design approach provides a further guarantee of the safety of structures sited in seismically active regions.

BIBLIOGRAPHY

Richardson G.N., Feger D., Fong A., Lee K., Seismic testing of Reinforced Earth walls, Jour. of Geot. Eng. Div. ASCE, pp 1-17, G11, January 1977.

Seed H.B., Mitchell J.K., Earthquake resistant design of Reinforced Earth walls (Internal study, progress report), Berkeley CA, U.S.A., 1981.

Udaka T., A method for soil - structure interaction analysis. Proceedings of the fourth symposium on the use of computers in Buildings engineering, Japan, March 1982.

Chida S., Minami K., Adachi K., Test de stabilité de remblais en Terre Armée (traduit du Japonais), 1982.

Bastick M., Schlosser F., Comportement et Dimensionnement dynamique des ouvrages en Terre Armée, 1er Coll. Nat. de Génie parasismique, Saint-Rémy-lès-Chevreuse (France), janvier 1986.

Elms DG, Nagel RB - University of Canterbury (New Zealand) and Ministry of Works and Development (NZ). Dynamic model testing of Reinforced Earth walls 1987.

International Geotechnical Symposium on Theory and Practice of Earth Reinforcement / Fukuoka Japan / 5-7 October 1988
© 1988 Balkema, Rotterdam. ISBN 90 6191 820 0

Experimental study of reinforced-earth wall

Zhonglun Shi & Mingde Cheng
Fourth Survey and Design Institute, Ministry of Railways, Wuhan, People's Republic of China

ABSTRACT: From a large number of model and full-scale testing data, this paper has analyzed the force acting on the reinforced-earth wall and deformation features and also has presented a design and analysis method of reinforced-earth wall.

reinforced-earth wall has several advantages : light weight, good appearance , flexible to deformation and wide application , also its construction is simple and efficient, taking less land and costs.So it develops rapidly and tends to replace gravity retaining wall . However, the functional mechanism, the load state of structure and design and analysis method etc.are still under study. For further understanding to this structure,based on the actual condition in china , full-scale tests and laboratorial model tests have been carried out. A large number of actual data has been obtained. Through comprehensive analysis, a conclusion could be drawn as follows. (Owing to the limitation of the testing data , the conclusion is only appropriate to the retaining wall less than 10 m high.)

1 DETERMINATION OF SOIL COLUMN STRESS IN REINFORCED-EARTH WALL MASS

It could be seen from Fig 1, the actual soil column stress varies in direct proportion with the fill height, and it approaches the theoretical value rh, so rh is adopted.

The supplemental soil column stress due to the load could be seen in Fig 2. Comparatively the actual value approaches the elastic theoretical value. So supplemental influence of the load could be suggested to be considered with distributed angle method.

By the superpose theory,the soil column stress in the reinforced-earth wall mass is the sum of soil column stress caused by soil mass and supplemental stress caused by load.

2 LATERAL PRESSURE ACTING ON REINFORCED-EARTH WALL FRONT PLATES

Different testing conditions produce different results (Fig 3) . However , according to the actual data from the full-scale wall, the lateral pressure distribution pattern is approximately consistent: the distribution of the upper part of the retaining wall is approximately linear , and it approaches the static soil pressure; the distribution of the lower part is in broken line . Because of the loose backfill and the difference of construction conditions between the model wall and the full-scale wall , while the supports being demolished,the deformation occurs in the whole wall mass.the lateral pressure is approximately linear along the wall height , and it is higher than the active soil pressure. Besides, the wider the strips are,the heavier the front plates will undertake.

According to the above-mentioned data , to the full-scale wall, the lateral pressure acting on the front plates without loading could be simplified for calculation by Fig 4.

The composite force of the simplified lateral pressure is larger than that of the actual lateral pressure,only the actual stress of the local site in lower part is larger than the simplified value, and it is also larger than the Coulomb's active soil pressure (1.18 times).

The comparative curve of the load to the lateral pressure is illustrated in Fig 5.

The actual result approaches the theoretical value of the elastic subgrade,its distribution pattern is also similar.Considering the insuf-

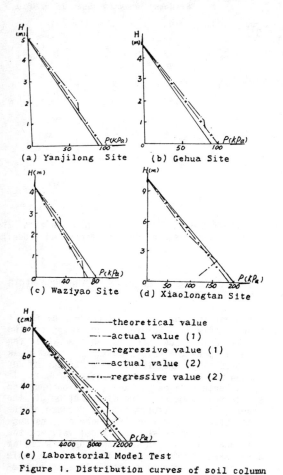

(a) Yanjilong Site   (b) Gehua Site

(c) Waziyao Site   (d) Xiaolongtan Site

————theoretical value
—·—·actual value (1)
—··—·regressive value (1)
—·—·actual value (2)
—··—·regressive value (2)

(e) Laboratorial Model Test
Figure 1. Distribution curves of soil column

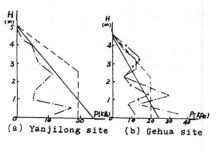

(a) Yanjilong site   (b) Gehua site

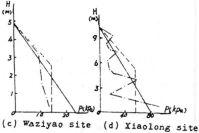

(c) Waziyao site   (d) Xiaolong site

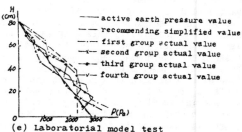

————active earth pressure value
———recommending simplified value
—·—·first group actual value
—×—second group actual value
—●—third group actual value
—v—fourth group actual value

(e) Laboratorial model test
Fig. 3. Distribution curves of lateral
pressure on the front plates.

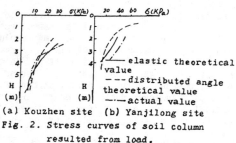

————elastic theoretical
value
———distributed angle
theoretical value
—·—·actual value

(a) Kouzhen site (b) Yanjilong site
Fig. 2. Stress curves of soil column
resulted from load.

Fig. 4. Simplified distribution curve of
lateral pressure due to backfill.

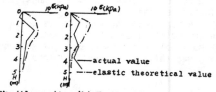

————actual value
———elastic theoretical value

(a) Yanjilong site (b) Waziyao site
Fig. 5. Additional lateral pressure and
stress curve due to load.

ficient data of this respect , a suggestion
would be to calculate the supplemental influence of loading to the lateral pressure
acting on the front plates with the elastic
subgrade theory. So the design load of the
front plates on the reinforced-earth wall would

order	Model test										Waziyao site					Yanjilong site		
	20x2x450mm strip					50x2x450mm strip												
	E	T₀	Tmax	T₀/E	Tmax/E	E	T₀	Tmax	T₀/E	Tmax/E	E	T₀	Tmax	T₀/E	Tmax/E	E	Tmax	Tmax/E
1	7.8	15.0	22.9	1.92	2.94	20.6	20.0	24.1	0.97	1.17	3800	3500	3700	0.92	0.97	4390	16490	3.83
2	21.3	21.0	24.9	0.99	1.17	25.3	27.0	30.7	1.16	1.32	5500	2500	6200	0.42	1.13	5850	16260	2.78
3	24.6	21.0	32.3	0.86	1.32	38.5	40.0	42.2	1.04	1.10	7000	3000	7000	0.43	1.0	3430	21850	6.37
4	37.8	43.0	55.4	1.14	1.47	34.3	45.0	56.6	1.31	1.65	5500	6300	8500	1.15	1.51			
5	43.8	67.0	70.3	1.44	1.61	52.0		70.2	1.21	1.35	9000	4300	5000	0.48	0.56			
6	57.2	62.0	69.6	1.08	1.22	55.4	80.0	101.2	1.44	1.83								
7	50.4	59.0	64.9	1.17	1.29	69.3	85.0	106.1	1.23	1.53								
Σ	262.7	284.0	340.5	1.17	1.60	283.4	346.0	451.1	1.25	1.47	30800	19600	52000	0.63	0.96	13580	54600	4.04

Note: E is the lateral pressure (actual value) (N) applied
to the plates.
T_0 is the tension of the connecting place between
plates and strips (actual value) (N)
T_{max} is the max. tension (actual value) (N)

Table 2. A comparative table of actual max. tension in model test to the designing lateral pressure on the corresponding layer of the plates.

order	12x8x450mm strip			20x2x450mm strip			30x2x450mm strip			50x2x450mm strip		
	Ed	Tmax	Tmax/Ed	Ed	Tmax	Tmax/Ed	Ed	Tmax	Tmax/Ed	Ed	Tmax	Tmax/Ed
1	17.3	15.4	0.89	17.3	22.9	1.32	17.3	15.6	0.90	17.3	24.1	1.39
2	28.8	20.9	0.73	28.8	24.9	0.89	28.8	32.6	1.13	28.8	30.7	1.07
3	40.4	29.5	0.73	40.4	32.3	0.80	40.4	36.4	0.90	40.4	42.2	1.05
4	46.1	43.6	0.95	46.1	55.4	1.20	46.1	55.3	1.20	46.1	56.6	1.23
5	46.1	47.5	1.03	46.1	70.3	1.53	46.1	57.4	1.25	46.1	70.2	1.53
6	46.1	54.2	1.18	46.1	69.6	1.51	46.1	69.5	1.51	46.1	101.2	2.20
7	46.1	63.8	1.38	46.1	64.9	1.41	46.1	69.3	1.50	46.1	106.1	2.30
Σ	270.9	274.9	1.02	270.9	340.3	1.26	270.9	336.1	1.24	270.9	431.1	1.59

Note: 1. E_d is the simplified value.
2. Unit: N.

be considered as the sum of the lateral pressure caused by the backfill itself (simplified value) and supplemental lateral pressure caused by load(an elastic subgrade theoretical value).

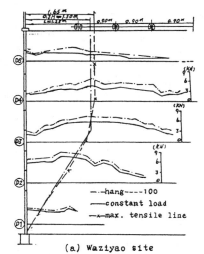

(a) Waziyao site

--hang----100
constant load
-x-max. tensile line

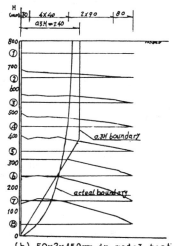

(b) 50x2x450mm in model test
Fig. 6. Tension distribution

3 DISTRIBUTION AND MAGNITUDE OF STRIP TENSION

The distribution of the strip tension is in curves along the length direction. The distribution curve of the tension of the full-scale wall presents waved due to the order of compaction by rolling and the soil nature. In the model wall, because of the even backfill , its distribution curve is parabolical. To the same strip, the increase scale varies in inverse proportion with the fill height.while with the backfill on the top, the tension increases accordingly. The tension varies in inverse proportion with the baried depth, in direct proportion with the load.

4 BOUNDARY DETERMINATION OF STRIPS IN ACTIVE AND PASSIVE AREA

According to the actual data , smoothly connecting sections with the maximum tension on each layers,then "a potential failure surface" could be obtained as shown in Fig 6.Inside the failure surface is defined as a passive area, and outside as an active area . Its shape is similar to the common " 0.3H method ".

5 DETERMINATION OF DESIGN EXTERNAL TENSION OF STRIPS

By the current design method , the tension T_0 of strips at the joint of the front plates should be equal to the lateral pressure E acting on the corresponding wall plates.maximum tension T_{max} should be greater than E . The actual results are shown in Tab 1. In the full-

Table 3.

order	Yanjilong site			Gehua site		
	Ed	$Tmax$	$Tmax/Ed$	Ed	$Tmax$	$Tmax/Ed$
1	6.7	12.5	1.87	5.4	4.8	0.89
2	10.5	14.3	1.36	9.5	4.5	0.47
3	10.5	18.9	1.80	9.7	10.2	1.05
Σ	27.7	45.7	1.65	24.6	19.5	0.79

Note: Unit is KN

Table 4. Actual displacement at plates on the retaining wall

Site name	H	Δmax	$\Delta max/H$
Gehua	4.5	1.57	0.35‰
Yanjilong	5.0	1.1	0.22‰
Kouzhen	7.0	4.0	0.57‰
Waziyao	5.0	5.66	1.13‰
Daqin bridge	3.5	0.42	0.12‰
Model test	0.8	0.5 (50x2x450) / 1.28 (20x2x450)	0.63‰ / 1.6‰

scale wall,owing to the errors of the measurements and the insufficiency of the actual data, its behavior is not very obvious , additional data are required. In the model wall, its behavior is clear. Although the corresponding E and T. have differences due to the interactions between the plates,comparatively speaking,they are almost equal. To the whole wall, the ΣT. $/\Sigma E$ is approximately 1.1 to 1.2 , from the viewpoint of the entire wall the T_{max} is larger than the corresponding E , $\Sigma T_{max}/\Sigma E$ of the whole wall is approximately 1.2 to 1.5. Therefore, a conclution could be drawn that the current design method has been verified by the model.

A comparative value about the actual maximum tension and the corresponding design lateral pressure of the plates in the model test and in site is listed in Tab 2 and 3.
In the model test, almost all their rates are larger than 1.0, only some values are between 1.02 to 1.59. But in the full-scale test,there are great differences. Considering the limited number of the full-scale test sites and the inconsistency of conditions on each site, further information is to be obtained.
According to a comprehensive analysis to the above-mentioned data for safty, two times of the lateral pressure of the plates would be suggested to be the design load of strips in the calculation.

6 DISPLACEMENT OF WALL MASS

The displacement of the wall mass in site and model test is shown in Tab 4.
The horizontal displacement of plates is very small after the completion of the reinforced-earth wall with backfill. Besides the filling soil nature and the density of compaction by rolling, factors influencing the displacement of the reinforced-earth wall have direct relation with the number of strips.
According to the distribution of plate displacement scale on each layer along the wall height,the upper displacement scale is usually larger than the lower, and the upper is mainly a horizontal displacement . The lower is very small,and could be considered as a rotation.

7 DETERMINATION OF TENSILE STRENGTH OF STRIPS

The tensile strength is determined by the tensile testing. As strips are undertaking an upright pressure from the filling around . the tensile strength has a relation with the friction between the strips and the filling. Besides that , owing to the tamping and compaction by rolling to the filling around , a meshing force is added to strips under the horizontal tension condition. So the tensile strength T is a combination of friction and the meshing force. The frictional resistance only varies with the fill height h,the meshing force is a constant.Then

$$T = ah + b$$

In this formula, T could be expressed as the linear function of h.So by means of the linear regression and the testing data, the tensile strength of strips could be determined.

8 DAMAGES OF REINFORCED-EARTH WALL

Through a large amount of loading stability tests in model wall, the damages of the model wall could be classified in three catagories as follows.

8.1 Damages due to the excess deformation

This kind of damage mainly appears in narrow strips or in the loose filling model wall or when the backfill is on the end of the wide strips. The damage features are: the displacement of the wall surface is convex. The place with loading has a considerable settlement.The boundary of the damage area is not clear . The sliding surface is approximately determined by the displacement trace of the filling: it is a logarithmic spiral passing through the rear edge of loading and the wall bottom.

8.2 The landsliding damage of the wall mass

This damage only appears in the model wall

with loading acting within the strip length
and the filling at high density.Under the condition of loading , a sudden landslide will
occur in the wall . The whole wall only has a
little deformation and could bear much backfill
before the damage.
It could be known from the tests that the
wall's bearing capacity is in direct proportion
to the filling density as well as the dimension
of strips, and it is in inverse proportion to
the wall height.Simulating tests have verified
that the ultimate bearing capacity of the reinforced-earth wall designed by the current design method is several times greater than the
design load.

8.3 Damages due to the insufficient strength
of units

The damage appears when the section of a strip
is wide, the filling is loose and the backfill
is acting on the top of the wall . The tensile
strength of the strips of the reinforced-earth
wall is larger than the bearing capacity of
the unit itself.When the unit breaks,the plate
falls off, as a result of that, plenty of filling runs off from the place with plates fal-
len off, hence damages the wall. these damage
types are, of course, obtained from the model
tests.Being confined by conditions, there probably exist some differences from the full-
scale wall damages. However, these tests present such a regular pattern:
when other conditions are invariable, the more
compact the filling is,the greater the bearing
capacity of the wall will be, and the less the
deformation will be . So suggest that: in the
construction of the reinforced-earth wall,
strictly control the quality of compaction by
rolling and ramming for the filling, so as to
meet the design requirements with a desired
density.

International Geotechnical Symposium on Theory and Practice of Earth Reinforcement / Fukuoka Japan / 5-7 October 1988
© *1988 Balkema, Rotterdam. ISBN 90 6191 820 0*

Geotextiles earth reinforced retaining wall tests

I. Yamaoka
Kansai University, Osaka, Japan

T. Yamashiro & K. Mondori
Kinki Engineering Office of the Construction Ministry, Osaka, Japan

ABSTRACT: The Kinki Construction Bureau decided to utilize geotextile reinforced soil retaining wall to investigate the wall mechanic and monitor the wall behavior. This paper describes the geotextile reinforced retaining wall tests by static and dynamic loads and provides the design, construction and instrumentation details. The wall has a gunite facing after wall behavior had become stable condition in march 1987. The wall was designed by conventional method and was assigned a lower than usual factor of safety to provide a more critical condition for static and dynamic load test.
Since construction, the wall has settled and it's face displaced foreword by the lack of compaction during construction time, and by static and dynamic loads tests. The wall performance, however, has been satisfactory and has not exhibited distress. After the project, the Kinki Construction Bureau constructed other larger retaining wall at Nomura tunnel entrance access road on national road No.175 in order to investigate their cost comparison and monitor the movement.

1 INTRODUCTION

The wall's general scope is as follows: height 5.4m except foundation drainage layer 0.4m, cover layer 0.4m, and pavement layer 0.1m height, top length 31.5m, bottom length 6.0m.(shown in Fig.1(a), and 1(b)).
From No.1 layer(h=0m) to layer 8 (h=3.0m), combination fabric with polyester needle punched felt white nonwoven fabric and polypropylene monofilament black woven fabric were utilized. And from layer 9(h=3.4m) through layer 13 (h=5.0m), single polyester needle punched fabric was utilized only.
Instrumentation at the site was designed to provide both qualitative and quantitative information on settlement and displacement in vicinity of the wall and within backfill soil mass and to identify specific layers or zones of settlements.
Information on horizontal deflection in the foundation soils and vertical deflection of the wall face and the assumed and separated face from the wall face (d=0.5m, 1.75m, 3.5m) was obtained. Measurement survey of the deflection of the wall face and the surface above the wall top were conducted to indicate their settlement.
Movements within the backfill soil mass

were also monitored.
These measurements were taken with 6 horizontal inclinometer casings (No.1 h=0m, No.2 h=1.4m, No.3,4,5 h=3.0m, No.6 h=4.6m) for settlement in the wall soil mass and foundation soil spaced vertically at the center section of the wall; 9 pressure gauge extensomater(h=1.4m, h=3.0m, h=4.6m, d=0.5m, 1.75m, 3.5m); 9 horizontal length gauge (h=1.4m, 3.5m, 4.6m, d=0.5m, 1.75m, 3.5m from wall face) for each their deflection of inclined face spaced horizontally 0.5m, 1.75m, and 3.5m in parallel with the wall faces; 18 direct measurement survey points for the wall face's displacement survey and 27 printed mark scale on the wall face fabric to investigate the occurrence of fabric creep on their layer lift fabric.
Most of these instruments can be identified in Fig.1(a),(b). After construction, 9 direct measurement survey points on the pavement surface above top of wall for the settlement above the embankment was installed.
To facilitate construction, geotextile layer spacing was 0.3m for lower layers (No.1 through No.3) and 0.4m for upper layers (No.4 through No.13) except 0.4m for cover layer and drainage layer. Therefore since the geotextile in upper layer

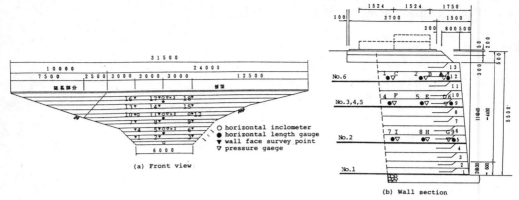

(a) Front view

○ horizontal inclometer
● horizontal length gauge
▼ wall face survey point
▽ pressure gaege

(b) Wall section

Fig.1 Profile of the retaining wall

and lower layer have different strength, the factor of safety are different for upper layer and lower layer. Safety considerations dictated that the wall not fail rapidly by static and dynamic load tests except excessive static loads test's pattern Ⅳ. Therefore some conservation was retained in this design method. However in static load pattern Ⅳ test, it is evident that the wall was falling forward rapidly and would exhibit significant strains in some fabric layers.

The wall was faced with a gunite after the wall's settlement was confirmed to be stable in March 1987. The project was performed by hanging the wire mesh on the end of folded rebars (ϕ =9mm, l=1.2m) buried in the wall soil and guniting about 11.5m³ of gunite were required for the approximately 100m² of wall face.

2 STATIC AND DYNAMIC LOADS TESTS

After the completion of the wall in Feb. 1981, the static loads and dynamic load tests were conducted from Aug. 4th through 7th and after that, the continuous applied static load condition was applied until Aug. 23. This paper describes the wall behavior from this period and to present.

2.1 Test methods

In this static load test, the steel plate (l=6m, b=1.5m, t=25mm, w=1.8tf), 30 sheets were placed in various pattern shown in Fig.2, in one line or 2 lines. The static load's application is listed in Table 1. At first in Aug. 5th, static applied loads

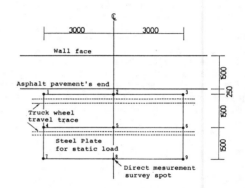

Fig.2 Loading positions

Table 1 The applications of static loads

	Front load		Back load	
	Number of plate	q (tf/m²)	Number of plate	q (tf/m²)
Pattern Ⅰ	1 0	1.96	———	———
Pattern Ⅱ	1 0	1.96	1 0	1.96
Pattern Ⅲ	2 0	3.92	1 0	1.96
Pattern Ⅳ	3 0	5.88	———	———
Pattern Ⅴ	———	———	3 0	5.88
Pattern Ⅵ	1 5	2.94	1 5	2.94

pattern Ⅰ , Ⅱ , Ⅲ , Ⅳ test by static loads was conducted. Next, dynamic loads test was performed. After dynamic load tests, static loads test Ⅴ , Ⅵ were done. The static load pattern Ⅵ load test was continuously applied for 16 days.

The dynamic load test was carried out by applying the truck loads (total weight 19.6tf). Every 50, 100, 200, 400, 600, and

Table 2 Pavement surface settlement

Load pattern	Survey point								
	1	2	3	4	5	6	7	8	9
Pattern III	0.003	0.004	0.002	0.002	0.003	0.0	0.0	0.0	0.0
Pattern IV	0.011	0.019	0.011	0.006	0.010	0.002	0.001	0.001	-0.001
50 passes	-0.008	0.0	0.001	0.003	0.001	0.0	0.001	0.001	-0.001
100 passes	-0.006	0.0	0.002	0.003	0.002	-0.001	0.0	0.0	-0.001
200 passes	0.006	0.002	0.005	0.003	0.001	0.0	0.0	0.0	-0.002
400 passes	0.015	0.003	0.007	0.003	0.001	0.002	0.0	0.001	-0.002
600 passes	0.010	0.008	0.010	0.006	0.0	0.002	0.0	0.001	-0.002
1000 passes	0.014	0.012	0.014	0.008	-0.001	0.002	0.003	0.002	0.002
Pattern V	0.014	0.013	0.014	0.008	-0.004	0.002	—	—	0.001
Pattern VI	0.011	0.012	0.017	0.011	-0.005	0.005	0.006	0.010	0.005

Table 3 Horizontal displacement
in wall mass

Height(m)	Distance from wall face(m)	Date						
		2/16	7/17	7/30	8/4	8/7	8/20	3/27
h=0.0	Wall face D=0.0	0	-0.3	-0.9	-1.2	-1.2	-1.1	-0.8
h=1.4	D=0.0	0	-0.1	0.3	0.3	0.3	0.7	1.2
h=3.0	D=0.0	0	1.4	1.5	1.8	3.7	3.9	4.6
h=4.6	D=0.0	0	1.9	3.3	4.9	7.1	7.2	7.5
h=1.4	No.3 D=0.50	0	0.2	-0.3	-0.3	-0.2	-0.2	0.6
h=3.0	No.5 D=0.50	0	1.1	1.0	1.5	3.1	3.3	4.2
h=4.6	No.9 D=0.50	0	1.8	2.7	4.4	6.5	6.5	7.6
h=1.4	No.2 D=1.75	0	-0.95	-0.85	0.75	-0.8	-0.5	-0.1
h=3.0	No.6 D=1.75	0	0.45	0.05	0.05	0.85	0.95	1.75
h=4.6	No.8 D=1.75	0	0.85	1.55	3.05	6.5	3.35	4.25
h=1.4	No.1 D=3.50	0	0.2	-0.9	-1.0	-1.2	-1.5	-0.5
h=3.0	No.4 D=3.50	0	1.1	-1.0	-1.0	-1.3	-1.4	-0.6
h=4.6	No.7 D=3.50	0	1.8	0.25	1.05	3.1	0.25	1.05

Table 4 Horizontal displacement
in wall by load

Height(m)	Distance from wall face(m)	static load test			deloading	dynamic load test						static load test		
		8/4	III	IV		50	100	200	400	600	1000	V	VI	8/26
h=0.0	Wall face D=0.0	0	0.1	0.2	0.2	0.0	0.0	-0.1	-0.2	-0.2	-0.2	-0.1	-0.1	-0.1
h=1.4	Wall face D=0.0	0	0.4	0.6	0.6	0.4	0.4	0.4	0.3	0.3	0.2	0.3	0.4	0.4
h=3.0	Wall face D=0.0	0	0.7	1.4	1.6	1.5	1.5	1.5	1.5	1.6	1.7	1.6	1.8	1.9
h=4.6	Wall face D=0.0	0	0.6	1.5	1.8	1.9	2.0	2.4	2.7	3.0	3.2	3.3	3.4	2.2
h=1.4	No.3 D=0.50	0	0.2	0.5	0.4	0.0	0.2	0.2	0.1	0.1	0.0	0.1	0.0	0.1
h=3.0	No.5 D=0.50	0	0.2	1.2	1.4	1.3	1.4	1.4	1.2	1.3	1.4	1.5	1.3	1.8
h=4.6	No.9 D=0.50	0	0.6	1.6	1.9	1.8	1.9	2.3	2.6	3.0	3.5	3.2	3.3	2.0
h=1.4	No.2 D=1.75	0	0.3	0.4	0.5	-0.1	-0.1	-0.1	-0.1	-0.1	-0.3	-0.2	-0.2	-0.2
h=3.0	No.6 D=1.75	0	0.2	0.5	0.7	0.8	0.7	0.7	0.6	0.7	0.7	0.6	0.3	0.6
h=4.6	No.8 D=1.75	0	0.4	0.7	0.8	0.9	0.9	0.9	1.1	1.2	1.4	1.4	1.4	0.7
h=1.4	No.1 D=3.50	0	0.1	0.3	0.3	-0.1	-0.1	-0.2	-0.2	-0.3	-0.2	-0.2	-0.2	-0.5
h=3.0	No.4 D=3.50	0	0.0	1.3	0.1	-0.3	-0.2	-0.3	-0.4	-0.4	-0.4	-0.5	-0.6	-0.8
h=4.6	No.7 D=3.50	0	0.0	1.3	0.3	0.1	0.0	1.4	0.2	0.3	0.3	0.4	0.4	-0.9

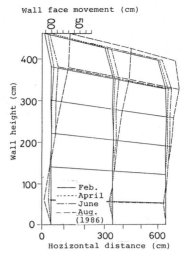

Fig.3 Wall face movement

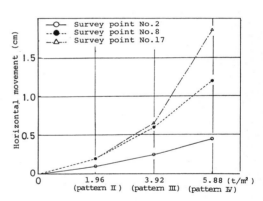

Fig.4 Horizontal movement on No.2, No.8,
No.17 under the pattern IV load

1000 passes the rut depth and settlement
at the direct measurement survey point on
the pavement surface were surveyed and
their result are listed in Table 2.
The largest direct settlement on the pavement above the top wall was about 2cm at
pt.2 and about 1cm at pt.1,3 and 5 in the
case of loads pattern . At that time,
cracking grew through pt.4 and 6 as a
circle which pt.2 is the center. As for
dynamic load test, about 4p rut depth was
produced after 1000 passes because of new
pavement roads (Table 2).

2.2 Measurement of the wall face behavior

The wall face movement by the static and
dynamic load test was shown in Fig.3,4 and
Table 3,4. The displacement and movement
became rapidly larger near the wall top.

In static load pattern IV, the top of the
wall moved about 2cm forward but at foundation ground, less displacement was
produced at any loads pattern. For load
application pattern IV, the displacement was
the largest and in very short time, the
wall top started to incline forward
rapidly. So this test was stopped. We
could recognize this displacement to be
the critical condition.
In the dynamic load test, as a number of
passes increase, the displacement grew
larger at h=4.6m survey point, however,
their values were less than 1cm. Under
h=3.0m, layers 8, the movement disappeared
gradually.

2.3 Horizontal distance variations measurement in the wall mass

On layer 4(h=1.4m), layer 8(h=3.0m) and layer 12(h=4.6m) at the center section, the displacement at pt.A-I which are at 0.5m, 1.75m and 3.5m from the wall face on layer 4,layer 8 and 12 of the wall center section were surveyed and calculated by observation of horizontal length gauge's relative and differential displacement. Their movement at each point looked similarly to the wall face movement but their values reduced as the distance from wall increased and the distance from the wall crest increased. The result was accurate and useful. Moreover their movement has a tended to displace toward back side under layer 4(h=1.4m) and beyond D=1.75m from the wall face. The wall top's movement was about 2cm forward in the case of

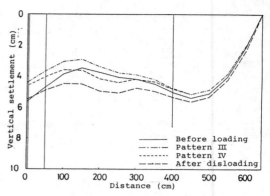

(c) Vertical settlement of No.6 inclinometer under the static load

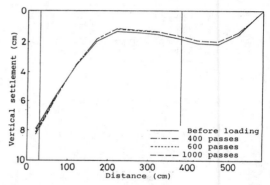

(d) Vertical settlement of No.4 inclinometer under the dynamic load

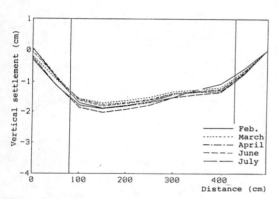

(a) Vertical settlement of No.1 inclinometer as the time elapsed

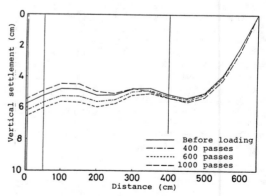

(e) Vertical settlement of No.6 inclinometer under the dynamic load

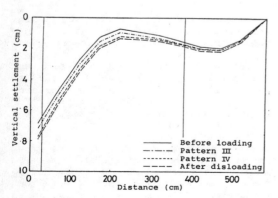

(b) Vertical settlement of No.4 inclinometer under the static load

Fig.5 Vertical settlement curves by inclinometer

pattern and about 1.4cm at 1000 pass cycle run time.

2.4 Settlement measurement by horizontal inclinometer

The settlement curves measured by the horizontal inclinometer installed in foundation ground (h=0.0m), layer 4, layer 8 and layer 12 was shown Fig. 5(a),(b),(c), (d),(e).
The settlement curves on foundation ground at construction completion was about 1.5cm the maximum settlement in vicinity of 1.0m from the wall face looked like downward bow. Their ground had been almost in stationary up to date, except critical situation by excessive loads because their curves were unchangeable up to date. The distance length between the left fine line, the wall face and the right fine line, the end of fabric are fabric installation length and the right ends of curves is the fixed point of the inclinometer casing. The vertical axis indicate the settlement and horizontal axis indicate the horizontal length. From Feb. to 4th Aug. the settlement on layer 8 and 12 inclined to increase rapidly in parts of beyond 3.4m length from the wall face, no fabric areas and front parts of less 2.0m from the wall face, the fabric folding back sections because lack of the compaction. Their shape were laterally adverse S character shape toward the fixed point from Feb. to 4th Aug.. In static and dynamic loads test the settlement curves on layer 12 (No.7) and 8 (No.3,4,5 h=3.0m) increases in proportion with the load's increasing and as shown in Fig.5(c),(e) the settlement curves in back fill soil just under the application loads is producing a very interesting second curves by loads.
The maximum settlements on layer 12 and 8 were about 1.0cm and less 0.5cm respectively in the case of loads pattern Ⅳ. At 1000 passes of dynamic load test, about 1.0cm settlement was observed on layer 12 while it disappeared on layer 8.
In the parts beyond the application load point, the settlements were not visible apparently and conversed toward a fixed point. It is found that the movement in the wall soil look like one on the pavement surface and the wall face.

2.5 Checking of safety factors

As for safety factors of pullout resistance and rupture resistance, the checking

Table 5 Calculation of safety factor under the pattern Ⅳ load

Layer	Depth to bottom of layer (m)	Depth to middle of layer (m)	Fhd (1/m)	Fhl (1/m)	Fh (1/m)	Le (m)	P (1/m)	F.S. (Pullout)	F.S. (Rupture)
13	0.95	0.75	0.211	0.517	0.728	1.392	1.882	2.586	1.551
12	1.35	1.15	0.324	0.546	0.869	1.600	3.075	3.537	1.265
11	1.75	1.55	0.436	0.084	0.520	1.808	4.504	8.662	2.115
10	2.15	1.95	0.549	0.127	0.676	2.016	6.171	9.128	1.627
9	2.55	2.35	0.661	0.083	0.744	2.225	8.075	10.849	1.478
8	2.95	2.75	0.774	0.091	0.865	2.433	10.216	11.806	5.431
7	3.35	3.15	0.887	0.095	0.982	2.641	12.594	12.828	4.787
6	3.75	3.55	0.999	0.096	1.095	2.849	15.210	13.894	4.293
5	4.15	3.95	1.112	0.094	1.205	3.058	18.062	14.986	3.900
4	4.55	4.35	1.224	0.090	1.314	3.266	21.152	16.094	3.576
3	4.95	4.75	1.337	0.086	1.422	3.474	24.419	17.209	3.304
2	5.25	5.10	1.077	0.061	1.138	3.630	27.129	23.849	4.132
1	5.50	5.40	1.140	0.058	1.198	3.786	29.913	24.968	3.923

Table 6 Calculation of safety factor under the dynamic load

Layer	Depth to bottom of layer (m)	Depth to middle of layer (m)	Fhd (1/m)	Sld (1/m)	Fh (1/m)	Le (m)	P (1/m)	F.S. (Pullout)	F.S. (Rupture)
13	0.95	0.75	0.211	0.144	0.355	1.392	1.882	5.295	3.095
12	1.35	1.15	0.324	0.108	0.432	1.600	3.075	7.118	2.547
11	1.75	1.55	0.436	0.071	0.508	1.808	4.504	8.874	2.167
10	2.15	1.95	0.549	0.046	0.595	2.015	6.171	10.375	1.849
9	2.55	2.35	0.661	0.030	0.692	2.225	8.075	11.677	1.591
8	2.95	2.75	0.774	0.020	0.794	2.433	10.216	12.862	5.917
7	3.35	3.15	0.887	0.014	0.901	2.641	12.594	13.983	5.218
6	3.75	3.55	0.999	0.010	1.009	2.849	15.210	15.070	4.657
5	4.15	3.95	1.112	0.007	1.119	3.058	18.062	16.139	4.200
4	4.55	4.35	1.224	0.005	1.230	3.266	21.152	17.198	3.821
3	4.95	4.75	1.337	0.004	1.341	3.474	24.419	18.251	3.504
2	5.25	5.10	1.077	0.003	1.079	3.630	27.129	25.139	4.355
1	5.50	5.40	1.140	0.002	1.142	3.786	29.913	26.192	4.115

was conducted, ignoring the fabric's elongation because the time from construction to test is very short, only 5 months.
Table 5,6 showed the safety factor calculation by static and dynamic loads. The calculation was conducted by New mark for uniform steel plates loads and Boussinesq's formula by the truck loads. The rupture resistance was a little short at some layers. For example, safety factor on layer 8 is 1.47 for rapture in the case of pattern Ⅳ and 1.59 for pullout in dynamic test. So it is safer for pullout resistance but is insufficient for rupture resistance. Thinking of long term creep criteria, the rupture resistance becomes a little dangerous. The present design method which takes long term creep criteria and safety factor 1.5 for both pullout and rupture resistance is believed to be conservative.

2.6 The wall's behavior with time and rainfall

The measurement was conducted twice or once month from 28th Feb. 1986, the day which is the construction completed to 4th Aug. and after that was continuously performed once a month. The wall's face horizontal movement on layer 8 exposed face from 27th Feb. to 4th Aug. on layer 12 was 4.4cm(5 month), caused by soil consolidation and compaction. The displacement by static and dynamic load test was 2.2cm while the movement from 20th Aug. to 27th Mar. 1987 was only 0.3cm during 7 months. The back fill soil was consolidated by excessive (static and dynamic) loads. So it is very important to compact the soil for geotextile reinforced wall engineering. There were total amount rainfalls 1046mm from Mar. to Aug. and over 200mm in May, June and July (shown in Fig.6). Especially in 20th July, there were 137mm rainfall and 31mm one hour rainfall (Yamashiro heavy rain disaster) about 0.9o horizontal displacement was observed on layer 12 face from 17th July to 30th July. However in 1986 there were less displacement.

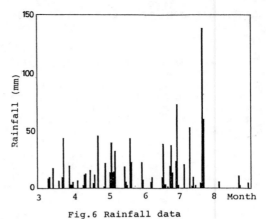

Fig.6 Rainfall data

2.7 Facing and cost consideration

The wall was faced with a gunite in March 1986 after the wall stabilization was confirmed.
At Nomura tunnel entrance approach road, the geotextile reinforced retaining wall was constructed. This wall's scope is following. Height 3.9m, wall top length 159.4m, surface area 217u. At this project, the comparative checking was made for cost and construction time on the geotextile wall, reinforced concrete wall (inverse T type) and Terre-arme type (metal crib or concrete crib). It is found that the geotextile reinforced wall is the most economical and construction time is the shortest. The geotextile reinforced wall was adopted because of this reason and limited right of way.
The behavior survey was performed but was not accurate because of the project work's vandalism.

3 CONCLUSION

As stated above, the static loads test to maximum loads intensity $5.88tf/m^2$ was performed. The maximum settlement was 2.0cm at pt.2, and less than 1.0cm at other point on the pavement and as for horizontal displacement the movement on the wall top was about 3.0cm forward ,but there were less than 1.0cm displacement backward displacement on the foundation ground as the consolidation progress. In dynamic loads test, the truck 20ton travel forth and back 1000 times and the settlement was less than 1.5cm at their maximum settlement point on the pavement surface and the settlement in proportion with numbers of pass increasing had not been observed.

As for safety factor, there are no problems as long as granular materials is utilized as backfill. Values of 35° and 1.6 t/m^3 are assumed for $\emptyset$ and unit weight of backfill. This value of $\emptyset$ may be low. The analysis is assumed the at rest earth pressure (K_O) rather than the active pressure (Ka). The use of the at rest pressure was reasonable when the heavy traffic loads were expected but may be excessively conservative for the dead loads only. This could result in over estimation of the total load to be resisted by the geotextiles. The theory is tie back analysis but wall is composite reinforced materials but how conservative is not known. At the present conventional design, safety factor values 1.5 for rupture resistance was appropriate. This wall was made of the granular material. The study for cohesion material will be done in the future.

REFERENCE

1) F.H.W.A. Geotextile Engineering Manual: federal high way Administration, Washington D.C.,1984.
2) T.A.Haliburton, J.D.Lawmaster: Use of Engineering Fabrics in Transportation related application, Oct.,1981,F.H.W.A. U.S.
3) J.Steward, R.Williamson and J.Mohney: Guidelines for use of Fabrics in construction and maintenance of Low volume loads, June,1977, Forest service U.S.

International Geotechnical Symposium on Theory and Practice of Earth Reinforcement / Fukuoka Japan / 5-7 October 1988
© 1988 Balkema, Rotterdam. ISBN 90 6191 820 0

The limit equilibrium of geotextile reinforced structures

Y. Zhang
Ministry of Water Resources, People's Republic of China

E. Schwing
University of Karlsruhe, FR Germany

ABSTRACT: The global stability of geotextile reinforced retaining structures is usually determined without any consideration of the deformation properties of the composite building material soil-geotextile. For practical design the orientation of the geotextile reinforcement towards potential slip surfaces is investigated using a simple equilibrium analysis. It is shown that safety increases reaching the limit state with the appearence of local deformations.

1 INTRODUCTION

The stability analysis for geotextile reinforced walls is still done the same way as for conventional retaining walls. A substantial difference however is that the slip surfaces can penetrate the supporting body of geotextile in the limit state. More accurate studies on stability analysis of geotextile reinforced retaining structures are presented in [1]. It is shown in the analysis of stability that one important difference between geotextile reinforced walls and classical retaining walls (such as gravity walls) is, that tensile forces are variable in direction and magnitude. Under these circumstances the classical elastic theory can not be directly used.

In practice, changes occur in relation to the direction and magnitude of tensile forces to the effect that the stability tends to increase at the initial state of slip in geotextile reinforced walls. Such tendency stops only when slip deformation reaches a certain degree (figure 1c). This phenomenon is a result of the limit state of geotextile reinforced walls.

The present paper aims at finding out the limit state of geotextile reinforced walls through a theoretical investigation of the influence of the tensile force inclinations on geotextile reinforced wall stability. The limit state can be expressed in a simple equation of limit equilibrium which therefore is very practical.

The adopted approach and part of the drawn conclusions are considered to be of general significance.

2 LIMIT STATE AND LIMIT EQUILIBRIUM

For inner stability of geotextile reinforced walls the factor of safety is defined as:

$$F = \frac{\sum_{i=1}^{n} \vec{R}_i}{\sum_{i=1}^{n} \vec{T}_i} = \frac{R}{E} \qquad (1)$$

with (figure 1)

$\vec{R}_i$: pull out resistance force of the i − layer,

$R : \sum_{i=1}^{n} \vec{R}_i$

$\vec{T}_i$: actual tensile force of the i − layer geotextile
    acting on the slip surface, $E = \sum_{i=1}^{n} \vec{T}_i$

E : active earth pressure

n : number of layers

The tensile forces are always horizontal, providing that no deformation occurs in the reinforced soil (geotextiles are built in horizontally). When slip occurs inside geotextile reinforced walls, $\vec{R}_i$ and $\vec{T}_i$ are sure to incline, leading to a corresponding change of F (figure 1 a,b,c).

As safety factors always refer to limit state equilibrium it may seem conservative to obtain a safety factor with $\theta=0$ as limit state (all forms of safety expressions lead to the same conclusion). Therefore it is of primary importance to determine the limit state.

2.1 Equation of equilibrium for local forces

The forces acting in the i-layer geotextile around the slip surface (figure 1d) can be expressed as follows:
a) from equilibrium of the horizontal forces

$$q_i \tan\beta r_i \sin\theta_i - q_i r_i (1 - \cos\theta_i) + T_i' - T_i \cos\theta_i = 0 \quad (2)$$

b) from equilibrium of the vertical forces

$$q_i \tan\beta r_i (1 - \cos\theta_i) + q_i r_i \sin\theta_i - T_i \sin\theta_i = 0 \quad (3)$$

with

T_i', T_i : tensile forces; $T_i' = T_i'' + 2\,l_i'\,\sigma_i \tan\beta$

$\tan\beta$: coefficient of friction between earth and geotextile.

Thus (2) and (3) yield

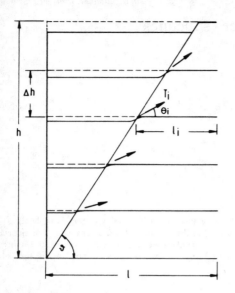

a) System

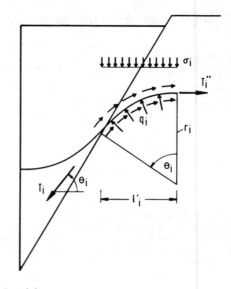

d) Local forces

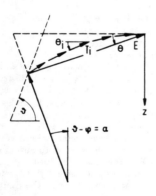

b) Polygon of forces

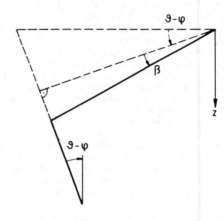

e) Significance of limit state

Fig.1. General description

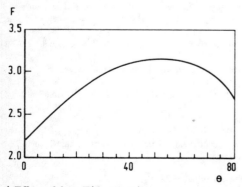

c) Effect of θ on $F(\vartheta=\text{const.})$

$$T_i = T_i' \frac{\cos(\beta - \theta_i/2)}{\cos(\beta + \theta_i/2)} \qquad (4)$$

Equ.(4) denotes the increase of tensile forces due to local pressures.

2.2 Limit state

It is shown in figure 1c, that in the course of deformation the tensile force inclines with the result of F,

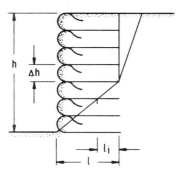

a) System

b) Model tests

Fig.2. Two-wedge failure system

increasing at the initial stage of deformation till θ reaches a certain degree. In other words, the maximum value of $F(\theta)$ corresponds to the limit state of geotextile reinforced walls. In this paper the cause of the inclination of θ (or θ_i) is of little importance. That geotextile reinforced soil is able to deform is all that matters.

From formula (1) ensues:

$$F = \frac{R}{E} = \frac{\sum_{i=1}^{n} R_i \cos(\theta - \theta_i)}{E_h \cos\alpha} \cos(\alpha - \theta) \quad (5)$$

with

$$E_h = E|_{\theta=0} = E\frac{\cos(\alpha - \theta)}{\cos\alpha}$$

(See figure 1 for characters used)

From $\delta F/\delta\theta_i = 0$, $\quad R = R_i' \frac{\cos(\beta - \theta_i/2)}{\cos(\beta + \theta_i/2)}$,

$\Delta\theta/\Delta\theta_i = R_i/R$:

$$R_i \sin(\alpha - \theta) + \cos(\alpha - \theta)[R_i \sin(\theta - \theta_i)$$
$$+ R_i \frac{\sin 2\beta \cos(\theta - \theta_i)}{\cos 2\beta + \cos\theta_i}] = 0 \quad (6)$$

Adding all $\delta F/\delta\theta_i = 0$:

$$\sum_{i=1}^{n} R_i \sin(\alpha - \theta) + \sin 2\beta \cos(\alpha - \theta)$$
$$\sum_{i=1}^{n} R_i \frac{\cos(\theta - \theta_i)}{\cos 2\beta + \cos\theta_i} = 0 \quad (7)$$

Within the range of $0 \leq \theta \leq \vartheta$, the solution of formula (7) is always equal to or slightly larger then $\alpha + \beta$. Therefore it is on the safe side. Limit state is expressed as:

$$\theta = \alpha + \beta \qquad (0 \leq \theta \leq \vartheta) \qquad (8)$$

with $\alpha = \vartheta - \varphi \qquad$; (φ : angle of internal friction)

2.3 Equation for limit state equilibrium

From (1) and (8), the equation for limit state equilibrium is :

$$F = \frac{\sum_{i=1}^{n} R_i' \cos(\beta - \theta_i/2)/\cos(\beta + \theta_i/2)}{E_h \cos\alpha} \cdot$$
$$\cos(\alpha - \theta) \quad (9)$$

with
$R_i' = 2\sigma_i\, l_i \tan\beta$: pull out resistance
l_i : anchor length behind the slip surface
σ_i : normal pressure on the geotextile

Though θ is uncertain in formula (9), the following formula is obtained from $0 \leq \theta \leq \vartheta$:

$$\frac{\sum_{i=1}^{n} R_i' \cos(\beta - \vartheta/2)\cos(\alpha - \theta)}{E_h \cos(\theta + \beta - \vartheta/2)\cos\alpha} \leq F \leq$$
$$\frac{\sum_{i=1}^{n} R_i' \cos(\beta - \theta/2)\cos(\alpha - \theta)}{E_h \cos(\beta + \theta/2)\cos\alpha} \quad (10)$$

with, if $\theta_i = \theta$ (i=1,2,...n) is true

$$F = \frac{\sum_{i=1}^{n} R_i'(\cos\beta - \theta/2)\cos(\alpha - \theta)}{Z_h \cos(\beta + \theta/2)\cos\alpha} \quad (11)$$

Generally, biased towards the lower safety bound of F, it is suggested, that :

$$F = \frac{\sum_{i=1}^{n} R_i' \cos(\beta - \vartheta/2)\cos(\alpha - \theta)}{Z_h \cos(\theta + \beta - \vartheta/2)\cos\alpha} \quad (12)$$

where ϑ is obtained corresponding to the minimum value of F.

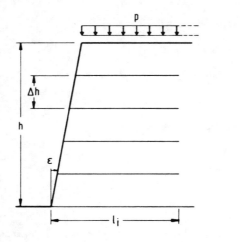

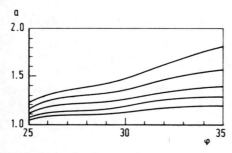

a) System; $\varepsilon=10$, $1/h=1.0$, $\Delta h/h=0.2$, $p/\gamma_h=0.4$

b) Inclination of $a=F_{\theta=\alpha+\beta}$ $|_{F_{\theta=0}}$

Fig.3. Effect of tensile forces

3 DISCUSSION

As limit state refers to the whole structure, resultant θ is used in formulas (8), (11), and (12) rather than θ_i for the purpose of clearness. It is true that θ_i sometimes provides more accurate results, but it complicates the expression of formulas (θ_i, q_i and r_i result from the distribution of earth pressure, mechanical characteristics of soil and geotextile and other factors). Actually the difference between F's upper and lower bound ((11),(12)) is relatively very small, so a somewhat accurate solution is of little significance.
. If for example friction by lower pressure is neglected ($R_i = R_i'$ with $\beta = 0$ (4)), maximum F is conditioned by $\theta = \alpha$ (E=mini., R=const.). It can be seen in figure 1e, where E, vertical to a side of the polygon, has the shortest distance. Correspondingly:

$$F = \frac{\sum_{i=1}^{n} R_i' \cos(\alpha - \theta)}{E_h \cos\alpha} = \frac{\sum_{i=1}^{n} R_i'}{E_h \cos\alpha} \qquad (13)$$

Obviously θ_i is irrelevant for the results, here. Formula (13) may also be considered as lower bound of F, when β/φ is smaller (eg. $\beta/\varphi < 0.4$).

Model tests (IBF, Univ. Karlsruhe) show that, if the slip surface extends into non-reinforced soil, two slip bodies emerge (figure 2a,b). Calculations in this paper are all done along this line. Calculation results for figure 3 indicate the influence of tensile force inclinations on the increase of the safety factor, which (though not always so obvious) is comparatively observable when θ becomes larger. It is demonstrated through calculation, that θ seems to be little affected by surcharge.

In general ϑ is smaller when tensile force inclination is considered.

So far, attention has been directed at a plain slip surface; formulas (8) and (12) can also be applied to a curved slip surface and other geo/soil structures, providing that the polygon of forces is in analogy to the one indicated in figure 1b.

4 CONCLUSION

A certain degree of slip deformation is favorable to the stability of geotextile reinforced walls, with the result that F increases.
Affected by tensile force inclination, the limit state of a geotextile reinforced wall is:

$$\theta = \alpha + \beta \qquad (0 \leq \theta \leq \vartheta),$$

which is corresponding to the maximum F (for a given ϑ)

The difference between the lower and the upper bound solution of F is relatively very small.

REFERENCES

Gudehus, G. & Schwing, E. (1986) : Standsicherheit kunststoffbewehrter Erdbauwerke an Geländesprüngen. Baugrundtagung, Nürnberg 1986, 129 - 147.

Proceedings of the 3rd symposium of the international conference on geotextiles. Vol. I – V, Vienna, 1986.

International Geotechnical Symposium on Theory and Practice of Earth Reinforcement / Fukuoka Japan / 5-7 October 1988
© 1988 Balkema, Rotterdam. ISBN 90 6191 820 0

26m height reinforced soil quaywall: Design and full-scale test

Zhongchun Ouyang
Chongqing Institute of Communication, People's Republic of China

ABSTRACT: In this paper the highlights of the design and main results of on-site test on 26m height geosynthetic-fiber reinforced soil quaywall are presented. Author proposes an experimental equation to calculate the tension in the extensible inclusions (geosynthetic-fiber). A reinforcing-tieback system failure plane is also discussed.

1 INTRODUCTION

Chongqing located by the Yangtze river in China is a beautiful mountain city, its topographic form is narrow, and its banks are high and precipitous. There are many tall retaining structures in the city. However what are the design outlines of the geosynthetic-fiber reinforced soil structures in this kind of city? And how to consider the calculating method? Our testing project, Bai Sha Wan reinforced soil quaywall, is an attempt.

Bai Sha Wan reinforced soil wharf is a coal export one. It is located on the north bank of the Yangtze river in Chongqing. The quaywall is 137m in length, the part of reinforced soil is 90m in length, and from 18m to 26m in height. The loading area is 2200m². The quaywall was finished in July, 1985 (Fig. 1), and the loading test has been done (Fig. 2).

2 THE UNITS OF STRUCTURE

Bai Sha Wan wharf is a vertical quay. The units of the structure are as follows (Fig. 3).

1. The facing made of precast reinforced concrete panels measuring 150*50*25cm.
2. The fill consists of sand and gravel. The particle size is shown in Table 1. The unit weight of the fill is 19.6 KN/m³. The angle of internal friction is 36°.
3. The reinforcements are polypropylene strips, 20mm in width, 1.2mm in thickness. The ultimate tension in a piece of strip is greater than 4KN. The elongation at break is 19.2%, and 5% was used in design.
4. The tieback parapet. In order to stabilize the tall reinforced soil structure we have to increase the length of reinforcements, i.e., we have to increase the width of reinforced soil mass. But it is impossible to do on site where the topographic form is narrow and the digging is difficult. And the costs will be increased also. Therefore we set the girt-tieback-parapets separately at the middle and the top of the quay, so that a stability of reinforced soil mass is increased but the length of the reinforcements is not increased. The parapet is 1.5m in height, 0.7m in thickness and 30m in length of segmentation. The parapets are connected with deadman using steel bars. A horizontal space between the bars is 1.0m. The deadman is 50*60*30cm. All deadmans are laid behind the reinforced soil mass.
5. Filter is close to the facing, 1.2m in width. The fill of the filter consists of sand and gravel with elements from 0 to 20mm.
6. Foundation was made of stone masonry or stone concrete, which

599

Fig. 1 Bai Sha Wan reinforced
soil wharf

Fig. 2   Loading condition in
test

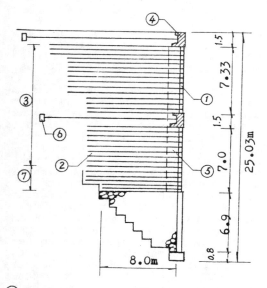

① Facing   ④ Parapet
② Fill   ⑤ Filter
③ Polypropylence ⑥ Deadman
  strips   ⑦ Metal strips

Fig.3 Section of Bai Sha Wan
reinforced soil quaywall

Table 1. Particle size of the fill
in Bai Sha Wan quaywall

Size (mm)	2	2-20	21-60	60
Content (%)	17.1	21.4	54.7	6.8

was built along the bank slope.
The top of the foundation is 8m in
width.

3 CONSIDERATION OF THE TENSION IN THE REINFORCEMENT

The calculation of the tension in
reinforcement relates not only to
the number and the length of reinforcement but also to the internal
stability of the reinforced soil.
The present methods to calculate
the tension in the reinforcement
are not suitable for greater than
15m height quaywall reinforced with
lower-modulus-extensible inclusions
(geosynthetic fiber etc.), e.g., if
we adopted the Rankine's equation
or the Mayehef's equation to calculate the tension in the course of
designing the Bai Sha Wan quaywall,
the tension at a node can be more
than 40KN, i.e., more than 30 pieces
of the geosynthetic strips are laid
within 0.5m, which makes it difficult not only to lay the strips but
also to limit mobilizing the friction between the strips and the
fill. Fig.4 shows that the tensions
calculated by Coulomb's equation or
by Rankine's equation are less. Because the above equations allowed
the fill is at active state of limiting equilibrium in full, which
is inclined to disadvantage for the
reinforced soil structures over 15m
height. Although Fig. 4 shows that
the tension diagram calculated by
Osman's equation could be approximately accordant with the result in
centrifugal model test (Fig. 5),
the Osman's tension will be reduced
to zero at the foot of wall, which
brings about blindness to lay the
reinforcements at the foot. In
theory it is possible that the lateral earth pressure is equal to
zero, in this case $\phi = 90$, the cohesion must be large enough. In
fact the reinforced soil are mostly
cohesionless and the "cohesion" increased by reinforcement is limit-

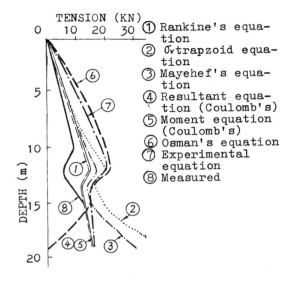

① Rankine's equa-
   tion
② σᵥtrapzoid equa-
   tion
③ Mayehef's equa-
   tion
④ Resultant equa-
   tion (Coulomb's)
⑤ Moment equation
   (Coulomb's)
⑥ Osman's equation
⑦ Experimental
   equation
⑧ Measured

Fig.4 Tension measurements and
    comparison

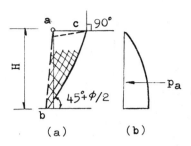

Fig.6 When a wall rotates around
    the top (a) failure plane
    (b) earth pressure distri-
    bution

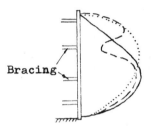

Fig.7 Earth pressure measurements
    with bracing in excavation

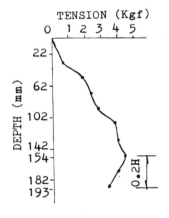

Fig.5 Peak tensions at 63g in
    centrifugal model

4 THE EXPERIMENTAL EQUATION

The tension in the reinforcement is
determined in terms of the lateral
earth pressure. The tension distribution along the height of the wall
should be fit in with the lateral
earth pressure. A lot of measurement as well as investigation of
Terzaghi and Peck showed that the
earth pressure distribution is such
as shown in Fig. 6(b) and Fig. 7
when the wall rotates around the
top. The earth pressure distribution measured from installation of
bracing as the excavation proceeded
had the shape of curve. In general
the Terzaghi's and Peck's "apparent-
earth-pressure" diagram was used to
design a retaining wall and the
Schnabel's "apparent-earth-pressure" diagram was used to design a
tied-back wall, which have performed successfully also (Fig. 8).

ed. Thus it is impossible that the
lateral earth pressure is zero at
the toe, then the tension can't be
also reduced to zero. Based on the
practice of the Bai Sha Wan quaywall the author suggested an experimental equation to calculate the
tension in geosynthetic fiber reinforcements.

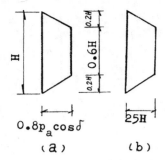

Fig.8 "Apparent-earth-pressure" envelope: (a) Terzaghi and Peck (b) Schnabel

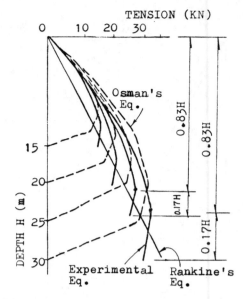

Fig.9 Experimental equation is compared with other equation

The acting force of the reinforced soil with geosynthetic-fiber strips is similar to that of the bracing wall and the tied-back wall in some ways. The bracing apply a thrust on the wall along the depth of the excavation, and the reinforced soil or tied-back wall apply an anchoring force on the wall behind the facing. Both of them are the horizontal supporting force. Therefore the diagram of the tension in reinforcements should be also similar to earth pressure distribution of the bracing wall and the Schnabel "tied-back wall", i.e. it has the shape of firstly-increased and laterly-reduced curve which has been demonstrated by Bolton's results from the centrifugal model test as shown in Fig. 5.

Based on the above considerations the following conclusions can be drawn:

1. The tensile distribution in the reinforcements has the shape of firstly-increased and laterly-reduced curve as shown in Fig. 5.

2. The maximum tension acts at an approximate distance of $0.2H$ above the bottom of the wall.

Therefore the author suggested an experimental equation to calculate the tension by means of the curve fitting as follow:

$$T_i = \sqrt{6K_a^{2.8}(H - 0.8h)/L} \quad rhS_xS_y$$

where $K = K_o(1 - h/6) + K_a$ when $h \leqslant 6m$

$K = K_a$ when $h > 6m$

K_o = coefficient of earth pressure at rest; $K_a = \tan(45° - \Phi/2)$; h=the depth at the ith layer; L= a strip length; r=density of soil; S_x = a horizontal spacing between the joint, and S_y = a vertical spacing; H=wall height; T_i = a tension in the reinforcement one joint at i layer.

Fig. 9 shows the comparison between the experimental equation and other equations. By comparison it is found that experimental equation diagram is close to the Osman's equation diagram above distance of $0.17H$ from the bottom of the wall, but within $0.17H$ distance the tension is slightly less than that of Rankine's equation, and it is not zero at the bottom, which just satisfied designing reinforced soil wall. The maximum tension of the experimental equation is at a distance of $0.17H$ from the bottom. The experimental equation diagram is also similar to Fig. 5, and it is also satisfactory for designing Bai Sha Wan quaywall.

In addition the experimental equation diagram was similar to the measurement of the earth pressure in Bai Sha Wan quaywall, which demonstrates the tension distribution

602

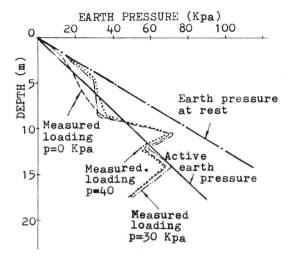

Fig.10 Measurement of the earth
pressure in Bai Sha Wan
quaywall

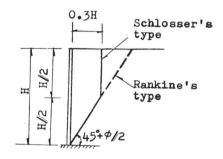

Fig.11 Typical failure planes of
reinforced soil

along the height of the wall calculated by the experimental equation
coincides with the earth pressure
distribution (Fig. 10). Therefore
the experimental equation is also
satisfactory for designing Bai Sha
Wan quaywall.

5 THE FAILURE PLANE

The present typical failure plane
of the reinforced earth are two
types, i.e., the Rankine's type
and the Schlosser's type (Fig. 11).

The Rankine's type required the
length of the reinforcement to be
0.8H. It is obvious that this
length may be impossible for the
great height reinforced earth wall
in the mountain area; and the
Schlosser's type required the
length of the reinforcement to be
0.3H enough, it is obvious that
this length may be too shorter for
the great high wall. From this reason we set separately the tieback
parapet which is like the girt at
the middle and the top of the quaywall to increase the stability of
the quaywall rather than to increase the length of reinforcement.
Owing to the restrained action of
the tieback parapet a lateral displacement of the upper part of the
quaywall might be developed diffi-
cultly, which forms the condition
to turn about the top of the wall.
Terzaghi and Peck considered that
the failure plane corresponding to
the above condition should be like
Fig. 6(a), i.e., the failure plane
which starts from b made an angle
of $(45° + \phi/2)$ with the horizontal
(Fig. 6), and then the failure
plane is orthogonal gradually to
the fill surface. That means before
the foot of the failure wedge arrives at the state of plastic equilibrium the upper part of the wedge
is still at the state of elastic
equilibrium, i.e., the upper fill
in the wedge is settlement rather
than shear failure. Thus when the
lower part of the reinforced fill
is firstly unsteady, the failure
plane will be developed, the fill
behind the parapet will be still at
state of the elastic equilibrium
(vertical settlement). In consequence of settlement the point of
the maximum tension in reinforcement behind the parapet moves towards to the facing and the reinforcement must be cut at M (Fig.12).
Therefore the failure plane of the
reinforced soil quaywall which has
a tieback parapet may be simplified
as shown in Fig. 13.

6 CONCLUSION

1. For the reinforced soil structure over 15m height the experimental equation presented in this paper might be adopted to calculate
the tension in the geosynthetic-

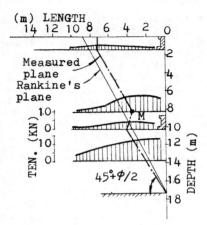

Fig.12 Measurements of the tension along the length of the strip

REFERENCES

Bolton, M.D. & Pang, P.L.R. 1982.
Collapse limit states of rein-
forced earth retaining walls.
Geotechnique 4: 349-367.
Harry Schnabel, Jr. 1982. Tiebacks
in foundation engineering and
construction: 129-149, Mcgraw-
Hill, Inc.
Lee, Kenneth L., Adams, Bobby Dean
and Vagneron, Jean-Marie. 1973.
Reinforced earth retaining walls.
Journal of the Soil Mechanics and
Foundation Division, Vol. 99, No.
SM10:745-764.
Ouyang Zhongchun, 1986. Design and
test of reinforced soil structure
over 15m height. Proceeding of
CCES 3rd convention: 387-398.
Terzaghi and Peck, 1948. Soil mech-
anics in engineering practice.
New York.

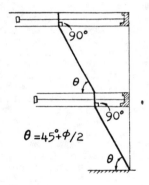

Fig.13 Reinforcing-tieback system failure plane

fiber reinforcement, which can
obtain the satisfactory results.
   2. For reinforced soil structure
with a tieback parapet, the simplified failure plane presented in
the paper can be used to determine
the length of the reinforcement.
The girt-tieback-parapet is set in
the great height reinforced soil
structures, which can reduce the
length of the reinforcement as well
as increase the stability of a
structure.

International Geotechnical Symposium on Theory and Practice of Earth Reinforcement / Fukuoka Japan / 5-7 October 1988
© 1988 Balkema, Rotterdam. ISBN 90 6191 820 0

Soil-nailing – Design and application to modern and ancient retaining walls

E.Schwing & G.Gudehus
Institute of Soil Mechanics and Rock Mechanics, University of Karlsruhe, FR Germany

ABSTRACT: Two practical projects of nailed retaining walls performed in West–Germany are presented. The first one is a steep cut; forces at the nail heads and displacements of the wall surface were measured during a period of about three years to investigate the influence of freezing. The second one is an ancient masonry retaining wall. A simplified limit state equation, based upon a two-body translatory mechanism and ground bearing capacity is formulated. By means of soil nailing the safety and reliability of the old structure increases to a sufficient level. A kind of observational method is applied.

1 INTRODUCTION

The technique of soil nailing has a wide range of applications for temporary and permanent structures. It is mainly used for constructing excavations and stabilizing slopes (Gässler & Gudehus 1981; Stocker et. al. 1979). Numerous projects have been carried out (Bruce &Jewell 1986). The theoretical investigations of the relevant failure mechanisms are summarized by Gässler (1987).

This paper deals first with the influence of frost. Measurement results of nail forces and displacements are presented. Second, the stability analysis of nailed masonry retaining walls on the basis of a consistent failure model is employed. The relevant failure mechanism is confirmed by model tests. By means of probability theory the risk of failure of the structure can be calculated.

2 STEEP CUT

2.1 General description

In the course of the construction of a new road a cut with a height of about 7m and an average inclination of 70° had to be executed. To stabilize the cut, soil-nailing was used with shotcrete and earth terraces. (Fig.1). The terraces were filled up with top soil and planted to cover the shotcrete face.

In the area of the cut weathered debris of Keuper marl (Gipskeuper) is covered by a layer of loam up to several meters thickness. Ground water was not detected. Following our experience, an angle of internal friction of $\varphi'=35°$ could be assumed. The unit weight amounts to 20 kN/m³. For the stability analysis, a two-body translation mechanism with partial safety factors for the relevant parameters was used, without further investigations (Gässler 1987). The par-

tial safety factors are: $\gamma_\varphi=1.20$ for the angle of internal friction, $\gamma_p=1.30$ for the live load and $\gamma_T=1.6$ for the pull-out resistance applied to the respective mean values. With these parameters, a failure probability of $p_f < 10^{-6}$ is obtained following the Level II approach (Gässler & Gudehus 1983, Hasofer & Lind 1974).

Loading tests were carried starting after the end of the construction in order to determine the pull-out resistance of the soil-nail system. The necessary values to prove the safety of the wall were reached. Loading was halted before the limit state was reached. Therefore a prediction of the precise limits value can not be made.

2.2 Measurements

The main attention was on the developement of forces and displacement after execution of the cut. Especially, the effects of repeated freeze-thaw changes were to be observed.

In a part of the nailed wall, the following instruments

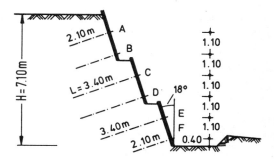

Fig.1 Cross section of cut

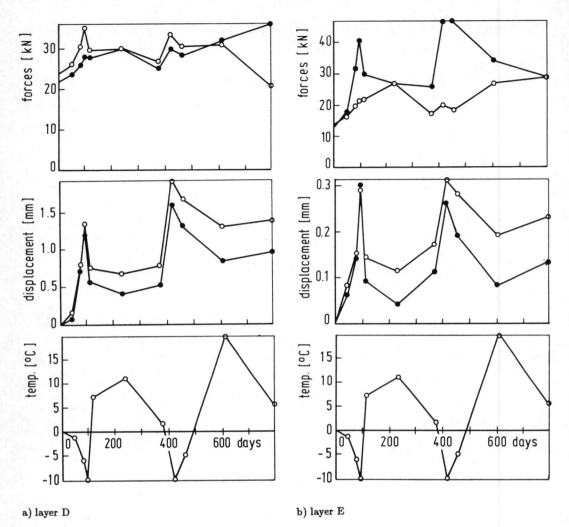

a) layer D

b) layer E

Fig.2: Temperatures, forces and displacements

were installed:
– 4 force meters to measure the nail-head forces,
– 4 extensiometers (length: 1.5m) to measure the displacement of the shotcrete wall (Fig.1); two of them in layer D and two in layer E with a horizontal distance of about 2m.

After execution of the cut the first (zero) measurement was made. During a time span of about 2.5 years 11 measurements were taken. Figs.2a,b contain the results.

2.3 Interpretation of the results

It can be seen that frost has a remarkable influence on the nail-head forces and on the displacement of the wall; both increase with frost-duration.

The nail-head forces amount, with low temperatures, to about twice the values of the zero measurement, on both the upper and the lower row of points of measurement. The scatter of results of measurements in the upper layer is remarkably bigger than in the upper layers, which indicates a heterogeneous structure of the soil in this area.

The forces built up in the first frost period decreased to a value greater than the one of the zero measurement. During the second frost period, with about the same low temperatures, they increased again to the values reached before.

The displacement of the wall showed the same changes with temperature; they also increase with frost duration.

A final evaluation of the development of forces and

606

displacements is not jet possible. It is planned to make more measurements, twice a year to the minimum, once during the frost period and once in summer.

The measurements taken at the foot of the wall (layer E) show greater values for displacement than the ones taken in the upper part. As a result of the construction technique, the nails take up smaller tensile forces. Because of the creep and settlement behaviour of the soil they asymtotically reach the values calculated in the final state.

3 ANCIENT RETAINING WALLS

3.1 General description

A lot of old masonry gravity-retaining walls in West–Germany , made out of natural stone with a height of 2m up to more than 8m, are not safe enough. Cracks and bulging witness the inadequate bearing capacity of the wall foundation. The removal of the unsafe constructions would not be adequate. A stabilisation was repeatedly carried out using soil nailing in order to reduce the active earth pressure.

In contrast to conventional soil nailing the ground under the wall can -according to its bearing capacity- still carry an amount of the earth pressure. A mechanism was found to describe this nail-gravity-wall system in the limit state. A statically and kinematically consistent failure model, correctly describing the system in limit state is employed.

A concrete face of the wall could not be applied out of esthetical reasons (preservation of a historical monument). The nail heads, therefore, had to be hidden. This paper shows how to hide the load bearing nail-heads behind the wall.

The stability analysis of the nail-retaining wall-system was completed by means of statistical methods.

3.2 Stability analysis

The basic failure mechanisms for nailed walls are presented in detail by Gässler (1987). The limit state equation can be simplified using a two-body translation mechanism. In our case, the necessary outer concrete face is replaced by the existing gravity wall. Depending on the width b at the foot, the wall can carry a certain part of the earth pressure. Recent investigations of more than 20 ancient retaining walls has shown that they have very shallow foundations, that their back side is nearly vertical, and that they have an almost constant thickness. Only walls of this type are investigated here. Fig. 3 shows the failure mechanism relevant after nailing.

Equation of limit state

The global system can be subdivieded into two bodies. The forces acting upon the retainig wall and the nailed body are shown in Fig.4 a,b. After some algebraic transformations an expression for the earth pressure is

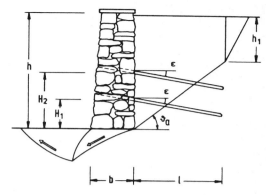

Fig.3 Nailed retaining wall with failure mechanism

obtained:

$$E = \frac{\tan(\vartheta_a - \varphi)}{\sin\delta + \cos\delta} \cdot \left[\sum \frac{T}{a} \sin\epsilon + E_a \sin\varphi + W_1 \right.$$
$$\left. - C \frac{\sin\vartheta_a \, 1}{\tan(\vartheta_a - \varphi)} \left(\sum T\cot\epsilon + E_a \cos\varphi - C\cos\vartheta_a \right) \right]$$

(1)

$\delta = \varphi$ can be assumed because the back side of the wall is rough and a slip plane developes there. Herein $W_1 = 1/2\gamma l(2h - l\tan\vartheta_a)$ denotes the weight of the nailed body, E_a the earth pressure acting on the vertical intermediate slip surface with $h_1 = h - l\tan\vartheta_a$ (for $E_a \leq 0$, it is substituted by $E_a = 0$), and the resultant cohesion $C = cl/\cos\vartheta_a$. ϑ_a is the inclination of the slip surface with the maximum of E, found by variation of the slip surface inclination.

$\sum T = \sum_{i=1}^{2} \frac{1 - H_i}{\tan\epsilon + \tan\vartheta_a} \frac{1}{a} \frac{1}{\cos\vartheta_a}$ the resultant nail forces per unit length (with a: horizontal distance of nails; ϵ: inclination of the nails).

The vertical forces at the bottom of the wall are limited by the bearing capacity of the soil.

$$V_b = b' \cdot (c \cdot N_c \cdot i_c + \gamma \cdot b' \cdot N_\gamma \cdot i_\gamma) \qquad (2)$$

with the bearing capacity factors values (DIN 4017 (1979))

$N_q = e^{\pi \cdot \tan\varphi} \cdot \tan^2(45 + \varphi/2)$

$N_\gamma = (N_q - 1)\tan\varphi \qquad N_c = (N_q - 1) \cdot \cot\varphi$

$i_\gamma = \left(1 - \frac{\sum H}{\sum V + b' \cdot c \cot\varphi} \right)^3$

$i_q = \left(1 - 0.7 \cdot \frac{\sum H}{\sum V + b' \cdot c \cot\varphi} \right)^3$

$i_c = i_q - \frac{1 - i_q}{N_q - 1}$

Herein, $\sum H = E\cos\varphi$: the resultant horizontal forces $\sum V = W_W + E\sin\varphi$: the resultant vertical forces and $W_W = \gamma_w bh$,the weight of the wall.

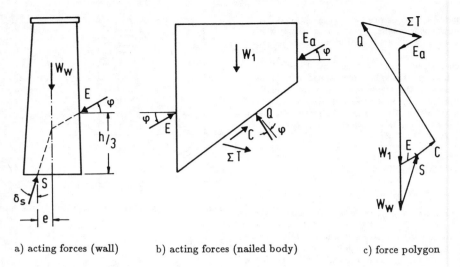

| a) acting forces (wall) | b) acting forces (nailed body) | c) force polygon |

Fig.4 System

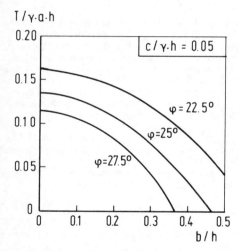

Fig.5 Design chart for $H_1/h=0.25$; $H_2/h=0.45$; $\gamma_w/\gamma=1.1$; $c/\gamma h=0.05$; $l/h=0.8$

The reduced width is calculated as

$b' = b - 2 \cdot e$

with the eccentricity $e = \dfrac{\sum M}{\sum V} = \dfrac{\sum H \cdot h/3 - E \cdot \sin\varphi \cdot b/2}{\sum V}$.

The earth-pressure resultant is assumed to have the height of h/3 above the level of foundation

Based on this failure mechanism, the limit state equation can now be formulated as:

$$V_b - \sum V = 0 \qquad (3)$$

It can be represented in diagrams with dimensionless parameters l/h, b/h, $c/\gamma h$, H_1/h, H_2/h, $T/\gamma ah$, γ_w/γ and φ and can be used for stability analysis.

3.3 Construction of nail heads

Nail heads can be anchored in front or behind the wall. Because of optical reasons and monument preservation, nail heads in front of the wall are not to be used. Anchoring in the body of the wall is problematic. The quality of masonry is not known and cavities are -in spite of injection- still present.

Thus a construction detail was designed to fit between wall and backfill. Using a water-jet injection the soil behind the wall is liquified within a diameter of 0.7 m, and sucked off. The cavity thus created is filled with cement grout after the placement of the nail. Experiments with this detail show a sufficient bearing capacity, much higher than the ones obtained with anchoring in the masonry of the wall. Moreover, the resulting displacements are much smaller.

3.4 Model Tests

Model tests were carried out according to the laws of similarity mechanics. Model walls were built out of hewn sandstone without mortar. The pull-out resistance of the model nails and their bending rigidity were correctly represented by using 1mm thick pipe cleaners. The length of the model nails was 10cm, and they were built in two layers. The nail head details were represented by thin rectangular sheets of wood. A contact to the wall was prevented. Dense dry sand (γ=16.3 kN/m^3, φ ~39°) was pluviated with thin black marker layers of 4cm distance.

Other tests showed that the failure mechanisms are

disturbed along the side walls of the model bin. Therefore, the earth bodies were cut vertically in the center after failure. This procedure was enabled by submerging and draining and thus producing sufficient capillary cohesion.

Model tests for unreinforced gravity walls showed that the wall should be safe up to a height of about 20cm. Knowing the pull-out resistance of the model nails and in consideration of the friction along the side walls the limit height of the reinforced wall at the point of collapse was predicted to 26cm using the limit state equation(3).

The sandstone wall was built up to the limit hight; collapse was triggered by pluviating sand behind the wall. Sufficient kinematic liberty was guaranteed, and no possible mechanism was favored beforehand. A combined mechanism as in Fig.6 with $\vartheta = 54°$ was observed. The calculated limit state mechanism is the one of Fig.4a with $\vartheta = 52.5°$ and is nearly the same as the one observed. Therefore the limit state function is sufficiently verified; especially the assumptions for the inclination and the acting height of the resultant earth pressure are confirmed.

3.5 Probability of failure

In the state limit function (equ.3) the quantities φ, c and T are scattering; they are so-called basic variables in the sense of the statistic safety theory (Hashofer and Lind 1974). The distributions of the angle of internal friction and the cohesion c are taken here from Walz and Genske (1987).They are for φ: log–normal,truncated at X_{φ_L}, coefficient of variation V_φ=0.05–0.075 and for c log–normal, coefficient of variation V_c=0.2. As shown by Gässler and Gudehus(1983), the unit weight of soil γ can be considered as non-scattering; likewise the unit weight of the wall γ_W. Pull–out tests to determine the distribu-

tion, the mean values and the coefficient of variation of the static friction nail/soil were carried out in different fields. The distribution can be fitted with a log–normal distribution. Typical test results of one field in silty soil are given in Fig. 7 a,b. It was found that the mean values were nearly independent of the nail length (Fig.7a). The coefficient of variation decreases with increasing length (Fig. 7b) as Gudehus (1987) proposes. The mean value of 5 fields (more than 100 tests) in silty soil is $T_0 = 34kN/m$ with a standard deviation $\sigma_{T_0} = 2.7$ kN/m and, therefore, a coefficient of variation $V_{T_0} \approx 8\%$. In the sense of the Bayesian approach this can be taken as prior information. This prior information can be combined with n observed data of pull-out test in a real project(sample mean $\bar{T}_x$ and sample standard deviation σ_{T_x}) to estimate the mean value m_T(Ang and Tang(1975)):

$$m_T = \frac{(\sigma_{T_x}^2/n)T_0 + \sigma_{T_0}^2 \bar{T}_x}{(\sigma_{T_x}^2/n) + \sigma_{T_0}} \qquad (4).$$

m_T is an average of the prior mean T_0 and the sample mean $\bar{T}_x$, weighted inversily by the respective variances.

The probability of failure, p_f, is estimated via the safety index β. In the Level II approach (Hasofer and Lind 1974) probability distributions are replaced by Gaussian ones. The limit state equation is linearized in the vicinity of the so-called design point with the

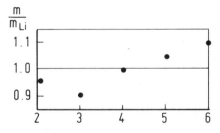

a) mean values

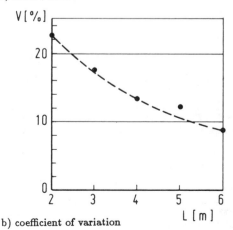

b) coefficient of variation

Fig.7 Results of pull-out tests

Fig.6 : Model test failure mechanism

design values X_i^*. β is the minimal distance of this point from the limit state function (Smith 1981).

The design values depend on the safety index β, the respective distributions, coefficients of variation, weight coefficient α_i and the limit state function. With the simplifying assumption that the weight coefficients α_i for the three basic variables φ, c and T are equal to $1/\sqrt{3}$ (Gudehus 1987), the design values can be derived for log–normal distributions as $X_i^* = m_i/\gamma_i$ with the partial safety factors

$$\gamma_i = \frac{\sqrt{1 + V_i^2}}{\exp(-\beta\sqrt{3}\sqrt{\ln(1 + V_i^2)})} \qquad (5)$$

respectively for at X_{L_i} truncated log-normal basic variables

$$\tilde{V} = \frac{V_i}{1 - X_{L_i}/m_i} \qquad (6)$$

$$\eta_i = \frac{\sqrt{1 + \tilde{V_i}^2}}{\exp(-\beta/\sqrt{3}\sqrt{\ln(1 + \tilde{V_i}^2)})} \qquad (7)$$

$$\gamma_i = \frac{\eta_i}{1 + X_{L_i}/m_i(\eta_i - 1)} \qquad (8)$$

3.6 Design example

A more than 200 year old masonry retaining wall with a height of 8m and a nearly constant width of 1.6m showed a bulged profile. Cracks and forward movements were observed. Bearing capacity calculations by the Level II approach showed a safety index β of 0.5 (i.e. probability of failure $p_f \approx 0.3$), whereas a safety index $\beta = 4.7$ (i.e. $p_f \approx 10^{-6}$) is required. In order to increase the safety to the sufficient level, the soil behind the wall was reinforced by nails. Three pull–out tests were performed.
Data:
walls: h=8m, b=1.6m, b/h=0.2, $\gamma_w = 22 kN/m^3$
silty soils: $m_\varphi = 30.5°$, $V_\varphi = 0.05$, $\varphi_L = 20°$,
$\quad m_c = 16 kN/m^2, V_c = 0.20, \gamma = 19 kN/m^3$
nails: sample mean $\tilde{T}_x = 30$ kN/m, standard deviation $\sigma_{T_x} = 3.6$ kN/m, n=3, L=6m, $\epsilon = 10°$, H_1=2m, H_2=3.6m
The horizontal nail distance a is to be determined. The design value of the cohesion is obtained (Eqn.5) as $c^* = 16/1.75 = 9.2$ kN/m^2 , for the angle of internal friction it can be found (Eqn.6–8) $V_\varphi = 0.145, \eta_\varphi = 1.5, \gamma_\varphi = 1.13$ and $\varphi^* = 30.5/1.13 = 27°$.

With the prior information $T_0 = 34$ kN/m, $\sigma_{T_0} = 2.7$kN/m and the observed data one obtained from Eqn.4 and 5 $m_T = 31.5$ kN/m, $\gamma_T = 1.4$ and $T^* = 31.5/1.4 = 22.5$ kN/m (mean length behind the slip surface ca. 4m, $V_T = 0.12$). The limit state function (Eqn. 3) requires a pull–out resistance of $T_1 = 13.3$kN/m per unit wall length. With the existing allowable value of $T^* = 22.5$ kN/m, the horizontal distance a has to be smaller than $T^*/T_1 = 22.5/13.3 = 1.67$ m. By means of the design chart (Fig. 5) one easily obtains, with $c^*/(\gamma h)$=0.06 and $\varphi = 27°$, $T/(\gamma ah) \approx 0.09$; a is given immediately by $a = T^*/(0.09 \cdot 20 \cdot 8) = 1.56$ m. The results of a Le-

vel II approach calculation for a choosen horizontal distance a=1.50m yield the sensitivity coefficient α_φ=0.545, α_c=0.694, φ_T=0.479 and the safety index β=5.2.

4 CONCLUSIONS

The conventional method of soil nailing requires a closed shotcrete face. The arrangement of berms in a steep cut leads to a stepped wall with open planes. Taking the frost action into account this wall is as safe as a conventional permanent nailed wall; the outer concrete face can be hidden by planting the earth terraces.

A failure model based on the kinematic failure mechanism of rigid bodies can be developed for nailed masonry retaining walls. The translation mechanism of two bodies combined with the ground bearing capacity is found as the relevant failure mode. It is found by variation of slip surfaces depending on all input data. The results of theoretical investigations can be verified by small model tests. By means of the new statistic–probabilistic theory partial safety factors for the relevant basic variables can be derived. Observed data of field tests combined with prior information yield a reliable structure. The proposed design procedure is simple and supported by more than a dozen successful applications.

5 REFERENCES

Ang, A. H-S. & Tang, W.H. (1975): Probability concepts in engineering planning and design. Vol. 1+2, John Wiley & Sons, Inc., New York.
Bruce,D.A & Jewell,R.A. (1986/87): Soil nailing: Application and practice. Part 1 and 2, Ground Engineering.
DIN 4017 (1979): Grundbruchberechnungen, Teil 1+2. ed. Deutsches Institut für Normung e.V., Beuth Verlag GmbH, Berlin.
Gässler, G. (1987): Vernagelte Geländesprünge - Tragverhalten und Standsicherheit (Doctor thesis). Veröff. d. Inst. f. Bodenmech. u. Felsmech., University of Karlsruhe, FRG, Heft 108.
Gässler, G. & Gudehus, G. (1981): Soil nailing - some aspects of a new technique. Proc 10th Int. Conf. Soil. Mech. and Found. Eng., Stockholm, Vol. 3: 665-670.
Gässler, G. & Gudehus,G. (1983): Soil nailing - Statistical design. Proc. 8th Eur. Conf. Soil Mech. and Found. Eng., Helsinki, Vol.2.
Gudehus,G. (1987): Sicherheitsnachweise für Grundbauwerke. Geotechnik, Heft 1, 4-34.
Hasofer, A.M. & Lind, N.C. (1974): Exact and invariant second moment code format. Journ. Eng. Mech. Div. ASCE, Vol 100, EM1: 111.
Smith, G.N. (1981): Probability theory in Geotechnics - An introduction. Ground engineering.
Stocker, M. Körber,G. Gässler, G. & Gudehus, G. (1979): Soil nailing. Proc. Colloque Int. sur le Renforcement des sols, 2, Paris, 469 -474.
Walz, D. & Genske, B. (1987): Anwendung der probabilistischen Sicherheitstheorie auf Grundbruchberechnungen nach DIN 4017. Geotechnik, Heft 2.

International Geotechnical Symposium on Theory and Practice of Earth Reinforcement / Fukuoka Japan / 5-7 October 1988
© 1988 Balkema, Rotterdam. ISBN 90 6191 820 0

Compatibility, serviceability and design factors for reinforced soil walls

R.A.Jewell
University of Oxford, UK

ABSTRACT: An approach to the design of reinforced soil walls is summarised. By considering the compatibility of strain in reinforced soil, the approach allows the design forces and strains in the soil and the reinforcement to be estimated, and the expected deformation of the reinforced soil wall to be calculated. Non-dimensional charts are presented for the calculation of outward movements at the wall face both at the end of construction and due to creep of the reinforcement material.

1 INTRODUCTION

This paper summarises in one location a general approach to the design of reinforced soil structures, the concepts of which have been described in separate papers, Jewell (1985, 1987a, 1987b), and Jewell and Greenwood (1988). Although the method is a general one, it is described here as it would apply to vertical walls.

The method has been developed with extensible reinforcement materials in mind, for which the prediction of structural movements is important. Compatibility of strain in the reinforced soil is considered directly, and influences the choice of design values for the soil shearing resistance and the reinforcement force.

The concepts of *equilibrium*, *compatibility of deformation* and *yield* are introduced into the analysis. These are fundamental concepts in the plastic analysis of structures, which, in turn, is fundamental to the limit state design of structures, CIRIA (1977). These concepts are likely to be needed for the limit state design of reinforced soil.

2 EQUILIBRIUM

Equilibrium is considered in terms of *required forces* and *available forces*. The soil geometry (wall height etc), loadings (external, self weight, porewater pressure) and mobilised soil shearing resistance (friction) determine the *required forces*. The aim is to calculate both the magnitude and the distribution of the reinforcement forces *required* to maintain equilibrium everywhere in the soil. Thus the *required forces* for equilibrium depend only on the soil properties and the loadings.

Possible reinforcement layouts (reinforcement type, spacing, length) are then considered. A satisfactory reinforcement layout is one in which the *available forces* from the reinforcement exceed the *required forces* everywhere in the soil. The calculation of the *available forces* involves consideration of the reinforcement material properties (strength and stiffness) and the available bond between the reinforcement and the soil. Clearly the *available force* in the reinforcement cannot exceed the strength of the material, and the magnitude of force selected governs the corresponding tensile strain in the reinforcement.

3 COMPATIBILITY

The link which determines whether a pair of *required* and *available force* distributions are indeed compatible is the mobilised strain in the soil and the reinforcement. The assumption is made for geotextile and polymer grid reinforcement that the tensile strain in both the soil and the reinforcement is equal in the direction of the reinforcement. Any slippage between the soil and the reinforcement would enable the soil to strain more than the reinforcement. The assumption of equal strain allows the assumed mobilised soil shearing resistance and the reinforcement force to be checked against the respective properties of the materials (stress-strain relationship) to ensure that the strains are indeed *compatible*.

Finally, the distribution of the *required force* along a reinforcement layer implies a distribution of mobilised tensile strain along the reinforcement and hence a net elongation. The elongation in the reinforcement is a major component of the horizontal deflexion at the face of the reinforced soil structure. The magnitude of this deflexion can be checked for the structure to see if it is acceptable.

4 REINFORCED SOIL WALLS

Two idealised equilibrium stress distributions for a vertical wall with horizontal reinforcement have been derived by Jewell (1987a) with the aim of **bounding** the likely equilibrium in a reinforced soil wall. The usefulness of such idealised solutions resides in the simple results for the required reinforcement forces and the expected horizontal deflexions.

The first stress distribution is called the *ideal reinforcement length layout* because the reinforcement extends into the soil to provide forces locally everywhere they are required, Fig. 1. No reinforcement force is required in the soil beyond the line defining the natural angle of repose. The required reinforcement force is constant in the uniform Rankine active zone behind the wall face, and this zone determines the maximum required horizontal stress in the soil and force in the reinforcement.

The distribution of the maximum required stress for a wall with uniform vertical surcharge is shown in Fig. 2. To match the required and available stresses exactly (to minimise the number of reinforcement layers) the spacing between the reinforcement layers would vary, reducing with depth. This is an *ideal spacing arrangement*, Fig. 2a. For practical reasons the reinforcement is usually arranged in zones of constant spacing, Fig. 2b. Note, the force in a reinforcement layer is assumed to be equivalent to a stress in the soil (in the direction of the reinforcement) calculated from the force acting on the soil extending to half the spacing to the reinforcement layer above and below.

The horizontal deflexion at the wall face δ_h due to the reinforcement extension may be represented in terms of a non-dimensional parameter

$$\frac{\delta_h K}{HP} \tag{1}$$

where H is the wall height, K the reinforcement stiffness and P the mobilised reinforcement force in any layer. The variation of horizontal deflexion with depth below the crest of the wall z/H for the *ideal length layout* is shown in Fig. 3.

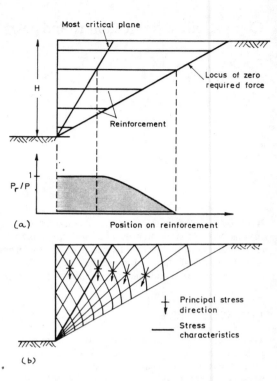

Fig. 1. *Ideal reinforcement length* stress field solution for a reinforced soil wall showing (a) reinforcement force distribution and (b) overall stresses in the soil. (Jewell, 1987a)

The second stress distribution is for a *truncated reinforcement length layout*. The solution gives the shortest length for the reinforcement able to maintain equilibrium in the soil with the magnitude of available reinforcement force mobilised in the uniform Rankine active zone. Two simplifying factors have been assumed in the analysis: (1) the unreinforced soil behind the reinforced zone is in an active state (smooth retaining wall) and (2) the available reinforcement force is constant along the full length of the reinforcement (no bond length at the free end). The *truncated length layout* is shown in Fig. 4.

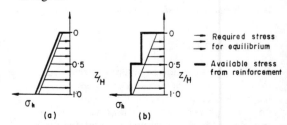

Fig. 2. Comparison of the *required* and *available stresses* for equilibrium. (a) *Ideal reinforcement spacing* matching exactly the *required stress*, (b) a typical reinforcement layout in two zones of constant spacing.

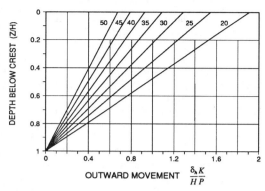

Fig. 3. Non-dimensional outward movement at the face due to deformation in the reinforced zone - *ideal reinforcement length case*. (Jewell, 1987a)

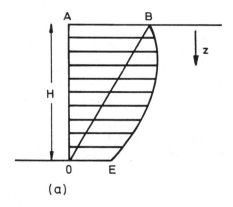

(a)

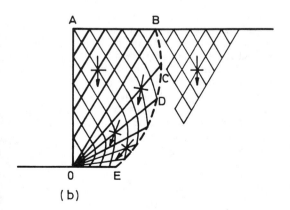

(b)

Fig. 4. *Truncated reinforcement length* stress field solution for a reinforced soil wall. Idealised case with reinforcement force constant along full length. (Jewell, 1987a)

The required reinforcement length and the horizontal deflexion at the wall face due to the reinforcement extension are both shown in Fig. 5. for the *truncated length layout*.

4.1 Wall deflexion

There are three main sources of outward movement at the face of a wall. These are (1) deformation in the reinforced zone, (2) deformation caused by the unreinforced soil behind the reinforced zone and (3) apparent deformation caused by incremental construction. This movement applies at the end of construction. When the reinforcement stiffness changes with time after construction there is additional deformation caused by (1) above.

If the wall face is fully propped until the wall has been built, then the deformation caused by incremental construction is eliminated, (3) above. For walls built incrementally, the additional movement can only be calculated approximately, and one such analysis has been described by Jewell (1987a).

For extensible reinforcement materials the main component of face deflexion comes from the deformation in the reinforced zone, (1) above. For a *uniformly spaced reinforcement* layout the **maximum horizontal deflexion** is found to be similar for the two idealised equilibrium states. This is illustrated in Fig. 6. for soil with a mobilised shearing resistance $\phi = 35°$. Fig. 6 has been derived directly from Figs. 3 and 5, but the uniform spacing reduces the mobilised reinforcement force P higher in the wall. The outward movement in Fig. 6 is expressed

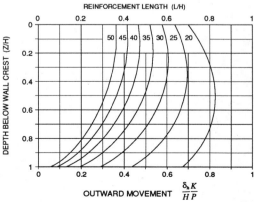

Fig. 5. Minimum required reinforcement length and non-dimensional outward movement - *truncated reinforcement length case*. (Jewell, 1987a)

613

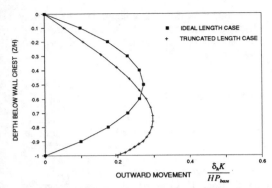

Fig. 6. Non-dimensional outward movement when the reinforcement is *uniformly spaced*. Comparison for *ideal* and *truncated reinforcement length* cases with $\phi = 35°$.

in terms of the reinforcement force at the base of the wall

$$\frac{\delta_h K}{HP_{base}} = \frac{\delta_h K}{K_a \gamma H^2 S_v} \qquad (2)$$

where S_v is the vertical reinforcement spacing and K_a the coefficient of active earth pressure. Calculations with the non-dimensional charts are fully described by Jewell (1987b).

The results (Fig. 6) predict the *maximum horizontal deflexion* δ_{max} in a wall with uniformly spaced reinforcement rather precisely, Fig. 7. The maximum deflexion at the end of construction should be calculated with an appropriate value for the reinforcement stiffness K. The maximum incremental deflexion due to reinforcement creep is given by the difference between the deflexion calculated with a new (lower) stiffness K_t and the original end of construction deflexion.

A similar analysis using Figs. 3 & 5 may be carried out for other reinforcement spacing arrangements.

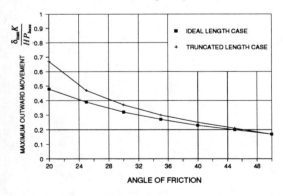

Fig. 7. Maximum outward movement for *uniformly spaced* reinforcement layouts. Comparison for *ideal* and *truncated reinforcement length* cases.

5 DESIGN VALUES AND SAFETY MARGINS

For geotextile and polymer grid reinforcement it is the material properties (strength and stiffness) which usually govern the maximum available reinforcement force. The length of the reinforcement layers also need to be chosen to ensure sufficient bond with the soil to allow the available force to be mobilised. For example, a bond length could be added to the required minimum reinforcement length of the *truncated length layout* (Fig. 5) to determine a satisfactory total reinforcement length. The choice of reinforcement length is not considered further.

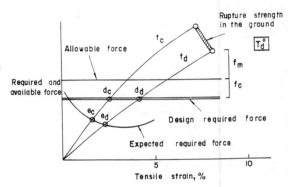

Fig. 8. Compatibility curve showing *required* and *available reinforcement forces*, and the **design** and **expected equilibrium** at the end of construction (d_c, e_c) and at the end of the design life (d_d, e_d). (Jewell and Greenwood, 1988)

5.1 Design required force

The *compatibility* for a reinforced soil wall is most conveniently represented in terms of the overall **maximum** *required* and *available force* for the wall plotted against the principal tensile strain in the soil and the tensile strain in the reinforcement (which are equal for horizontal reinforcement), Jewell (1985). The resulting **compatibility curve** is also most simply drawn as if for a propped wall construction: initially both the soil and the reinforcement have zero tensile strain, and the soil is under *at rest* conditions, ie the *required force* for equilibrium (prop force) is due to the *at rest* earth pressure. Releasing the prop force disturbs the equilibrium and tensile strain develops in the soil and the reinforcement. This increases the soil shearing resistance thus reducing the maximum *required force*, and increases the reinforcement force thus increasing the maximum *available force*, Fig. 8.

The proposal for design with extensible reinforcement is to calculate the maximum *design required force* from the active earth pressure with the **critical state shearing resistance** for the soil, Fig. 8. This is because the critical state value is the lowest expected shearing resistance for compact soil over the range of strains likely in a reinforced soil wall. The **maximum expected** soil unit weight and applied external surcharge would also be used, Table 1.

5.2 Allowable reinforcement force

The case of woven geotextiles, polymer grids and other polymer reinforcement materials whose mechanical properties are not influenced by confinement pressure in the soil are examined below.

The first criterion for the reinforcement is to select an *allowable force* which provides a margin of safety on the expected rupture strength which is consistent with existing structural codes. The material properties of strength and stiffness for polymer reinforcement depend on the *time* under load and the operational *temperature*. Installation of the reinforcement in the ground can reduce the reinforcement properties measured on *ex works* material due to *mechanical damage* and the *chemical environment* in the soil. For structures with long design lives there is likely to be a need for *extrapolation of data* to determine the expected reinforcement stength *in the ground at the end of the design life*.

The variation of reinforcement strength with time measured at the design temperature T_d is illustrated in Fig. 9. The datum is the behaviour of the *ex works* material. The material strength is likely to be reduced by *mechanical damage* and the *soil environment* and the maximum expected reduction due to these causes at the end of the design life t_d is represented by the partial factors f_d and f_{env} respectively, Fig. 9. Although represented by partial factors, the magnitude of the reduction in strength can only be assessed directly from test data, and cannot be prescribed by standard partial factor values.

The material partial factor f_m which determines the margin between the **expected strength** in the ground and the **allowable reinforcement force** must take into account the amount and quality of the data on the reinforcement material properties and any extrapolation of data.

Finally, allowance is made for an overall margin f_c between the maximum *required force* and the maximum *available* force (**allowable force from above**), Figs. 8 and 9. This factor enables the

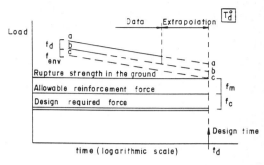

Fig. 9. Summary for polymer reinforcement material strength, illustrating the *strength in the ground* and the *allowable reinforcement force* which apply at the end of the design life *(t_d,T_d)*. (Jewell and Greenwood, 1988)

consequences and seriousness of attaining the limit state to be taken into account.

The recommended design values and partial factors are summarised in **Table 1**. The approach to design, the required data for polymer reinforcement materials, and suggested magnitudes for the partial factors are discussed by Jewell and Greenwood, (1988).

5.3 Design and expected equilibrium

The reinforcement stiffness (the relationship between the tensile force and the reinforcement extension) also varies with *time* and *temperature* for polymer materials. By working with the *isochronous curves* derived from creep test data, the total available reinforcement force-extension relationship (for the number of layers used) at the end of construction t_c and at the end of the design life t_d may be drawn on the **compatibility curve**, Fig. 8.

The **compatibility curve** indicates the *design equilibrium* point at the end of construction d_c and at the end of the design life d_d, Fig. 8. If there is data on the expected relationship between the soil shearing resistance and principal tensile strain, then the **compatibility curve** also indicates the *expected equilibrium* tensile strain in the soil and the reinforcement, Fig. 8.

The magnitude of the horizontal face deflexion implied by the calculated equilibrium points, both at the end of construction and during the design life, can be assessed as described in section 4. If found to be too great, increasing the number of reinforcement layers directly increases the total available reinforcement force at a given tensile strain. Alternatively, a stiffer or less creep susceptible reinforcement could be adopted.

Table 1. Proposed partial factors and design values for reinforced soil wall design strength calculations (Jewell and Greenwood, 1988).

	FACTOR	COMMENT
REINFORCEMENT		
Start with the **ideal strength** at (t_d, T_d)		Properties based on long term test data
	f_d	Factor based on test data to allow for **mechanical damage**
	f_{env}	Factor based on test data to allow for the **chemical and biological environment**
Based on **expected strength in the ground** at (t_d, T_d) from above, calculate the **allowable design force**	f_m	Specified **material factor** allowing for poorer performance in the structure and for extrapolation of data
SOIL		
Choose the **critical state shearing resistance** for design		Provides a safety margin over the expected shearing resistance
LOADINGS		
Soil self weight: choose the **maximum expected unit weight**		
Applied surcharge load: choose the **worst expected surcharge load**		Design surcharge loading may be specified in related codes
OVERALL FACTOR		
Overall margin between the **required** and **available** reinforcement force	f_c	Allows for seriousness and consequences of attaining the limit state

6 CONCLUSIONS

An approach to the design of reinforced soil has been summarised which provides a well defined basis for the selection of safety margins and design values for the soil and reinforcement, and which examines the compatibility of strain in the reinforced soil, allowing the deflexion in the reinforced soil to be estimated. The approach builds on the following concepts:

- clear separation of the *required forces* for equilibrium which depend on the soil, and the *available forces* which depend on the reinforcement.
- recognition that the properties of polymeric reinforcement not only depend on *time* and *temperature*, but also on *mechanical damage* and the *environmental conditions* in the soil.
- linking *compatibility* between the reinforcement and the soil through equality of tensile strain in both materials in the direction of the reinforcement.
- allowing the *likely deformation* in the structure to be estimated from idealised equilibrium stress distributions in the reinforced soil structure which are expected to bound the actual equilibrium.

7 REFERENCES

CIRIA 1977. *Rationalisation of safety and serviceability factors in structural codes.* Construction Industry Research and Information Association, London.

Jewell R.A. 1985. Limit equilibrium analysis of reinforced soil walls. *Proc. 11 Int. Conf. Soil Mechs. and Fndn. Engng,* San Francisco, **Vol 3,** 1705-1708, Balkema, Holland.

Jewell R.A. 1987a. Reinforced soil wall analysis and behaviour. Proc. NATO Advanced Scientific Workshop, *Application of polymeric reinforcement in soil retaining structures,* Martinus Nijhoff, Holland (in press).

Jewell R.A. 1987b. Analysis and predicted behaviour for the Royal Military College trial wall. Proc. NATO Advanced Scientific Workshop, *Application of polymeric reinforcement in soil retaining structures,* Martinus Nijhoff, Holland (in press).

Jewell R.A. and Greenwood J.H. 1988. Long term strength and safety in steep soil slopes reinforced by polymer materials. *Geotextiles and Geomembranes,* **Vol 7, No 1** (in press), Elsevier, London.

Authors' index